JUNQUEIRA'S
Basic
Histology
TEXT & ATLAS

Anthony L. Mescher, PhD
Professor of Anatomy and Cell Biology
Indiana University School of Medicine
Bloomington, Indiana

Medical

New York   Chicago   San Francisco   Lisbon   London
Madrid   Mexico City   Milan   New Delhi   San Juan   Seoul
Singapore   Sydney   Toronto

Junqueira's Basic Histology, Twelfth Edition

Copyright © 2010 by The McGraw-Hill Companies, Inc. All rights reserved. Printed in the United States of America. Except as permitted under the United States copyright Act of 1976, no part of this publication may be reproduced or distributed in any form or by any means, or stored in a data base or retrieval system, without the prior written permission of the publisher.

1 2 3 4 5 6 7 8 9 0   BAN/BAN   14 13 12 11 10 9

Set ISBN 978-0-07-163020-7; MHID 0-07-163020-1
Book ISBN 978-0-07-160431-4; MHID 0-07-160431-6
CD ISBN 978-0-07-160432-1; MHID 0-07-160432-4

ISSN 0891-2106

Notice

Medicine is an ever-changing science. As new research and clinical experience broaden our knowledge, changes in treatment and drug therapy are required. The author and the publisher of this work have checked with sources believed to be reliable in their efforts to provide information that is complete and generally in accord with the standards accepted at the time of publication. However, in view of the possibility of human error or changes in medical sciences, neither the author nor the publisher nor any other party who has been involved in the preparation or publication of this work warrants that the information contained herein is in every respect accurate or complete, and they disclaim all responsibility for any errors or omissions or for the results obtained from use of the information contained in this work. Readers are encouraged to confirm the information contained herein with other sources. For example and in particular, readers are advised to check the product information sheet included in the package of each drug they plan to administer to be certain that the information contained in this work is accurate and that changes have not been made in the recommended dose or in the contraindications for administration. This recommendation is of particular importance in connection with new or infrequently used drugs.

This book was set in Adobe Garamond by International Typesetting and Composition.
The editors were Michael Weitz and Karen Davis.
The production supervisor was Catherine H. Saggese.
Production management was provided by Harleen Chopra, International Typesetting and Composition.
The illustration manager was Armen Ovsepyan.
The artwork was done by Electronic Publishing Services, Inc.
The book designer was Eve Siegel.
The cover design was by Perhsson Design.
The index was prepared by BIM Indexing & Proofreading Services.
RR Donnelley was the printer and binder.

This book is printed on acid-free paper.

The Figure Credit section for this book begins on page 441 and is considered an extension of the copyright page.

Contents

CONTENTS / **v**

Key Features of
Junqueira's Basic Histology,
Twelfth Edition

- **Recognized for more than three decades** as the most authoritative, comprehensive and effective approach to understanding medical histology

- **Unmatched** in its ability to explain the relationship between cell and tissue structure and their function in the human body

- **Updated** to reflect the latest research and developments in the field

- **Medical Applications** are incorporated throughout every chapter to give the content greater clinical relevance

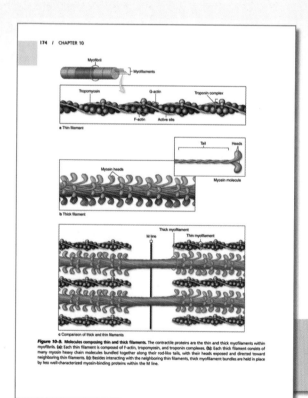

- **Full-color design** includes more than 1,000 state-of-the-art photomicrographs and drawings

- **Expanded legends** eliminate the need to jump from image to text

- **New author Anthony L. Mescher, PhD,** has more than 30 years' teaching experience

1,000 illustrations bring important concepts to life

Student-suggested detailed legends

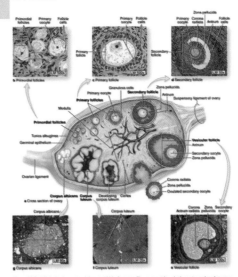

State-of-the art micrographs are the core of an all-new art program

Medical Applications are incorporated into every chapter

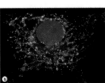

Preface

Since its inception, *Junqueira's Basic Histology* has set the standard for a concise yet thorough presentation of tissue structure and function for students in the health professions and advanced undergraduates. Junqueira treated histology, the study of cellular and tissue biology, not as microscopic anatomy, but as the key to integrating all of anatomy with physiology, cell and molecular biology, and biochemistry and as the foundation of pathology. Translated into many languages, *Basic Histology* is used worldwide and its concise, integrated style has been imitated in most subsequent histology texts.

As editor of the twelfth edition of *Junqueira's Basic Histology,* I undertook to maintain or improve three key features of the book. First, the **written text** itself has been upgraded in many areas while keeping its clear expository style and its integration with related subjects. All topics have been re-examined in the light of current literature and updated or refined if necessary. The result is a thoroughly modern treatment of cell and tissue biology, focused on the needs of students in the health professions. Students desiring additional information or greater detail on a topic can place the bold terms or other key words into any web-based search engine or into PubMed to access the most recent reviews on that topic. To simplify a preliminary overview or rapid review of chapters, main points for each subject are included in the expanded legends that accompany each figure.

Second, **micrographs** have been replaced as needed and now comprise a complete color atlas of tissue sections that include the important features of every tissue and organ in the human organism stained by standard methods. The light microscope photos are supplemented by electron micrographs and specifically stained preparations when these are useful in explaining unusual cells and tissues and their functional significance. Students purchasing the text can also now be linked for the first time to a virtual microscope and a complete collection of normal tissue specimens, most of which were used for the new micrographs in this atlas.

Finally, new **art** has been introduced throughout the text in a comprehensive set of modern, full color, three-dimensional drawings prepared by a certified team of medical illustrators. Figures chosen for this new edition include introductory material for each chapter that allows for rapid comprehension of an organ system's basic anatomy. Other illustrations highlight key features of each tissue and organ, along with their functional significance. Especially useful classic illustrations used in previous editions are still present, usually with more color or other new features. For each figure the goals are complete accuracy and sufficient detail to clarify the accompanying text and make learning easier. The result is a comprehensive program of art that strikes a balance between earlier simplistic diagrams and traditional medical illustrations with excessive details.

The overall organization of the highly successful eleventh edition has been retained. Unlike most histology texts, this includes an introductory chapter on laboratory methods used for the study of tissues, including the most important types of microscopy. Separate chapters focus on the cytoplasmic and nuclear compartments of the cell, followed by chapters on the four basic tissues that form the organs. Individual chapters are then devoted to each of the organ systems. Each chapter utilizes Junqueira's cell biological approach, emphasizing the specialized properties and activities of the basic tissue components as the key to understanding the functions of each organ. Also included is a chapter on the eye and ear, with thorough treatments of the structure and function of these organs at the cellular level. With minor changes in the placement of certain topics, the text covers every tissue of the body and is as up-to-date as possible.

KEY FEATURES

- Each topic is covered concisely yet thoroughly and includes all histological information required by students in the health professions.
- The number of illustrations have been doubled, to over 1000, with expanded figure legends.
- Figures and their legends include key points to facilitate preview or review study of a chapter.
- New light micrographs provide a complete atlas of human tissues and organs in standard preparations.
- Light micrographs are supplemented as needed by useful but not excessive electron micrographs and other microscopic preparations.
- A comprehensive program of new, modern illustrations facilitates comprehension of the micrographs without unnecessary detail.

- Each topic has been revised as needed to reflect new findings or interpretations of cell and tissue structure.
- Terms used are fully consistent with the new *Terminologia Histologica: International Terms for Human Cytology and Histology* and with standard usage in both clinical and basic sciences.
- Additional "Medical Applications" concisely state aspects of most topics' clinical relevance.
- The importance of stem cells in organ renewal or repair is emphasized as needed for each organ.

I am confident that *Junqueira's Basic Histology* will continue to be one of the most useful histology texts available. Comments and suggestions for further improving the next edition are welcome!

Anthony L. Mescher, PhD
Bloomington, Indiana
mescher@indiana.edu

Acknowledgments

I wish to thank the editors and staff of McGraw-Hill, especially Michael Weitz and Karen Davis, who helped me immensely in this thorough revision of *Junqueira's Basic Histology*. Those colleagues of Dr. Junqueira who helped write and review previous editions and those medical scientists, including Dr. James C. Williams, Jr., who provided me with additional suggestions, are also gratefully acknowledged. I also thank my family and my research colleagues for bearing with me during this undertaking and acknowledge finally the invaluable help of the medical, graduate, and undergraduate students with whom I have studied histology and cell biology for over 27 years at Indiana University Bloomington. All these groups have helped with this new edition of *Junqueira's Basic Histology*.

Histology & Its Methods of Study

1

Histology is the study of the tissues of the body and how these tissues are arranged to constitute organs. The Greek root *histo* can be translated as either "tissue" or "web" and both translations are appropriate because most tissues are webs of interwoven filaments and fibers, both cellular and noncellular, with membranous linings. Histology involves all aspects of tissue biology, with the focus on how cells' structure and arrangement optimize functions specific to each organ.

Tissues are made of two interacting components: cells and extracellular matrix. The extracellular matrix consists of many kinds of molecules, most of which are highly organized and form complex structures, such as collagen fibrils and basement membranes. The main functions once attributed to the extracellular matrix were to furnish mechanical support for the cells, to transport nutrients to the cells, and to carry away catabolites and secretory products. We now know that, although the cells produce the extracellular matrix, they are also influenced and sometimes controlled by molecules of the matrix. There is, thus, an intense interaction between cells and matrix, with many components of the matrix recognized by and attaching to receptors present on cell surfaces. Most of these receptors are molecules that cross the cell membranes and connect to structural components of the intracellular cytoplasm. Thus, cells and extracellular matrix form a continuum that functions together and reacts to stimuli and inhibitors together.

Each of the fundamental tissues is formed by several types of cells and typically by specific associations of cells and extracellular matrix. These characteristic associations facilitate the recognition of the many subtypes of tissues by students. Most organs are formed by an orderly combination of several tissues, except the central nervous system, which is formed almost solely by nervous tissue. The precise combination of these tissues allows the functioning of each organ and of the organism as a whole.

The small size of cells and matrix components makes histology dependent on the use of microscopes. Advances in chemistry, molecular biology, physiology, immunology, and pathology—and the interactions among these fields—are essential for a better knowledge of tissue biology. Familiarity with the tools and methods of any branch of science is essential for a proper understanding of the subject. This chapter reviews several of the more common methods used to study cells and tissues and the principles involved in these methods.

PREPARATION OF TISSUES FOR STUDY

The most common procedure used in the study of tissues is the preparation of histological sections or tissue slices that can be studied with the aid of the light microscope. Under the light microscope, tissues are examined via a light beam that is transmitted through the tissue. Because tissues and organs are usually too thick for light to pass through them, they must be sectioned to obtain thin, translucent sections and then attached to glass slides before they can be examined.

The ideal microscope tissue preparation should be preserved so that the tissue on the slide has the same structure and molecular composition as it had in the body. However, as a practical matter this is seldom feasible and artifacts, distortions, and loss of components due to the preparation process are almost always present. The basic steps used in tissue preparation for histology are shown in Figure 1–1.

Fixation

If a permanent section is desired, tissues must be fixed. To avoid tissue digestion by enzymes present within the cells (autolysis) or by bacteria and to preserve the structure and molecular composition, pieces of organs should be promptly and adequately treated before, or as soon as possible after, removal from the animal's body. This treatment—**fixation**—can be done by chemical or, less frequently, physical methods. In chemical fixation, the tissues are usually immersed in solutions of stabilizing or cross-linking agents called **fixatives.** Because the fixative needs some time to fully diffuse into the tissues, the tissues are usually cut into small fragments before fixation to facilitate the penetration of the fixative and to guarantee preservation of the tissue. Intravascular perfusion of fixatives can be used. Because the fixative in this case rapidly reaches the tissues through the blood vessels, fixation is greatly improved.

One of the best fixatives for routine light microscopy is formalin, a buffered isotonic solution of 37% formaldehyde.

The chemistry of the process involved in fixation is complex and not always well understood. Formaldehyde and glutaraldehyde, another widely used fixative, are known to react with the amine groups (NH_2) of tissue proteins. In the case of glutaraldehyde, the fixing action is reinforced by virtue of its being a dialdehyde, which can cross-link proteins.

In view of the high resolution afforded by the electron microscope, greater care in fixation is necessary to preserve ultrastructural detail. Toward that end, a double fixation procedure, using a buffered glutaraldehyde solution followed by a second fixation in buffered osmium tetroxide, is a standard procedure in preparations for fine structural studies. The effect of osmium tetroxide is to preserve and stain lipids and proteins.

Embedding & Sectioning

Tissues are usually embedded in a solid medium to facilitate sectioning. To obtain thin sections with the microtome, tissues must

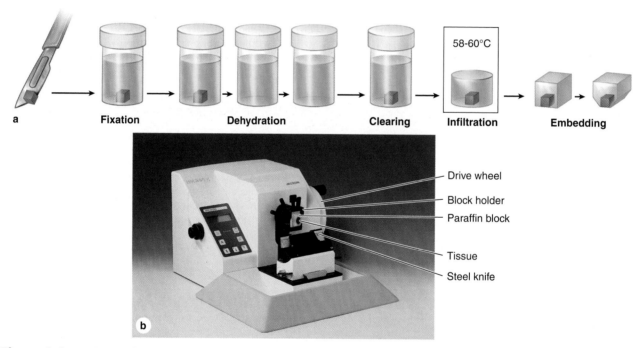

Figure 1–1. **Sectioning fixed and embedded tissue.** Most tissues studied histologically are prepared as shown. **(a):** Small pieces of fresh tissue are placed in **fixative** solutions which generally cross-link proteins, inactivating degradative enzymes and preserving cell structures. The fixed pieces then undergo "**dehydration**" by being transferred through a series of increasingly more concentrated alcohol solutions, ending in 100% which effectively removes all water from the tissue. The alcohol is then removed in a **clearing** solution miscible in both alcohol and melted paraffin. When the tissue is then placed in melted paraffin at 58°C it becomes completely infiltrated with this substance. All steps to this point are commonly done today by robotic devices in active histology or pathology laboratories. After **infiltration** the tissue is placed in a small mold containing melted paraffin, which is then allowed to harden. The resulting paraffin block is trimmed to expose the tissue for sectioning (slicing). Similar steps are used in preparing tissue for transmission electron microscopy, except that smaller tissue samples are fixed in special fixatives and dehydrating solutions are used that are appropriate for embedding in epoxy resins which become much harder than paraffin to allow very thin sectioning. **(b):** A **microtome** is used for sectioning paraffin-embedded tissues for light microscopy. After mounting a trimmed block with the tissue specimen, rotating the drive wheel moves the tissue-block holder up and down. Each turn of the drive wheel advances the specimen holder a controlled distance, generally between 1 and 10 μm, and after each forward move the tissue block passes over the steel knife edge, which cuts the sections at a thickness equal to the distance the block advanced. Paraffin sections are then adhered to glass slides, deparaffinized, and stained for microscopic examination. For transmission electron microscopy sections less than 1 μm thick are prepared from resin-embedded cells using an ultramicrotome with a glass or diamond knife.

be infiltrated after fixation with embedding substances that impart a rigid consistency to the tissue. Embedding materials include paraffin and plastic resins. Paraffin is used routinely for light microscopy; resins are used for both light and electron microscopy.

The process of paraffin embedding, or tissue impregnation, is ordinarily preceded by two main steps: **dehydration** and **clearing.** The water is first extracted from the fragments to be embedded by bathing them successively in a graded series of mixtures of ethanol and water, usually from 70% to 100% ethanol (dehydration). The ethanol is then replaced with a solvent miscible with both alcohol and the embedding medium. As the tissues are infiltrated with this solvent, they generally become transparent (clearing). Once the tissue is impregnated with the solvent, it is placed in melted paraffin in an oven, typically at 52–60°C. The heat causes the solvent to evaporate, and the spaces within the tissues become filled with paraffin. The tissue together with its impregnating paraffin hardens after removal from the oven. Tissues to be embedded with plastic resin are also dehydrated in ethanol and—depending on the kind of resin used—subsequently infiltrated with plastic solvents. The ethanol or the solvents are later replaced by plastic solutions that are hardened by means of cross-linking polymerizers. Plastic embedding prevents the shrinking effect of the high temperatures needed for paraffin embedding and gives little or no distortion to the cells.

The hard blocks containing the tissues are then placed in an instrument called a **microtome** (Figure 1–1) and are sliced by the microtome's steel or glass blade into sections 1 to10 micrometers thick. Remember that one micrometer (1 μm) equals 1/1,000 of a millimeter (mm) = 10^{-6} m. Other units of distance commonly used in histology are the nanometer (1 nm = 0.001 μm = 10^{-6} mm = 10^{-9} m) and angstrom (1 Å = 0.1 nm or 10^{-4} μm). The sections are floated on water and then transferred to glass slides to be stained.

An alternate way to prepare tissue sections is to submit the tissues to rapid freezing. In this process, the tissues are fixed by freezing (physical, not chemical fixation) and at the same time become hard and thus ready to be sectioned. A freezing microtome—the **cryostat**—is then used to section the frozen block with tissue. Because this method allows the rapid preparation of sections without going through the long embedding procedure described above, it is routinely used in hospitals to study specimens during surgical procedures. Freezing of tissues is also effective in the histochemical study of very sensitive enzymes or small molecules, since freezing, unlike fixation, does not inactivate most enzymes. Finally, because immersion in solvents such as xylene dissolves cell lipids in fixed tissues, frozen sections are also useful when structures containing lipids are to be studied.

Staining

To be studied microscopically sections must typically be stained or dyed because most tissues are colorless. Methods of staining tissues have therefore been devised that not only make the various tissue components conspicuous but also permit distinctions to be made between them. The dyes stain tissue components more or less selectively. Most of these dyes behave like acidic or basic compounds and have a tendency to form electrostatic (salt) linkages with ionizable radicals of the tissues. Tissue components with a net negative charge (anionic) stain more readily with basic dyes and are termed **basophilic**; cationic components, such as proteins with many ionized amino groups, have affinity for acidic dyes and are termed **acidophilic.**

Examples of basic dyes are toluidine blue, alcian blue, and methylene blue. Hematoxylin behaves like a basic dye, that is, it stains the basophilic tissue components. The main tissue components that ionize and react with basic dyes do so because of acids in their composition (nucleic acids, glycosaminoglycans, and acid glycoproteins). Acid dyes (eg, orange G, eosin, acid fuchsin) stain the acidophilic components of tissues such as mitochondria, secretory granules, and collagen.

Of all dyes, the simple combination of **hematoxylin** and **eosin (H&E)** is used most commonly. Hematoxylin stains DNA of the cell nucleus and other acidic structures (such as RNA-rich portions of the cytoplasm and the matrix of cartilage) blue. In contrast, eosin stains other cytoplasmic components and collagen pink (Figure 1–2). Many other dyes, such as the **trichromes** (eg, Mallory stain, Masson stain), are used in different histologic procedures. The trichromes, besides showing the nuclei and cytoplasm very well, help to distinguish extracellular tissue components better than H&E. A good technique for differentiating collagen is the use of picrosirius, especially when associated with polarized light (see Polarizing Microscopy).

The chemical basis of other staining procedures is more complicated than the electrostatic interactions underlying basophilia and acidophilia. DNA can be specifically identified and quantified in nuclei using the Feulgen reaction, in which deoxyribose sugars are hydrolyzed by mild hydrochloric acid, followed by treatment with **periodic acid** and **Schiff reagent (PAS).** The PAS technique is based on the transformation of 1,2-glycol groups present in the sugars into aldehyde residues, which then react with Schiff reagent to produce a purple or magenta color.

Polysaccharides constitute an extremely heterogeneous group in tissues and occur either in a free state or combined with proteins and lipids. Because of their hexose sugar content, many polysaccharides can also be demonstrated by the PAS reaction. A ubiquitous free polysaccharide in animal cells is **glycogen**, which can be demonstrated by PAS in liver, striated muscle, and other tissues where it accumulates.

Short branched chains of sugars (oligosaccharides) are attached to specific amino acids of **glycoproteins**, making most glycoproteins PAS-positive. Figure 1–2b shows an example of cells stained by the PAS reaction. **Glycosaminoglycans** (GAGs) are anionic, unbranched long-chain polysaccharides containing aminated sugars. Many glycosaminoglycans are synthesized while attached to a core protein and constitute a class of macromolecules called **proteoglycans**, which upon secretion make up important parts of the extracellular matrix (ECM) (see Chapters 5 and 7). Unlike a glycoprotein, a proteoglycan's carbohydrate chains are greater in weight and volume than the protein core of the molecule. GAGs and many acidic glycoproteins do not undergo the PAS reaction, but because of their high content of anionic carboxyl and sulfate groups show a strong electrostatic interaction with alcian blue and other basic stains.

Basophilic or PAS-positive material can be further identified by **enzyme digestion** pretreatment of a tissue section with an enzyme that specifically digests one substrate, leaving other adjacent sections untreated. For example, pretreatment with ribonuclease will greatly reduce cytoplasmic basophilia with little effect on chromosomes, indicating the importance of RNA for the cytoplasmic staining. Similarly, free polysaccharides are digested by amylase, which can therefore be used to distinguish glycogen from glycoproteins in PAS-positive material.

In many staining procedures certain structures such nuclei become labeled, but other parts of cells are often not visible. In this case a **counterstain** is used to give additional information. A counterstain is usually a single stain that is applied to a section by another method to allow better recognition of nuclei or other structures.

Lipid-rich structures are best revealed with **lipid-soluble dyes** to avoid the steps of slide preparation that remove lipids such as treatment with heat, xylene, or paraffin. Typically frozen sections are stained in alcohol solutions saturated with a lipophilic dye such as Sudan black. The stain dissolves in cellular lipid droplets and other lipid-rich structures, which become stained in black. Specialized methods for the localization of cholesterol, phospholipids, and glycolipids are useful in diagnosis of metabolic diseases in which there are intracellular accumulations of different kinds of lipids. In addition to tissue staining with dyes, **metal impregnation techniques** usually using silver salts are a common method of visualizing certain ECM fibers and specific cellular elements in nervous tissue.

The whole procedure, from fixation to observing a tissue in a light microscope, may take from 12 hours to $2^1/_2$ days, depending on the size of the tissue, the fixative, the embedding medium, and the method of staining. The final step before observation is mounting a protective glass coverslip on the slide with adhesive mounting media.

LIGHT MICROSCOPY

Conventional bright-field microscopy, as well as fluorescence, phase-contrast, differential interference, confocal, and polarizing microscopy are all based on the interaction of light and tissue components and can be used to reveal and study tissue features.

Bright-field Microscopy

With the **bright-field microscope**, widely used by students of histology, stained preparations are examined by means of ordinary light that passes through the specimen. The microscope is composed of mechanical and optical parts (Figure 1–3). The optical components consist of three systems of lenses. The **condenser** collects and focuses light, producing a cone of light that illuminates the object to be observed. The **objective** lenses enlarge and project the illuminated image of the object in the direction of the eyepiece. The **eyepiece** or ocular lens further magnifies this image and projects it onto the viewer's retina, photographic film, or (to obtain a digital image) a detector such as a charge-coupled device (CCD) camera. The total magnification is obtained by multiplying the magnifying power of the objective and ocular lenses.

The critical factor in obtaining a crisp, detailed image with a light microscope is its **resolving power,** defined as the smallest distance between two particles at which they can be seen as

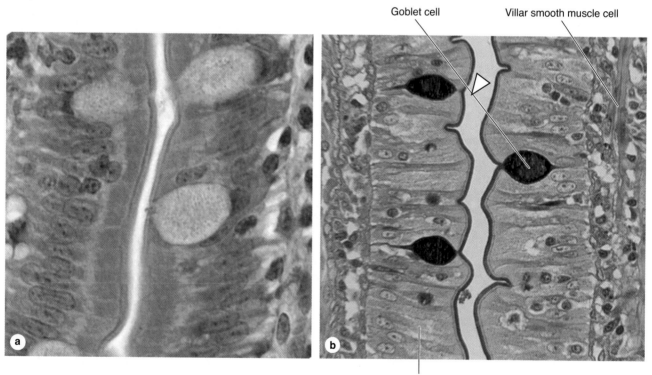

Figure 1–2. **Hematoxylin & Eosin (H&E) and Periodic acid-Schiff (PAS) staining.** Micrographs of the columnar epithelium lining the small intestine. **(a):** Micrograph stained with hematoxylin and eosin (H&E). **(b):** Micrograph stained by the periodic acid-Schiff (PAS) reaction for glycoproteins. With H&E, basophilic cell nuclei are stained purple while cytoplasm stains pink. Cell regions with abundant oligosaccharides on glycoproteins, such as the apical ends of the cells or the scattered mucus-secreting goblet cells in the layer are poorly stained. With PAS, staining is most intense at the cell surface, where projecting microvilli have a prominent layer of glycoproteins (arrow head) and in the mucin-rich secretory granules of goblet cells. Cell surface glycoproteins and mucin are PAS-positive due to their high content of oligosaccharides and polysaccharides. The PAS-stained tissue was counterstained with hematoxylin to show the cell nuclei. Both X300.

separate objects. The maximal resolving power of the light microscope is approximately 0.2 μm; this power permits good images magnified 1000–1500 times. Objects smaller or thinner than 0.2 μm (such as a ribosome, a membrane, or a filament of actin) cannot be distinguished with this instrument. Likewise, two objects such as mitochondria will be seen as only one object if they are separated by less than 0.2 μm. The quality of the image—its clarity and richness of detail—depends on the microscope's resolving power. The magnification is of value only when accompanied by high resolution. The resolving power of a microscope depends mainly on the quality of its objective lens. The eyepiece lens enlarges only the image obtained by the objective; it does not improve resolution. For this reason, when comparing objectives of different magnifications, those that provide higher magnification also have higher resolving power.

Video cameras highly sensitive to light enhance the power of the bright-field and other light microscopes and allow the capture of digitized images suitable for computerized image analysis and printing. The frontiers of light microscopy have been redefined by the use of such cameras. With digital cameras and image-enhancement programs (to enhance contrast, for example), objects that may not be visible when viewed directly through the ocular may be made visible in the video screen. These video systems are also useful for studying living cells for long periods of time, because they use low-intensity light and thus avoid the cellular damage from heat that can result from intense illumination. Moreover, software developed for image analysis allows rapid measurements and quantitative study of microscopic structures.

Fluorescence Microscopy

When certain substances are irradiated by light of a proper wavelength, they emit light with a longer wavelength. This phenomenon is called fluorescence. In **fluorescence microscopy**, tissue sections are usually irradiated with ultraviolet (UV) light and the emission is in the visible portion of the spectrum. The fluorescent substances appear brilliant on a dark background. For this method, the microscope has a strong UV light source and special filters that select rays of different wavelengths emitted by the substances.

Fluorescent compounds with affinity for specific cell macromolecules may be used as fluorescent stains. Acridine orange, which binds both DNA and RNA, is an example. When observed in the fluorescence microscope, these nucleic acids emit slightly different fluorescence, allowing them to be localized separately in cells (Figure 1–4a). Other compounds such as Hoechst stain and DAPI specifically bind DNA and are used to stain cell nuclei, emitting a characteristic blue fluorescence under UV. Another important application of fluorescence microscopy is achieved by coupling fluorescent compounds to molecules that will specifically bind to certain cellular components and thus allow the identification of these structures under the microscope (Figure 1–4b). Antibodies labeled with fluorescent compounds are extremely important in immunohistological staining. (See Detection Methods Using Specific Interactions Between Molecules).

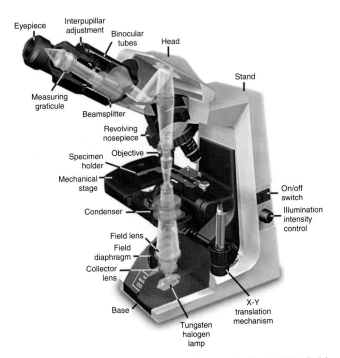

Figure 1–3. **Components and light path of a bright-field microscope.** Photograph of a bright-field light microscope showing its main components and the pathway of light from the substage lamp to the eye of the observer. The optical system has three sets of lenses: a condenser, a set of objectives, and either one or two eyepieces. The **condenser** collects and focuses light, producing a cone of light that illuminates the tissue slide on the stage. **Objective** lenses enlarge and project the illuminated image of the object in the direction of the eyepiece. For routine histological studies objectives having three different magnifications are generally used: **X4** for low magnification observations of a large area (field) of the tissue; **X10** for medium magnification of a smaller field; and **X40** for high magnification of more detailed areas. The **eyepiece** or ocular further magnifies this image another X10 and projects it onto the viewer's retina, yielding a total magnification of X40, X100, or X400. (With permission, from Nikon Instruments.)

Phase-Contrast Microscopy & Differential Interference Microscopy

Some optical arrangements allow the observation of unstained cells and tissue sections. Unstained biological specimens are usually transparent and difficult to view in detail, because all parts of the specimen have almost the same optical density. **Phase-contrast microscopy**, however, uses a lens system that produces visible images from transparent objects (Figure 1–5).

Phase-contrast microscopy is based on the principle that light changes its speed when passing through cellular and extracellular structures with different refractive indices. These changes are used by the phase-contrast system to cause the structures to appear lighter or darker in relation to each other. Because it does not require fixation or staining, phase-contrast microscopy allows observation of living cells and tissue cultures, and such microscopes are prominent tools in all cell culture labs. A related method of observing unstained cells or tissue sections is the Nomarski **differential interference microscopy**, which produces an image with a more apparent three-dimensional aspect than in routine phase-contrast microscopy (Figure 1–5).

Confocal Microscopy

With a regular bright-field microscope the beam of light is relatively large and fills the specimen. Stray light reduces contrast within the image and compromises the resolving power of the objective lens. Confocal microscopy avoids stray light and achieves greater resolution by using (1) a small point of high-intensity light provided by a laser and (2) a plate with a pinhole aperture in front of the image detector. The point light source, the focal point of the lens, and the detector's pinpoint aperture are all optically conjugated or aligned to each other in the focal plane (confocal) and unfocused light does not pass through the pinhole. This greatly improves resolution of the object in focus and allows the localization of specimen components with much greater precision than with the bright-field microscope.

Most confocal microscopes include a computer-driven mirror system (the beam splitter) to move the point of illumination across the specimen automatically and rapidly. Digital images captured at many individual spots in a very thin plane-of-focus are used to produce an "optical section" of that plane. Moreover, creating optical sections at a series of focal planes through the specimen allows them to be digitally reconstructed into a three-dimensional image. Important features of confocal microscopes are shown in Figure 1–6.

Polarizing Microscopy

Polarizing microscopy allows the recognition of structures made of highly organized molecules. When normal light passes through a **polarizing filter** (such as a Polaroid), it exits vibrating in only one direction. If a second filter is placed in the microscope above the first one, with its main axis perpendicular to the first filter, no light passes through. If, however, tissue structures containing oriented macromolecules are located between the two polarizing filters, their repetitive structure rotates the axis of the light emerging from the polarizer and they appear as bright structures against a dark background (Figure 1–7). The ability to rotate the direction of vibration of polarized light is called **birefringence** and is a feature of crystalline substances or substances containing highly oriented molecules, such as cellulose, collagen, microtubules, and microfilaments.

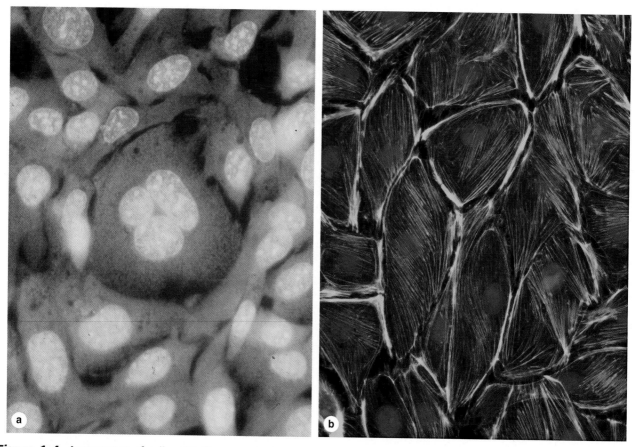

Figure 1–4. **Appearance of cells with fluorescent microscopy.** Components of cells in culture are often stained with compounds visible by fluorescence microscopy. **(a):** Kidney cells stained with acridine orange, which binds nucleic acids. Under a fluorescence microscope, nuclear DNA emits yellow light and the RNA-rich cytoplasm appears reddish or orange. **(b):** The less dense culture of kidney cells stained with DAPI (4′,6-diamino-2-phenylindole) which binds DNA, and with phalloidin, which binds actin filaments. Nuclei of these cells show a blue fluorescence and actin filaments appear green. Important information such as the greater density of microfilaments at the cell periphery is readily apparent. (Figure 1–4b, with permission, from Drs. Claire E. Walczak and Rania Risk, Indiana University School of Medicine, Bloomington.)

ELECTRON MICROSCOPY

Transmission and scanning electron microscopes are based on the interaction of electrons and tissue components. The wavelength in the electron beam is much shorter than of light, allowing a thousand-fold increase in resolution.

Transmission Electron Microscopy

The transmission electron microscope (TEM) is an imaging system that permits resolution around 3 mm (Figure 1–8a). This high resolution allows magnifications of up to 400,000 times to be viewed with details. Unfortunately, this level of magnification applies only to isolated molecules or particles. Very thin tissue sections can be observed with details at magnifications of up to about 120,000 times.

The TEM functions on the principle that a beam of electrons can be deflected by electromagnetic fields in a manner similar to light deflection in glass lenses. The beam is produced by a cathode at the top of the instrument and passes down through the chamber in a vacuum. Because electrons change their path when submitted to electromagnetic fields, the beam can be focused by passing through electric coils which can be considered electromagnetic lenses.

The first lens is a condenser focusing the beam of electrons on the specimen section. Some electrons interact with atoms in the section and their course is modified, while others simply cross the specimen without interacting. Electrons passing through the specimen reach the objective lens, which forms a focused, magnified image that is then magnified further through other lenses and captured on a viewing screen. The image of the specimen shows areas of white, black, and shades of gray corresponding to areas through which electrons readily passed (appearing brighter or electron lucent) and areas where electrons were absorbed or deflected (appearing darker or more electron dense).

To provide a useful interaction between the specimen and the electrons, TEM requires very thin sections (40–90 nm); therefore, embedding is performed with a hard epoxy and sectioning is done with a glass or diamond knife. The extremely thin sections are collected on small metal grids and transferred to the interior of the microscope to be analyzed.

Freezing techniques (**freeze fracture, cryofracture, freeze etched**) combined with electron microscopy have been very useful for examining membrane structure. Very small tissue specimens are rapidly frozen in liquid nitrogen and fractured in a vacuum with a knife. A replica of the still frozen exposed surface is produced by applying thin coats of vaporized carbon, platinum, or other atoms. Tissue is then dissolved away and the replica of the surface is examined by SEM. The random fracture planes often split the lipid bilayers of membranes, exposing protein components whose size, shape, and distribution can then be studied.

Scanning Electron Microscopy

Scanning electron microscopy (SEM) permits pseudo–three-dimensional views of the surfaces of cells, tissues, and organs. Like the TEM this microscope produces and focuses a very narrow beam of electrons, but in this instrument the beam does not pass through the specimen (Figure 1-8b). Instead the surface of the specimen is first dried and coated with a very thin layer of metal atoms through which electrons do not pass readily. When the beam is scanned from point to point across the specimen it interacts with the metal atoms and produces reflected electrons or secondary electrons emitted from the metal. These are captured by a detector and the resulting signal is processed to produce a black-and-white image on a monitor. SEM images are usually easily understood, because they present a view that appears to be illuminated from above, just as our ordinary macroscopic world is filled with highlights and shadows caused by illumination from above.

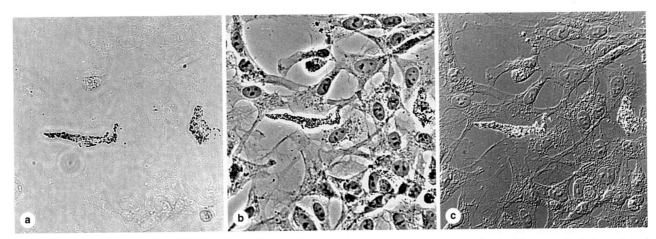

Figure 1–5. **Unstained cells' appearance in three types of light microscopy.** Neural crest cells growing as a single layer in culture appear differently with various techniques of light microscopy. These cells are unstained and the same field of cells, including two differentiating pigment cells, is shown in each photo. **(a): Bright-field microscopy**: without fixation and staining, only the two pigment cells can be seen. **(b): Phase-contrast microscopy**: cell boundaries, nuclei, and cytoplasmic structures with different refractive indices affect in-phase light differently and produce an image of these features in all the cells. **(c): Differential interference microscopy**: cellular details are highlighted in a different manner using Nomarski optics. Phase-contrast microscopy, with or without differential interference, is widely used to observe live cells grown in tissue culture. All X200. (With permission, from Sherry Rogers, Department of Cell Biology and Physiology, University of New Mexico.)

AUTORADIOGRAPHY

Autoradiography is a method of localizing newly synthesized macromolecules (DNA, RNA, protein, glycoproteins, and polysaccharides) in cells or tissue sections. Radioactively labeled metabolites (nucleotides, amino acids) incorporated into the macromolecules emit weak radiation that is restricted to the cellular regions where the molecules are located. Radiolabeled cells or mounted tissue sections are coated in a darkroom with photographic emulsion containing silver bromide crystals, which act as microdetectors of this radiation in the same way that they respond to light in common photographic film. After an adequate exposure time in lightproof boxes the slides are developed photographically. The silver bromide crystals reduced by the radiation are reduced to small black grains of metallic silver,

indicating locations of radiolabeled macromolecules in the tissue. This general procedure can be used in preparations for both light microscopy and TEM (Figure 1–9).

Much information becomes available by autoradiography of cells or tissues. Thus, if a radioactive amino acid is used, it is possible to know which cells in a tissue produce more protein and which cells produce less, because the number of silver grains formed over the cells is proportional to the intensity of protein synthesis. If a radioactive precursor of DNA (such as tritium-labeled thymidine) is used, it is possible to know which cells in a tissue (and how many) are preparing to divide. Dynamic events may also be analyzed. For example, if one wishes to know where in the cell protein is produced, if it is secreted, and which path it follows in the cell before being secreted, several animals are injected with a radioactive amino acid and tissues collected at

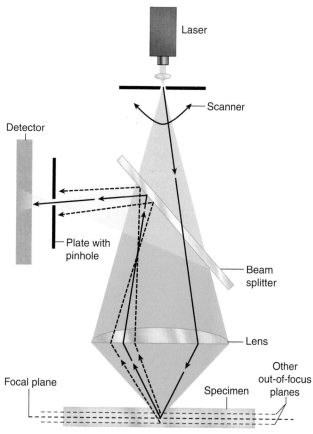

Figure 1–6. Principle of confocal microscopy. Although a very small spot of light originating from one plane of the section crosses the pinhole and reaches the detector, rays originating from other planes are blocked by the blind. Thus, only one very thin plane of the specimen is focused at a time. The diagram shows the practical arrangement of a confocal microscope. Light from a laser source hits the specimen and is reflected. A beam splitter directs the reflected light to a pinhole and a detector. Light from components of the specimen that are above or below the focused plane is blocked by the blind. The laser scans the specimen so that a larger area of the specimen can be observed.

Figure 1–7. Tissue appearance with bright-field and polarizing microscopy. Polarizing light microscopy produces an image only of material having repetitive, periodic macromolecular structure; features without such structure are not seen. Shown here is a piece of thin mesentery that was stained with red picrosirius, orcein, and hematoxylin, and was then placed directly on a slide and observed by bright-field and polarizing microscopy. **(a):** Under routine bright-field microscopy collagen fibers appear red, along with thin dark elastic fibers and cell nuclei. **(b):** Under polarizing light microscopy, only collagen fibers are visible and these exhibit intense birefringence and appear bright red or yellow; elastic fibers and nuclei lack oriented macromolecular structure and are not visible.

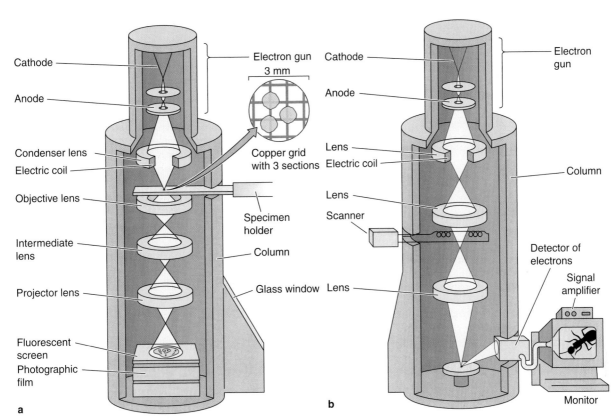

Figure 1–8. **Electron microscopes.** Electron microscopes are large instruments generally housed in a specialized EM facility. **(a):** Schematic view of a transmission electron microscope (TEM) with its lenses and the pathway of the electrons. With the microscope's entire column in a vacuum, electrons are released by heating a very thin metallic (usually tungsten) filament (cathode). The released electrons are then submitted to a voltage difference of 60–120 kV between the cathode and the anode, which is a metallic plate with a hole in its center. Electrons are thus attracted to the anode, accelerated to high speeds, and form a beam of electrons as they pass through the central opening in the anode. Passing through electric coils the beam is deflected in a way roughly analogous to the effect of optical lenses on light because electrons change their path when submitted to electromagnetic fields.

The configuration of the TEM is similar to that of an upside-down light microscope. The first lens is a condenser that focuses the beam of electrons on the section. Some electrons interact with atoms of the section and continue their course, while others simply cross the specimen without interacting. Most electrons reach the objective lens, which forms a magnified image that is then projected through other magnifying lenses. Because the human eye is not sensitive to electrons, the image is finally projected on a fluorescent screen or is registered by photographic plates or a CCD camera.

In a TEM image areas of the specimen through which electrons passed appear bright (electron lucent), while those areas which are naturally dense or which bind heavy metals during specimen preparation or "staining" absorb or deflect electrons and appear dark (electron dense). Such images are therefore always black, white, and shades of gray.

(b): Schematic view of a scanning electron microscope (SEM) with many similarities to a TEM. However, here the electron beam focused by electromagnetic lenses does not pass through the specimen, but rather is moved sequentially (scanned) from point to point across its surface similar to the way an electron beam is scanned across a television tube. The specimen was coated previously with a very thin coating of metal atoms and the beam interacts with these atoms, and produces reflected electrons and newly emitted secondary electrons. All of these are captured by a detector and transmitted to amplifiers and other devices which produce a signal to a cathode ray tube monitor, resulting in a black-and-white image. The SEM shows only surface views of the coated specimen but with a striking three-dimensional quality. The inside of organs or cells can be analyzed by sectioning them to expose their internal surfaces.

different times after the injections. Autoradiography of the tissues representing the various times throughout the experiment will indicate the migration of the radioactive proteins. If one wishes to know where new cells are produced in an organ and where they migrate, several animals are injected with radioactive thymidine and tissues collected at different times after the injection. Autoradiographs of the sections will show the location of the dividing cells and where they migrate.

CELL & TISSUE CULTURE

Live cells and tissues can be maintained and studied outside the body. In a complex organism, tissues and organs are formed by several kinds of cells. These cells are bathed in fluid derived from blood plasma, which contains many different molecules required for growth. Cell culture has been very helpful in isolating the effects of single molecules on specific types of cells. It also allows the direct observation of the behavior of living cells under a phase contrast microscope. Many experiments that cannot be performed in the living animal can be accomplished *in vitro.*

The cells and tissues are grown in complex solutions of known composition (salts, amino acids, vitamins) to which serum components or specific growth factors are added. In preparing cultures from a tissue or organ, cells must be initially dispersed mechanically or enzymatically. Once isolated, the cells can be cultivated in a clear dish to which they adhere, usually as a single layer of cells (Figure 1–5). Cultures of cells that are isolated in this way are called **primary cell cultures.** Many cell types

once isolated from normal or pathologic tissue have been maintained *in vitro* ever since because they have been immortalized and now constitute a permanent **cell line.** Most cells obtained from normal tissues have a finite, genetically programmed life span. Certain changes, however (some related to oncogenes; see Chapter 3), can promote cell immortality, a process called **transformation,** which are similar to the initial changes in a normal cell's becoming a cancer cell. Because of improvements in culture technology, most cell types can now be maintained in the laboratory. All procedures with living cells and tissues must be performed in a sterile area, using sterile solutions and equipment, to avoid contamination with microorganisms.

As shown in the next chapter, incubation of living cells *in vitro* with a variety of new fluorescent compounds that are sequestered and metabolized in specific compartments of the cell provides a new approach to understanding these compartments both structurally and physiologically. Other histological techniques applied to cultured cells have been particularly important for understanding the locations and functions of microtubules, microfilaments, and other components of the cytoskeleton.

MEDICAL APPLICATION

Cell culture has been widely used for the study of the metabolism of normal and cancerous cells and for the development of new drugs. This technique is

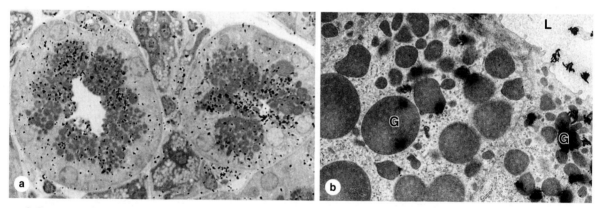

***Figure 1–9.* Autoradiography.** Autoradiographs are tissue preparations in which particles called **silver grains** indicate the regions of cells in which specific macromolecules were synthesized just prior to fixation. Precursors such as nucleotides, amino acids, or sugars with isotopes substituted for specific atoms are provided to the tissues and after a period of incorporation, tissues are fixed, sectioned, and mounted on slides or TEM grids as usual. This processing removes all radiolabeled precursors, leaving only the isotope in the fixed macromolecules. In a darkroom the slides are coated with a thin layer of chemicals like those in photographic film and dried. In a black box the isotope in newly synthesized macromolecules emits radiation exposing the layer of photographic chemicals immediately adjacent to the isotopes' location. The minute regions of exposed chemicals in the photographic layer are revealed as silver grains by "developing" the preparation as if it were film, followed by microscopic examination. Shown here are autoradiographs from the salivary gland of a mouse injected with ^{3}H-fucose 8 h before tissue fixation. Fucose is incorporated into oligosaccharides and the results reveal location of newly synthesized glycoproteins containing such sugars. **(a):** Black "silver grains" are visible over regions with secretory granules and the duct indicating glycoprotein locations. X1500. **(b):** The same tissue prepared for TEM autoradiography shows silver grains with a coiled or amorphous appearance again localized mainly over the granules (G) and in the gland lumen (L). X7500. (Figure 1–9b, with permission, from Ticiano G. Lima and A. Antonio Haddad, School of Medicine, Ribeirão Preto, Brazil.)

also useful in the study of parasites that grow only within cells, such as viruses, mycoplasma, and some protozoa. In cytogenetic research, determination of human karyotypes (the number and morphology of an individual's chromosomes) is accomplished by short-term cultivation of blood cells or fibroblasts and by examining the chromosomes during mitotic division. In addition, cell culture is central to contemporary techniques of molecular biology and recombinant DNA technology.

Because peroxidase is extremely active and rapidly produces an appreciable amount of insoluble precipitate, it is also widely used for an important practical application: tagging other proteins as described in the next section.

HISTOCHEMISTRY & CYTOCHEMISTRY

The terms **histochemistry** and **cytochemistry** indicate methods for localizing cellular structures in tissue sections using unique enzymatic activity present in those structures. To preserve these enzymes histochemical procedures are usually applied to unfixed or mildly fixed tissue, often sectioned on a cryostat to avoid adverse effects of heat and paraffin on enzymatic activity. Enzyme histochemistry usually works in the following way: (1) tissue sections are immersed in a solution that contains the substrate of the enzyme to be localized; (2) the enzyme is allowed to act on its substrate; (3) at this stage or later, the section is put in contact with a marker compound; (4) this compound reacts with a molecule produced by enzymatic action on the substrate; (5) the final reaction product, which must be insoluble and which is visible by light or electron microscopy only if it is colored or electron-dense, precipitates over the site that contains the enzyme. When examining such a section in the microscope, one can see the cell regions (or organelles) covered with a colored or electron-dense material.

Examples of enzymes that can be detected histochemically include the following:

Phosphatases split the bond between a phosphate group and an alcohol residue of phosphorylated molecules. The visible, insoluble reaction product of phosphatases is usually lead phosphate or lead sulfide. Both alkaline phosphatases which have their maximum activity at an alkaline pH and acid phosphatases can be detected (Figure 1–10).

Dehydrogenases remove hydrogen from one substrate and transfer it to another. Like phosphatases, dehydrogenases play an important role in several metabolic processes. They are detected histochemically by incubating nonfixed tissue sections in a substrate solution containing a molecule that receives hydrogen and precipitates as an insoluble colored compound. Mitochondria can be specifically identified by this method, since dehydrogenases are key enzymes in the citric acid (Krebs) cycle of this organelle.

Peroxidase, which is present in several types of cells, promotes the oxidation of certain substrates with the transfer of hydrogen ions to hydrogen peroxide, forming molecules of water. In this method, sections of adequately fixed tissue are incubated in a solution containing hydrogen peroxide and 3,3'-diaminoazobenzidine (DAB). The latter compound is oxidized in the presence of peroxidase, resulting in an insoluble, brown, electron-dense precipitate that permits the localization of peroxidase activity by light and electron microscopy. Peroxidase staining in white blood cells is important in the diagnosis of certain leukemias.

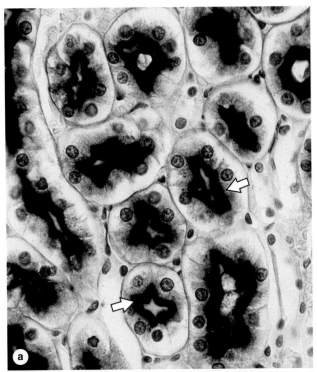

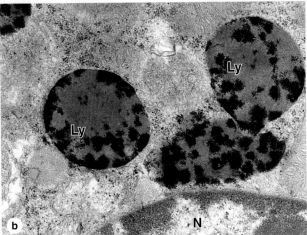

Figure 1–10. Enzyme histochemistry. **(a):** Micrograph of cross sections of kidney tubules treated histochemically by the Gomori method for alkaline phosphatases show strong activity of this enzyme at the apical surfaces of the cells at the lumen of the tubules (arrows). **(b):** TEM image of a kidney cell in which acid phosphatase has been localized histochemically in three lysosomes (Ly) near the nucleus (N). The dark material within these structures is lead phosphate that precipitated in places with acid phosphatase activity. X25,000. (Figure 1–10b, with permission, from Eduardo Katchburian, Department of Morphology, Federal University of Sao Paulo, Brazil.)

DETECTION METHODS USING SPECIFIC INTERACTIONS BETWEEN MOLECULES

A specific macromolecule present in a tissue section may sometimes be identified by using tagged compounds or macromolecules that specifically interact with the material of interest (Figure 1–11). The compounds that will interact with the molecule must be tagged with a label that can be detected under the light or electron microscope. The most commonly used labels are fluorescent compounds (which can be seen with a fluorescence or laser microscope), radioactive atoms (which can be detected with autoradiography), molecules of peroxidase or other enzymes (which can be detected with histochemistry), and metal (usually gold) particles that can be observed with light and electron microscopy. These methods can be used for detecting and localizing specific sugars, proteins, and nucleic acids.

Examples of molecules that interact specifically with other molecules include the following:

Phalloidin is a compound extracted from the mushroom *Amanita phalloides* and interacts strongly with actin. Tagged with fluorescent dyes, phalloidin is commonly used to demonstrate actin filaments in cells.

Protein A is obtained from *Staphylococcus aureus* and binds to the Fc region of immunoglobulin (antibody) molecules. Labeled protein A can therefore be used to localize naturally occurring or applied antibodies bound to cell structures.

Lectins are proteins or glycoproteins, derived mainly from plant seeds and that bind to carbohydrates with high affinity and specificity. Different lectins bind to specific sugars or sequences of sugar residues. Fluorescently labeled lectins are used to stain specific glycoproteins, proteoglycans, and glycolipids and are used to characterize membrane components with specific sequences of sugar residues.

Immunohistochemistry

A highly specific interaction between molecules is that between an antigen and its antibody. For this reason, methods using labeled antibodies have become extremely useful in identifying and localizing many specific proteins, not just those with enzymatic activity that can be demonstrated by histochemistry.

The body's immune cells are able to discriminate its own molecules (self) from foreign ones. When exposed to foreign molecules—called **antigens**—the body responds by producing **antibodies** that react specifically and bind to the antigen, thus helping to eliminate the foreign substance. Antibodies belong to the **immunoglobulin** family of glycoproteins, produced by lymphocytes.

In immunohistochemistry, a tissue section (or cells in culture) that one believes contains the protein of interest is incubated in a solution containing an antibody to this protein. The antibody binds specifically to the protein, whose location in the tissue or cell can then be seen with either the light or electron microscope, depending on the type of compound used to label the antibody. Antibodies are commonly tagged with fluorescent

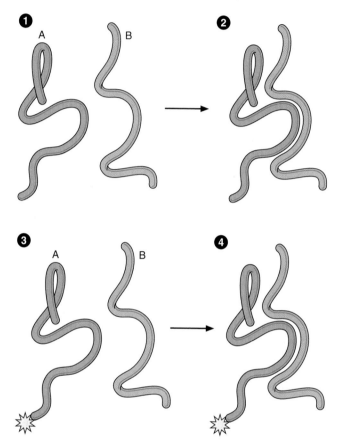

Figure 1–11. **Labeling by specific, high-affinity interactions.** Compounds or macromolecules that have specific affinity toward certain cell or tissue macromolecules can be tagged with a label and used to identify that component and determine its location in cells and tissues. **(1)** Molecule A has a high and specific affinity toward a portion of molecule B. Examples of such interacting macromolecules are an antibody that recognizes specific antigens, usually proteins, or a segment of single-stranded DNA with sequence-specific complementarity to RNA molecules in a cell. Molecule A can also be a small compound like phalloidin, which specifically binds actin filaments, or a protein such as "protein A" which binds all immunoglobulins. **(2)** When A and B are mixed, A binds to the portion of B it recognizes. **(3)** Molecule A may be tagged with a label that can be visualized with a light or electron microscope. The label can be a fluorescent compound, an enzyme such as peroxidase, an electron-dense particle, or a radioisotope. **(4)** If molecule B is present in a cell or extracellular matrix that is incubated with labeled molecule A, molecule B can be detected and localized by visualizing the labeled molecule A bound to it.

compounds, with peroxidase or alkaline phosphatase for histochemical detection, or with electron-dense gold particles.

For immunocytochemistry one must have an antibody against the protein that is to be detected. This means that the protein must have been previously purified using biochemical or molecular approaches so that antibodies against it can be produced. To produce antibodies against protein *x* of a certain animal species (eg, a human or rat), the protein is first isolated and then injected into an animal of another species (eg, a rabbit or a goat). If the protein's amino acid sequence is sufficiently different for this animal to recognize it as foreign—that is, as an antigen—the animal will produce antibodies against the protein.

Different groups (clones) of lymphocytes in the animal that was injected recognize different parts of protein *x* and each clone produces an antibody against that part. These antibodies are collected from the animal's plasma and constitute a mixture of **polyclonal antibodies**, each capable of binding a different region of protein *x*.

It is also possible, however, to inject protein *x* into a mouse and then days later to isolate the activated lymphocytes and place them into culture. Growth and activity of these cells can be prolonged indefinitely by fusing them with lymphocytic tumor cells to produce **hybridoma cells.** Different hybridoma clones produce different antibodies against the several parts of protein *x* and each clone can be isolated and cultured separately so that the different antibodies against protein *x* can be collected separately. Each of these antibodies is a **monoclonal antibody.** An advantage to using a monoclonal antibody rather than polyclonal antibodies is that it can be selected to be highly specific and to bind strongly to the protein to be detected, producing less non-specific binding to other proteins similar to the one of interest.

In the **direct method of immunocytochemistry,** the antibody (either monoclonal or polyclonal) is tagged itself with an appropriate label. A tissue section is incubated with the antibody for some time so that the antibody interacts with and binds to protein *x*. The section is then washed to remove the unbound antibody, processed by the appropriate method and examined microscopically to study the location or other aspects of protein *x* (Figure 1–12).

The **indirect method of immunocytochemistry** is more sensitive but requires two antibodies and additional steps. Instead of labeling the (primary) antibody specific for protein *x*, the detectible tag is conjugated to a **secondary antibody** made in a different "foreign" species against the immunoglobulin class to which the primary antibody belongs. For example, primary antibodies made by mouse lymphocytes (such as most monoclonal antibodies) are specifically bound by rabbit anti-mouse antibodies.

The indirect immunocytochemical detection is performed by initially incubating a section of a human tissue believed to contain protein *x* with mouse anti-*x* antibody. After washing, the tissue sections are incubated with labeled rabbit or goat antibody against mouse antibodies. These secondary antibodies will recognize the mouse antibody that had recognized protein *x* (Figure 1–12). Protein *x* can then be detected by using a microscopic technique appropriate for the label used for the secondary antibody. There are other indirect methods that involve the use of other intermediate molecules, such as the biotin-avidin technique.

Examples of indirect immunocytochemistry are shown in Figure 1–13, demonstrating the use of labeling methods with cells in culture or after sectioning for both light microscopy and TEM.

MEDICAL APPLICATION

Immunocytochemistry has contributed significantly to research in cell biology and to the improvement of medical diagnostic procedures. Table 1–1 shows some of the routine applications of immunocytochemical procedures in clinical practice.

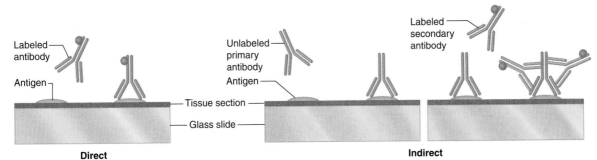

Figure 1–12. **Immunocytochemistry.** Immunocytochemistry (or immunohistochemistry) can be direct or indirect. **Direct immunocytochemistry** uses an antibody made against the tissue protein of interest and tagged directly with a label such as a fluorescent compound or peroxidase. When placed with the tissue section on a slide, these labeled antibodies bind specifically to the protein (antigen) against which they were produced and can be visualized by the appropriate method. The more widely used technique of **indirect immunocytochemistry** uses two different antibodies. A **primary antibody** is made against the protein (antigen) of interest and applied to the tissue section first to bind its specific antigen. Then a **labeled secondary antibody** is obtained that was (1) made in another vertebrate species against immunoglobulin proteins (antibodies) from the species in which the primary antibodies were made and then (2) labeled with a fluorescent compound or peroxidase. When this labeled secondary antibody is applied to the tissue section it specifically binds the primary antibodies, indirectly labeling the protein of interest on the slide. Since more than one labeled secondary antibody can bind each primary antibody molecule, labeling of the protein of interest is amplified by the indirect method.

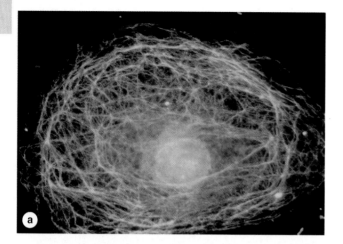

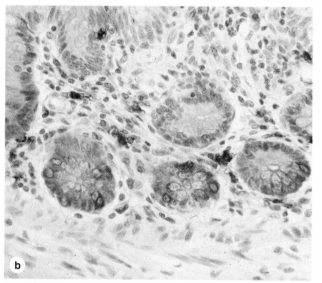

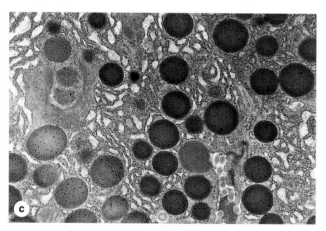

Figure 1–13. Cells and tissues stained by immunohisto-chemistry. Immunocytochemical methods to localize specific proteins in cells can be applied to either light microscopic or TEM preparations using a variety of labels. **(a):** A decidual cell grown *in vitro* stained to reveal a mesh of intermediate filaments throughout the cytoplasm. Primary antibodies against the protein desmin, which forms these intermediate filaments, and FITC-labeled secondary antibodies were used in an indirect immunofluorescence technique. The nucleus is

Table 1–1. Many pathologic conditions are diagnosed by localizing specific markers of the disorder using antibodies against those antigens in immuno-histochemical staining.

Antigens	Diagnosis
Specific cytokeratins	Tumors of epithelial origin
Protein and polypeptide hormones	Protein or polypeptide hormone–producing endocrine tumors
Carcinoembryonic antigen (CEA)	Glandular tumors, mainly of the digestive tract and breast
Steroid hormone receptors	Breast duct cell tumors
Antigens produced by viruses	Specific virus infections

Hybridization Techniques

The central challenge in modern cell biology is to understand the workings of the cell in molecular detail. This goal requires techniques that permit analysis of the molecules involved in the process of information flow from DNA to protein. Many techniques are based on **hybridization.** Hybridization is the binding between two single strands of nucleic acids (DNA with DNA, RNA with RNA, or RNA with DNA) that recognize each other if the strands are complementary. The greater the similarities of the sequences, the more readily complementary strands form "hybrid" double-strand molecules. Hybridization thus allows the specific identification of sequences of DNA or RNA. This is commonly performed with nucleic acids in solution, but hybridization also occurs when solution of nucleic acid are applied directly to cells and tissue sections, a procedure called ***in situ* hybridization** (ISH).

This technique is ideal for (1) determining if a cell has a specific sequence of DNA (such as a gene or part of a gene), (2) identifying the cells containing specific mRNAs (in which the corresponding gene is being transcribed), or (3) determining the localization of a gene in a specific chromosome. DNA and RNA

counterstained light blue with DAPI. **(b):** A section of small intestine stained with an antibody against the enzyme lysozyme. The secondary antibody labeled with peroxidase was then applied and the localized brown color produced histochemically with the peroxidase substrate DAB. The method demonstrates lysozyme-containing structures in scattered macrophages and in the clustered Paneth cells. Nuclei were counterstained with hematoxylin. **(c):** A section of pancreatic acinar cells in a TEM preparation incubated with an antibody against the enzyme amylase antibody and then with protein A coupled with gold particles. Protein A has high affinity toward antibody molecules and the resulting image reveals the presence of amylase with the gold particles localized as very small black dots over dense secretory granules and developing granules (left). With specificity for immunoglobulin molecules, labeled protein A can be used to localize any primary antibody. (Figure 1–13c, with permission, from Moise Bendayan, Departments of Pathology and Cell Biology, University of Montreal.)

of the cells must be initially denatured by heat or other agents to become completely single-stranded. They are then ready to be hybridized with a segment of single-stranded DNA or RNA (called a **probe**) that is complementary to the sequence one wishes to detect. The probe may be obtained by cloning, by PCR amplification of the target sequence, or by chemical synthesis if the desired sequence is short. The probe is tagged with nucleotides containing a radioactive isotope (which can be localized by autoradiography) or modified with a small compound such as digoxygenin (which can be identified by immunocytochemistry). A solution containing the probe is placed over the specimen for a period of time necessary for hybridization. After washing off the excess unbound probe, the localization of the hybridized probe is revealed through its label (Figure 1–14).

PROBLEMS IN THE STUDY OF TISSUE SECTIONS

A key point to be remembered in studying and interpreting stained tissue sections is that microscope preparations are the end result of a series of processes that began with collecting the tissue and ended with mounting a coverslip on the slide. Several

steps of this procedure may distort the tissues, producing minor structural abnormalities called **artifacts.** Structures seen microscopically then may differ slightly from the structures present when they were alive.

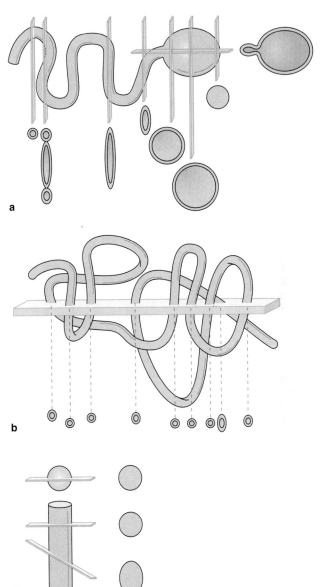

Figure 1–15. Interpretation of 3-D structures in 2-D tissue sections. Three-dimensional structures appear to have only two dimensions in thin sections. **(a):** Sections through a hollow swelling on a tube produce large and small circles, oblique sections through bent regions of the tube produce ovals of various dimensions. **(b):** A single section through a highly coiled tube shows many small, separate round or oval sections. On first observation it may be difficult to realize that these represent a coiled tube, but it is important to develop such interpretive skill in understanding histological preparations. **(c):** Round structures in sections may be portions of either spheres or cylinders. Additional sections or the appearance of similar nearby structures help reveal a more complete picture.

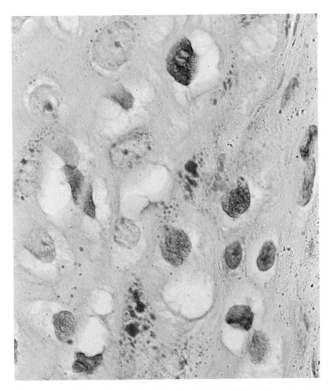

Figure 1–14. Cells stained by *in situ* **hybridization.** *In situ* hybridization shows that many of the epithelial cells in this section of a genital wart contain the human papillomavirus (HPV), which causes this benign proliferative condition. The section was incubated with a solution containing a digoxygenin-labeled cDNA probe for the HPV DNA. The probe was then visualized by direct immunohistochemistry using peroxidase-labeled antibodies against digoxygenin. This procedure stains brown only those cells containing HPV. X400. H&E counterstain. (With permission, from Jose E. Levi, Virology Lab, Institute of Tropical Medicine, University of Sao Pāulo, Brazil.)

One such distortion is minor shrinkage of cells or tissue regions produced by the fixative, by the ethanol, or by the heat needed for paraffin embedding. Shrinkage can produce the appearance of artificial spaces between cells and other tissue components. Another source of artificial spaces is the loss of molecules such as lipids, glycogen, or low molecular weight substances that are not kept in the tissues by the fixative or removed by the dehydrating and clearing fluids. Slight cracks in sections also appear as large spaces in the tissues.

Other artifacts may include wrinkles of the section (which may be confused with linear structures such as blood capillaries) and precipitates of stain (which may be confused with cellular structures such as cytoplasmic granules). Students must be aware of the existence of artifacts and able to recognize them.

Another point to remember in studying histological sections is the impossibility of differentially staining all tissue components on a slide stained by a single procedure. With the light microscope it is necessary to examine several preparations stained by different methods to obtain an idea of the tissue's complete composition and structure. The TEM, on the other hand, allows the observation of cells with all organelles and inclusions, surrounded by the components of the ECM.

Finally, when a **three-dimensional tissue** volume is cut into very thin sections, the sections appear microscopically to have only two dimensions: length and width. When examining a section under the microscope, one must always keep in mind that something may be missing in front of or behind that section because many tissue structures are thicker than the section. Round structures seen microscopically may be sections through spheres or cylinders and tubes in cross-section look like rings (Figure 1–15). Also since structures within a tissue have different orientations, their two-dimensional appearance will vary depending on the plane of section. A single convoluted tube will appear histologically as several rounded structures.

To understand the architecture of an organ, one often must study sections made in different planes. Examining many parallel sections (**serial sections**) and reconstructing the images three-dimensionally provides better understanding of a complex organ or organism.

The Cytoplasm

CELL DIFFERENTIATION
CYTOPLASMIC ORGANELLES
    Plasma Membrane
    Mitochondria
    Ribosomes
    Endoplasmic Reticulum
    Golgi Apparatus
    Secretory Vesicles or Granules

    Lysosomes
    Proteasomes
    Peroxisomes or Microbodies
THE CYTOSKELETON
    Microtubules
    Microfilaments (Actin Filaments)
    Intermediate Filaments
INCLUSIONS

Cells and extracellular material together comprise all the tissues that make up the organs of multicellular animals. In all tissues, cells themselves are the basic structural and functional units, the smallest living parts of the body. Animal cells are **eukaryotic** (Gr. *eu*, good, + *karyon*, nucleus), with distinct membrane-limited nuclei surrounded by cytoplasm containing many varied membrane-limited organelles. In contrast the small prokaryotic cells of bacteria typically have a cell wall around the plasmalemma, lack other membranous structures including an envelope around the genetic material (DNA). Different cells of the animal become specialized by concentrating specific organelles and greatly developing specific cellular activities which can generally be found to more limited extents in all animal cells.

CELL DIFFERENTIATION

The human organism presents about 200 different cell types, all derived from the zygote, the single cell formed by fertilization of an oocyte with a spermatozoon. The first cellular divisions of the zygote produce cells called **blastomeres** and as part of the **inner cell mass** blastomeres give rise to all tissue types of the adult. Explanted to tissue culture such cells have been termed **embryonic stem cells**. During their specialization process, called **cell differentiation,** the cells synthesize specific proteins, change their shape, and become very efficient in specialized functions. For example, muscle cell precursors elongate into fiber-like cells that synthesize and accumulate large arrays of actin and myosin. The resulting cell is specialized to efficiently convert chemical energy into contractile force.

The main cellular functions performed by specialized cells in the body are listed in Table 2–1. It is important to understand that the functions listed there can be performed by most cells of the body; specialized cells have greatly expanded their capacity for one or more functions during differentiation.

The body's cells can experience various environments under both normal and pathological conditions and the same cell type can exhibit different characteristics and behaviors in different regions and circumstances. Thus, macrophages and neutrophils (both of which are phagocytic defense cells) will shift from oxidative metabolism to glycolysis in an anoxic, inflammatory environment. Cells that appear to be structurally similar may react in different ways because they have different families of receptors for signaling molecules such as hormones and extracellular matrix macromolecules. For example, because of their diverse library of receptors, breast fibroblasts and uterine smooth muscle cells are exceptionally sensitive to female sex hormones while most other fibroblasts and smooth muscle cells are insensitive.

CYTOPLASMIC ORGANELLES

The cell is composed of two basic parts: **cytoplasm** (Gr. *kytos*, cell, + *plasma,* thing formed) and **nucleus** (L. *nux*, nut). Individual cytoplasmic components are usually not clearly distinguishable in common hematoxylin-and-eosin–stained preparations; the nucleus, however, appears intensely stained dark blue or black.

The outermost component of the cell, separating the cytoplasm from its extracellular environment, is the **plasma membrane (plasmalemma)**. However, although the plasma membrane defines the external limit of the cell, a continuum exists between the interior of the cell and extracellular macromolecules. The plasma membrane contains proteins called **integrins** that are linked to both cytoplasmic cytoskeletal filaments and extracellular matrix components. Through these linkages there is a constant exchange of influences, in both directions, between the extracellular matrix and the cytoplasm. The cytoplasm itself is composed of a fluid component, or **cytosol,** in which are contained metabolically active structures, the **organelles**, which can

Table 2–1. Cellular functions in some specialized cells.

Function	Specialized Cell(s)
Movement	Muscle and other contractile cells
Form adhesive and tight junctions between cells	Epithelial cells
Synthesize and secrete components of the extracellular matrix	Fibroblasts, cells of bone and cartilage
Convert physical and chemical stimuli into action potentials	Neurons and sensory cells
Synthesis and secretion of enzymes	Cells of digestive glands
Synthesis and secretion of mucous substances	Mucous-gland cells
Synthesis and secretion of steroids	Some adrenal gland, testis, and ovary cells
Ion transport	Cells of the kidney and salivary gland ducts
Intracellular digestion	Macrophages and some white blood cells
Lipid storage	Fat cells
Metabolite absorption	Cells lining the intestine

be membranous (such as mitochondria) or non-membranous protein complexes (such as ribosomes and proteasomes). The shape and motility of eukaryotic cells are determined by components of the **cytoskeleton**. Other minor cytoplasmic structures are **inclusions** which are generally deposits of carbohydrates, lipids, or pigments.

The **cytosol** contains hundreds of enzymes, such as those of the glycolytic pathway, that produce building blocks for larger molecules and break down small molecules to liberate energy. All the machinery converging on the ribosomes for protein synthesis (mRNA, transfer RNA, enzymes, and other factors) is also contained within the cytosol. Oxygen, CO_2, electrolytic ions, low molecular weight substrates, metabolites, waste products, etc all diffuse through the cytosol, either freely or bound to proteins, passing to or leaving the organelles where they are used or produced.

Plasma Membrane

All eukaryotic cells are enveloped by a limiting membrane composed of phospholipids, cholesterol, proteins, and chains of oligosaccharides covalently linked to phospholipid and protein molecules. The plasma, or cell, membrane functions as a selective barrier that regulates the passage of certain materials into and out of the cell and facilitates the transport of specific molecules. One important role of the cell membrane is to keep constant the ion content of cytoplasm, which is different from that of extracellular fluid. Membranes also carry out a number of specific recognition and regulatory functions (discussed later in this section), playing an important role in the interactions of the cell with its environment.

Membranes range from 7.5 to 10 nm in thickness and consequently are visible only in the electron microscope. The line between adjacent cells sometimes seen with the light microscope is formed by plasma membrane proteins of the cells plus extracellular material, which together can reach a dimension visible by light microscopy. Electron micrographs reveal that the plasmalemma—and, for that matter, all other organellar membranes—exhibit a trilaminar structure after fixation in osmium tetroxide (Figure 2–1). Because all membranes have this appearance, the 3-layered structure was designated the **unit membrane** (Figure 2–1).

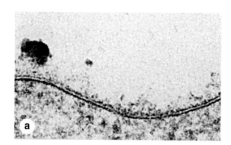
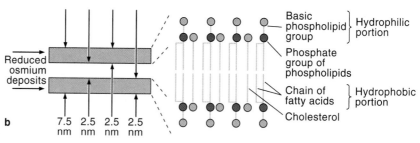

Figure 2–1. Membrane structure. **(a):** A TEM of a sectioned cell surface shows the trilaminar **unit membrane** with two dark (electron-dense) lines enclosing a clear (electron-lucent) band. These three layers of the unit membrane correspond to reduced osmium deposited on the hydrophilic phosphate groups present on each side of the internal bilayer of fatty acids where osmium is not deposited. The "fuzzy" material on the outer surface of the membrane represents the **glycocalyx** of oligosaccharides attached to phospholipids and proteins. Components of the glycocalyx are important for cell-cell recognition in many biological processes and for adsorption and uptake of many molecules by cells. X100,000. **(b):** Schematic drawing depicts the trilaminar ultrastructure (left) and molecular organization (right) of the lipid bilayer in a cell membrane. The shaded bands at left represent the two dense layers observed in the TEM caused by the deposit of osmium in the hydrophilic portions of the phospholipid molecules. The right side of the diagram shows the orientation of the phospholipids that form the bilayer of biological membranes. The **hydrophilic polar heads** of the phospholipids are directed toward each surface of the membrane, in direct contact with water, and the **hydrophobic nonpolar fatty acid chains** of the phospholipids are buried in the middle, away from water. Cholesterol molecules are interspersed throughout the lipid bilayer, affecting the packing and fluidity of the fatty acid chains.

Membrane phospholipids, such as phosphatidylcholine (lecithin), consist of two non-polar (hydrophobic or water-repelling) long-chain fatty acids linked to a charged polar (hydrophilic or water-attracting) head group. Cholesterol is also present, often at nearly a 1:1 ratio with the phospholipids in plasma membranes. Membrane phospholipids are most stable when organized into a double layer (bilayer) with their hydrophobic fatty acid chains directed toward the middle away from water and their hydrophilic polar heads directed outward to contact water on both sides (Figure 2–1). Cholesterol molecules insert among the close packed the phospholipid fatty acids, restricting their movement, and thus modulate the fluidity and movement of all membrane components. The lipid composition of each half of the bilayer is different. For example, in red blood cells phosphatidylcholine and sphingomyelin are more abundant in the outer half of the membrane, whereas phosphatidylserine and phosphatidylethanolamine are more concentrated in the inner half. Some of the lipids, known as glycolipids, possess oligosaccharide chains that extend outward from the surface of the cell membrane and thus contribute to the lipid asymmetry (Figures 2–2a and 2–3).

Proteins, which are a major molecular constituent of membranes (~50% w/w in the plasma membrane), can be divided into two groups. **Integral proteins** are directly incorporated

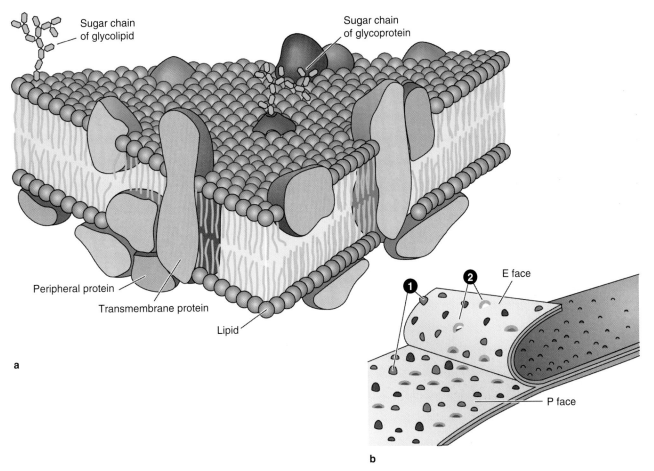

Sugar chain of glycolipid

Sugar chain of glycoprotein

Peripheral protein

Transmembrane protein

Lipid

E face

P face

a

b

Figure 2–2. **The fluid mosaic model of membrane structure. (a):** The fluid mosaic model emphasizes that a membrane consisting of a phospholipid bilayer also contains proteins inserted in it or bound to the cytoplasmic surface (peripheral proteins) and that many of these proteins move within the fluid lipid phase. **Integral proteins** are firmly embedded in the lipid layers. Other proteins completely span the bilayer and are called **transmembrane proteins.** Hydrophobic amino acids of the integral membrane protein interact with the hydrophobic fatty acid portions of the membrane. Both the proteins and lipids may have externally exposed oligosaccharide chains. When cells are frozen and fractured (**cryofracture**), the lipid bilayer of membranes is often cleaved along the hydrophobic center.

(b): Membrane splitting occurs along the line of weakness formed by the fatty acid tails of membrane phospholipids, since only weak hydrophobic interactions bind the halves of the membrane along this line. Electron microscopy of cryofracture preparation replicas is a useful method of studying membranous structures. Most of the protruding membrane particles seen (**1**) are proteins or aggregates of proteins that remain attached to the half of the membrane adjacent to the cytoplasm (the P or protoplasmic face). Fewer particles are found attached to the outer half of the membrane (E or extracellular face). For every protein particle that bulges on one surface, a corresponding depression (**2**) appears in the opposite surface.

within the lipid bilayer itself, whereas **peripheral proteins** exhibit a looser association with one of the two membrane surfaces. The loosely bound peripheral proteins can be easily extracted from cell membranes with salt solutions, whereas integral proteins can be extracted only by drastic methods using detergents to disrupt lipids. Some integral proteins span the membrane one or more times, from one side to the other. Accordingly, they are called **one-pass** or **multipass transmembrane proteins** (Figure 2–3). Many integral and peripheral proteins that function as components of large enzyme complexes are located in specialized patches of membrane having higher concentrations of cholesterol. Within these regions called **lipid rafts** membrane fluidity is reduced, allowing the associated proteins to remain in closer proximity and interact more efficiently.

Freeze-fracture electron microscope studies of membranes show that many integral proteins are only partially embedded in the lipid bilayer and protrude from either the outer or inner surface (Figure 2–2b). Transmembrane proteins are large enough to extend across the two lipid layers and may protrude from both

membrane surfaces. The carbohydrate moieties of the glycoproteins and glycolipids project from the external surface of the plasma membrane; they are important components of specific molecules called **receptors** that participate in important interactions such as cell adhesion, recognition, and response to protein hormones. As with lipids, the distribution of membrane proteins is different in the two surfaces of the cell membranes. Therefore, all membranes in the cell are asymmetric.

Integration of the proteins within the lipid bilayer is mainly the result of hydrophobic interactions between the lipids and nonpolar amino acids present on the outer region of the proteins. Some membrane proteins are not bound rigidly in place and are able to move within the plane of the cell membrane (Figure 2–4). However, unlike lipids, most membrane proteins are restricted in their lateral diffusion by attachment to cytoskeletal components. In most epithelial cells, tight junctions (see Chapter 4) also restrict lateral diffusion of unattached transmembrane proteins and outer layer lipids to specific membrane domains.

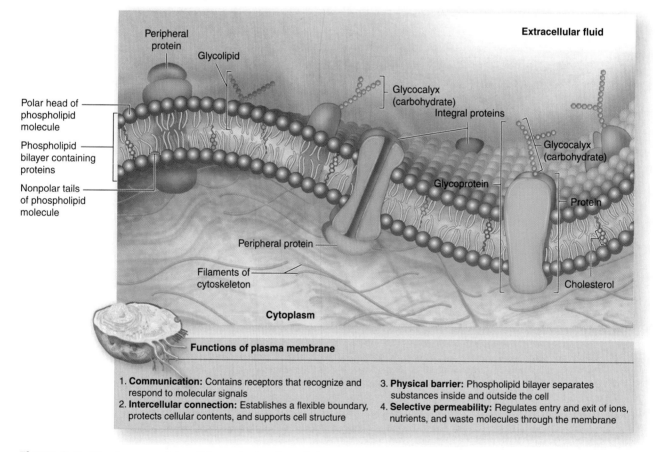

Extracellular fluid

Peripheral protein

Glycolipid

Glycocalyx (carbohydrate)

Integral proteins

Polar head of phospholipid molecule

Phospholipid bilayer containing proteins

Nonpolar tails of phospholipid molecule

Glycocalyx (carbohydrate)

Glycoprotein

Protein

Peripheral protein

Filaments of cytoskeleton

Cholesterol

Cytoplasm

Functions of plasma membrane

1. **Communication:** Contains receptors that recognize and respond to molecular signals
2. **Intercellular connection:** Establishes a flexible boundary, protects cellular contents, and supports cell structure
3. **Physical barrier:** Phospholipid bilayer separates substances inside and outside the cell
4. **Selective permeability:** Regulates entry and exit of ions, nutrients, and waste molecules through the membrane

Figure 2–3. **Membrane proteins.** Schematic drawing of plasma membrane structure shows a globular peripheral protein on the external face of the membrane and two integral transmembrane proteins. **One-pass** transmembrane proteins have single hydrophobic regions along the length of amino acids and for maximal stability this becomes buried in the internal region of the lipid bilayer. **Multipass** transmembrane proteins have several hydrophobic amino acid sequences all buried in the bilayer, with terminal and intervening hydrophilic sequences exposed at either the external or cytoplasmic face of the membrane. Many physiologically important membrane proteins, including ion pumps and channels, are multipass proteins.

Such observations and data from biochemical, electron microscopic and other studies showed both the mosaic disposition of membrane proteins and the fluid nature of the lipid bilayer and led to the well-established **fluid mosaic model** for membrane structure (Figure 2–2a). Membrane proteins are synthesized in the rough endoplasm reticulum, are modified and completed in the Golgi apparatus, and transported in vesicles to the cell surface (Figure 2–5).

In the TEM the external surface of the cell shows a fuzzy carbohydrate-rich region called the **glycocalyx** (Figure 2–1). This layer is made of carbohydrate chains linked to membrane proteins and lipids and of cell-secreted glycoproteins and proteoglycans. The glycocalyx has a role in cell recognition and attachment to other cells and to extracellular molecules. The plasma membrane is the site where materials are exchanged between the cell and its environment. Some ions, such as Na^+, K^+, and Ca^{2+}, cross the cell membrane by passing through integral membrane proteins. This can involve **passive diffusion** through **ion channels** or **active transport** via **ion pumps** using energy from the breakdown of adenosine triphosphate (ATP).

ENDOCYTOSIS

Bulk uptake of material also occurs across the plasma membrane in a general process called **endocytosis**, which involves folding and fusion of this membrane to form **vesicles** which enclose the material transported. Cells show three general types of endocytosis (Figure 2-6).

1. Phagocytosis. Phagocytosis literally means "cell eating." Certain white blood cells, such as macrophages and neutrophils, are specialized for engulfing and removing particulate matter such as bacteria, protozoa, dead cells, and unneeded extracellular constituents. When a bacterium becomes bound to the surface of a neutrophil, cytoplasmic processes of the cell are extended and ultimately surround the bacterium. The membranes of these processes meet and fuse, enclosing the bacterium in an intracellular vacuole, a **phagosome.**

2. Fluid-phase Endocytosis. In fluid-phase pinocytosis ("cell drinking"), with a mechanism comparable to that of phagocytosis, smaller invaginations of the cell membrane form and entrap extracellular fluid and anything it has in solution. **Pinocytotic vesicles** (about 80 nm in diameter) pinch off inwardly from the cell surface. In most cells such vesicles usually fuse with lysosomes (see the section on lysosomes later in this chapter). In the lining cells of capillaries (endothelial cells), however, pinocytotic vesicles may move to the cell surface opposite their origin. There they fuse with the plasma membrane and release their contents outside the cell, thus accomplishing bulk transfer of material across the cell. This process is termed **transcytosis.**

3. Receptor-mediated Endocytosis. Receptors for many substances, such as low-density lipoproteins and protein hormones, are integral proteins of the cell membrane. Binding of the ligand (a molecule with high affinity for a receptor) to its receptor causes widely dispersed receptors to aggregate in special membrane regions called **coated pits**. The electron-dense coating on the cytoplasmic surface of the membrane is composed of several polypeptides, the major one being clathrin. In a developing coated pit clathrin molecules interact like the struts in a geodesic dome, forming that region of cell membrane into a cage-like invagination that is pinched off into the cytoplasm, forming a **coated vesicle** (Figure 2–7) carrying the ligand and its receptor.

In all these endocytotic processes, the vesicles or vacuoles produced quickly enter and fuse with the **endosomal compartment**, a dynamic system of membranous vesicles (Figure 2–7) and tubules located in the cytoplasm near the cell surface (**early endosomes**) or deeper in the cytoplasm (**late endosomes**). The clathrin molecules separated from the coated vesicles recycle to the cell membrane to participate in the formation of new coated pits.

Figure 2–4. **Experiment demonstrating the fluidity of membrane proteins. (a, b):** Two types of cells were grown in tissue cultures, one with fluorescently labeled transmembrane proteins in the plasmalemma (right) and one without. Cells of each type were fused together into hybrid cells through the action of Sendai virus. **(c):** Minutes after the fusion of the cell membranes, the fluorescent proteins of the labeled cell spreads to the entire surface of the hybrid cells. Such experiments provide important data in support of the fluid mosaic model. However, in many cells most transmembrane proteins show very restricted lateral movements along the cell membrane and are anchored in place by other proteins linking them to the cytoskeleton.

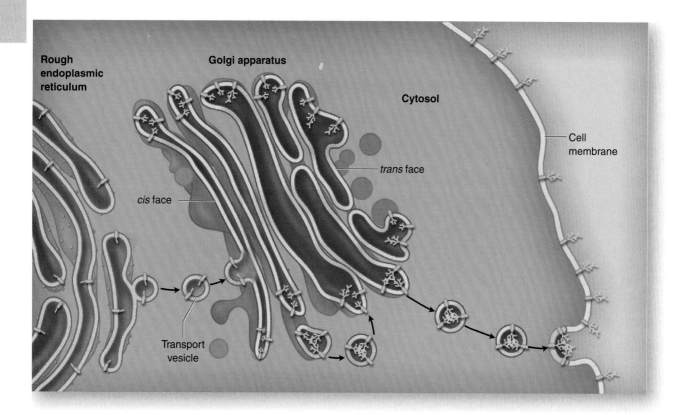

Figure 2–5. Formation and maturation of cell membrane proteins. Membrane proteins of the plasmalemma are synthesized in the rough endoplasmic reticulum and then move in **transport vesicles** to a **Golgi apparatus**, another cytoplasmic structure with several flattened membrane saccules or cisternae. While in the Golgi apparatus, the oligosaccharide chains are added (glycosylation) to many membrane proteins by enzymes in the Golgi saccules. When glycosylation and other posttranslational modifications are complete, the mature membrane proteins are isolated within vesicles that leave the Golgi apparatus. These vesicles move to the cell membrane and fuse with it, thus incorporating the new membrane proteins (along with the lipid bilayer of the vesicle) into the cell membrane.

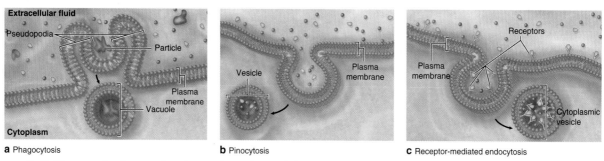

Figure 2–6. Three major forms of endocytosis. Endocytosis is a process in which a cell takes in material from the extracellular fluid using dynamic movements and fusion of the cell membrane to form cytoplasmic, membrane-enclosed structures containing the material. Such cytoplasmic structures formed during endocytosis fall into the general category of **vesicles** or **vacuoles**. **(a): Phagocytosis** involves the extension from the cell of large folds called pseudopodia which engulf particles, for example bacteria, and then internalize this material into a cytoplasmic vacuole or **phagosome**. **(b):** In **pinocytosis** the cell membrane invaginates (dimples inward) to form a pit containing a drop of extracellular fluid. The pit pinches off inside the cell when the cell membrane fuses and forms a pinocytotic vesicle containing the fluid. **(c): Receptor-mediated endocytosis** includes membrane proteins called receptors which bind specific molecules (ligands). When many such receptors are bound by their ligands, they aggregate in one membrane region which then invaginates and pinches off to create vesicle or **endosome** containing both the receptors and the bound ligands.

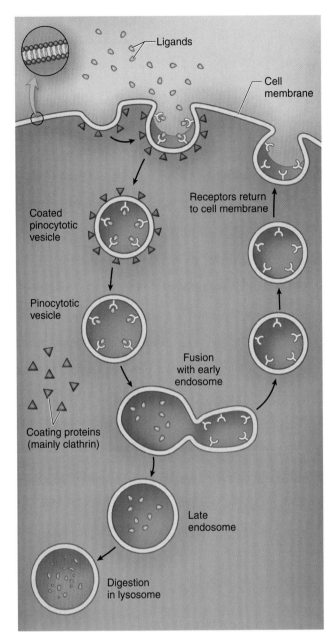

Figure 2–7. Endocytosis and membrane trafficking. Ligands, such as hormones and growth factors, are internalized by receptor-mediated endocytosis, which is mediated by the cytoplasmic peripheral membrane protein **clathrin** or other proteins which promote invagination and temporarily coat the newly formed vesicles. Such **coated vesicles** can be identified by TEM. After detachment of the coating molecules, the vesicle fuses with one or more vesicles of the endosomal compartment, where the ligands detach from their receptors and are sorted into other vesicles. Vesicles of membrane with empty receptors return to the cell surface and after fusion the receptors are ready for reuse. Vesicles containing the free ligands typically fuse with lysosomes, as discussed below. The cytoskeleton with associated motor proteins is responsible for all such directional movements of vesicles.

The membrane of endosomes contains ATP-driven H^+ pumps that acidify their interior.

While phagosomes and pinocytotic vesicles soon fuse with lysosomes, molecules penetrating the endosomal compartment after receptor-mediated endocytosis may take more than one pathway (Figure 2–7). The acidic pH of early endosomes causes many ligands to uncouple from their receptors, after which the two molecules are sorted into separate vesicles. The receptors may be returned to the cell membrane to be reused. Low-density lipoprotein receptors for example (Figure 2–8) are recycled several times. The ligands typically are transferred to late endosomes. However, some ligands are returned to the extracellular milieu with their receptors and both are used again. An example of this activity is the iron-transport protein transferrin: ferric atoms dissociate from the carrier at low endosomal pH and both apotransferrin and the receptor return to the cell surface. Late endosomes most commonly fuse with lysosomes for degradation of their contents.

EXOCYTOSIS

In **exocytosis** a membrane-limited cytoplasmic vesicle fuses with the plasma membrane, resulting in the release of its contents into the extracellular space without compromising the integrity of the plasma membrane (Figure 2–6). Often exocytosis of stored products from epithelial cells occurs specifically at the apical domains of cells, such as in the exocrine pancreas and the salivary glands (see Chapter 4). The fusion of membranes during exocytosis is a highly regulated process involving interactions between several specific membrane proteins. Exocytosis is triggered in many cells by transient increase in cytosolic Ca^{2+}.

During endocytosis, portions of the cell membrane become endocytotic vesicles; during exocytosis, the membrane is returned to the cell surface. This process of membrane movement and recycling is called **membrane trafficking** (Figures 2–7 and 2–8). Trafficking and sorting of membrane components occur continuously in most cells and are not only crucial for cell maintenance but also physiologically important in processes such as reducing blood lipid levels.

SIGNAL RECEPTION AND TRANSDUCTION

Cells in a multicellular organism need to communicate with one another to regulate their development into tissues, to control their growth and division, and to coordinate their functions. Many cells form communicating junctions that couple adjacent cells and allow the exchange of ions and small molecules (see Chapter 4). Through these channels, also called gap junctions, signals may pass directly from cell to cell without reaching the extracellular fluid.

Soluble extracellular signaling molecules bind receptor proteins only found on their **target cells**. Each cell type in the body contains a distinctive set of receptor proteins that enable it to respond to a complementary set of signaling molecules in a specific, programmed way (Figure 2–9). Such signaling can take different routes:

- In **endocrine signaling,** the signal molecules (called hormones) are carried in the blood to target cells throughout the body.
- In **paracrine signaling,** the chemical mediators are rapidly metabolized so that they act only on local cells very close to the source.
- In **synaptic signaling,** a special kind of paracrine interaction, neurotransmitters act only on adjacent cells through special contact areas called **synapses** (see Chapter 9).

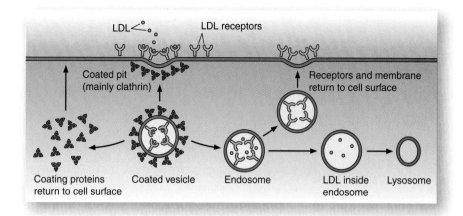

Figure 2–8. Internalization of low-density lipoproteins. Endocytosis of low-density lipoproteins (LDL) is an important mechanism that keeps the concentration of LDL in extracellular body fluids low and is a well-studied example of endocytosis and membrane trafficking. LDL, which is often rich in cholesterol, binds with high affinity to its specific receptors in the cell membranes. This binding activates the formation of clathrin-coated endocytotic pits that form coated vesicles. The vesicles soon lose their coat proteins, which return to the inner surface of the plasmalemma. The uncoated vesicles fuse with endosomes and the free LDL and the receptors are sorted into separate vesicles. Receptors are returned to the cell surface and the LDL is transferred to lysosomes for digestion and separation of their components to be utilized by the cell.

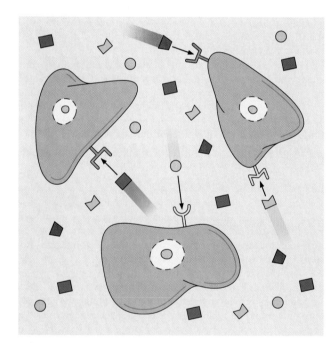

Figure 2–9. Receptors and their ligands. Cells respond to certain external chemical signals that act as ligands, such as hormones and lipoproteins, according to the library of receptors they have. Such receptors are always proteins, typically transmembrane proteins. In this schematic representation, three cells appear with different receptors. The extracellular environment is shown to contain several ligands, which can interact only with the appropriate specific receptors. Considering that the extracellular environment contains a multitude of molecules, it is important that ligands and the respective receptors exhibit complementary morphology and great binding affinity.

- In **autocrine signaling**, signals bind receptors on the same cell type that produced the messenger molecule.
- In **juxtacrine signaling**, important in early embryonic tissue interactions, signaling molecules remain part of a cell's surface and bind surface receptors of the target cell when the two cells make direct physical contact.

Hydrophilic signaling molecules, including most hormones, local chemical mediators (paracrine signals), and neurotransmitters activate receptor proteins on the surface of target cells. These receptors, often transmembrane proteins, relay information to a series of intracellular intermediaries that ultimately pass the signal (**first messenger**) to its final destination in either the cytoplasm or the nucleus in a process called **signal transduction**. One of the best-studied classes of such intermediary proteins, the **G proteins,** binds guanine nucleotides and acts on other membrane-bound intermediaries called **effectors** which propagate the signal further into the cell (Figure 2–10). Effector proteins are usually ion channels or enzymes that generate large quantities of small **second messenger** molecules, such as 1,2-diacyglycerol (DAG), cyclic adenosine monophosphate (cAMP), and inositol 1,4,5-triphosphate (IP_3). The ions or second messengers diffuse through the cytoplasm, amplifying the first signal and triggering a cascade of molecular reactions that lead to changes in gene expression or cell behavior.

MEDICAL APPLICATION

Several diseases have been shown to be caused by defective receptors. For example, pseudohypoparathyroidism and a type of dwarfism are caused by nonfunctioning parathyroid and growth hormone receptors. In these two conditions the glands produce the respective hormones, but the target cells do not respond because they lack normal receptors.

Signaling molecules differ in their water solubility. **Hydrophobic signaling molecules,** such as small steroid and thyroid hormones, bind reversibly to carrier proteins in the plasma for transport through the body. Such hormones are lipophilic and once released from their carrier proteins, they diffuse directly through the plasma membrane lipid bilayer of the target cell and bind to specific **intracellular receptor** proteins. With many steroid hormones, receptor binding activates that protein, enabling the complex to move into the nucleus and bind with high affinity to specific DNA sequences. This generally increases the level of transcription from specific genes. Each steroid hormone is recognized by a different member of a family of homologous receptor proteins.

Mitochondria

Mitochondria (Gr. *mitos*, thread, + *chondros*, granule) are membrane-enclosed organelles with enzyme arrays specialized for aerobic respiration and production of **adenosine triphosphate (ATP)**, which contains energy stored in high-energy phosphate bonds and is used in most energy-requiring cellular activities. Glycolysis converts glucose anaerobically to pyruvate in the cytoplasm, releasing some energy. The rest of the energy is captured when pyruvate is imported into mitochondria and oxidized to CO_2 and H_2O. Mitochondrial enzymes yield 15 times more ATP than is produced by glycolysis alone. Some of the energy released in mitochondria is not stored in ATP but is dissipated as heat which maintains body temperature.

Mitochondria are usually elongated structures 0.5–1 μm in diameter and lengths up to ten times greater (Figure 2–11). They are highly plastic, rapidly changing shape, fusing with one another and dividing, and are moved through the cytoplasm along microtubules. The number of mitochondria is related to

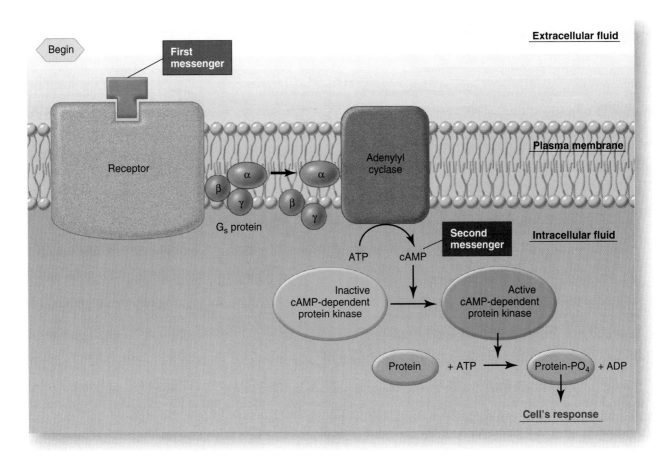

Figure 2–10. G proteins and initiation of signal transduction. When a hormone or other signal binds to a membrane receptor, the hormone can begin to cause changes in the cell's activities after a signal transduction process initiated by the bound receptor. The first step in receptor signaling often involves **G proteins** which bind guanosine diphosphate (GDP) when inactive and are activated when GDP is exchanged for GTP. A simplified version of G protein activity is shown here. Conformational changes occur in the receptor when it binds its ligand and the changed receptor activates the G protein–GDP complex. A GDP-GTP exchange releases the α subunit of the G protein, which then moves laterally to bind with a transmembrane effector protein, activating it to propagate the signal further by various mechanisms. The α subunit GTP is rapidly converted back to GDP, allowing the polypeptide to reassociate with the rest of the G protein complex, ready to be activated again when the receptor is again bound by hormone.

the cell's energy needs. Thus, cells with a high-energy metabolism (eg, cardiac muscle, cells of some kidney tubules) have abundant mitochondria, whereas cells with a low-energy metabolism have few mitochondria. Mitochondria also accumulate in cytoplasmic regions where energy utilization is more intense.

Mitochondria are often large enough to be visible with the light microscope as numerous discrete eosinophilic organelles. Under the TEM each mitochondrion is seen to have two separated and very different membranes which together create two compartments: the innermost **matrix** and a narrow **intermembrane space** (Figure 2–12). Both mitochondrial membranes contain a large number of protein molecules compared with other membranes in the cell and have reduced fluidity. The **outer membrane** is sieve-like, containing many transmembrane proteins called **porins** that form channels through which small molecules (<5000 daltons) readily pass to enter the intermembrane space from the cytoplasm.

The **inner membrane** is folded to form a series of long infoldings called **cristae**, which project into the matrix and greatly increase the membrane's surface area (Figure 2–12). The number of cristae in mitochondria also corresponds to the energy needs of the cell. The lipid bilayer of the inner membrane contains unusual phospholipids and is highly impermeable to ions (Figure 2–13). Integral proteins include various transport proteins that make the inner membrane selectively permeable to the small molecules required by mitochondrial enzymes in the matrix. Matrix enzymes include those that oxidize pyruvate and fatty acids to form acetyl coenzyme A (CoA) and those of the citric acid cycle that oxidize acetyl CoA, releasing CO_2 as waste and small energy-rich molecules which provide electrons for transport along the **respiratory chain** or **electron transport chain**. Enzymes and other components of this chain are embedded in the inner membrane and allow oxidative phosphorylation, which produces most of the ATP in animal cells.

Formation of ATP by oxidative phosphorylation enzymes of the respiratory chain occurs in a **chemiosmotic process**. Membrane proteins guide the small electron carrier molecules through closely packed enzyme complexes so that the electrons move sequentially along the chain. Electron transfer is coupled with oriented proton uptake and release which causes protons to accumulate in the intermembrane space (Figure 2–13). This produces an **electrochemical gradient** across the inner membrane. Other membrane-associated proteins make up the **ATP synthase** system, forming 10 nm, multisubunit globular complexes on stalk-like structures densely packed on the matrix side of the inner membrane (Figure 2–12). Through this enzyme complex runs a hydrophilic pathway that allows protons to flow down the electrochemical gradient, crossing the membrane back

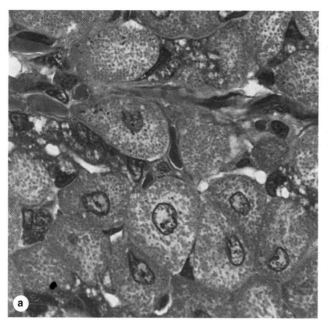
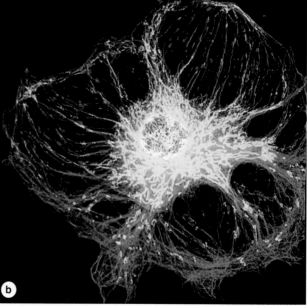

Figure 2–11. Mitochondria in the light microscope. (a): In sectioned cells stained with H&E, such as certain cells of the stomach inner lining, mitochondria typically appear as numerous eosinophilic structures throughout the cytoplasm. The mitochondria usually appear round or slightly elongated and are more numerous in cytoplasmic regions with higher energy demands, such as near the cell membrane in cells undergoing much active transport. The central nuclei are also clearly seen in these cells. **(b):** Entire mitochondria can be shown in cultured cells, such as the endothelial cells shown here and often appear as the elongated structures (shown in yellow or orange here), usually arrayed in parallel along microtubules. These preparations along with TEM studies indicate that the elongated shape is typical of mitochondria and that their shape can be quite plastic and variable. Specific mitochondrial staining such as that shown here involves incubating living cells with specific fluorescent compounds that are specifically sequestered into these organelles, followed by fixation and immunocytochemical staining of the microtubules. In this preparation, microtubules are stained green and mitochondria appear yellow or orange, depending on their association with the green microtubules. The cell nucleus was stained with DAPI. (Figure 2–11b, with permission, from Invitrogen.)

into the matrix. Passage of protons through this narrow channel causes rapid spinning of specific polypeptides in the globular ATP synthase complex, converting the energy of proton flow into the mechanical energy of protein movement. Mechanical energy is stored in the new phosphate bond of ATP by other subunit polypeptides that bind ADP and inorganic phosphate (Figure 2–13). A steady torrent of protons along the gradient allows each of these remarkable synthase complexes to produce more than 100 molecules of ATP per second.

The mitochondrial matrix also contains a small circular chromosome of DNA (like that of prokaryotic organisms), ribosomes, messenger RNA (mRNA) and transfer RNA, all with similarities to the corresponding bacterial components. Protein synthesis occurs in mitochondria, but because of the reduced amount of mitochondrial DNA, only a small proportion of the mitochondrial proteins is produced locally. Most are encoded by nuclear DNA and synthesized in the cytoplasm. These proteins

have a small amino acid sequence that is a signal for their uptake across the mitochondrial membranes. The fact that mitochondria have certain bacterial characteristics has led to the hypothesis that mitochondria originated from an ancestral aerobic prokaryote that adapted to a symbiotic life within an ancestral eukaryotic host cell.

During cell mitosis each daughter cell receives approximately half the mitochondria in the parent cell. New mitochondria originate from preexisting mitochondria by growth and subsequent division (fission) of the organelle itself.

Ribosomes

Ribosomes are small electron-dense particles, about 20×30 nm in size. Ribosomes found in the cytosol are composed of four segments of rRNA and approximately 80 different proteins. Those of the mitochondria (and chloroplasts), like prokaryotic ribosomes, are somewhat smaller with fewer constituents. All ribosomes are composed of two different-sized subunits.

In eukaryotic cells, the RNA molecules of both subunits are synthesized within the nucleus. Their numerous proteins are synthesized in the cytoplasm but then enter the nucleus and associate with rRNAs. The assembled large and small subunits then leave the nucleus and enter the cytoplasm to participate in protein synthesis.

Ribosomes are intensely basophilic because of the numerous phosphate groups of the constituent rRNAs which act as polyanions. Thus, sites in the cytoplasm rich in ribosomes stain intensely with hematoxylin and basic dyes, such as methylene and toluidine blue.

The large and small ribosomal subunits come together by binding an mRNA strand (Figure 2–14a) and typically numerous ribosomes are present on an mRNA as **polyribosomes** (or **polysomes**). The nucleotide sequence of the mRNA determines the amino acid sequence of the protein synthesized, with ribosomes assembling the polypeptide from amino acids ferried in by transfer RNA (tRNA). The compact core of each ribosome contains the rRNA molecules which not only provide structural support but also position tRNAs in the correct "reading frame" and as ribozymes catalyze the formation of the covalent peptide

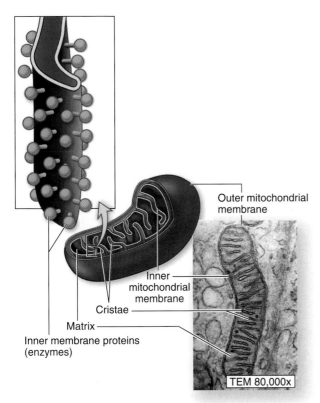

Energy synthesis: Produce ATP by cellular respiration for energy needs of the cell; called the "powerhouses" of the cell

Figure 2–12. **Mitochondria.** The two mitochondrial membranes and central matrix can be seen here in the diagram and the TEM. The **outer membrane** is smooth and the **inner membrane**, shown at left, has many sharp folds called **cristae** which increase its surface area greatly. Cristae are most numerous in mitochondria of highly active cells. The **matrix** is a gel containing numerous enzymes. The inner membrane surface in contact with the matrix is studded with many multimeric protein complexes resembling globular units on short stalks. These contain the ATP synthase complexes that generate most of the cell's ATP.

bonds. The more peripheral proteins of the ribosome seem to function primarily to stabilize the catalytic RNA core.

Proteins synthesized for use within the cell cytosol (eg, glycolytic enzymes) are synthesized on polyribosomes existing as isolated clusters within the cytoplasm. Polyribosomes that are attached to the membranes of the endoplasmic reticulum (via their large subunits) translate mRNAs that code for proteins that are sequestered across the membranes of this organelle (Figure 2–14b).

Endoplasmic Reticulum

The cytoplasm of eukaryotic cells contains an anastomosing network of intercommunicating channels and sacs formed by a continuous membrane which encloses a space called a **cisterna.** In sections cisternae appear separated, but high-resolution microscopy of whole cells reveals that they are continuous. This membrane

system is called the **endoplasmic reticulum** (ER) (Figure 2–15). In many places the cytosolic side of the membrane is covered by polyribosomes synthesizing protein molecules which are injected into the cisternae. This permits the distinction between the two types of endoplasmic reticulum: **rough** and **smooth.**

ROUGH ENDOPLASMIC RETICULUM

Rough endoplasmic reticulum (RER) is prominent in cells specialized for protein secretion, such as pancreatic acinar cells (digestive enzymes), fibroblasts (collagen), and plasma cells (immunoglobulins). The RER consists of saclike as well as parallel stacks of flattened cisternae (Figure 2–15), limited by membranes that are continuous with the outer membrane of the nuclear envelope. The name "rough endoplasmic reticulum" refers to the presence of polyribosomes on the cytosolic surface of this structure's membrane (Figures 2–15 and 2–16). The presence

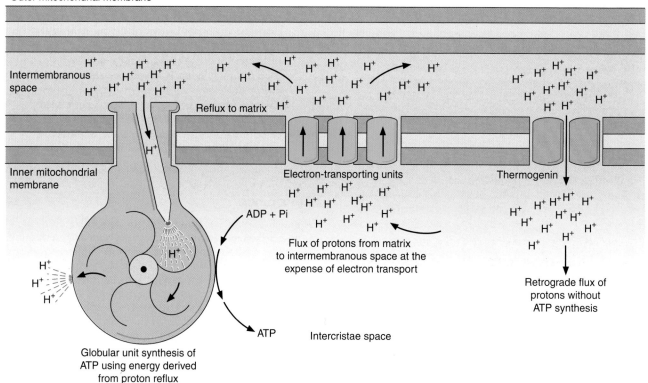

Figure 2–13. The chemiosmotic process of energy transduction. The movement of electrons along the units of the inner mitochondrial membrane **electron transport system (middle)** is accompanied by the directed movement of protons (H⁺) from the matrix into the intermembranous space. The inner membrane is highly impermeable to protons and the result is an electrochemical gradient across the membrane. The other membrane-associated proteins **(left)** make up the **ATP synthase** systems, each of which forms a 10 nm, multisubunit globular complex on a stalk-like structure projecting from the matrix side of the inner membrane (Figure 2–12). A channel runs through this enzyme complex and specifically allows protons to flow through it, down the electrochemical gradient and across the membrane back into the matrix. Passage of protons through this narrow path causes rapid spinning of specific polypeptides in the globular ATP synthase complex. In this manner the energy of proton flow is converted into the mechanical energy of protein movement. Other subunit proteins of the complex store this energy in the new phosphate bond of ATP which leaves the mitochondrion for use throughout the cell. It is estimated that each ATP synthase complex produces more than 100 molecules of ATP per second. In some mitochondria, particularly those in cells of multilocular adipose tissue, another inner membrane protein called **thermogenin** forms a shunt for the return of protons into the matrix **(right).** This reflux of protons does not produce ATP, but instead dissipates energy as heat which warms blood flowing through the tissue (see Chapter 6).

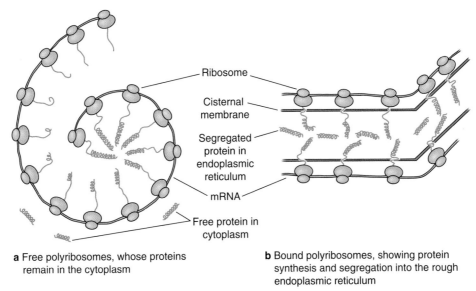

a Free polyribosomes, whose proteins remain in the cytoplasm

b Bound polyribosomes, showing protein synthesis and segregation into the rough endoplasmic reticulum

Figure 2–14. **Polyribosomes.** Proteins that are to remain freely soluble within the cytoplasm are synthesized on (free) polyribosomes (ie, not attached to the endoplasmic reticulum). **(a):** Many ribosomes attach to the same mRNA and move along it during translation, with each ribosome producing and at the end of the mRNA releasing one copy of the protein encoded by that message.

(b): Proteins that are to be incorporated into membranes, or eventually to be extruded from the cytoplasm (secreted proteins) or sequestered into lysosomes, are made on polysomes that adhere to the membranes of endoplasmic reticulum. The proteins produced by these ribosomes are segregated during translation into the interior of the endoplasmic reticulum's membrane cisternae.

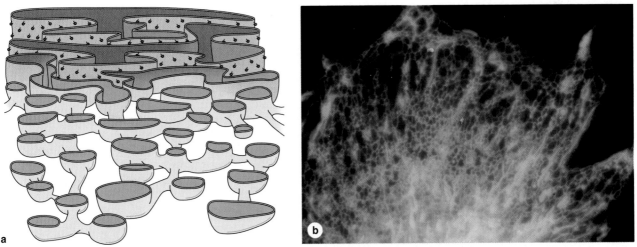

Figure 2–15. **Endoplasmic reticulum.** The **endoplasmic reticulum** is an anastomosing network of intercommunicating channels or **cisternae** formed by a continuous membrane. **(a):** Electron microscopy shows that some regions of endoplasmic reticulum, called smooth ER (foreground), are devoid of ribosomes, the small granules that are present in the rough ER (background). Both types of ER are continuous with one another. The interconnected membranous cisternae of the smooth ER are frequently tubular, whereas those in the rough ER are flattened sacs. **(b):** In a very thin cultured endothelial cell, both ER (green) and mitochondria (orange) can be visualized with vital fluorescent dyes that are sequestered specifically into those organelles. This staining method with intact cells clearly reveals the continuous, lacelike ER present in all regions of the cytoplasm. (Figure 2-15b, with permission, from Invitrogen.)

of polyribosomes also confers basophilic staining properties on this organelle when viewed with the light microscope.

The principal function of the RER is to segregate proteins not destined for the cytosol. Additional functions include the initial (core) glycosylation of glycoproteins, the synthesis of phospholipids, the assembly of multichain proteins, and certain posttranslational modifications of newly formed polypeptides. All protein synthesis begins on polyribosomes that are not attached to the ER. Messenger RNA for proteins destined to be segregated in the ER contain an additional sequence of bases at their 5′ end that codes for 20–50 mainly hydrophobic amino acids comprising the protein's **ER signal sequence** (Figure 2–17). Upon translation, the signal sequence interacts with a complex of six different polypeptides bound to a small RNA molecule, a complex referred to as the **signal-recognition particle** (**SRP**). The SRP inhibits further polypeptide elongation until the SRP-polyribosome complex binds to receptors in the ER membrane. When the complex is bound, the SRP is released from the polyribosomes, allowing translation to continue (Figure 2–17). The growing polypeptide chain is translocated across the membrane through a pore formed by another protein complex.

Once inside the lumen of the RER, the signal sequence is removed by an enzyme, signal peptidase. Translation of the protein continues, accompanied by intracisternal secondary and tertiary structural changes as well as certain posttranslational modifications of the polypeptides.

Proteins synthesized in the RER can have several destinations: intracellular storage (eg, in lysosomes and specific granules of leukocytes), provisional intracellular storage of proteins before exocytosis (eg, in the pancreas, some endocrine cells), and as integral membrane proteins. Figure 2–18 shows several cell types with distinct differences in the destinations of their major protein products and how these differences can determine histological features of the cells.

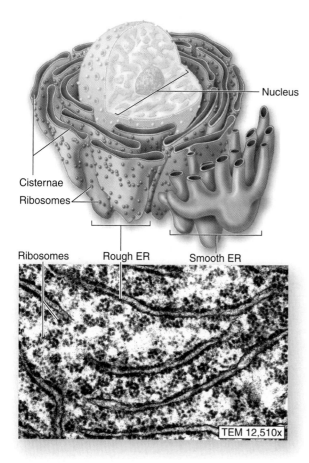

SMOOTH ENDOPLASMIC RETICULUM

Regions of ER that lack bound polyribosomes make up the smooth endoplasmic reticulum (SER), which in most cells is less abundant that RER but is continuous with it (Figures 2–15 and 2–16). SER cisternae are often more tubular and more likely to appear as a profusion of interconnected channels of various shapes and sizes than as stacks of flattened cisternae (Figure 2–15).

SER contains enzymes associated with a wide variety of specialized functions. A major role of SER is the synthesis of the various phospholipid molecules that constitute all cellular membranes. The phospholipids are transferred to other membranes from the SER (1) by direct communication with the RER allowing lateral diffusion, (2) by vesicles that detach, move to and fuse with other membranous organelles, or (3) by being carried individually by **phospholipid transfer proteins** (Figure 2–19).

In cells that synthesize steroid hormones (eg, cells of the adrenal cortex), SER occupies a large portion of the cytoplasm and contains some of the enzymes required for steroid synthesis. SER is abundant in liver cells, where it contains enzymes responsible for the oxidation, conjugation, and methylation processes that degrade certain hormones and neutralize noxious substances such as alcohol and barbiturates. An important example of such detoxification reactions are those catalyzed by the cytochrome P-450 family of enzymes. The SER of liver cells also contains the enzyme glucose-6-phosphatase, which is involved in the utilization of glucose originating from glycogen. This enzyme is also found in the RER, an example of the lack of absolute partitioning of functions between these regions.

Another function of the SER is to sequester and release Ca²⁺ in a controlled manner, which is part of the rapid response of cells to various external stimuli. This function is well-developed in muscle cells, where the SER participates in the contraction process and assumes a specialized form called the **sarcoplasmic reticulum** (see Chapter 10).

Figure 2–16. Functions of rough and smooth ER. As seen with the TEM the cisternae of **rough ER** are flattened, with polyribosomes on their outer surfaces and concentrated material in their lumens. Such cisternae appear separated in sections made for electron microscopy, but they actually form a continuous channel or compartment in the cytoplasm. **Smooth ER** is continuous with rough ER but is involved with a much more diverse range of functions. Three major activities associated with smooth ER are (1) lipid biosynthesis, (2) detoxification of potentially harmful compounds, and (3) sequestration of Ca⁺⁺ ions. Specific cell types with well-developed smooth ER are usually specialized for one of these functions.

Golgi Apparatus

The highly dynamic **Golgi apparatus,** or Golgi complex, completes posttranslational modifications and then packages and addresses proteins synthesized in the RER. This organelle, named for histologist Camillo Golgi who discovered it in 1898, is

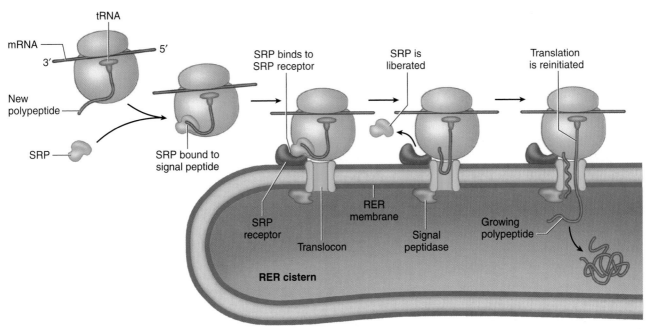

Figure 2–17. Movement of polypeptides into the RER. Proteins to be incorporated into membranes or sequestered into vesicles contain 20 to 25 hydrophobic amino acids comprising the **signal sequence** or signal peptide in the region translated first. This sequence is bound by a cytoplasmic signal-recognition particle (SRP). The bound SRP then recognizes and binds to a receptor on the ER. Another receptor in the ER membrane binds a structural protein of the large ribosomal subunit, more firmly attaching the ribosome to the ER. The hydrophobic signal peptide is translocated through a protein pore (translocon) in the ER membrane and the SRP is freed for reuse. The signal peptide is removed from the growing protein by a peptidase and translocation of the growing polypeptide continues until it is completely segregated into the ER cisterna.

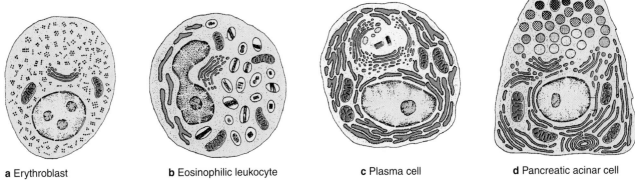

a Erythroblast   **b** Eosinophilic leukocyte   **c** Plasma cell   **d** Pancreatic acinar cell

Figure 2–18. Protein synthesis and cell morphology. The ultrastructure and many general histological aspects of a cell are determined by the nature of the most prominent proteins the cell is making. Schematic representations show cell types that illustrate this idea. **(a):** Cells that make few or no proteins for secretion have very little rough ER, with essentially all polyribosomes free in the cytoplasm. **(b):** Cells that synthesize, segregate, and store various proteins in specific secretory granules or vesicles always have rough ER, a Golgi apparatus, and a supply of granules containing the proteins ready to be secreted. **(c):** Cells with extensive RER and a well-developed Golgi apparatus show few secretory granules because the proteins undergo exocytosis immediately after Golgi processing is complete. Many cells, especially those of epithelia, are *polarized*, meaning that the distribution of RER and secretory vesicles is different in various regions or poles of the cell. **(d):** Epithelial cells specialized for secretion have distinct polarity, with RER abundant at their basal ends and mature secretory granules at the apical poles undergoing exocytosis into an enclosed extracellular compartment, the lumen of a gland.

composed of smooth membranous **saccules** in which these functions occur (Figures 2–20, 2–21, and 2–22). In polarized secretory cells with apical and basal ends, such as mucus-secreting goblet cells, the Golgi apparatus occupies a characteristic position between the nucleus and the apical plasma membrane.

The Golgi apparatus generally shows two distinct sides structurally and functionally, which reflects the complex traffic of vesicles within cells. Near the Golgi, the RER can be seen budding off small **transport vesicles** that shuttle newly synthesized proteins to the Golgi apparatus for further processing. The Golgi saccules nearest this point make up the entry or *cis* face. On the opposite side of the Golgi network, which is the exit or *trans* face, larger saccules sometimes called **condensing vacuoles** can be seen to accumulate (Figure 2–20). These structures bud from the maturing saccules and generate vesicles that carry completed protein products to organelles away from the Golgi. Vesicle formation is driven by assembly of various coat proteins (including clathrin). Such proteins help regulate vesicular traffic to, through, and beyond the Golgi apparatus in conjunction with specific receptor and fusion-promoting proteins that mark the membrane at the vesicles' destinations.

The TEM and cytochemical methods have shown that Golgi saccules contain different enzymes at different *cis-trans* levels and that the Golgi apparatus is important for glycosylation, sulfation, phosphorylation, and limited proteolysis of proteins. Furthermore, the Golgi apparatus initiates packing, concentration,

and storage of secretory products. Figure 2–22 gives an overall view of the transit of material through this organelle.

Secretory Vesicles or Granules

Originating in the Golgi apparatus, secretory vesicles are found in those cells that store a product until its release by exocytosis is signaled by a metabolic, hormonal, or neural message (regulated secretion). These vesicles are surrounded by a membrane and contain a concentrated form of the secretory product (Figure 2–23). The contents of some secretory vesicles may be up to 200 times more concentrated than those in the cisternae of the RER. Secretory vesicles with dense contents of digestive enzymes are referred to as **zymogen granules.**

Lysosomes

Lysosomes are sites of intracellular digestion and turnover of cellular components. Lysosomes (Gr. *lysis,* solution, + *soma,* body) are membrane-limited vesicles that contain about 40 different hydrolytic enzymes and are particularly abundant in cells with great phagocytic activity (eg, macrophages, neutrophils). Although the nature and activity of lysosomal enzymes vary depending on the cell type, the most common are acid hydrolyases such as proteases, nucleases, phosphatase, phospholipases, sulfatases, and β-glucuronidase. As can be seen from this list, lysosomal enzymes are capable of breaking down most macromolecules.

Cytosolic components are protected from these enzymes by the membrane surrounding lysosomes and because the enzymes have optimal activity at an acidic pH (~5.0). Any leaked lysosomal enzymes are practically inactive at the pH of cytosol (~7.2) and harmless to the cell.

Lysosomes, which are usually spherical, range in diameter from 0.05 to 0.5 μm and present a uniformly granular, electron-dense appearance in the transmission electron microscope (TEM) (Figure 2–24). In macrophages and neutrophils, lysosomes are slightly larger and thus visible with the light microscope.

Lysosomal hydrolases are synthesized and segregated in the RER and subsequently transferred to the Golgi apparatus, where the enzymes are further modified and packaged in vacuoles that form lysosomes. The marker mannose-6-phosphate (M6P) is added by a phosphotransferase in the *cis* Golgi only to the N-linked oligosaccharides of the hydrolases destined for lysosomes. Membrane receptors for M6P-containing proteins in the *trans* Golgi network then bind these proteins and divert them from the main secretory pathway for segregation into lysosomes.

Material taken from the cellular environment by endocytosis is digested when lysosomes fuse with the membrane of the phagosome or pinocytotic vesicle. The endocytosed material mixes with the hydrolytic enzymes, a proton pump in the lysosomal membrane is activated to lower the internal pH, and digestion follows. The composite structure is now termed a secondary or **heterolysosome.** Heterolysosomes are generally 0.2–2 μm in diameter and present a heterogeneous appearance in the TEM because of the wide variety of materials they may be digesting (Figure 2–24c).

During this digestion of macromolecules, released nutrients diffuse into the cytosol through the lysosomal membrane. Indigestible material is retained within the vacuoles, which are now called **residual bodies** or telolysosomes (Figure 2–25). In some long-lived cells (eg, neurons, heart muscle), residual bodies can accumulate and are referred to as **lipofuscin granules.**

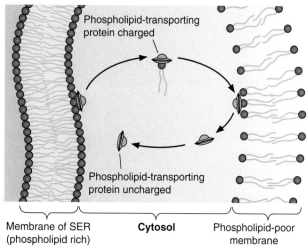

Figure 2–19. Phospholipid transport. Phospholipids or more complex lipids such as cholesterol are generally synthesized by enzymes located in smooth ER. The products are inserted into the lipid bilayers of that organelle and are distributed in membranes throughout the cell by movement through the ER, Golgi apparatus, secretory vesicles, and other organelles. As shown here however, individual phospholipids can also be transported from smooth ER directly to membranes elsewhere in the cell but only after binding a water-soluble transporting protein. There is a specific transfer protein (also called exchange proteins) for each specific type of phospholipid and each can be reused many times. **Phospholipid transfer proteins** are an important mechanism for redistributing lipids between different membrane-enclosed compartments or organelles, such as the ER and mitochondria.

Within the figure: Phospholipid-transporting protein charged; Phospholipid-transporting protein uncharged; Membrane of SER (phospholipid rich); **Cytosol**; Phospholipid-poor membrane

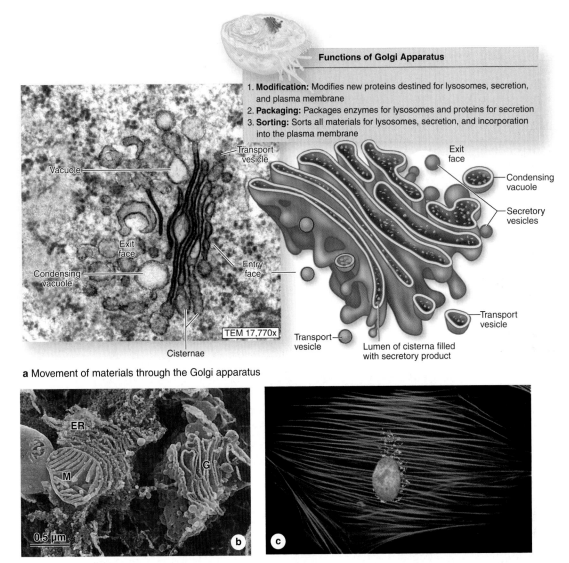

Functions of Golgi Apparatus

1. **Modification:** Modifies new proteins destined for lysosomes, secretion, and plasma membrane
2. **Packaging:** Packages enzymes for lysosomes and proteins for secretion
3. **Sorting:** Sorts all materials for lysosomes, secretion, and incorporation into the plasma membrane

a Movement of materials through the Golgi apparatus

Figure 2–20. **Golgi apparatus.** The **Golgi apparatus** is a highly plastic, morphologically complex system of membrane vesicles and cisternae in which proteins and other molecules made in the ER undergo modification and maturation and then are sorted into specific vesicles destined for different roles in the cell. **(a):** Transport vesicles emerging from the RER move toward and fuse at the *cis*, entry, or forming face of the Golgi, merging with the first of several flattened Golgi cisternae. Movement through the Golgi remains a subject of intense investigation, but data suggest that other transport vesicles move proteins serially through the cisternae until at the *trans*, exit, or maturing face larger vesicles and vacuoles emerge to carry fully modified proteins elsewhere in the cell. Formation and fusion of the vesicles through the Golgi apparatus is controlled by specific membrane proteins. Depending on their protein contents, vesicles are directed toward different regions of the Golgi by specific interactions of these proteins with other membrane proteins. Peripheral membrane proteins important for directed vesicle fusion are the **golgins**. These are an important Golgi-specific family of proteins, characterized by central coiled-coil domains, which interact with GTPases and many other binding proteins to organize, shape, and specify Golgi membranes. Golgi vesicles may become lysosomes, secretory vesicles that undergo exocytosis, and portions of the plasma membrane. **(b):** Morphological aspects of the Golgi apparatus are revealed by the SEM, which shows a three-dimensional snapshot of the region between RER and the Golgi membrane compartments. Cells may have multiple Golgi apparatuses, each with stacks of cisternae and dynamic *cis* and *trans* faces, and these typically are situated near the cell nucleus. This has been shown in careful TEM studies but is also clearly seen in intact cultured cells. **(c):** The fibroblast was processed by immunocytochemistry using an antibody against golgin-97 to show many complexes of Golgi vesicles (green), all near the nucleus, against a background of microfilaments organized as stress fibers and stained with fluorescent phalloidin (violet). Because of the abundance of lipids in its many membranes, the Golgi apparatus is difficult to visualize by light microscopy in typical paraffin-embedded, H&E stained sections. In cells with very active Golgi complexes however, such as developing white blood cells, the organelle can sometimes be seen as a faint unstained juxtanuclear region (sometimes called a "Golgi ghost") surrounded by basophilic cytoplasm. (Figure 2–20b reproduced, with permission, from T. Naguro and A. Iino: *Prog. Clin. Biol.* Res. 1989;295:250. Copyright ©1989 by Wiley-Liss, Inc., a subsidiary of John Wiley & Sons, Inc. Figure 2–20c, with permission, from Invitrogen.)

Lysosomes also function in the removal of nonfunctional organelles or excess cytoplasmic structures, a process called **autophagy** (Figure 2–26). A membrane forms around those organelles or portions of cytoplasm that are to be removed, producing an **autophagosome** (Gr. *autos*, self, + *phagein*, to eat, + *soma*). These fuse with lysosomes which initiate lysis of the enclosed cytoplasm. Autophagy is enhanced in secretory cells that have accumulated excess secretory granules. Digested products from autophagosomes are reutilized in the cytoplasm.

In some cases, lysosomes release their contents extracellularly, and their enzymes act in the extracellular milieu. An example is the destruction of bone matrix by the collagenases synthesized and released by osteoclasts during normal bone tissue formation (see Chapter 8). Lysosomal enzymes acting in the extracellular milieu also play a significant role in the response to inflammation or injury. Several possible pathways relating to lysosome activities are schematically illustrated in Figure 2–25.

Lysosomes play an important role in the metabolism of several substances in the human body, and consequently many diseases have been ascribed to deficiencies of lysosomal enzymes. In metachromatic leukodystrophy, there is an intracellular accumulation of sulfated cerebrosides caused by lack of lysosomal sulfatases. In most of these diseases, a specific lysosomal enzyme is absent or inactive, and certain molecules (eg, glycogen, cerebrosides, gangliosides, sphingomyelin, glycosaminoglycans) are not digested. As a result, these substances accumulate in the cells, interfering with their normal functions. This diversity of affected cell types explains the variety of clinical symptoms observed in lysosomal diseases (Table 2–2).

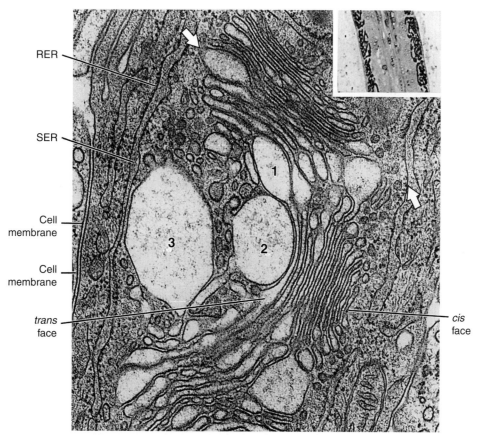

Figure 2–21. **Golgi apparatus.** Though only snapshots of this highly dynamic organelle, electron micrographs of the Golgi apparatus provided early evidence about how this organelle functions, evidence that has now been strengthened by biochemical and other studies. To the right is a cisterna (arrow) of the rough ER containing granular material. Close to it are small vesicles containing apparently similar material. These are very close to the *cis* face of the Golgi apparatus. In the center are the characteristic flattened, curved, and stacked medial cisternae of the complex. Dilatations (upper left arrow) are seen extending from the ends of the cisternae. Similar dilatations gradually detach themselves from the cisternae and fuse at the *trans* face, forming the secretory granules (**1, 2,** and **3**). Near the plasma membranes of two neighboring cells is more rough ER and smooth ER. X30,000. **Inset:** a small region of a Golgi apparatus in a 1-μm section impregnated with silver, which demonstrates the abundance of glycoproteins within some cisternae. X1200.

I-cell disease (inclusion cell disease) is a rare inherited condition clinically characterized by defective physical growth and mental retardation and is caused by a deficiency in a phosphorylating enzyme normally present in the Golgi apparatus. Lysosomal enzymes coming from the RER are not phosphorylated in the Golgi apparatus. Nonphosphorylated protein molecules are not separated to form *lysosomes, but instead follow the main secretory pathway. The secreted lysosomal enzymes are present in the blood of patients with I-cell disease, whereas their lysosomes are empty. Cells of these patients show large inclusion granules that interfere with normal cellular metabolism.*

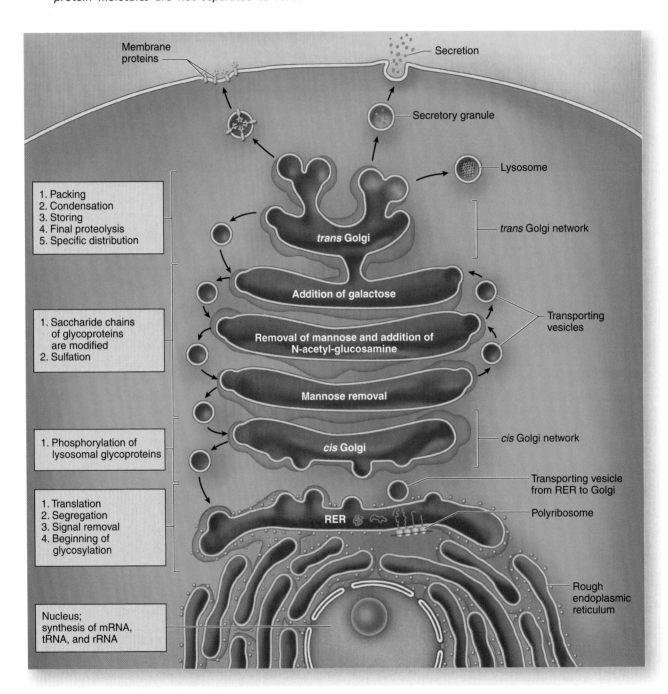

Figure 2–22. Summary of Golgi apparatus structure and function. Summary of the main events occurring during protein trafficking and sorting from the rough ER through the Golgi complex. Numbered at the left are the main molecular processes that take place in the compartments shown. In the *trans* Golgi network, the proteins and glycoproteins combine with specific receptors that guide them to the next stages toward their destinations. On the left side of the drawing is the returning flux of membrane, from the Golgi to the endoplasmic reticulum.

Proteasomes

Proteasomes are abundant cytoplasmic protein complexes not associated with membrane, each approximately the size of the small ribosomal subunit. They function to degrade denatured or otherwise nonfunctional polypeptides. Proteasomes also remove proteins no longer needed by the cell and provide an important mechanism for restricting activity of a specific protein to a certain window of time. Whereas lysosomes digest bulk material introduced into the cell, or whole organelles and vesicles, proteasomes deal primarily with proteins as individual molecules.

The proteasome is a cylindrical structure made of four stacked rings, each composed of seven proteins including proteases. At each end of the cylinder is a regulatory particle that contains ATPase and recognizes proteins with ubiquitin molecules attached. **Ubiquitin** is an abundant cytosolic 76-amino acid protein found in all cells and is highly conserved during evolution—it has virtually the same structure from bacteria to humans. Denatured proteins or proteins with oxidized amino acids are targeted for destruction after recognition by enzyme complexes which conjugate a molecule of ubiquitin to a lysine residue in the protein, followed by formation of a multiubiquitin chain.

A ubiquinated protein is recognized by the regulatory particle of proteasomes, unfolded by the ATPase using energy from ATP, and then translocated into the core particle, where it is broken into short peptides. These peptides are transferred to the cytosol and the ubiquitin molecules are released by the regulatory particles for reuse.

The peptides may be broken down further to amino acids or they may have other specialized destinations, such as the antigen-presenting complexes of cells activating an immune response.

MEDICAL APPLICATION

Failure of proteasomes or other aspects of a cell's protein quality control can allow large aggregates of protein to accumulate in affected cells. Such aggregates may adsorb other macromolecules to them and damage or kill cells. Aggregates released from dead cells can accumulate in the extracellular matrix

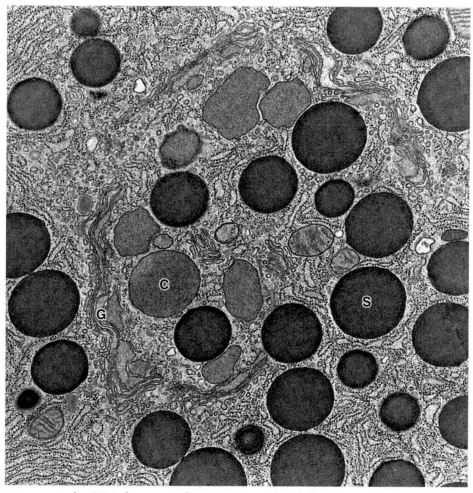

Figure 2–23. **Secretory granules.** TEM of one area of a pancreatic acinar cell shows numerous mature, electron-dense **secretory granules** (S) in association with condensing vacuoles (C) of the Golgi apparatus (G). Such granules form as the contents of the Golgi vacuoles becomes more condensed. In H&E stained sections secretory granules are often shown as intensely eosinophilic structures, which in polarized epithelial cells are concentrated at the apical region prior to exocytosis. X18,900.

of the tissue. In the brain this can interfere directly with cell function and lead to neurodegeneration. Alzheimer disease and Huntington disease are two neurologic disorders caused initially by such protein aggregates.

Peroxisomes or Microbodies

Peroxisomes (peroxide + *soma*) are spherical membrane-limited organelles approximately 0.5 μm in diameter (Figure 2–27). They utilize oxygen but do not produce ATP and do not participate directly in cellular metabolism. Peroxisomes oxidize specific organic substrates by removing hydrogen atoms that are

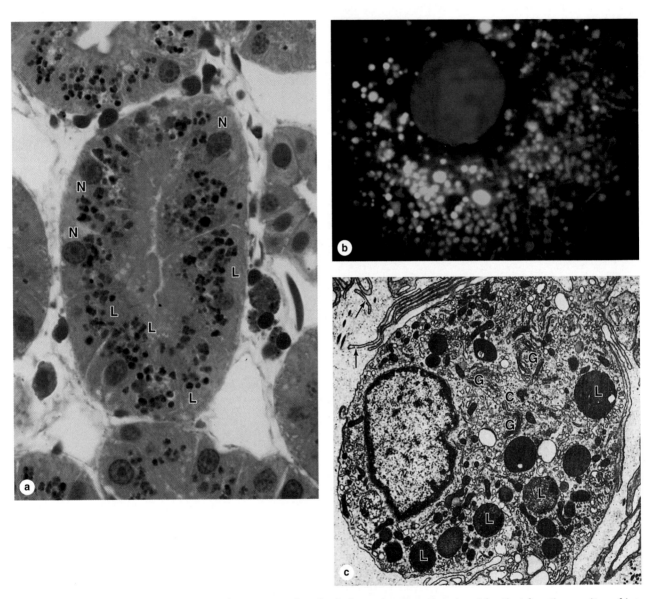

Figure 2–24. **Lysosomes. Lysosomes** are large, generally spherical membrane-enclosed vesicles that function as sites of intracellular digestion and are particularly numerous in cells active in various types of endocytosis. Lysosomes are not well-shown on H&E stained cells, but can be visualized by light microscopy after staining with toluidine blue. **(a):** Cells in a kidney tubule show numerous purple lysosomes (L) in the cytoplasmic area between the basally located nuclei (N) and apical ends of the cells at the center of the tubule. Using endocytosis, these cells actively take up small proteins in the lumen of the tubule, degrade the proteins in lysosomes, and then release the resulting amino acids for reuse. X300 toluidine blue. **(b):** Lysosomes in cultured vascular endothelial cells can be specifically stained using fluorescent dyes sequestered into these organelles (green), which are abundant around the blue Hoechst-stained nucleus. Mitochondria (red) are scattered among the lysosomes. **(c):** In the TEM lysosomes (L) have a characteristic very electron-dense appearance and are shown here near groups of Golgi cisternae (G) and a centriole (C). Less electron-dense lysosomes represent heterolysosomes in which digestion of the contents is underway. The cell is a macrophage with numerous fine cytoplasmic extensions (arrows). X15,000. (Figure 2–24b, with permission, from Invitrogen.)

transferred to molecular oxygen (O₂). This produces hydrogen peroxide (H_2O_2), a substance potentially damaging to the cell which is immediately broken down by **catalase**, another enzyme in all peroxisomes. The transfer by catalase of oxygen atoms from H_2O_2 to other compounds has clinical implications: it oxidizes various potentially toxic molecules as well as prescription drugs, particularly in the large and abundant peroxisomes of liver and kidney cells. For example, 50% of ingested ethyl alcohol is degraded to acetic aldehyde in peroxisomes of these cells. Peroxisomes of these cells contain additional enzymes, including D- and L-amino acid oxidases, and hydroxyacid oxidase. In most animals except humans, urate oxidase is also present and can become highly concentrated, appearing ultrastructurally as a crystalloid core in an otherwise homogeneous matrix.

Peroxisomes also contain enzymes involved in lipid metabolism. Thus, the β-oxidation of long-chain fatty acids (18 carbons and longer) is preferentially accomplished by peroxisomal enzymes that differ from their mitochondrial counterparts. Certain reactions leading to the formation of bile acids and cholesterol also have been localized in highly purified peroxisomal fractions.

Peroxisome formation is not well understood, but involves precursor vesicles that appear to bud off the ER. Many peroxisomal enzymes are synthesized on free cytosolic polyribosomes, with a small sequence of amino acids near the carboxyl terminus that functions as a specific import signal. Proteins with this signal are recognized by receptors located in the membrane of peroxisomes and internalized by the organelle.

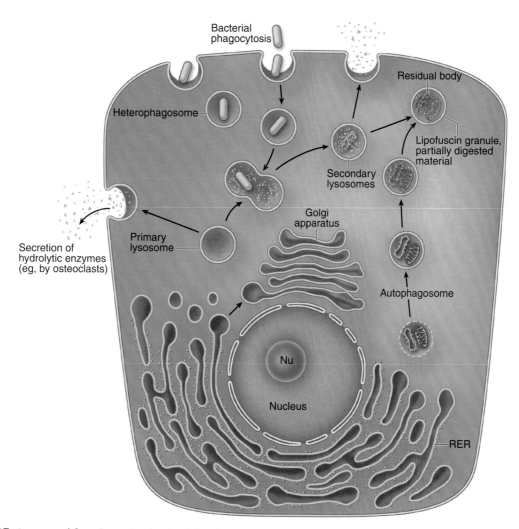

Figure 2–25. **Lysosomal functions.** Synthesis of the digestive enzymes occurs in the rough ER, and the enzymes are packaged in the Golgi apparatus. **Heterophagosomes,** in which bacteria are being destroyed, are formed by the fusion of the phagosomes and lysosomes. **Autophagosomes,** such as those depicted here with ER and mitochondria in the process of digestion, are formed after nonfunctional or surplus organelles become enclosed with membrane and the resulting structure fuses with a lysosome. The products of digestion can be excreted from the cell by exocytosis, but may remain in a membrane-enclosed **residual body,** containing remnants of indigestible molecules. Residual bodies can accumulate in long-lived cells and be visualized as lipofuscin granules. In some cells, such as osteoclasts, the lysosomal enzymes are secreted to a restricted extracellular compartment.

A large number of disorders arise from defective peroxisomal proteins, because this organelle is involved in several metabolic pathways. Probably the most common peroxisomal disorder is X-chromosome-linked adrenoleukodystrophy, caused by a defective integral membrane protein that participates in transporting very long-chain fatty acids into the peroxisome for β-oxidation. Accumulation of these fatty acids in body fluids destroys the myelin sheaths in nerve tissue, causing severe neurologic symptoms. Deficiency in peroxisomal enzymes causes the fatal Zellweger syndrome, with severe muscular impairment, liver and kidney lesions, and disorganization of the central and peripheral nervous systems. Electron microscopy reveals empty peroxisomes in liver and kidney cells of these patients.

THE CYTOSKELETON

The cytoplasmic cytoskeleton is a complex network of (1) microtubules, (2) microfilaments (actin filaments), and (3) intermediate filaments. These protein structures determine the shape of cells, play an important role in the movements of organelles and cytoplasmic vesicles, and also allow the movement of entire cells.

Table 2–2. Examples of diseases caused by lysosomal enzyme failure and accumulation of undigested material in different cell types.

Disease	Faulty Enzyme	Main Organs Affected
Hurler	α-L-Iduronidase	Skeleton and nervous system
Sanfilippo syndrome A	Heparan sulfate sulfamidase	Skeleton and nervous system
Tay-Sachs	Hexosaminidase-A	Nervous system
Gaucher	β-D-glycosidase	Liver and spleen
I-cell disease	Phosphotransferase	Skeleton and nervous system

Microtubules

Within the cytoplasmic matrix of eukaryotic, cells are fine tubular structures known as **microtubules** (Figures 2–28 and 2–29). Microtubules are also found in cytoplasmic processes called cilia (Figure 2–30) and flagella. They have an outer diameter of 24 nm, with a dense wall 5 nm thick and a hollow lumen. Microtubules are variable in length, but they can become many micrometers long. Occasionally, two or more microtubules are linked by protein arms or bridges, which are particularly important in cilia and flagella (Figure 2–31).

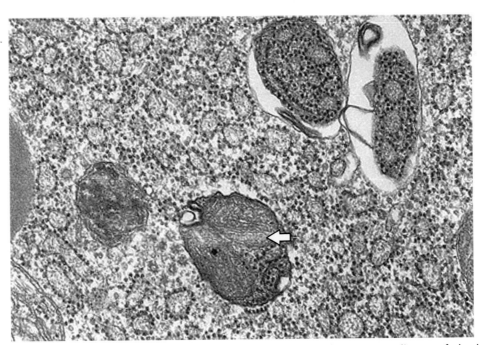

Figure 2–26. Autophagosomes. Autophagy is a process in which the cell uses lysosomes to dispose of obsolete or non-functioning organelles or membranes. Details of the process are highly regulated but not well-understood. Membrane of unknown origin encloses the organelles to be destroyed, forming an autophagosome which then fuses with a lysosome for digestion of the contents. In the TEM autophagosomes can sometimes be recognized by their contents, as shown here. **Upper right:** Two autophagosomes containing portions of the RER that are slightly more electron-dense than neighboring normal RER. **Center:** An autophagosome containing what may be mitochondrial membranes (arrow) plus RER. **Left:** A vesicle that may represent a residual body with indigestible material. X20,000.

The protein subunit of a microtubule is a heterodimer composed of α and β **tubulin** molecules of closely related amino acid composition, each with a molecular mass of about 50 kDa.

Under appropriate conditions (*in vivo* or *in vitro*), tubulin heterodimers polymerize to form microtubules, which have a slight spiral organization visible with special EM preparations. A total of 13 units is present in one complete turn of the spiral (Figure 2–28). Longitudinally aligned subunits make up protofilaments and 13 parallel protofilaments constitute a microtubule.

Polymerization of tubulins to form microtubules *in vivo* is directed by **microtubule organizing centers (MTOCs),** which contain γ-tubulin ring complexes that act as nucleating sites for polymerization. MTOCs include centrosomes and the basal bodies of cilia. Microtubules are polarized structures and growth, via tubulin polymerization, occurs more rapidly at one end of existing microtubules (Figure 2-31a). This end is referred to as the plus (+) end, and the other is the minus (–) end. Microtubules show dynamic instability, with tubulin polymerization and depolymerization dependent on concentrations of Ca^{2+}, Mg^{2+}, GTP and specific **microtubule-associated proteins (MAPs).** Microtubule stability is variable; for example, microtubules of cilia are very stable, whereas microtubules of the mitotic spindle have a short duration. The antimitotic alkaloid colchicine binds specifically to tubulin, and when the complex tubulin-colchicine binds to microtubules, it prevents the addition of more tubulin in the plus (+) extremity. Mitotic microtubules are broken down because the depolymerization continues, mainly at the minus (–) end, and the lost tubulin units are not replaced.

Cytoplasmic microtubules are stiff structures that play a significant role in the formation and maintenance of cell shape. Procedures that disrupt microtubules result in the loss of cellular asymmetry.

Complex microtubule networks also participate in the intracellular transport of organelles and vesicles. Examples include axoplasmic transport in neurons, melanin transport in pigment cells, chromosome movements by the mitotic spindle, and vesicle movements among different cell compartments. In each of these examples, movement is suspended if microtubules are disrupted. Transport along microtubules is under the control of special MAPs called **motor proteins,** which use ATP to move molecules and vesicles. **Kinesins** carry organelles away from the MTOC toward the plus end of microtubules; **cytoplasmic dyneins** carry vesicles in the opposite direction.

Microtubules provide the basis for several complex cytoplasmic components, including centrioles, basal bodies, cilia, and flagella (Figure 2–31b and c). **Centrioles** are cylindrical structures (0.15 μm in diameter and 0.3–0.5 μm in length) composed primarily of short, highly organized microtubules (Figure 2–31c). Each centriole has nine microtubular triplets and adjacent microtubules share some protofilaments. A pair of centrioles surrounded by a matrix of tubulin subunits close to the nucleus of nondividing cells constitutes a **centrosome** (Figure 2–32).

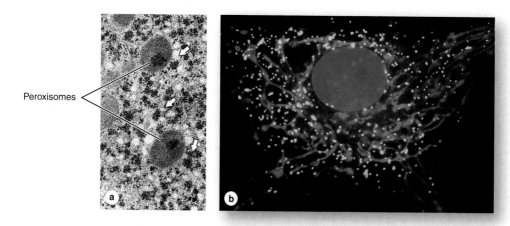

Peroxisomes

Figure 2–27. Peroxisomes. **Peroxisomes** (or microbodies) are small spherical, membranous organelles, containing enzymes that use O_2 to remove hydrogen atoms from substrates, typically fatty acids, in a reaction that produces hydrogen peroxide (H_2O_2) which must be broken down to water and O_2 by another enzyme, **catalase. (a):** By TEM peroxisomes generally show a homogenous matrix of moderate electron-density, but may include darker crystalloid internal structures representing very dense concentrations of enzymes. The arrows indicate small aggregates of glycogen. (x30,000) **(b):** A cultured endothelial cell processed by immunocytochemistry shows many peroxisomes (green) distributed throughout the cytoplasm among the vitally stained elongate mitochondria (red) around the DAPI-stained nucleus (blue). Peroxisomes shown here were specifically stained using an antibody against the membrane protein PMP70. (Figure 2–27b, with permission, from Invitrogen.)

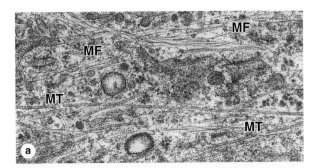

Figure 2–29. Microtubules and actin filaments in cytoplasm. **(a):** Actin micro filaments (MF) and **microtubules** (MT) can both be clearly distinguished in this TEM photo of fibroblast cytoplasm. The image also provides a good comparison of the relative diameters of these two cytoskeletal components. X60,000.

(b): The ultrastructural view can be compared to the appearance of microfilaments and microtubules in a cultured cell stained by immunocytochemistry. Actin filaments (red) are most concentrated at the cell periphery, forming prominent circumferential bundles from which finer filaments project into the transient cellular extensions at the edge of the cell and push against the cell membrane. Such an arrangement of actin filaments forms a dynamic network important for cell shape changes such as those during cell division, locomotion, and formation of cellular processes, folds, pseudopodia, lamellipodia, veils, microvilli, etc. which serve to change a cell's surface area or give direction to a cell's crawling movements.

Microtubules (green/yellow) are present throughout the cytoplasm and are oriented in arrays which generally extend from the area around the nucleus into the most peripheral extensions. Besides serving to stabilize cell shape, microtubules form the tracks for kinesin-based transport of vesicles and organelles into the cell periphery and dynein-based transport toward the cell nucleus. Variations of these arrangements of microfilaments and microtubules can be seen in Figure 2–20c and Figure 2–11b, respectively. (Figure 2–29b, with permission, from Albert Tousson, University of Alabama—Birmingham High Resolution Imaging Facility.)

Protofilament

Figure 2–28. Molecular organization of a microtubule. **Microtubules** are rigid structures which assemble from heterodimers of α and β **tubulin**. Microtubules have an outer diameter of 24 nm and a hollow lumen 14 nm wide. Tubulin molecules are arranged to form 13 **protofilaments**, as seen in the cross section in the upper part of the drawing. The specific orientation of the tubulin dimers results in structural polarity of the microtubule. Microtubules elongate or rapidly shorten by the addition or removal of tubulin at the ends of individual protofilaments. The lengths and locations of cytoplasmic microtubules vary greatly during different phases of cell activity, with assembly dependent on shifting balances between polymerized and unpolymerized tubulin and other factors in "dynamic instability."

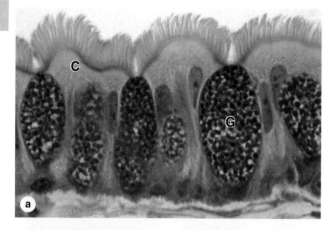

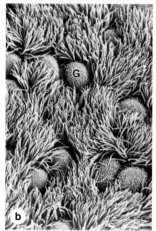
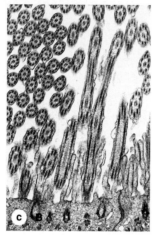

In each pair the long axes of the centrioles are at right angles to each other. Before cell division, more specifically during the S period of the interphase, each centrosome duplicates itself so that now each centrosome has two pairs of centrioles. During mitosis, the centrosome divide into halves, which move to opposite poles of the cell, and become organizing centers for the microtubules of the mitotic spindle.

Cilia and **flagella** (singular: cilium, flagellum) are motile processes, covered by cell membrane, with a highly organized microtubule core. Ciliated cells typically possess a large number of cilia, each about 2–3 μm in length. The main function of cilia is to sweep fluid along the surface of cell sheets. In humans, the spermatozoa are the only cell type with a flagellum, with a length close to 100 μm, used for motility.

Both cilia and flagella possess the same core structure, consisting of nine peripheral microtubular doublets surrounding two central microtubules. This assembly of microtubules with the **9 + 2 pattern** is called an **axoneme** (Gr. *axon*, axis, + *nema*, thread). Microtubules of the nine peripheral doublets each share a few protofilaments (Figure 2–31b). The microtubules of the peripheral doublets are identified as A (complete with 13 protofilaments), and B (with only 10 protofilaments). Adjacent peripheral doublets are linked to each other by protein bridges called **nexins** and each doublet has a **radial spoke** projecting toward the center. Extending from the surface of microtubule A are inner and outer arms of **axonemal dynein,** which project toward the B microtubule of the next doublet. ATP-dependent interactions of the dyneins with the neighboring microtubule cause repetitive conformational changes that are coordinated to produce a repeated beating motion of the entire axoneme. At the base of each cilium or flagellum is a **basal body,** essentially similar to a centriole, which controls the assembly of the axoneme.

Figure 2–30. Cilia. Cilia are motile structures projecting from a cell, typically the apical end of epithelial cells. Each cilium is covered by the cell membrane and contains cytoplasm dominated by a specialized assembly of unusually stable microtubules, the **axoneme.** Shifting movements between microtubules of an axoneme produce whip-like motions of the cilia. Most epithelial cells lining the respiratory tract, such as those shown in the three micrographs here, have numerous cilia which move to propel mucus along the tract toward the pharynx. Between the ciliated cells are mucus-producing, non-ciliated goblet cells (G) with basal nuclei and apical cytoplasm filled with mucus granules. The relative size and spacing of the ciliated cells and goblet cells is seen in micrographs. **(a):** Light micrograph. X400. Pararosaniline-toluidine blue, PT. **(b):** SEM. X300. **(c):** TEM shows the axonemes of cilia cut in different orientations and their basal bodies in the apical cytoplasm. X9200. (Figure 2–30b reproduced, with permission from P. Andrews: *Am J Anat* 1974; 139:421. Copyright ©1974 by Wiley-Liss, Inc., a subsidiary of John Wiley & Sons, Inc.)

In addition to the numerous cilia on specialized cells such as these, many (perhaps most) other cell types have a single, short *primary cilium* with similar axoneme structure. Primary cilia lack dynein and are nonmobile, but serve as sensory structures receiving mechanical and chemical signals which are transduced by the cell to generate an appropriate response. Many signaling proteins, including those of developmentally important pathways, are concentrated in primary cilia which have various functions, including specific cell interactions during embryonic development.

MEDICAL APPLICATION

Several mutations have been described in the proteins of the cilia and flagella. They are responsible for the immotile cilia syndrome, the symptoms of which are immotile spermatozoa, male infertility, and chronic respiratory infections caused by the lack of the cleansing action of cilia in the respiratory tract.

Microfilaments (Actin Filaments)

Contractile activity in cells results primarily from an interaction between **actin** and its associated protein, **myosin.** Actin is present as thin (5–7 nm diameter) polarized **microfilaments** composed of globular subunits organized into a double-stranded helix (Figures 2–33 and 2–29). There are several types of actin and this protein is present in all cells. Actin is usually found in cells as polymerized filaments of F-actin mingled with free globular G-actin subunits.

Within cells, actin microfilaments (F-actin) can be organized in several forms.

1. In skeletal muscle, they assume a stable array integrated with thick (16-nm) myosin filaments.

2. In most cells, microfilaments form a thin sheath or network just beneath the plasmalemma. These filaments are involved

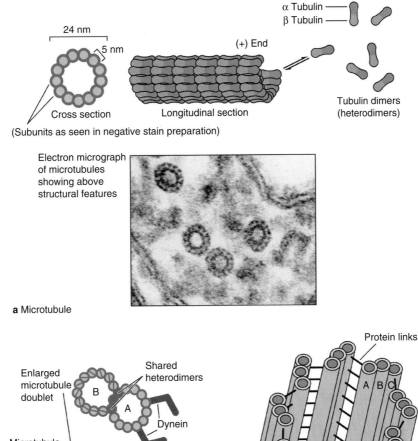

24 nm

5 nm

α Tubulin
β Tubulin

(+) End

Cross section

Longitudinal section

Tubulin dimers
(heterodimers)

(Subunits as seen in negative stain preparation)

Electron micrograph
of microtubules
showing above
structural features

a Microtubule

Protein links

Enlarged
microtubule
doublet

B

A

Shared
heterodimers

Dynein

A B C

Microtubule
doublet

Plasmalemma

Nexin

Central
sheath

Radial
spokes

Axoneme (with 9 + 2 pattern)

b Cilium

c Centriole

Figure 2–31. Microtubules, cilia, and centrioles. Microtubules are seen **(a)**: cross-section by TEM after fixation with tannic acid in glutaraldehyde, which leaves the unstained tubulin subunits delineated by the dense tannic acid. Cross sections of tubules reveal the ring of 13 subunits of dimeric tubulin which are arranged lengthwise as protofilaments. Changes in microtubule length are caused by the addition or loss of individual tubulin subunits from protofilaments. **(b)**: A diagrammatic cross-section through a cilium reveals a cytoplasmic core of microtubules called an **axoneme**. The axoneme consists of **two central microtubules** surrounded by **nine peripheral microtubular doublets** associated with several other proteins. In the doublets, microtubule A is complete, consisting of 13 protofilaments, whereas microtubule B shares some of A's protofilament heterodimers. A series of protein complexes containing ciliary dynein, the **inner and outer dynein arms**, are bound to microtubule A along its length. When activated by ATP, the dynein arms briefly link microtubule B of the adjacent doublet and provide for slight sliding of the doublets against each other, which is then immediately reversed. This rapid back-and-forth shift between adjacent doublets, produced by the ciliary dynein motors, causes the rhythmic changes of axonemal shape that bring about the flailing motion of the entire cilium.

Each axoneme is continuous with a **basal body** located at the base of the cilium. Basal bodies are structurally very similar to centrioles, which nucleate and organize the growth of microtubules during formation of the mitotic spindle. **(c)**: Each centriole consists of nine relatively short **microtubular triplets** linked together in a pinwheel-like arrangement. In the triplets, microtubule A is complete and consists of 13 protofilaments, whereas microtubules B and C share protofilaments. Under normal circumstances, these organelles are found in pairs and are oriented at right angles to one another. The *pair* of centrioles is called a **centrosome**.

in all cell shape changes such as those during endocytosis, exocytosis, and cell locomotion.

3. Microfilaments are intimately associated with several cytoplasmic organelles, vesicles, and granules and play a role in moving or shifting cytoplasmic components (cytoplasmic streaming).

4. Microfilaments are associated with myosin and form a "purse-string" ring of filaments whose constriction results in the cleavage of mitotic cells.

5. In crawling cells actin filaments are organized into parallel contractile bundles called **stress fibers** (Figure 2–20C).

Although actin filaments in muscle cells are structurally stable, in nonmuscle cells they readily dissociate and reassemble. Actin filament polymerization appears to be under the direct control of minute changes in Ca^{2+} and cyclic AMP levels. A large number of **actin-binding proteins** with different activities have been demonstrated in various cells and include:

- actin motor proteins such as the *myosins*, which carry other molecules or vesicles along microfilaments,
- actin-capping proteins such as *tropomyosin*, which bind the free end and stabilize microfilaments,
- actin filament-severing proteins such as *gelsolin*, which break microfilaments into short pieces,
- actin-bundling proteins such as *fimbrin, villin,* and *α-actinin,* which crosslink microfilaments, and
- actin-branching proteins such as *formin*, which produce branch points along a microfilament.

Intermediate Filaments

In addition to microtubules and the thin actin filaments, eukaryotic cells contain a class of filaments intermediate in size between the other two cytoskeletal components and with a more variable

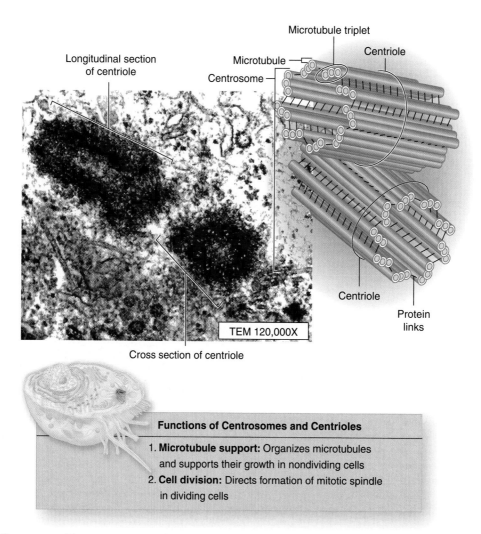

Functions of Centrosomes and Centrioles

1. **Microtubule support:** Organizes microtubules and supports their growth in nondividing cells
2. **Cell division:** Directs formation of mitotic spindle in dividing cells

***Figure 2–32.* Centrosome.** The **centrosome** is the microtubule-organizing center for the mitotic spindle and consists of paired centrioles. The TEM reveals that the two centrioles in a centrosome exist at right angles to one another in a dense matrix of free tubulin subunits and other proteins. Each centriole consists of **nine microtubular triplets**. In a poorly understood process, the centrosome duplicates itself and is divided equally during a cell's interphase, each half having a duplicated centriole pair. At the onset of mitosis, the two daughter centrosomes move to opposite sides of the nucleus and become the two poles of the mitotic spindle of microtubules attaching to chromosomes.

diameter averaging 10–12 nm (Figure 2–34). In comparison with microtubules and actin filaments, **intermediate filaments** are much more stable and vary in their protein subunit structure in different cell types. A dozen or more heterogeneous protein classes that form such intermediate filaments have been identified and localized immunocytochemically, some of which are listed in Table 2–3. The size of these intermediate filament subunits ranges from 40 to 240 kDa. All are essentially rod-like rather than globular proteins that form coiled tetramers which self-assemble into large cable-like arrays stabilized by further interactions laterally.

Intermediate filament proteins have been organized chemically and genetically into four major groups:

- **Keratins** (Gr. *keras*, horn) or cytokeratins are a diverse family of more than 20 proteins found in all epithelial cells and in the hard structures produced by epidermal cells (eg, nails, horns, feathers and scales). They are encoded by related genes but have different chemical and immunologic properties and play various roles. In epidermal cells (Figure 2–35) keratins strengthen the tissue and provide protection against abrasion and water loss.

Table 2–3. Examples of intermediate filaments found in eukaryotic cells.

Filament Type	Cell Type	Examples
Cytokeratins	Epithelium	Both keratinizing and nonkeratinizing epithelia
Vimentin	Mesenchymal cells	Fibroblasts, chondroblasts, macrophages, endothelial cells, vascular smooth muscle
Desmin	Muscle	Striated and smooth muscle (except vascular smooth muscle)
Glial fibrillary acidic proteins	Glial cells	Astrocytes
Neurofilaments	Neurons	Nerve cell body and processes

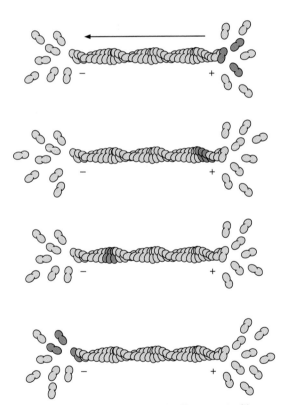

Figure 2–33. Actin filament treadmilling. Actin filaments or microfilaments are helical two-stranded polymers assembled from **globular actin subunits**. The filaments are flexible structures, with diameters in various cells of 5-9 nm, depending on associated proteins. Assembly of actin filaments (F-actin) results in their polarity, with actin subunits (G-actin) added to the plus (+) end and removed at the minus (−) end. Even actin filaments of a constant length are highly dynamic structures, balancing G-actin assembly and disassembly at the opposite ends, with a net movement or flow along the polymer known as treadmilling.

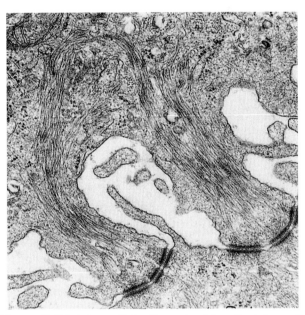

Figure 2–34. Intermediate filaments of keratin. Intermediate filaments display an average diameter of 10-12 nm, between that of actin filaments and microtubules, and serve to provide mechanical strength or stability to cells. Unlike the other two cytoskeletal polymers, intermediate filaments are composed of various protein subunits in different types of cells. All such subunits appear to be rodlike rather than globular and undergo step-wise assembly into a structure resembling a cable with many strands. A large and important class of intermediate filaments is composed of **keratin** subunits, which are prominent in epithelial cells. Bundles of keratin filaments associate with certain classes of intercellular junctions common in epithelial cells and are easily seen with the TEM, as shown here in two extensions in an epidermal cell bound to a neighboring cell.

Table 2–4. Some human and animal diseases related to specific alterations of organelles.

Organelle	Disease	Molecular Defect	Morphologic Change	Clinical Consequence
Mitochondrion	Mitochondrial cytopathy	Defect of oxidative phosphorylation	Increase in size and number of muscle mitochondria	High basal metabolism without hyperthyroidism
Microtubule	Immotile cilia syndrome	Lack of dynein in cilia and flagella	Lack of arms of the doublet microtubules	Immotile cilia and flagella with male sterility and chronic respiratory infection
	Mouse (*Acomys*) diabetes	Reduction of tubulin in pancreatic β cells	Reduction of microtubules in β cells	High blood sugar content (diabetes)
Lysosome	Metachromatic leukodystrophy	Lack of lysosomal sulfatase	Accumulation of lipid (cerebroside) in tissues	Motor and mental impairment
	Hurler disease	Lack of lysosomal α-L-iduronidase	Accumulation of dermatan sulfate in tissues	Growth and mental retardation
Golgi apparatus	I-cell disease	Phosphotransferase deficiency	Inclusion-particle storage in several cells	Psychomotor retardation, bone abnormalities

MEDICAL APPLICATION

As discussed throughout this chapter, many diseases are related to molecular alterations in organelles or other specific cytoplasmic components. In several of these diseases, structural changes can be detected by light or electron microscopy or by cytochemical techniques. Table 2–4 lists a few more such diseases and emphasizes the importance of understanding the many cell components in pathobiology.

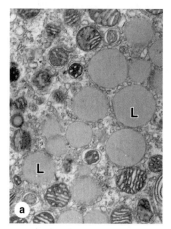 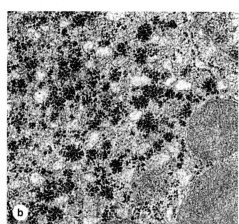 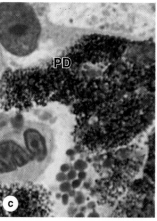

Figure 2–35. Cellular inclusions. **Inclusions** are cytoplasmic structures or deposits filled with stored macromolecules and are not present in all cells. **(a):** **Lipid droplets** are abundant in cells of the adrenal cortex, and appear with the TEM as small spherical structures with homogenous matrices (L). Mitochondria are also seen here. As aggregates of hydrophobic lipid molecules these inclusions are enclosed by a single monolayer of phospholipids with various peripheral proteins, including enzymes for lipid metabolism. In routine processing of tissue for paraffin sections fat droplets are generally removed, leaving empty spaces in the cells. Common fat cells have cytoplasm essentially filled with one large lipid droplet. X19,000.

(b): TEM of a liver cell cytoplasm shows numerous individual or clustered electron-dense particles representing **glycogen granules**, although these granules lack membrane. Glycogen granules usually form characteristic aggregates such as those shown. Glycogen is a ready source of energy and such granules are often abundant in cells with high metabolic activity. X30,000. **(c):** **Pigment deposits** (PD) occur in many cell types and may contain various complex substances, such as **lipofuscin** or **melanin**. Lipofuscin granules represent an accumulating by-product of lysosomal digestion in long-lived cells, but melanin granules serve to protect cell nuclei from damage to DNA caused by light. Many cells contain pigmented deposits of **hemosiderin granules** containing the protein ferritin, which forms a storage complex for iron. Hemosiderin granules are very electron-dense, but with the light microscope appear brownish and resemble lipofuscin. The liver cells shown have large cytoplasmic regions filled with pigment deposits which probably represent iron-containing hemosiderin. X400. Giemsa.

- **Vimentin** is a single protein (56–58 kDa) and is the most common intermediate filament protein in mesenchymal cells derived from the middle layer of the early embryo. Important vimentin-like proteins are **desmin** found in almost all muscle cells and **glial fibrillar acidic protein (GFAP)** found in astrocytes, supporting cells of the central nervous system tissues. The desmin filaments of a cultured cell are shown after immunocytochemistry in Figure 1–13.
- **Neurofilaments** consist of at least three high-molecular-weight polypeptides (68, 140, and 210 kDa) with different chemical structures and different roles. All are restricted to neurons.
- **Lamins** consist of three proteins averaging about 70 kDa in size present in the nucleus of animal cells. They form a structural framework just inside the nuclear envelope.

MEDICAL APPLICATION

The presence of a specific type of intermediate filament in tumors can reveal which cell originated the tumor, information important for diagnosis and treatment of the cancer. Identification of intermediate filament proteins by means of immunocytochemical methods is a routine procedure.

INCLUSIONS

Unlike organelles, cytoplasmic **inclusions** are composed mainly of accumulated metabolites or other substances and are often transitory components of the cytoplasm. Nonmotile and with little or no metabolic activity, inclusions are not considered organelles. Important and commonly seen inclusions include:

- **Fat droplets**, accumulations of lipid molecules that are prominent in adipocytes (fat cells), adrenal cortex cells, liver and other cells (Figure 2–35).
- **Glycogen granules**, aggregates of a carbohydrate polymer in which glucose is stored and are also visible in several cell types, mainly liver cells, in the form of irregular clumps of PAS-positive or electron-dense material (Figure 2–35). They are not enclosed with membrane.
- **Lipofuscin granules**, small pigmented (golden-brown) bodies present in many cells, but which accumulate with age in stable nondividing cells (eg, neurons, cardiac muscle). Lipofuscin granules contain a complex mixture of material derived from residual bodies after lysosomal digestion.

The Cell Nucleus

The nucleus contains a blueprint for all cell structures and activities encoded in the DNA of the chromosomes. It also contains the molecular machinery to replicate its DNA and to synthesize and process all types of RNA. Macromolecular transfer between the nuclear and cytoplasmic compartments is regulated. Because functional ribosomes do not occur in the nucleus, no proteins are produced there. The numerous protein molecules needed for the activities of the nucleus are imported from the cytoplasm.

COMPONENTS OF THE NUCLEUS

The nucleus frequently appears as a rounded or oval structure, usually in the center of the cell (Figure 3–1). Its main components are the **nuclear envelope, chromatin** consisting of DNA and associated proteins, and a specialized region of chromatin called the **nucleolus** (Figures 3–2 and 3–3). The size and morphologic features of nuclei in a specific normal tissue tend to be uniform. In contrast, the nuclei in cancer cells often have irregular shapes, variable sizes, and atypical chromatin patterns.

Nuclear Envelope

Electron microscopy shows that the nucleus is surrounded by two parallel unit membranes separated by a narrow (30–50 nm) **perinuclear space** (Figure 3–2). Together, the paired membranes and the intervening space make up the **nuclear envelope**. Polyribosomes are attached to the outer nuclear membrane, indicating continuity of the nuclear envelope with the endoplasmic reticulum. Closely associated with the inner nuclear membrane is a meshwork of fibrous proteins called the **nuclear lamina** (Figure 3–4), which helps to stabilize the nuclear envelope. Major components of this lamina are intermediate filament proteins called **lamins** which bind to membrane proteins and associate with chromatin in nondividing cells. The pattern of association is regular from cell to cell within a tissue, supporting the conclusion that chromosomes have a definite localization within the nucleus. (Whether nuclei contain a matrix of proteins in

addition to lamins for organizing and moving chromatin, proteins and ribonucleoproteins remains an area of dispute among cell biologists.)

At sites where the inner and outer membranes of the nuclear envelope fuse, the resulting lipid-free spaces contain **nuclear pore complexes** or NPCs (Figures 3–5, 3–6, and 3–7), which contain the machinery that regulates most bidirectional transport between the nucleus and the cytoplasm. The nucleus of a typical mammalian cell contains 3000–4000 such pore complexes, each composed of subunits with some 30 different NPC proteins or **nucleoporins** (Figure 3–7).

Chromatin

In nondividing nuclei, chromatin is the chromosomal material in a largely uncoiled state. Two types of chromatin can be distinguished with both the light and electron microscopes, which reflect the degree of chromosomal condensation (Figures 3–2 and 3–3). **Heterochromatin** (Gr. *heteros*, other, + *chroma*, color), which is electron dense, appears as coarse granules in the electron microscope and as basophilic clumps in the light microscope. **Euchromatin** is the less coiled portion of the chromosomes, visible as finely dispersed granular material in the electron microscope and as lightly stained basophilic areas in the light microscope. The regions of heterochromatin and euchromatin account for the patchy light-and-dark appearance of nuclei in tissue sections as seen by both light and electron microscopy. The intensity of nuclear staining of the chromatin is frequently used to distinguish and identify different tissues and cell types in the light microscope.

Chromatin is composed mainly of coiled strands of DNA bound to basic proteins called **histones** and to various nonhistone proteins. The basic structural unit of chromatin and histones is the **nucleosome** (Figure 3–8), which has a core of eight small histones (two copies each of histones H2A, H2B, H3, and H4), around which is wrapped DNA with about 150 base pairs. Each nucleosome also has a larger linker histone (H1) that binds

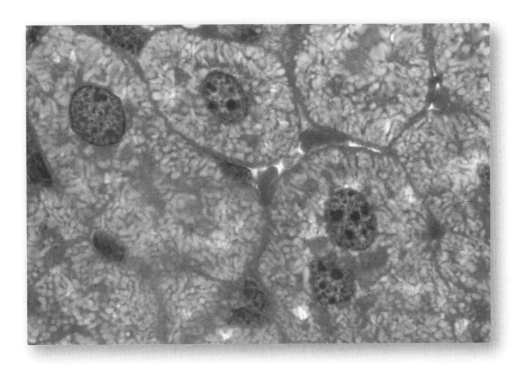

Figure 3–1. **Nuclei of large, active cells.** Liver cells (hepatocytes) have large, well-stained nuclei located in the center of the cytoplasm. One or more nucleoli are seen inside each nucleus, indicating intense protein synthesis by these cells. Most of the chromatin is light-staining or euchromatic, with small areas of more darkly stained heterochromatin scattered throughout the nucleus and just inside the nuclear envelope. This superficial heterochromatin allows the boundary of the organelle to be seen more easily by light microscopy. One cell here has two nuclei, which is fairly common in the liver. X500. Pararosaniline–toluidine blue.

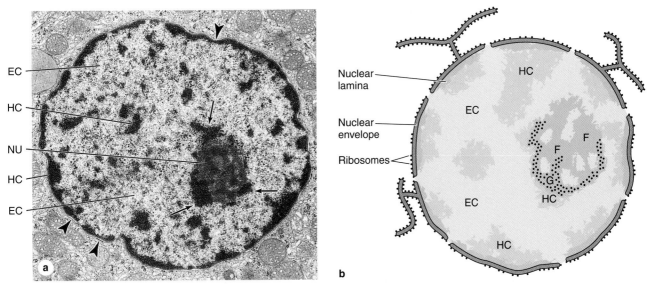

Figure 3–2. **Structural components of the nucleus. (a):** TEM of a typical cell nucleus clearly shows the electron-dense heterochromatin (HC) and the more diffuse euchromatin (EC). The arrows indicate the nucleolus-associated heterochromatin around the nucleolus (NU). Arrowheads indicate areas where the perinuclear space between the two membranes of the nuclear envelope is clearly seen. Just inside the nuclear envelope is a thin electron-dense region containing the nuclear lamina and more heterochromatin. X26,000. **(b):** Schematic representation of a cell nucleus shows that the nuclear envelope is made of two membranes separated by the perinuclear space. The outer membrane has ribosomes bound to it and is continuous with the ER. The two membranes fuse at many places to form nuclear pores. Heterochromatin clumps (HC) are associated with the meshwork of the nuclear lamina just inside the nuclear envelope, whereas the euchromatin (EC) appears dispersed in the interior of the nucleus. The nucleolus contains distinct regions called the pars granulosa (G) and the pars fibrosa (F).

both wrapped DNA and the surface of the core. The series of nucleosomes in chromatin is also associated with many diverse nonhistone proteins with a wide variety of enzymatic functions.

DNA bound to nucleosomes is then folded further in the next order of chromatin organization which is the 30-nm fiber, but the mechanism of this folding is less well understood. Higher orders of chromatin coiling into microscopically visible stained structures, the **chromosomes**, also occur, which are especially important during the condensation of chromatin for mitosis and meiosis (Figure 3–9).

The chromatin pattern of a nucleus is a guide to the cell's activity. Generally cells with lightly stained nuclei are more active in protein synthesis than those with condensed, dark nuclei. In light-stained nuclei with much euchromatin and few heterochromatic clumps, more DNA surface is available for the transcription of RNA. In dark-stained nuclei rich in highly condensed heterochromatin, the tightly coiled DNA is less accessible for transcription.

Careful study of the chromatin of mammalian cell nuclei reveals a mass of heterochromatin that is frequently observed in somatic cells of females but not males. This chromatin clump is the **sex chromatin** and is one of the two X chromosomes present in female cells (Figure 3–10). The X chromosome that constitutes the sex chromatin remains tightly coiled and visible between mitotic cycles, whereas the other X chromosome is uncoiled and not visible. The heterochromatic sex chromatin is transcriptionally inactive. The male cell has one X chromosome and one Y chromosome; like the other chromosomes the interphase X chromosome is uncoiled and therefore no sex chromatin is visible in males. X chromosome inactivation involves a number of specific chemical modifications of its histones.

MEDICAL APPLICATION

The study of sex chromatin discloses the genetic sex in patients whose external sex organs do not permit determination of gender, as in hermaphroditism and pseudohermaphroditism. Sex chromatin analysis also helps the study of other anomalies involving the sex chromosomes—eg, Klinefelter syndrome, in which testicular abnormalities, azoospermia (absence of spermatozoa), and other symptoms are associated with the presence of XXY chromosomes.

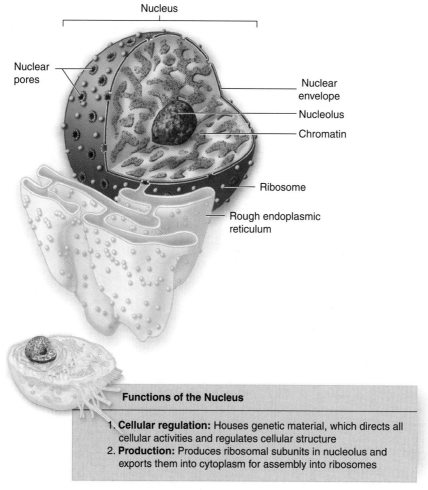

Nucleus

Nuclear pores

Nuclear envelope

Nucleolus

Chromatin

Ribosome

Rough endoplasmic reticulum

Functions of the Nucleus

1. **Cellular regulation:** Houses genetic material, which directs all cellular activities and regulates cellular structure
2. **Production:** Produces ribosomal subunits in nucleolus and exports them into cytoplasm for assembly into ribosomes

Figure 3–3. **Relationship of nuclear envelope to the rough ER.** Three-dimensional representation of a cell nucleus shows a single large nucleolus and the distribution of the nuclear pores in the envelope. The number of nuclear pores varies greatly from cell to cell, increasing in cells actively involved in protein synthesis.

The X and Y chromosomes contain genes which determine whether an individual will develop as a female or a male. In humans most cells of the body, the somatic cells, contain 22 pairs of **autosomes** in addition to the pair of sex chromosomes. Each of these 23 pairs of chromosomes contains one chromosome originally derived from the mother and one derived from the father. The members of each chromosomal pair are called **homologous** because although from different parents they contain forms (alleles) of the same genes. Somatic cells are considered **diploid** because they contain paired chromosomes. Geneticists refer to diploid cells as $2n$, where n is the number of unique chromosomes in cells of a species, 23 in humans. Sperm cells and mature oocytes are **haploid**, with half the diploid number of chromosomes (n), each pair of chromosomes having been separated during meiosis (described below).

Study of chromosomes themselves usually uses cells grown *in vitro* and the arrest of mitotic cells during metaphase using colchicine which binds tubulin and disrupts microtubules. Arrested cells are then immersed in a hypotonic solution, which causes swelling, stained in various ways, and then flattened between a glass slide and a coverslip. The mitotic chromosomes from one nucleus are then photographed under the light microscope, cut individually from the photograph, and arranged to produce a **karyotype** in which the stained chromosomal bands can be analyzed (Figure 3–11).

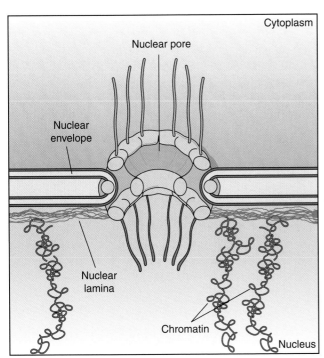

***Figure 3–4.* Nuclear lamina.** The **nuclear lamina** is formed from a class of intermediate filaments proteins, the **lamins**, which assemble as a lattice adjacent to the inner nuclear membrane. When the nuclear envelope disperses during early prophase of cell division, at least some lamin proteins remain attached to the membrane fragments and reassembly of the nuclear lamina immediately after cell division facilitates reformation of the nuclear envelopes of the two new nuclei. The nuclear lamina also contains binding sites for **chromatin**, helping to organize this material in the nucleus. Chromatin is not present at the openings through the nuclear envelope called nuclear pore complexes.

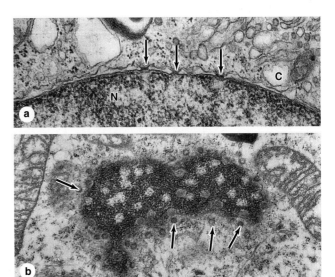

***Figure 3–5.* Nuclear pores.** TEM micrographs show nuclear envelopes and nuclear pores between nucleus (N) and cytoplasm (C). **(a):** Section through the nuclear envelope and the two-membrane structure of the nuclear envelope clearly. The electron-dense proteins that make up the nuclear pore complexes can also be seen (arrows). Immediately beneath the nuclear envelope is the nuclear lamina and heterochromatin, material that is not present however at the nuclear pores. **(b):** Tangential section through a nuclear envelope shows the electron-dense nuclear pore complexes (arrows) and the electron-lucent patches in the peripheral heterochromatin which represent the areas just inside the pores. X80,000.

Nucleolus

The nucleolus is a generally spherical, highly basophilic structure present in the nuclei of cells active in protein synthesis (Figure 3–12). The intense basophilia of nucleoli is due not to heterochromatin, but to the presence of densely concentrated rRNA which is transcribed, processed, and complexed into ribosomal subunits in that nuclear region. Nucleoli are always associated with nuclei of cells that are intensely synthesizing proteins for growth or secretion. As seen with the transmission electron microscope (TEM), the nucleolus consists of distinct subregions with different staining characteristics (Figure 3–13). The molecules of rRNA synthesized and modified in the nucleolus very quickly associate with the many ribosomal proteins which are imported from the cytoplasm through the nuclear pore complexes. The newly organized small and large ribosomal subunits are then exported back to the cytoplasm through those same nuclear pores.

MEDICAL APPLICATION

Large nucleoli are encountered in cells that are actively synthesizing proteins and in cells of rapidly growing malignant tumors. The nucleolus disperses during the prophase of cell division but reappears in the telophase stage of mitosis.

CELL DIVISION

Cell division, or mitosis (Gr. *mitos,* a thread), can be observed with the light microscope. During this process, the parent cell divides, and each of the daughter cells receives a chromosomal set identical to that of the parent cell. Essentially, a longitudinal duplication of the chromosomes takes place, and these chromosomes are distributed to the daughter cells. The period between mitoses is called **interphase,** during which the DNA is replicated and the nucleus appears as it is most commonly seen in histological preparations. The process of mitosis is subdivided into four phases (Figures 3–14 and 3–15).

In **prophase** of mitosis the replicated chromatin condenses into discrete rod-shaped bodies, the chromosomes, each consisting of duplicate sister chromatids closely associated longitudinally. Outside the nucleus, the centrosomes with their centrioles separate and migrate to opposite poles of the cell. The duplication of the centrosomes and centrioles occurs during interphase. Simultaneously with the centrosome migration, the microtubules of the mitotic spindle appear between the two centrosomes and the nucleolus disappears as transcriptional activity there stops. Late in prophase, the nuclear envelope breaks down when proteins of the nuclear lamina and inner membrane are phosphorylated (PO_4^{3-} groups added). The nuclear lamina and pore complexes disassemble and these proteins along with membrane vesicles disperse in local cytosol and ER.

During **metaphase,** the condensed chromosomes attach to microtubules of the mitotic spindle (Figures 3–14 and 3–16) at large electron-dense protein complexes called **kinetochores**

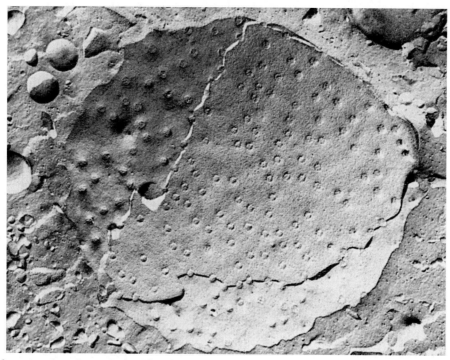

Figure 3–6. **Cryofracture of nuclear envelop showing nuclear pores.** Electron micrograph obtained by freeze-fracture of an intestinal cell shows the two components of the nuclear envelope and the nuclear pores. The fracture plane occurs partly *between* the two nuclear envelope membranes (left) but mostly just inside the envelope with the chromatin falling away. The size and distribution of the nuclear pore complexes are clearly seen. The same nuclear pore complexes can mediate both the import and export of macromolecules between the nucleus and cytoplasm using tightly controlled processes in each direction.

(Gr. *kinetos,* moving, + *chora,* central region), which are located at a constricted region of each chromatid called the **centromere** (Gr. *kentron,* center, + *meros,* part). The chromosomes are moved to the equatorial plane of the now more spherical cell. Kinetochore microtubules bound to sister chromatids are continuous with centrosomes at opposite poles of the mitotic spindle.

In **anaphase,** the sister chromatids separate from each other and are slowly pulled at their kinetochores toward opposite spindle poles by kinesin motors moving along the microtubules. During this time the spindle poles also move farther apart.

At **telophase** the two sets of chromosomes are at the spindle poles and begin reverting to their decondensed state. Microtubules of the spindle depolymerize and the nuclear envelope begins to reassemble around each set of daughter chromosomes. A belt-like **contractile ring**, containing actin filaments associated with myosins, develops in the peripheral cytoplasm at the

equator of the parent cell. During **cytokinesis** at the end of telophase, constriction of this ring produces a **cleavage furrow** and progresses until the cytoplasm and its organelles are divided in two daughter cells, each with one nucleus.

Most tissues undergo constant cell turnover because of continuous cell division and the ongoing death of cells. Nerve and cardiac muscle cells are exceptions, because they do not multiply postnatally and therefore have greatly reduced potential for regeneration. The turnover rate of cells varies greatly from one tissue to another—rapid in the epithelium of the digestive tract and the epidermis, slow in the pancreas and the thyroid gland. Mitotic cells are often difficult to identify conclusively in sectioned adult organs, but can be recognized in rapidly growing tissues by their condensed chromatin (Figure 3–17).

THE CELL CYCLE

Mitosis is the visible manifestation of cell division, but other processes, less easily observed with the microscope, play fundamental roles in cell multiplication. Principal among these is the phase in which DNA is replicated. This process can be analyzed by introducing labeled DNA precursors (eg, [^{3}H] thymidine or thymidine analogs) into the cell and tracing them by means of biochemical, autoradiographic, or immunocytochemical methods. DNA replication occurs during interphase. The cyclic alternation between mitosis and interphase, known as the **cell cycle,** occurs in all tissues with cell turnover.

The cell cycle has four distinct phases: mitosis, and three interphase periods termed G_1 (the time gap between mitosis and DNA replication), S (the period of DNA synthesis), and G_2 (the gap between DNA duplication and the next mitosis). The

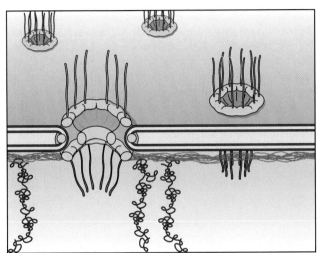

Figure 3–7. Nuclear pore complexes. A nuclear pore complex (NPC) is made of transmembrane proteins and other proteins which form an octagonal annulus or ring, with filaments extending into both the cytoplasm and the nucleus. Each complex contains about 30 different proteins, which have been referred to as **nucleoporins.** Multiple copies of many nucleoporins are assembled to form each octagonal NPC. The nuclear envelope is impermeable to ions and molecules of all sizes and the exchange of substances between the nucleus and the cytoplasm occur only through the nuclear pores. Ions and small molecules pass through the nuclear pore by passive diffusion. Larger molecules and molecular complexes are imported by a two-stage process. First proteins with certain amino acid sequences called **nuclear localization signals** are bound by soluble import receptor proteins and the resulting complexes then attach to filaments of nucleoporins on the cytoplasmic face of the NPC. Translocation of the protein across the envelop appears to occur by repeated low-affinity interactions with a series of discrete binding sites along these nucleoporin filaments, initially on the cytoplasmic face, then in the pore itself, and finally on the nucleoplasmic side of the NPC. Release of the protein cargo from nucleoporins inside the nucleus requires energy from GTP hydrolysis. Export of RNA and ribosomal subunits from the nucleus depends on a similar system of nuclear export signals and export receptor proteins binding the nucleoporins.

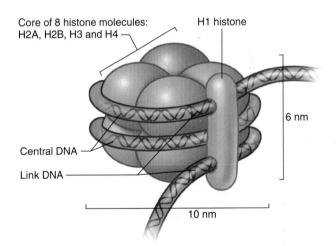

Core of 8 histone molecules: H2A, H2B, H3 and H4

H1 histone

Central DNA

Link DNA

6 nm

10 nm

Figure 3–8. Components of a nucleosome. Nucleosome is a structure that produces the initial organization of free double-stranded DNA into chromatin. Each nucleosome has an octomeric core complex made up of four types of **histones,** two copies each of H2A, H2B, H3, and H4. Around this core is wound DNA approximately 150 base pairs in length. One H1 histone is located outside the DNA on the surface of each nucleosome. DNA associated with nucleosomes *in vivo* thus resembles a long string of beads. Nucleosomes are very dynamic structures, with H1 loosening and DNA unwrapping at least once every second to allow other proteins, including transcription factors and enzymes, access to the DNA.

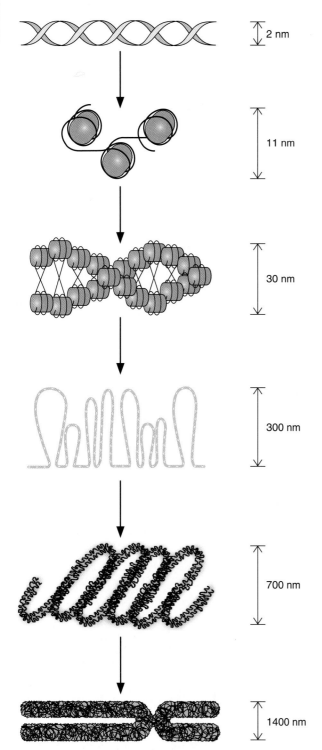

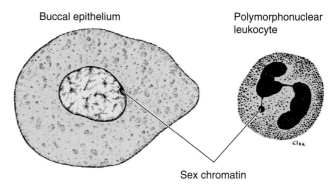

Buccal epithelium   Polymorphonuclear leukocyte

Sex chromatin

Figure 3–10. **Sex chromatin.** Either X chromosome in cells from females can undergo inactivation and clumping to form heterochromatic **sex chromatin.** Morphologic features of sex chromatin can be seen in human female epithelial cells lining the mouth and neutrophils. **Left:** In the oral epithelial cells, heterochromatic sex chromatin appears as a small granule adhering to the nuclear envelope. These superficial buccal cells lining the cheeks are frequently used to study sex chromatin or as a very convenient source of nucleated cells for DNA analyses. **Right:** In neutrophils, chromatin often has the shape of a drumstick projecting from the multilobed nucleus that is unique to these cells. The genetically inactive, heterochromatic X chromosome comprising the sex chromatin is sometimes called a **Barr body,** after the cytologist who first discovered it in the cells of females.

approximate times of these phases in rapidly dividing human cells are illustrated in Figures 3–18 and 3–19. During the G_1 phase there is active synthesis of RNA and proteins, including proteins that control the cell cycle, and the cell volume, reduced to one-half by mitosis, grows to its previous size. The S phase is characterized by the synthesis of DNA and histones and by the beginning of centrosome duplication. In the relatively short G_2 phase, proteins required for mitosis accumulate. As postmitotic cells begin to specialize and differentiate, cell cycle activities may be temporarily or permanently suspended and the cells are referred to as being in the G_0 phase. Some differentiated cells, such as those of the liver, renew cycling under certain conditions; others, including most muscle and nerve cells, are *terminally differentiated.*

Cycling in postmitotic cells (bypassing the G_0 state) is triggered by protein signals from the extracellular environment called mitogens or **growth factors,** which activate cell surface receptors. Nutrients and proteins required for DNA replication

2 nm

11 nm

30 nm

300 nm

700 nm

1400 nm

Figure 3–9. **From DNA to chromatin.** Several orders of **chromatin packing** are believed to occur during condensation of chromatin during mitotic prophase, although the protein associations involved at each stage are not well understood. Genetic activity is almost completely shut down during this process and histones are chemically modified in various ways. The top drawing shows the 2-nm DNA double helix, followed by the association of DNA with histones to form 11-nm filaments of nucleosomes connected by the DNA ("beads on a string"). Nucleosomes on the DNA then interact in a manner

not well understood to form a more compact 30-nm fiber. Through further condensation, filaments with diameters of 300 nm and 700 nm are formed. The highly folded loops of chromatin at these stages are stabilized by interactions with protein complexes made of **condensins** which eventually make up a central framework at the long axis of each chromatid. The bottom drawing shows a metaphase chromosome, which exhibits the maximum packing of DNA. The chromosome consists of two chromatids held together at a narrow point called the centromere.

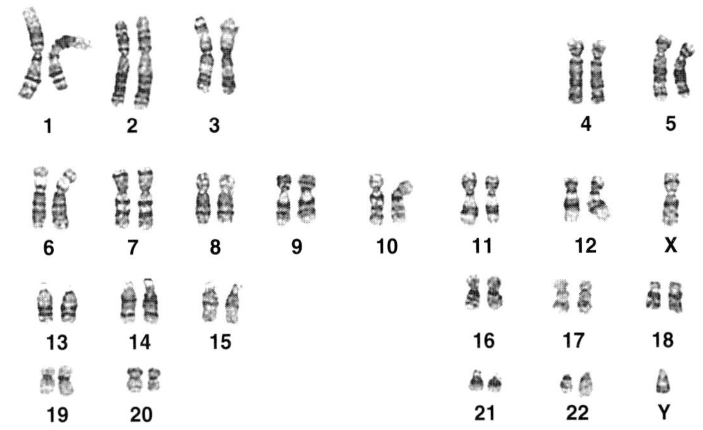

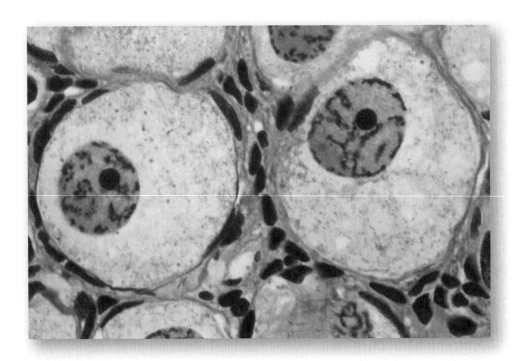

Figure 3–11. Karyotyping. Human karyotype preparations are made by staining and then photographing the chromosomes of cells disrupted after mitotic arrest with colchicine. Nuclei are chosen for analysis in which the individual chromosomes are maximally condensed. The individual chromosomes are cut from the photograph and pasted together in various ways for study. With certain stains each chromosome has a particular pattern of banding that facilitates its identification and shows the relationship of the banding pattern to genetic anomalies. The 22 pairs of autosomes are numbered in order of decreasing size; the pair of X and Y chromosomes differ in both size and morphology.

Figure 3–12. Nucleoli. Primary oocytes are very large cells with very large round euchromatic nuclei. The cells are actively increasing in volume, synthesizing much protein and many ribosomes, and each nucleus has one well-developed, intensely basophilic nucleolus. The strong basophilia reflects the high concentration of rRNA being processes in this small region of nucleoplasm. Other cells may each have one large nucleolus or a few smaller nucleoli. All are involved in transcription and processing of rRNA. Primary oocytes arrest for a prolonged period during prophase of the first meiotic division, when the chromosomes have already begun to condense. Parts of the condensed chromosomes are seen as the stained material in the sectioned nuclei shown here. Meiosis in oocytes will proceed just before they are ovulated (extruded from the ovary; see Chapter 22).

accumulate and when all is ready (at the *restriction point*) DNA synthesis begins. Entry or progression through each phase of the cycle is controlled by specific sets of proteins, the **cyclins** and **cyclin-dependent kinases (CDKs),** each of which phosphorylates proteins in various other complexes (such as the nuclear lamins at the beginning of mitosis). In this way diverse cellular activities are coordinated with specific phases of the cell cycle.

MEDICAL APPLICATION

Some growth factors are being used in medicine. One example is erythropoietin, which stimulates proliferation, differentiation, and survival of red blood cell precursors in the bone marrow.

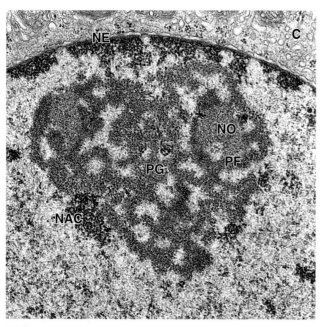

***Figure 3–13.* Regions within a nucleolus.** Different regions of the nucleolus can often be seen in sections of cell nuclei examined by TEM. The major parts of the nucleolus identified in this way are one or more pale-staining regions containing **nucleolar organizer (NO)** DNA—sequences of bases coding for rRNA. In the human genome, five pairs of chromosomes contain nucleolar organizers. Closely associated with the nucleolar organizers are densely packed 5- to 10-nm ribonucleoprotein fibers of the **pars fibrosa (PF),** which consists of primary rRNA transcripts. The **pars granulose (PG)** consists of 15- to 20-nm granules that represent maturing ribosomal subunits. Proteins, synthesized in the cytoplasm, become associated with rRNAs in the nucleolus. The resulting ribosomal subunits are then exported to the cytoplasm. A small amount of heterochromatic **nucleolus-associated chromatin (NAC)** is also sometimes part of the nucleolus, but its functional significance is unknown. X30,000.

Progression through the cell cycle is also regulated by various signals which halt cycling under adverse conditions. DNA damage can arrest the cell cycle not only at the G_1 restriction point, but also during S or at a checkpoint in G_2 (Figure 3–19). G_1 arrest may permit repair of the damage to occur before the cell enters S phase, so that the damaged DNA is not replicated. If the problem encountered at any checkpoint cannot be corrected while cycling is halted, **tumor suppressor** genes or proteins (such as p53) are activated and the cell's activity is redirected toward cell suicide or apoptosis. The gene encoding p53 is often mutated in cancer cells, thus reducing the cell's ability to detect and repair damaged DNA. Inheritance of damaged DNA by daughter cells results in a greater frequency of mutations and general instability of the genome, which may contribute to the development of the cancer.

MEDICAL APPLICATION

*Rapidly growing tissues (eg, intestinal epithelium) frequently contain cells in mitosis, whereas slowly growing tissues do not. The increased number of mitotic figures and abnormal mitoses in tumors are important characteristics that distinguish rapidly growing malignant tumors from benign tumors. Cell proliferation and differentiation are controlled by a group of genes called **proto-oncogenes**; altering the structure or expression of these genes promotes the production of tumors. Proto-oncogenes can be changed into **oncogenes** by a mutation in their DNA sequences or by DNA translocations that move genes to active promoter sites causing them to be inappropriately or permanently expressed. Altered protooncogenes have been associated with several tumors and hematologic cancers. Proto-oncogenes encode almost any protein involved in the control of mitotic activity, including various specific growth factors, the receptors for growth factors, and various kinases and other proteins involved in intracellular signaling of growth factors. There is an extensive and growing list of such proto-oncogenes.*

Various cancer-causing factors (eg, certain chemical substances, certain types of radiation, and certain viral infections) can induce DNA damage or mutations which can lead to abnormal cell proliferation that bypasses normal regulation mechanisms for controlled growth and results in the formation of tumors.

*The term **tumor**, initially used to denote any localized swelling in the body caused by inflammation or abnormal cell proliferation, is now usually used as a synonym for **neoplasm** (Gr. neos, new, + plasma, thing formed). Neoplasm can be defined as an abnormal mass of tissue formed by uncoordinated cell proliferation. Neoplasms are either benign or malignant according to their characteristics of slow growth and no invasiveness (benign) or rapid growth and great capacity to invade other tissues and organs (malignant). **Cancer** is the common term for all malignant tumors.*

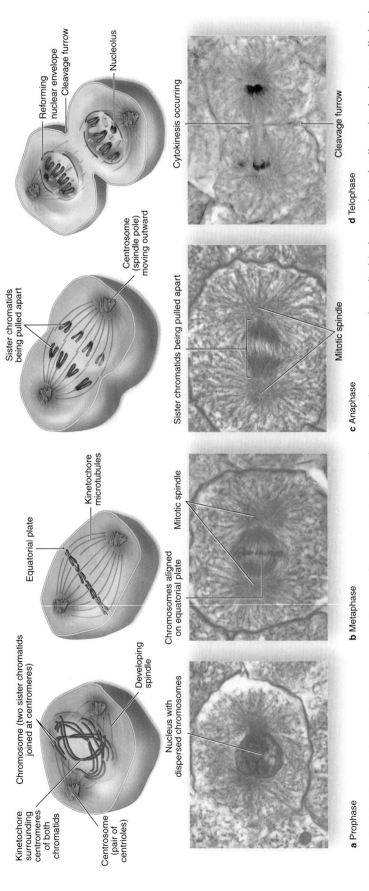

Kinetochore surrounding centromeres of both chromatids

Chromosome (two sister chromatids joined at centromeres)

Equatorial plate

Kinetochore microtubules

Reforming nuclear envelope

Cleavage furrow

Nucleolus

Centrosome (pair of centrioles)

Developing spindle

Nucleus with dispersed chromosomes

Sister chromatids being pulled apart

Centrosome (spindle pole) moving outward

Chromosomes aligned on equatorial plate

Mitotic spindle

Sister chromatids being pulled apart

Mitotic spindle

Cytokinesis occurring

Cleavage furrow

a Prophase

b Metaphase

c Anaphase

d Telophase

Figure 3–14. **Phases of mitosis.** Chromosomal changes during mitosis are easily seen and most commonly studied in large cultured cells or in the large cells in the very early embryos of invertebrates or primitive vertebrates after sectioning. Shown here are cells in sections of a fish blastodisc. **a.** During the relatively long **prophase** the centrosomes move to opposite poles, the nuclear envelope fragments, and chromosomes condense and become visible. Having undergone DNA replication, each chromosome consists of two chromatids joined at their centromere regions by a kinetochore protein complex. **b.** At the short **metaphase** the chromosomes have become aligned at the equatorial plate as a result of their attachments to the dynamic microtubules which run from the kinetochores to the two centrosomes. **c.** During **anaphase** the kinetochores come apart and the chromatids (now called chromosomes themselves) are pulled on microtubules toward the two centrosomes. **d.** In **telophase** the cell pinches in two by constriction of bundled actin filaments in the cell cortex and the chromosomes decondense, transcription resumes, nucleoli reappear, and the nuclear lamina and nuclear envelopes reassemble. X600. H&E.

57

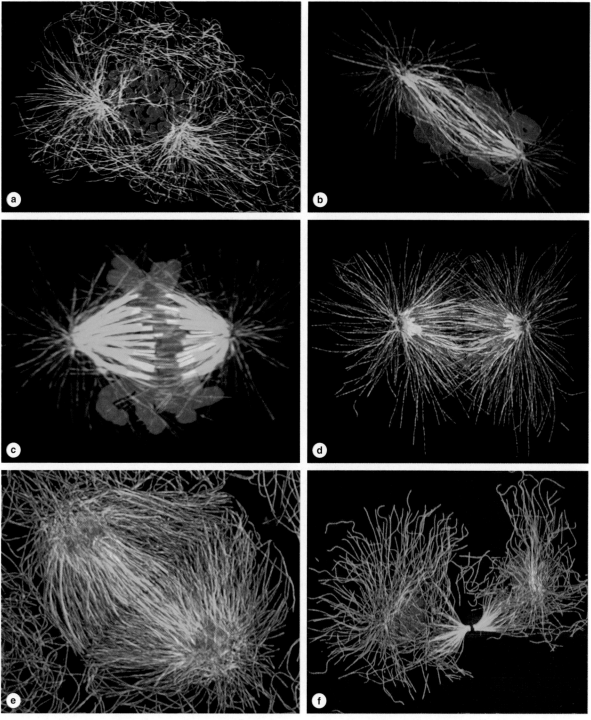

***Figure 3–15.* Confocal immunofluorescent images of mitotic cells.** Images obtained with a confocal laser scanning microscope from cultured cells in various phases of mitosis. Chromosomes are stained orange and microtubules, green. **(a):** Prophase: The chromosomes have undergone DNA replication and each consists of two very close sister chromatids. Two microtubule-organizing centers, the centrosomes, have moved apart and each is associated with microtubules forming the mitotic spindle. **(b):** Prometaphase: Chromosomes attach to spindle microtubules at their kinetochores and begin to be moved. **(c):** Metaphase: Chromosomes have become aligned at the middle of the spindle, near the cell equator. Kinetochore microtubules attach to each sister chromatid and to opposite poles of the spindle. **(d):** Anaphase: Sister chromatids separate from each other to become individual chromosomes which are pulled toward the spindle poles. The poles move apart and the kinetochore microtubules get shorter. **(e):** Telophase: The two sets of daughter chromosomes arrive at the spindle poles. **(f):** Late telophase and cytokinesis: A contractile ring of myosin-associated actin filaments forms a cleavage furrow that pinches the cell into two daughter cells, each with one nucleus and a complete set of chromosomes ready to undergo another round of DNA replication. (With permission, from Julie C. Canman and Ted Salmon, Department of Biology, University of North Carolina at Chapel Hill.)

STEM CELLS AND TISSUE RENEWAL

Throughout an individual's lifetime many tissues and organs contain a small population of undifferentiated **stem cells** whose cycling serves to renew the differentiated cells of the tissues as needed. Many stem cells divide infrequently but the divisions are always asymmetric, that is, one daughter cell remains as a stem cell while the other becomes committed to a path that leads to differentiation. Stem cells of many tissues are found in specific locations or niches where the microenvironment helps maintain their uniquely undifferentiated properties; they are often rare and inconspicuous by routine histological methods.

The commitment to differentiation may quickly yield a new specialized cell that is functionally integrated into the tissue or organ. This seems to be the case in tissues with highly *stable or static cell populations* which normally show little or no mitotic activity. Many tissues are maintained by occasional mitotic activity among the functionally differentiated cells. In other tissues,

such as the blood and epidermis, cells become *terminally differentiated*, meaning they cannot renew cycling and exist for a short period of time. Such tissues have *rapidly renewing cell populations* and many more cells with mitotic activity. Most of these dividing cells are not the stem cells but the more rapidly dividing progeny of the cells committed to differentiation (Figure 3–20). They are commonly called **progenitor cells** or **transit amplifying cells** because they are in transit along the path from the stem cell niche to a differentiated state, while still amplifying by mitosis the number of new cells available for the differentiated tissue.

MEIOSIS

Meiosis is a specialized process involving two closely associated cell divisions that occurs only in the cells that will form sperm and egg cells in the gonads. Differentiation of these two forms of "germ cells" or **gametes** is discussed fully in Chapters 21 and 22, but the chromosomal aspects of meiosis are mentioned here for better comparison with the events of mitosis. Two key features characterize meiosis. (1) The cells produced are **haploid**, with just one chromosome from each pair present in the rest of the body's (somatic) cells. Union of haploid eggs and sperm cells at fertilization forms a new diploid cell (the zygote) which can develop into a new individual. (2) Early in the process the homologous chromosomes of each pair (one from the mother, one from the father) physically associate along their lengths in an activity termed **synapsis**. During synapsis double-stranded breaks and repairs occur in the DNA, some of which result in reciprocal DNA exchanges called **crossovers** between the aligned maternal and paternal chromosomes. Crossing over produces new combinations of genes in the chromosomes in the germ cells so that few if any chromosomes are exactly the same as those from the mother and father.

As shown in Figure 3–21, the important events of meiosis unfold as follows:

- The cell entering meiosis has just completed DNA replication in a typical S phase so that each of its chromosomes contains the two identical copies referred to as **sister chromatids.**

- During a greatly elongated prophase of the first meiotic division (prophase I) chromatin condenses as usual, but early in condensation homologous chromosomes begin to come together physically in synapsis. Because each chromosome has two sister chromatids at this point geneticists refer to paired synaptic chromosomes as **tetrads**, emphasizing that the structures have four copies of each genetic sequence. During synapsis crossing over occurs among these DNA filaments which mixes up the genes inherited from each parent and yields a new and different set of genes to be passed on to the next generation. Though not well-understood at the molecular level, the events of synapsis and crossing over are obviously under tight controls. Prophase I is normally extended for 3 weeks during male gametogenesis in humans, whereas oocytes arrest in this meiotic phase from the time of their formation in the fetal ovary through the woman's reproductive maturity, that is, for about 12 years to 5 decades!

- When synapsis and crossing over are completed, the chromosomes condense further and undergo microscopically typical metaphase, anaphase, and telophase events as the cell divides in two. Importantly, the anaphase I separation involves the homologous chromosomes that came together during synapsis. Each of the separated chromosomes still contains two chromatids held together at the centromere.

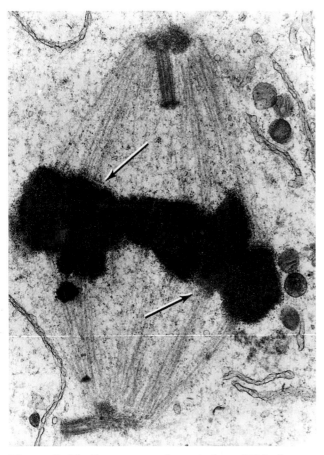

Figure 3–16. Chromosomes in metaphase. TEM of a sectioned metaphase cell shows several features of the mitotic apparatus, including the very electron-dense chromosomes bound at their kinetochores (arrows) to microtubules of the spindle. The microtubules are seen to converge on the centrosomes, in each of which centriole-like structures are found. Large flattened membrane vesicles near the mitotic spindle may represent the fragmented nuclear envelope which will begin to reform during late telophase. X19,000. (With permission, from Richard McIntosh, Department of Molecular, Cellular and Developmental Biology, University of Colorado at Boulder.)

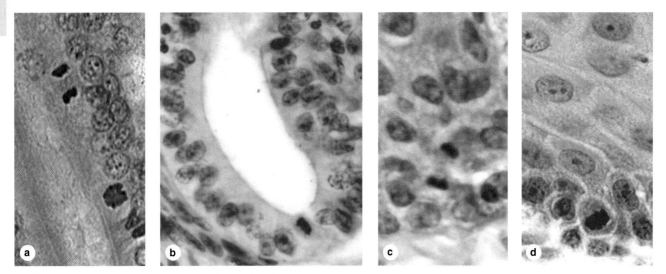

Figure 3–17. **Mitotic cells in adult tissues.** Dividing cells in recognizable stages of mitosis are rarely observed in adult tissues but can sometimes be identified as *mitotic figures* are shown here in various rapidly renewing tissues. **(a):** In the lining of the small intestine, many mitotic *transit amplifying cells*, progeny of nearby stem cells that are not yet fully differentiated, can be found in the area above the most basal region of the intestinal crypts. Condensed chromosomes of cells in late anaphase and telophase phase can be distinguished. **(b):** Metaphase cells in a gland of proliferating uterine endometrium. **(c):** Telophase cells in the esophagus lining. **(d):** Metaphase in the basal layer of epidermis. Mitotic figures are normally difficult to identify in most animal tissues, both because they are rare and because the various cell shapes and locations seldom allow specific phases of mitosis to be seen clearly. Most commonly, mitotic figures in organs appear simply as nuclei with clumped, darkly stained chromatin. X400. H&E.

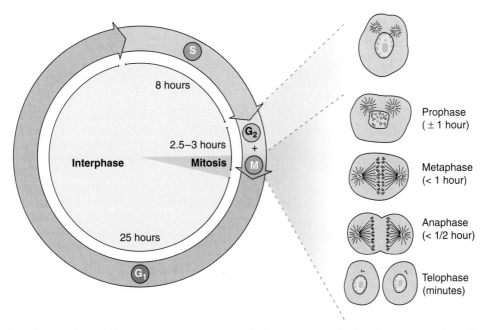

Figure 3–18. **The cell cycle.** The ability to recognize microscopically cells during both mitosis and DNA replication (by autoradiography after administering radiolabeled thymidine) led to the concept of the cell "cycle." In this concept new interphase cells undergo a period after leaving mitosis and before starting DNA synthesis called the first gap or G_1. Another gap, G_2, occurs after DNA replication and before the next mitotic prophase. After mitosis both new cells repeat this cycle. In rapidly dividing cells, G_1 is a period in which cells accumulate the enzymes and nucleotides required for DNA replication, S is the period devoted primarily to DNA replication, G_2 is a usually short period of preparation for mitosis, and M includes all phases of mitosis itself. In rapidly growing human tissues the cell cycle varies from 24 to 36 hours. The length of G_1 depends on many factors and is usually the longest and most variable period; the length of S is largely a function of the genome size. G_2 and mitosis together normally last only 2-3 hours.

- Each of the two new cells now divides again, much more rapidly and without a new phase of DNA replication. In this division the chromatids now separate at the centromere and are pulled to the opposite poles as individual chromosomes. In each new cell a nuclear envelope forms around this new haploid set of chromosomes.

In summary, meiosis and mitosis share many aspects of chromatin condensation and separation (Figure 3–21), but differ in key ways:

- Mitosis is a cell division that produces *two diploid cells*. Meiosis consists of two connected cell divisions and produces *four haploid cells*.

- During meiotic crossing over, new combinations of gene alleles are produced and every haploid cell is genetically unique. Lacking synapsis and the opportunity for DNA recombination, mitosis yields two cells that are the same genetically.

APOPTOSIS

Less evident, but no less important than cell proliferation for body functions, is the process of cell suicide or programmed cell death called **apoptosis** (Gr. *apo*, off + *ptosis*, a falling). Apoptosis is a highly regulated cellular activity that occurs rapidly and produces small membrane-enclosed *apoptotic bodies,* which quickly undergo phagocytosis by neighboring cells or macrophages specialized for debris removal. Unlike cells undergoing **necrosis** as a result of accidental injury, apoptotic cells do not rupture and release none of their contents. This difference is highly significant because release of cellular components causes a rapid series of local reactions and immigration of leukocytes in an elaborate reaction called an inflammatory response. Such

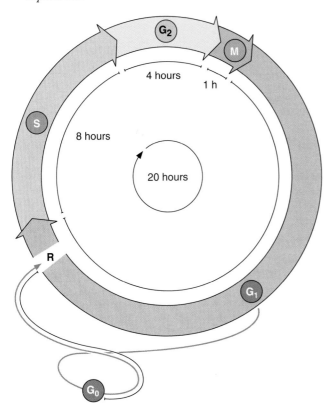

Figure 3–19. Control of the cell cycle. One factor determining the time a cell spends in G₁ is the cell's state of differentiation, or how much time it spends expressing gene products specific to its cell type before resuming DNA replication. Differentiating cells in growing tissues may have very long G₁ periods and such cells are often said to be "in a G₀ phase" of the cell cycle. From this phase many differentiated cells can return to the cycle, but some stay in G₀ for a long time or even for their entire lifetime. Entry into each phase of the cell cycle is controlled by proteins called **cyclins** and **cyclin-dependent kinases** which phosphorylate/activate many proteins needed for phase-specific functions. Cyclin activity produces an important **restriction point (R)** late in G₁ and a similar **G₂/M checkpoint** which are important for the maintenance of chromosome stability and cell viability. These control points stop the cycle under conditions unfavorable to the cell and help insure that neither the DNA replication nor the mitotic phases occurs prematurely. For example, at the G₂/M checkpoint the cell pauses while enzymes insure that all DNA has been replicated properly.

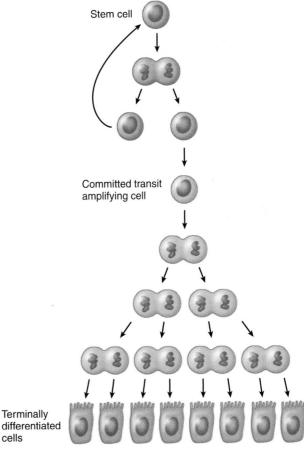

Figure 3–20. Stem cells. In rapidly growing adult tissues and perhaps in other tissues there are slowly dividing populations of **stem cells**. Stem cells divide asymmetrically, producing one cell that remains as a stem cell and another which becomes committed to a differentiative pathway but divides a few more times at a more rapid rate. Such cells have been termed "transit amplifying cells," each of which eventually stops dividing and becomes fully differentiated.

a response is unwanted when cells are routinely eliminated following DNA damage or as part of a normal development process. These routine cell eliminations therefore occur rapidly and without repercussions by apoptosis.

A few examples of apoptosis will illustrate its significance. Inside the thymus, T lymphocytes with the potential to react against self-antigens receive signals that activate the apoptotic program and they die before leaving the thymus (see Chapter 14). In the mature ovary, apoptosis is the mechanism of both the monthly loss of luteal cells and the removal of excess oocytes and their follicles. Programmed cell death was first discovered in developing embryos, where apoptosis is an essential process for shaping various developing organs or body regions (morphogenesis), such as the tissue between the digits on a developing limb bud. Apoptosis also plays an important role in formation of the central nervous system.

Apoptosis is an important means of eliminating cells whose survival is blocked by lack of nutrients, by damage caused by

MITOSIS

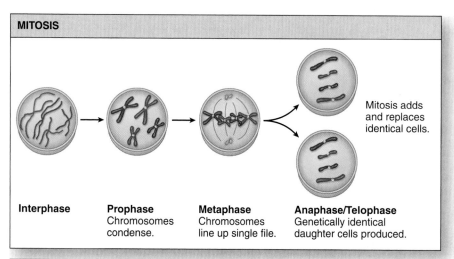

| Interphase | **Prophase** Chromosomes condense. | **Metaphase** Chromosomes line up single file. | **Anaphase/Telophase** Genetically identical daughter cells produced. |

Mitosis adds and replaces identical cells.

MEIOSIS

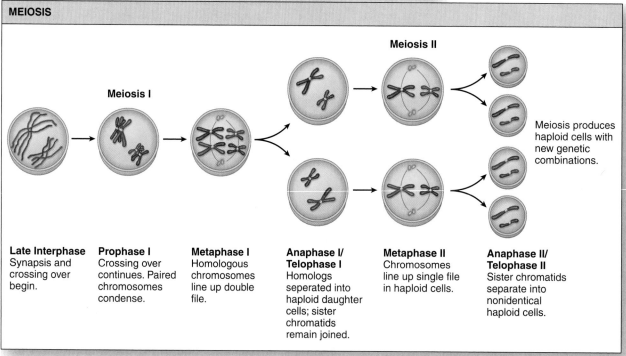

Meiosis I

Meiosis II

Meiosis produces haploid cells with new genetic combinations.

| **Late Interphase** Synapsis and crossing over begin. | **Prophase I** Crossing over continues. Paired chromosomes condense. | **Metaphase I** Homologous chromosomes line up double file. | **Anaphase I/ Telophase I** Homologs seperated into haploid daughter cells; sister chromatids remain joined. | **Metaphase II** Chromosomes line up single file in haploid cells. | **Anaphase II/ Telophase II** Sister chromatids separate into nonidentical haploid cells. |

Figure 3–21. **Mitosis and meiosis.** Mitosis and meiosis share many aspects of chromatin condensation and separation, but differ in various key ways. As chromosomal condensation begins in meiosis, the two homologous maternal and paternal chromosomes physically align in **synapsis** and regions are exchanged during **crossing over** or recombination. This is followed by **two meiotic divisions** with no intervening S phase. Mitosis produces *two diploid cells* which are the same genetically. Meiosis with its two successive cell divisions produces *four haploid cells*. During meiotic crossing over, new combinations of genes arise so that every haploid cell is genetically unique.

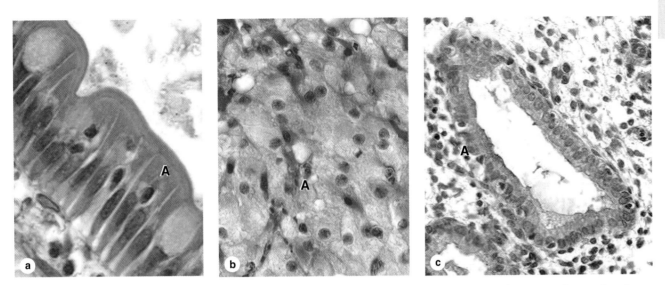

Figure 3–22. Apoptotic cells. Apoptotic cells in adult tissues are also rarely observed because the process is completed very rapidly. Moreover, with their condensed nuclear chromatin, such cells may superficially resemble some mitotic cells. Shown here are apoptotic cells (A) in epithelium of a villus from the lining of the small intestine **(a)**, in a corpus luteum beginning to undergo involution **(b)**, and epithelium of a uterine endometrial gland at the onset of menstruation **(c)**. X400. H&E.

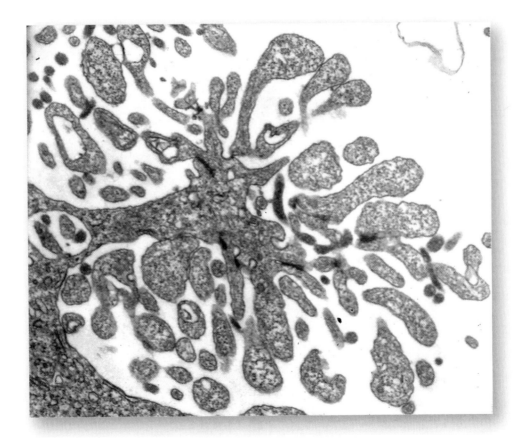

Figure 3–23. Late apoptosis—formation of apoptotic bodies. TEM of a cell in late apoptosis shows that during this process the cell's shape changes radically and large cytoplasmic vesicles (blebs) are formed. These detach from the cell and often separate one from another, but remain contained within plasma membrane so that no cytoplasmic contents are released into the extracellular space. The membrane surrounding such apoptotic bodies is changed in such a way that the blebs are recognized by neighboring cells or macrophages and are very rapidly phagocytosed. The rapid formation and engulfment of such blebs without their disruption allow apoptosis to occur without eliciting an inflammatory reaction. X10,000.

free radicals or radiation, or by the action of tumor suppressor proteins. In all examples studied apoptosis occurs very rapidly, in less time than required for mitosis, and the affected cells are removed without a trace.

MEDICAL APPLICATION

Most cells of the body can activate their apoptotic program when major changes occur in their DNA— eg, when a number of mutations accumulate in the DNA. In this way, apoptosis prevents the proliferation of such cells to form a clone and develop into a tumor. Malignant cells sometimes deactivate the genes that control the apoptotic process, thus avoiding death and allowing cancer progression.

Whether apoptosis is induced by external signals or by adverse internal conditions that cannot be remedied, the process involves the following features:

- *Loss of mitochondrial function*: Mitochondrial membrane integrity is not maintained, causing the end of normal activity and release of cytochrome c into the cytoplasm where it activates proteolytic enzymes called **caspases**. The initial caspases activate a cascade of other caspases, resulting in protein degradation throughout the cell.

- *Fragmentation of DNA*: **Endonucleases** are activated which cleave DNA between nucleosomes into small fragments. (The new ends produced in the fragmented DNA allow specific histochemical staining of apoptotic cells using an appropriate enzyme that adds labeled nucleotides at these sites.)
- *Shrinkage of nuclear and cell volumes*: Small dark-stained (pyknotic) nuclei can sometimes be identified with the light microscope (Figure 3–22).
- *Cell membrane changes*: The integrity of the plasmalemma is maintained, but the cell undergoes dramatic shape changes, such as "blebbing" (Figure 3–23), as membrane proteins and cytoskeleton are degraded. Phospholipids normally found only in the inner layer move to the outer layer, serving as signals to induce phagocytosis.
- *Formation and phagocytic removal of these* **apoptotic bodies**.

MEDICAL APPLICATION

*The accidental death of cells, a pathologic process, is called **necrosis**. Necrosis can be caused by microorganisms, viruses, chemicals, and other harmful agents. Necrotic cells swell; their organelles increase in volume; and finally they burst, releasing their contents into the extracellular space. Macrophages engulf the debris of necrotic cells by phagocytosis and then secrete molecules that activate other immunodefensive cells to promote inflammation.*

Epithelial Tissue

4

CHARACTERISTIC FEATURES OF EPITHELIAL CELLS
    Basal Laminae & Basement Membranes
    Intercellular Adhesion & Other Junctions
SPECIALIZATIONS OF THE APICAL CELL SURFACE
    Microvilli
    Stereocilia
    Cilia

TYPES OF EPITHELIA
    Covering or Lining Epithelia
    Glandular Epithelia
TRANSPORT ACROSS EPITHELIA
RENEWAL OF EPITHELIAL CELLS

Despite its complexity, the human body is composed of only **four basic types of tissue:** epithelial, connective, muscular, and nervous. These tissues, which are formed by cells and molecules of the **extracellular matrix,** exist not as isolated units but rather in association with one another and in variable proportions, forming different organs and systems of the body. The main characteristics of these basic types of tissue are shown in Table 4–1. Also of great functional importance are the free cells found in body fluids such as blood and lymph.

Connective tissue is characterized by the abundance of extracellular material produced by its cells; muscle tissue is composed of elongated cells specialized for contraction and movement; and nerve tissue is composed of cells with elongated processes extending from the cell body that have the specialized functions of receiving, generating, and transmitting nerve impulses. Organs can be divided into **parenchyma,** which is composed of the cells responsible for the main functions typical of the organ, and **stroma,** which is the supporting tissue. Except in the brain and spinal cord, the stroma is made of connective tissue.

Epithelial tissues are composed of closely aggregated polyhedral cells with very little extracellular substance. These cells have strong adhesion and form cellular sheets that cover the surface of the body and line its cavities.

The principal functions of epithelial (Gr. *epi,* upon, + *thele,* nipple) tissues are:

- Covering, lining, and protecting surfaces (eg, skin)
- Absorption (eg, the intestines)
- Secretion (eg, the epithelial cells of glands)
- Contractility (eg, myoepithelial cells).

Specific cells of certain epithelia are also highly specialized sensory cells, such as those of taste buds or the olfactory epithelium. Because epithelial cells line all external and internal surfaces of the body, everything that enters or leaves the body must cross an epithelial sheet.

CHARACTERISTIC FEATURES OF EPITHELIAL CELLS

The forms and dimensions of epithelial cells range from high **columnar** to **cuboidal** to low **squamous** cells. Their common polyhedral form results from their close juxtaposition in cellular layers or masses and is similar to what would be observed if a large number of inflated balloons were compressed into a limited space. Epithelial cell nuclei have a distinctive shape, varying from spherical to elongated or elliptic. The nuclear form often corresponds roughly to the cell shape; thus, cuboidal cells have spherical nuclei, and squamous cells have flattened nuclei. The long axis of the nucleus is always parallel to the main axis of the cell.

Because the lipid-rich membranes between cells are frequently indistinguishable with the light microscope, the stained cell nucleus is a clue to the shape and number of cells. Nuclear form is also useful to determine whether the cells are arranged in layers, a primary morphologic criterion for classifying epithelia.

Most epithelia rest on connective tissue. In the case of epithelia lining the cavity of internal organs (especially in the digestive, respiratory, and urinary systems) this layer of connective tissue is often called the **lamina propria.** The lamina propria not only serves to support the epithelium but also provides nutrition and binds it to underlying structures. The area of contact between epithelium and lamina propria is increased by irregularities in the connective tissue surface in the form of small evaginations called **papillae** (L. diminutive of *papula,* nipple; singular **papilla**). Papillae occur most frequently in epithelial tissues subject to friction, such as the covering of the skin or tongue.

Table 4–1. Main characteristics of the four basic types of tissues.

Tissue	Cells	Extracellular Matrix	Main Functions
Nervous	Intertwining elongated processes	None	Transmission of nervous impulses
Epithelial	Aggregated polyhedral cells	Small amount	Lining of surface or body cavities, glandular secretion
Muscle	Elongated contractile cells	Moderate amount	Movement
Connective	Several types of fixed and wandering cells	Abundant amount	Support and protection

Epithelial cells generally show **polarity**, with organelles and membrane proteins distributed unevenly in different parts of the cell. The region of the cell that faces the connective tissue is called the **basal pole**, whereas the opposite pole, usually facing a space, is the **apical pole** and the intervening sides apposed in neighboring cells are the **lateral surfaces**. The membranes on the lateral surfaces of adjoining cells often have numerous infoldings to increase the area of that surface, increasing its functional capacity. The different regions of polarized cells may have different functions.

Basal Laminae & Basement Membranes

All epithelial cells in contact with subjacent connective tissue have at their basal surfaces a felt-like sheet of extracellular material called the **basal lamina** (Figure 4–1). This structure is visible only with the electron microscope, where it appears as an electron-dense layer, 20–100 nm thick, consisting of a network of fine fibrils, the **dense layer** or **lamina densa** (Figure 4–2). In addition, basal laminae may have electron-lucent layers on one or both sides of the dense layer, called **clear layers** or **laminae lucida.** Between epithelia with no intervening connective tissue, such as in lung alveoli and renal glomeruli, the basal lamina is often thicker due to the fusion of the basal laminae from each epithelial layer.

The macromolecular components of basal laminae form precise three-dimensional arrays and are described individually in the next chapter. The best known of these include:

- **Laminin**: These are large glycoprotein molecules that self-assemble to form a lace-like sheet immediately below the cells' basal poles where they are held in place by the transmembrane integrins.
- **Type IV collagen**: Monomers of type IV collagen contain three polypeptide chains and self-assemble further to form a felt-like sheet associated with the laminin layer.
- **Entactin (nidogen)**, a glycoprotein, and **perlecan**, a proteoglycan with heparan sulfate side chains: these glycosylated proteins and others serve to link together the laminin and type IV collagen sheets.

All these components are secreted at the basal poles of the epithelial cells. Their precise proportions in basal laminae vary between and within tissues. Basal laminae are attached to **reticular fibers** made of **type III collagen** in the underlying connective tissues by **anchoring fibrils** of **type VII collagen**. These proteins are produced by cells of the connective tissue and form a layer below the basal lamina called the **reticular lamina** that is also visible by TEM (Figure 4–2).

Basal laminae are found not only in epithelial tissues but also where other cell types come into contact with connective tissue. Muscle cells, adipocytes, and Schwann cells secrete laminin, type IV collagen, and other components that provide a barrier limiting or regulating exchanges of macromolecules between these cells and connective tissue.

Basal laminae have many functions. In addition to simple structural and filtering functions, they are also able to influence cell polarity; regulate cell proliferation and differentiation by binding and concentrating growth factors; influence cell metabolism and survival; organize the proteins in the adjacent plasma membrane (affecting signal transduction); and serve as pathways for cell migration. The basal lamina seems to contain the information necessary for many cell-to-cell interactions, such as the reinnervation of denervated muscle cells. The presence of the basal lamina around a muscle cell is necessary for the establishment of new neuromuscular junctions.

The term **basement membrane** is used to specify a periodic acid–Schiff (PAS)-positive layer, visible with the light microscope beneath epithelia (Figure 4–3). The basement membrane is formed by the combination of a basal lamina and a reticular lamina and is therefore thicker. The terms basement membrane and basal lamina are often used indiscriminately, causing confusion.

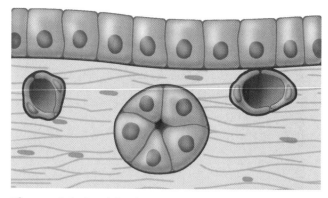

Figure 4–1. Basal laminae. An extracellular **basal lamina** always lies at the interface of epithelial cells and connective tissue. The basal laminae to two neighboring epithelia can fuse or appear to fuse in places where there is no intervening connective tissue. Nutrients for epithelial cells must diffuse across the basal lamina. Nerve fibers normally penetrate this structure, but small blood capillaries (being epithelial themselves) never enter an epithelium across a basal lamina. When components of a basal lamina are resolved with the light microscope, the structure is often called a **basement membrane**.

In this book, "basal lamina" is used to denote the lamina densa and its adjacent layers and structures seen with the TEM. "Basement membrane" is used to denote the structures seen with the light microscope.

Intercellular Adhesion & Other Junctions

Several membrane-associated structures contribute to adhesion and communication between cells. They are present in most tissues but are particularly numerous and prominent in epithelia and will be described here. Epithelial cells are extremely cohesive and relatively strong mechanical forces are necessary to separate them. Intercellular adhesion is especially marked in epithelial tissues that are subjected to traction and pressure (eg, in the skin).

The lateral membranes of epithelial cells exhibit several specialized **intercellular junctions**. Various junctions serve to function as:

- Seals to prevent the flow of materials between the cells (**occluding junctions**)
- Sites of adhesion (**adhesive** or **anchoring junctions**)

- Channels for communication between adjacent cells (**gap junctions**).

In several epithelia such junctions are present in a definite order from the apical to the basal ends of the cells. **Tight junctions,** or **zonulae occludens** (singular, **zonula occludens**), are the most apical of the junctions. The Latin terminology gives important information about the geometry of the junction. "Zonula" indicates that the junctions form bands completely encircling each cell, and "occludens" refers to the membrane fusions that close off the space between the cells. In properly stained thin sections viewed in the TEM, the adjacent membranes appear tightly apposed or fused (Figures 4–4 and 4–5). The seal between the membranes is due primarily to direct interactions between the transmembrane protein **claudin** on each cell. After cryofracture (Figure 4–6), the replicas show these fusion sites as a band of branching strands around each cell. The number of these sealing strands or fusion sites is inversely correlated with the leakiness of the epithelium. Epithelia with one or very few fusion sites (eg, proximal renal tubule) are more permeable to water and solutes than are epithelia with numerous fusion sites (eg, the lining of the urinary bladder). Thus, the

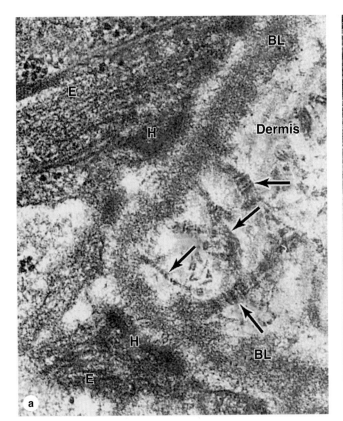 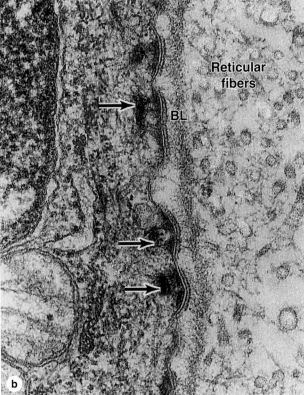

Figure 4-2. **Ultrastructural components of the basal lamina.** Details of the basal lamina are revealed by two TEM of sectioned human skin. **(a):** The basal lamina (BL) is shown to have a dense layer with a clear layer on each side. The underlying dermis contains **anchoring fibrils** (arrows) of collagen which help anchor the epithelium to the underlying connective tissue. Hemidesmosomes (H) occur at the epithelial–connective tissue junction. X54,000. **(b):** The basal lamina, hemidesmosomes (arrows), and underlying **reticular fibers** of the reticular lamina typically comprise a basement membrane sometimes visible with the light microscope. X80,000.

principal function of the tight junction is to form a seal that prevents the flow of materials between epithelial cells (the paracellular pathway) in either direction. In this way, zonulae occludens in sheets of epithelial cells help form two functional compartments: an apical compartment that is composed of an organ cavity (such as the lumen of a secretory unit or the gut) and a basal compartment that begins at the junctions and encompasses the underlying tissue.

Besides forming a seal between compartments on either side of an epithelium, the zonulae occludens of epithelial cells help prevent the integral membrane proteins of the apical surface from being transferred to the basolateral surface and vice versa. This allows the two sides of the epithelium to maintain different receptors and function differently.

The next type of junction is the **adherent junction** or **zonula adherens** (Figures 4–4 and 4–5). This junction also encircles the cell, usually immediately below the zonula occludens, and provides for the firm adhesion of one cell to its neighbors. Adhesion is mediated by transmembrane glycoproteins of each cell, the **cadherins**, which lose their adhesive properties in the absence of Ca^{2+}. Inside the cell, cadherins bind the protein catenin which is linked by means of actin-binding proteins to actin filaments, all of which produce electron-dense plaques of material on the cytoplasmic surfaces of

adherent junctions. The numerous actin filaments form part of the **terminal web,** a cytoskeletal feature at the apical pole in many epithelial cells with a role in cytoplasmic motility and other functions.

Another junction specialized for adhesion is the **desmosome** or **macula adherens** (L. *macula,* spot). As the names imply, this junctional type resembles a single "spot-weld" and does not form a belt around the cell. The desmosome is a disk-shaped structure at the surface of one cell that is matched with an identical structure at the surface of an adjacent cell (Figures 4–4 and 4–5). Between cell membranes at a desmosome are variable amounts of electron-dense material, principally larger members of the cadherin family. On the cytoplasmic side of each cell membrane these cadherin-type proteins inset into a dense **attachment plaque** of anchoring proteins (**plakophilin, plakoglobin,** and **desmoplakin**) which bind intermediate filaments rather than actin filaments. Cable-like filaments of **cytokeratin** are most common in desmosomes of epithelia. Because intermediate filaments of the cytoskeleton are very strong, desmosomes provide firm adhesion among the cells. In nonepithelial cells, the intermediate filaments attached to desmosomes are composed of other proteins, such as desmin or vimentin.

Gap or **communicating junctions** can occur almost anywhere along the lateral membranes of epithelial cells, but are also found between cells in nearly all mammalian tissues. With conventional TEM, gap junctions appear as regions where adjacent cell membranes are closely apposed (Figure 4–7a). After cryofracture, these junctions are seen as aggregated transmembrane protein complexes that form circular patches in the plasma membrane (Figure 4–7b).

The proteins of gap junctions, called **connexins,** form hexameric complexes called **connexons,** each of which has a central hydrophilic pore about 1.5 nm in diameter. When two cells attach, connexins in the adjacent cell membranes move laterally and align to form connexons between the two cells (Figure 4–4), with each gap junction having dozens or hundreds of aligned pairs of connexons. Gap junctions permit the rapid exchange between cells of molecules with small (<1.5 nm) diameters. Some molecules mediating signal transduction, such as cyclic AMP, cyclic GMP, and ions, move readily through gap junctions, allowing cells in many tissues to act in a coordinated manner rather than as independent units. A good example is heart muscle, where abundant gap junctions are greatly responsible for the heart's coordinated beat.

In the contact area between epithelial cells and the subjacent basal lamina, **hemidesmosomes** (Gr. *hemi,* half, + *desmos* + *soma*) can often be observed ultrastructurally. These adhesive structures resemble a half-desmosome and bind the cell to the basal lamina (Figure 4–2). However, while in desmosomes the attachment plaques contain cadherins, in hemidesmosomes the plaques contain abundant **integrins,** transmembrane proteins that are receptor sites for the extracellular macromolecules laminin and collagen type IV.

Blood vessels do not normally penetrate an epithelium and nutrients for the epithelial cells must pass out of the capillaries in the underlying lamina propria. These nutrients then diffuse across the basal lamina and are taken up through the basolateral surfaces of the epithelial cell, usually by an energy-dependent process. Receptors for chemical messengers (eg, hormones, neurotransmitters) that influence the activity of epithelial cells are localized in the basolateral membranes. In absorptive epithelial cells, the apical cell membrane contains, as integral membrane proteins, enzymes such as disaccharidases

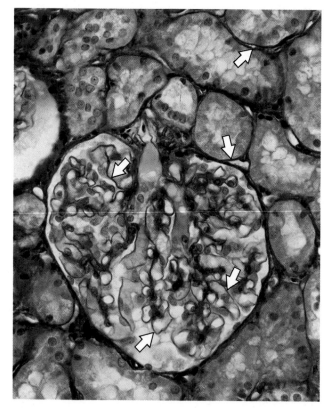

***Figure 4–3.* Basement membranes.** This section of kidney shows the typical basement membranes (arrows) of several tubules and of structures within the single glomerulus included here. In renal glomeruli the basement membrane, besides having a supporting function, has an important role as a filter. X100. Picrosirius-hematoxylin (PSH).

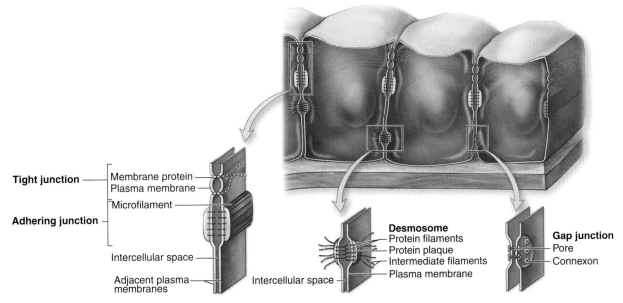

Tight junction — Membrane protein
— Plasma membrane

Adhering junction — Microfilament

Intercellular space —

Adjacent plasma — membranes

Intercellular space —

Desmosome
— Protein filaments
— Protein plaque
— Intermediate filaments
— Plasma membrane

Gap junction
— Pore
— Connexon

Types of intercellular junctions

Figure 4–4. Junctional complexes of epithelial cells. Three cuboidal epithelial cells, emptied of their contents, show the four major types of junctional complexes between cells. The **tight junction** (zonula occludens) and **adherent junction** (zonula adherens) are typically close together and each forms a continuous ribbon around the cell's apical end. Multiple ridges of the tight, occluding junctions prevent passive flow of material between the cells, but are not very strong; the adhering junctions immediately below them serve to stabilize and strengthen these circular bands around the cells and help hold the layer of cells together. Both desmosomes and gap junctions make spotlike plaques between two cells. Bound to intermediate filaments inside the cells, **desmosomes** form very strong attachment points which supplement the role of the zonulae adherens and play a major role to maintain the integrity of an epithelium. **Gap junctions**, each a patch of many **connexons** in the adjacent cell membranes, have little strength but serve as intercellular channels for flow of molecules. All of these junctional types are also found in certain other cell types besides epithelia.

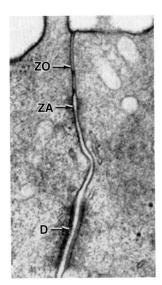

Figure 4–5. Junctional complex as seen in the TEM. A section showing the apical regions of two epithelial cells reveals a junctional complex with its zonula occludens (ZO), zonula adherens (ZA), and a desmosome (D). The major components of zonula occludens are each cell's transmembrane proteins called claudins which make tight contact across the intercellular space, creating a seal. The cytoplasmic electron-dense material at the zonula adherens includes cadherins, catenin, actin-binding proteins and actin filaments, but that of the desmosomes consists of a plaque of "anchoring proteins," such as plakophilin, plakoglobin, and desmoplakin, which are bound by intermediate filaments primarily those composed of keratins. X80,000.

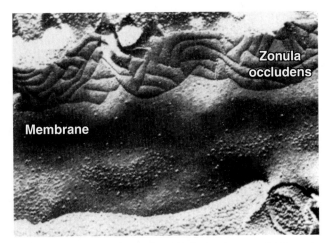

Figure 4–6. View of zonula occludens after cryofracture. In this electron micrograph of an epithelial cell after cryofracture, the fracture crosses through the cytoplasm in the lower portion, then shows a region of relatively smooth cell membrane, above which are the ridges and grooves of the zonula occludens. The membranes of adjoining cells basically fuse in the zonula occludens caused by tight interaction between claudins. X100,000.

and peptidases, which complete the digestion of molecules to be absorbed.

SPECIALIZATIONS OF THE APICAL CELL SURFACE

The free or apical surface of many types of epithelial cells has specialized structures to increase the cell surface area or to move substances or particles bound to the epithelium.

Microvilli

When viewed in the electron microscope, many cells are seen to have cytoplasmic projections. These projections may be short or long fingerlike extensions or folds that pursue a sinuous course, and they range in number from a few to many. Most are temporary, reflecting cytoplasmic movements and the activity of actin filaments.

In absorptive cells, such as the lining epithelium of the small intestine, the apical surface presents orderly arrays of many

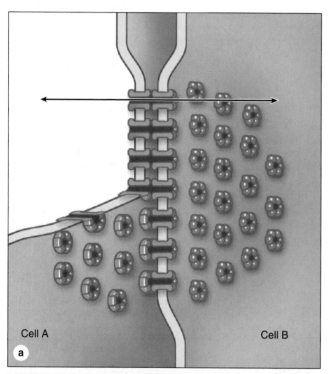

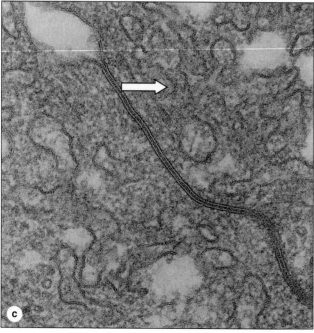

Figure 4–7. Gap junctions. **(a):** The diagram of a gap junction (oblique view) depicts the structural elements that allow the exchange of nutrients and signal molecules between cells without loss of material into the intercellular space. The communicating channels are formed by pairs of abutting particles (**connexons**), which are in turn composed of six dumbbell-shaped protein subunits (connexins) that span the lipid bilayer of each cell membrane. The channel passing through the cylindrical bridges (arrow) is about 1.5 nm in diameter, limiting the size of the molecules that can pass through it. **(b):** A cryofracture preparation shows a gap junction between epithelial cells. The junction appears as a plaquelike agglomeration of intramembrane protein particles, the connexons. X45,000. **(c):** A section through a gap junction between two cells shows that the two cell membranes are very closely apposed, separated only by a 2-nm-wide electron-dense space. Individual connexons are not resolved in cell sections. X193,000. (Figure 4–7c, with permission, from Mary C. Williams, Pulmonary Center, Boston University School of Medicine.)

hundreds of more permanent **microvilli** (L. *villus*, tuft) (Figure 4–8). The average microvillus is only about 1 μm high and 0.08 μm wide, but with hundreds or thousands present on the end of each absorptive cell, the total surface area can be increased as much as 20- or 30-fold. In these absorptive cells the glycocalyx is thicker than that of most cells and includes enzymes for the final stages of certain macromolecules' breakdown. The complex

of microvilli and glycocalyx is easily seen in the light microscope and is called the **brush** or **striated border.**

Within each microvillus are bundles of actin filaments (Figure 4–8c,d) cross-linked to each other and to the surrounding plasma membrane by other proteins. These filaments insert into the actin filaments of the terminal web. The array of microfilaments stabilizes the microvillus and allows it to contract

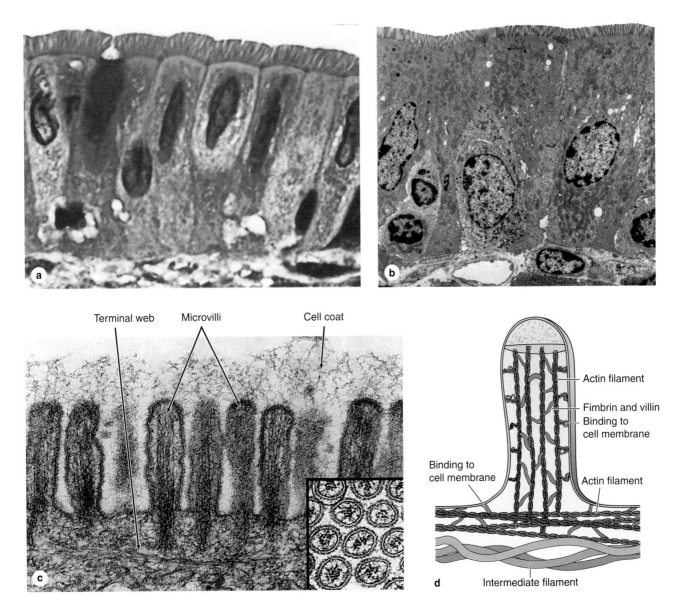

Figure 4–8. **Microvilli.** Absorptive cells lining the small intestine demonstrate microvilli particularly well. **(a):** With the light microscope microvilli at the apical side of the epithelium are usually faintly visible and make up the so-called **striated border** of the cells. **(b):** Individual microvilli are better seen by the TEM with a slightly higher magnification. Scattered endocrine cells (E) in this epithelium do not extend to the apical surface and lack microvilli. **(c):** At higher magnification the bundles of vertical microfilaments constituting the core of each microvillus are clearly seen. Below the microvilli is the terminal web, a horizontal network of actin microfilaments and associated proteins including myosins. On the plasmalemma of the microvilli is a thick extracellular cell coat (glycocalyx) containing glycoproteins and enzymes that allow the final stages of digestion to be linked to the uptake of digestion products across the cell membrane. The inset of cross-sectioned microvilli shows the internal disposition of the bundled actin filaments, the surrounding cell membrane, and the glycocalyx. X45,000. **(d):** The diagram indicates important proteins in a microvillus: the **actin filaments** cross-linked to one another by proteins such as **fimbrin** and **villin** and bound to the plasma membrane by proteins such as myosin I. The actin filaments are oriented in the same direction, with their plus ends associated with amorphous material at the tip of the microvillus.

slightly and intermittently which helps maintain optimal conditions for absorption across its plasmalemma.

Stereocilia

Stereocilia are long apical processes of cells in other absorptive epithelia such as that lining the epididymis (Figure 4–9) and

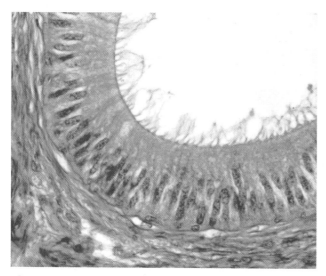

Figure 4–9. **Stereocilia.** At the apical ends of the tall epithelial cells lining organs such as the epididymis (shown here) are numerous very long stereocilia, which increase the surface area available for cellular absorption. Each stereocilium is typically much longer than a microvillus and may show a branching structure. Stereocilia have cytoplasmic actin filament bundles and external cell coats similar to those of microvilli. X400. H&E.

ductus deferens. These structures are much longer and less motile than microvilli, are branched, and should not be confused with true cilia. Like microvilli, stereocilia also increase the cells' surface area, facilitating the movement of molecules into and out of the cell.

Cilia

Cilia are elongated, highly motile structures on the surface of some epithelial cells, 5–10 μm long and 0.2 μm in diameter, which is much longer and two times wider than a typical microvillus. As discussed in Chapter 2, each cilium is bounded by the cell membrane and contains an axoneme with a central pair of microtubules surrounded by nine peripheral microtubular pairs (Figure 4–10). Cilia are inserted into **basal bodies,** which are electron-dense structures at the apical pole just below the cell membrane (Figure 4–10). Basal bodies have a structure similar to that of centrioles. In living organisms, cilia exhibit rapid back-and-forth movements coordinated to propel a current of fluid and suspended matter in one direction over the ciliated epithelium. The motion occurs due to activity of **ciliary dynein** present on the peripheral microtubular doublets of the axoneme, with adenosine triphosphate (ATP) as the energy source. A ciliated cell of the trachea lining is estimated to have about 250 cilia. Flagella, present in the human body only in spermatozoa (Chapter 21), are similar in structure to cilia but are much longer and are normally limited to one flagellum per cell.

TYPES OF EPITHELIA

Epithelia can be divided into two main groups according to their structure and function: **covering (or lining) epithelia** and **glandular epithelia.** This is an arbitrary division, for there are lining epithelia in which all the cells secrete (eg, the lining of the stomach)

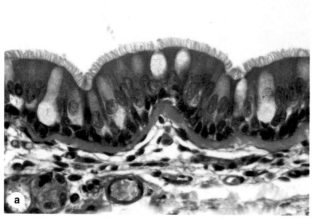

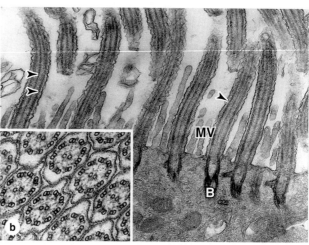

Figure 4-10. **Cilia.** TEMs of the apical portions of cells lining the respiratory tract show very well-developed cilia. **(a):** By light microscopy cilia usually appear as long, somewhat tangled projections. X400. Mallory trichrome. **(b):** TEM of cilia sectioned longitudinally reveals the axoneme of each, with arrowheads on the left side showing the central and peripheral microtubules. The arrowhead at right indicates the plasma membrane surrounding a cilium. At the base of each cilium is a basal body (B) from which it grows. Much shorter microvilli (MV) can be seen between the cilia. X59,000. **Inset:** Cilia seen in cross section clearly show the 9 + 2 array of the axoneme microtubules in each cilium. X80,000.

or in which glandular cells are distributed among the lining cells (eg, mucous cells in the small intestine or trachea).

Covering or Lining Epithelia

Covering epithelia are tissues in which the cells are organized in layers that cover the external surface or line the cavities of the body. They are classified according to the number of cell layers and the morphologic features of the cells in the surface layer (Table 4–2). **Simple epithelia** contain only one layer of cells and **stratified epithelia** contain more than one layer.

Based on cell shape, simple epithelia are classified as **squamous** (thin cells), **cuboidal** (cells roughly as thick as they are wide) or **columnar** (cells taller than they are wide) Examples of simple epithelia are shown in Figures 4–11, 4–12, and 4–13.

Stratified epithelia are classified according to the cell shape of the *superficial* layer(s): **squamous, cuboidal, columnar,** and **transitional.**

The very thin surface cells of stratified squamous epithelia can be "keratinized" (rich in keratin intermediate filaments) or "nonkeratinized" (with relatively sparse amounts of keratin). **Stratified squamous keratinized epithelium** is found mainly in the epidermis of skin. Its cells form many layers, and the cells closer to the underlying connective tissue are usually cuboidal or low columnar. The cells become irregular in shape and flatten as they accumulate keratin in the process of **keratinization** and are moved progressively closer to the surface, where they become thin, metabolically inactive packets (**squames**) of keratin lacking nuclei. This surface layer of cells helps protect against water loss across this epithelium. (See Chapter 18 for more detailed information on skin.) **Stratified squamous nonkeratinized epithelium** (Figure 4–14) lines wet cavities (eg, mouth, esophagus, and vagina). In such areas where water loss is not a problem, the flattened cells of the epithelial surface layer are living cells containing much less keratin and retaining their nuclei.

Stratified cuboidal and **stratified columnar epithelia** are rare. Stratified columnar epithelium can be found in the conjunctiva lining the eyelids, where it is both protective and mucus secreting. Stratified cuboidal epithelium is restricted to large excretory ducts of sweat and salivary glands, where it apparently provides a lining more robust than that of a simple epithelium.

Transitional epithelium or **urothelium**, which lines only the urinary bladder, the ureter, and the upper part of the urethra, is characterized by a superficial layer of domelike cells that are neither squamous nor columnar (Figure 4–15). These cells, sometimes called umbrella cells, are essentially protective against the hypertonic and potentially cytotoxic effects of urine. Importantly, the form of the surface cells changes according to the degree of distention of the bladder wall. This type of epithelium is discussed in detail in Chapter 19.

In addition to these various stratified epithelia, there is another type classified as **pseudostratified columnar epithelium**, so called because all cells are attached to the basal lamina even though their nuclei lie at different levels in the epithelium and the height of some cells does not extend to the surface. The best-known example of pseudostratified columnar epithelium is that lining the passages of the upper respiratory tract (Figure 4–16). The columnar cells of this epithelium are also heavily ciliated.

Table 4–2. Common types of covering epithelia in the human body.

Number of Cell Layers	Cell Form	Examples of Distribution	Main Function
Simple (one layer)	Squamous	Lining of vessels (endothelium). Serous lining of cavities; pericardium, pleura, peritoneum (mesothelium).	Facilitates the movement of the viscera (mesothelium), active transport by pinocytosis (mesothelium and endothelium), secretion of biologically active molecules (mesothelium).
	Cuboidal	Covering the ovary, thyroid.	Covering, secretion.
	Columnar	Lining of intestine, gallbladder.	Protection, lubrication, absorption, secretion.
Pseudostratified (layers of cells with nuclei at different levels; not all cells reach surface but all adhere to basal lamina)		Lining of trachea, bronchi, nasal cavity.	Protection, secretion; cilia-mediated transport of particles trapped in mucus out of the air passages.
Stratified (two or more layers)	Squamous keratinized (dry)	Epidermis.	Protection; prevents water loss.
	Squamous nonkeratinized (moist)	Mouth, esophagus, larynx, vagina, anal canal.	Protection, secretion; prevents water loss.
	Cuboidal	Sweat glands, developing ovarian follicles.	Protection, secretion.
	Transitional	Bladder, ureters, renal calyces.	Protection, distensibility.
	Columnar	Conjunctiva.	Protection.

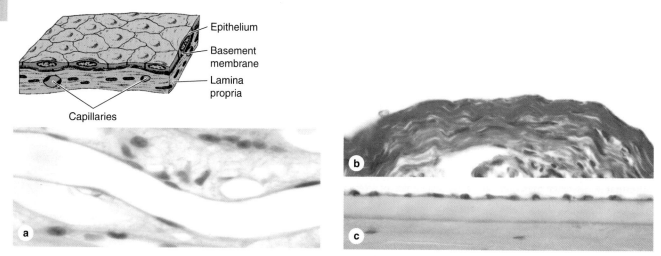

Figure 4–11. Simple squamous epithelia. In simple squamous epithelium, cells of the single layer are flat and usually very thin, with only the thicker cell nucleus appearing as a bulge to denote the cell. Simple epithelia are typically specialized as lining of vessels and cavities and regulate substances which can enter underlying tissue from the vessel or cavity. The thin cells often exhibit transcytosis. Examples shown here are those lining the renal loops of Henle **(a)**, the mesothelium lining a mesentery **(b)**, and the endothelium lining the inner surface of the cornea **(c)**. Endothelium and mesothelium are nearly always simple squamous. All X400. H&E.

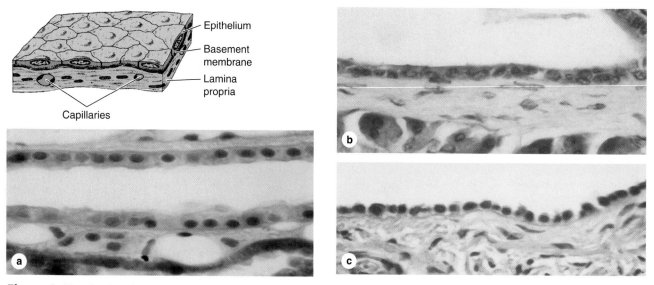

Figure 4–12. Simple cuboidal epithelium. Cells of simple cuboidal epithelia vary in their height but are roughly as tall as they are wide. Their greater thickness often includes cytoplasm rich in mitochondria providing energy for a high level of active transport of substances across the epithelium. Examples of simple cuboidal epithelia shown here are from a renal collecting tubule **(a)**, a pancreatic duct **(b)**, and the mesothelium covering an ovary **(c)**. All X400. H&E.

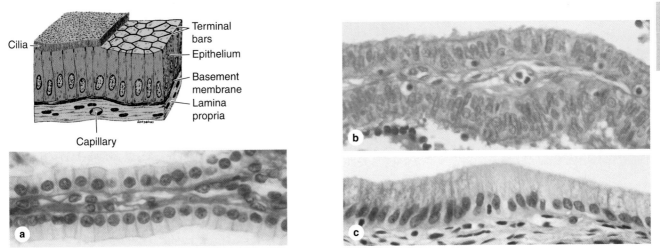

Figure 4–13. **Simple columnar epithelium.** Cells of simple columnar epithelia are taller than they are wide. Such cells are usually highly specialized for absorption, with microvilli, and often have interspersed secretory cells or ciliated cells. Such epithelial cells always have tight and adherent junctional complexes at their apical ends, but are often loosely associated in more basolateral areas. This allows for rapid transfer of absorbed material to the space between the cells rather than transport the full length of the cells. The additional cytoplasm in columnar cells allows additional mitochondria and other organelles needed for absorption and processing. The examples shown here are from a renal collecting duct **(a)**, the oviduct lining, with both secretory and ciliated cells **(b)**, and the lining of the gall bladder **(c)**. All X400. H&E.

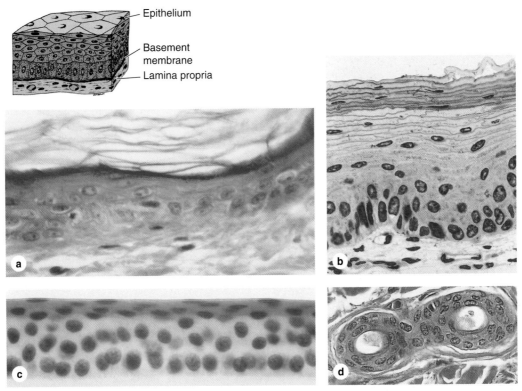

Figure 4–14. **Stratified epithelia.** Stratified squamous epithelia have protective functions: protection against easy invasion of underlying tissue by microorganisms and protection against water loss. In the skin, protection against water loss and desiccation is particularly important and the epithelium is **keratinized**. As epidermal cells of the skin **(a)** differentiate they become filled with keratin and other substances and eventually lose their nuclei and other organelles. The superficial flattened squames form a layer which impedes water loss and eventually slough off and are replaced from below. Keratinization will be discussed fully in Chapter 18. Epithelia lining many internal surfaces such as the esophagus **(b),** or covering the cornea **(c)** are considered **nonkeratinized** because the differentiating cells accumulate much less keratin and retain their nuclei. Such epithelia still provide protection against microorganisms, but do not fill with keratin because water loss is less of an issue. Stratified cuboidal or columnar epithelia are fairly rare, but are found in excretory ducts of some glands **(d)** where the double layer of cells apparently provides a more robust lining than a simple epithelium would. All X400; (b) PT, (a, c, and d) H&E.

Glandular Epithelia

Glandular epithelia are formed by cells specialized to secrete. The molecules to be secreted are generally stored in the cells in small membrane-bound vesicles called **secretory granules.**

Glandular epithelial cells may synthesize, store, and secrete proteins (eg, in the pancreas), lipids (eg, adrenal, sebaceous glands), or complexes of carbohydrates and proteins (eg, salivary glands). Mammary glands secrete all three substances. The cells of some glands have low synthetic activity (eg, sweat glands) and secrete mostly water and electrolytes transferred into the gland from the blood.

The epithelia that form glands can be classified according to various criteria. Unicellular glands consist of large isolated secretory cells and multicellular glands have clusters of cells. The classic unicellular gland is the **goblet cell** in the lining of the small intestine (Figure 4–17) or respiratory tract. The term "gland," however, is usually used to designate large aggregates of secretory epithelial cells, such as in the salivary glands and the pancreas.

Glands develop during fetal life from covering epithelia by means of cell proliferation and invasion of the subjacent connective tissue, followed by further differentiation (Figure 4–18).

Exocrine glands retain their connection with the surface epithelium, the connection taking the form of tubular ducts lined with epithelial cells through which the secretions pass to the surface. **Endocrine glands** have lost their connection to the surface from which they originated during development. These glands are therefore ductless and their secretions are picked up and transported to their sites of action by the bloodstream rather than by a duct system. Multicellular glands, whether exocrine or endocrine, also have connective tissue in a surrounding capsule and in septa that divide the gland into lobules. These lobules then subdivide, and in this way the connective tissue separates and binds the glandular components together (Figure 4–19).

Exocrine glands have a **secretory portion,** which contains the cells specialized for secretion, and **ducts,** which transport the secretion out of the gland. The morphology of these components allows the glands to be classified according to the scheme shown in Figure 4–20 and summarized as follows:

- Ducts can be **simple** (unbranched) or **compound** (with two or more branches).
- Secretory portions can be **tubular** (either short or long and **coiled**) or **acinar** (round or globular).
- Either type of secretory portion may be **branched.**
- Compound glands can have tubular, acinar, or tubuloacinar secretory portions.

Exocrine glands are also classified functionally according to the way the secretory products leave the cell (Figure 4–21):

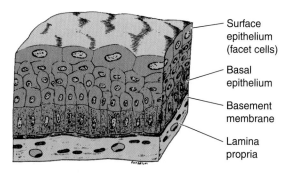

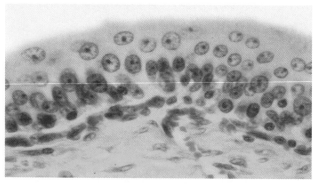

Figure 4–15. Transitional epithelium or urothelium. Stratified transitional epithelium lining the urinary bladder has rounded or dome-shaped superficial cells with two unusual features. The surface cells have specialized membranes and are able to withstand the hypertonic effects of urine and protect underlying cells from this toxic solution. Cells of transitional epithelium are also able to adjust their relationships with one another as the bladder fills and the wall is stretched, so that the transitional epithelium of a full, distended bladder seems to have fewer cell layers than that of an empty bladder. These unique features of urothelium will be discussed more fully in Chapter 19. X400. H&E.

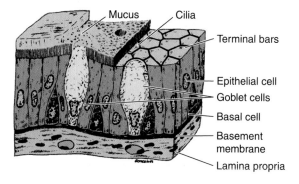

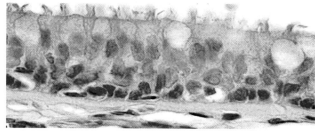

Figure 4–16. Pseudostratified epithelium. Cells of pseudostratified epithelia appear to be in layers, but the basal ends of the cells are all in contact with the basement membrane, which is often very thick in these epithelia. The best example of this epithelial type is the pseudostratified ciliated columnar epithelium of the upper respiratory tract, which contains cell types with their nuclei at different levels that give the false appearance of cellular stratification. This epithelium is discussed in detail in Chapter 17. X400. H&E.

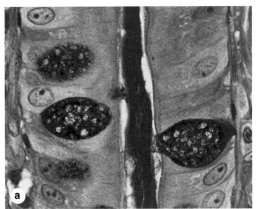
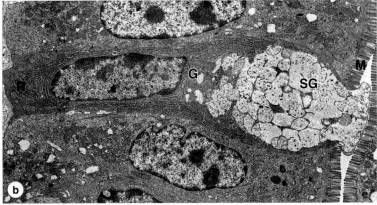

Figure 4–17. Goblet cells: unicellular glands. A section of epithelial lining of the large intestine shows scattered goblet cells secreting mucus to the extracellular space **(a):** With the stain for glycoproteins used here, both the mucus precursor stored in cytoplasmic graules of the goblet cells as well as the secreted mucus are stained dark blue. X400. PAS-PT. **(b):** Ultrastructurally a goblet cell shows a basal nucleus surrounded by RER (R), a large Golgi complex (G) just above the nucleus, and an apical end filed with large secretory granules (SG) containing mucins. This highly viscous material is secreted by exocytosis and is then hydrated to form mucus in the lumen lined by microvilli (M). X17,000.

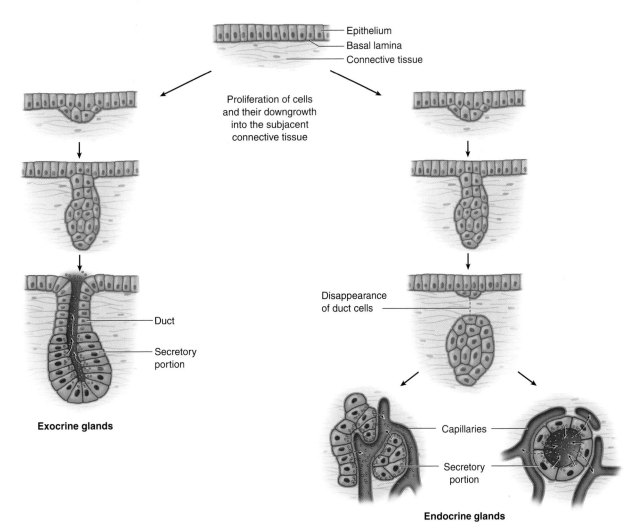

Figure 4–18. Formation of glands from covering epithelia. During fetal development epithelial cells proliferate and penetrate the underlying connective tissue. They may—or may not—maintain a connection with the surface epithelium. When the connection is maintained, exocrine glands are formed; with the connection lost, endocrine glands are formed. Exocrine glands secrete to the body surface or gut via duct systems formed from the epithelial connection. The cells of endocrine glands, which secrete hormones (see Chapter 20) can be arranged in cords or in follicles with lumens for storing the secretory product. From either the cords (left) or follicles (right) of endocrine cells, the secretory product is released outside the cells and picked up by the blood vessels for distribution throughout the body.

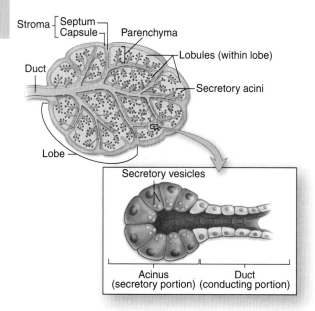

Stroma {Septum Capsule} — Parenchyma

Duct — Lobules (within lobe)

Secretory acini

Lobe

Secretory vesicles

Acinus (secretory portion) — Duct (conducting portion)

Figure 4–19. General structure of exocrine glands. Exocrine glands by definition have ducts that lead to an organ or body surface. Inside the gland the duct runs through connecting septa and branches repeatedly, until its smallest branches end in the secretory portions of the gland.

- **Merocrine secretion** (sometimes called eccrine) involves typical exocytosis of proteins or glycoproteins. This is the most common mode of secretion.
- **Holocrine secretion** involves the cell filling with secretory product and then the whole cell being disrupted and shed. This is best seen in the sebaceous glands of skin (Figure 4–22).
- In an intermediate type, **apocrine secretion**, the secretory product is typically a large lipid droplet and is discharged together with some of the apical cytoplasm and plasmalemma (Figure 4–23).

Exocrine glands with merocrine secretion can be further categorized as either **serous** or **mucous** according to the nature of the proteins or glycoproteins secreted and the resulting staining properties of the secretory cells. The acinar cells of the pancreas and parotid salivary glands are examples of the serous type which secrete **digestive enzymes**. The basal ends of serous cells have well-developed RER and Golgi complexes and the cells are filled apically with secretory granules in different stages of maturation (Figure 4–24). Serous cells therefore stain intensely with any basophilic or acidophilic stain.

Mucous cells, such as goblet cells, while also rich in RER and Golgi complexes are filled apically with secretory granules containing strongly hydrophilic glycoproteins called **mucins**. When mucins are released from the cell, they become hydrated and form **mucus**, a viscous, elastic, protective lubricant material. Mucin-containing granules stain well with the periodic

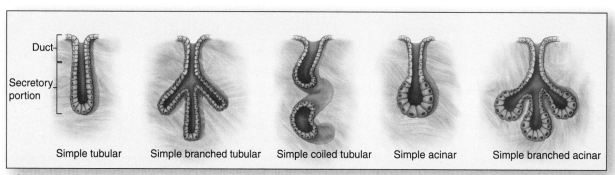

Duct

Secretory portion

Simple tubular · Simple branched tubular · Simple coiled tubular · Simple acinar · Simple branched acinar

a Simple glands

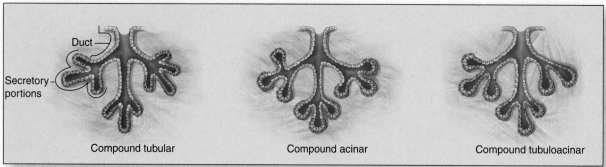

Duct

Secretory portions

Compound tubular · Compound acinar · Compound tubuloacinar

b Compound glands

Figure 4–20. Structural classes of exocrine glands. (a): Simple glands have unbranched ducts, although the ducts may be short or long and coiled. The secretory portions attached to these ducts may themselves be branched. The secretory portions are either **tubular**, if more or less cylindrical in shape, or **acinar**, if bulbous or saclike. **(b):** If the ducts branch to serve multiple secretory units, the gland is **compound**. On compound glands, the secretory units may be all tubular, all acinar, or a combination of the two shapes.

acid-Schiff (PAS) method for glycoproteins (Figure 4–17a), but are not intensely acidophilic like zymogen granules of serous cells (Figure 4–25). Mucous cells of large glands are organized as secretory tubules and in mixed seromucous salivary glands crescent-shaped clumps of serous cells frequently share the ends of the tubules as serous demilunes (Figure 4–26).

Several exocrine glands (eg, sweat, lachrymal, salivary, and mammary glands) contain stellate or spindle-shaped **myoepithelial cells** located between the basal lamina and the basal pole of secretory or duct cells (Figure 4–27). Long processes of these cells embrace an acinus as an octopus might embrace a rounded boulder. Along ducts they are more longitudinally arranged.

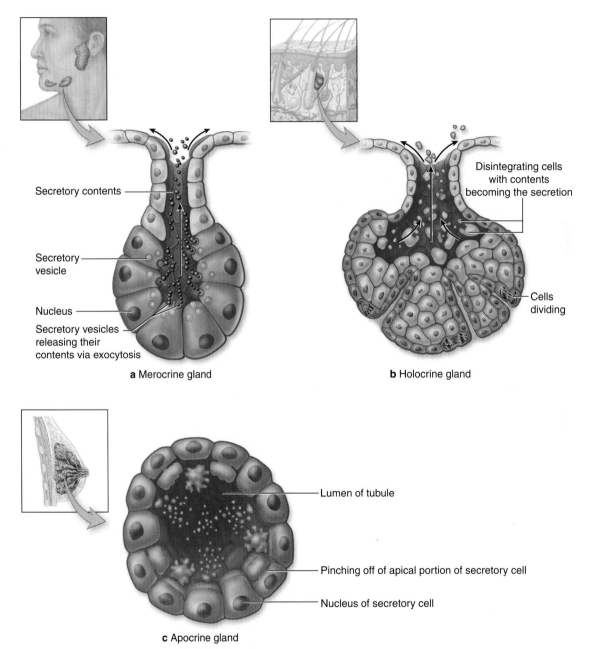

Figure 4–21. Functional classification of exocrine glands. Different cellular secretion processes are used in exocrine glands, depending on what substance is being secreted. **(a):** Merocrine glands secrete products, usually containing proteins, by means of exocytosis at the apical end of the secretory cells. Most exocrine glands are merocrine. **(b):** Holocrine gland secretion is produced by the disintegration of the secretory cells themselves as they complete differentiation which involves becoming filled with product. Sebaceous glands of hair follicles are the best examples of holocrine glands. **(c):** Apocrine gland secretion involves loss of a large membrane-enclosed portion of apical cytoplasm, usually containing one or more lipid droplets. This apical portion of the cell may subsequently break down to release its contents during passage into the duct. Apocrine secretion, along with merocrine secretion, is seen in mammary glands.

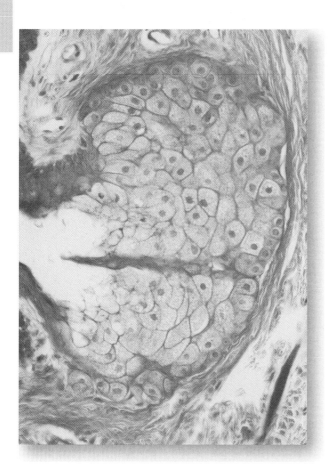

Figure 4–22. Holocrine secretion in a sebaceous gland. In holocrine secretion, best seen in the sebaceous gland adjacent to hair follicles, entire cells fill with a product and are released during secretion. Undifferentiated cells deep and peripheral in the gland fill with lipid-rich granules and become metabolically inactive as they mature and move upward and toward the gland's center. When terminally differentiated, the cells separate and quickly disintegrate to form the secretion which serves to protect and lubricate adjacent skin and hair. Sebaceous glands lack myoepithelial cells; cell proliferation inside a dense, inelastic connective tissue capsule continuously forces product into the duct. X200. H&E.

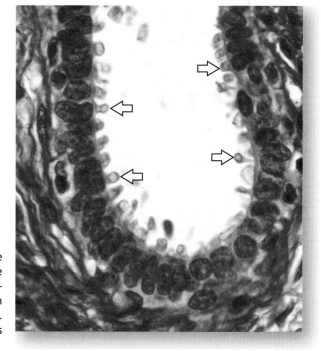

Figure 4–23. Apocrine secretion in the mammary gland. The secreting portions of a mammary gland demonstrate apocrine secretion and are characterized by the discharge of the secretion product with a pinched off portion of the apical cytoplasm (arrows). The released portion of cell contains lipid droplet(s). Merocrine secretion also occurs from the same and other cells of the secretory units. X400. PSH.

Myoepithelial cells are connected to each other and to the epithelial cells by both gap junctions and desmosomes. These cells are specialized for contraction, containing myosin and a large number of actin filaments. Their major function is to contract around the secretory or conducting portion of the gland and thus help propel secretory products into the duct.

Endocrine glands are the producers of **hormones**, which are generally polypeptide or lipid-derived factors that are released into the interstitial fluid. Hormones diffuse into the blood for circulation and bind specific receptors on target cells elsewhere in the body, often within other endocrine glands. The receptors may also be on cells very close to the hormone-secreting cells or on the secreting cell itself; in these cases the cellular signaling is termed **paracrine** or **autocrine**, respectively. Hormones can be secreted from single cells that are sparsely distributed or from cells with other major functions, such as certain cardiac muscle cells. In the large endocrine glands the parenchymal cells form strands or cords interspersed between dilated capillaries (eg, the adrenal cortex; see Figure 4–18) or can line a follicle filled with stored secretory product (eg, the thyroid gland; Figure 4–18). Some endocrine glands have cells releasing more than one hormone.

Some organs such as the pancreas have both endocrine and exocrine functions, and in the liver one cell type may function both ways, secreting bile components into a duct system, as well as releasing other products into the bloodstream.

TRANSPORT ACROSS EPITHELIA

As discussed in Chapter 2, all cells have the ability to actively transport certain ions against a concentration and electrical-potential gradient. An important example is the active extrusion of Na^+ by means of Mg^{2+}-activated Na^+/K^+-ATPase (**sodium pump**), by which cells maintain the required low intracellular sodium concentration (5-15 mmol/L vs. ~140 mmol/L in extracellular fluid).

Some epithelial cells actively transfer ions and fluid across the epithelium, from its apex to its base or vice-versa; this is known as **transcellular transport** (Figure 4–28). For transport in either direction, the tight junctions play an important role in the transport process, sealing the apical portions of the epithelium and preventing back-diffusion of materials already transported across the epithelium. A well-studied site of epithelial transport is the proximal renal tubule cell, where the apical surface is freely permeable to Na^+ in the lumen. To maintain electrical and osmotic balance, equimolar amounts of chloride and water follow the Na^+ ion into the cell. The basal surfaces of these cells are elaborately folded and many long invaginations of the basolateral membrane are seen in electron micrographs (Figure 4–29). In addition, there is interdigitation of membrane folds between adjacent cells, all of which increase the surface area for transport. Sodium pumps are localized in both the basal and the lateral plasma membranes and located between the folds are vertically oriented mitochondria

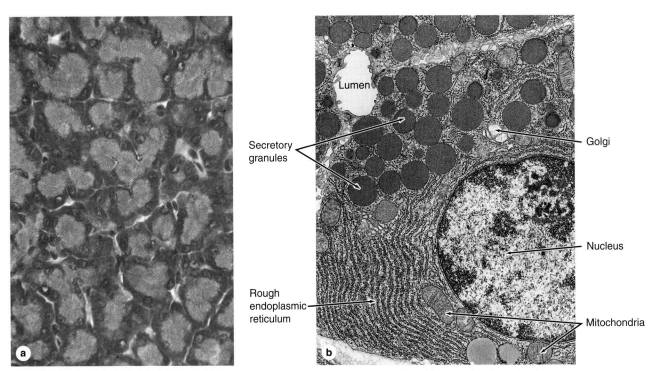

Figure 4–24. Serous cells. Serous acinar cells of the exocrine pancreas are arranged in small acini of 5-10 cells with a very small central lumen. Each acinar cell is roughly pyramid-shaped, with its apex at the lumen. **(a):** As seen by light microscopy, the apical ends are eosinophilic due to the abundant immature and mature secretory granules present there. The cells' basal ends contain the large rounded nuclei and an abundance of rough ER, making the cells highly basophilic basally. X200. PT. **(b):** A portion of one acinar cell is shown ultrastructurally, indicating the abundant RER, Golgi complexes, and secretory granules and the very small size of the acinus lumen. X13,000. Secretion here is merocrine and typically the mature **zymogen granules**, filled with digestive enzymes, remain in the apical cell region until the cell is stimulated to secrete. Other cells secrete constitutively, with small granules undergoing exocytosis as soon as they emerge fully formed from the Golgi apparatus.

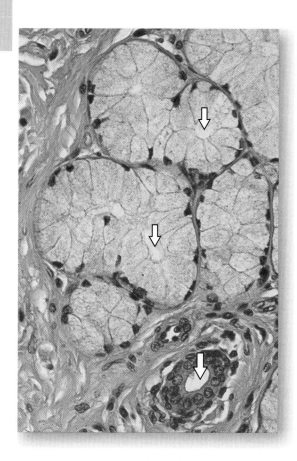

Figure 4–25 **Mucous cells.** Mucous cells are typically larger than serous cells, with more flattened basal nuclei. The apical region and most of the other cytoplasm of each mucous cell is filled with secretory granules containing mucin like that of goblet cells. The basal region contains the RER, nucleus, and a well-developed Golgi apparatus. The RER and Golgi are very rich in enzymes called glycosyltransferases, which attach sugars to polypeptide chains to make glycoproteins. Mucus contains many glycoproteins with important water-binding properties. The lumens (small arrows) of mucous tubules are larger than those of serous acini. The large arrow indicates a secretory duct. X200. PT. Other types of mucous cells are found in the stomach, the various salivary glands, the respiratory tract, and the genital tract. These cells show great variability in both their morphologic features and in the chemical nature of their secretions.

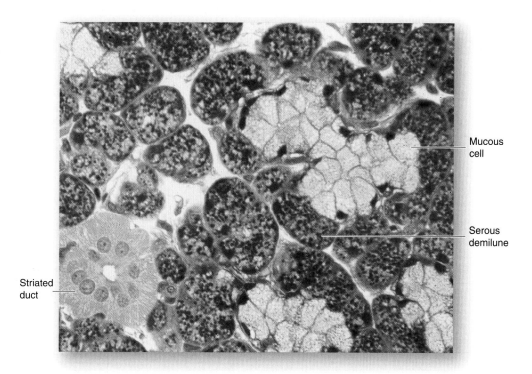

Mucous cell

Serous demilune

Striated duct

Figure 4–26. **Seromucous, compound tubuloacinar gland.** The submandibular salivary glands have both mucous and serous secretory units, typically shaped as acini and tubules respectively. Clumps of serous cells at the ends of some mucous tubules appear as crescent-shaped structures called **serous demilunes.** At the left is seen a **striated duct** whose cells' basal membranes are folded into long folds with many mitochondria, an arrangement specialized for ion transport across the epithelium. X400. PT.

that supply the ATP for the active extrusion of Na⁺ from the cell basally. Chloride and water again follow passively. In this way, sodium is returned to the circulation and is not lost in massive amounts in the urine.

Extracellular molecules and fluid are also internalized in the cytoplasm of most cells by pinocytotic vesicles that form abundantly at the plasmalemma. This activity is clearly observed in the simple squamous epithelia that line the blood and lymphatic capillaries (endothelia) or the body cavities (mesothelia). These cells have few organelles other than the abundant pinocytotic vesicles, which cross the thin cells in both directions and secrete their contents on the opposite side by exocytosis. This process, termed **transcytosis**, is not restricted to simple squamous epithelia. Uptake of material at the apical epithelial pole followed by exocytosis at the basolateral surface occurs actively in many simple cuboidal and columnar epithelia and is important in various physiological processes.

RENEWAL OF EPITHELIAL CELLS

Epithelial tissues are relatively labile structures whose cells are renewed continuously by mitotic activity. The renewal rate is variable; it can be fast in tissues such as the intestinal epithelium, which is replaced every week, or slow, as in the large glands. In stratified epithelial tissues, mitosis only occurs within the basal layer in contact with the basal lamina. In some functionally complex epithelia, stem cells have been identified only in restricted

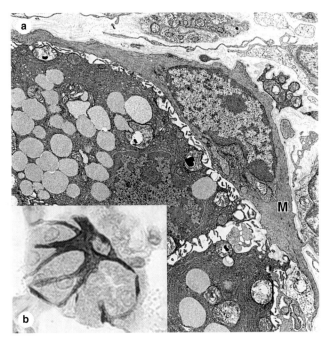

Figure 4–27. **Myoepithelial cells. (a):** Portion of a salivary gland acinus shows two secretory cells with secretory granules. A myoepithelial cell (M) embraces the acinus with contractile processes. X20,000. **(b):** A myoepithelial cell immunostained against smooth muscle actin shows its association with an entire acinus. Contraction of the myoepithelial cell compresses the acinus and aids in the expulsion of secretory products into the duct. X200. H&E counterstain.

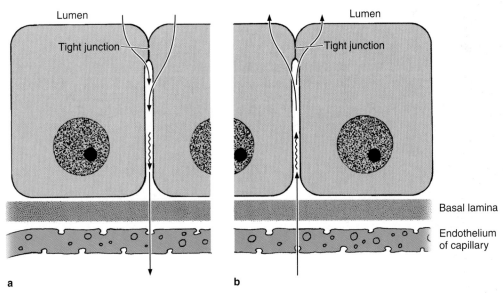

Figure 4–28. **Ion and water absorption and secretion.** Ion and water transport across epithelia can occur in different directions, depending on which tissue is involved. **(a):** The direction of transport is from the lumen to the blood vessel, as in the gallbladder and intestine. This process is called **absorption**, and serves to concentrate bile and obtain water and ions in these organs. **(b):** Transport in the opposite direction, as in the choroid plexus, ciliary body, and sweat glands, is called **secretion** and serves to expel water from the interstitial fluid into specialized aqueous fluids in these tissues. Whether the epithelia are absorbing or secreting water, the presence of apical occluding junctions is necessary to maintain tight compartmentalization and consequent control over ion distribution.

niches some distance from the transit amplifying cells and differentiating cells. For example, the epithelium lining the small intestine is derived completely from stem cells found in the simple glands between the intestinal villi. In the epidermis, stem cells are located at a characteristic position along the wall of hair follicles.

MEDICAL APPLICATION

Both benign and malignant tumors can arise from most types of epithelial cells. A **carcinoma** *(Gr. karkinos, cancer, + oma, tumor) is a malignant tumor of epithelial cell origin. Malignant tumors derived from glandular epithelial tissue are usually called* **adenocarcinomas** *(Gr. adenos, gland, + karkinos); these are by far the most common tumors in adults. In children up to age 10 years, most tumors develop (in decreasing order) from hematopoietic organs, nerve tissues, connective tissues, and epithelial tissues. This proportion gradually changes, and after age 45 years, more than 90% of all tumors are of epithelial origin.*

Carcinomas composed of differentiated cells reflect cell-specific morphologic features and behaviors (eg, the production of keratins, mucins, and hormones). Undifferentiated carcinomas are often difficult to diagnose by morphologic analysis alone. Since these carcinomas usually contain keratins, the detection of keratins by immunocytochemistry often helps to determine the diagnosis and treatment of these tumors.

Epithelia are normally capable of rapid repair and replacement of apoptotic or damaged cells. In some large glands, most notably the liver, mitotic activity is normally rare but is actively renewed following major damage to the organ. When a portion of liver tissue is removed surgically or lost by the acute effects of toxic substances, cells of undamaged regions quickly begin active proliferation and normal functional mass of liver tissue is soon regenerated.

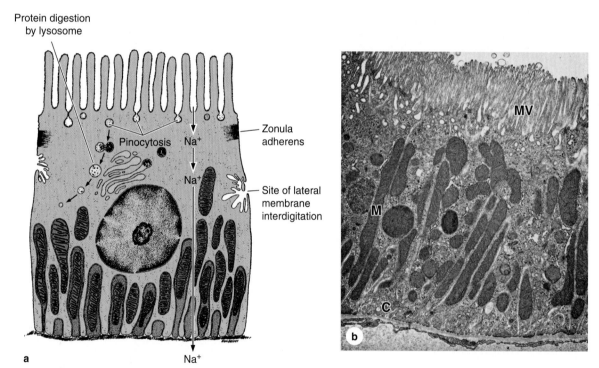

Figure 4–29. Absorptive cells. An ultrastructural diagram and TEM of epithelial cells highly specialized for absorption: cells of proximal convoluted tubule of the kidney. Long invaginations of the basal cell membrane outline regions filled with vertically oriented mitochondria, a typical disposition present in ion-transporting cells. Interdigitations from neighboring cells interlock with those of this cell. Immediately below the microvilli are junctional complexes between individual cells. The basolateral membranes can be discerned in continuity with the junctional complexes. Apically are vesicles that have undergone pinocytosis, soon to fuse with lysosomes, as shown in the upper left portion of the diagram. Sodium ions diffuse passively through the apical membranes of renal epithelial cells and are then actively transported out of the cells by Na^+/K^+ -ATPase located in the basolateral membranes of the cells. Energy for this sodium pump is supplied by the nearby mitochondria. Immediately below the basal lamina is a capillary for removal of the water absorbed across this part of the epithelium. X9600.

MEDICAL APPLICATION

Some epithelial cells are prone to abnormal growth called neoplasia that may lead to cancers. Neoplastic growth is reversible and does not always result in cancer.

Under certain abnormal conditions, one type of epithelial tissue may undergo transformation into another type in another reversible process called **metaplasia,** *which is illustrated by the following examples.*

In heavy cigarette smokers, the ciliated pseudostratified epithelium lining the bronchi can be transformed into stratified squamous epithelium.

In individuals with chronic vitamin A deficiency, epithelial tissues of the type found in the bronchi and urinary bladder are gradually replaced by stratified squamous epithelium.

Metaplasia is not restricted to epithelial tissue; it may also occur in connective tissue.

CELLS OF CONNECTIVE TISSUE
  Fibroblasts
  Adipocytes
  Macrophages & the Mononuclear
    Phagocyte System
  Mast Cells
  Plasma Cells
  Leukocytes

FIBERS
  Collagen
  Reticular Fibers
  Elastic Fibers
GROUND SUBSTANCE
TYPES OF CONNECTIVE TISSUE
  Connective Tissue Proper
  Reticular Tissue
  Mucous Tissue

The different types of connective tissue are responsible for providing and maintaining the form of organs throughout the body. Functioning in a mechanical role, they provide a matrix that connects and binds other tissues and cells in organs and gives metabolic support to cells as the medium for diffusion of nutrients and waste products.

Structurally, connective tissue is formed by three classes of components: cells, fibers, and ground substance. Unlike the other tissue types (epithelium, muscle, and nerve), which consist mainly of cells, the major constituent of connective tissue is the **extracellular matrix (ECM).** Extracellular matrices consist of different combinations of **protein fibers** (collagen, reticular, and elastic fibers) and **ground substance.** Ground substance is a highly hydrophilic, viscous complex of anionic macromolecules (glycosaminoglycans and proteoglycans) and multiadhesive glycoproteins (laminin, fibronectin, and others) that stabilizes the ECM by binding to receptor proteins (**integrins**) on the surface of cells and to the other matrix components. In addition to its major structural role, molecules of connective tissue serve other important biological functions, such as forming a reservoir of factors controlling cell growth and differentiation. The hydrated nature of much connective tissue provides the medium through which nutrients and metabolic wastes are exchanged between cells and their blood supply.

The wide variety of connective tissue types in the body reflects variations in the composition and amount of the cells, fibers, and ground substance which together are responsible for the remarkable structural, functional, and pathologic diversity of connective tissue.

The connective tissues originate from the **mesenchyme,** an embryonic tissue formed by elongated undifferentiated cells, the **mesenchymal cells** (Figure 5–1). These cells are characterized by oval nuclei with prominent nucleoli and fine chromatin. They possess many thin cytoplasmic processes and are immersed in an abundant and viscous extracellular substance containing few fibers. The mesenchyme develops mainly from the middle layer of the embryo, the **mesoderm.** Mesodermal cells migrate from their site of origin in the embryo, surrounding and penetrating developing organs. In addition to being the point of origin of all types of connective tissue cells, mesenchyme develops into other types of structures, such as blood cells, endothelial cells, and muscle cells.

CELLS OF CONNECTIVE TISSUE

A variety of cells with different origins and functions are present in connective tissue (Figure 5–2 and Table 5–1). **Fibroblasts** originate locally from undifferentiated mesenchymal cells and spend all their life in connective tissue; other cells such as **mast cells**, **macrophages**, and **plasma cells** originate from hematopoietic stem cells in bone marrow, circulate in the blood, and then move into connective tissue where they remain and execute their functions. White blood cells (leukocytes) are transient cells of most connective tissues; they also originate in the bone marrow and move to the connective tissue where they reside for a few days, then usually die by apoptosis.

Fibroblasts

Fibroblasts synthesize collagen, elastin, glycosaminoglycans, proteoglycans and multiadhesive glycoproteins. Fibroblasts are the most common cells in connective tissue (Figure 5–3) and are responsible for the synthesis of extracellular matrix components. Two stages of activity—active and quiescent—are often observed in these cells (Figure 5–3b). Cells with intense synthetic activity are morphologically distinct from the quiescent fibroblasts that are scattered within the matrix they have already synthesized. Some histologists reserve the term **fibroblast** to denote the active cell and **fibrocyte** to denote the quiescent cell.

The active fibroblast has an abundant and irregularly branched cytoplasm. Its nucleus is ovoid, large, and pale-staining, with fine chromatin and a prominent nucleolus. The cytoplasm is rich in rough ER, and the Golgi apparatus is well developed. The quiescent fibroblast or fibrocyte is smaller than the active fibroblast and is usually spindle-shaped. It has fewer processes; a smaller, darker, elongated nucleus; and more acidophilic cytoplasm with much less RER.

Fibroblasts synthesize most components of connective tissue ECM, including proteins, such as collagen and elastin, which upon secretion form collagen, reticular, and elastic fibers, and the glycosaminoglycans, proteoglycans, and glycoproteins of the ground substance. Fibroblasts are targets of various **growth factors** that influence cell growth and differentiation. In adults, fibroblasts in connective tissue rarely undergo division; mitosis can resume when the organ requires additional fibroblasts as in wound healing.

by inflammation or traumatic injury. In these cases, the spaces left after injury to tissues whose cells do not divide (eg, cardiac muscle) are filled by connective tissue, which forms a scar. The healing of surgical incisions depends on the reparative capacity of connective tissue. The main cell type involved in repair is the fibroblast.

*When it is adequately stimulated, such as during wound healing, the fibrocyte reverts to the fibroblast state, and its synthetic activities are reactivated. In such instances the cell reassumes the form and appearance of a fibroblast. The **myofibroblast**, a cell with features of both fibroblasts and smooth muscle cells, is also observed during wound healing. These cells have most of the morphological characteristics of fibroblasts but contain increased amounts of actin microfilaments and myosin and behave much like smooth muscle cells. Their activity is responsible for wound closure after tissue injury, a process called **wound contraction**.*

Adipocytes

Adipocytes (L. *adeps*, fat, + Gr. *kytos*, cell) are connective tissue cells that have become specialized for storage of neutral fats or for the production of heat. Often called **fat cells**, they have considerable metabolic significance and are discussed in detail in Chapter 6.

Macrophages & the Mononuclear Phagocyte System

Macrophages were discovered and initially characterized by their phagocytic ability. They have a wide spectrum of morphologic features that correspond to their state of functional activity and to the tissue they inhabit.

In the electron microscope, macrophages are characterized by an irregular surface with pleats, protrusions, and indentations, a morphologic expression of their active pinocytotic and phagocytic activities. They generally have a well-developed Golgi apparatus, many lysosomes, and rough ER (Figure 5–4).

Macrophages derive from bone marrow precursor cells that divide, producing **monocytes** which circulate in the blood. These cells cross the wall of venules and capillaries to penetrate the connective tissue, where they mature and acquire the morphologic features of **macrophages.** Therefore, monocytes and macrophages are the same cell in different stages of maturation. Macrophages are also sometimes referred to as "histiocytes."

Macrophages are distributed throughout the body and are present in most organs. Along with other monocyte-derived cells, they comprise a family of cells called the **mononuclear phagocyte system** (Table 5–2). All are long-living cells and may survive for months in the tissues. In most organs these cells are highly important for the up-take, processing, and presentation of antigens for lymphocyte activation. The macrophage-like cells have been given different names in different organs, eg, Kupffer cells in the liver, microglial cells in the central nervous system, Langerhans cells in the skin, and osteoclasts in bone tissue. However, all are derived from monocytes. The transformation from monocytes to macrophages in connective tissue involves increases in cell size, increased protein synthesis, and increases in the

MEDICAL APPLICATION

The regenerative capacity of the connective tissue is clearly observed when tissues are destroyed

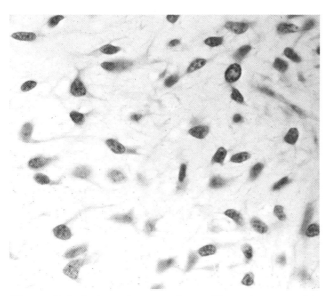

Figure 5–1. Embryonic mesenchyme. Mesenchyme consists of a population of undifferentiated cells, generally elongated but with many shapes, having large euchromatic nuclei and prominent nucleoli which indicate high levels of synthetic activity. These cells are called **mesenchymal cells.** Mesenchymal cells are surrounded by an extracellular matrix which they produced and which consists largely of a simple ground substance rich in hyaluronan (hyaluronic acid). This section is stained with Masson trichrome which stains collagen fibers blue and the lack of collagen in mesenchyme is apparent. X200.

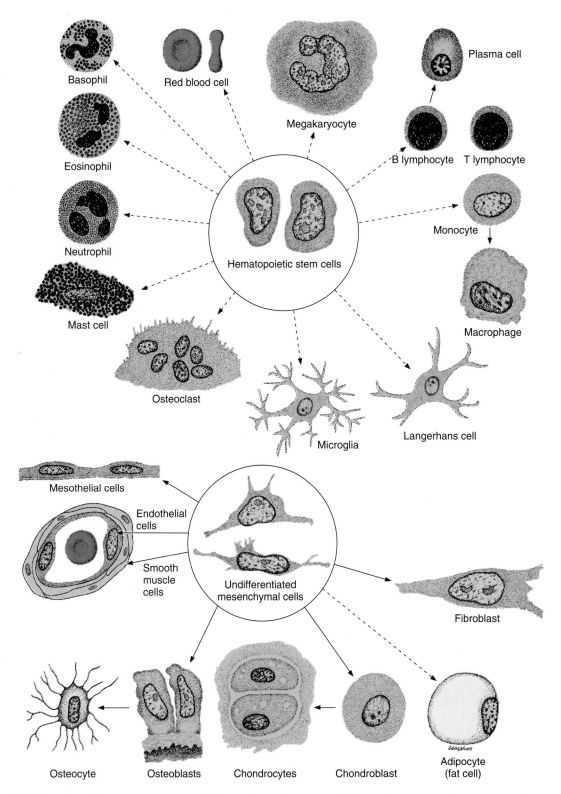

Figure 5–2. **Lineages of connective tissue cells.** This simplified representation of the connective tissue cell lineage includes cells from the multipotential embryonic **mesenchyme cells** and **hematopoietic stem cells** of bone marrow. Dotted arrows indicate that one or more intermediate cell types exist between the examples illustrated. The cells are not drawn in proportion to actual sizes, eg, adipocyte, megakaryocyte, and osteoclast cells are significantly larger than the other cells illustrated.

number of Golgi complexes and lysosomes. A typical macrophage measures between 10 and 30 μm in diameter and has an oval or kidney-shaped nucleus located eccentrically.

MEDICAL APPLICATION

*When adequately stimulated, macrophages may increase in size and are arranged in clusters forming **epithelioid** cells (named for their vague resemblance to epithelial cells), or several may fuse to form **multinuclear giant cells**. Both cell types are usually found only in pathological conditions.*

Macrophages act as defense elements. They phagocytose cell debris, abnormal extracellular matrix elements, neoplastic cells, bacteria, and inert elements that penetrate the organism. Macrophages are also antigen-presenting cells that participate in the processes of partial digestion and presentation of antigen to other cells (see Chapter 14). A typical example of an antigen-processing cell is the macrophage present in the skin epidermis, called the Langerhans cell (see Chapter 18). Although macrophages are the main antigen presenting cells, under certain circumstances many other cell types, such as fibroblasts, endothelial cells, astrocytes, and thyroid epithelial cells, are also able to perform this function. Macrophages also participate in cell-mediated resistance to infection by bacteria, viruses, protozoans, fungi, and metazoans (eg, parasitic worms); in cell-mediated resistance to tumors; and in extrahepatic bile production, iron and fat metabolism, and the destruction of aged erythrocytes.

*When macrophages are stimulated (by injection of foreign substances or by infection), they change their morphological characteristics and metabolism. They are then called **activated macrophages** and acquire characteristics not present in their nonactivated state. These activated macrophages, in addition to showing an increase in their capacity for phagocytosis and intracellular digestion, exhibit enhanced metabolic and lysosomal enzyme activity. Macrophages also have an important role in removing cell debris and damaged extracellular components formed during the physiological involution process.*

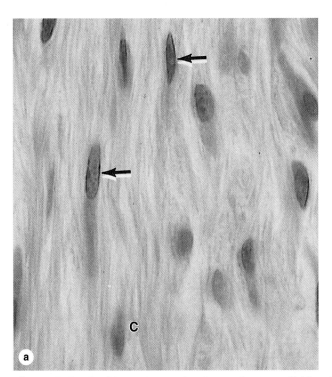
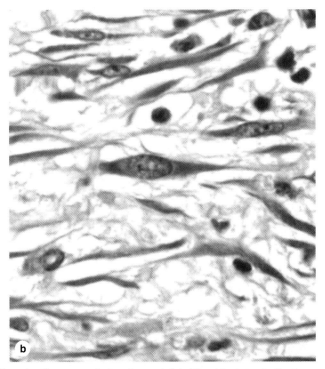

Figure 5–3. **Fibroblasts.** Connective tissue where parallel bundles of collagen are being formed. **(a):** Fibroblasts typically show large active nuclei and eosinophilic cytoplasm tapering off in both directions along the axis of the nucleus, a morphology usually called "spindle-shaped." The nuclei (arrows) are clearly seen, but the cytoplasmic processes resemble the collagen bundles (C) that fill the extracellular matrix and are difficult to distinguish in H&E-stained sections. **(b):** Both active and quiescent fibroblasts may sometimes be distinguished, as in this section of dermis. Active fibroblasts are large cells with large, euchromatic nuclei and basophilic cytoplasm, whereas inactive fibroblast or fibrocytes are smaller with less prominent, heterochromatic nuclei. The very basophilic round cells in (b) are leukocytes. Both X400. H&E.

For example, during pregnancy the uterus increases in size. Immediately after parturition, the uterus suffers an involution during which some of its tissues are destroyed by the action of macrophages. Macrophages are also secretory cells that produce an impressive array of substances, including enzymes (eg, collagenase) and cytokines that participate in defensive and reparative functions, and they exhibit increased tumor cell–killing capacity.

Mast Cells

Mast cells are large, oval or round connective tissue cells, 20–30 μm in diameter, whose cytoplasm is filled with basophilic secretory granules. The rather small, spherical nucleus is centrally situated and may be obscured by the cytoplasmic granules (Figure 5–5).

The secretory granules are 0.3–2.0 μm in diameter. Their interior is electron-dense and heterogeneous. Mast cells function in the localized release of many bioactive substances with roles in the inflammatory response, innate immunity, and tissue repair.

Because of their high content of acidic radicals in their sulfated glycosaminoglycans, mast cell granules display **metachromasia**, which means that they can change the color of some basic dyes (eg, toluidine blue) from blue to purple or red. The granules are poorly preserved by common fixatives, so that mast cells are frequently difficult to identify. Mast cell granules contain a wide variety of paracrine compounds that promote different aspects of a local inflammatory response. A partial list of important molecules released from these granules includes:

- **Heparin**, a sulfated glycosaminoglycan that acts locally as an anticoagulant
- **Histamine**, which promotes increased vascular permeability and smooth muscle contraction

- **Serine proteases**, which activate various mediators of inflammation
- **Eosinophil** and **neutrophil chemotactic factors** which attract those leukocytes
- **Leukotrienes C$_4$, D$_4$, and E$_4$** (or the slow-reacting substance of anaphylaxis, SRS-A) which also trigger smooth muscle contraction.

Mast cells occur in many connective tissues, but are especially numerous near small blood vessels in skin and mesenteries (**perivascular mast cells**) and in the mucosa lining digestive and respiratory tracts (**mucosal mast cells**). The average size and granular content of these two populations differ somewhat.

Mast cells originate from progenitor cells in the bone marrow. These progenitor cells circulate in the blood, cross the wall of venules and capillaries, and penetrate connective tissues, where they proliferate and differentiate. Although they are in many respects similar to basophilic leukocytes, they have a separate stem cell.

Release of the chemical mediators stored in mast cells promotes the allergic reactions known as **immediate hypersensitivity reactions,** because they occur within a few minutes after penetration by an antigen of an individual previously sensitized to the same or a very similar antigen. There are many examples of immediate hypersensitivity reaction; a dramatic one is anaphylactic shock, a potentially fatal condition. The process of anaphylaxis consists of the following sequential events: The first exposure to an antigen (allergen), such as bee venom, results in production of the IgE class of immunoglobulins (antibodies) by plasma cells. IgE is avidly bound to the surface of mast cells. A second exposure to the antigen results in binding of the antigen to IgE on the mast cells. This event triggers release of the mast cell granules, liberating histamine, leukotrienes, ECF-A, and heparin (Figure 5–6). Degranulation of mast cells also occurs as a result of the action of the complement molecules that participate in the immunological reaction cited in Chapter 14. Histamine causes contraction of smooth muscle (mainly of the

Table 5–1. Functions of connective tissue cells.

Cell Type	Representative Product or Activity	Representative Function
Fibroblast, chondroblast, osteoblast, odontoblast	Production of fibers and ground substance	Structural
Plasma cell	Production of antibodies	Immunologic (defense)
Lymphocyte (several types)	Production of immunocompetent cells	Immunologic (defense)
Eosinophilic leukocyte	Participation in allergic and vasoactive reactions, modulation of mast cell activities and the inflammatory process	Immunologic (defense)
Neutrophilic leukocyte	Phagocytosis of foreign substances, bacteria	Defense
Macrophage	Secretion of cytokines and other molecules, phagocytosis of foreign substances and bacteria, antigen processing and presentation to other cells	Defense
Mast cell and basophilic leukocyte	Liberation of pharmacologically active molecules (eg, histamine)	Defense (participate in allergic reactions)
Adipocyte	Storage of neutral fats	Energy reservoir, heat production

Table 5–2. Distribution and main functions of the cells of the mononuclear phagocyte system.

Cell Type	Location	Main Function
Monocyte	Blood	Precursor of macrophages
Macrophage	Connective tissue, lymphoid organs, lungs, bone marrow	Production of cytokines, chemotactic factors, and several other molecules that participate in inflammation (defense), antigen processing and presentation
Kupffer cell	Liver	Same as macrophages
Microglia cell	Nerve tissue of the central nervous system	Same as macrophages
Langerhans cell	Skin	Antigen processing and presentation
Dendritic cell	Lymph nodes	Antigen processing and presentation
Osteoclast	Bone (fusion of several macrophages)	Digestion of bone
Multinuclear giant cell	Connective tissue (fusion of several macrophages)	Segregation and digestion of foreign bodies

bronchioles) and dilates and increases permeability (mainly in postcapillary venules). Any liberated histamine is inactivated immediately after release. Leukotrienes produce slow contractions in smooth muscle, and ECF-A attracts blood eosinophils.

Heparin is a blood anticoagulant, but blood clotting remains normal in humans during anaphylactic shock. Mast cells are widespread in the human body but are particularly abundant in the dermis and in the digestive and respiratory tracts.

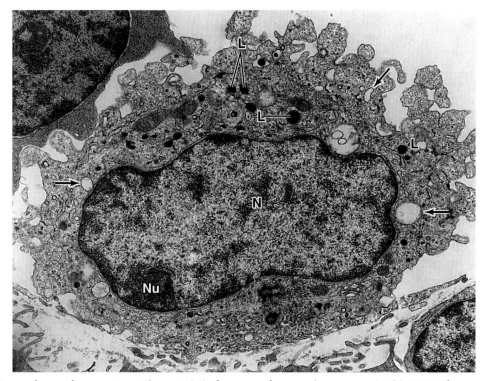

Figure 5–4. Macrophage ultrastructure. Characteristic features of macrophages seen in this TEM of one such cell are the prominent nucleus (N) and the nucleolus (Nu) and the numerous secondary lysosomes (L). The arrows indicate phagocytic vacuoles near the protrusions and indentations of the cell surface. X10,000.

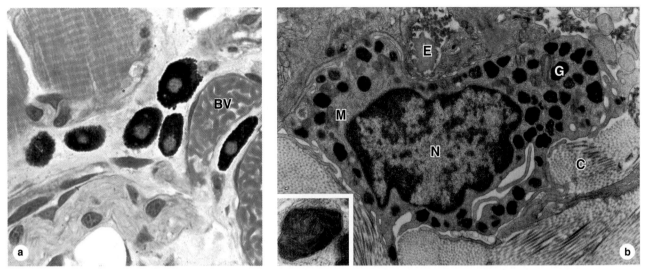

Figure 5–5. Mast cells. Mast cells are components of loose connective tissues, often located near small blood vessels (BV). **(a):** They are typically oval-shaped, with cytoplasm filled with strongly basophilic granules. X400. PT. **(b):** Ultrastructurally mast cells show little else around the nucleus (N) besides these cytoplasmic granules (G), except for occasional mitochondria (M). The granule staining in the TEM is heterogeneous and variable in mast cells from different tissues; at higher magnifications some granules may show a characteristic scroll-like substructure (inset) that contains preformed mediators such as histamine and proteoglycans. The ECM near this mast cell includes elastic fibers (E) and bundles of collagen fibers (C).

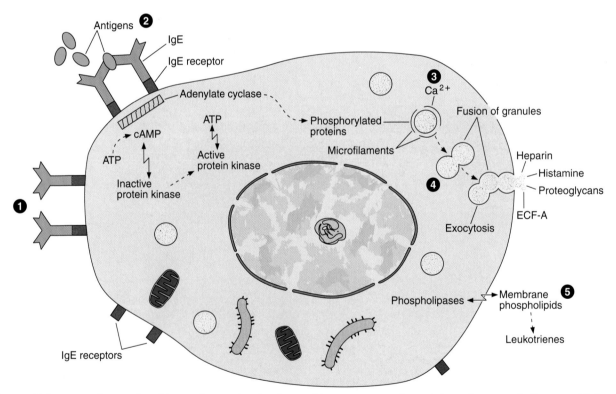

Figure 5–6 Mast cell secretion. Mast cell secretion is triggered by re-exposure to certain antigens and allergens. Molecules of IgE antibody produced in an initial response to an allergen such as pollen or bee venom are bound to surface receptors for IgE (**1**), of which 300,000 are present per mast cell. When a second exposure to the allergen occurs, IgE molecules bind this antigen and a few IgE receptors very rapidly become cross-linked (**2**). This activates adenylate cyclase, leading to phosphorylation of specific proteins and (**3**) entry of Ca^{2+} and rapid exocytosis of some granules (**4**). In addition, phospholipases act on specific membrane phospholipids, leading to production and release of leukotrienes (**5**). The components released from granules, as well as the leukotrienes, are immediately active in the local microenvironment and promote a variety of controlled local reactions which together normally comprise part of the inflammatory process called the immediate hypersensitivity reaction. ECF-A, eosinophil chemotactic factor of anaphylaxis.

Plasma Cells

Plasma cells are large, ovoid cells that have a basophilic cytoplasm due to their richness in rough ER. The juxtanuclear Golgi apparatus and the centrioles occupy a region that appears pale in regular histologic preparations (Figure 5–7).

The nucleus of the plasma cell is generally spherical but eccentrically placed. Many of these nuclei contain compact, peripheral regions of heterochromatin alternating with lighter areas of euchromatin, a configuration that can give the nucleus of a plasma cell the appearance of a clock-face. There are few plasma cells in most connective tissues. Their average lifespan is short, 10–20 days.

MEDICAL APPLICATION

Plasma cells are derived from B lymphocytes and are responsible for the synthesis of antibodies. Antibodies are immunoglobulins produced in response to penetration by antigens. Each antibody is specific for the one antigen that gave rise to its production and reacts specifically with molecules possessing similar epitopes (see Chapter 14). The results of the antibody-antigen reaction are variable. The capacity of the reaction to neutralize harmful effects caused by antigens is important. An antigen that is a toxin (eg, tetanus, diphtheria) may lose its capacity to do harm when it combines with its respective antibody.

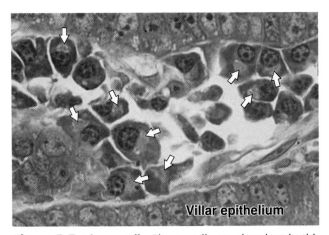

Figure 5–7. Plasma cells. Plasma cells are abundant in this portion of an inflamed intestinal villus. The plasma cells are characterized by their abundant basophilic cytoplasm involved in the synthesis of antibodies. A large pale Golgi apparatus (arrows) near each nucleus is the site of the terminal glycosylation of the antibodies (glycoproteins). Plasma cells can leave their sites of origin in lymphoid tissues, move to connective tissue, and produce the antibodies that mediate immunity. X400 PT.

Leukocytes

Connective tissue normally contains leukocytes that migrate from the blood vessels by diapedesis. Leukocytes (Gr. *leukos,* white, + *kytos*), or white blood corpuscles, are the wandering cells of the connective tissue. They leave blood by migrating between the endothelial cells lining capillaries and postcapillary venules to enter connective tissue by a process called **diapedesis**. This process increases greatly during inflammation, which is a vascular and cellular defensive reaction against foreign substances, in most cases pathogenic bacteria or irritating chemical substances. The classic signs of inflammation were first described by Celsus in the first century as redness and swelling with heat and pain (*rubor et tumor cum calore et dolore*).

Inflammation begins with the local release of **chemical mediators of inflammation,** substances of various origin (mainly from local cells and blood plasma proteins) that induce some of the events characteristic of inflammation, eg, **increase of blood flow** and **vascular permeability, chemotaxis,** and **phagocytosis.**

MEDICAL APPLICATION

Increased vascular permeability is caused by the action of vasoactive substances; an example is histamine, which is liberated from mast cells and basophilic leukocytes. Increases in blood flow and vascular permeability are responsible for local swelling (edema), redness, and heat. Pain is due mainly to the action of chemical mediators on nerve endings. Chemotaxis (Gr. chemeia, alchemy, + taxis, orderly arrangement), the phenomenon by which specific cell types are attracted by some molecules, is responsible for the migration of large quantities of specific cell types to regions of inflammation. As a consequence of chemotaxis, leukocytes cross the walls of venules and capillaries by diapedesis, invading the inflamed areas.

Leukocytes do not return to the blood after arriving in connective tissue except for the lymphocytes. These cells circulate continuously in various compartments of the body: blood, lymph, lymphatic organs, and the interstitial fluid of connective tissue. Lymphocytes are particularly abundant in the connective tissue of the digestive tract. A detailed analysis of the structure and functions of leukocytes and lymphocytes is presented in Chapters 12 and 14.

FIBERS

The connective tissue fibers are formed by proteins that polymerize into elongated structures. The three main types of connective tissue fibers are **collagen, reticular,** and **elastic**

fibers. Collagen and reticular fibers are both formed by the protein **collagen,** and elastic fibers are composed mainly of the protein **elastin.** These fibers are distributed unequally among the types of connective tissue and the predominant fiber type is usually responsible for conferring specific properties on the tissue.

Collagen

The collagens constitute a family of proteins selected during evolution for the execution of several (mainly structural) functions. During the process of evolution of multicellular organisms, a family of structural proteins was selected by both environmental influences and the functional requirements of the animal organism and developed to acquire varying degrees of rigidity, elasticity, and strength. These proteins are known collectively as **collagen,** and the chief examples among its various types are present in the skin, bone, cartilage, smooth muscle, and basal lamina.

Collagen is the most abundant protein in the human body, representing 30% of its dry weight. The collagens are produced by several cell types and are distinguishable by their molecular compositions, morphologic characteristics, distribution, functions, and pathologies. More than 20 types of collagen have been identified and designated with Roman numerals; the most important of these are listed in Table 5–3. They are classified into the following four categories according to their structure and general functions.

COLLAGENS THAT FORM LONG FIBRILS

The molecules of long fibril–forming collagens aggregate to form fibrils clearly visible in the electron or light microscope (Figure 5–8). Collagen type I is the most abundant and has a widespread distribution. It occurs in tissues as structures that are classically designated as **collagen fibers** forming structures such as tendons, organ capsules, and dermis.

Table 5–3. Collagen types.

Type	Molecule Composition	Structure	Optical Microscopy	Representative Tissues	Main Function
Collagens that form fibrils					
I	$[\alpha 1\,(I)]_2[\alpha 2\,(I)]$	300-nm molecule, 67-nm banded fibrils	Thick, highly picrosirius birefringent, nonargyrophilic fibers	Skin, tendon, bone, dentin	Resistance to tension
II	$[\alpha 1\,(II)]_3$	300-nm molecule, 67-nm banded fibrils	Loose aggregates of fibrils, birefringent	Cartilage, vitreous body	Resistance to pressure
III	$[\alpha 1\,(III)]_3$	67-nm banded fibrils	Thin, weakly birefringent, argyrophilic fibers	Skin, muscle, blood vessels, frequently together with type I	Structural maintenance in expansible organs
V	$[\alpha 1\,(V)]_3$	390-nm molecule, N-terminal globular domain	Frequently forms fiber together with type I	Fetal tissues, skin, bone, placenta, most interstitial tissues	Participates in type I collagen function
XI	$[\alpha 1\,(XI)]\,[\alpha 2\,(XI)]$ $[\alpha 3\,(XI)]$	300-nm molecule	Small fibers	Cartilage	Participates in type II collagen function
Fibril-associated collagens					
IX	$[\alpha 1\,(IX)]\,[\alpha 2\,(IX)]$ $[\alpha 3\,(IX)]$	200-nm molecule	Not visible, detected by immunocytochemistry	Cartilage, vitreous body	Bound glycosaminoglycans; associated with type II collagen
XII	$[\alpha 1\,(XII)]_3$	Large N-terminal domain; interacts with type I collagen	Not visible, detected by immunocytochemistry	Embryonic tendon and skin	Interacts with type I collagen
XIV	$[\alpha 1\,(XIV)]_3$	Large N-terminal domain; cross-shaped molecule	Not visible; detected by immunocytochemistry	Fetal skin and tendon	
Collagen that forms anchoring fibrils					
VII	$[\alpha 1\,(VII)]_3$	450 nm, globular domain at each end	Not visible, detected by immunocytochemistry	Epithelia	Anchors skin epidermal basal lamina to underlying stroma
Collagen that forms networks					
IV	$[\alpha 1\,(VII)]_2$ $[\alpha 1\,(IV)]$	Two-dimensional cross-linked network	Not visible, detected by immunocytochemistry	All basement membranes	Support of delicate structures, filtration

FIBRIL-ASSOCIATED COLLAGENS

Fibril-associated collagens are short structures that bind the surfaces of collagen fibrils to one another and to other components of the ECM. Molecules in this category are also known as FACIT collagens, an acronym for "fibril-associated collagens with interrupted triple helices."

COLLAGENS THAT FORM ANCHORING FIBRILS

Anchoring collagen is type VII collagen, present in the anchoring fibrils that bind the basal lamina to reticular fibers in the underlying connective tissue (Figure 4–2).

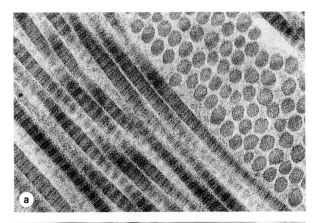

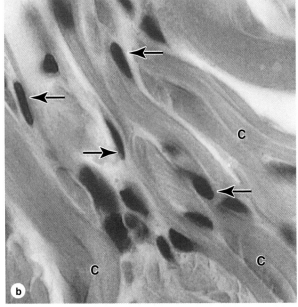

***Figure 5–8.* Type I collagen.** Molecules of type I collagen, the most abundant type, assemble to form much larger structures. **(a):** TEM shows fibrils cut longitudinally and transversely. In longitudinal sections the fibrils display alternating dark and light bands that are further divided by cross-striations and in cross-section the cut ends of individual collagen molecules can be seen. Ground substance completely surrounds the fibrils. X100,000. **(b):** In H&E stained tissues, type I collagen fibrils can often be seen to aggregate further into large collagen bundles (C) of very eosinophilic fibers. Subunits for these fibers were secreted by fibroblasts (arrows) associated with them. X 400.

COLLAGENS THAT FORM NETWORKS

An important network-forming collagen is type IV collagen, whose molecules assemble in a meshwork that constitutes a major structural component of the basal lamina.

Collagen synthesis, an activity once thought restricted to fibroblasts, chondroblasts, osteoblasts, and odontoblasts, has now been shown to occur in many cell types. The polypeptides initially formed on ribosomes of the rough ER are called **procollagen α chains,** which intertwine in ER cisternae to make triple helices. Every third amino acids in the α chains is glycine; two other small amino acids abundant in collagen are hydroxylated post-translationally to form **hydroxyproline** and **hydroxylysine.** Many different α chains have been identified, encoded by related genes and varying in length and amino acid sequence.

The triple helix of α chains forms a rod-like procollagen molecule, which in type I and II collagen measures 300 nm in length and 1.5 nm in width. Procollagen molecules may be homotrimeric, with all three chains identical, or heterotrimeric, with two or all three chains having different sequences (Figure 5–9). Different combinations of the many procollagen α chains in procollagen molecules are largely responsible for the various types of collagen with different structures and functional properties. In collagen types I, II, and III, collagen molecules aggregate and become packed together to form **fibrils.** Hydrogen bonds and hydrophobic interactions are important in the aggregation and packing of these subunits. In a subsequent step, this structure is reinforced by the formation of covalent cross-links, a process catalyzed by lysyl oxidases.

Collagen fibrils are thin, elongated structures with diameters ranging from 20 to 90 nm and can be several micrometers in length; they have transverse striations with a characteristic periodicity of 64-68 nm (Figure 5–10). The striations are caused by the regular, overlapping arrangement of the collagen molecule subunits (Figure 5–10). The dark (electron-dense) bands retain more of the lead-based stain used in TEM studies because their more numerous free chemical groups react more intensely with the lead solution than do the light bands. In some collagen types these fibrils associate further with FACIT collagens to form fibers. In collagen type I, the fibers can form large bundles (Figure 5–10). Collagen type II (present in cartilage) occurs as fibrils but does not form fibers or bundles. Collagen type IV, present in all basement membranes, assembles as a lattice-like network in the basal lamina.

Because collagen type I is so abundant, its synthesis has been studied most thoroughly. Synthesis of this important protein involves several steps, which are summarized in Figure 5–11:

1. Polypeptides called procollagen α chains are produced on polyribosomes bound to membranes of RER and translocated into the cisternae and the signal peptide is clipped off.

2. Hydroxylation of proline and lysine begins after the peptide chain has reached a certain minimum length and is still bound to the ribosomes. The enzymes involved are **prolyl hydroxylases** and **lysyl hydroxylase** and the reactions require O_2, Fe^{2+}, and ascorbic acid (vitamin C) as co-factors.

3. Glycosylation of some hydroxylysine residues occurs, with the various collagen types having different amounts of galactose linked to hydroxylysines.

4. Both the amino- and carboxyl-terminal ends of each α chain make up nonhelical portions of the polypeptides, sometimes called the extension propeptides, which may help ensure that the appropriate α chains (α1, α2) assemble in the correct

position as a triple helix. In addition, the nonhelical propeptides make the resulting **procollagen molecule** soluble and prevent its premature intracellular assembly and precipitation as collagen fibrils. Procollagen is transported through the Golgi network and undergoes exocytosis to the extracellular environment.

5. Outside the cell, specific proteases called **procollagen peptidases** remove the extending propeptides, converting the procollagen molecules to collagen molecules. These are now capable of self-assembly into polymeric collagen fibrils, usually in specialized niches near the cell surface.

6. In some collagen types, fibrils aggregate to form fibers. Certain proteoglycans and types of collagen (types V and XI) participate in the aggregation of collagen molecules to form fibrils and in the formation of fibers from fibrils. FACIT collagens help stabilize the molecules in collagen fibrils and fibers and bind these structures to other components of the ECM.

7. Fibrillar structure is reinforced further by the formation of covalent cross-links between assembled collagen molecules, a process catalyzed by the extracellular enzyme **lysyl oxidase**.

The other fibrillar collagens are formed in processes similar to that described for collagen type I. In summary, collagen synthesis involves a cascade of unique post-translational modifications of the original procollagen polypeptides. All these modifications are critical to the structure and function of normal mature collagen. Because there are so many steps in collagen biosynthesis, there are many points at which the process can be interrupted or changed by defective enzymes or by disease processes.

Although fresh collagen fibers are colorless strands, when present in large numbers (eg, in tendons) they appear white. The highly regular orientation of subunits in collagen fibers makes them birefringent under the polarizing microscope (Figure 1–7). In the light microscope, collagen fibers are acidophilic; they stain pink with eosin, blue with Mallory trichrome stain, green with Masson trichrome stain, and red with Sirius red. Because of the long and tortuous course of collagen bundles, their length and diameter are better studied in spread preparations than in histologic sections (Figure 1–7). Mesentery is frequently used for this purpose; when spread on a slide, this structure is sufficiently thin to let the light pass through; it can be stained and examined directly under the

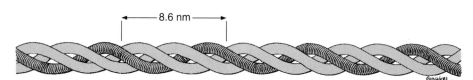

Figure 5–9. Procollagen. In the most abundant form of collagen, type I, each procollagen molecule is composed of two α1 and one α2 peptide chains, each with a molecular mass of approximately 100 kDa, intertwined in a right-handed helix and held together by hydrogen bonds and hydrophobic interactions. Each complete turn of the helix spans a distance of 8.6 nm. The length of each tropocollagen molecule is 300 nm, and its width is 1.5 nm.

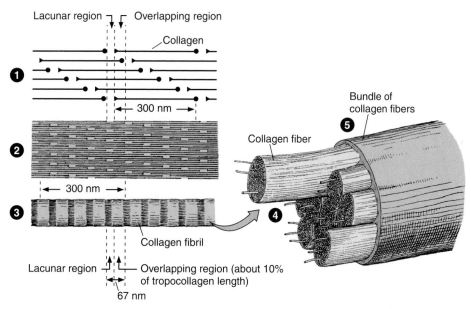

Figure 5–10. Assembly of collagen molecules into collagen fibers. This diagram shows an aggregate of collagen molecules, fibrils, fibers, and bundles. There is a stepwise overlapping arrangement of rodlike collagen molecules, each measuring 300 nm (**1**). This arrangement results in the production of alternating spaces and overlapping regions (**2**), which cause the cross-striations characteristic of collagen fibrils and confer a 67-nm periodicity of dark and light bands when the fibril is observed in the electron microscope (**3**). Fibrils aggregate and are covalently cross-linked to form fibers (**4**), which in collagen type I aggregate further to form bundles (**5**) routinely called collagen fibers when seen by light microscopy.

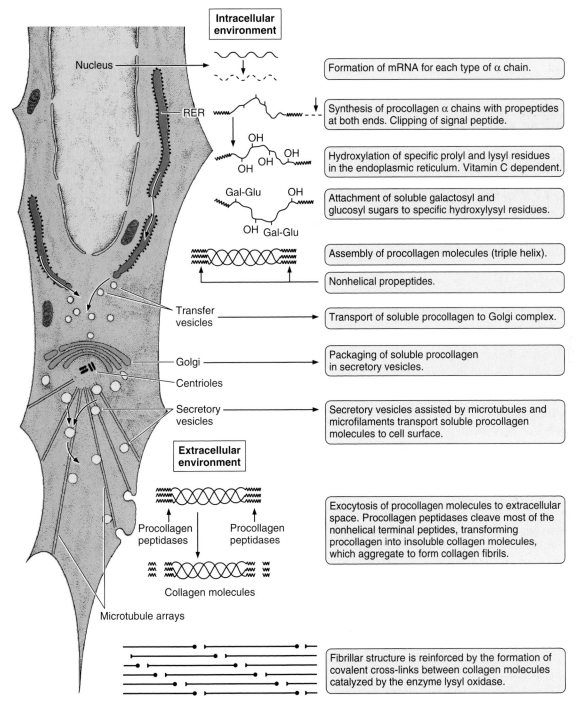

Figure 5–11. **Collagen synthesis.** Hydroxylation and glycosylation of procollagen α chains and their assembly into triple helices occurs in the RER and further assembly into fibrils occurs in the ECM after secretion of procollagen. Because there are many slightly different genes for procollagen α chains and collagen production depends on several post-translational events involving several other enzymes, many diseases involving defective collagen synthesis have been described.

microscope. Mesentery consists of a central portion of connective tissue lined on both surfaces by a simple squamous epithelium, the mesothelium. The collagen fibers in a spread preparation appear as elongated and tortuous cylindrical structures of indefinite length, with a diameter that varies from 1 to 20 μm.

MEDICAL APPLICATION

Collagen synthesis depends on the expression of several genes and several posttranslational events. It should not be surprising, therefore, that a large number of pathological conditions are directly attributable to insufficient or abnormal collagen synthesis.

Certain mutations in the α1 (I) or α2 (I) genes lead to **osteogenesis imperfecta**. *Many cases of osteogenesis imperfecta are due to deletions of all or part of the α1 (I) gene. However, a single amino acid change is sufficient to cause certain forms of this disease, particularly mutations involving glycine. Glycine must be at every third position for the collagen triple helix to form.*

In addition to these disorders, several diseases result from an over-accumulation of collagen. In **progressive systemic sclerosis**, *almost all organs may present an excessive accumulation of collagen* **(fibrosis)**. *This occurs mainly in the skin, digestive tract, muscles, and kidneys, causing hardening and functional impairment of the implicated organs.*

Keloid *is a local swelling caused by abnormal amounts of collagen that form in scars of the skin. Keloids, which occur most often in individuals of black African descent, can be a troublesome clinical problem to manage; not only can they be disfiguring, but excision is almost always followed by recurrence.*

Vitamin C (ascorbic acid) deficiency leads to **scurvy**, *a disease characterized by the degeneration of connective tissue. Without this vitamin, fibroblasts synthesize defective collagen, and the defective fibers are not replaced. This process leads to a general degeneration of connective tissue that becomes more pronounced in areas in which collagen renewal takes place at a faster rate. The periodontal ligament that holds teeth in their sockets has a relatively high collagen turnover; consequently, this ligament is markedly affected by scurvy, which leads to a loss of teeth. Ascorbic acid is a cofactor for proline hydroxylase, which is essential for the normal synthesis of collagen. Table 5–4 lists a few examples of the many disorders caused by failure of collagen biosynthesis.*

Collagen turnover and renewal in normal connective tissue is generally a very slow process. In some organs, such as tendons and ligaments, the collagen is very stable, whereas in others, as in the periodontal ligament surrounding teeth, the collagen turnover rate is very high. To be renewed, the collagen must first be degraded. Degradation is initiated by specific enzymes called **collagenases**, which are members of an enzyme class called matrix metalloproteinases or MMPs. Collagenases clip collagen molecules in such a way that they are then susceptible to further degradation by nonspecific proteases.

Reticular Fibers

Once thought to be uniquely distinct from collagen, **reticular fibers** are now known to consist mainly of collagen type III, which forms extensive networks of extremely thin (diameters 0.5–2 μm) and heavily glycosylated fibers in certain organs. They are not visible in hematoxylin-and-eosin (H&E) preparations but can be easily stained black by impregnation with silver salts (Figure 5–12). Because of their affinity for silver salts, these fibers are called **argyrophilic** (Gr. *argyros*, silver). Reticular fibers are also periodic acid–Schiff (PAS)-positive, which like argyrophilia is due to the high content of sugar chains associated with these fibers. Reticular fibers contain 6–12% hexoses as opposed to 1% in most collagen fibers.

Reticular fibers constitute a network around the parenchymal cells of various organs (eg, liver, endocrine glands) and are particularly abundant in the framework of hematopoietic organs (eg, spleen, lymph nodes, red bone marrow). In the latter sites the network is produced by fibroblast-like cells called **reticular cells**. The loose disposition of reticular fibers creates a flexible network in these organs and others that are subject to changes in form or volume, such as the arteries, uterus, and intestinal muscle layers.

MEDICAL APPLICATION

Ehlers–Danlos type IV disease, *a deficiency of collagen type III, is characterized by ruptures in arteries and the intestine (Table 5–4), both structures rich in reticular fibers.*

Elastic Fibers

Elastic fibers are also thinner than the average collagen fiber and form sparse networks interspersed with collagen bundles in many organs subject to much bending or stretching, such as the wall of large arteries. The name indicates the major functional property such fibers impart to these resilient organs (Figure 5–13).

Elastic fibers develop through successive stages. In the first stage, a core of 10-nm microfibrils forms from several different glycoproteins, notably the large glycoprotein called **fibrillin** (350kDa). Fibrillin binds elastin and forms the scaffolding necessary for the deposition of elastin. Defective fibrillin results in the formation of fragmented elastic fibrils. In the second stage of development, the protein **elastin** is deposited between the microfibrils, forming larger fibers. During the third stage, elastin gradually accumulates until it comprises most of the fiber bundles, which are further surrounded by a thin sheath of microfibrils. These are the mature **elastic fibers,** the most numerous component of the elastic fiber system. Stages of elastic fiber formation are shown in Figure 5–14. In the wall of large blood vessels, especially arteries, elastin also occurs as fenestrated sheets called **elastic lamellae**.

Microfibrils of fibrillin alone are used in some organs, such as to hold in place the lens of the eye. Such microfibrils are not elastic but are highly resistant to pulling forces, whereas the mature elastic fibers stretch easily in response to tension. By using different proportions of fibrillin and elastin, a family of fibers can be formed whose variable functional characteristics are adapted to local tissue requirements.

Like collagen elastin matures in the ECM. Elastin molecules are globular (molecular mass 70 kDa) and are secreted by fibroblasts in connective tissue and by smooth muscle cells in the walls of blood vessels. Elastin molecules are rich in glycine and proline, with many regions having random-coil conformations (like that of natural rubber). Elastin molecules polymerize to form fibers or sheet-like structures, both of which can be stretched by external forces. Elastin contains two unusual amino acids, **desmosine** and **isodesmosine,** which are produced when covalent cross-links are formed among four lysine residues in different elastin molecules (Figure 5–15). These effectively cross-link the subunits of elastin and help account for the rubberlike qualities of this protein. Elastin is resistant to digestion by most proteases, but is easily hydrolyzed by pancreatic **elastase**.

Table 5–4. Examples of clinical disorders resulting from defects in collagen synthesis.

Disorder	Defect	Symptoms
Ehlers-Danlos type IV	Faulty transcription or translation of collagen type III	Aortic and/or intestinal rupture
Ehlers-Danlos type VI	Faulty lysine hydroxylation	Augmented skin elasticity, rupture of eyeball
Ehlers-Danlos type VII	Decrease in procollagen peptidase activity	Increased articular mobility, frequent luxation
Scurvy	Lack of vitamin C (cofactor for prolyl hydroxylase)	Ulceration of gums, hemorrhages
Osteogenesis imperfecta	Change of one nucleotide in genes for collagen type I	Spontaneous fractures, cardiac insufficiency

MEDICAL APPLICATION

*Fibrillin is a family of proteins related to the scaffolding necessary for the deposition of elastin. Mutations in the fibrillin gene result in **Marfan syndrome**, a disease characterized by a lack of resistance in the tissues rich in elastic fibers.*

Because the large arteries are rich in components of the elastic system and because the blood pressure is high in the aorta, patients with this disease often experience aortic swellings called aneurysms, a life-threatening condition.

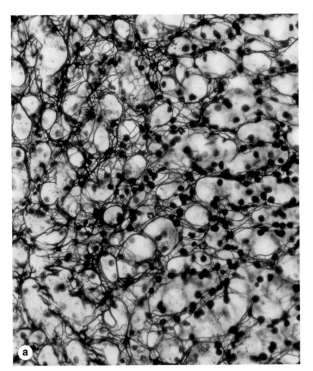 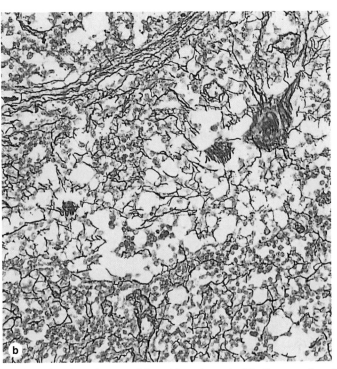

***Figure 5–12.* Reticular fibers.** In these silver-stained sections of both adrenal cortex **(a)** and lymph node **(b)**, the prominent feature is a network of reticular fibers which provides a framework for cell attachment. Reticular fibers contain type III collagen that is heavily glycosylated, which produces the argyrophilia. Cell nuclei are also dark but cytoplasm is unstained. X100.

GROUND SUBSTANCE

The ground substance of the ECM is a highly hydrated, transparent, complex mixture of macromolecules, principally in three classes: **glycosaminoglycans** (or GAGs), **proteoglycans**, and **multiadhesive glycoproteins**. The complex molecular mixture of the ground substance is transparent and rich in bound water. It fills the space between cells and fibers of connective tissue and, because it is viscous, acts as both a lubricant and a barrier to the penetration of invaders. When adequately fixed for histologic analysis, its components aggregate and precipitate in the tissues as granular material that is observed in TEM preparations as electron-dense filaments or granules (Figure 5–16).

GAGs (originally called **mucopolysaccharides**) are linear polysaccharides formed by repeating disaccharide units usually composed of a uronic acid and a hexosamine. The hexosamine can be **glucosamine** or **galactosamine,** and the uronic acid can be **glucuronic** or **iduronic acid.** The largest, most unique, and most ubiquitous GAG is **hyaluronic acid** (or **hyaluronan**). With a molecular weight from 100s to 1000s kDa, hyaluronic acid is a long polymer of the disaccharide glucosamine – glucuronate. It is synthesized directly into the ECM by an enzyme complex, **hyaluronate synthase**, located in the cell membrane of many cells. Hyaluronic acid forms a dense, viscous network of polymers which binds a considerable amount of water, giving it an

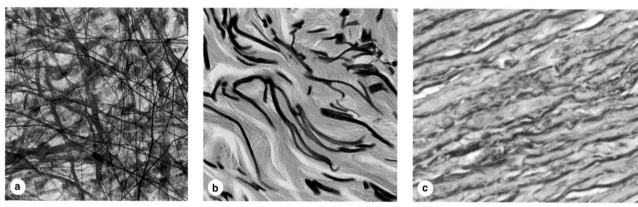

Figure 5-13. Elastic fibers. Elastic fibers or lamellae (sheets) add the resiliency to connective tissue. They are difficult to discern in H&E stained material and are usually demonstrated in preparations made using compounds such as aldehyde fuscin which stains elastin a dark magenta. **(a):** The length and density of fine elastic fibers is best seen in spread preparation of connective tissue in a thin mesentery. X200. Orcein-H&E. **(b):** At higher magnification, sectioned elastic fibers can be seen among the eosinophilic collagen bundles in dermis. X400. Aldehyde fuscin & eosin. **(c):** Elastic fibers and lamellae are abundant between layers of smooth muscle in the wall of elastic arteries such as the aorta. X200. Van Gieson-H&E.

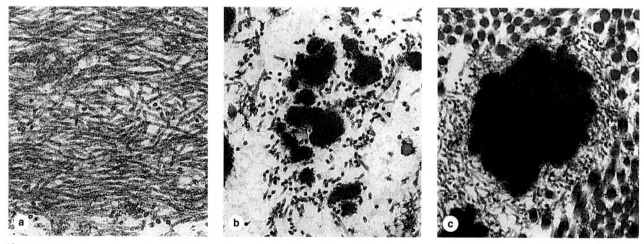

Figure 5–14. Formation of elastic fibers. Stages in the formation of elastic fibers can be seen by TEM. **(a):** Initially a developing fiber consists of many small microfibrils composed of the glycoprotein **fibrillin** secreted by fibroblasts, smooth muscle cells or other cells. **(b):** With further development, to the microfibrils are added amorphous deposits of **elastin**. Elastin is secreted by the cells and like procollagen molecules quickly polymerizes. **(c):** Elastin accumulates and ultimately occupies the center of an elastic fiber, which retains fibrillin microfibrils at the surface. Collagen fibrils, seen in cross section, are also present. X50,000.

important role in allowing diffusion of molecules in connective tissue and in lubricating various organs and joints.

All other GAGs are much smaller (10–40 kDa), are covalently attached to proteins (as parts of proteoglycans), are synthesized in Golgi complexes, and are rich in sulfate. The four main GAGs found in proteoglycans are **dermatan sulfate, chondroitin sulfates, keratan sulfate,** and **heparan sulfate** all of which have different disaccharide units and tissue distributions (Table 5–5). Like hyaluronic acid these GAGs are intensely hydrophilic, are highly viscous, and are polyanions, binding a great number of cations (usually sodium) by electrostatic (ionic) bonds.

Proteoglycans are composed of a core protein to which are covalently attached various numbers and combinations of the sulfated GAGs just mentioned. Like glycoproteins they are synthesized on RER, mature in the Golgi and secreted from cells by exocytosis. The main structural differences between proteoglycans and glycoproteins are shown in Figure 5–17. In cartilage, the core proteins of secreted proteoglycans are bound via small link proteins to a hyaluronic acid chain, forming much larger structures—proteoglycan aggregates. The acidic groups of proteoglycans cause these molecules to bind to the basic amino acid residues of collagen. Proteoglycans are distinguished for their diversity and include cell-surface and ECM families. A given matrix may contain several different types of core proteins, each with different numbers of GAGs of different lengths and composition. One of the most important ECM proteoglycans is **aggrecan,** the dominant proteoglycan in cartilage. In aggrecan the core protein bears several chondroitin sulfate and keratan sulfate chains and is in turn bound via a link protein to hyaluronic acid. Cell-surface proteoglycans such as **syndecan** are present on many types of cells, particularly epithelial cells. The core protein of cell-surface proteoglycans spans the plasma membrane, with a short cytoplasmic extension. A small number of

heparan sulfate chains are attached to the extracellular extension of the core protein.

Besides acting as structural components of the ECM and anchoring cells to the matrix, both extracellular and surface proteoglycans also bind and sequester certain signaling proteins eg, fibroblast growth factor (FGF). Degradation of proteoglycans releases these stored growth factors which then stimulate new cell growth and ECM synthesis.

MEDICAL APPLICATION

The degradation of proteoglycans is carried out by several cell types and depends on the presence of several lysosomal enzymes. Several disorders have been described in which a deficiency in lysosomal enzymes causes glycosaminoglycan degradation to be blocked, with the consequent accumulation of these compounds in tissues. The lack of specific hydrolases in the lysosomes has been found to be the cause of several disorders in humans, including Hurler, Hunter, sanfilippo, and Morquio syndromes.

*Because of their high viscosity, intercellular substances act as a barrier to the penetration of bacteria and other microorganisms. Bacteria that produce **hyaluronidase**, an enzyme that hydrolyzes hyaluronic acid and other glycosaminoglycans, have greater invasive power because they reduce the viscosity of the connective tissue ground substance.*

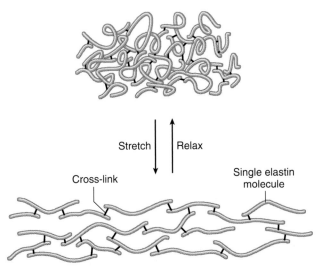

Figure 5–15. Molecular basis of elasticity. Subunits of the glycoprotein elastin are joined by covalent bonds formed among lysine residues of different subunits, catalyzed by lysyl oxidase. This produces an extensive and durable cross-linked network of elastin. (Such bonds give rise to the unusual amino acids desmosine and isodesmosine.) Each elastin molecule in the network has multiple random-coil domains which expand and contract; this allows the entire network to stretch and recoil like a rubber band.

Multiadhesive glycoproteins have carbohydrates attached, but in contrast to proteoglycans the protein moiety usually predominates. The carbohydrate moiety of glycoproteins is frequently a branched structure. Several such glycoproteins have important roles in the adhesion of cells to their substrate. **Fibronectin** (L. *fibra,* fiber, + *nexus,* interconnection) is an important example synthesized by fibroblasts and some epithelial cells. This dimeric molecule, with a molecular mass of 222–240 kDa, has binding sites for collagens, certain GAGs, and integrins of cell membranes, ie, it is multiadhesive. Interactions at these sites help to mediate normal cell adhesion and migration and cause fibronection to be distributed as a network in the intercellular spaces of many tissues (Figure 5–18a). Another multiadhesive glycoprotein, **laminin** is a larger, trimeric, cross-shaped glycoprotein that participates in the adhesion of epithelial cells to the basal lamina, with binding sites for type IV collagen, GAGs, and integrins. All basal laminae are rich in laminin (Figure 5–18b).

Cells interact with extracellular matrix components by using cell-surface molecules (**matrix receptors**) that bind to collagen, fibronectin, and laminin. These receptors are the **integrins,** a family of transmembrane linker proteins (Figures 5–19). Integrins bind their ligands in the ECM with relatively low affinity, allowing cells to explore their environment without losing attachment to it or becoming glued to it. Integrins also interact with the cytoskeleton, usually the actin microfilaments, an interaction mediated by several intracellular proteins, such as **talin** and **vinculin.** The interactions that integrins mediate between the

ECM and the cytoskeleton exert effects in both directions and play an important role in orienting both the cells and the ECM in tissues (Figure 5–19).

MEDICAL APPLICATION

The participation of fibronectin and laminin in both embryonic development and the increased ability of cancer cells to invade other tissues has been well-studied. The importance of fibronectin is shown by the fact that mice whose fibronectin gene has been inactivated die during early embryogenesis.

In connective tissue, in addition to the hydrated ground substance, there is a small quantity of free fluid—called **interstitial** or **tissue fluid**—that is similar to blood plasma in its content of ions and diffusible substances. Tissue fluid contains a small percentage of plasma proteins of low molecular weight that pass through the capillary walls as a result of the hydrostatic pressure of the blood. Although only a small proportion of connective tissue proteins are plasma proteins, it is estimated that because of its volume and wide distribution, as much as one third of the plasma proteins of the body are stored in the intercellular connective tissue matrix.

MEDICAL APPLICATION

Edema is promoted by the accumulation of water in the extracellular spaces. Water in the extracellular compartment of connective tissue comes from the blood, passing through the capillary walls into the extracellular compartment of the tissue. The capillary wall is only slightly permeable to macromolecules but permits the passage of water and small molecules, including low-molecular-weight proteins. In several pathologic conditions, the quantity of tissue fluid may increase considerably, causing edema.

Edema may result from venous or lymphatic obstruction or from a decrease in venous blood flow (eg, congestive heart failure). It may also be caused by the obstruction of lymphatic vessels due to parasitic plugs or tumor cells and chronic starvation; protein deficiency results in a lack of plasma proteins and a decrease in colloid osmotic pressure. Water therefore accumulates in the connective tissue and is not drawn back into the capillaries.

Another possible cause of edema is increased permeability of the blood capillary endothelium resulting from chemical or mechanical injury or the release of certain substances produced in the body (eg, histamine).

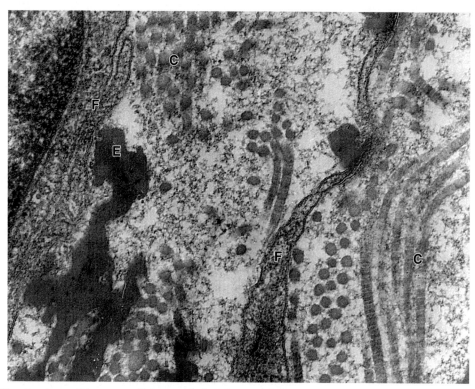

Figure 5–16. **Ultrastructure of the extracellular matrix (ECM).** TEM of the connective tissue extracellular matrix reveals ground substance as either empty or containing fine granular material that fills spaces between the collagen (C) and elastic (E) fibers and surrounds fibroblast cells and processes (F). The granularity of ground substance is an artifact of the glutaraldehyde–tannic acid fixation procedure. X100,000.

Table 5–5. Composition and distribution of glycosaminoglycans in connective tissue and their interactions with collagen fibers.

Glycosaminoglycan	Repeating Disaccharides		Distribution	Electrostatic Interaction with Collagen
	Hexuronic Acid	Hexosamine		
Hyaluronic acid	D-glucuronic acid	D-glucosamine	Umbilical cord, synovial fluid, vitreous humor, cartilage	
Chondroitin 4-sulfate	D-glucuronic acid	D-galactosamine	Cartilage, bone, cornea, skin, notochord, aorta	High levels of interaction, mainly with collagen type II
Chondroitin 6-sulfate	D-glucuronic acid	D-galactosamine	Cartilage, umbilical cord, skin, aorta (media)	High levels of interaction, mainly with collagen type II
Dermatan sulfate	L-iduronic acid or D-glucuronic acid	D-galactosamine	Skin, tendon, aorta (adventitia)	Low levels of interaction, mainly with collagen type I
Heparan sulfate	D-glucuronic acid or L-iduronic acid	D-galactosamine	Aorta, lung, liver, basal laminae	Intermediate levels of interaction, mainly with collagen types III and IV
Keratan sulfate (cornea)	D-galactose	D-galactosamine	Cornea	None
Keratan sulfate (skeleton)	D-galactose	D-glucosamine	Cartilage, nucleus pulposus, annulus fibrosus	None

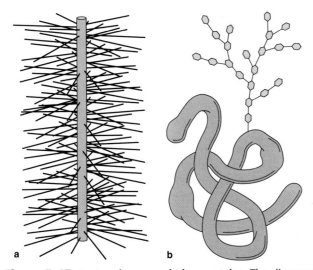

***Figure 5–17.* Proteoglycans and glycoproteins.** The diagram indicates the major structural features of proteoglycans and glycoproteins. **(a):** Proteoglycans contain a core of protein (vertical rod in drawing) to which molecules of sulfated glycosaminoglycans (GAGs) are covalently bound. A GAG is an unbranched polysaccharide made up of repeating disaccharides; one component is an amino sugar, and the other is uronic acid. Proteoglycans contain a greater amount of carbohydrate than do glycoproteins. In general the three-dimensional structure of proteoglycans can be pictured as a test tube brush, with the wire stem representing the core protein and the bristles representing the sulfated GAGs. **(b):** Glycoproteins are globular protein molecules to which branched chains of monosaccharides are covalently attached. Their polypeptide content is greater than their polysaccharide content. (Reproduced, with permission, from Junqueira LCU, Carneiro J: *Biologia Celular e Molecular,* 8th ed. Editora Guanabara Koogan. Rio de Janeiro, 2000.)

Blood vessels bring to connective tissue the various nutrients required by its cells and carry away metabolic waste products to the detoxifying and excretory organs, the liver and kidneys.

Two forces act on the water contained in the capillaries: the **hydrostatic pressure** of the blood caused by the pumping action of the heart, which forces water out across the capillary wall; and the colloid **osmotic pressure** of the blood plasma, which draws water back into the capillaries (Figure 5–20). Osmotic pressure is due mainly to plasma proteins. (Because the ions and low-molecular-weight compounds that pass easily through the capillary walls have approximately the same concentration inside and outside these blood vessels, the osmotic pressures they exert are approximately equal on either side of the capillaries and cancel each other.) The colloid osmotic pressure exerted by the blood proteins—which are unable to pass through the capillary walls—is not counterbalanced by outside pressure and tends to bring water back into the blood vessel (Figure 5–20).

The quantity of water drawn back is less than that which passes out through the capillaries. Rather than accumulating in connective tissue, this excess fluid is continuously drained by lymphatic capillaries and eventually returned to the blood. The smallest lymphatic vessels, the lymphatic capillaries originate in connective tissue as delicate, blind-ending, endothelium-lined tubes that join increasingly larger lymphatic vessels that drain into veins at the base of the neck (see Chapter 11).

TYPES OF CONNECTIVE TISSUE

Connective tissues composed of the cells, fibers, and ground substance components already described nevertheless are quite variable in histological structure. This has led to the use of descriptive names or classifications for various connective tissue types, denoting either the major component or a structural characteristic of the tissue. Table 5-6 shows one classification commonly used for the main types of connective tissue.

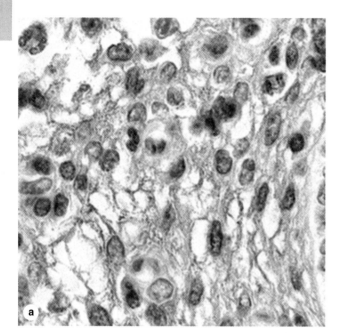

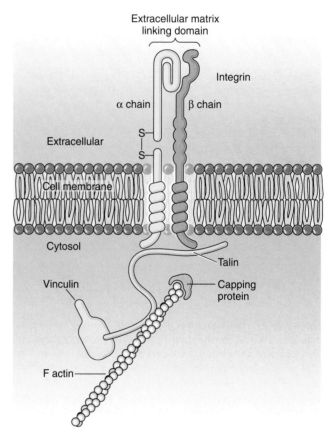

Connective Tissue Proper

There are two general classes of connective tissue proper: loose and dense (Figure 5–21).

Loose connective tissue is a very common type of connective tissue that supports many structures which are normally under some pressure and low friction. It usually supports epithelial tissue, forms a layer around small blood and lymphatic vessels, and fills the spaces between muscle and nerve fibers. Loose connective tissue is also found in the papillary layer of the dermis, in the hypodermis, in the linings of the peritoneal and pleural cavities, in glands, and in the mucous membranes (wet membranes that line the hollow organs) supporting the epithelial cells.

Loose connective tissue, sometimes called **areolar tissue**, has all the main components of connective tissue (cells, fibers, and ground substance) in roughly equal parts. The most numerous cells are fibroblasts and macrophages, but other types of connective tissue cells are also present. Collagen, elastic, and reticular fibers also appear in this tissue. With a moderate amount

Figure 5–18. **Fibronectin and laminin localization.** Immunohistochemistry of sections with connective tissue shows that **fibronectin (a)** is ubiquitous throughout the ECM, while **laminin (b)** is restricted to the basal lamina of the epithelium (top of the picture) and of cross-sectioned muscle fibers, nerves, and small blood vessels (lower half of picture). Both glycoproteins (and many other similar glycoproteins) are multiadhesive, with binding sites for collagens and other ECM components and for integrins of cell surfaces. They play important roles in cell migration, in embryonic tissue formation, and in maintaining tissue structure. **(a):** X400, **(b):** X200, both hematoxylin counterstain. (Reproduced, with permission, from Junqueira LCU, Carneiro J: *Biologia Celular e Molecular*, 8th ed. Editora Guanabara Koògan. Rio de Janeiro, 2000.)

Figure 5–19. **Integrin cell-surface matrix receptor.** By binding to a matrix protein and to the actin cytoskeleton (via talin) inside the cell, integrins serves as transmembrane links by which cells adhere to components of the ECM. The molecule is a heterodimer, with α and β chains. The head portion may protrude some 20 nm from the surface of the cell membrane into the ECM where it interacts with fibronectin, laminin, or collagens.

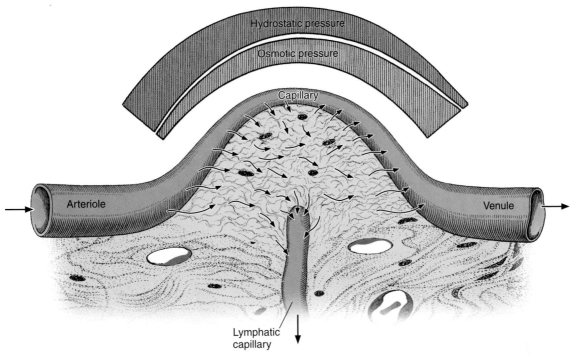

Figure 5–20. Movement of fluid in connective tissue. Normally, water passes through capillary walls to the ECM of the surrounding connective tissues primarily at the arterial end of a capillary, because the hydrostatic pressure there is greater than the colloid osmotic pressure. As indicated at the top of the figure, however, hydrostatic pressure decreases along the length of the capillary toward the venous end. As the hydrostatic pressure falls, osmotic pressure of the fluid in the capillary rises because the protein concentration is increasing due to the passage of water from the capillaries. As a result of the increased protein concentration and decreased hydrostatic pressure, osmotic pressure becomes greater than hydrostatic pressure at the venous end and water is drawn back into the capillary. In this way, metabolites circulate in the connective tissue, feeding its cells. Not all water that leaves capillaries by hydrostatic pressure is drawn back in by osmotic pressure. This excess tissue fluid is normally drained by the lymphatic capillaries, blind-ended vessels that arise in connective tissue and enter the one-way lymphatic system which eventually delivers the fluid (now called lymph) back to veins.

Table 5–6. Types of connective tissue.

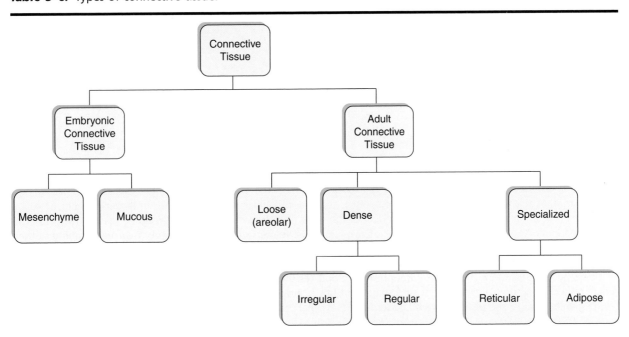

of ground substance, loose connective tissue has a delicate consistency; it is flexible, well vascularized, and not very resistant to stress.

Dense connective tissue is adapted to offer resistance and protection. It has the same components found in loose connective tissue, but there are fewer cells and a clear predominance of collagen fibers over ground substance. Dense connective tissue is less flexible and far more resistant to stress than is loose connective tissue. It is known as **dense irregular** connective tissue when the collagen fibers are arranged in bundles without a definite orientation. The collagen fibers form a 3-dimensional network in dense irregular tissue, providing resistance to stress from all directions. Dense irregular connective tissue is often found closely associated with loose connective tissue. The two types frequently grade into one another and distinctions between them are often arbitrary (Figure 5–21).

The collagen bundles of **dense regular** connective tissue are arranged according to a definite pattern, with collagen fibers aligned with the linear orientation of fibroblasts in response to prolonged stresses exerted in the same direction (Figure 5–22). This arrangement offers great resistance to traction forces.

Tendons and ligaments are the most common examples of dense regular connective tissue. These elongated cylindrical structures hold together components of the musculoskeletal system; by virtue of their richness in collagen fibers, they are white and inextensible. They have parallel, closely packed bundles of collagen separated by a very small quantity of ground substance (Figure 5–22a). Their fibrocytes contain elongated nuclei

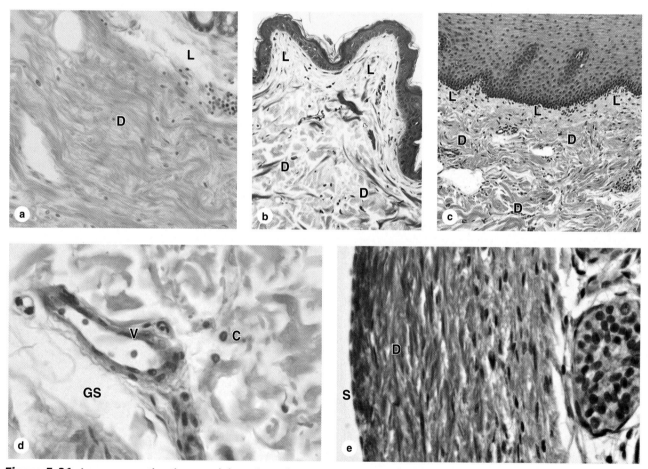

Figure 5–21. **Loose connective tissue and dense irregular connective tissue.** Shown here are three examples of connective tissue which display the types designated loose (or areolar) and dense irregular. Loose connective tissue (L) contains faint ground substance with fine fibers of collagen and is adjacent to epithelium in the examples shown here. Dense irregular connective tissue (D) underlies the thinner layer of loose connective tissue and is invariably much richer in larger bundles of collagen. **(a):** Micrograph of a mammary gland, showing a duct at the top of the figure. In the dense irregular connective tissue can be seen scattered leukocytes, and the irregular spaces of two lymphatic vessels (left). X100. H&E. **(b):** Trichrome staining of the skin demonstrates the blue staining of collagen with this method. X100. Mallory Trichrome. **(c):** Loose and dense irregular connective tissue within the esophagus is seen below the stratified squamous epithelium. X100. H&E. **(d):** At higher magnification ground substance (GS) is more clearly seen around small blood vessels (V) and collagen bundles (C). X200. H&E. **(e):** The dense irregular connective tissue (D) capsule that surrounds the testis is shown here. Similar capsules are found around many organs and large glands. That of the testis is covered by serous mesothelial cells (S), which produce a hyaluronate-rich lubricant around the organ. X200. H&E.

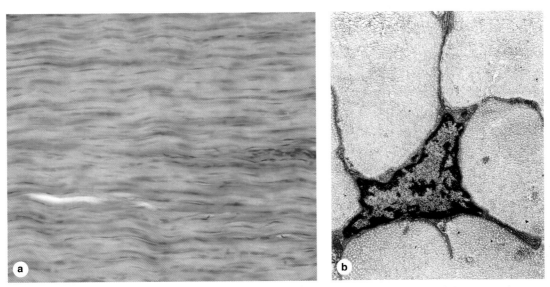

Figure 5–22. Dense regular connective tissue. (a): Micrograph shows a longitudinal section of dense regular connective tissue of a tendon. Long, parallel bundles of collagen fibers fill the spaces between the elongated nuclei of fibrocytes. X100. H&E stain. **(b):** The electron micrograph shows one fibrocyte in a cross-section of tendon, revealing that the sparse cytoplasm of the fibrocytes is divided into numerous thin cytoplasmic processes extending among adjacent collagen fibers. X25,000.

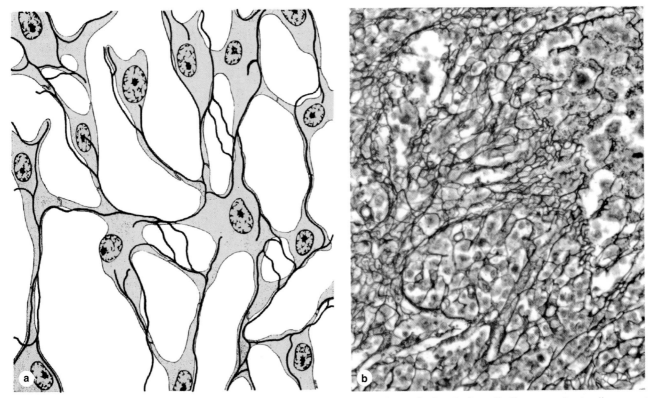

Figure 5–23. Reticular tissue. (a): The diagram shows only the fibers and attached reticular cells (free, transient cells are not represented). Reticular fibers of type III collagen are produced and enveloped by the reticular cells, forming an elaborate network through which interstitial fluid or lymph and wandering cells from blood pass continuously. **(b):** The micrograph shows a silver-stained section of lymph node in which reticular fibers are seen as irregular black lines. Reticular cells are also heavily stained and dark. Most of the smaller, more lightly stained cells are lymphocytes passing through the lymph node. X200. Silver.

parallel to the fibers and sparse cytoplasmic folds that envelop portions of the collagen bundles (Figure 5–22b). The cytoplasm of these fibrocytes is rarely revealed in H&E stains, not only because it is sparse but also because it stains the same color as the fibers. Some ligaments, such as the yellow ligaments of the vertebral column, also contain abundant parallel elastic fiber bundles.

The collagen bundles of tendons vary in size and are enveloped by small amounts of loose connective tissue containing small blood vessels and nerves. Overall, however, tendons are poorly vascularized and repair of damaged tendons is very slow.

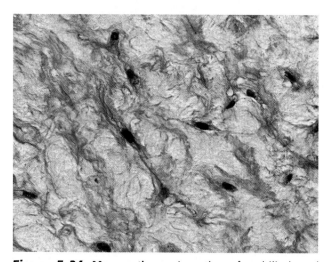

Figure 5–24. **Mucous tissue.** A section of umbilical cord show large fibroblasts surrounded by a large amount of very loose ECM containing mainly ground substances very rich in hyaluronan, with wisps of collagen. Histologically mucous connective tissue resembles embryonic mesenchyme in many respects and is rarely found in adult organs. X200. H&E

Externally, the tendon is surrounded by a sheath of dense irregular connective tissue. In some tendons, this sheath is made up of two layers, both lined by flattened **synovial cells** of mesenchymal origin. One layer is attached to the tendon, and the other lines neighboring structures. The space between these linings contains a viscous fluid (similar to the fluid of synovial joints) composed of water, proteins, hyaluronate and other GAGs. This synovial secretion acts as a lubricant permitting easy sliding movements of the tendon within its sheath.

Reticular Tissue

Individual reticular fibers form delicate three-dimensional networks that support cells in **reticular tissue**. This specialized connective tissue consists of reticular fibers of type III collagen produced by specialized fibroblasts called reticular cells (Figure 5–23). The heavily glycosylated reticular fibers provide the architectural framework that creates special microenvironments for hematopoietic organs and lymphoid organs (bone marrow, lymph nodes, and spleen). The reticular cells are dispersed along this framework and partially cover the reticular fibers and ground substance with cytoplasmic processes. The resulting cell-lined system creates a spongelike structure (Figure 5–12) within which cells and fluids are freely mobile.

In addition to the reticular cells, cells of the mononuclear phagocyte system are strategically dispersed along the trabeculae. These cells monitor the slow flow of materials through the sinuslike spaces and remove invaders by phagocytosis.

Mucous Tissue

Mucous tissue is found mainly in the umbilical cord and fetal tissues. Mucous tissue has an abundance of ground substance composed chiefly of hyaluronic acid, making it a jellylike tissue containing very few collagen fibers with scattered fibroblasts (Figure 5–24). Mucous tissue is the principal component of the umbilical cord, where it is referred to as **Wharton's jelly.** A similar form of connective tissue is also found in the pulp cavity of young teeth.

Adipose Tissue

WHITE ADIPOSE TISSUE	BROWN ADIPOSE TISSUE
Storage & Mobilization of Lipids	Function of Brown Adipocytes
Histogenesis of White Adipose Tissue	Histogenesis of Brown Adipose Tissue

Adipose tissue is a specialized type of connective tissue in which **adipocytes** or fat cells predominate. These cells can be found isolated or in groups within loose or irregular connective tissue, often in large aggregates where they are the major component of adipose tissue. Located in many areas throughout the body, adipose tissue represents 15–20% of the body weight in men of normal weight; in women of normal weight, 20–25% of body weight. Long considered little more than inert masses of energy stored as fat, adipocytes are now recognized as key regulators of the body's energy metabolism. Because of a growing worldwide epidemic of obesity and its associated problems, including diabetes and heart disease, adipocytes are now the most widely studied cell of connective tissue.

Adipose tissue is the largest repository of energy (in the form of triglycerides, the neutral fats) in the body. The other organs that store energy, notably the liver and skeletal muscle, do so in the form of glycogen. However the supply of glycogen is limited and a large store of calories must be mobilized between meals. Because triglycerides are of lower density than glycogen and have a higher caloric value (9.3 kcal/g for triglycerides versus 4.1 kcal/g for carbohydrates), adipose tissue has evolved as a very efficient storage tissue. It is in a state of continuous turnover and is sensitive to both nervous and hormonal stimuli. Moreover, adipocytes themselves release hormones and a number of important factors, and adipose tissue is now recognized as a major endocrine and signaling organ. With its unique physical properties, adipose tissue or fat is a poor heat conductor and it contributes to the thermal insulation of the body. Adipose tissue also fills up spaces between other tissues and helps to keep some organs in place. Subcutaneous layers of adipose tissue help to shape the surface of the body, whereas deposits in the form of pads act as shock absorbers, chiefly in the soles and palms.

There are two types of adipose tissue with different locations, structures, colors, and pathologic characteristics. **White adipose tissue**, the more common type, is composed of cells that, when completely developed, contain one large central droplet of whitish-yellow fat in their cytoplasm. **Brown adipose tissue** contains cells with multiple lipid droplets interspersed among abundant mitochondria, which give these cells the darker appearance. Both types of adipose tissue have a rich blood supply.

WHITE ADIPOSE TISSUE

Specialized for long-term energy storage, white adipose cells are spherical when isolated but are polyhedral when closely packed in adipose tissue. Each cell is very large, between 50 and 150 μm in diameter and contains one huge droplet of lipid that makes up 85% of the cell's weight. White adipocytes are called **unilocular** because triglycerides are stored in a single locus. Since lipid is removed from cells by the alcohol and xylene used in routine histological techniques, a unilocular adipocyte appears in standard microscope preparations as a thin ring of cytoplasm surrounding the empty vacuole left by the dissolved lipid droplet, sometimes referred to as the **signet ring cell.** The large droplet causes these cells to have eccentric and flattened nuclei (Figure 6–1). The rim of cytoplasm that remains after removal of the stored triglycerides may rupture and collapse, distorting the tissue structure.

The thickest portion of the cytoplasm surrounds the nucleus of these cells and contains a Golgi apparatus, mitochondria, poorly developed cisternae of the rough ER, and free polyribosomes. The rim of cytoplasm surrounding the lipid droplet contains cisternae of smooth ER and numerous pinocytotic vesicles. TEM studies reveal that each adipose cell usually possesses minute lipid droplets in addition to the single large droplet seen with the light microscope; the droplets are not enveloped by a membrane but show many vimentin intermediate filaments in their periphery. Each adipose cell is surrounded by a thin external or basal lamina.

White adipose tissue is subdivided into incomplete lobules by a partition of connective tissue containing a rich vascular bed and nerve network. Fibroblasts, macrophages, and other cells make up about half the total number of cells. Reticular fibers form a fine interwoven network that supports individual fat cells and binds them together. Although blood vessels are not always apparent in tissue sections, adipose tissue is richly vascularized.

The color of freshly dissected white adipose tissue depends on the diet and varies from white to bright yellow, due mainly to the presence of carotenoids dissolved in the fat droplets. Almost all adipose tissue in adults is of this type and it is found in many organs throughout the body. Age and gender determine the distribution and density of adipose deposits.

In the newborn, white adipose tissue has a more uniform thickness throughout the body. As the child matures, the tissue tends to disappear from some parts of the body and increase in others. Its distribution is partly regulated by sex hormones, which control adipose deposition in the breasts and thighs.

Storage & Mobilization of Lipids

The white adipose tissue is a large depot of energy for the organism. The lipids stored in adipose cells are chiefly triglycerides, ie, esters of fatty acids and glycerol. Triglycerides stored by these cells originate in dietary fats brought to adipocytes as circulating

chylomicrons, in triglycerides synthesized in the liver and transported to adipose tissue in the form of **very low-density lipoproteins (VLDLs)**, and by the local synthesis of free fatty acids and glycerol from glucose to form triglycerides.

Chylomicrons (Gr. *chylos*, juice, + *micros*, small) are particles up to 3 μm in diameter, formed in intestinal epithelial cells and transported in blood plasma and mesenteric lymph. They consist of a central core, composed mainly of triglycerides and a small quantity of cholesterol esters, surrounded by a stabilizing monolayer consisting of apolipoproteins, cholesterol, and phospholipids. VLDL are smaller than chylomicrons (providing a greater surface-to-volume ratio) and have proportionately more lipid in their surface layer. VLDL also have different apolipoproteins at the surface and contain a higher proportion of cholesterol esters

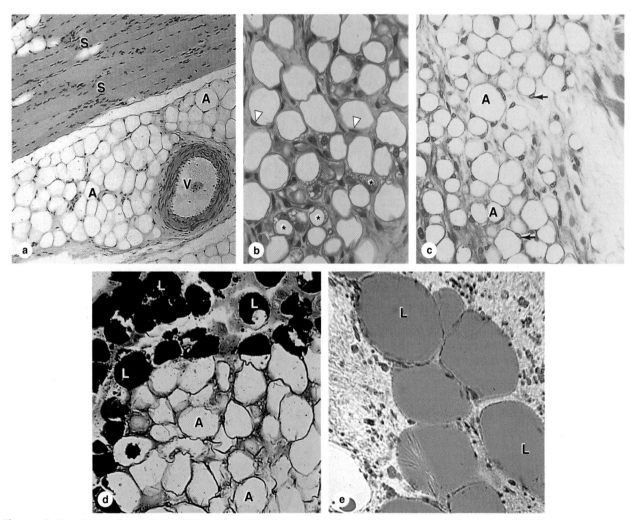

Figure 6–1. **White adipose tissue.** White or unilocular adipose tissue is commonly seen in sections of many human organs. **(a):** In this photomicrograph adipocytes (A) are seen in the connective tissue associated with a blood vessel (V) in striated muscle (S). The fat cells are very large, many with nuclei not present in the section, and are empty because lipid was dissolved away in slide preparation. X100. H&E. **(b):** The specimen here was from a young mammal and the adipocytes marked with **asterisks** are not yet unilocular, having many small lipid droplets in their cytoplasm, indicating that their differentiation is not yet complete. Nuclei of unilocular cells are indicated by arrowheads. X200. PT. **(c):** The rim of cytoplasm around the adipocyte (A) fat droplet can barely be detected, but arrows denote their nuclei. X200. Azure. **(d):** Tissue was fixed here with osmium tetroxide, which preserves lipid (L) and stains it black. Many adipocytes (A) in the preparation retain at least part of their lipid droplet. X440. Osmium tetroxide. **(e):** Another method of studying intact adipocytes of white fat is to prepare frozen sections on a cryostat and stain them with fat soluble dyes such as oil red O, as shown in the lipid-filled (L) cells in this micrograph. X450. Oil red O.

to triglycerides than do chylomicrons. Chylomicrons and VLDL are both hydrolyzed at the luminal surfaces of blood capillaries in adipose tissue by **lipoprotein lipase**, an enzyme synthesized by the adipocyte and transferred to the capillary cell membrane. Free fatty acids enter the adipocyte by both active transport and diffusion (Figure 6–2). Pinocytotic vesicles seen ultrastructurally at the surfaces of adipocytes are probably not involved. Within the adipocyte, the fatty acids combine with glycerol phosphate, supplied by glucose metabolism, to form triglycerides once again. These are then deposited in the triglyceride droplets. Mitochondria and smooth ER participate actively in the process of lipid uptake and storage.

Adipose cells can synthesize fatty acids from glucose, a process accelerated by insulin. Insulin also stimulates the uptake of glucose into adipocytes and increases the synthesis of lipoprotein lipase.

When adipocytes are stimulated by nerves or hormones, stored lipids are mobilized and fatty acids and glycerol are released into the blood. Norepinephrine is liberated at the endings of the postganglionic sympathetic nerves present in adipose tissue. This neurotransmitter triggers the activation of **hormone-sensitive lipase** which breaks down triglycerides at the surface of the stored lipid droplets. The free fatty acids diffuse across the membranes of the adipocyte and the capillary endothelium and then they bind the carrier protein albumin in blood for transport throughout the body. The more water-soluble glycerol remains free and is taken up by the liver. Several hormones are important in regulating lipid synthesis and mobilization in adipocytes. Insulin inhibits the hormone-sensitive lipase, reducing fatty acid release and also stimulates enzymes for lipid synthesis. Glucagon and growth hormone promote triglyceride breakdown and release of fatty acids.

Adipose tissue also functions as an important endocrine organ. Adipocytes are the sole source of the 16 kDa polypeptide hormone **leptin** (Gr. *leptos*, thin), a "satiety factor" with target cells in the hypothalamus and other organs, which regulates the appetite under normal conditions and participates in regulating the amount of adipose tissue.

Although all white adipose tissue appears histologically and physiologically similar, differences in gene expression have been noted between visceral deposits (in the abdomen) and subcutaneous deposits of white fat. Such differences may be important in the medical risks of obesity; it is well-established that increased visceral adipose tissue raises the risk of diabetes and cardiovascular disease whereas increased subcutaneous fat does not. The release of visceral fat products directly to the portal circulation and liver may also influence the medical importance of this form of obesity.

In response to body needs, lipids are mobilized uniformly in all parts of the body, although adipose tissue in the palms, soles, and retroorbital fat pads resists long periods of starvation. After periods of starvation, most unilocular adipocytes lose nearly all their fat and become polyhedral or spindle-shaped cells with very few lipid droplets.

Histogenesis of White Adipose Tissue

Like the fiber-producing cells of connective tissue, adipocytes undergo differentiation from embryonic mesenchymal cells. Such differentiation is first seen with the appearance of **lipoblasts**. Early lipoblasts have the appearance of fibroblasts but are able to accumulate fat in their cytoplasm. Lipid accumulations are isolated from one another at first but soon fuse to form the single larger droplet that is characteristic of unilocular adipose tissue cells (Figure 6–3).

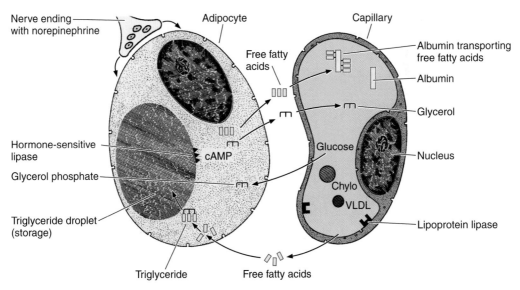

Figure 6–2. Lipid storage and mobilization from adipocytes. The main features of the process by which lipid is stored and released by the adipocyte are shown. Triglycerides are transported by lymph and blood from the intestine and liver in lipoprotein complexes known as **chylomicrons** (Chylo) and **very low-density lipoproteins** (VLDLs). In capillary endothelial cells of adipose tissue, these complexes are partly broken down by lipoprotein lipase, releasing free fatty acids and glycerol. The free fatty acids diffuse from the capillary into the adipocyte, where they are re-esterified to glycerol phosphate, forming triglycerides. These resulting triglycerides are stored in droplets until needed. Norepinephrine from nerve endings stimulates the cyclic AMP (cAMP) system, which activates hormone-sensitive lipase to hydrolyze the stored triglycerides to free fatty acids and glycerol. These substances diffuse into the capillary, where free fatty acids bind to albumin for transport to distant sites for use as an energy source.

Humans are one of the few mammals born with fat stores, which begin to accumulate at week 30 of gestation and are well-developed by birth in both the visceral and subcutaneous compartments. After birth, the development of new adipocytes is common around small blood vessels, where undifferentiated mesenchymal cells are fairly abundant.

Excessive formation of adipose tissue, or obesity, occurs when energy intake exceeds energy expenditure. Although fat cells can differentiate from mesenchymal stem cells throughout life, adult-onset obesity is generally believed to involve largely increased size or hypertrophy in existing adipocytes (hypertrophic obesity). Childhood obesity can involve both increased adipocyte size and formation of new adipocytes by differentiation and hyperplasia of preadipocytes from mesenchymal cells. This early increase in the number of adipocytes may predispose an individual to hyperplastic obesity in later life.

BROWN ADIPOSE TISSUE

The color of brown adipose tissue or **brown fat** is due to both the numerous mitochondria (containing colored cytochromes) scattered through the adipocytes and the large number of blood capillaries in this tissue. Adipocytes of brown fat contain many small lipid inclusions and are therefore called **multilocular** (Figure 6-3). The many small lipid droplets, abundant mitochondria, and rich vasculature all help mediate this tissue's principal function of **heat production**. In comparison with white adipose tissue, which is present throughout the body, brown adipose tissue has a much more limited distribution.

Cells of brown adipose tissue cells are polygonal and generally smaller than cells of white adipose tissue but their cytoplasm contains a great number of lipid droplets of various sizes (Figure 6–4). These adipocytes have spherical and central nuclei and the numerous mitochondria have abundant long cristae.

Brown adipose tissue resembles an endocrine gland in that its cells assume an almost epithelial arrangement closely associated with blood capillaries. The tissue is subdivided by partitions of connective tissue into lobules that are better delineated than the lobules of white adipose tissue. Cells of this tissue receive direct sympathetic innervation.

Function of Brown Adipocytes

The main function of the multilocular adipose cells is to produce heat by nonshivering thermogenesis. The physiology of multilocular adipose tissue is best understood in the study of hibernating species. Because it is more abundant in hibernating animals, it was at one time called the hibernating gland.

In animals ending their hibernation period, or in newborn mammals (including humans) that are exposed to an environment colder than the mother's uterus, nerve impulses liberate norepinephrine into brown adipose tissue. As in white fat, this neurotransmitter activates the hormone-sensitive lipase present in adipose cells, promoting hydrolysis of triglycerides to fatty acids and glycerol. However, unlike white fat, liberated fatty acids of multilocular adipocytes are quickly metabolized, with a consequent increase in oxygen consumption and heat production, elevating the temperature of the tissue and warming the blood passing through it. Heat production is increased in these cells because the mitochondria have in their inner membrane a transmembrane protein called **thermogenin** or **uncoupling protein (UCP-1)**, a marker unique to brown fat. Thermogenin permits the backflow of protons previously transported to the intermembranous space without passing through the ATP-synthetase system in the mitochondrial globular units (Figure 2–13). Consequently, the energy generated by proton flow is not used to synthesize ATP and is dissipated as heat. Warmed blood circulates throughout the body, distributing the heat and carrying fatty acids not metabolized in the adipose tissue for use elsewhere.

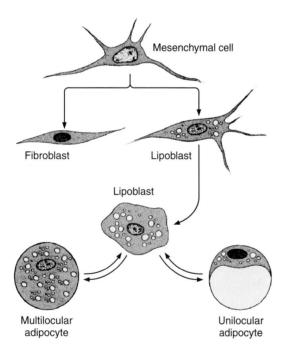

Figure 6–3. Development of white and brown fat cells. Undifferentiated mesenchymal cells differentiate as preadipocytes and are transformed into **lipoblasts** as they accumulate fat and thus give rise to mature fat cells. The mature fat cell is larger than that shown here in relation to the other cell types. Undifferentiated mesenchymal cells also give rise to a variety of other cell types, including fibroblasts. When a large amount of lipid is mobilized by the body, mature unilocular fat cells may return to the lipoblast stage.

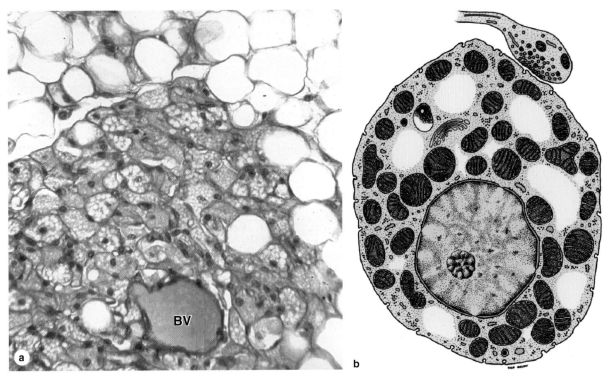

Figure 6–4. **Brown adipose tissue. (a):** The photomicrograph shows brown adipose tissue around a small blood vessel (BV) and adjacent white adipose tissue. Brown adipocytes are slightly smaller and characteristically contain many small lipid droplets and central spherical nuclei. If the lipid has been dissolved from the cells, as shown here, the many mitochondria among the lipid spaces are retained and can be easily discerned. X200. PT. **(b):** The diagram of a multilocular adipocyte shows the relationship between the multiple fat droplets and the mitochondria. Also shown is a sympathetic nerve ending which releases norepinephrine to induce the mitochondrial production of heat via thermogenin activity.

Histogenesis of Brown Adipose Tissue

Brown adipose tissue also develops from embryonic mesenchyme, that emerges earlier than white fat during fetal development. Its cells become arranged differently from white adipose tissue. The mesenchymal cells and lipoblasts that constitute this tissue may resemble epithelium (thus suggesting an endocrine gland) before accumulating much fat.

In humans the amount of brown fat is maximal relative to body weight at birth, when nonshivering thermogenesis is most needed. The tissue largely disappears (by involution) or is replaced by white fat during childhood. In adults it is found only in scattered areas, especially around the kidneys and adrenal glands, the aorta, and mediastinum.

The number of brown adipocytes increases again during cold adaptation, usually appearing as clusters of multilocular cells in white adipose tissue. This likely represents differentiation of mesenchymal stem cells within the white adipose tissue. Transformation of one type of adipose tissue directly to the other apparently does not occur. Besides stimulating thermogenic activity, autonomic nerves also promote brown adipocyte differentiation and prevent apoptosis in mature brown cells.

MEDICAL APPLICATION

*Unilocular adipocytes can generate very common benign tumors called **lipomas**. Malignant adipocyte-derived tumors (**liposarcomas**) are infrequent in humans.*

Cartilage

Cartilage is characterized by an extracellular matrix (ECM) enriched with glycosaminoglycans and proteoglycans, macromolecules that interact with collagen and elastic fibers. Variations in the composition of these matrix components produce three types of cartilage adapted to local biomechanical needs.

Cartilage is a specialized form of connective tissue in which the firm consistency of the ECM allows the tissue to bear mechanical stresses without permanent distortion. In the respiratory system cartilage forms a framework supporting soft tissues. Because it is smooth-surfaced and resilient, cartilage provides a shock-absorbing and sliding area for joints and facilitates bone movements. Cartilage is also essential for the development and growth of long bones, both before and after birth (see Chapter 8).

Cartilage consists of cells called **chondrocytes** (Gr. *chondros,* cartilage + *kytos,* cell) and an extensive **extracellular matrix** composed of fibers and ground substance. Chondrocytes synthesize and secrete the ECM and the cells themselves are located in matrix cavities called **lacunae.** Collagen, hyaluronic acid, proteoglycans, and small amounts of several glycoproteins are the principal macromolecules present in all types of cartilage matrix.

Because collagen and elastin are flexible, the firm gel-like consistency of cartilage depends on electrostatic bonds between collagen fibers and the glycosaminoglycan side chains of matrix proteoglycans. It also depends on the binding of water (solvation water) to the negatively charged glycosaminoglycan chains that extend from the proteoglycan core proteins.

As a consequence of different functional requirements, three forms of cartilage have evolved, each exhibiting variation in matrix composition. In the matrix of **hyaline cartilage,** the most common form, type II collagen is the principal collagen type (Figure 7–1). The more pliable and distensible **elastic cartilage** possesses, in addition to collagen type II, an abundance of elastic fibers within its matrix. **Fibrocartilage,** present in regions of the body subjected to pulling forces, is characterized by a matrix containing a dense network of coarse type I collagen fibers.

In all three forms, cartilage is avascular and is nourished by the diffusion of nutrients from capillaries in adjacent connective tissue (perichondrium) or from synovial fluid in joint cavities.

In some instances, large blood vessels traverse cartilage to nourish other tissues, but these vessels do not supply nutrients to the cartilage. As might be expected of cells in an avascular tissue, chondrocytes exhibit low metabolic activity. Cartilage also lacks lymphatic vessels and nerves.

The **perichondrium** (Figure 7–2) is a sheath of dense connective tissue that surrounds cartilage in most places, forming an interface between the cartilage and the tissue supported by the cartilage. The perichondrium harbors the vascular supply for the avascular cartilage and also contains nerves and lymphatic vessels. Articular cartilage, which covers the surfaces of the bones in movable joints, is devoid of perichondrium and is sustained by the diffusion of oxygen and nutrients from the synovial fluid.

HYALINE CARTILAGE

Hyaline cartilage (Figure 7–2) is the most common and best studied of the three forms. Fresh hyaline cartilage is bluish-white and translucent. In the embryo, it serves as a temporary skeleton until it is gradually replaced by bone.

In adult mammals, hyaline cartilage is located in the articular surfaces of the movable joints, in the walls of larger respiratory passages (nose, larynx, trachea, bronchi), in the ventral ends of ribs, where they articulate with the sternum, and in the **epiphyseal plate,** where it is responsible for the longitudinal growth of bone (see Chapter 8).

Matrix

Forty percent of the dry weight of hyaline cartilage consists of collagen embedded in a firm, hydrated gel of proteoglycans and structural glycoproteins. In routine histology preparations, the collagen is indiscernible for two reasons: the collagen is in the form of fibrils which have submicroscopic dimensions and the refractive index of the fibrils is almost the same as that of the surrounding substances. Hyaline cartilage contains primarily type II collagen, although small amounts of collagen types VI and IX are also present.

Cartilage proteoglycans contain chondroitin 4-sulfate, chondroitin 6-sulfate, and keratan sulfate, covalently linked to core proteins. Hundreds of these proteoglycans are bound noncovalently to long molecules of hyaluronic acid by link proteins, forming very large **proteoglycan aggregates** such as aggrecan that interact with collagen (Figure 7–3). Structurally, proteoglycans resemble bottle brushes, the protein core being the stem and the radiating glycosaminoglycan (GAG) chains the bristles.

The high content of solvation water bound to the negative charges of the GAGs acts as a shock absorber or biomechanical spring; this is of great functional importance, especially in articular cartilages (see Chapter 8).

In addition to type II collagen and proteoglycan, an important component of cartilage matrix is the structural multiadhesive glycoprotein **chondronectin.** Like fibronectin in connective tissue, this macromolecule binds specifically to GAGs, collagen type II and integrins, mediating the adherence of chondrocytes to the ECM.

Cartilage matrix is generally basophilic due to the high concentration of sulfated GAGs and staining variations within the matrix reflect differences in the molecular composition. Immediately surrounding each chondrocyte the ECM is richer in GAGs and poor in collagen. These areas comprise the **territorial matrix** and usually stain differently from the rest of the matrix (Figure 7–2).

Chondrocytes

At the periphery of hyaline cartilage, young chondrocytes have an elliptic shape, with the long axis parallel to the surface. Farther in, they are round and may appear in groups of up to eight cells originating from mitotic divisions of a single chondrocyte. These groups are called **isogenous aggregates** (Gr. *isos,* equal, + *genos,* family). Chondrocytes synthesize collagens and the other matrix molecules. As matrix is produced, cells in the aggregates are moved apart and occupy separate lacunae.

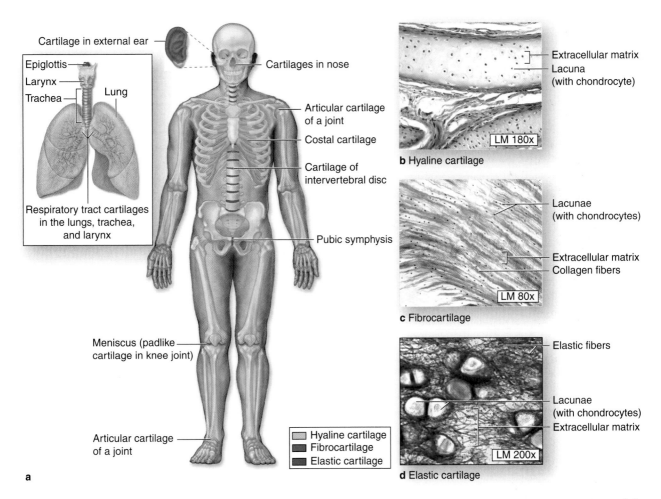

Figure 7–1. **Distribution of cartilage in adults. (a):** There are three types of adult cartilage distributed in many areas of the skeleton, particularly in joints and where pliable support is useful, as in the ribs, ears, and nose. Cartilage support of other tissues throughout the respiratory system is also prominent. The photomicrographs show the main features of **(b)** hyaline cartilage, **(c)** fibrocartilage, and **(d)** elastic cartilage.

Cartilage cells and the matrix often shrink during routine histologic preparation, resulting in both the irregular shape of the chondrocytes and their retraction from the matrix. In living tissue, and in properly prepared sections, the chondrocytes fill the lacunae completely.

Because cartilage is devoid of blood capillaries, chondrocytes respire under low oxygen tension. Hyaline cartilage cells metabolize glucose mainly by anaerobic glycolysis to produce lactic acid as the end product. Nutrients from the blood diffuse through the perichondrium to reach the more deeply placed cartilage cells. Transport of water and solutes is promoted by the pumping action of intermittent cartilage compression and decompression. Because of the limits of diffusion, the maximum width of the cartilage is limited and cartilage usually is found as small, thin plates of tissue.

Chondrocyte function is hormone dependent. Synthesis of sulfated GAGs is accelerated by growth hormone, thyroxin, and testosterone and is slowed by cortisone, hydrocortisone, and estradiol. Cartilage growth depends mainly on the pituitary-derived growth hormone **somatotropin.** This hormone does not act on cartilage cells directly but promotes the endocrine release in the liver of insulin-like growth factor-1 (IGF-1), sometimes called somatomedin C. IGF-1 acts directly on cartilage cells, promoting their growth.

Perichondrium

Except in the articular cartilage of joints, all hyaline cartilage is covered by a layer of dense connective tissue, the **perichondrium**, which is essential for the growth and maintenance of cartilage (Figure 7–2). It consists largely of collagen type I fibers and contains numerous fibroblasts. Although cells in the inner layer of the perichondrium resemble fibroblasts, they are precursors for chondroblasts which divide and differentiate into chondrocytes.

ELASTIC CARTILAGE

Elastic cartilage is essentially very similar to hyaline cartilage except that it contains an abundant network of fine elastic fibers in addition to collagen type II fibrils (Figure 7–4). Fresh elastic cartilage has a yellowish color owing to the presence of elastin in the elastic fibers.

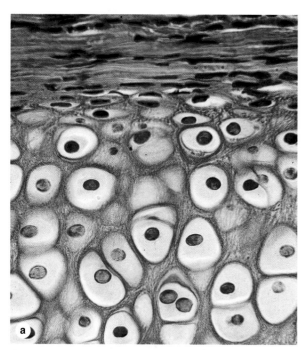
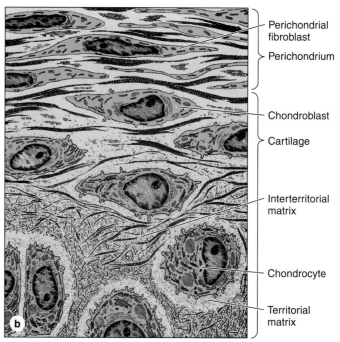

Figure 7–2. **Hyaline cartilage and perichondrium. (a):** A section of hyaline cartilage shows chondrocytes located in matrix lacunae. Preparation for sectioning usually causes shrinkage of the matrix which may cause the chondrocytes to pull away from the matrix and become distorted. The upper part of the figure shows the more eosinophilic perichondrium, an example of dense connective tissue consisting largely of type I collagen. There is a gradual transition and differentiation of cells from the perichondrium to the cartilage, with elongated fibroblastic cells becoming larger and more rounded chondrocytes with irregular surfaces contacting the matrix secreted by the cells. X200. H&E. **(b):** Diagram of the area of transition between the perichondrium and the hyaline cartilage. In living cartilage chondrocytes essentially fill their lacunae. Closely associated groups of two or four lacunae indicate isogenous groups, or clones of chondrocytes derived from the same cell. Staining differences are apparent between the matrix immediately around each lacuna, called the territorial matrix, and that more distant from lacunae, the interterritorial matrix. Collagen is more abundant in the interterritorial parts of the matrix.

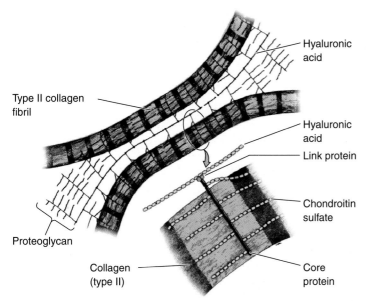

Figure 7–3. Molecular organization of hyaline cartilage ECM. Schematic representation of the most abundant molecules in cartilage matrix shows the interaction between type II collagen fibrils and proteoglycans linked to hyaluronic acid. Link proteins noncovalently bind the protein core of proteoglycans to the linear hyaluronic acid molecules. The chondroitin sulfate side chains of the proteoglycan electrostatically bind to the collagen fibrils, forming a cross-linked matrix. The oval outlines the area shown larger in the lower part of the figure. Physical properties of these matrix components produce a highly hydrated, pliable material with great strength. Approximately 75% of the wet weight of hyaline cartilage is water.

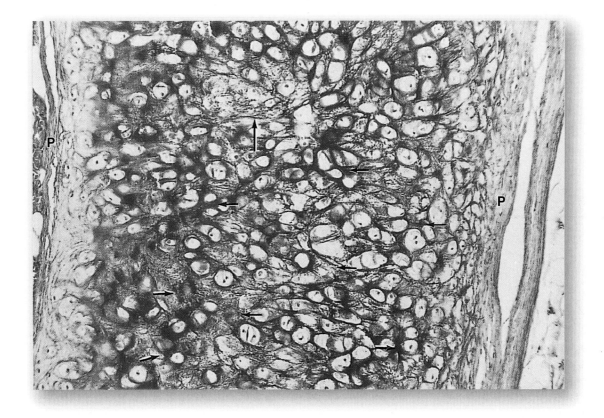

Figure 7–4. Elastic cartilage. Photomicrograph of elastic cartilage from the epiglottis shows perichondrium (P) on both surfaces. Cell size and distribution in elastic cartilage is very similar to that of hyaline cartilage. With special staining for elastic fibers however, the matrix is seen to be filled with this material (arrows), providing greater flexibility to this form of cartilage. X100. Weigert resorcin-fuscin.

Elastic cartilage is frequently found to be gradually continuous with hyaline cartilage. Like hyaline cartilage, elastic cartilage possesses a perichondrium.

Elastic cartilage is found in the auricle of the ear, the walls of the external auditory canals, the auditory (eustachian) tubes, the epiglottis, and the cuneiform cartilage in the larynx.

FIBROCARTILAGE

Fibrocartilage is a tissue intermediate between dense connective tissue and hyaline cartilage. It is found in intervertebral disks, in attachments of certain ligaments, and in the pubic symphysis (Figure 7–1). Fibrocartilage is always associated with dense connective tissue and the border between these two tissues is not clear-cut, showing a gradual transition.

Fibrocartilage contains chondrocytes, either singly or in isogenous aggregates, usually arranged axially, in long rows separated by coarse collagen type I fibers and less proteoglycans than other forms of cartilage (Figure 7–5). Because it is richer in collagen type I, the fibrocartilage matrix is more acidophilic.

In fibrocartilage dense collagen fibers can form either irregular or parallel bundles between the axial aggregates of chondrocytes (Figure 7–5). The general orientation of the collagen depends on the stresses on fibrocartilage, since the collagen bundles take up a direction parallel to those stresses. There is no distinct perichondrium in fibrocartilage.

Intervertebral disks are composed of fibrocartilage primarily. They are situated between the vertebrae and are held to them by ligaments. Each disk has two major histological components: the peripheral **annulus fibrosus** rich in bundles of type I collagen and the central **nucleus pulposus** with a gel-like matrix rich in hyaluronic acid (see Chapter 8). Intervertebral disks act as lubricated cushions and shock absorbers preventing adjacent vertebrae from being damaged by abrasive forces or impact during movement of the spinal column.

CARTILAGE FORMATION, GROWTH AND REPAIR

All cartilage derives from the embryonic mesenchyme in the process of **chondrogenesis** (Figure 7–6). The first indication of cell differentiation is the rounding up of the mesenchymal cells, which retract their extensions, multiply rapidly, and form cellular condensations. The cells formed by this direct differentiation of mesenchymal cells, now called **chondroblasts,** have a ribosome-rich basophilic cytoplasm. Synthesis and deposition of the matrix then begin to separate the chondroblasts from one another (Figure 7–7). During embryonic development, the differentiation of cartilage takes place primarily from the center outward; therefore, the more central cells have the characteristics of chondrocytes, whereas the peripheral cells are typical chondroblasts. The superficial mesenchyme develops into the perichondrium.

Further growth of cartilage is attributable to two processes: **interstitial growth,** resulting from the mitotic division of preexisting chondrocytes; and **appositional growth,** resulting from the differentiation of perichondrial cells. In both cases, the synthesis of matrix contributes greatly to the growth of the cartilage. Interstitial growth is the less important of the two processes postnatally. It occurs during the early phases of cartilage formation, when it increases tissue mass by expanding the cartilage matrix from within. Interstitial growth also occurs in the epiphyseal plates of long bones and within articular cartilage. In the epiphyseal plates, interstitial growth is important in increasing the length of long bones (see Chapter 8). In articular cartilage, as the cells and matrix near the articulating surface are gradually worn away, the cartilage must be replaced from within, since there is no perichondrium there to add cells by appositional growth. In cartilage found elsewhere in the body, interstitial growth becomes less pronounced, as the matrix becomes increasingly rigid from the cross-linking of matrix molecules. Cartilage

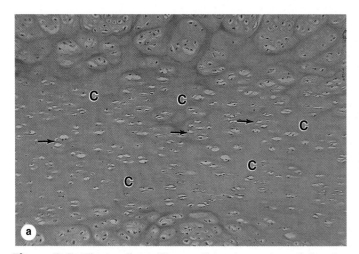 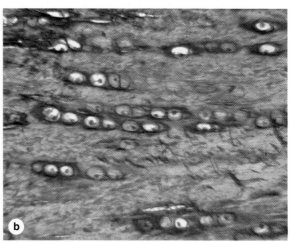

Figure 7–5. **Fibrocartilage.** Fibrocartilage shows rows of chondrocytes aligned parallel to the direction of greatest stress and separated by parallel or irregular bundles of type I collagen fibers. No separate perichondrium is present on fibrocartilage. **(a):** Micrograph of pubic symphysis shows staining variations in the matrix caused by varying concentrations of collagen (C). Lacunae (arrows) of chondrocytes are also seen. A section of intervertebral disk. X100. Masson trichrome. **(b):** The axial aggregates of chondrocytes are separated by collagen. Fibrocartilage is also frequently found in the insertion of tendons on the epiphyseal hyaline cartilage. X400. Picrosirius-hematoxylin.

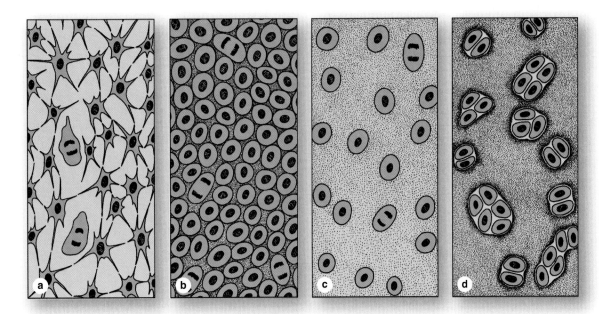

Figure 7–6. Chondrogenesis. Diagrams of the major stages by which cartilage is formed. **(a):** Embryonic mesenchyme is the precursor tissue of all types of cartilage. **(b):** Mitotic proliferation of mesenchymal cells and early differentiation gives rise to a tissue with condensations of rounded cells called chondroblasts. **(c):** Chondroblasts are separated from one another by their own production of various matrix components which collectively swell with water and form a great amount of ECM. **(d):** Multiplication of cartilage cells gives rise to isogenous aggregates, each surrounded by a condensation of territorial matrix. In mature cartilage this interstitial mitotic activity ceases and all chondrocytes typically become more widely separated by their production of matrix.

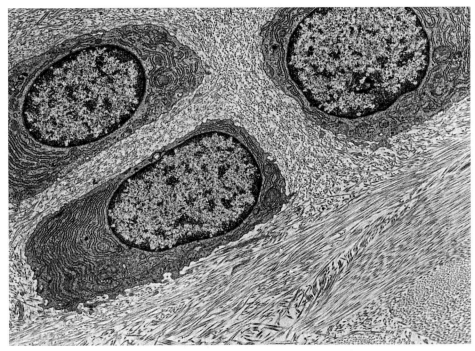

Figure 7–7. Chondrocytes in growing fibrocartilage. This TEM of fibrocartilage from a young animal shows three chondrocytes in their lacunae. RER is abundant in the cells, which are actively secreting their collagen-rich matrix. Fine collagen fibers, sectioned in several orientations, are prominent around the chondrocytes of fibrocartilage. Growing chondrocytes in hyaline and elastic cartilage have more prominent Golgi complexes and synthesize abundant proteoglycans in addition to collagens. X3750.

then increases in girth only by appositional growth. Chondroblasts differentiate in the inner layers of the perichondrium, proliferate, and become chondrocytes once they have surrounded themselves with cartilaginous matrix and are incorporated into the existing cartilage (Figure 7–2).

Except in young children, damaged cartilage undergoes slow and often incomplete **regeneration**, by activity of cells in the perichondrium which invade the injured area and generate new cartilage. In extensively damaged areas—and occasionally in small areas—the perichondrium produces a scar of dense connective tissue instead of forming new cartilage. The poor regenerative capacity of cartilage is due in part to the avascularity of this tissue.

MEDICAL APPLICATION

In contrast to other tissues, hyaline cartilage is more susceptible to degenerative aging processes. Calcification of the matrix, preceded by an increase in the size and volume of the chondrocytes and followed by their death, is a common process in some cartilage. "Asbestiform" degeneration, frequent in aged cartilage, is due to the formation of localized aggregates of thick, abnormal collagen fibrils

GLIAL CELLS & NEURONAL ACTIVITY

Glial cells are 10 times more abundant in the mammalian brain than neurons. In the CNS glial cells surround most of the neuronal cell bodies, which are usually much larger than glial cells, and the processes of axons and dendrites that occupy the spaces between neurons. Except around the larger blood vessels, the CNS has only a very small amount of connective tissue or ECM. Glial cells (Table 9–2) furnish a microenvironment ideal for neuronal activity. A dense network of fibers from processes of both neurons and glial cells fills the interneuronal space of the CNS and is called the **neuropil** (Figure 9–9).

Key facts for glial cells are summarized in Table 9–2 and shown schematically in Figure 9–10. There are six kinds of glial cells:

Oligodendrocytes

Oligodendrocytes (Gr. *oligos*, small, few + *dendron*, tree + *kytos*, cell) produce the myelin sheath that provides the electrical insulation for neurons in the CNS. Oligodendrocytes extend processes that wrap around parts of several axons, producing a myelin sheath as shown in Figure 9–10a. They are the predominant glial cell in CNS white matter. The processes are not visible by routine light microscope staining, in which oligodendrocytes usually appear as small cells with rounded, condensed nuclei and unstained cytoplasm (Figures 9–9a and 9–10a).

Astrocytes

Astrocytes (Gr. *astron*, star, + *kytos*) have a large number of radiating processes (Figures 9–10b and 9–11) and are unique to the CNS. Astrocytes with relatively few long processes are called **fibrous astrocytes** and are located in the white matter; **protoplasmic astrocytes,** with many short, branched processes, are found in the gray matter. Astrocytes have supportive roles for neurons and are very important for proper formation of the CNS during embryonic and fetal development. Located mainly in gray matter, astrocytes are by far the most numerous glial cells and exhibit considerable morphologic and functional diversity.

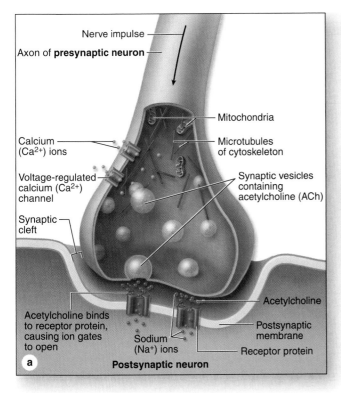

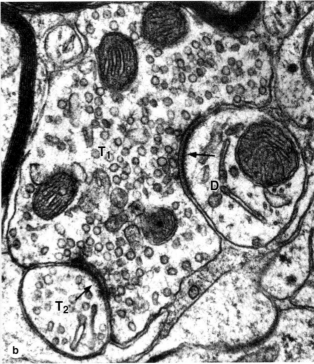

Figure 9–6. **Synapse. (a):** Diagram showing how neurotransmitters are released from the terminal bouton in a chemical synapse. Presynaptic terminals always contain a large number of **synaptic vesicles** containing neurotransmitters, numerous **mitochondria**, and smooth ER as a source of new membrane. Some neurotransmitters are synthesized in the cell body and then transported in vesicles to the presynaptic terminal. Upon arrival of a nerve impulse, voltage-regulated Ca^{2+} channels permit Ca^{2+} entry, which triggers exocytosis releasing neurotransmitter into the synaptic cleft. Excess membrane accumulating at the presynaptic region as a result of exocytosis is recycled by clathrin-mediated endocytosis, which is not depicted here. The retrieved membrane fuses with the SER in the presynaptic compartment for reuse in the formation of more synaptic vesicles. Some neurotransmitters are synthesized in the presynaptic compartment, using enzymes and precursors brought there by axonal transport. **(b):** TEM shows a large presynaptic terminal (T_1) filled with synaptic vesicles and asymmetric electron-dense regions around 20–30 nm wide synaptic clefts (arrows). The postsynaptic membrane contains the neurotransmitter receptors and mechanisms to initiate an impulse at the postsynaptic neuron. The postsynaptic membrane on the right is part of a dendrite (D), associated with fewer vesicles of any kind, showing this to be an axodendritic synapse. On the left is another presynaptic terminal (T_2), suggesting an axoaxonic synapse with a role in modulating activity of the other terminal. X35,000.

In addition to their supporting function, astrocytes have major roles in controlling the ionic environment of neurons. Some astrocytes develop processes with expanded **perivascular feet** that cover capillary endothelial cells and contribute to the blood-brain barrier. The perivascular feet are important for the ability of astrocytes to regulate vasodilation and transfer of O_2, ions and other substances from the blood to the neurons. Other expanded processes form a layer, the superficial **glial limiting membrane** which lines the pia mater, the innermost meningial layer at the external surface of the CNS. Furthermore, when the

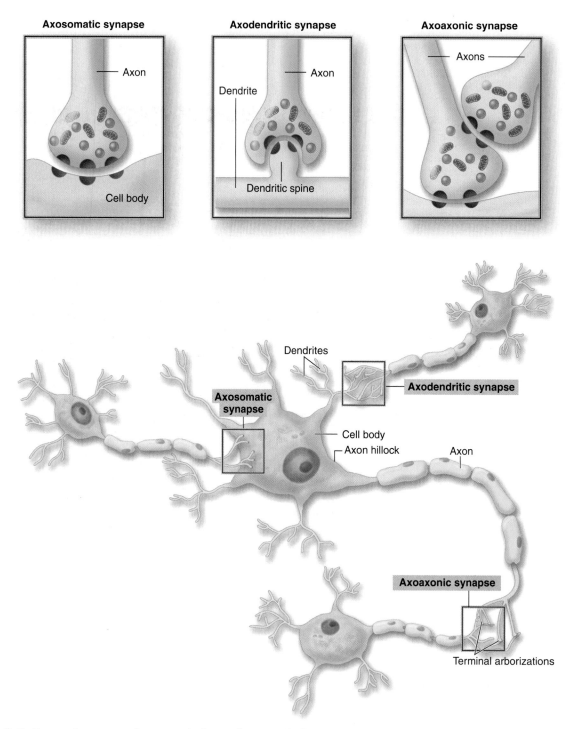

Figure 9–7. **Types of synapses.** Axon terminals usually transmit the nerve impulse to another neuron's cell body (or soma) or to its dendrites (usually at a dendritic spine). Less frequently, axon terminals form synapses with another axon terminal, an arrangement that appears to modulate synaptic activity. Features of these three common morphological types of synapses are shown at the top of the figure.

CNS is damaged, astrocytes proliferate to form cellular scar tissue (which often interferes with neuronal regeneration).

Astrocytic functions are essential for neuronal survival. They regulate constituents of the extracellular environment, absorb local excess of neurotransmitters, and secrete numerous metabolites and factors regulating neuronal activities. Finally, astrocytes are in direct communication with one another via gap junctions, forming a network through which information can flow from one point to another, reaching distant sites.

The processes of all astrocytes are reinforced with bundles of intermediate filaments made of **glial fibrillary acid protein (GFAP)**, which serves as a unique marker for astrocytes, the most common source of brain tumors.

Ependymal Cells

Ependymal cells are low columnar or cuboidal cells that line the ventricles of the brain and central canal of the spinal cord (Figures 9–10c and 9–12). In some CNS locations, the apical ends of ependymal cells have cilia, which facilitate the movement of cerebrospinal fluid (CSF), or long microvilli, which are likely involved in absorption.

Ependymal cells are joined apically by junctional complexes similar to those of epithelia. However, unlike a true epithelium there is no basal lamina. Instead, the basal ends of ependymal

cells are elongated and extend branching processes into the adjacent neuropil.

Microglia

Somewhat less numerous than oligodendrocytes or astrocytes but more evenly distributed throughout gray and white matter, **microglia** are small cells with short irregular processes (Figures 9–10d and 9–13). Unlike other glial cells microglia migrate through the neuropil, analyzing the tissue for damaged cells and invading microorganisms. They secrete a number of immunoregulatory cytokines and constitute the major mechanism of immune

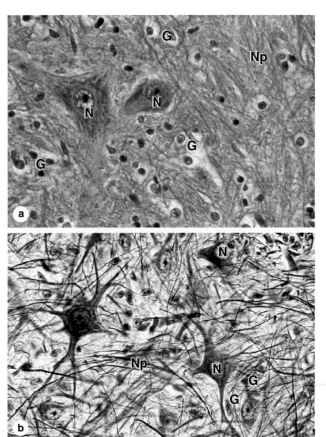

Figure 9–9. Neurons, neuropil, and the common glial cells of the CNS. (a): Most neuronal cell bodies (N) in the CNS are larger than the much more numerous glial cells (G) that surround them. The various types of glial cells and their relationships with neurons are difficult to distinguish by most routine light microscopic methods. However, **oligodendrocytes** have condensed, rounded nuclei and unstained cytoplasm due to very abundant Golgi complexes, which stain poorly and are very likely represented by the cells with those properties seen here. The other glial cells similar in overall size but with very little cytoplasm and more elongated or oval nuclei, are mostly **astrocytes**. Routine H&E staining does not allow **neuropil** (Np) to stand out well. X200. H&E. **(b):** With the use of gold staining for neurofibrils, neuropil is more apparent. X200. Gold & hematoxylin.

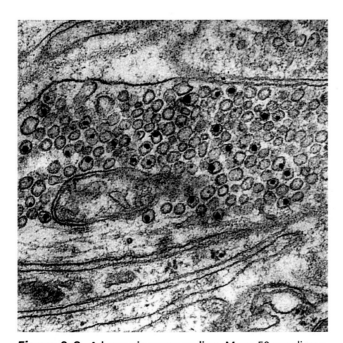

Figure 9–8. Adrenergic nerve ending. Many 50-nm-diameter vesicles, with electron-dense cores containing norepinephrine, fill the axon terminal shown here. X40,000. (Machado AB: Straight OsO₄ versus glutaraldehyde-OsO₄ in sequence as fixatives for the granular vesicles in sympathetic axons of the rat pineal body. Stain *Technology* 1967, 42(6):293. Reproduced, with permission, from Taylor & Francis Group, http:www.informaworld.com, and Angelo B. Machado, Department of Morphology, Federal University of Minas Gerais, Belo Horizonte, Brazil)

CNS Glial Cells

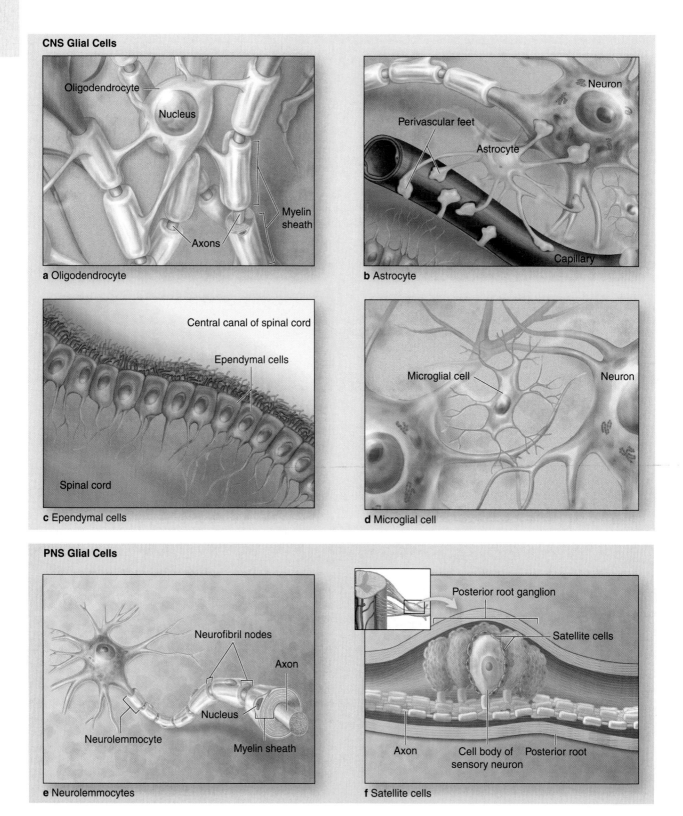

a Oligodendrocyte

Oligodendrocyte
Nucleus
Myelin sheath
Axons

b Astrocyte

Neuron
Perivascular feet
Astrocyte
Capillary

c Ependymal cells

Central canal of spinal cord
Ependymal cells
Spinal cord

d Microglial cell

Microglial cell
Neuron

PNS Glial Cells

e Neurolemmocytes

Neurofibril nodes
Axon
Nucleus
Neurolemmocyte
Myelin sheath

f Satellite cells

Posterior root ganglion
Satellite cells
Axon
Cell body of sensory neuron
Posterior root

Figure 9–10. **Glial cells of the CNS and PNS.** Glial cell in the CNS. **(a): Oligodendrocytes** myelinate parts of several axons. **(b): Astrocytes** have multiple processes and form perivascular feet that completely enclose all capillaries (only a few such feet are shown here to allow their morphology to be seen). **(c): Ependymal cells** are epithelial-like cells that line the ventricles and central canal. **(d): Microglial cells** have a protective, phagocytic, immune-related function. Glial cells in the PNS. **(e): Neurolemmocytes,** commonly called Schwann cells, form a series enclosing axons. **(f): Satellite cells** are restricted to ganglia where they cover and support the large neuronal cell bodies.

extracellular environment provided by insulating glial cells. This shift is the beginning of the **action potential** or **nerve impulse**. The +30 mV potential rapidly closes the sodium channels and opens the K^+ channels, allowing this ion to leave the axon by diffusion and returning the membrane potential to –65 mV. The duration of these local events is very short, only about 5 milliseconds.

However, the action potential propagates along the axonal membrane, producing the nerve impulse. The electrical disturbance opens neighboring sodium channels and, in sequence, potassium channels. In this way the action potential propagates at a high speed along the axon, with several occurring per second. When an action potential arrives at the nerve ending, it promotes discharge of stored neurotransmitter that stimulates or inhibits another neuron or a non-neural cell, such as a muscle or gland cell.

MEDICAL APPLICATION

Local anesthetics are hydrophobic molecules that bind to sodium channels, inhibiting sodium transport and, consequently, also the action potential responsible for the nerve impulse.

Synaptic Communication

The synapse (Gr. *synapsis*, union) is responsible for the transmission of nerve impulses from neuron to another cell and insures that transmission is unidirectional. Synapses are sites of

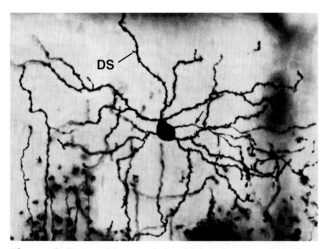

Figure 9–5. **Dendrites and dendritic spines.** In this silver-stained preparation of cells in a section of cerebellum, the many dendrites emerging from a single stellate neuron are clearly seen. Each dendrite can be seen further to have many dendritic spines (DS) along its surface. The dendritic spines are sites of synapses with other neurons. Their length and morphology are dependent on actin filaments and are highly plastic. Arrow indicates the cell's axon. X500. Golgi.

functional contact between neurons or between neurons and other effector cells. The function of the synapse is to convert an electrical signal (impulse) from the **presynaptic** cell into a chemical signal that acts on the **postsynaptic** cell. Most synapses transmit information by releasing **neurotransmitters** during this signaling process. Neurotransmitters are chemicals that bind specific receptor proteins to either open or closed ion channels or initiate second-messenger cascades. A synapse (Figure 9–6) has the following structure:

- Presynaptic axon terminal (**terminal bouton**) from which neurotransmitter is released,
- Postsynaptic cell membrane with receptors for the transmitter and ion channels or other mechanisms to initiate a new impulse,
- 20–30 nm wide intercellular space called the **synaptic cleft** separating the presynaptic and postsynaptic membranes.

Nerve impulses sweep rapidly (in milliseconds) along the axolemma as an explosive wave of electrical activity (depolarization). At the presynaptic region the nerve impulse briefly opens calcium channels, promoting a calcium influx that triggers the exocytosis of synaptic vesicles. The released neurotransmitters diffuse across the synaptic cleft and bind receptors at the postsynaptic region, promoting a transient electrical activity (depolarization) at the postsynaptic membrane. These synapses are called **excitatory,** because their activity promotes impulses in the postsynaptic cell membrane. In some synapses the neurotransmitter-receptor interaction has an opposite effect, promoting membrane **hyperpolarization** with no transmission of the nerve impulse. These are called **inhibitory** synapses. Thus, synapses can excite or inhibit impulse transmission and thereby regulate nerve activity.

Once used, neurotransmitters are removed quickly by enzymatic breakdown, diffusion, or endocytosis mediated by specific receptors on the presynaptic membrane. This removal of neurotransmitters is functionally important because it prevents an undesirable sustained stimulation of the postsynaptic neuron.

Morphologically, various types of synapses are seen between neurons. If an axon forms a synapse with a cell body, it is called an **axosomatic synapse;** with a dendrite, **axodendritic;** or with an axon, **axoaxonic** (Figure 9–7). The axoaxonic synapse is less common and is used to modulate synaptic activity.

The first neurotransmitters to be described were acetylcholine and norepinephrine. A norepinephrine-releasing axon terminal is shown in Figure 9–8. Most neurotransmitters are amines, amino acids, or small peptides (neuropeptides). Inorganic substances such as nitric oxide can also act as neurotransmitters. Several peptides that can act as neurotransmitters are used as paracrine hormones elsewhere in the body eg, in the digestive tract. Neuropeptides are involved in regulating feelings and drives, such as pain, pleasure, hunger, thirst, and sex.

Neuromodulators are chemical messengers that modify neuron sensitivity to synaptic stimulation or inhibition, without acting directly on synapses. Some neuromodulators are neuropeptides or steroids produced in the nerve tissue, others are circulating steroids.

Although most synapses are **chemical synapses** and use chemical neurotransmitters, some synapses transmit ionic signals through gap junctions between the pre- and postsynaptic membranes, thereby conducting neuronal signals directly. These synapses are **electrical synapses** and are prominent in cardiac and smooth muscle.

synthesized in the cell body move by **anterograde transport** along the axon from the perikaryon to the synaptic terminals. **Retrograde transport** in the opposite direction carries certain other macromolecules, such as material taken up by endocytosis (including viruses and toxins), from the periphery to the cell body. Retrograde transport can be used to study the pathways of neurons: if peroxidase or another marker is injected into regions with axon terminals, its distribution along the entire axon after a period of time can be followed histochemically.

Axonal transport in both directions utilizes motor proteins attached to microtubules, as discussed in Chapter 2. **Kinesin,** a microtubule-activated ATPase, attaches to vesicles and allows them to move along microtubules in axons away from the perikarya. **Dynein** is a similar ATPase that allows retrograde transport in axons, toward the cell bodies.

Anterograde and retrograde transport both occur fairly rapidly, at rates of 50 to 400 mm/day. A much slower anterograde stream (only a few millimeters per day) involves movement of the axonal cytoskeleton itself. This slow transport system corresponds roughly to the rate of axon growth.

Membrane Potentials

Many integral membrane proteins of neuronal cell membrane act as pumps and channels that transport or allow diffusion of ions into and out of the cytoplasm. The axolemma or limiting membrane of the axon pumps Na^+ out of the axoplasm, maintaining a concentration of Na^+ that is only a tenth of that in the extracellular fluid. In contrast, the concentration of K^+ is maintained at a level many times greater than that in the extracellular environment. This produces a potential difference across the axolemma of about –65 mV with the inside negative to the outside: the **resting membrane potential.** When a neuron is stimulated, ion channels open and there is a sudden influx of extracellular Na^+ that changes the resting potential from –65 mV to +30 mV and makes the cell interior positive in relation to the

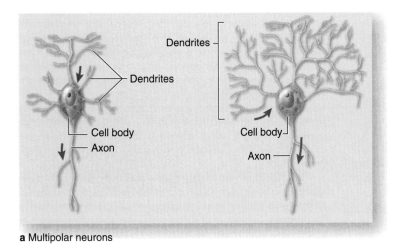

a Multipolar neurons

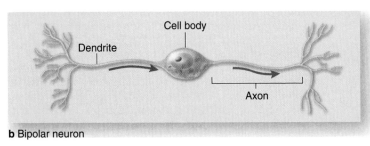

b Bipolar neuron

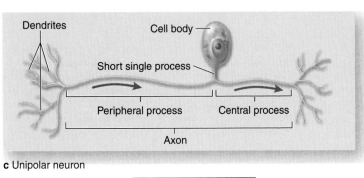

c Unipolar neuron

→ Input → Output

Figure 9–4. Structural classes of neurons. Simplified views of the three main types of neurons, according to their number of processes. **(a): Multipolar neurons** have one axon but a large number of branching dendrites. **(b): Bipolar neurons** have one axon and one dendrite emerging from the cell body. **(c): (Pseudo)unipolar neurons** have one short process coming off the perikaryon, but this immediately bifurcates into a long process extending peripherally and a shorter branch extending toward the CNS. Both of these processes have terminal arborizations (branches) and those of the peripheral process serve as dendrites, receiving stimuli that travel directly to the terminals at the other end of the axon without passing through the perikaryon. Pseudounipolar neurons form when two initial processes move together and fuse, becoming one single fiber. In these neurons, the cell body does not seem to be involved in impulse conduction, but remains as the synthetic center for the entire cell.

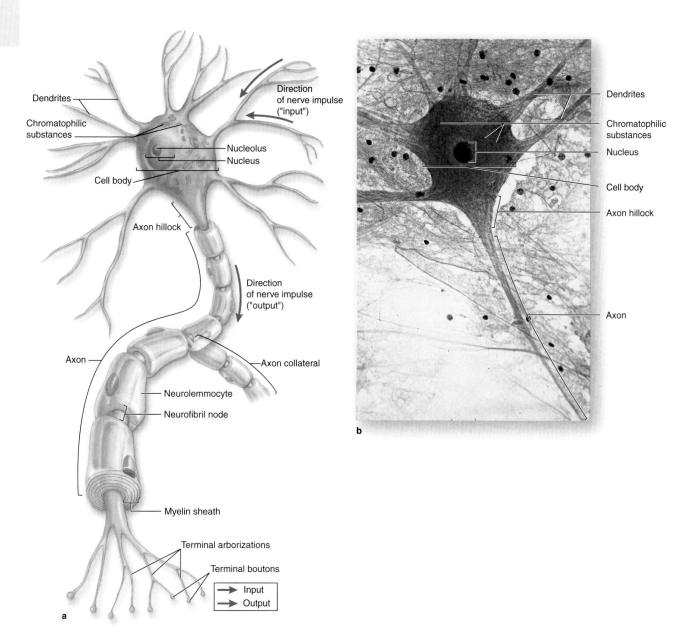

Figure 9–3. **Structures of neuron. (a):** The diagram of a "typical" neuron has many features of a motor neuron, but shows the three major parts of every neuron. The **cell body** is large and has a large, euchromatic nucleus with a well-developed nucleolus. The perikaryon also contains **chromatophilic substance** or Nissl bodies, which are large masses of free polysomes and RER and indicate the cell's rate of protein synthesis. Numerous short **dendrites** extend from the perikaryon, carrying input from other neurons. A long **axon** carries impulses from the cell body and is covered by a myelin sheath composed of other cells. Arrows show the direction of the nerve impulse. The ends of axons usually have many small branches called **terminal arborizations**, each of which usually has a swollen end called bouton which forms a functional connection (synapse) with another neuron or other cell. Axons can also branch closer to the cell bodies and form **collateral branches** that connect to other groups of cells. **(b):** Micrograph of a large motor neuron showing the large cell body with a long axon and several dendrites emerging from it. Evenly dispersed chromatophilic substance can be seen throughout the cell body and cytoskeletal elements can be detected in the processes. X100. H&E.

action potential. In contrast to dendrites, axons have a constant diameter and do not branch profusely. Occasionally, the axon, shortly after its departure from the cell body, gives rise to a branch that returns to the area of the nerve cell body. All axon branches are known as **collateral branches** (Figure 9–3). Axoplasm contains mitochondria, microtubules, neurofilaments, and some

cisternae of smooth ER. The absence of polyribosomes and rough ER emphasizes the dependence of the axon on the perikaryon for its maintenance. If an axon is severed, its peripheral parts quickly degenerate.

There is a lively bidirectional transport of small and large molecules along the axon. Organelles and macromolecules

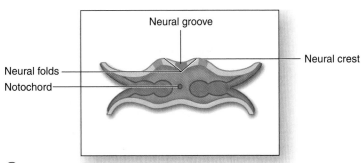

① Neural folds and neural groove form from the neural plate.

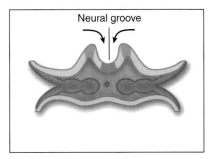

② Neural folds elevate and approach one another.

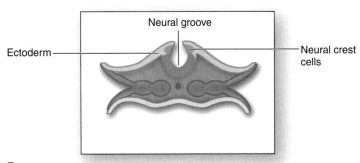

③ As neural folds prepare to fuse and form the neural tube and dorsal epidermis, neural crest cells loosen and become mesenchymal.

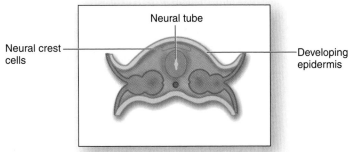

④ The mass of neural crest cells initially lies atop the newly formed neural tube.

Figure 9–2. **Neurulation in the early embryo.** The top diagram shows a cross-section of a 21-day human embryo, when it is approximately 1 mm in length, with the surrounding amnion membrane removed. Cross-sections through the embryo at four times during the following week show stages in the process of neurulation, the embryonic process by which the CNS and PNS initially form. Under an inductive influence from the axial notochord, the overlying layer of ectodermal cells thickens to become the **neural plate**. All other ectoderm will become epidermis.

The neural plate forms two lateral **folds**, separated by the **neural groove** (1). These folds rise and fuse at the midline (2), converting the neural groove into the **neural tube**. The neural tube, which is large at the cranial end of the embryo and much narrower caudally, will give rise to the CNS. As the neural folds fuse and the resulting tube detaches from the now overlying ectoderm (3), a population of neural cells separates and becomes a mass of mesenchymal cells called the **neural crest**.

Located initially between the neural tube and the epidermis (4), neural crest cells immediately begin migrating laterally. Neural crest cells form the sensory ganglia and all other cells of the PNS, as well as contribute to many other developing structures, many non-neural, including melanocytes, meningeal layers around the brain, the adrenal medulla, cells of the teeth, and cartilage of the head.

that these cells synthesize both structural proteins and proteins for transport and secretion. When appropriate stains are used, RER and free ribosomes appear under the light microscope as clumps of basophilic material called **chromatophilic substance** (often called **Nissl bodies**) (Figure 9–3). The amount of chromatophilic substance varies according to the type and functional state of the neuron and is particularly abundant in large nerve cells such as motor neurons (Figure 9–3b). The Golgi apparatus is located only in the cell body, but mitochondria can be found throughout the cell and are usually abundant in the axon terminals.

Intermediate filaments are abundant both in perikarya and processes and are called **neurofilaments** in this cell. Neurofilaments become cross-linked with certain fixatives and when impregnated with silver stains, they form neurofibrils visible with the light microscope. The neurons also contain microtubules identical to those found in other cells. Nerve cells occasionally contain inclusions of pigmented material, such as **lipofuscin,** which consists of residual bodies left from lysosomal digestion.

Dendrites

Dendrites (Gr. *dendron,* tree) are usually short and divide like the branches of a tree (Figure 9–3). They are often covered with many synapses and are the principal signal reception and processing sites on neurons. Most nerve cells have numerous dendrites, which considerably increase the receptive area of the cell. The arborization of dendrites makes it possible for one neuron to receive and integrate a great number of axon terminals from other nerve cells. It has been estimated that up to 200,000 axonal terminations establish functional contact with the dendrites of a single large Purkinje cell of the cerebellum.

Unlike axons, which maintain a constant diameter, dendrites become much thinner as they subdivide into branches. The cytoplasmic composition of the dendrite base, close to the neuron body, is similar to that of the perikaryon but is devoid of Golgi complexes. Most synapses impinging on neurons are located on **dendritic spines,** which are usually short blunt structures 1 to 3 μm long projecting from dendrites, visible with silver staining methods (Figure 9–5). These spines occur in vast numbers, estimated to be on the order of 10^{14} for cells of the human cerebral cortex, and serve as the first processing locale for synaptic signals arriving on a neuron. The processing apparatus is contained in a complex of proteins attached to the cytosolic surface of the postsynaptic membrane, which is visible under the transmission electron microscope (TEM). The morphology of such spines is based on actin filaments and can be highly plastic; dendritic spines participate widely in the constant changes that make up **neuronal plasticity** which underlies adaptation, learning, and memory.

Axons

Most neurons have only one axon, a cylindrical process that varies in length and diameter according to the type of neuron. Axons are usually very long processes. For example, axons of the motor cells of the spinal cord that innervate the foot muscles may have a length of up to 100 cm (~40 inches). All axons originate from a pyramid-shaped region, the **axon hillock,** arising from the perikaryon (Figure 9–3). The plasma membrane of the axon is often called the **axolemma** and its contents are known as **axoplasm.**

Just beyond the axon hillock, at an area called the **initial segment**, is the site where various excitatory and inhibitory stimuli impinging on the neuron are algebraically summed, resulting in the decision to propagate—or not to propagate—a nerve impulse. Several types of ion channels are localized in the initial segment and these channels are important in generating the

Table 9–1. Structural and functional divisions of the nervous system.

Organization	Components	General description
Structural divisions:		
Central nervous system (CNS)	Brain and spinal cord	Overall "command center," processing and integrating information
Peripheral nervous system (PNS)	Nerves and ganglia	Receives and projects information to and from the CNS; mediates some reflexes
Functional divisions:		
Sensory nervous system	Some CNS and PNS components	Includes all axons that transmit impulses from a peripheral structure to the CNS
	Somatic sensory	Transmits input from skin, fascia, joints, and skeletal muscles
	Visceral sensory	Transmits input from stomach and intestines (viscera)
Motor nervous system	Some CNS and PNS components	Includes all axons that transmit nerve impulses from the CNS to a muscle or gland
	Somatic motor (somatic nervous system)	Voluntary control of skeletal muscle
	Autonomic motor (autonomic nervous system)	Involuntary control of smooth muscle, cardiac muscle, and glands

migrate extensively and differentiate as all the cells of the PNS, as well as a number of other nonneuronal cell types.

NEURONS

The functional unit in both the CNS and PNS is the neuron or nerve cell. Most neurons consist of three parts (Figure 9–3): the **cell body,** or **perikaryon,** which is the synthetic or trophic center for the entire nerve cell and is receptive to stimuli; the **dendrites,** many elongated processes specialized to receive stimuli from the environment, sensory epithelial cells, or other neurons; and the **axon** (Gr. *axon*, axis), which is a single process specialized in generating and conducting nerve impulses to other cells (nerve, muscle, and gland cells). Axons may also receive information from other neurons, information that mainly modifies the transmission of action potentials to those neurons. The distal portion of the axon is usually branched as the **terminal arborization.** Each branch terminates on the next cell

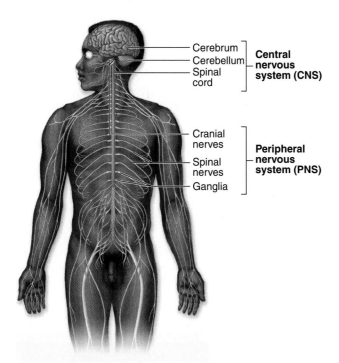

Figure 9–1. **The general organization of the nervous system.** The major anatomical division of nervous tissue is as components of the central and peripheral nervous systems. The CNS includes all parts of the brain, including the large cerebrum and cerebellum seen here, and the spinal cord, which are protected by the bones of the cranium and spinal column respectively. The PNS includes **nerves,** which contain both long nerve processes (fibers) growing from motor neurons whose cell bodies are within the spinal cord and processes of sensory neurons which are grouped in a series of ganglia outside the spinal cord. Motor nerves are efferent, carrying impulses away from the CNS; sensory fibers are afferent, carrying impulses to the CNS. Another group of peripheral neurons, fibers, and ganglia comprises the more dispersed autonomic nervous system, not shown here.

in dilatations called **end bulbs** (**boutons**), which interact with other neurons or nonnerve cells at structures called **synapses.** Synapses initiate impulses in the next cell of the circuit.

Neurons and their processes are extremely variable in size and shape. Cell bodies can be very large, measuring up to 150 μm in diameter. Other cells are among the smallest cells in the body; for example, the cell bodies of granule cells of the cerebellum are only 4–5 μm in diameter.

Neurons can be classified according to the number of processes extending from the cell body (Figure 9–4):

- **Multipolar neurons,** which have one axon and two or many dendrites;
- **Bipolar neurons,** with one dendrite and one axon; and
- **Unipolar** or **pseudounipolar neurons,** which have a single process that bifurcates close to the perikaryon, with the longer branch extending to a peripheral ending and the other toward the CNS.

Most neurons are multipolar. Bipolar neurons are found in the retina, olfactory mucosa, and the (inner ear) cochlear and vestibular ganglia, where they serve the senses of sight, smell and balance respectively. Pseudounipolar neurons are found in the spinal ganglia (the sensory ganglia found with the spinal nerves) and in most cranial ganglia. Since processes emerging from perikarya are seldom seen in sections of nervous tissue, neurons cannot be classified by visual inspection and it is simpler to remember the major locations of these structural types.

Neurons can also be subdivided according to their functional roles (Table 9-1). **Motor** (**efferent**) **neurons** control effector organs such as muscle fibers and exocrine and endocrine glands. **Sensory** (**afferent**) **neurons** are involved in the reception of sensory stimuli from the environment and from within the body.

Interneurons establish relationships among other neurons, forming complex functional networks or circuits (as in the retina). During mammalian evolution the number and complexity of interneurons have increased greatly. Highly developed functions of the nervous system cannot be ascribed to simple circuits of two or three neurons; rather, they depend on complex interactions established by the integrated functions of many neurons.

In the CNS nerve cell bodies are present only in the gray matter; neuronal processes but no cell bodies are found in the white matter. These names refer to the appearance of dissected but unstained tissue. In the PNS cell bodies are found in ganglia and in some sensory regions, such as the olfactory mucosa.

Cell Body (Perikaryon)

The cell body, or **perikaryon,** is the part of the neuron that contains the nucleus and surrounding cytoplasm, exclusive of the cell processes (Figure 9–3). It is primarily a trophic center, although most neurons perikarya also receive a great number of nerve endings that convey excitatory or inhibitory stimuli generated in other nerve cells.

Most nerve cells have a spherical, unusually large, euchromatic (pale-staining) nucleus with a prominent nucleolus. Binuclear nerve cells are sometimes seen in sympathetic and sensory ganglia. The chromatin is finely dispersed, reflecting the intense synthetic activity of these cells.

Cell bodies often contain a highly developed rough ER organized into aggregates of parallel cisternae. In the cytoplasm between the cisternae are numerous polyribosomes, suggesting

9 Nerve Tissue & the Nervous System

The human nervous system is by far the most complex system in the body histologically and physiologically and is formed by a network of many billion nerve cells (**neurons**), all assisted by many more supporting **glial cells**. Each neuron has hundreds of interconnections with other neurons, forming a very complex system for processing information and generating responses.

Nerve tissue is distributed throughout the body as an integrated communications network. Anatomists divide the nervous system into the following:

- **Central nervous system (CNS)**, consisting of the brain and spinal cord
- **Peripheral nervous system (PNS)**, composed of the cranial, spinal, and peripheral nerves conducting impulses to and from the CNS (motor and sensory nerves respectively) and **ganglia** which are small groups of nerve cells outside the CNS (Figure 9–1; Table 9–1).

Both central and peripheral nerve tissue consists of two cell types: **nerve cells,** or **neurons,** which usually show numerous long processes; and various **glial cells** (Gr. *glia,* glue), which have short processes, support and protect neurons, and participate in neural activity, neural nutrition, and defense of cells in the central nervous system.

Neurons respond to environmental changes (**stimuli**) by altering the ionic gradient that exists between the inner and outer surfaces of their membranes. All cells maintain such a gradient, also called an electrical potential, but cells that can rapidly change this potential in response to stimuli (eg, neurons, muscle cells, some gland cells) are said to be **excitable** or **irritable.** Neurons react promptly to stimuli with a reversal of the ionic gradient (**membrane depolarization**) that generally spreads from the place that received the stimulus and is propagated across the neuron's entire plasma membrane. This propagation, called the **action potential**, the **depolarization wave**, or the **nerve impulse,** is capable of traveling long distances along neuronal processes, transmitting such signals to other neurons, muscles, and glands.

By creating, analyzing, identifying, and integrating information in such signals, the nervous system continuously stabilizes the intrinsic conditions of the body (eg, blood pressure, O_2 and CO_2 content, pH, blood glucose levels, and hormone levels) within normal ranges and maintains behavioral patterns (eg, feeding, reproduction, defense, interaction with other living creatures).

DEVELOPMENT OF NERVE TISSUE

The nervous system develops from the outer embryonic layer, the ectoderm, beginning in the third week of human embryonic life (Figure 9–2). With signals from the notochord, the underlying axial structure, ectoderm along the mid-dorsal side of the embryo thickens to form the epithelial **neural plate**. The lateral sides of this plate fold upward, bend and grow toward each other medially and within a few days fuse to form the **neural tube**. Cells of this tube give rise to the entire CNS, including neurons, most glial cells, ependymal cells, and the epithelial cells of the choroid plexus.

As the folds fuse and the neural tube separates from the now overlying ectoderm that will form epidermis, a large population of important cells called the **neural crest** separates from the neuroepithelium and becomes mesenchymal. Neural crest cells tissue, but is mainly composed of overlapping laminae of fibrocartilage in which collagen bundles are orthogonally arranged in adjacent layers. The multiple lamellae, with the 90-degree registration of type I collagen fibers in adjacent layers, provide the disk with unusual resilience that enables it to withstand the pressures generated by the impinging vertebrae.

The **nucleus pulposus** is situated in the center of the annulus fibrosus. It may contain a few scattered cells derived from the embryonic notochord, but is largely composed of viscous, gel-like matrix rich in hyaluronic acid and fibers of type II collagen. The nucleus pulposus is large in children, but gradually becomes smaller with age and is partially replaced by fibrocartilage. The nucleus pulposus allows each intervertebral disk to function as a shock absorber within the spinal column.

MEDICAL APPLICATION

Rupture of the annulus fibrosus, which most frequently occurs in the posterior region where there are fewer collagen bundles, results in expulsion of the nucleus pulposus and a concomitant flattening of the disk. As a consequence, the disk frequently dislocates or slips from its position between the vertebrae. If it moves toward the spinal cord, it can compress the nerves and result in severe pain and neurologic disturbances. The pain accompanying a slipped disk may be perceived in areas innervated by the compressed nerve fibers—usually the lower lumbar region.

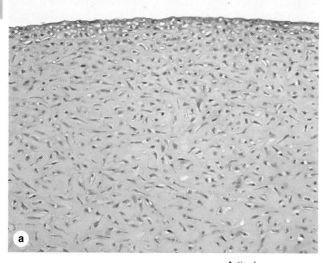

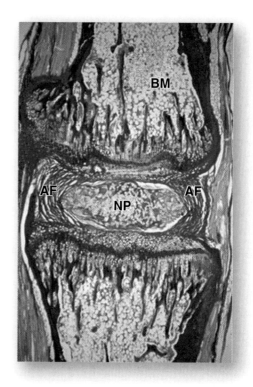

Figure 8–22. Intervertebral disk. Section of a rat tail showing a small intervertebral disk and two adjacent vertebrae. The disk consists of concentric layers of fibrocartilage, comprising the **annulus fibrosus**, which surrounds the **nucleus pulposus**. The nucleus pulposus contains scattered residual cells of the embryonic notochord embedded in abundant gel-like matrix. In humans the intervertebral disks have similar components but function primarily as shock absorbers within the spinal column, as well as making possible greater mobility of the spinal column. X40. PSH.

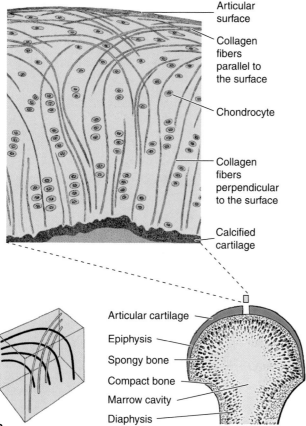

Articular surface

Collagen fibers parallel to the surface

Chondrocyte

Collagen fibers perpendicular to the surface

Calcified cartilage

Articular cartilage
Epiphysis
Spongy bone
Compact bone
Marrow cavity
Diaphysis

b

Figure 8–21. Articular cartilage. (a): Articular surfaces of a diarthrosis are made of hyaline cartilage that lacks the usual perichondrium covering. X40. H&E. **(b):** The upper drawing shows that in such cartilage, collagen fibers are first perpendicular and then bend gradually, forming a broad arch parallel to the cartilage surface. More deeply located chondrocytes are arranged in vertical rows. Superficially placed chondrocytes are flattened and are no longer organized in aggregates. The lower left drawing shows a more three-dimensional view of collagen fibers in articular cartilage. Proteoglycan aggregates bound to hyaluronic acid and collagen fill the space among the collagen fiber network and bind a large amount of water.

The large assembly of GAGs functions as a biomechanical spring in articular cartilage. When pressure is applied, water is forced out of the cartilage matrix into the synovial fluid. When water is expelled, electrostatic repulsion of the negatively charged carboxyl and sulfate groups in the GAGs occurs, separating the GAGs again and thus creating spaces for the return of water. When pressure is released, water is attracted back into the interstices of the GAG matrix. These water movements are brought about constantly by using the joint and are essential for nutrition of the cartilage and for facilitating the interchange of O_2, CO_2, and other molecules between the synovial fluid and the articular cartilage.

skull bones, which, in children and young adults, are united only by dense connective tissue

- **Synchondroses**, in which the bones are joined by hyaline cartilage. The epiphyseal plates of growing bones are one example and in adults a synchondrosis unites the first rib to the sternum, with little movement
- **Syndesmoses**, in which bones are joined by an interosseous ligament of dense connective tissue or fibrocartilage (eg, the pubic symphysis, Figure 7–1), again with very limited movement.

Diarthroses (Figure 8–19) are joints that generally unite long bones and have great mobility, such as the elbow and knee joints. In a diarthrosis, ligaments and a capsule of dense connective tissue maintain proper alignment of the bones. The capsule encloses a sealed **joint cavity** that contains **synovial fluid,** a colorless, transparent, viscous fluid. The joint cavity is not lined by epithelium, but by a specialized connective tissue called the **synovial membrane** which extends folds and villi into the cavity and secretes the lubricant synovial fluid. Synovial fluid is derived from blood plasma, but with a high concentration of hyaluronic acid produced by cells of the synovial membrane.

The synovial membrane or layer may have prominent regions with various types of connective tissue (areolar, fibrous, or adipose) in different diarthrotic joints. At the surface contacting the synovial fluid the tissue is usually well-vascularized, with many porous (fenestrated) capillaries, and contains two specialized cells (synoviocytes) with distinctly different functions and origins (Figure 8–20). Rounded synoviocytes in contact with the synovial cavity are phagocytic and remove wear-and-tear debris from the synovial fluid. Among the capillaries are many more fibroblastic synoviocytes specialized to produce the long, nonsulfated glycosaminoglycan (GAG) hyaluronic acid and secrete other components of ground substance. These GAGs along with plasma from the capillaries leave the synovial membrane, oozing into the synovial fluid. This viscous, gel-like fluid lubricates the joint, reducing friction on all internal surfaces, and supplies nutrients and oxygen to the articular cartilage.

The collagen fibers of the hyaline articular cartilage are disposed as arches with their tops at the exposed surface, which unlike most cartilage is not covered by perichondrium (Figure 8–21). This arrangement of collagen helps to distribute more evenly the forces generated by pressure on joints. The resilient articular cartilage is also an efficient absorber of the intermittent mechanical pressures to which many joints are subjected.

A similar mechanism is seen in **intervertebral disks** (Figure 8–22) which are thick disks of fibrocartilage between the articular surfaces of successive bony vertebrae. The **annulus fibrosus** of each disk has an external layer of dense connective

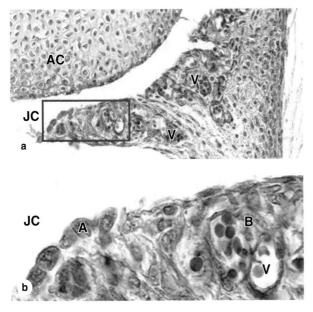
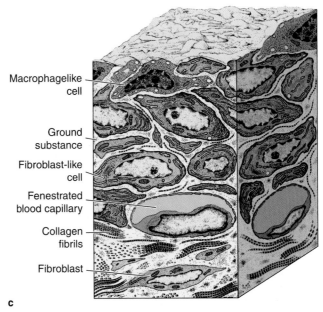

Figure 8–20. **Synovial membrane.** The synovial membrane is a specialized connective tissue that lines capsules of synovial joints and contacts the synovial fluid lubricant, which it is primarily responsible for maintaining. **(a):** Micrograph showing this membrane projects folds into the joint cavity (*JC*) and is highly vascularized (V). The joint cavity surrounds the articular cartilage (AC). X100. Mallory trichrome. **(b):** Higher magnification of the fold showing a high density of capillaries and two specialized types of cells called synoviocytes. Contacting the synovial fluid at the tissue surface are many rounded **macrophage-like synoviocytes (type A)** derived from blood monocytes. These cells bind, engulf, and remove tissue debris from synovial fluid. These cells often form a layer at the tissue surface (A) and can superficially resemble an epithelium, but there is no basal lamina and the cells are not joined together by cell junctions. **Fibroblast-like (type B) synoviocytes** (B) are mesenchymally derived and specialized for synthesis of hyaluronic acid and other components of ground substance. Together with plasma leaking from the many porous fenestrated capillaries, this GAG-rich ground substance enters the synovial fluid, replenishing it. X400. **(c):** Schematic representation of ultrastructural features of the synovial membrane. Among the macrophage-like and fibroblast-like synoviocytes, are collagen fibers and other typical components of connective tissue. There is no basal lamina or other ultrastructural features of epithelium, despite the superficial resemblance. Blood capillaries are of the fenestrated type, which facilitates exchange of substances between blood and synovial fluid.

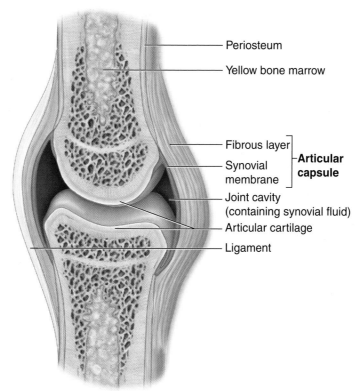

- Periosteum
- Yellow bone marrow

- Fibrous layer ⎤
- Synovial membrane ⎦ **Articular capsule**
- Joint cavity (containing synovial fluid)
- Articular cartilage
- Ligament

a Typical synovial joint

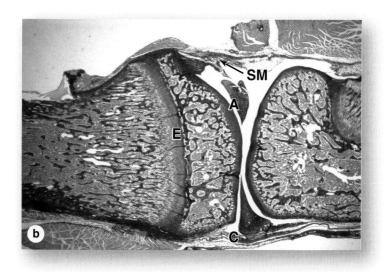

Figure 8–19. Diarthroses, or synovial joints. Diarthroses are joints that allow free movement of the attached bones, such as knuckles, knees, and elbows. **(a):** Diagram shows the components of a diarthrosis that include (1) a capsule continuous with a covering ligament which inserts into the periosteum of both bones, (2) a synovial or joint cavity lined by synovial membrane and containing synovial fluid as a lubricant, and (3) the ends of epiphyses covered by articular cartilage. **(b):** Longitudinal section through a diarthrosis of growing long bones shows the position near the boundaries of the capsule (C) of the epiphyseal growth plates (E) where endochondral ossification occurs. Also shown here are the articular cartilage (A) and the folds of synovial membrane (SM) which extend prominently into the joint cavity from connective tissue of the capsule for production of synovial fluid. X10. PSH stain.

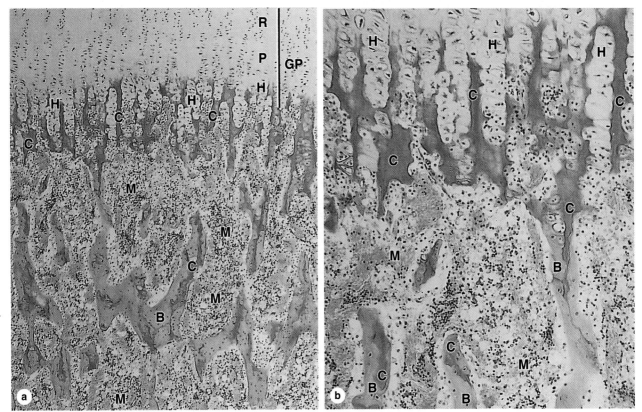

Figure 8–17. Cells and matrices of the epiphyseal growth plate. (a): At the top of the micrograph the growth plate (GP) shows its zones of hyaline cartilage with cells undergoing rest (R), proliferation (P), and hypertrophy (H). As the chondrocytes swell and degenerate they release phosphatase, activities which compress the matrix and cause an initial deposition of $CaPO_4$. This produces calcified spicules (C) in the former cartilage matrix. The tunnel-like lacunae in which the chondrocytes have undergone apoptosis are invaded from the diaphysis by large, thin-walled blood vessels which begin to convert these spaces into marrow (M) cavities. Endosteum with osteoblasts also moves in from the diaphyseal primary ossification center and these cells cover the spicules of calcified cartilage and lay down layers of osteoid, forming a supportive matrix that is now largely primary woven bone (B). X40. H&E. **(b):** Higher magnification shows more detail of the cells and matrix spicules in the zones undergoing hypertrophy (H) and ossification. Staining properties of the matrix clearly change in this process: first when it is compressed and begins to calcify (C), and then when osteoid and bone (B) are laid down. The large spaces between the ossifying matrix spicules become the marrow cavity (M), in which sinuses of eosinophilic red blood cells and aggregates of basophilic white blood cell precursors can be distinguished. The marrow is the major site of blood cell formation in adults. X100. H&E.

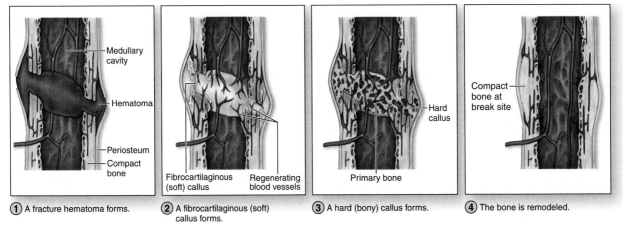

① A fracture hematoma forms. ② A fibrocartilaginous (soft) callus forms. ③ A hard (bony) callus forms. ④ The bone is remodeled.

Figure 8–18. Main features of bone fracture repair. Repair of a fractured bone occurs through several stages, but utilizes mechanisms already in place for bone remodeling. **(1):** Blood vessels torn within the fracture release blood which clots to produce a large fracture hematoma. **(2):** This is gradually removed by macrophages and replaced by a soft fibrocartilage-like mass of procallus tissue rich in collagen and fibroblasts. If broken, the periosteum re-establishes continuity over this tissue. **(3):** This soft procallus is invaded by regrowing blood vessels and osteoblasts. In the next few weeks the fibrocartilage is gradually replaced by trabeculae of primary bone, forming a hard callus throughout the original area of fracture. **(4):** The primary bone is then remodeled as compact and cancellous bone in continuity with the adjacent uninjured areas and fully functional vasculature is re-established.

found in immobilized patients and in post-menopausal women, is an imbalance in skeletal turnover so that bone resorption exceeds bone formation.

Hormones Acting on Bone Tissue

In addition to PTH and calcitonin, several other hormones act on bone. The anterior lobe of the pituitary synthesizes growth hormone (GH or somatotropin), which stimulates the liver to produce insulin-like growth factor-1 (IGF-1 or somatomedin). IGF has an overall growth effect, especially on the epiphyseal cartilage. Consequently, lack of growth hormone during the growing years causes **pituitary dwarfism**; *an excess of growth hormone causes excessive growth of the long bones, resulting in* **gigantism***. Adult bones cannot increase in length even with excess IGF because they lack epiphyseal cartilage, but they do increase in width by periosteal growth. In adults, an increase in GH causes* **acromegaly***, a disease in which the bones—mainly the long ones—become very thick.*

The sex hormones, both male (androgens) and female (estrogens), have a complex effect on bones and are, in a general way, stimulators of bone formation. They influence the time of appearance and development of ossification centers and accelerate the closure of epiphyses.

Bone Tumors

Cancer originating directly from bone cells is fairly uncommon (0.5% of all cancer deaths) but a form called **osteosarcoma** *can arise in osteoblasts. The skeleton is often the site of metastases from tumors originating from malignancies in other organs, most commonly from breast, lung, prostate, kidney, and thyroid tumors.*

JOINTS

Joints are regions where bones are capped and surrounded by connective tissues that firmly hold the bones together and determine the type and degree of movement between them. Joints may be classified as **diarthroses**, which permit free bone movement, and **synarthroses** (Gr. *syn*, together, + *arthrosis*, articulation), in which very limited or no movement occurs. There are three types of synarthroses, based on the type of tissue uniting the bone surfaces:

- **Synostoses**, in which bones are united by bone tissue and no movement takes place. In older adults, synostoses unite the

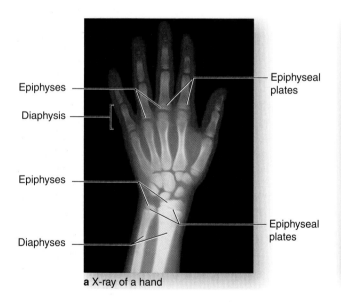
a X-ray of a hand

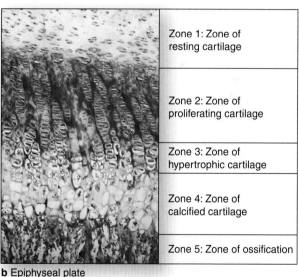
b Epiphyseal plate

Zone 1: Zone of resting cartilage

Zone 2: Zone of proliferating cartilage

Zone 3: Zone of hypertrophic cartilage

Zone 4: Zone of calcified cartilage

Zone 5: Zone of ossification

Epiphyses — Diaphysis — Epiphyses — Diaphyses — Epiphyseal plates

Figure 8–16. Epiphyseal growth plate: locations and zones of activity. The large and growing primary ossification center in long bone diaphyses and the secondary ossification centers in epiphyses are separated in each developing bone by a plate of cartilage called the epiphyseal plate. **(a):** Epiphyseal plates can be identified in an x-ray of a child's hand as marrow regions of lower density between the denser ossification centers. Cells in epiphyseal growth plates are responsible for continued elongation of bones until the body's full size is reached. Developmental activities in the epiphyseal growth plate occur in overlapping zones with distinct histological appearances. **(b):** Moving from the epiphysis to the diaphysis, these zones include cells specialized for the following: (1) normally appearing hyaline cartilage, (2) cartilage with proliferating chondroblasts aligned in lacunae as axial aggregates, (3) degenerating cartilage in which the aligned cells are hypertrophic and the matrix condensed, (4) an area in which the chondrocytes have disappeared and the matrix is undergoing calcification, and (5) a zone in which blood vessels and osteoblasts have invaded the lacunae of the old cartilage, producing marrow cavities and osteoid for new bone. X100. H&E.

METABOLIC ROLE OF BONE

Calcium ions are required for the activity of many enzymes and other proteins mediating cell adhesion, cytoskeletal movements, exocytosis, membrane permeability, and other functions in cells throughout the body. The skeleton serves as the calcium reservoir and contains 99% of the body's total calcium in crystals of hydroxyapatite. The concentration of calcium in the blood and tissues is generally quite stable because of a continuous interchange between blood calcium and bone calcium.

The principal mechanism for raising blood calcium levels is the mobilization of ions from hydroxyapatite crystals to interstitial fluid. This takes place mainly in cancellous bone. The younger, more lightly calcified lamellae that exist even in adult bone (because of continuous remodeling) receive and lose calcium more readily. These lamellae are more important for the maintenance of calcium concentration in the blood than are the older, more densely calcified lamellae, whose role is mainly that of support and protection.

The action of two key hormones on cells in bone regulates the process of calcium mobilization from hydroxyapatite. **Parathyroid hormone** (PTH) from the parathyroid glands raises low blood calcium levels. The principal target cells of this polypeptide are osteoblasts, which stop production of osteoid and matrix vesicles and instead secrete a paracrine protein, osteoclast stimulating factor. This factor promotes osteoclastic resorption of the bone matrix, liberating calcium. Osteoclast activity is inhibited by another hormone, **calcitonin,** which is synthesized by the parafollicular cells of the thyroid gland. This slows matrix resorption and thereby gradually lowers blood calcium levels.

MEDICAL APPLICATION

Because the concentration of calcium in tissues and blood must be kept constant, nutritional deficiency of calcium results in decalcification of bones. Severely decalcified bones are more likely to fracture.

Decalcification of bone may also be caused by excessive production of PTH (hyperparathyroidism), which can cause increased osteoclastic activity, intense resorption of bone, elevation of blood Ca^{2+} and PO^{3-}_4 levels, and abnormal deposits of calcium in the kidneys and arterial walls.

The opposite occurs in **osteopetrosis** *(L. petra, stone), a disease caused by defective osteoclast function that results in overgrowth, thickening, and hardening of bones. This process can obliterate the bone marrow cavities, depressing blood cell formation and causing anemia and the loss of white blood cells.*

Nutritional Deficiencies and Bone Remodeling

Especially during growth, bone is sensitive to nutritional factors. Calcium deficiency, which leads to incomplete calcification of the organic bone matrix, can be due either to a lack of calcium in the diet or a failure to produce the steroid prohormone vitamin D, which is important for the absorption of Ca^{2+} and PO^{3-}_4 by the small intestine.

Calcium deficiency in children causes **rickets,** *a disease in which the bone matrix does not calcify normally and the epiphyseal plate becomes distorted by the normal strains of body weight and muscular activity. Ossification processes at this level are consequently hindered, and the bones not only grow more slowly but also become deformed.*

Calcium deficiency in adults gives rise to **osteomalacia** *(osteon + Gr. malakia, softness), which is characterized by deficient calcification of recently formed bone and partial decalcification of already calcified matrix. Osteomalacia should not be confused with* **osteoporosis.** *In osteomalacia, there is a decrease in the amount of calcium per unit of bone matrix. Osteoporosis, frequently*

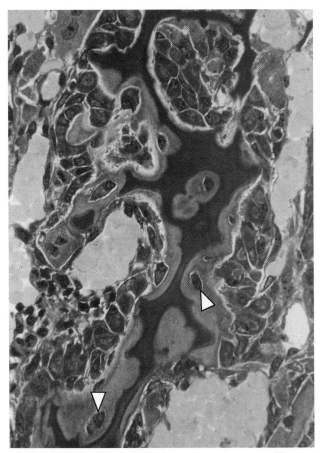

Figure 8–15. Cells and matrices of a primary ossification center. A small region of a primary ossification center showing key features of endochondral ossification. Compressed remnants of calcified cartilage matrix (dark purple), now devoid of chondrocytes, are enclosed by more lightly stained osteoid or bone matrix. This newly formed bone is surrounded by a layer of large, active osteoblasts. Some osteoblasts that were captured by the matrix have become smaller osteocytes (arrowheads). X200. Pararosaniline–toluidine blue.

has an excellent capacity for repair and regeneration. Bone fractures and other damage are repaired efficiently using cells and processes already active in bone remodeling. Surgically created gaps in bone can be filled with new bone, especially when periosteal tissue remains nearby.

MEDICAL APPLICATION

When a bone is fractured, blood vessels are disrupted and bone cells adjoining the fracture die. The damaged blood vessels produce a localized hemorrhage and form a blood clot.

Soon the blood clot is removed by macrophages and the adjacent matrix of bone is resorbed by osteoclasts. The periosteum and the endosteum at the site of the fracture respond with intense proliferation producing a soft callus of fibrocartilage-like tissue that surrounds the fracture and covers the extremities of the fractured bone (Figure 8–18).

Primary bone is then formed by a combination of endochondral and intramembranous ossification. Further repair produces irregularly formed trabeculae of primary bone that temporarily unite the extremities of the fractured bone, forming a hard bone callus (Figure 8–18).

Stresses imposed on the bone during repair and during the patient's gradual return to activity serve to remodel the bone callus. The primary bone of the callus is gradually resorbed and replaced by secondary bone, remodeling and restoring the original bone structure. Unlike other connective tissues, bone tissue heals without forming a scar.

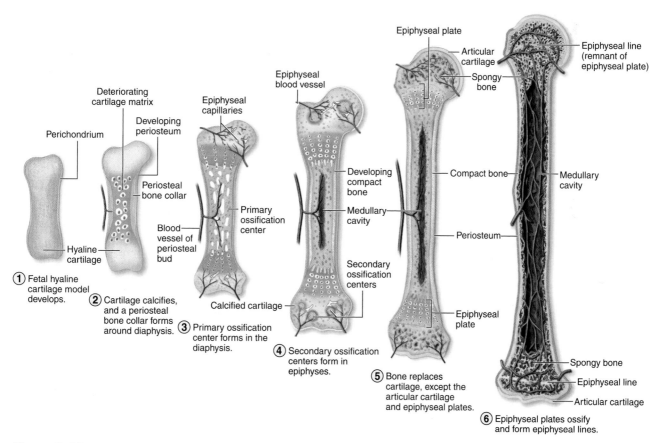

Figure 8–14. Osteogenesis of long bones by endochondral ossification. Endochondral ossification forms most bones of the skeleton and occurs in the fetus in models made of hyaline cartilage **(1)**. The process takes many weeks and major developmental stages include: formation of a bone collar around the middle of the cartilage model and degeneration of the underlying cartilage **(2)**, followed by invasion of the resulting ossification center by capillaries and osteoprogenitor cells from the periosteum **(3)**, osteoid deposition by the new osteoblasts, calcification of woven bone, and its remodeling as compact bone **(4)**. This **primary ossification center** develops in the diaphysis, along the middle of each developing bone. **Secondary ossification centers** develop somewhat later by a similar process in the epiphyses. The primary and secondary ossification centers are separated by the **epiphyseal plate (5)** which provides for continued bone elongation. The two ossification centers do not merge until the epiphyseal plate disappears **(6)** when full stature is achieved.

closed. Once the epiphyses have closed, growth in length of bones becomes impossible, although bone widening may still occur.

A plate of epiphyseal cartilage is divided into five zones (Figure 8–16), starting from the epiphyseal side of cartilage:

1. The **resting zone** consists of hyaline cartilage with typical chondrocytes.

2. In the **proliferative zone,** chondrocytes begin to divide rapidly and form columns of stacked cells parallel to the long axis of the bone.

3. The **hypertrophic cartilage zone** contains swollen chondrocytes whose cytoplasm has accumulated glycogen. Hypertrophy compresses the matrix into thin septa between the chondrocytes.

4. In the **calcified cartilage zone,** loss of the chondrocytes by apoptosis is accompanied by calcification of the septa of cartilage matrix by the formation of hydroxyapatite crystals (Figure 8–17).

5. In the **ossification zone,** bone tissue first appears. Capillaries and osteoprogenitor cells originating from the periosteum invade the cavities left by the chondrocytes. Many of these cavities will be merged and become the marrow cavity. The osteoprogenitor cells form osteoblasts, which settle in a discontinuous layer over the septa of calcified cartilage matrix. The osteoblasts deposit osteoid over the spicules of calcified cartilage matrix, forming woven bone (Figure 8–17).

In summary, growth in length of a long bone occurs by proliferation of chondrocytes in the epiphyseal plate adjacent to the epiphysis. At the same time, chondrocytes in the diaphyseal side of the plate hypertrophy, their matrix becomes calcified, and the cells die. Osteoblasts lay down a layer of primary bone on the calcified cartilage matrix. Because the rates of these two opposing events (proliferation and destruction) are approximately equal,

the epiphyseal plate does not change thickness. Instead, it is displaced away from the middle of the diaphysis, resulting in growth in length of the bone.

BONE GROWTH, REMODELING, & REPAIR

Bone growth is generally associated with partial resorption of preformed tissue and the simultaneous laying down of new bone (exceeding the rate of bone loss). This process permits the shape of the bone to be maintained as it grows. The rate of bone remodeling (**bone turnover**) is very active in young children, where it can be 200 times faster than that in adults. Bone remodeling in adults is a dynamic physiologic process that occurs simultaneously in multiple locations of the skeleton and is not always related to bone growth.

Despite its hardness, the constant remodeling makes bone very plastic and capable of internal structural changes according to the various stresses to which it is subjected. A well-known example of bone plasticity is the ability of the positions of teeth in the jawbone to be modified by the lateral pressures produced by orthodontic appliances. Bone is formed on the side where traction is applied and is resorbed on the opposite side where pressure is exerted. In this way, teeth are moved within the jaw while the bone is being remodeled.

Cranial bones grow mainly because of the formation of bone tissue by the periosteum between the sutures and on the external bone surface. At the same time, resorption takes place on the internal surface. The plasticity of bone allows it to respond to the growth of the brain and form a skull of adequate size. The skull will be small if the brain does not develop completely and will be larger than normal in a person suffering from hydrocephalus, a disorder characterized by abnormal accumulation of spinal fluid and dilatation of the cerebral ventricles.

Because it contains osteoprogenitor stem cells throughout the endosteum and periosteum and has an extensive blood supply, bone

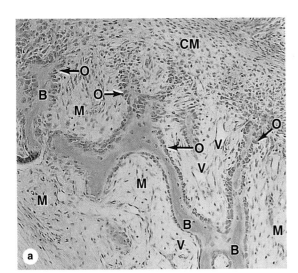
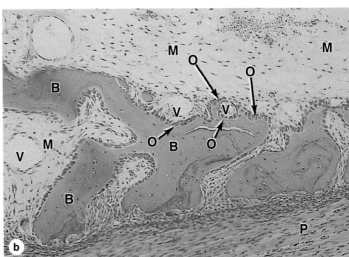

Figure 8–13. Intramembranous ossification. A section of jaw from a fetal pig undergoing intramembranous ossification. **(a):** Areas of typical mesenchyme (M), condensed mesenchyme (CM) adjacent to aggregates of new osteoblasts (O). Some osteoblasts have secreted matrices of bone (B) which remain covered by osteoblasts. Between these trabeculae of newly formed primary bone are vascularized areas (V) that will form marrow cavities. X40. H&E. **(b):** Higher magnification shows the developing periosteum (P) that covers masses primary bone that will soon merge to form a continuous plate of bone. The larger mesenchyme-filled region at the top is the developing marrow cavity. X100. H&E.

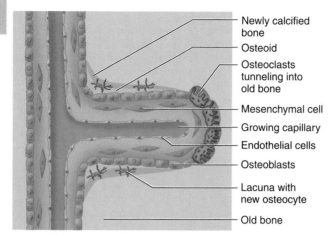

Newly calcified bone
Osteoid
Osteoclasts tunneling into old bone
Mesenchymal cell
Growing capillary
Endothelial cells
Osteoblasts
Lacuna with new osteocyte
Old bone

Figure 8–11. Development of osteons. During remodeling of compact bone, a group of osteoclasts acts as a boring cone to make a tunnel into bone matrix. Behind these cells a population of osteoblasts enters the tunnel and lines its walls. As the osteoblasts secrete osteoid in a cyclic manner, they produce layers of new matrix with cells trapped in lacunae. The cells in lacunae are now osteocytes. The tunnel becomes constricted with multiple concentric layers of new matrix and its lumen finally exists as only a narrow central canal with small blood vessels.

to the porous central region. Next, osteoblasts adhere to the calcified cartilage matrix and produce continuous layers of primary bone that surround the cartilaginous matrix remnants. At this stage, the calcified cartilage appears basophilic, and the primary bone is eosinophilic (Figure 8–15).

This process in the diaphysis forms the **primary ossification center** (Figure 8–14). **Secondary ossification centers** appear slightly later at the epiphyses of the cartilage model and develop in a similar manner. During their expansion and remodeling, the primary and secondary ossification centers produce cavities that are gradually filled with bone marrow.

In the secondary ossification centers, cartilage remains in two regions: the **articular cartilage** (Figure 8–14), which persists throughout adult life and does not contribute to bone growth in length, and the **epiphyseal cartilage** (also called **epiphyseal plate** or **growth plate**), which connects each epiphysis to the diaphysis (Figures 8–16 and 8–17). The epiphyseal cartilage is responsible for the growth in length of the bone and disappears in adults, which is why bone growth ceases in adulthood. Elimination of the epiphyseal plates ("epiphyseal closure") occurs at different times with different bones and is complete in all bones by about age twenty. In forensics or through X-ray examination of the growing skeleton, it is possible to determine the "bone age" of a young person, noting which epiphyses are open and which are

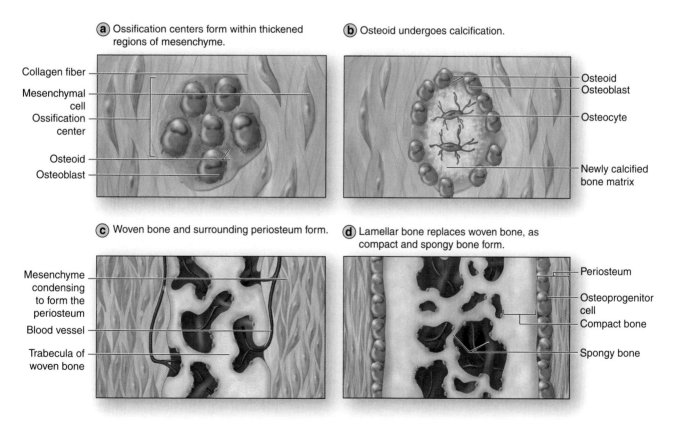

(a) Ossification centers form within thickened regions of mesenchyme.

Collagen fiber
Mesenchymal cell
Ossification center
Osteoid
Osteoblast

(b) Osteoid undergoes calcification.

Osteoid
Osteoblast
Osteocyte
Newly calcified bone matrix

(c) Woven bone and surrounding periosteum form.

Mesenchyme condensing to form the periosteum
Blood vessel
Trabecula of woven bone

(d) Lamellar bone replaces woven bone, as compact and spongy bone form.

Periosteum
Osteoprogenitor cell
Compact bone
Spongy bone

Figure 8–12. Intramembranous ossification. Developmental process by which most bones of the skull are formed. **(a):** Groups of mesenchymal cells in a "membrane" or sheet of this embryonic tissue, round up and differentiate as osteoblasts producing osteoid. **(b):** Cells trapped in the calcifying matrix differentiate as osteocytes. **(c):** Woven bone is produced in this manner, with vascularized internal spaces that will form the marrow cavity and surrounded on both sides by developing periosteum. **(d):** Remodeling of the woven bone produces the two layers of compact lamellar bone with cancellous bone in between, which is characteristic of these flat bones.

- **Endochondral ossification**, in which the matrix of preexisting hyaline cartilage is eroded and replaced by osteoblasts producing osteoid.

In both processes, the bone tissue that appears first is primary or woven. Primary bone is a temporary and is soon replaced by the definitive secondary lamellar bone. During bone growth, areas of primary bone, areas of resorption, and areas of secondary bone all appear side by side.

Intramembranous Ossification

Intramembranous ossification, by which most flat bones are produced, is so called because it takes place within condensations of embryonic mesenchymal tissue. The frontal and parietal bones of the skull—as well as parts of the occipital and temporal bones and the mandible and maxilla—are formed by intramembranous ossification. The process is summarized in Figure 8–12.

In the mesenchymal condensation layer or "membrane," the starting point for bone formation is called an **ossification center.** The process begins when groups of mesenchymal cells differentiate into osteoblasts. Osteoblasts produce osteoid matrix and calcification follows, resulting in the encapsulation of some osteoblasts, which then become osteocytes. These islands of developing bone form walls that delineate elongated cavities containing capillaries, bone marrow cells, and undifferentiated cells. Several such groups arise almost simultaneously at the ossification center, and their fusion between the walls gives the bone a spongy appearance. The connective tissue that remains among the bone walls is penetrated by growing blood vessels and additional undifferentiated mesenchymal cells, giving rise to the bone marrow. The ossification centers of a bone grow radially and finally fuse together, replacing the original connective tissue (Figures 8–12 and 8–13).

In cranial flat bones there is a marked predominance of bone formation over bone resorption at both the internal and external surfaces. Thus, two layers of compact bone (internal and external plates) arise, while the central portion (diploë) maintains its spongy nature. The fontanelles or "soft spots" on the heads of newborn infants are areas in the skull that correspond to parts of the connective tissue that are not yet ossified. The portions of the connective tissue layer that do not undergo ossification give rise to the endosteum and the periosteum of the new bone.

Endochondral Ossification

Endochondral (Gr. *endon*, within, + *chondros*, cartilage) ossification takes place within a piece of hyaline cartilage whose shape resembles a small version, or model, of the bone to be formed. This type of ossification is principally responsible for the formation of short and long bones.

Endochondral ossification of a long bone consists of the sequence of events shown schematically in Figure 8–14. Initially, the first bone tissue appears as a collar surrounding the diaphysis of the cartilage model. This **bone collar** is produced by local osteoblast activity within the surrounding perichondrium. The collar now impedes diffusion of oxygen and nutrients into the underlying cartilage, promoting degenerative changes there. The chondrocytes begin to produce alkaline phosphatase and swell up (hypertrophy), enlarging their lacunae. These changes both compress the matrix into narrower trabeculae and lead to calcification in these structures. Death of the chondrocytes results in a porous three-dimensional structure formed by the remnants of the calcified cartilage matrix (Figure 8–15). Blood vessels from the former perichondrium now the periosteum penetrate through the bone collar previously perforated by osteoclasts, bringing osteoprogenitor cells

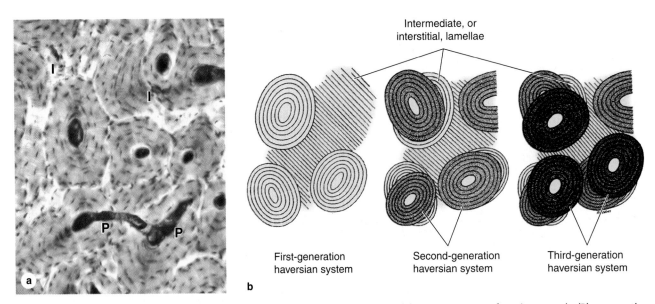

Intermediate, or interstitial, lamellae

First-generation haversian system

Second-generation haversian system

Third-generation haversian system

a

b

Figure 8–10. **Lamellar bone: Perforating canals and interstitial lamellae. (a):** Transverse perforating canals (P) connecting adjacent osteons are shown at the left side of the micrograph. Such canals "perforate" lamellae and provide another source of microvasculature for the central canals of osteons. Among the intact osteons are also found remnants of eroded osteons, seen as irregular interstitial or intermediate lamellae (I). X100. **(b):** Schematic diagram shows remodeling of compact lamellar bone showing three generations of osteonic haversian systems and their successive contributions to the formation of interstitial lamellae. Remodeling is a continuous process that involves the coordinated activity of osteoblasts and osteoclasts, and is responsible for adaptation of bone to changes in stress, especially during the body's growth.

is replaced in adults by secondary bone tissue except in a very few places in the body, eg, near the sutures of the calvaria, in tooth sockets, and in the insertions of some tendons.

In addition to the irregular array of collagen fibers, other characteristics of primary bone tissue are a lower mineral content (it is more easily penetrated by x-rays) and a higher proportion of osteocytes than that in secondary bones.

Secondary Bone Tissue

Secondary bone tissue is the type usually found in adults. It characteristically shows multiple layers of calcified matrix (each 3–7 μm thick) and is often referred to as **lamellar bone.** The lamellae are quite organized, either parallel to each other or concentrically around a vascular canal. Each complex of concentric bony lamellae surrounding a small canal containing blood vessels, nerves,

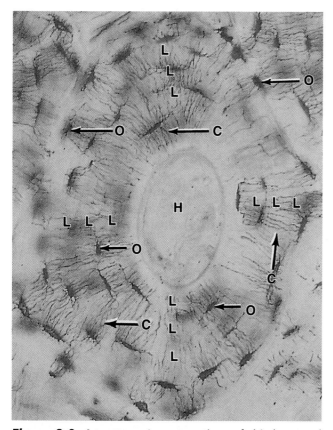

Figure 8–9. An osteon. In preparations of dried, ground bone osteons can be seen with lacunae (L) situated between concentric lamellae and interconnected by fine canaliculi (C). Although it is not apparent by light microscopy, each lamella consists of multiple parallel arrays of collagen fibers. In adjacent lamellae, the collagen fibers are oriented in different directions. The presence of large numbers of lamellae with differing fiber orientations provides the bone with great strength, despite its light weight. Only remnants of the osteocytes (O) in some lacunae and of the osteonic canal's contents are seen in ground bone. In living tissue osteocytic processes connected via gap junctions are present in successive canaliculi, making cells in all the lamellae in communication with the blood vessels in the central canal. X500.

and loose connective tissue is called an **osteon** (formerly known as an **haversian system**) (Figures 8–1 and 8–9). Lacunae with osteocytes are found between the lamellae, interconnected by canaliculi which allow all cells to be in contact with the source of nutrients and oxygen in the osteonic canal (Figure 8–9). The outer boundary of each osteon is a more collagen-rich layer called the **cement line.**

In each lamella, type I collagen fibers are aligned in parallel and follow a helical course. The pitch of the helix is, however, different for different lamellae, so that at any given point, fibers from adjacent lamellae intersect at approximately right angles (Figure 8–1). The specific organization of collagen fibers in successive lamellae of each osteon is highly important for the great strength of secondary bone.

In compact bone (eg, the diaphysis of long bones) besides forming osteons, the lamellae also exhibit a typical organization consisting of multiple **external circumferential lamellae** (Figure 8–1) and often some **inner circumferential lamellae.** Inner circumferential lamellae are located around the marrow cavity and external circumferential lamellae are located immediately beneath the periosteum.

Each osteon is a long, often bifurcated cylinder generally parallel to the long axis of the diaphysis. It consists of a central canal surrounded by 4–10 concentric lamellae. Each endosteum-lined canal contains blood vessels, nerves, and loose connective tissue. The central canals communicate with the marrow cavity and the periosteum and with one another through transverse or oblique **perforating canals** (formerly known as Volkmann canals) (Figures 8–1 and 8–10). The transverse canals do not have concentric lamellae; instead, they perforate the lamellae. All osteonic and perforating canals in bone tissue come into existence when matrix is laid down around preexisting blood vessels.

Among the osteons between the two circumferential systems are numerous irregularly shaped groups of parallel lamellae called **interstitial lamellae.** These structures are lamellae remaining from osteons partially destroyed by osteoclasts during growth and remodeling of bone (Figure 8–10).

Bone **remodeling** is continuous throughout life and involves a combination of bone synthesis and removal. In compact bone, remodeling resorbs parts of old osteons and produces new ones. Resorption involves the actions of osteoclasts, often working in groups to remove old bone in tunnel-like cavities having the approximate diameter of new osteons. Such tunnels are quickly invaded by many osteoprogenitor cells and sprouting loops of blood capillaries, both derived from the endosteum or periosteum. Osteoblasts develop, line the wall of the tunnels, and begin to secrete osteoid in a cyclic manner, forming concentric lamellae of bone with trapped osteocytes (Figure 8–11). In healthy adults 5–10% of the bone turns over annually.

Variations in the remodeling activity produce great variability in the sizes of osteons, osteonic canals, and interstitial lamellae. As osteons form by successive deposition of lamellae by osteoblasts, moving inward from the periphery, younger osteons usually have larger canals. In mature osteons the most recently formed lamella is the one closest to the central canal.

OSTEOGENESIS

Bone can be formed initially by either of two ways:

- **Intramembranous ossification,** in which osteoblasts differentiate directly from mesenchyme and begin secreting osteoid

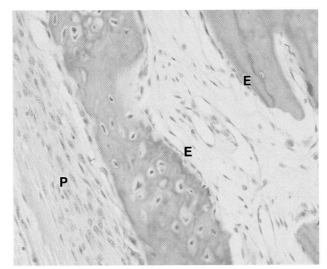

Figure 8–6. Periosteum and endosteum. Section through a thin portion of the wall of a long-bone diaphysis showing both periosteum (P) and endosteum (E). The periosteum covers bone and provides a supply of osteoprogenitor cells which become osteoblasts for new bone formation. These cells are located in the inner, more cellular layers of the periosteum, next to the bone matrix. Externally the periosteum consists of a thick layer of more fibrous, dense connective tissue, which merges with ligaments and other connective tissues. Perforating fibers fastening the periosteum to the bone matrix are not seen in routine light microscope preparations. The periosteum has a good blood supply, but the cavities lined by endosteum, the marrow cavities, are very rich in blood sinuses and blood-forming tissue. X100. H&E.

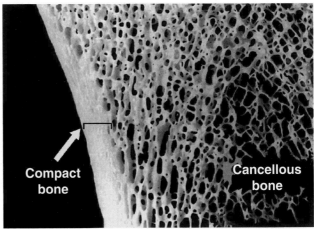

Figure 8–7. Compact and cancellous (spongy or trabecular) bone. Close gross examination of a thick section of dried bone illustrating the cortical compact bone and the lattice of trabeculae in cancellous bone at the bone's interior. In living tissue the compact bone is covered externally with periosteum and all surfaces of cancellous bone are covered with endosteum.

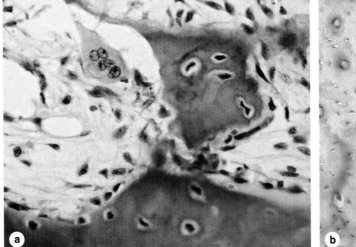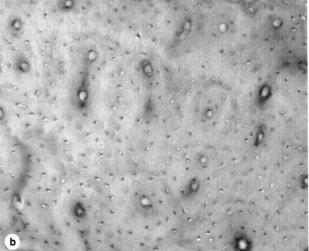

Figure 8–8. Primary (woven) bone and secondary (lamellar) bone. (a): Micrograph of a fractured bone undergoing repair. Primary bone is newly formed, immature bone, rich in osteocytes, with randomly arranged bundles of calcified collagen. Osteoclasts and osteoblasts are numerous in the surrounding endosteum. X200. H&E. **(b):** Secondary or mature bone shows matrix organized as lamellae, seen faintly here as concentric lines surrounding osteonic canals. X100. H&E.

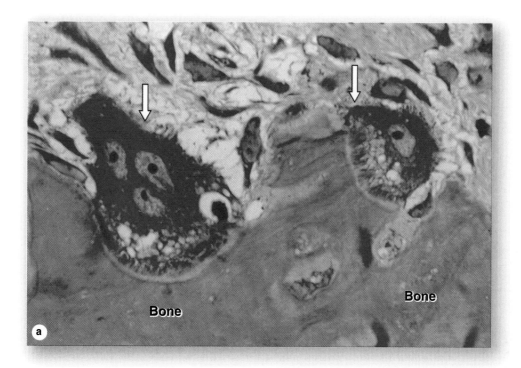

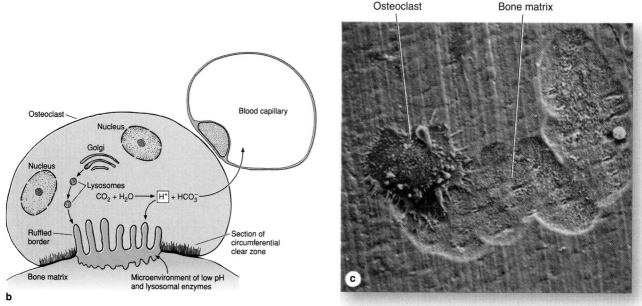

Figure 8–5. **Osteoclasts and their activity.** The osteoclast is a large cell with several nuclei derived by the fusion in bone of several blood-derived monocytes. **(a):** Microscopic section showing two osteoclasts (arrows) digesting or resorbing bone matrix in resorption bays on the matrix surface. X400. H&E. **(b):** Diagram showing each osteoclast has a **circumferential zone** where integrins tightly bind the matrix and surround a **ruffled border** of cytoplasmic projections close to this matrix. The sealed space between the cell and the matrix is acidified by a proton pump localized in the osteoclast membrane and receives hydrolytic enzymes secreted by the cell. It is the place of decalcification and matrix digestion and can be compared to a giant extracellular lysosome. Acidification of this confined space facilitates the dissolution of $CaPO_4$ from bone and creates the optimal pH for activity of the lysosomal hydrolases. Bone matrix is thus resorbed and ions and products of matrix digestion are released for re-use. **(c):** SEM showing an active osteoclast cultured on a flat substrate of bone. A trench is formed on the bone surface as the osteoclast crawls along. X5000. (Figure 8–5c, with permission, from Alan Boyde, Centre for Oral Growth and Development, University of London.)

*the two fluorescent layers is proportional to the rate of bone apposition. This procedure is of diagnostic importance in such diseases as **osteomalacia,** in which mineralization is impaired, and **osteitis fibrosa cystica,** in which increased osteoclast activity results in removal of bone matrix and fibrous degeneration.*

Osteoclasts

Osteoclasts are very large, motile cells with multiple nuclei (Figure 8–5). The large size and multinucleated condition of osteoclasts is due to their origin from the fusion of bone marrow-derived cells. In areas of bone undergoing resorption, osteoclasts lie within enzymatically etched depressions or crypts in the matrix known as **resorption bays** (formerly called **Howship lacunae**).

In active osteoclasts, the surface against the bone matrix is folded into irregular projections, which form a **ruffled border.** Formation of the ruffled borders is related to the activity of osteoclasts. Surrounding the ruffled border is a clear cytoplasmic zone rich in actin filaments which is the site of adhesion to the bone matrix. This circumferential adhesion zone creates a microenvironment between the osteoclast and the matrix in which bone resorption occurs (Figure 8–5).

Into this subcellular pocket the osteoclast secretes collagenase and other enzymes and pumps protons, forming an acidic environment locally for dissolving hydroxyapatite and promoting the localized digestion of collagen. Osteoclast activity is controlled by local signaling factors and hormones. Osteoclasts have receptors for calcitonin, a thyroid hormone, but not for parathyroid hormone. Osteoblasts activated by PTH produce a cytokine called osteoclast stimulating factor. Thus, activity of these two cells is coordinated and both are essential in bone remodeling.

BONE MATRIX

Inorganic material represents about 50% of the dry weight of bone matrix. Hydroxyapatite is most abundant, but bicarbonate, citrate, magnesium, potassium, and sodium are also found. Significant quantities of amorphous (noncrystalline) $CaPO_4$ are also present. The surface ions of hydroxyapatite are hydrated and a layer of water and ions forms around this crystal. This layer, the **hydration shell,** facilitates the exchange of ions between the crystal and the body fluids.

MEDICAL APPLICATION

*In the genetic disease **osteopetrosis,** which is characterized by dense, heavy bones ("marble bones"), the osteoclasts lack ruffled borders, and bone resorption is defective.*

The organic matter embedded in the calcified matrix is type I collagen and ground substance, which contains proteoglycan aggregates and several specific multiadhesive glycoproteins, including **osteonectin.** Calcium-binding glycoproteins, notably osteocalcin, and the phosphatases released in matrix vesicles by osteoblasts promote calcification of the matrix. Other tissues containing type I collagen do not contain these glycoproteins or matrix vesicles and are not normally calcified. Because of its high collagen content, decalcified bone matrix is usually acidophilic.

The association of minerals with collagen fibers is responsible for the hardness and resistance of bone tissue. After a bone is decalcified, its shape is preserved, but it becomes as flexible as a tendon. Removal of the organic part of the matrix—which is mainly collagenous—also leaves the bone with its original shape; however, it becomes fragile, breaking and crumbling easily when handled.

PERIOSTEUM & ENDOSTEUM

External and internal surfaces of bone are covered by layers of bone-forming cells and vascularized connective tissue called periosteum and endosteum.

The **periosteum** consists of a dense fibrous outer layer of collagen bundles and fibroblasts (Figures 8–1 and 8–6). Bundles of periosteal collagen fibers, called **perforating (or Sharpey's) fibers,** penetrate the bone matrix, binding the periosteum to bone. The innermost cellular layer of the periosteum contains mesenchymal stem cells called **osteoprogenitor cells,** with the potential to divide by mitosis and differentiate into osteoblasts. Osteoprogenitor cells play a prominent role in bone growth and repair.

The large internal marrow cavities of bone are lined by **endosteum** (Figures 8–1 and 8-6). Endosteum is a single very thin layer of connective tissue, containing flattened osteoprogenitor cells and osteoblasts, which covers the small spicules or trabeculae of bone that project into these cavities. The endosteum is therefore considerably thinner than the periosteum.

The principal functions of periosteum and endosteum are nutrition of osseous tissue and provision of a continuous supply of new osteoblasts for repair or growth of bone.

TYPES OF BONE

Gross observation of bone in cross section shows dense areas generally without cavities—corresponding to **compact bone**—and areas with numerous interconnecting cavities—corresponding to **cancellous (spongy) bone** (Figure 8–7). Under the microscope, however, both compact bone and the trabeculae separating the cavities of cancellous bone have the same basic histologic structure.

In long bones, the bulbous ends—called **epiphyses** (Gr. *epiphysis,* an excrescence)—are composed of spongy bone covered by a thin layer of compact bone. The cylindrical part—the **diaphysis** (Gr. *diaphysis,* a growing between)—is almost totally composed of compact bone, with a thin component of spongy bone on its inner surface around the bone marrow cavity. Short bones usually have a core of spongy bone surrounded completely by compact bone. The flat bones that form the calvaria (skull-cap) have two layers of compact bone called **plates** (tables), separated by a thicker layer of spongy bone called the **diploë.**

Microscopic examination of bone shows two types: immature **primary bone** and mature **secondary bone** (Figure 8–8).

Primary Bone Tissue

Primary bone is the first bone tissue to appear in embryonic development and in fracture repair. It is characterized by random disposition of fine collagen fibers and is therefore often called **woven bone** (Figure 8–8). Primary bone tissue is usually temporary and among the noncollagen proteins secreted by osteoblasts is the small, vitamin K-dependent polypeptide **osteocalcin**, which together with various glycoproteins binds Ca^{2+} ions and raises their concentration locally. Osteoblasts also release membrane-enclosed vesicles rich in alkaline phosphatase and other enzymes whose activity raises the local concentration of PO_4^- ions. With high concentrations of both ions, these **matrix vesicles** serve as foci for the formation of hydroxyapatite $[Ca_{10}(PO_4)_6(OH)_2]$ crystals, the first visible step in calcification. These crystals grow rapidly by accretion of more mineral and eventually form a confluent mass of calcified material embedding the collagen fibers and proteoglycans.

Osteocytes

Individual osteoblasts are gradually surrounded by their own secretion and become **osteocytes** enclosed singly within spaces called **lacunae**. In the transition from osteoblasts to osteocytes the cells extend many long cytoplasmic processes, which also become surrounded by calcifying matrix. An osteocyte and its processes occupy each lacuna and the canaliculi radiating from it (Figures 8–4 and 8–1).

Processes of adjacent cells make contact via gap junctions, and molecules are passed via these structures from cell to cell. The exchange via gap junctions can provide nourishment for a chain of about ten cells. Some molecular exchange between

osteocytes and blood vessels also takes place through the small amount of extracellular fluid located between osteocytes and the bone matrix.

When compared with osteoblasts, the flat, almond-shaped osteocytes exhibit a significantly reduced RER and Golgi apparatus and more condensed nuclear chromatin (Figure 8–4a). These cells are involved in maintaining the bony matrix and their death is followed by resorption of this matrix.

MEDICAL APPLICATION

The fluorescent antibiotic tetracycline interacts with great affinity with recently deposited mineralized bone matrix. Based on this interaction, a method was developed to measure the rate of bone apposition—an important parameter in the study of bone growth and the diagnosis of bone growth diseases. Tetracycline is administered twice to patients, with an interval of 5 days between injections. A bone biopsy is then performed, and the sections are studied by means of fluorescence microscopy. The distance between

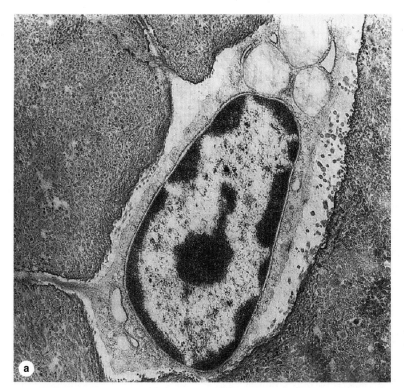 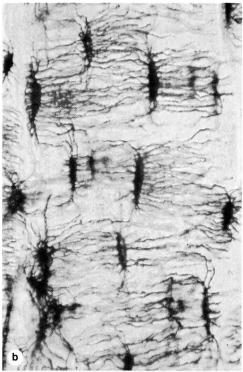

***Figure 8–4.* Osteocytes in lacunae. (a):** TEM section of bone showing an osteocyte with its cytoplasmic processes surrounded by matrix. Such processes are extended as osteoid is being secreted and this material calcifies around the processes giving rise to canaliculi in the bony matrix. The ultrastructure of the cell nucleus and cytoplasm is that of a cell no longer active in matrix synthesis. **(b):** Photomicrograph of bone, not decalcified and sectioned, but dried and ground very thin for demonstration of lacunae and canaliculi, but not cells. The lacunae and canaliculi appear dark and show the communication between these structures through which nutrients derived from blood vessels diffuse and are passed from cell to cell in living bone. X400. Ground bone.

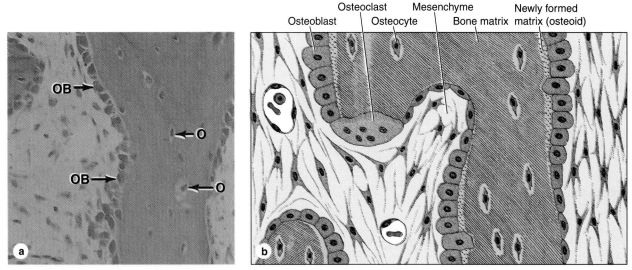

Figure 8–2. Osteoblasts and osteocytes. (a): The photomicrograph of developing bone shows the location and morphological differences between osteoblasts (OB) and osteocytes (O). Rounded osteoblasts, derived from the mesenchymal cells nearby, appear as a simple row of cells adjacent to a very thin layer of lightly stained matrix covering the more heavily stained matrix. The lightly stained matrix is osteoid. Osteocytes are less rounded and located within lacunae. In thin spicules of bone like those seen here, canaliculi are usually not present. X300. H&E. **(b):** Schematic diagram shows the relationship of osteoblasts to osteoid, bone matrix, and osteocytes.

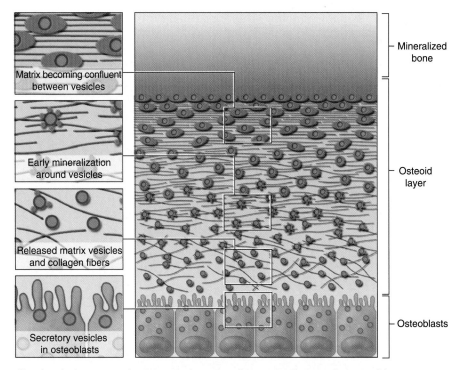

Figure 8–3. Mineralization in bone matrix. From their ends adjacent to the matrix, osteoblasts secrete type I collagen, several glycoproteins, and proteoglycans. Some of these factors, notably osteocalcin and certain glycoproteins, bind Ca^{2+} with high affinity, thus raising the local concentration of these ions. Osteoblasts also release very small membrane-enclosed **matrix vesicles** with which alkaline phosphatase and other enzymes are associated. These enzymes hydrolyze PO_4^- ions from various macromolecules, creating a high concentration of these ions locally. The high ion concentrations cause crystals of $CaPO_4$ to form on the matrix vesicles. The crystals grow and mineralize further with formation of small growing masses of hydroxyapatite [$Ca_{10}(PO_4)_6(OH)_2$] which surround the collagen fibers and all other macromolecules. Eventually the masses of hydroxyapatite merge as a confluent solid bony matrix as calcification of the matrix is completed.

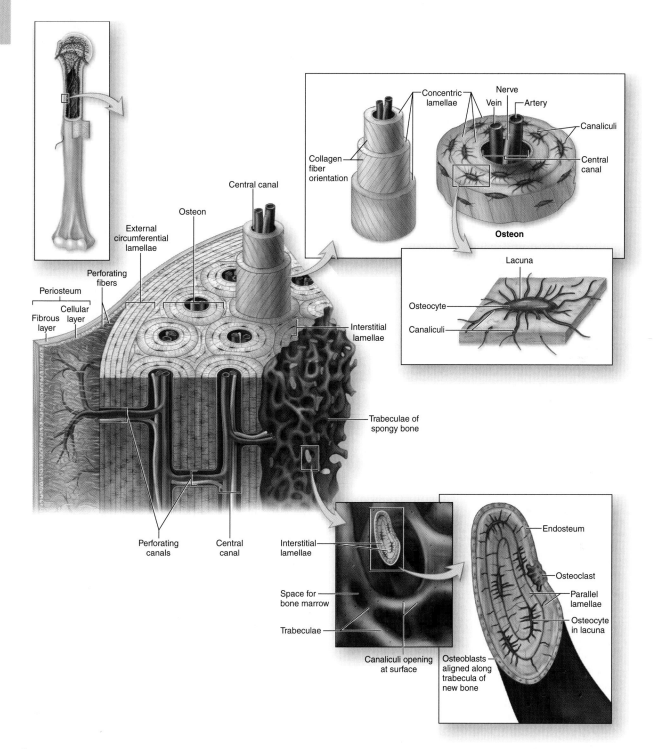

Figure 8–1. Components of bone. The diagram shows an overview of the basic features of bone, including the three key cell types osteocytes, osteoblasts, and osteoclasts; their usual locations; and the typical lamellar organizations of bone. Osteoblasts secrete the matrix which then hardens by calcification, trapping the differentiating cells now called osteocytes in individual lacunae. Osteocytes maintain the calcified matrix and receive nutrients from blood vessels via very small channels through the matrix called canaliculi. Osteoclasts are monocyte-derived cells in bone which are important in bone remodeling.

The periosteum consists of dense connective tissue, with a primarily fibrous layer covering a more cellular layer. Bone is vascularized by small vessels that penetrate the matrix from the periosteum.

As the main constituent of the adult skeleton, bone tissue supports fleshy structures, protects vital organs such as those in the cranial and thoracic cavities, and harbors the bone marrow, where blood cells are formed. Bone also serves as a reservoir of calcium, phosphate, and other ions that can be released or stored in a controlled fashion to maintain constant concentrations of these important ions in body fluids.

In addition, bones form a system of levers that multiply the forces generated during skeletal muscle contraction and transform them into bodily movements. This mineralized tissue therefore confers mechanical and metabolic functions to the skeleton.

Bone is a specialized connective tissue composed of calcified intercellular material, the **bone matrix,** and three cell types:

- **Osteocytes** (Gr. *osteon,* bone + *kytos,* cell), which are found in cavities (**lacunae**) between layers (lamellae) of bone matrix (Figure 8–1)
- **Osteoblasts** (*osteon* + Gr. *blastos,* germ), which synthesize the organic components of the matrix
- **Osteoclasts** (*osteon* + Gr. *klastos,* broken), which are multinucleated giant cells involved in the resorption and remodeling of bone tissue.

Because metabolites are unable to diffuse through the calcified matrix of bone, the exchanges between osteocytes and blood capillaries depend on communication through the **canaliculi** (L. *canalis,* canal), which are very thin, cylindrical spaces that perforate the matrix (Figure 8–1).

All bones are lined on both internal and external surfaces by layers of connective tissue containing osteogenic cells—**endosteum** on the internal surface and **periosteum** on the external surface.

Because of its hardness, bone cannot be sectioned with the microtome, and special steps must be taken to study it histologically.

A common technique that permits the observation of the cells and the organic part of the matrix is based on the decalcification of bone preserved by standard fixatives. The mineral is removed by immersion in a solution containing a calcium-chelating substance such as ethylenediaminetetraacetic acid (EDTA). The decalcified tissue is then embedded, sectioned, and stained as usual.

BONE CELLS

Osteoblasts

Osteoblasts are responsible for the synthesis of the organic components of bone matrix, consisting of type I collagen fibers, proteoglycans, and several glycoproteins including osteonectin. Deposition of the inorganic components of bone also depends on viable osteoblasts. Osteoblasts are located exclusively at the surfaces of bone matrix, usually side by side in a layer somewhat resembling a simple epithelium (Figure 8–2). When they are actively engaged in matrix synthesis, osteoblasts have a cuboidal to columnar shape and basophilic cytoplasm. When their synthesizing activity declines, they flatten and cytoplasmic basophilia is reduced. Osteoblast activity is stimulated by parathyroid hormone (PTH).

During matrix synthesis, osteoblasts have the ultrastructure of cells actively synthesizing proteins for secretion. Osteoblasts are polarized cells: matrix components are secreted at the cell surface in contact with older bone matrix, producing a layer of new (but not yet calcified) material called **osteoid** between the osteoblast layer and the bone formed earlier (Figure 8–2). This process of bone appositional growth is completed by subsequent deposition of calcium salts into the newly formed matrix.

Calcification of the matrix is not completely understood, but basic aspects of the process are shown in Figure 8–3. Prominent

then increases in girth only by appositional growth. Chondroblasts differentiate in the inner layers of the perichondrium, proliferate, and become chondrocytes once they have surrounded themselves with cartilaginous matrix and are incorporated into the existing cartilage (Figure 7–2).

Except in young children, damaged cartilage undergoes slow and often incomplete **regeneration**, by activity of cells in the perichondrium which invade the injured area and generate new cartilage. In extensively damaged areas—and occasionally in small areas—the perichondrium produces a scar of dense connective tissue instead of forming new cartilage. The poor regenerative capacity of cartilage is due in part to the avascularity of this tissue.

MEDICAL APPLICATION

In contrast to other tissues, hyaline cartilage is more susceptible to degenerative aging processes. Calcification of the matrix, preceded by an increase in the size and volume of the chondrocytes and followed by their death, is a common process in some cartilage. "Asbestiform" degeneration, frequent in aged cartilage, is due to the formation of localized aggregates of thick, abnormal collagen fibrils

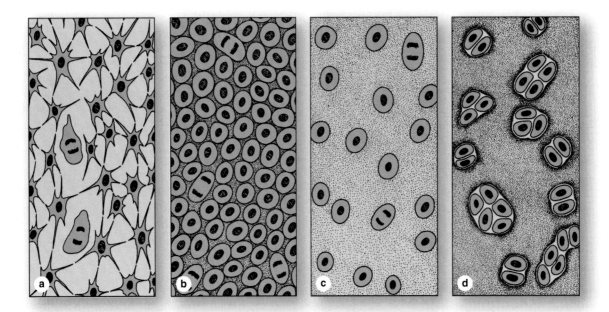

Figure 7–6. Chondrogenesis. Diagrams of the major stages by which cartilage is formed. **(a):** Embryonic mesenchyme is the precursor tissue of all types of cartilage. **(b):** Mitotic proliferation of mesenchymal cells and early differentiation gives rise to a tissue with condensations of rounded cells called chondroblasts. **(c):** Chondroblasts are separated from one another by their own production of various matrix components which collectively swell with water and form a great amount of ECM. **(d):** Multiplication of cartilage cells gives rise to isogenous aggregates, each surrounded by a condensation of territorial matrix. In mature cartilage this interstitial mitotic activity ceases and all chondrocytes typically become more widely separated by their production of matrix.

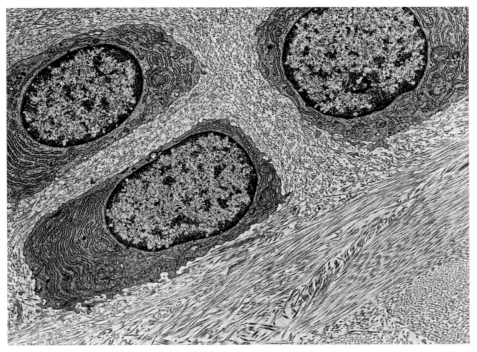

Figure 7–7. Chondrocytes in growing fibrocartilage. This TEM of fibrocartilage from a young animal shows three chondrocytes in their lacunae. RER is abundant in the cells, which are actively secreting their collagen-rich matrix. Fine collagen fibers, sectioned in several orientations, are prominent around the chondrocytes of fibrocartilage. Growing chondrocytes in hyaline and elastic cartilage have more prominent Golgi complexes and synthesize abundant proteoglycans in addition to collagens. X3750.

Table 9–2. Origin and principal functions of neuroglial cells.

Glial Cell Type	Origin	Location	Main Functions
Oligodendrocyte	Neural tube	Central nervous system	Myelin production, electric insulation
Neurolemmocyte	Neural crest	Peripheral nerves	Myelin production, electric insulation
Astrocyte	Neural tube	Central nervous system	Structural support, repair processes
			Blood-brain barrier, metabolic exchanges
Ependymal cell	Neural tube	Central nervous system	Lining cavities of central nervous system
Microglia	Bone marrow	Central nervous system	Immune-related activity

defense in CNS tissues. Microglia originate not from the embryonic neural tube but from circulating blood monocytes, belonging to the same family as macrophages and other antigen-presenting cells.

Nuclei of microglial cells can be recognized in routine H&E preparations by their dense elongated structure, which contrasts with the spherical, more lightly stained nuclei of other glial cells. Immunohistochemistry using antibodies against cell surface antigens of immune cells demonstrates microglial processes. When activated, microglia retract their processes and assume the morphologic characteristics of macrophages, become phagocytic and act as antigen-presenting cells (see Chapter 14).

MEDICAL APPLICATION

In multiple sclerosis, the myelin sheath is damaged by an autoimmune mechanism with various neurologic consequences. In this disease, microglia phagocytose and degrade myelin debris by receptor-mediated phagocytosis and lysosomal activity. In addition, AIDS dementia complex is caused by HIV-1 infection of the central nervous system. Overwhelming experimental evidence

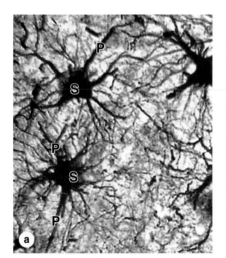
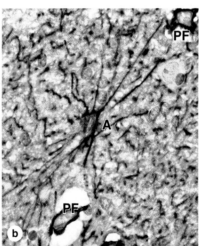
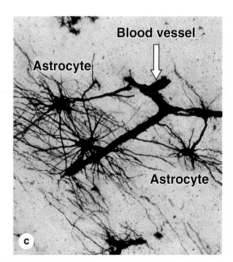

Figure 9–11. **Astrocytes. (a):** Astrocytes are the most numerous glial cells of the CNS and are characterized by numerous cytoplasmic processes (P) radiating from the glial cell body or soma (S). Astrocytic processes are not seen with routine light microscope staining, but are easily seen after gold staining. Morphology of the processes allows astrocytes to be classified as fibrous (relatively few and straight processes) or protoplasmic (numerous branching processes), but functional differences between these types are not clear. X500. Gold chloride. **(b):** All astrocytic processes contain intermediate filaments of glial fibrillary acidic protein (GFAP) and antibodies against this protein provide a simple method to stain these cells, as seen here in a fibrous astrocyte (A) and its processes. The small pieces of other GFAP-positive processes in the neuropil around this cell give an idea of the density of this glial cell and its processes in the CNS. Astrocytes are an important part of the blood-brain barrier regulating entry of molecules and ions from blood into CNS tissue. Capillaries at the extreme upper right and lower left corners of (b) are enclosed by GFAP-positive perivascular feet (PF) at the ends of numerous astrocytic processes. X500. Anti-GFAP immunoperoxidase and hematoxylin counterstain. **(c):** A length of capillary is shown here completely enclosed within stained processes of astrocytes. X400. Rio Hortega silver.

indicates that microglia are infected by HIV-1. A number of cytokines, such as interleukin-1 and tumor necrosis factor-α, activate and enhance HIV replication in microglia.

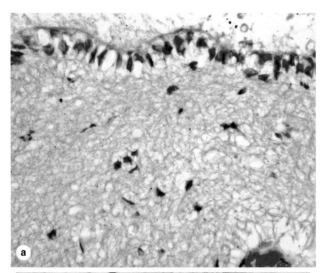

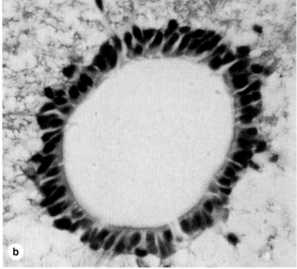

Figure 9–12. Ependymal cells. Ependymal cells are epithelial-like cells that form a single layer lining the fluid-filled ventricles of **(a)** the cerebrum and **(b)** the central canal of the spinal cord. Cuboidal or low columnar, ependymal cells in various CNS regions may be clearly ciliated or have long microvilli at the apical surfaces. These modifications help circulate the CSF and monitor its contents. Also at their apical ends ependymal cells have junctional complexes like those of epithelial cells. However, they lack a basal lamina. The cells' basal ends are elongated and tapered, extending processes that branch and penetrate some distance into the adjacent neuropil. Ependymal cells not only help move and absorb CSF, they are responsible for production of this fluid in specialized ventricular tissues of the choroid plexus. Both: X200. H&E.

Schwann Cells (Neurolemmocytes)

Schwann cells, also called **neurolemmocytes**, are found only in the PNS and have trophic interactions with axons and allow for their myelination like the oligodendrocytes of the CNS. One neurolemmocyte forms myelin around a segment of one axon, in contrast to the ability of oligodendrocytes to branch and sheath parts of more than one axon. Figure 9–10e shows how a series of Schwann cells covers the full length of an axon.

Satellite Cells of Ganglia

Derived from the embryonic neural crest like neurolemmocytes, small **satellite cells** form a covering layer over the large neuronal cell bodies in PNS ganglia (Figure 9–10f). Closely associated with the neurons, the satellite cells exert a trophic or supportive role, but the molecular basis of their support is poorly understood.

CENTRAL NERVOUS SYSTEM

The principal structures of the CNS are the **cerebrum, cerebellum,** and **spinal cord.** It has virtually no connective tissue and is therefore a relatively soft, gel-like organ.

When sectioned, the cerebrum, cerebellum, and spinal cord show regions of white (**white matter**) and gray (**gray matter**), differences caused by the differential distribution of myelin. The main components of white matter are myelinated axons (Figure 9–14) and the myelin-producing oligodendrocytes. White matter does not contain neuronal cell bodies, but microglia are present.

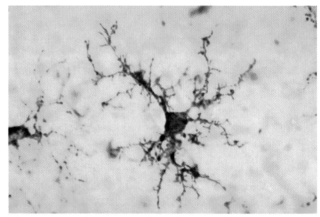

Figure 9–13. Microglial cells. Microglia are the monocyte-derived, antigen-presenting immune cells of the CNS and are evenly distributed in both gray and white matter. By immunohistochemistry, here using a monoclonal antibody against HLA antigens found on many immune-related cells, the short branching processes of microglia can be seen. Routine staining does not demonstrate the processes, but only the small dark nuclei of the cells. Microglia move around and are constantly employed in immune surveillance of CNS tissues. When activated by products of cell damage or microorganisms, the cells retract their processes, begin phagocytosing the damage- or danger-related material, and behave as antigen-presenting cells. X500. Antibody against HLA-DR and peroxidase. (Used, with permission, from Wolfgang Streit, Department of Neuroscience, University of Florida College of Medicine, Gainesville.)

Gray matter contains abundant neuronal cell bodies, dendrites, the initial unmyelinated portions of axons, astrocytes, and microglial cells. This is the region where synapses occur. Gray matter is prevalent at the surface or cortex of the cerebrum and cerebellum, whereas white matter is present in more central regions. Aggregates of neuronal cell bodies forming islands of gray matter embedded in the white matter are called **nuclei.** Neuroscientists recognize six layers in the **cerebral cortex** with most neurons arranged vertically. The most abundant neurons are the efferent **pyramidal neurons** which come in many sizes (Figure 9–15). Cells of the cerebral cortex function in the integration of sensory information and the initiation of voluntary motor responses.

The **cerebellar cortex**, which coordinates muscular activity throughout the body, has three layers (Figure 9–16): an outer **molecular layer**, a central layer of very large neurons called **Purkinje cells**, and an inner **granule layer**. The Purkinje cell bodies are conspicuous even in H&E stained material and their dendrites extend throughout the molecular layer as a branching basket of nerve fibers (Figure 9-16). The granule layer is formed by very small neurons (the smallest in the body), which are packed together densely, in contrast to the neuronal cell bodies in the molecular layer which are sparse (Figure 9–16).

In cross sections of the **spinal cord,** white matter is peripheral and gray matter is internal and has the general shape of an H (Figure 9–17). In the center is an opening, the **central canal,** which develops from the lumen of the embryonic neural tube and is lined by ependymal cells. The gray matter forms the **anterior** horns, which contain motor neurons whose axons make up the ventral roots of spinal nerves, and the posterior horns, which receive sensory fibers from neurons in the spinal ganglia (dorsal roots). Spinal cord neurons are large and multipolar, especially the motor neurons in the anterior horns (Figure 9–17).

Meninges

The skull and the vertebral column protect the CNS. Between the bone and nervous tissue are membranes of connective tissue called the **meninges** (Figures 9–18 and 9–19). Three meningial layers are distinguished:

DURA MATER

The dura mater is the thick external layer consisting of dense, fibroelastic connective tissue continuous with the periosteum of the skull. Around the spinal cord the dura mater is separated from the periosteum of the vertebrae by the epidural space, which contains a plexus of thin-walled veins and areolar connective tissue.

The dura mater is always separated from the arachnoid by the thin **subdural space**. The internal surface of all dura mater, as well as its external surface in the spinal cord, is covered by simple squamous epithelium of mesenchymal origin (Figure 9–18).

ARACHNOID

The arachnoid (Gr. *arachnoeides,* spiderweblike) has two components: (1) a sheet of connective tissue in contact with the dura mater and (2) a system of loosely arranged trabeculae containing fibroblasts and collagen. This trabecular system is continuous with the deeper pia mater. Surrounding the trabeculae is a large, sponge-like cavity, the **subarachnoid space,** filled with CSF. This space forms a hydraulic cushion that protects the CNS from trauma. The subarachnoid space communicates with the ventricles of the brain.

The connective tissue of the arachnoid is said to be avascular because it lacks nutritive capillaries, but larger blood vessels run through it (Figure 9–18). Because the arachnoid has fewer trabeculae in the spinal cord, it can be more clearly distinguished from the pia mater in that area. The arachnoid and the pia mater are intimately associated and are often considered a single membrane called the **pia-arachnoid.**

In some areas, the arachnoid perforates the dura mater and protrudes into blood-filled venous sinuses within the dura mater. These CSF-filled protrusions, which are covered by vascular endothelial cells, are called **arachnoid villi.** Their function is to transport CSF from the subarachnoid space into venous sinuses.

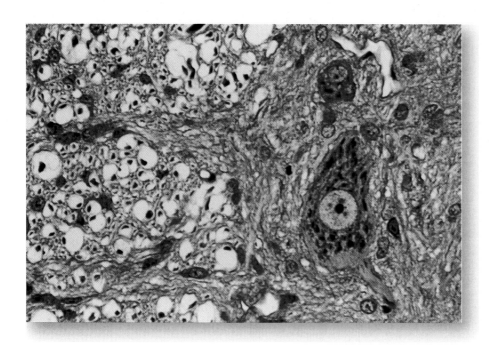

Figure 9–14. **White versus gray matter, stained.** A cross section of spinal cord shows the transition between white matter (left) and gray matter (right). The white matter consists mainly of nerve fibers whose myelin sheaths were dissolved in the preparation procedure, leaving the round empty spaces shown. Each such space surrounds a dark-stained spot which is the axon. Neuronal cell bodies, astrocytes, and abundant cell processes predominate in the gray matter. X400. PT.

PIA MATER

The innermost pia mater is lined internally by flattened, mesenchymally derived cells closely applied to the entire surface of the CNS tissue, but this layer does not directly contact nerve cells or fibers. Between the pia mater and the neural elements is a thin limiting layer of astrocytic processes, which adheres firmly to the pia mater. Together the pia mater and glial layer form a physical barrier at the CNS periphery. This barrier separates the CNS tissue from the CSF in the subarachnoid space (Figure 9–19).

Blood vessels penetrate the CNS through tunnels covered by pia mater—the **perivascular spaces.** The pia mater disappears when the blood vessels branch to the smallest capillaries. However, these capillaries remain completely covered by expanded perivascular processes of astrocytes (Figure 9–11).

Blood-Brain Barrier

The blood-brain barrier (BBB) is a functional barrier that allows much tighter control than that in most tissues over the passage of substances moving from blood into the CNS tissue, protecting the nature of the neuronal microenvironment. The main structural component of the BBB is the **capillary endothelium**, in which the cells are tightly sealed together with well-developed occluding junctions and show little or no transcytosis. Moreover, the basal lamina of capillaries in most CNS regions is enveloped by the **perivascular feet of astrocytes** (Figure 9–11) which further regulate passage of molecules and ions from blood to brain.

The BBB allows the stable composition and constant balance of ions in the interstitial fluid surrounding neurons and glial cells that is required for their function and protects these cells from potential toxins and infectious agents. The components of the BBB are not found in the choroid plexus where CSF is produced, in the posterior pituitary which releases hormones, or in regions of the hypothalamus where plasma components are monitored.

Choroid Plexus

The **choroid plexus** is a highly specialized tissue that projects as elaborate folds with many villi into the four large ventricles of the brain (Figure 9–20). It is found in the roofs of the third and fourth ventricles and in parts of the walls of the two lateral ventricles, all regions in which the ependymal lining directly contacts the pia mater.

Each villus of the choroid plexus contains a thin layer of well-vascularized pia mater covered by cuboidal ependymal cells. The main function of the choroid plexus is to remove water from blood and release it as cerebrospinal fluid (CSF). This fluid completely fills the ventricles, the central canal of the spinal cord, the subarachnoid space, and the perivascular spaces. It is

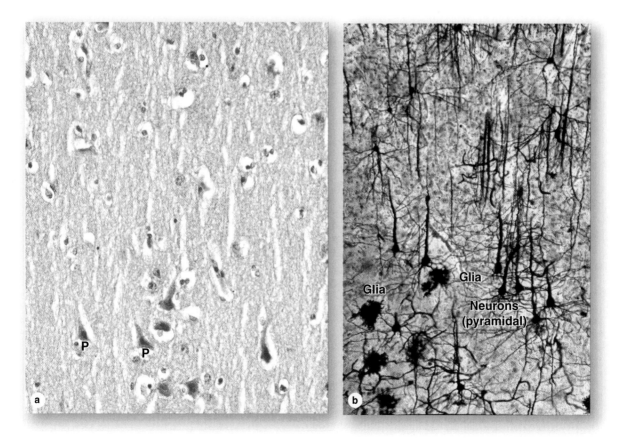

Figure 9–15. Cerebral cortex. (a): Important neurons of the cerebrum are pyramidal neurons (P), which are arranged vertically and interspersed with numerous glial cells in the eosinophilic neuropil. X200. H&E. (b): From the apical ends of pyramidal neuron, long dendrites extend in the direction of the cortical surface, which can be best seen in thick silver-stained sections in which only a few other protoplasmic glial cells are seen. X200. Silver.

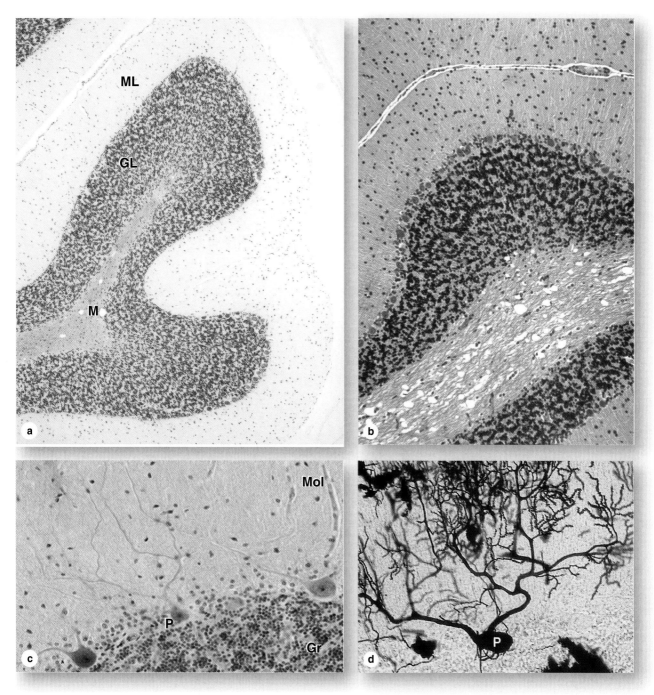

Figure 9–16. Cerebellum. **(a):** The cerebellar cortex is convoluted with many distinctive small folds, each supported at its center by cerebellar medulla (M), which is white matter consisting of large tracts of axons. X6. Cresyl violet. **(b):** Immediately surrounding the white matter of the medulla is the granular layer (GL) of the cortex, which is densely packed with very small, rounded neuronal cell bodies. The outer, "molecular layer" (ML) consists of neuropil with fewer, more scattered small neurons. X20. H&E. **(c):** At the interface between the granular and molecular layers is a single layer with very large neuronal cell bodies of unique Purkinje cells (P), whose axons pass through the granular layer (Gr) to join tracts in the medulla and whose multiple branching dendrites ramify throughout the molecular layer (Mol). X40. H&E. **(d):** Although not seen until well after H&E staining, dendrites of Purkinje cells have hundreds of small branches, each covered with dendritic spines, which can be demonstrated with silver stains. Axons from the small neurons of the granular layer are unmyelinated and run together into the molecular layer where they form synapses with the dendrites spines of Purkinje cells. The molecular layer of the cerebellar cortex contains relatively few neurons or other cells. X40. Silver.

Central canal and
gray commissure

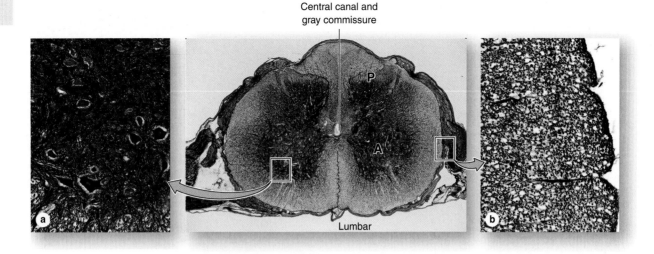

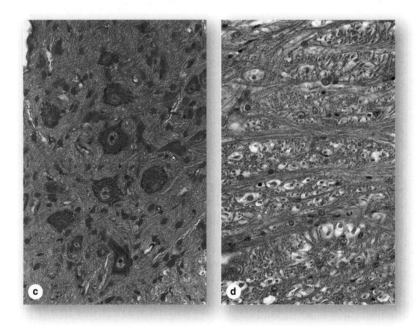

Figure 9–17. Spinal cord. The spinal cord varies slightly in diameter along its length, but in cross-section always shows bilateral symmetry around the small, CSF-filled central canal. Unlike the cerebrum and cerebellum, in the spinal cord the gray matter is internal, forming a roughly H-shaped structure that consists of two posterior (P) horns (sensory) and two anterior (A) (motor) horns all joined by the gray commissure around the central canal. **(a):** The gray matter contains abundant astrocytes and large neuronal cell bodies, especially those of motor neurons in the ventral horns. **(b):** The white matter surrounds the gray matter and contains primarily oligodendrocytes and tracts of myelinated axons running along the length of the cord.

(c): Micrograph of the large motor neurons of the ventral horns show large nuclei, prominent nucleoli, and cytoplasm rich in chromatophilic substance (Nissl substance), all of which indicate extensive protein synthesis to maintain the axons of these cells which extend great distances. **(d):** In the white commissure ventral to the central canal, tracts run lengthwise along the cord, seen here in cross-section with empty myelin sheaths surrounding axons, as well as tracts running from one side of the cord to the other, seen here as several longitudinally sectioned tracts of eosinophilic axons. Center: X5; a–d: X200.; Center, a, b: silver; c, d: H&E.

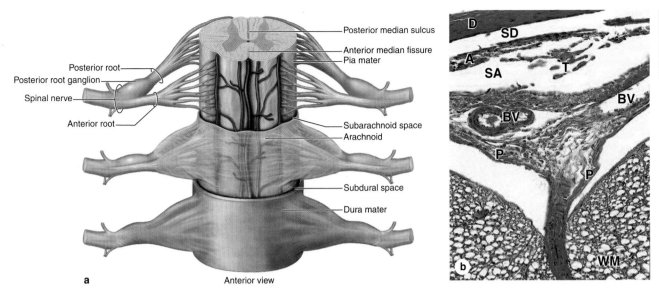

Figure 9–18. Spinal cord and meninges. (a): Diagram of spinal cord indicates the relationship of the three meningeal layers of connective tissue: the innermost **pia mater**, the **arachnoid**, and the **dura mater**. The dura fuses partially with the periosteum of the protective vertebrae, which are not shown. Also depicted are the blood vessels coursing through the subarachnoid space and the nerve rootlets that fuse to form the posterior and anterior roots of the spinal nerves. The posterior root ganglia contain the cell bodies of sensory nerve fibers and are located in intervertebral formamina. **(b):** Section of an area near the anterior median fissure showing the tough dura mater (D) and subdural space (SD) lined by flattened epithelial-like cells. The middle meningeal layer is the thicker weblike arachnoid mater (A) containing the large subarachnoid space (SA) and connective tissue trabecular (T). The subarachnoid space is filled with CSF and the arachnoid acts as a shock absorbing pad between the brain and skull. Fairly large blood vessels (BV) course through the arachnoid mater. The innermost pia mater (P) is thin and is not clearly separate from the arachnoid; together they are sometimes referred to as the pia-arachnoid or the leptomeninges. The space between the pia and the white matter (WM) of the spinal cord here is an artifact created during dissection; normally the pia is very closely applied to a layer of astrocytic processes at the surface of the CNS tissue. X100. H&E.

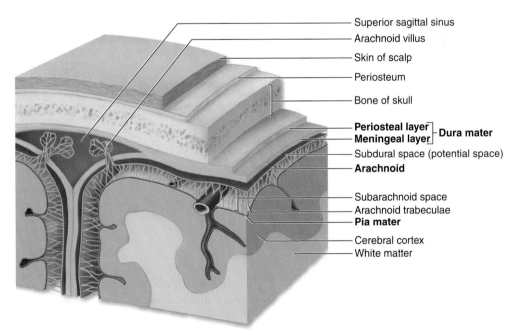

Figure 9–19. Meninges around the brain. The dura, arachnoid, and pia mater also cover the entire surface of the brain, but the periosteal dura often adheres to the cranium when the brain is removed. The relationships among the cranial meninges are similar to that described for those of the spinal cord. The diagram includes arachnoid villi, which are outpocketings of arachnoid away from the brain which penetrate the dura mater and enter blood-filled venous sinuses found within the vasculature of the periosteum. The arachnoid villi function in releasing excess CSF into the blood. Blood vessels from the arachnoid mater branch into smaller arteries and veins that enter brain tissue carrying oxygen and nutrients. These small vessels are initially covered with pia mater, but as capillaries are covered only by the perivascular feet of astrocytes.

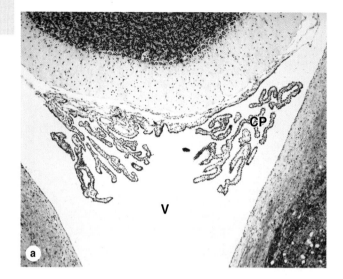

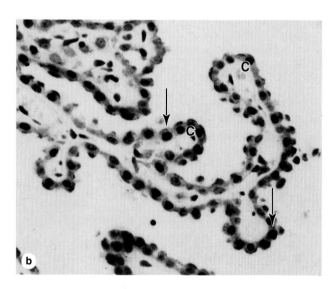

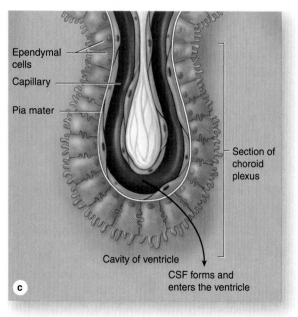

Ependymal cells

Capillary

Pia mater

Section of choroid plexus

Cavity of ventricle

CSF forms and enters the ventricle

important for metabolism within the CNS and acts to absorb mechanical shocks.

CSF is clear, has a low density, contains Na^+, K^+, and Cl^- ions but very little protein, and its only cells are normally very sparse lymphocytes. It is produced continuously across the walls of the choroid plexus villi and circulates through the ventricles and central canal, from which it passes into the subarachnoid space. There, arachnoid villi provide the main pathway for absorption of CSF into the venous circulation since there are no lymphatic vessels in CNS tissue.

MEDICAL APPLICATION

A decrease in the absorption of CSF or a blockage of outflow from the ventricles during fetal or postnatal development results in the condition known as **hydrocephalus** *(Gr. hydro, water, + kephale, head), which promotes a progressive enlargement of the head followed by mental impairment.*

PERIPHERAL NERVOUS SYSTEM

The main components of the peripheral nervous system are the **nerves, ganglia,** and **nerve endings.** Nerves are bundles of nerve fibers (axons) surrounded by glial cells and connective tissue.

Nerve Fibers

Nerve fibers consist of axons enclosed within a special sheath of cells derived from the embryonic neural crest. Like tracts within the CNS, peripheral nerves contain groups of nerve fibers. In peripheral nerve fibers, axons are sheathed by **Schwann cells,** also called **neurolemmocytes** (Figure 9–10e). The sheath may or may not form myelin around the axons, depending on their diameter.

Axons of small diameter are usually **unmyelinated nerve fibers** (Figures 9–22 and 9–25). Progressively thicker axons are generally sheathed by increasingly numerous concentric wrappings of the enveloping cell, forming the **myelin sheaths.** These fibers are known as **myelinated nerve fibers** (Figures 9–21, 9–22, and 9–23).

Figure 9–20. **Choroid plexus.** The choroid plexus consists of highly specialized regions of CNS tissue containing ependyma cells and vascularized pia mater that project from specific walls of the ventricles. **(a):** Section of the bilateral choroid plexus (CP) projecting into the fourth ventricle (V) near the cerebrum and cerebellum. It is elaborately folded with many finger-like villi. X12. H&E. **(b):** At higher magnification, each villus is seen to be well-vascularized with capillaries (C) and covered by a continuous layer of ependymal cells (arrow). X100. H&E. **(c):** The choroid plexus is specialized for transport of water and ions across the capillary endothelium and ependymal layer and the elaboration of these as CSF.

Myelinated Fibers

As axons of large diameter grow in the PNS, they are engulfed along their length by many undifferentiated neurolemmocytes and become **myelinated nerve fibers**. The plasma membrane of the covering neurolemmocyte (Schwann cell) fuses around the axon and becomes wrapped around the nerve fiber as the glial cell body moves around and around the axon many times (Figure 9–21). The multiple layers of Schwann cell membrane unite as a layer **myelin,** a whitish lipoprotein complex whose abundant lipid component is partly removed by standard histologic procedures, as in all cell membranes (Figures 9–14 and 9–17). With the TEM the myelin sheath can appear as a thick electron-dense cover in which individual membrane layers are seen (Figure 9–22).

Membranes of Schwann cells have a higher proportion of lipids than do other cell membranes and the myelin sheath serves to protect axons and maintain a constant ionic microenvironment required for action potentials. Between adjacent Schwann cells the myelin sheath shows small **nodal gaps** along the axon, also called **nodes of Ranvier** (Figures 9–10e and 9–23). Interdigitating processes of Schwann cells partially cover each node (Figure 9–24). The length of axon covered by one Schwann cell is called the **internodal segment** and may be more than 1 millimeter.

Unlike oligodendrocytes of the CNS, Schwann cells only form myelin around a portion of one axon.

Unmyelinated Fibers

The CNS is rich in unmyelinated axons which are not sheathed at all but run free among the other neuronal and glial processes. However in the PNS, even all unmyelinated axons are enveloped within simple folds of Schwann cells (Figure 9–25). In this situation the glial cell does not form multiple wrapping of itself as myelin. Unlike their association with individual myelinated axons, each Schwann cell can enclose portions of many unmyelinated axons with small diameters. Adjacent Schwann cells along unmyelinated nerve fibers do not form nodes of Ranvier.

Nerves

In the PNS nerve fibers are grouped into bundles to form nerves. Except for very thin nerves containing only unmyelinated fibers, nerves have a whitish, glistening appearance because of their myelin and collagen content.

Axons and Schwann cells of nerves are enclosed within connective tissue layers (Figures 9–26 and 9–27). Externally is a dense, irregular fibrous coat called **epineurium,** which continues more

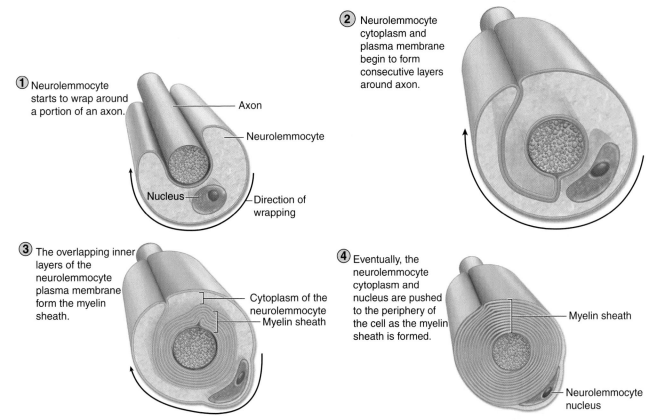

Figure 9–21. Myelination of large diameter PNS axons. A neurolemmocyte (Schwann cell) engulfs a portion along the axon. The Schwann cell membrane fuses around the axon and elongates as it becomes wrapped around the axon while the cell body moves around the axon many times. The neurolemmocyte membrane wrappings constitute the myelin sheath, with the cell body on its outer surface. The myelin layers are very rich in lipid and provide insulation and facilitate formation of action potentials along the axolemma.

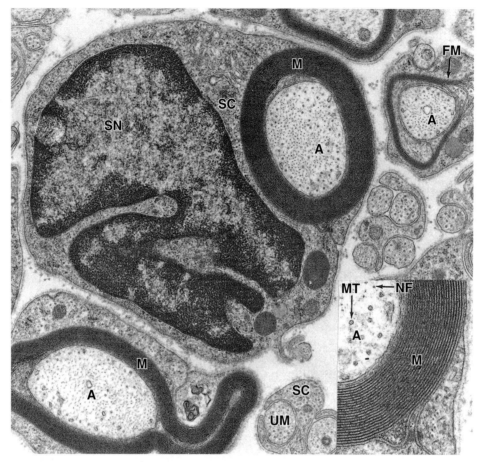

Figure 9–22. Ultrastructure of myelinated and unmyelinated fibers. Cross-section of PNS fibers in the TEM reveals differences between myelinated and unmyelinated axons. Large axons are wrapped in a thick myelin sheath (M) of multiple layers of Schwann cell membrane. The inset shows a portion of myelin in which individual membrane layers can be distinguished easily, as well as the neurofilaments (NF) and microtubules (MT) in the axoplasm (A). At the center of the photo is a Schwann cell showing its active nucleus (SN) and Golgi-rich cytoplasm (SC). At the right is an axon around which myelin is still forming (FM). Unmyelinated axons (UM) are much smaller in diameter and many such fibers may be engulfed by a single Schwann cell (SC). The glial cell does not form myelin wrappings around such small axons, but simply encloses them. Whether it forms myelin or not, each Schwann cell is surrounded, as shown, by an external lamina containing type IV collagen and laminin like the basal laminae of epithelial cells. X70,000. (Used, with permission, from Mary Bartlett Bunge, The Miami Project to Cure Paralysis, University of Miami Miller School of Medicine.)

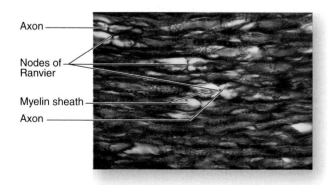

Axon —
Nodes of Ranvier —
Myelin sheath —
Axon —

Figure 9–23. Nodes of Ranvier and endoneurium. A longitudinal section of myelinated nerve fibers reveals the nodes of Ranvier, or nodal gaps, which are small, physiologically important gaps in the myelin sheath between adjacent Schwann cells. The axon can be seen spanning each nodal gap. Schwann cells produce an external lamina that surrounded their external surface. Like the basal laminae of epithelia, this structure contains type IV collagen and is continuous with the surrounding connective tissue rich in reticular fibers. This connective tissue makes up the endoneurium around the Schwann cells of all peripheral nerve fibers and is stained blue in this preparation. X400. Mallory trichrome.

deeply to also fill the space between bundles of nerve fibers. Each such bundle or **fascicle** is surrounded by the **perineurium,** a sleeve of specialized connective tissue formed by layers of flattened epithelial-like cells. The cells of each layer of the perineurium are joined at their edges by tight junctions, an arrangement that makes the perineurium a barrier to the passage of most macromolecules and has the important function of protecting the nerve fibers and helping maintain the internal microenvironment. Within the perineurial sheath run the Schwann cell–covered axons and their enveloping connective tissue, the **endoneurium** (Figure 9–27). The endoneurium consists of a sparse layer of loose connective tissue that merges with an external lamina of type IV collagen, laminin, and other proteins produced by the Schwann cells.

Very small nerves consist of one fascicle. Small nerves can be found in sections of many organs and often show a winding disposition in connective tissue (Figure 9–28).

The nerves establish communication between centers in the brain and spinal cord and the sense organs and effectors (muscles, glands, etc). They generally contain both afferent and efferent fibers. **Afferent** fibers carry information from the interior of the body and the environment to the CNS. **Efferent** fibers carry impulses from the CNS to effector organs commanded by these centers. Nerves possessing only sensory fibers are called **sensory nerves;** those composed only of fibers carrying impulses to the effectors are called **motor nerves.** Most nerves have both sensory and motor fibers and are called **mixed nerves** which usually have both myelinated and unmyelinated axons (Figure 9–27b).

Ganglia

Ganglia are typically ovoid structures containing neuronal cell bodies and glial cells supported by connective tissue. Because

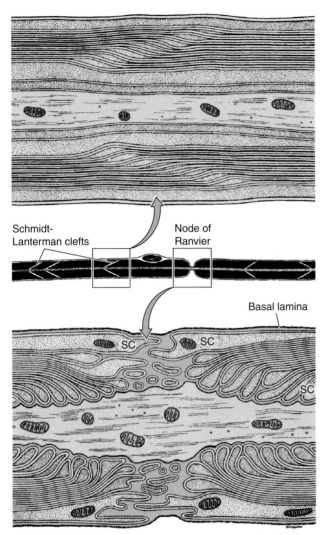

Figure 9–24. Myelin maintenance and nodal gaps (of Ranvier). The center drawing shows a myelinated peripheral nerve fiber as seen under the light microscope. The axon is enveloped by the myelin sheath, which, in addition to membrane, contains some Schwann cell cytoplasm in spaces between the membranes called **myelin clefts** (Schmidt-Lanterman clefts). The upper drawing shows one set of clefts ultrastructurally. The clefts contain Schwann cell cytoplasm that was not displaced to the cell body during myelin formation. This cytoplasm moves slowly along the myelin sheath, opening temporary spaces (the clefts) between the membrane layers, which allows renewal of some membrane components as needed and maintenance of the sheath.

The lower drawing shows the ultrastructure of a **nodal gap** or **node of Ranvier.** Interdigitating processes extending from the outer layers of the Schwann cells (SC) partly cover and contact the axolemma at the nodal gap. This contact acts as a partial barrier to the movement of materials in and out of the periaxonal space between the axolemma and the Schwann sheath. The basal or external lamina around Schwann cells is continuous over the nodal gap. Covering the nerve fiber is a thin connective tissue layer that belongs to the endoneurial sheath of the peripheral nerve fibers.

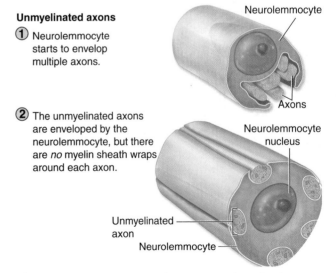

Figure 9–25. Unmyelinated nerves. During development portions of several small diameter axons are engulfed by one neurolemmocyte (Schwann cell). Subsequently the axons are separated and each becomes enclosed within its own fold or pocket of Schwann cell surface. No myelin is formed. Small diameter axons utilize action potentials whose formation and maintenance do not depend on the insulation provided by the myelin sheath required by large diameter axons.

they serve as relay stations to transmit nerve impulses, one nerve enters and another exits from each ganglion. The direction of the nerve impulse determines whether the ganglion will be a **sensory** or an **autonomic** ganglion.

SENSORY GANGLIA

Sensory ganglia receive afferent impulses that go to the CNS. Sensory ganglia are associated with both cranial nerves (cranial ganglia) and the dorsal root of the spinal nerves (spinal ganglia). The large neuronal cell bodies of ganglia (Figure 9–29) are associated with thin, sheet-like extensions of small glial cells called **satellite cells** (Figure 9–10f). These neural crest-derived

cells create the microenvironments of the perikarya, allowing the production of membrane action potentials and regulating metabolic exchange.

Sensory ganglia are supported by a distinct connective tissue capsule and framework continuous with the connective tissue layers of the nerves. The neurons of these ganglia are pseudounipolar and relay information from the ganglion's nerve endings to the gray matter of the spinal cord via synapses with local neurons.

AUTONOMIC GANGLIA

Autonomic (Gr. *autos,* self, + *nomos,* law) nerves effect the activity of smooth muscle, the secretion of some glands, modulate

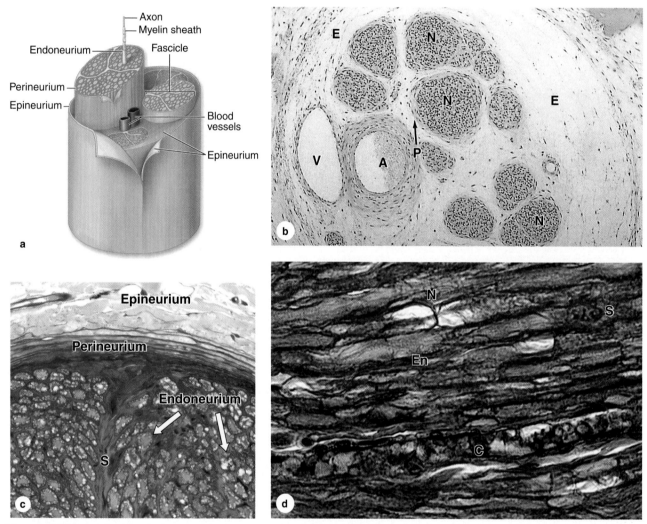

Figure 9–26. Peripheral nerve connective tissue. Peripheral nerves are protected by three layers of connective tissue, as depicted in the diagram **(a)**. **(b):** The outer **epineurium** (E) consists of a dense superficial region and a looser deep region that contains large blood vessels (A,V) and fascicles in which nerve fibers (N) are bundled. Each fascicle is surrounded by the **perineurium** (P), consisting of a few layers of unusual epithelial-like fibroblastic cells which are all joined at the peripheries by tight junctions to form a blood-nerve barrier that helps regulate the microenvironment inside the fascicle. Axons and Schwann cells are in turn surrounded by a thin layer of endoneurium. X140. H&E. **(c):** As shown here the perineurium can extend as septa (S) into larger fascicles. X200. PT. **(d):** This micrograph shows a longitudinally oriented nerve. Within fascicles is the **endoneurium** (En) which surrounds capillaries (C) and is continuous with the external lamina produced by the Schwann cells. Collagen of the endoneurium is stained blue and a node of Ranvier (N) and a Schwann cell nucleus (S) are also clearly seen. X400. Mallory trichome.

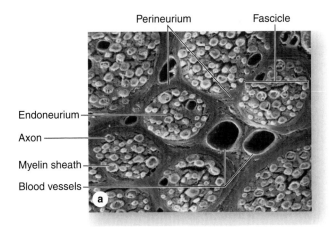

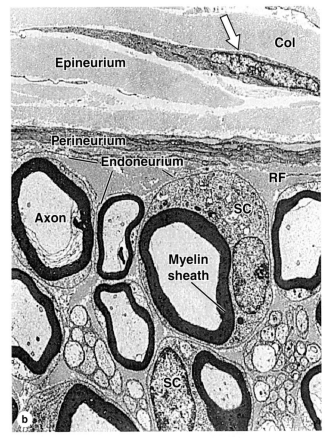

***Figure 9–27.* Peripheral nerve ultrastructure. (a):** SEM of transverse sections of a large peripheral nerve showing several fascicles, each surrounded by perineurium and packed with endoneurium which surrounds the individual myelin sheaths. Each fascicle contains at least one capillary. Endothelial cells of these capillaries are tightly joined as part of the blood-nerve barrier and regulate the kinds of plasma substances released to the endoneurium. Larger blood vessels course through the deep epineurium which fills the space around the perineurium and fascicles. X450. **(b):** TEM shows a fibroblast (arrow) surrounded by collagen in the epineurium and three or four layers of highly flattened cells in the perineurium which form another part of the blood-nerve barrier. Inside the perineurium the endoneurium is rich in reticulin fibers (RF) which surround all Schwann cells. Nuclei of two

cardiac rhythm and other involuntary activities by which the body maintains a constant internal environment (**homeostasis**).

Autonomic ganglia are small bulbous dilatations in autonomic nerves. Some are located within certain organs, especially in the walls of the digestive tract, where they constitute the **intramural ganglia** (Chapter 15). The capsules of these ganglia are less well-defined. Autonomic ganglia usually have multipolar neurons. A layer of satellite cells also envelops the neurons of autonomic ganglia (Figure 9-29), although in intramural ganglia, only a few satellite cells may be seen around each neuron.

Autonomic nerves use two-neuron circuits. The first neuron of the chain, with the **preganglionic fiber,** is located in the CNS. Its axon forms a synapse with postganglionic fibers of the second multipolar neuron in the chain located in a peripheral ganglion system. The chemical mediator present in the synaptic vesicles of all preganglionic axons is **acetylcholine**.

Autonomic nerves comprise an autonomic nervous system with two parts, called the **sympathetic** and the **parasympathetic divisions**. Neuronal cell bodies of preganglionic sympathetic nerves are located in the thoracic and lumbar segments of the spinal cord and those of the parasympathetic division are in the medulla and midbrain and in the sacral portion of the spinal cord. Sympathetic second neurons are located in small ganglia along the vertebral column, while second neurons of the parasympathetic series are found in very small ganglia always located near or within the effector organs, for example in the walls of the stomach and intestines. Parasympathetic ganglia may lack distinct capsules altogether, perikarya and associated satellite cells simply forming a loosely organized plexus within the surrounding connective tissue.

NEURAL PLASTICITY & REGENERATION

Despite its general stability, the nervous system exhibits plasticity even in adults. Plasticity is very high during embryonic development, when an excess of nerve cells is formed, and the cells that do not establish correct synapses with other neurons are eliminated by apoptosis. In adult mammals after an injury, the neuronal circuits may be reorganized by the growth of neuronal processes, forming new synapses to replace the ones lost by injury. Thus, new communications are established with some degree of functional recovery. This **neural plasticity** and reformation of processes are controlled by several growth factors produced by both neurons and glial cells in a family of growth factors called **neurotrophins.**

Neuronal stem cells are present in adult CNS, located in part among the cells of the ependyma, which can supply new neurons, astrocytes and oligodendrocytes. Because neurons cannot divide to replace those lost by injury or disease, the potential of neural stem cells to allow regeneration of CNS components is a subject of intense investigation.

Injured fibers in peripheral nerves have a good capacity for regeneration and return of function. In a wounded nerve fiber, it is important to distinguish changes occurring proximal to the injury from those in the distal segment. The proximal segment

Schwann cells (SC) of myelinated fibers are visible. Many unmyelinated axons within two Schwann cells are also present. X1200.

maintains its continuity with the trophic center in the perikaryon and can regenerate, while the distal segment, separated from the nerve cell body, degenerates (Figure 9–30). The onset of regeneration is accompanied by changes in the perikaryon: **chromatolysis** or dissolution of the RER and a consequent decrease in cytoplasmic basophilia; an increase in the volume of the perikaryon; and migration of the nucleus to a peripheral position in the perikaryon. The proximal segment of the axon degenerates close to the wound for a short distance, but growth starts as soon as debris is removed by macrophages. Macrophages produce cytokines which stimulate Schwann cells to secrete neurotrophins.

In the nerve segment distal to the injury the axon and myelin, but not the connective tissue, degenerate completely and are removed by macrophages. While these regressive changes take place, Schwann cells proliferate within the connective tissue sleeve, giving rise to rows of cells that serve as guides for the sprouting axons formed during the reparative phase.

MEDICAL APPLICATION

When there is an extensive gap between the distal and proximal segments of cut or injured peripheral nerves, or when the distal segment

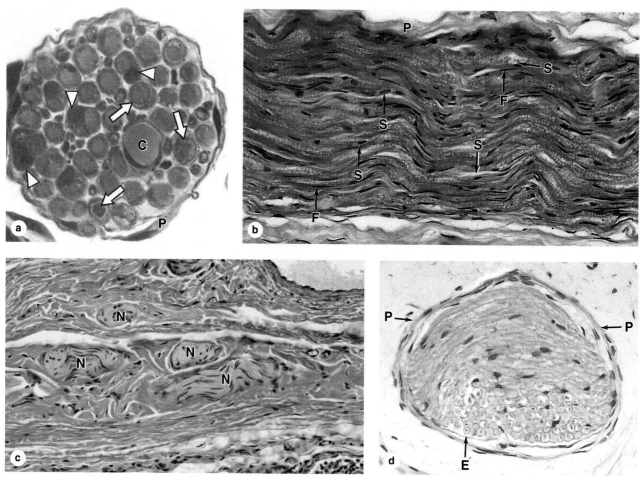

Figure 9–28. **Small nerves.** Small nerves can be seen in sections from most organs. **(a):** In cross-section an isolated, resin-embedded nerve is seen to have a thin perineurium, one capillary (C), and many large axons (arrows) associated with Schwann cells (arrowheads). A few nuclei of fibroblasts can be seen in the endoneurium between the myelinated fibers. X400. PT. **(b):** In longitudinal sections the flattened nuclei of endoneurial fibroblasts (F) and more oval nuclei of Schwann cells (S) can be distinguished. Nerve fibers are held rather loosely in the endoneurium and in low-magnification longitudinal section are seen to be wavy rather than straight. This indicates a slackness of fibers within the nerve which allows nerves to stretch slightly during body movements with no potentially damaging tension on the fibers. X200. H&E. **(c):** In sections of mesentery and other tissues, a highly wavy or tortuous disposition of a single small nerve (N) will be seen as multiple oblique or transverse pieces as the nerve enters and leaves the area in the section. X200. H&E. **(d):** Often a section of small nerve will have some fibers cut transversely and others cut obliquely within the same fascicle, again suggesting the relatively unrestrained nature of the fibers within the endoneurium (E) and perineurium (P). X300. H&E.

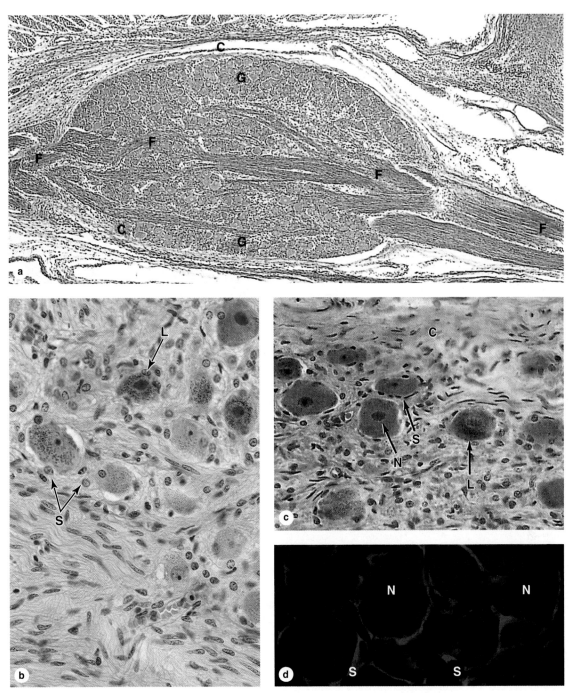

***Figure 9–29.* Ganglia. (a):** A **sensory ganglion** (G) has a distinct connective tissue capsule (C) and internal framework continuous with the epineurium and other components of peripheral nerves, except that no perineurium is present and there is no blood-nerve barrier function. Fascicles of nerve fibers (F) enter and leave these ganglia. X56. Luxol fast blue. **(b):** Higher magnification shows the small, rounded nuclei of glia cells called **satellite cells** (S) which produce thin, sheet-like cytoplasmic extensions that completely envelope each large neuronal perikaryon, some containing lipofuscin (L). X400. H&E. **(c):** Ganglia of sympathetic nerves are smaller than most sensory ganglia, but similar in having large neuronal cell bodies (N), some containing lipofuscin (L). Sheets from satellite cells (S) enclose each neuronal cell body with morphology slightly different from that of sensory ganglia. Autonomic ganglia generally have less well-developed connective tissue capsules (C) than sensory ganglia. X400. H&E. **(d):** Immunostained satellite cells form thin sheets (S) surrounding neuronal cell bodies (N). Like the effect of Schwann cells on axons, satellite glial cells insulate, nourish, and regulate the microenvironment of the neuronal cell bodies. X1000. Rhodamine red-labeled antibody against glutamine synthetase. (Figure 9–29d, with permission, from Menachem Hanani, Laboratory of Experimental Surgery, Hadassah University Hospital, Jerusalem.)

disappears altogether (as in the case of amputation of a limb), the newly growing axons may form a swelling, or **neuroma**, that can be the source of spontaneous pain.

Regeneration is functionally efficient only when the fibers and the columns of Schwann cells are directed to the correct place. In an injured mixed nerve, if regenerating sensory fibers grow into columns connected to motor end-plates that were occupied by motor fibers, the function of the muscle will not be reestablished.

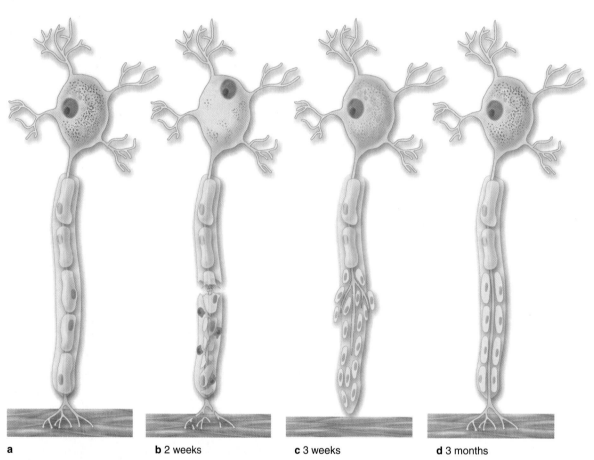

a　　　　　　**b** 2 weeks　　　　　　**c** 3 weeks　　　　　　**d** 3 months

Figure 9–30. **Regeneration in peripheral nerves.** In an injured or cut peripheral nerve, segments of axons distal to the injury lose their support from the cell bodies and degenerate completely. The proximal segments can regenerate from their cut ends after a delay. The main changes that take place in an injured nerve fiber are shown here. **(a):** Normal nerve fiber, with its perikaryon and effector cell (skeletal muscle). The cell body has much well-developed RER. **(b):** When the axon is injured, the neuronal nucleus moves to the cell periphery, and the RER is greatly reduced. The nerve fiber distal to the injury degenerates along with its myelin sheath. Debris is phagocytosed by macrophages. **(c):** The muscle fiber shows denervation atrophy. Schwann cells proliferate, forming a compact cord penetrated by the regrowing axon. The axon grows at the rate of 0.5–3 mm/day. **(d):** Here, the nerve fiber regeneration was successful and the muscle fiber was also regenerated after receiving nerve stimuli.

Muscle Tissue

Muscle tissue is composed of cells differentiated for optimal use of the universal cell property termed contractility. Microfilaments and associated proteins together generate the forces necessary for cellular contraction, which drives movement within certain organs and the body as a whole. Nearly all muscle cells are of mesodermal origin and they differentiate mainly by a gradual process of cell lengthening with simultaneous synthesis of myofibrillar proteins.

Three types of muscle tissue can be distinguished on the basis of morphologic and functional characteristics (Figure 10–1) and the structure of each type is adapted to its physiologic role. **Skeletal muscle** is composed of bundles of very long, cylindrical, multinucleated cells that show cross-striations. Their contraction is quick, forceful, and usually under voluntary control. It is caused by the interaction of thin actin filaments and thick myosin filaments whose molecular configuration allows them to slide upon one another. The forces necessary for sliding are generated by weak interactions in the bridges between actin and myosin. **Cardiac muscle** also has cross-striations and is composed of elongated, branched individual cells that lie parallel to each other. At sites of end-to-end contact are the **intercalated disks,** structures found only in cardiac muscle. Contraction of cardiac muscle is involuntary, vigorous, and rhythmic. **Smooth muscle** consists of collections of fusiform cells that do not show striations. Their contraction process is slow and not subject to voluntary control.

Some muscle cell organelles have names that differ from their counterparts in other cells. The cytoplasm of muscle cells is called **sarcoplasm** (Gr. *sarkos,* flesh, + *plasma,* thing formed) and the smooth ER is called **sarcoplasmic reticulum.** The **sarcolemma** (*sarkos* + Gr. *lemma,* husk) is the cell membrane, or plasmalemma.

SKELETAL MUSCLE

Skeletal muscle consists of **muscle fibers,** which are long, cylindrical multinucleated cells with diameters of 10–100 μm. Multinucleation results from the fusion of embryonic mesenchymal cells called **myoblasts** (Figure 10–2). The long oval nuclei are usually found at the periphery of the cell under the cell membrane. This characteristic nuclear location is helpful in discriminating skeletal muscle from cardiac and smooth muscle, both of which have centrally located nuclei.

MEDICAL APPLICATION

The variation in diameter of skeletal muscle fibers depends on factors such as the specific muscle and the age and sex, state of nutrition, and physical training of the individual. It is a common observation that exercise enlarges the musculature and decreases fat depots. The increase in muscle thus obtained is caused by formation of new myofibrils and a pronounced growth in the diameter of individual muscle fibers. This process, characterized by increased of cell volume, is called **hypertrophy** *(Gr. hyper, above, + trophe, nourishment). Tissue growth by an increase in the number of cells is termed* **hyperplasia** *(hyper + Gr. plasis, molding), which takes place most readily in smooth muscle, whose cells have not lost the capacity to divide by mitosis.*

Organization

The masses of fibers that make up the various types of muscle are arranged in regular bundles surrounded by the **epimysium,** an external sheath of dense connective tissue surrounding the entire muscle (Figures 10–3 and 10–4). From the epimysium, thin septa of connective tissue extend inward, surrounding the **fascicles** or bundles of fibers within a muscle. The connective tissue around each fascicle is called the **perimysium**. Each muscle fiber is itself surrounded by a more delicate connective tissue, the **endomysium**, composed of a basal lamina synthesized by the multinucleated fibers themselves as well as reticular fibers and fibroblasts. Within each fiber the nuclei are displaced peripherally against the sarcolemma.

One of the most important roles of this connective tissue is to transmit the mechanical forces generated by the contracting muscle cell/fibers because individual muscle cells seldom extend from one end of a muscle to the other.

Blood vessels penetrate the muscle within the connective tissue septa and form a rich capillary network in the endomysium (Figure 10–5). Lymphatic vessels and larger blood vessels are found in the other connective tissue layers.

Most muscles taper off at their extremities and connective tissue components of the epimysium show continuity with tendons through **myotendinous junctions** (Figure 10–6). TEM studies show that in these transitional regions, collagen fibers of the tendon insert themselves among muscle fibers and associate with complex infoldings of sarcolemma.

Muscle Fibers

As observed with the light microscope, longitudinally sectioned skeletal muscle fibers show cross-striations of alternating light and dark bands (Figures 10–7). The darker bands are called **A bands** (*ani*sotropic or birefringent in polarized light); the lighter bands are called **I bands** (*i*sotropic, do not alter polarized light).

In the TEM each I band is seen to be bisected by a dark transverse line, the **Z line** (Ger. *Zwischenscheibe*, between the discs). The repetitive functional subunit of the contractile apparatus, the **sarcomere**, extends from Z line to Z line (Figure 10–8) and is about 2.5 μm long in resting muscle.

The sarcoplasm has little RER or free ribosomes and is filled primarily with long cylindrical filamentous bundles called **myofibrils** running parallel to the long axis of the fiber. The myofibrils have a diameter of 1–2 μm and consist of an end-to-end repetitive arrangement of sarcomeres (Figure 10–8). The lateral registration of sarcomeres in adjacent myofibrils causes the entire muscle fiber to exhibit a characteristic pattern of transverse striations.

The A and I banding pattern in sarcomeres is due mainly to the regular arrangement of two types of **myofilaments**—thick and thin—that lie parallel to the long axis of the myofibrils in a symmetric pattern.

The thick filaments are 1.6 μm long and 15 nm wide; they occupy the A band, the central portion of the sarcomere. The thin filaments run between and parallel to the thick filaments and have one end attached to the Z line (Figure 10–8). Thin filaments are 1.0 μm long and 8 nm wide. As a result of this arrangement, the I bands consist of the portions of the thin filaments that do not overlap the thick filaments (which is why they are lighter staining). The A bands are composed mainly of thick filaments in addition to overlapping portions of thin filaments. Close observation of the A band shows the presence of a lighter zone in its center, the **H zone**, that corresponds to a region consisting only of the rod-like portions of the myosin molecule with no thin filaments present (Figure 10–8).

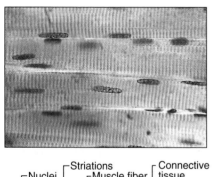
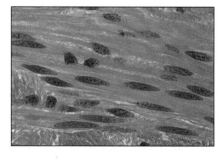

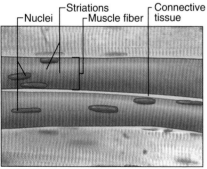
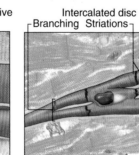
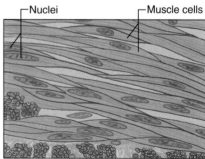

a Skeletal muscle   **b** Cardiac muscle   **c** Smooth muscle

Figure 10–1. **The three types of muscle.** Light micrographs of each type, accompanied by labeled drawings. **(a): Skeletal muscle** is composed of large, elongated, multinucleated fibers that show strong, quick, voluntary contractions. **(b): Cardiac muscle** is composed of irregular branched cells bound together longitudinally by intercalated disks and shows strong, involuntary contractions. **(c): Smooth muscle** is composed of grouped, fusiform cells with weak, involuntary contractions. The density of intercellular packing seen reflects the small amount of extracellular connective tissue present. (a, b): X200. (c): X300. All H&E.

Bisecting the H zone is the **M line** (Ger. *Mitte,* middle), a region where lateral connections are made between adjacent thick filaments (Figure 10–8). Major proteins present in the M line region are **myomesin**, a myosin-binding protein which holds the thick filaments in place, and creatine kinase, which catalyzes the transfer of phosphate groups from phosphocreatine (a storage form of high-energy phosphate groups) to adenosine diphosphate (ADP), thus helping to supply adenosine triphosphate (ATP) for muscle contraction.

Thin and thick filaments overlap for some distance within the A band. As a consequence, a cross section in the region of filament overlap shows each thick filament surrounded by six thin filaments in the form of a hexagon (Figure 10–8).

Thin filaments are composed of **F-actin**, associated with **tropomyosin**, which also forms a long fine polymer, and **troponin**, a globular complex of three subunits. Thick filaments consist primarily of **myosin**. Myosin and actin together represent 55% of the total protein of striated muscle.

F-actin consists of long filamentous polymers containing two strands of globular (G-actin) monomers, 5.6 nm in diameter, twisted around each other in a double helical formation (Figure 10–9). G-actin molecules are asymmetric and polymerize to produce a filament with polarity. Each G-actin monomer contains a binding site for myosin (Figure 10–9). Actin filaments, which are anchored perpendicularly on the Z line by the actin-binding protein α-actinin, exhibit opposite polarity on each side of the line (Figure 10–8c).

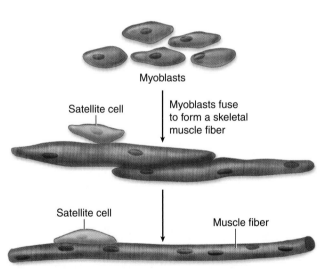

Figure 10–2. Development of skeletal muscle. Skeletal muscle begins to differentiate when mesenchymal cells called myoblasts align and fuse together to make longer, multinucleated tubes called myotubes. Myotubes synthesize the proteins to make up myofilaments and gradually begin to show cross striations by light microscopy. Myotubes continue differentiating to form functional myofilaments and the nuclei are displaced against the sarcolemma. Part of the myoblast population does not fuse and differentiate, but remains as a group of mesenchymal cells called muscle **satellite cells** located on the external surface of muscle fibers inside the developing external lamina. Satellite cells proliferate and produce new muscle fibers following muscle injury.

Each **tropomyosin** subunit is a long, thin molecule about 40 nm in length containing two polypeptide chains, which assembles to form a long polymer located in the groove between the two twisted actin strands (Figure 10–9).

Troponin is a complex of three subunits: TnT, which attaches to tropomyosin; TnC, which binds calcium ions; and TnI, which inhibits the actin-myosin interaction. Troponin complexes are attached at specific sites at regular intervals along each tropomyosin molecule (Figure 10–9).

Myosin is a much larger complex (molecular mass ~500 kDa). Myosin can be dissociated into two identical heavy chains and two pairs of light chains. Myosin heavy chains are thin, rod-like molecules (150 nm long and 2–3 nm thick) made up of two heavy chains twisted together as myosin tails. Small globular projections at one end of each heavy chain form the heads, which have ATP binding sites as well as the enzymatic capacity to hydrolyze ATP (ATPase activity) and the ability to bind actin. The four light chains are associated with the head. Several hundred myosin molecules are arranged within each thick filament with their rodlike portions overlapping and their globular heads directed toward either end (Figure 10–9).

Analysis of thin sections of striated muscle shows the presence of crossbridges between thin and thick filaments. These bridges are formed by the head of the myosin molecule plus a short part of its rodlike portion. These bridges are involved in the conversion of chemical energy into mechanical energy.

Sarcoplasmic Reticulum & Transverse Tubule System

In muscle the smooth endoplasmic reticulum (SER) is specialized for Ca^{2+} ion sequestration. The depolarization of this sarcoplasmic reticulum membrane, which results in the release of Ca^{2+} ions, is initiated at the specialized myoneural junction on the surface of the muscle cell. Surface-initiated depolarization signals would have to diffuse throughout the cell to produce Ca^{2+} release from internal sarcoplasmic reticulum cisternae. In larger muscle cells, such diffusion of the depolarization signal would lead to a wave of contraction with peripheral myofibrils contracting before those more centrally positioned. To provide for a uniform contraction, skeletal muscle fibers have a system of **transverse (T) tubules** (Figure 10–10). These fingerlike invaginations of the sarcolemma form a complex network of tubules that encircles every myofibril near the A-I band boundaries of each sarcomere (Figures 10–10 and 10–11).

Adjacent to opposite sides of each T tubule are expanded **terminal cisternae** of the sarcoplasmic reticulum. This specialized complex, consisting of a T tubule and usually two small cisternae of sarcoplasmic reticulum, is known as a **triad** (Figures 10–10 and 10–11). At the triad, depolarization of the sarcolemma-derived T tubules is transmitted to the sarcoplasmic reticulum membrane.

Muscle contraction depends on the availability of Ca^{2+} ions, and muscle relaxation is related to an absence of Ca^{2+}. The sarcoplasmic reticulum specifically regulates calcium flow, which is necessary for rapid contraction and relaxation cycles. The sarcoplasmic reticulum system consists of a branching network of small cisternae surrounding each myofibril (Figure 10–10). After a neurally mediated depolarization of the sarcoplasmic reticulum membrane, Ca^{2+} ions concentrated within these cisternae are passively released into the vicinity of the overlapping thick and thin filaments, whereupon they bind to troponin and allow bridging between actin and myosin molecules. When the

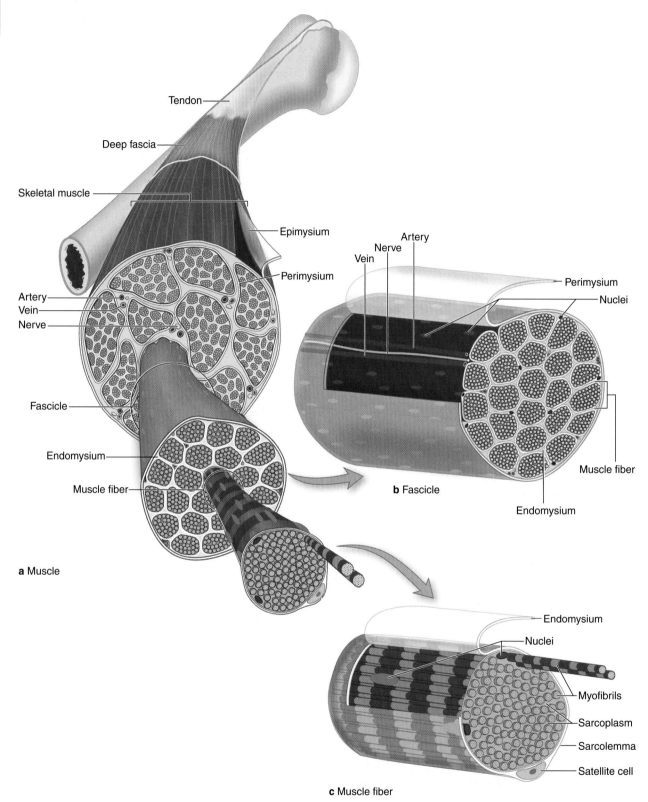

a Muscle

b Fascicle

c Muscle fiber

Figure 10–3. **Organization of skeletal muscle. (a):** An entire skeletal muscle is enclosed within a dense connective tissue layer called the **epimysium** continuous with the tendon binding it to bone. **(b):** Each fascicle of muscle fibers is wrapped in another connective tissue layer called the **perimysium**. **(c):** Individual muscle fibers (elongated multinuclear cells) is surrounded by a very delicate layer called the **endomysium**, which includes an external lamina produced by the muscle fiber (and enclosing the satellite cells) and ECM produced by fibroblasts.

membrane depolarization ends, the sarcoplasmic reticulum actively transports the Ca^{2+} back into the cisternae, ending contractile activity.

Mechanism of Contraction

Resting sarcomeres consist of partially overlapping thick and thin filaments. During contraction, neither the thick nor thin filaments changes their length. Contraction is the result of an increase in the amount of overlap between the filaments caused

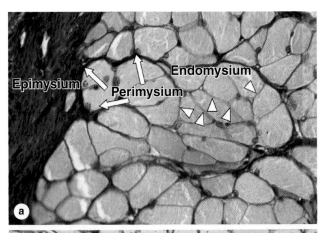

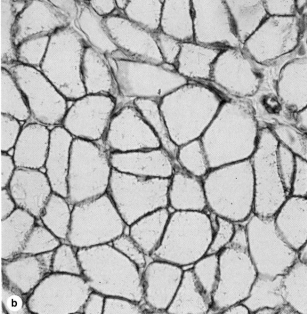

Figure 10–4. Skeletal muscle. (a): Micrograph shows a cross section of striated muscle demonstrating connective tissue and cell nuclei. The endomysium around individual muscle fibers is indicated by arrowheads. At left is a portion of the epimysium. All three of these tissues contain collagen types I and III (reticulin). X200. H&E. (b): Adjacent section immunohistochemically stained for laminin, which specifically stains the external lamina part of the endomysium produced by the muscle fibers themselves. X400. Anti-laminin.

by the sliding of thin and thick filaments past one another. Contraction is induced by an action potential produced at a synapse, the neuromuscular junction, between the muscle fiber and a terminus of a motor axon. Key molecular events in muscle contraction are summarized in Figure 10–11.

Although a large number of myosin heads extend from the thick filament, at any one time during the contraction only a small number of heads align with available actin-binding sites. As the bound myosin heads move the actin, however, they provide for alignment of new actin-myosin crossbridges. The old actin-myosin bridges detach only after the myosin binds a new ATP molecule; this action also resets the myosin head and prepares it for another contraction cycle. If no ATP is available, the actin-myosin complex becomes stable, which accounts for the extreme muscular rigidity (**rigor mortis**) that occurs after death. A single muscle contraction is the result of hundreds of bridge-forming and bridge-breaking cycles. The contraction activity that leads to a complete overlap between thin and thick filaments continues until Ca^{2+} ions are removed and the troponin–tropomyosin complex again covers the myosin-binding site.

During contraction, the I band decreases in size as thin filaments penetrate the A band. The H band—the part of the A band with only thick filaments—diminishes in width as the thin filaments completely overlap the thick filaments. A net result is that each sarcomere, and consequently the whole cell (fiber), is greatly shortened (Figure 10–12).

Innervation

Myelinated motor nerves branch out within the perimysium connective tissue, where each nerve gives rise to several terminal twigs. At the site of innervation, the axon loses its myelin sheath and forms a dilated termination situated within a trough on the muscle cell surface. This structure is called the **motor end-plate,**

Figure 10–5. Capillaries of skeletal muscle. The blood vessels were injected with plastic polymer before the muscle was collected and sectioned longitudinally. A rich network of capillaries in endomysium surrounding muscle fibers is revealed by this method. X200. Giemsa with polarized light.

or the **neuromuscular junction** (Figure 10–13). At this site, the axon is covered only by a thin cytoplasmic extension from a Schwann cell. Within the axon terminal are numerous mitochondria and synaptic vesicles, the latter containing the neurotransmitter **acetylcholine.** Between the axon and the muscle is a space, the **synaptic cleft,** in which lies an amorphous basal lamina matrix from the muscle fiber. At the junction, the sarcolemma is thrown into numerous deep **junctional folds,** which provide for greater surface area. In the sarcoplasm below the folds lie several nuclei and numerous mitochondria, ribosomes, and glycogen granules.

When an action potential reaches the motor end plate, acetylcholine is liberated from the axon terminal, diffuses across the cleft, and binds to acetylcholine receptors in the folded sarcolemma. Binding of the transmitter opens Na^+ channels in the sarcolemma, producing **membrane depolarization.** Excess acetylcholine is hydrolyzed by the enzyme cholinesterase bound to the synaptic cleft basal lamina. Acetylcholine breakdown is necessary to avoid prolonged contact of the transmitter with its receptors.

As shown in Figure 10–11, the depolarization initiated at the motor end-plate is propagated along the surface of the muscle cell and deep into the fibers via the transverse tubule system. At each triad, the depolarization signal is passed to the sarcoplasmic reticulum and results in the release of Ca^{2+}, which initiates the contraction cycle. When depolarization ceases, the Ca^{2+} is actively transported back into the sarcoplasmic reticulum cisternae, and the muscle relaxes.

A single nerve fiber (**axon**) can innervate one muscle fiber, or it may branch and be responsible for innervating 160 or more muscle fibers. In the case of multiple innervation, a single nerve fiber and all the muscle fibers it innervates are called a **motor unit.** Individual striated muscle fibers do not show graded contraction—they contract either all the way or not at all. To vary the force of contraction, the fibers within a muscle bundle do not all contract at the same time. Since muscles are composed of many motor units, the firing of a single motor axon will generate tension proportional to the number of muscle fibers innervated by that axon. Thus, the number of motor units and the

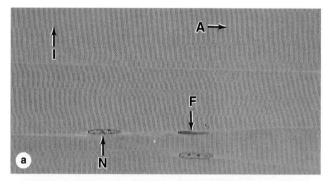

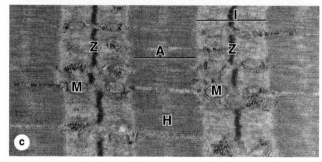

Figure 10–7. Striated skeletal muscle in longitudinal section. Longitudinal sections reveal the striations characteristic of skeletal muscle. **(a):** Parts of three muscle fibers separated by very small amounts of endomysium. One fibroblast nucleus (F) is shown. Muscle nuclei (N) are found against the sarcolemma. Along each fiber thousands of dark-staining A bands alternate with lighter I bands. X200. H&E. **(b):** At higher magnification, each fiber can be seen to have three or four myofibrils, with their striations slightly out-of-alignment with one another. Myofibrils are cylindrical bundles of thick and thin myofilaments which fill most of each muscle fiber. The middle of each I band can be seen to have a darker Z line (or disk). X500. Giemsa. **(c):** TEM showing the more electron-dense A bands bisected by a narrow, less electron-dense region called the H zone and in the I bands the presence of sarcoplasm with mitochondria (M), glycogen granules, and small cisternae of SER around the Z line. X24,000. (Figure 10–7c, with permission, from Mikel H. Snow, Department of Cell and Neurobiology, Keck School of Medicine at the University of Southern California.)

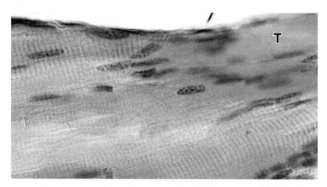

Figure 10–6. Myotendinous junction. Tendons develop together with skeletal muscles and join muscles to the periosteum of bones. The collagen fibers of tendons are continuous with those in the connective tissue layers in the muscle, forming a strong unit that allows muscle contraction to move the skeleton. The longitudinal section shows part of a tendon (T) inserted into the endomysium and perimysium of a muscle. X400. H&E.

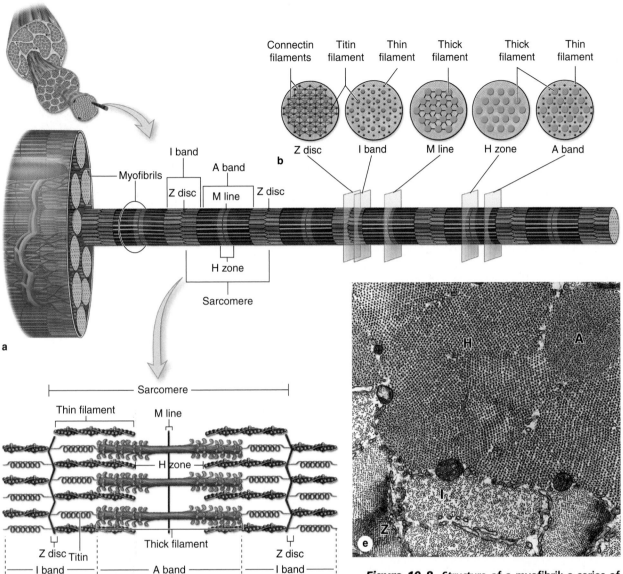

Connectin filaments | Titin filament | Thin filament | Thick filament | Thick filament | Thin filament

Z disc | I band | M line | H zone | A band

b

Myofibrils

I band

A band

Z disc | M line | Z disc

H zone

Sarcomere

a

Sarcomere

Thin filament | M line

H zone

Z disc | Titin | Thick filament | Z disc

I band | A band | I band

c

H | A

I

Z

e

Sarcomere

Z disc | M line | Z disc

H zone

I band | A band | I band

d

Figure 10–8. **Structure of a myofibril: a series of sarcomeres. (a):** Diagram indicates that each muscle fiber contains several parallel bundles called myofibrils. **(b):** Each myofibril consists of a long series of sarcomeres which contain thick and thin filaments and are separated from one another by Z discs. **(c):** Thin filaments are actin filaments with one end bound to **α-actinin**, the major protein of the Z disc. Thick filaments are bundles of myosin, which span the entire A band and are bound to proteins of the M line and to the Z disc across the I bands by a very large protein called **titin**, which has spring-like domains. **(d):** The molecular organization of the sarcomeres has bands of greater and lesser protein density, resulting in staining differences that produce the dark and light-staining bands seen by light microscopy and TEM. **(e):** TEM cross-sections through different regions of the sarcomere, as shown here, were useful in determining the relationships between thin and thick myofilaments and other proteins, as shown in part b of this figure. Thin and thick filaments are arranged so that each myosin bundle contacts six actin filaments.

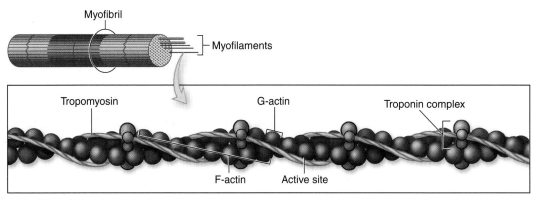

a Thin filament

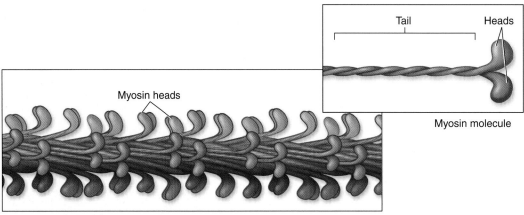

b Thick filament

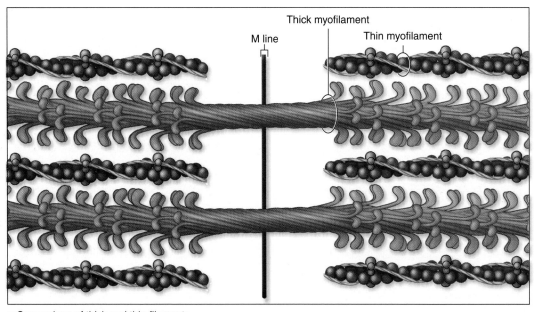

c Comparison of thick and thin filaments

Figure 10–9. **Molecules composing thin and thick filaments.** The contractile proteins are the thin and thick myofilaments within myofibrils. **(a):** Each thin filament is composed of F-actin, tropomyosin, and troponin complexes. **(b):** Each thick filament consists of many myosin heavy chain molecules bundled together along their rod-like tails, with their heads exposed and directed toward neighboring thin filaments. **(c):** Besides interacting with the neighboring thin filaments, thick myofilament bundles are held in place by less well-characterized myosin-binding proteins within the M line.

variable size of each unit can control the intensity of a muscle contraction. The ability of a muscle to perform delicate movements depends on the size of its motor units. For example, because of the fine control required by eye muscles, each of their fibers is innervated by a different nerve fiber. In larger muscles exhibiting coarser movements, such as those of the limb, a single, profusely branched axon innervates a motor unit that consists of more than 100 individual muscle fibers.

MEDICAL APPLICATION

Myasthenia gravis is an autoimmune disorder characterized by progressive muscular weakness caused by a reduction in the number of functionally active acetylcholine receptors in the sarcolemma of the myoneural junction. This reduction is caused by circulating antibodies that bind to the acetylcholine receptors in the junctional folds and inhibit normal nerve-muscle communication. As the body attempts to correct the condition, membrane segments with affected receptors are internalized, digested by lysosomes, and replaced by newly formed receptors. These receptors, however, are again made unresponsive to acetylcholine by similar antibodies, and the disease follows its progressive course.

Muscle Spindles & Tendon Organs

Striated muscles and myotendinous junctions contain sensory receptors that are encapsulated proprioceptors (L. *proprius,* one's own, + *capio,* to take). Among the muscle fascicles are stretch detectors known as **muscle spindles** (Figure 10–14). These structures consist of a connective tissue capsule surrounding a fluid-filled space that contains a few thin, nonstriated muscle fibers densely filled with nuclei and called **intrafusal fibers**. Several sensory nerve axons penetrate each muscle spindle and wrap around individual intrafusal fibers. Changes in length (usually stretch) of the surrounding striated (extrafusal) fibers caused by body movements are detected by the muscle spindles and the sensory nerves relay this information to the spinal cord. Different types of sensory and intrafusal fibers mediate reflexes of varying complexity to maintain posture and to regulate the activity of opposing muscle groups involved in motor activities such as walking.

In tendons, near the insertion sites of muscle fibers, a connective tissue sheath encapsulates the large collagen bundles of the myotendinous junction. Sensory nerves penetrate this capsule and form another sensory receptor known as (Golgi) **tendon organs** (Figure 10–14). Tendon organs detect changes in tension within tendons produced by muscle contraction and act to inhibit motor nerve activity if tension becomes excessive.

Because both of these sensory receptors detect increases in tension, they help to regulate the amount of effort required to perform movements that call for variable amounts of muscular force.

Muscle Fiber Types

Skeletal muscle cells are highly adapted for discontinuous production of intense work through the release of chemical energy.

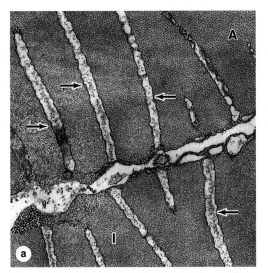
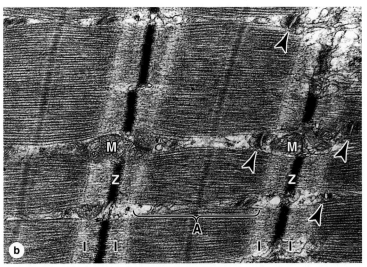

Figure 10–10. **Transverse tubule system.** Transverse tubules are invaginations of the sarcolemma that penetrate deeply into the muscle fiber around all myofibrils. **(a):** TEM shows portions of two fibers in cross-section and the intercellular space, and includes several transverse or T tubules cut lengthwise (arrows). X60,000. **(b):** TEM of a longitudinal section of skeletal muscle shows T tubules cut transversely (arrowheads) near the A-I interface, the most common location of T tubules in muscles of primates. Between the three myofibrils seen shown here is sarcoplasm containing mitochondria (M) and sarcoplasmic reticulum. Cisternae of this reticulum usually lie on each side of the transverse tubules, forming the triad of structures responsible for the cyclic release of Ca^{2+} from the cisternae and its sequestration again which occurs during muscle contraction and relaxation. The association between SR cisternae and T tubules is shown diagrammatically in the next figure. X40,000.

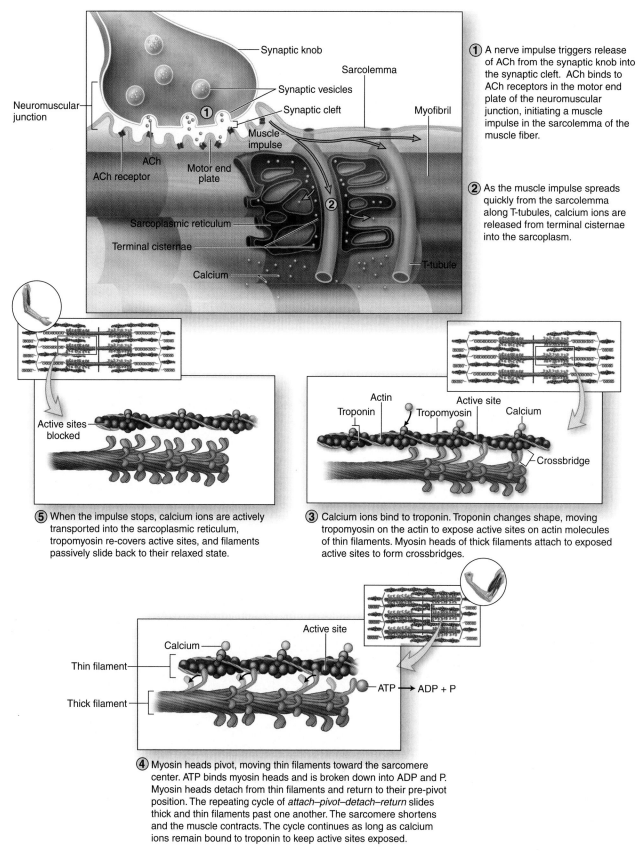

① A nerve impulse triggers release of ACh from the synaptic knob into the synaptic cleft. ACh binds to ACh receptors in the motor end plate of the neuromuscular junction, initiating a muscle impulse in the sarcolemma of the muscle fiber.

② As the muscle impulse spreads quickly from the sarcolemma along T-tubules, calcium ions are released from terminal cisternae into the sarcoplasm.

⑤ When the impulse stops, calcium ions are actively transported into the sarcoplasmic reticulum, tropomyosin re-covers active sites, and filaments passively slide back to their relaxed state.

③ Calcium ions bind to troponin. Troponin changes shape, moving tropomyosin on the actin to expose active sites on actin molecules of thin filaments. Myosin heads of thick filaments attach to exposed active sites to form crossbridges.

④ Myosin heads pivot, moving thin filaments toward the sarcomere center. ATP binds myosin heads and is broken down into ADP and P. Myosin heads detach from thin filaments and return to their pre-pivot position. The repeating cycle of *attach–pivot–detach–return* slides thick and thin filaments past one another. The sarcomere shortens and the muscle contracts. The cycle continues as long as calcium ions remain bound to troponin to keep active sites exposed.

Figure 10–11. **Events of muscle contraction.**

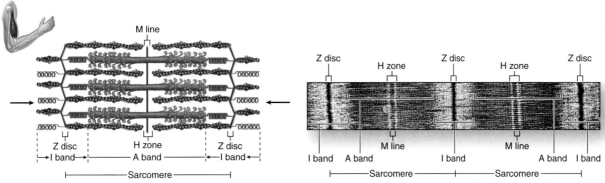

a Relaxed muscle
Sarcomere, I band, and H zone at a relaxed length.

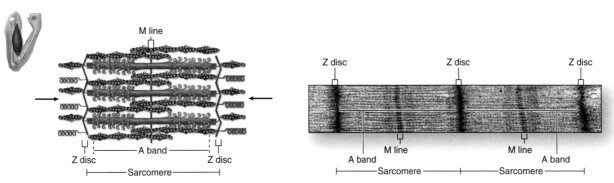

b Partially contracted muscle
Thick and thin filaments start to slide past one another. The sarcomere, I band, and H zone are narrower and shorter.

c Fully contracted muscle
The H zone and I band disappear, and the sarcomere is at its shortest length. Remember the lengths of the thick and thin filaments do not change.

***Figure 10–12.* Sliding filaments and sarcomere shortening in contraction.** Diagrams and TEM micrographs compare changes in the striations of skeletal muscle fibers according to the sliding filament mechanism. **(a):** In their relaxed state the sarcomere, I band and H zone are at their expanded length. The spring-like action of titin molecules, which span the I band, help pull thin and thick filaments past one another in relaxed muscle. **(b):** The Z discs at the sarcomere boundaries are drawn closer together during contraction as they move toward the ends of thick filaments in the A band. Titin molecules are compressed during contraction. **(c):** At maximal contraction, the H zone and I bands narrow and may disappear altogether.

Muscle fibers have depots of energy to cope with bursts of activity. The most readily available forms of energy are ATP and phosphocreatine, both of which are energy-rich phosphate compounds. Chemical energy is also stored in glycogen particles, which constitute about 0.5–1% of muscle weight (Figure 10–7c). Muscle tissue obtains energy as ATP and phosphocreatine from the aerobic metabolism of fatty acids and glucose. Fatty acids are broken down to acetate by the enzymes of β-oxidation in the mitochondrial matrix. Acetate is then further oxidized by the citric acid cycle, with the resulting energy being conserved in the form of ATP. When skeletal muscles are subjected to a short-term (sprint) exercise, they use anaerobic metabolism of glucose (coming mainly from glycogen stores), producing lactate and causing an oxygen debt that is repaid during the recovery period. The lactate formed during this type of exercise is the cause of cramping and pain in skeletal muscles.

Skeletal muscle fibers of humans are classified into three types based on their physiological, biochemical, and histochemical characteristics (Figure 10–15). All three fiber types are normally found throughout most muscles.

- **Type I** or **slow, red oxidative fibers** contain many mitochondria and abundant **myoglobin**, a protein with iron groups that bind O_2 and produce a dark red color. Red fibers derive energy primarily from aerobic oxidative phosphorylation of fatty acids and are adapted for slow, continuous contractions over prolonged periods, as required for example in the postural muscles of the back.

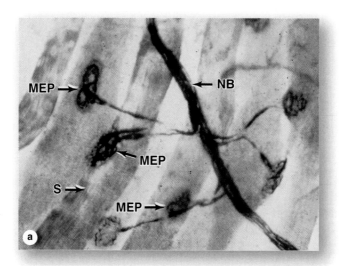

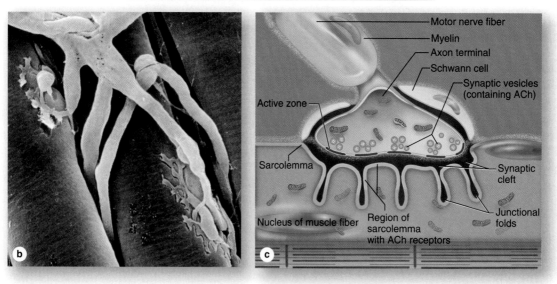

Figure 10–13. The neuromuscular junction (NMJ). Before it terminates in a skeletal muscle, each motor axon bundled in the nerve forms many branches, each of which forms a synapse with a muscle fiber. **(a):** Silver staining can reveal the nerve bundle (NB), the terminal axonal twigs, and the motor end plates (MEP) on striated muscle fibers (S). X1200. **(b):** A SEM shows the branching ends of a motor axon, each covered by an extension of the last Schwann cell and expanded terminally as a motor end plate embedded in a groove in the external lamina of the muscle fiber. **(c):** Diagram indicating key features of a typical neuromuscular junction: synaptic vesicles of acetylcholine (ACh), a synaptic cleft, and a postsynaptic membrane. This membrane, the sarcolemma, is highly folded to increase the number of Ach receptors at the NMJ. Receptor binding initiates muscle fiber depolarization, which is carried to the deeper myofibrils by the T tubules.

- **Type IIa** or **fast, intermediate oxidative-glycolytic fibers** have many mitochondria and much myoglobin, but also have considerable glycogen. They utilize both oxidative metabolism and anaerobic glycolysis and are intermediate between the other fiber types both in color and in energy metabolism. They are adapted for rapid contractions and short bursts of activity, such as those required for athletics.

- **Type IIb** or **fast, white glycolytic fibers** have fewer mitochondria and less myoglobin, but abundant glycogen, making them very pale in color. They depend largely on glycolysis for energy and are adapted for rapid contractions, but fatigue quickly. They are typically small muscles with a relatively large number of neuromuscular junctions, such as the muscles that move the eyes and digits.

The classification of fiber types in muscle biopsies has clinical significance for the diagnosis of muscle diseases, or myopathies (*myo* + Gr. *pathos,* suffering).

The differentiation of muscle into red, white, and intermediate fiber types is controlled by the frequency of impulses from its motor innervations, and fibers of a single motor unit are of the same type. If nerves to red and white fibers are exchanged experimentally, the fibers change their morphologic and physiologic characteristics to conform to the innervating nerve. Simple denervation of muscle leads to fiber atrophy and paralysis.

CARDIAC MUSCLE

During embryonic development, the mesoderm cells of the primitive heart tube align into chainlike arrays. Rather than fusing into multinucleated cells, as in the development of skeletal muscle fibers, cardiac muscle cells form complex junctions between extended processes (Figure 10–16). Cells within a fiber often branch and bind to cells in adjacent fibers. Consequently, the heart consists of tightly knit bundles of cells, interwoven in a fashion that provides for a characteristic wave of contraction that leads to a wringing out of the heart ventricles.

Mature cardiac muscle cells are approximately 15 μm in diameter and from 85 to 100 μm in length. They exhibit a cross-striated banding pattern comparable to that of skeletal muscle. Unlike multinucleated skeletal muscle, however, each cardiac muscle cell possesses only one or two centrally located pale-staining nuclei. Surrounding the muscle cells is a delicate sheath of endomysium containing a rich capillary network.

A unique and distinguishing characteristic of cardiac muscle is the presence of dark-staining transverse lines that cross the chains of cardiac cells at irregular intervals (Figures 10–16 and 10–17). These **intercalated discs** represent the interface between adjacent muscle cells where many junctional complexes are

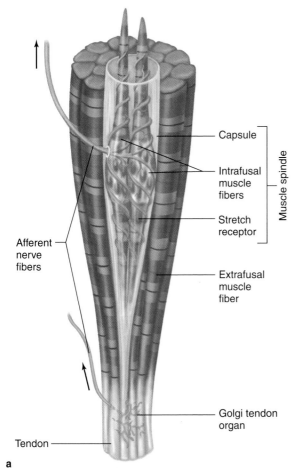

a

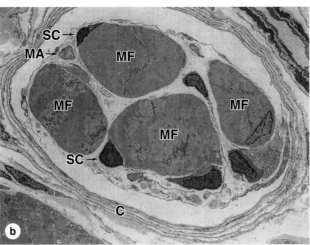

b

Figure 10–14. Sensory receptors associated with skeletal muscle. **(a):** Diagram shows both a **muscle spindle** and a **tendon organ.** Muscle spindles have afferent sensory and efferent motor nerve fibers associated with the intrafusal fibers, which are modified muscle fibers. The size of the spindle is exaggerated relative to the extrafusal fibers to show better the nuclei in the intrafusal fibers. **(b):** TEM cross-section near the end of a muscle spindle shows the capsule (C), sensory myelinated axons (MA), and the intrafusal muscle fibers (MF). These thin fibers differ from the ordinary skeletal muscle fibers in having essentially no myofibrils. Their many nuclei can either be closely aligned (nuclear chain fibers) or piled in a central dilatation (nuclear bag fibers). Satellite cells (SC) are also present within the external lamina of intrafusal fibers. Muscle spindles detect contraction of neighboring (extrafusal) muscle fibers during body movement and participate in the nervous control of body posture and the coordinate action of opposing muscles. The tendon organ collects information about the degree of tension among tendons and relays this data to the CNS, where the information is processed with that from muscle spindles to protect myotendinous junctions and help coordinate fine muscular contractions.

present (Figures 10–16 and 10–17). Transverse regions of these steplike discs have many **desmosomes** and **fascia adherentes** (which resemble the zonula adherentes between epithelial cells) and together these serve to bind cardiac cells firmly together to prevent their pulling apart under constant contractile activity. The more longitudinal portions of each disc have multiple **gap junctions**, which provide ionic continuity between adjacent cells. These act as "electrical synapses" and allow cells of cardiac muscle to act as in a multinucleated syncytium, with contraction signals passing in a wave from cell to cell.

The structure and function of the contractile proteins in cardiac cells are essentially the same as in skeletal muscle. The T tubule system and sarcoplasmic reticulum, however, are not as regularly arranged in cardiac fiber. The T tubules are more numerous and larger in cardiac muscle than in skeletal muscle and the sarcoplasmic reticulum is less well developed (Figure 10–18). Cardiac muscle cells contain numerous mitochondria, which occupy 40% or more of the cytoplasmic volume (Figure 10–18), reflecting the need for continuous aerobic metabolism in heart muscle. By comparison, only about 2% of skeletal muscle fiber is occupied by mitochondria. Fatty acids, transported to cardiac muscle cells by lipoproteins, are the major fuel of the heart and are stored as triglycerides in numerous lipid droplets seen in many cardiac muscle cells. Glycogen particles may also be present. Lipofuscin pigment granules are often found near the nuclei of cardiac muscle cells.

A few differences in structure exist between atrial and ventricular muscle. The arrangement of myofilaments is the same in both, but atrial muscle has markedly fewer T tubules, and the cells are somewhat smaller. Membrane-limited granules, each about 0.2–0.3 μm in diameter, are found at the poles of atrial muscle nuclei and are associated with Golgi complexes in this region (Figure 10–18). These granules release the peptide hormone atrial natriuretic factor (ANF) which acts on target cells in the kidney to affect Na^+ excretion and water balance. The contractile cells of the heart's atria thus also serve an endocrine function.

The rich autonomic nerve supply to the heart and the rhythmic impulse-generating and conducting structures are discussed in Chapter 11.

SMOOTH MUSCLE

Smooth muscle fibers are elongated, tapering, and nonstriated cells, each of which is enclosed by a thin basal lamina and a fine network of reticular fibers (Figure 10–19). The connective tissues serve to combine the forces generated by each smooth muscle fiber into a concerted action, eg, peristalsis in the intestine.

Smooth muscle cells may range in length from 20 μm in small blood vessels to 500 μm in the pregnant uterus. Each cell has a single nucleus located in the center of the cell's broadest part. To achieve the tightest packing, the narrow part of one cell lies adjacent to the broad parts of neighboring cells. Such an arrangement viewed in cross section shows a range of diameters, with only the largest profiles containing a nucleus (Figure 10–20a). The borders of the cell become scalloped when smooth muscle contracts and the nucleus becomes distorted.

Concentrated near the nucleus are mitochondria, polyribosomes, cisternae of rough ER, and the Golgi apparatus. Pinocytotic vesicles are frequent near the cell surface.

A rudimentary sarcoplasmic reticulum is present in smooth muscle cells, but T tubules are not. The characteristic contractile activity of smooth muscle is related to the structure and organization of its actin and myosin filaments, which do not exhibit the organization present in striated muscles. In smooth muscle cells, bundles of thin and thick myofilaments crisscross obliquely through the cell, forming a latticelike network. Smooth muscle actin and myosin contract by a sliding filament mechanism similar to that in striated muscles. However, myosin proteins are bundled differently and the cross-bridges interact with fewer F-actin filaments.

The thin filaments of smooth muscle cells lack troponin complexes and instead utilize **calmodulin,** a calcium-binding protein that is also involved in the contraction of non-muscle cells. As in all muscle, an influx of Ca^{2+} is involved in initiating contraction in smooth muscle cells. However in these cells the Ca^{2+} calmodulin complex activates **myosin light chain kinase (MLCK)**, the enzyme that phosphorylates myosin, which is required for myosin's interaction with F-actin. A number of hormones and other factors affect the activity of MLCK and thus influence the degree of contraction of smooth muscle cells.

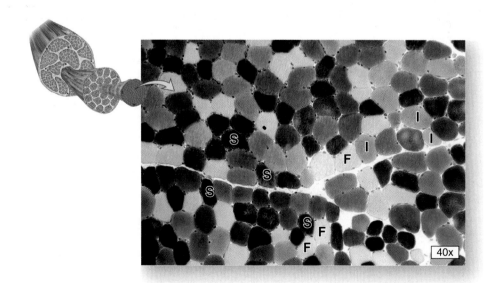

Figure 10–15. Skeletal muscle fiber types. Cross-section of skeletal muscle stained histochemically to detect the density of myofibrillar myosin-ATPase can be used to demonstrate the distribution of slow (S) type I fibers, intermediate (I) type IIa fibers, and fast (F) type IIb fibers.

Smooth muscle cells have an elaborate array of 10-nm intermediate filaments. **Desmin** is the major intermediate filament protein in all smooth muscles and **vimentin** is an additional component in vascular smooth muscle. Both intermediate filaments and F-actin filaments insert into **dense bodies** (Figure 10–20) which can be membrane-associated or cytoplasmic. Dense bodies contain α-actinin and are thus functionally similar to the Z discs of striated and cardiac muscles. The attachments of thin and intermediate filaments to the dense bodies helps transmit contractile force to adjacent smooth muscle cells and their surrounding network of reticular fibers (Figure 10–21).

Contraction of smooth muscle is not under voluntary control, but is regulated by autonomic nerves, certain hormones, and local physiological conditions such as the degree of stretch. The cells occur either as **multiunit smooth muscle**, in which each cell is innervated and can contract independently, or more commonly as **unitary smooth muscle**, in which only a few cells are innervated but all cells are interconnected by gap junctions. Gap junctions allow the stimulus for contraction to spread as a synchronized wave among adjacent cells. Smooth muscle lacks neuromuscular junctions like those in skeletal muscle. Instead axonal swellings with synaptic vesicles simply lie in close contact with the sarcolemma, with little or no specialized structure to the junctions.

Because smooth muscle is usually spontaneously active without nervous stimuli, its nerve supply serves primarily to modify activity rather than initiate it. Smooth muscle receives both adrenergic and cholinergic nerve endings that act antagonistically, stimulating or depressing its activity. In some organs, the cholinergic endings activate and the adrenergic nerves depress; in others, the reverse occurs.

In addition to contractile activity, smooth muscle cells also synthesize collagen, elastin, and proteoglycans, extracellular matrix (ECM) components normally synthesized by fibroblasts.

REGENERATION OF MUSCLE TISSUE

The three types of adult muscle have different potentials for regeneration after injury.

In skeletal muscle, although the nuclei are incapable of undergoing mitosis, the tissue can undergo limited regeneration. The source of regenerating cells is the sparse population of mesenchymal **satellite cells** that lies within the external lamina of

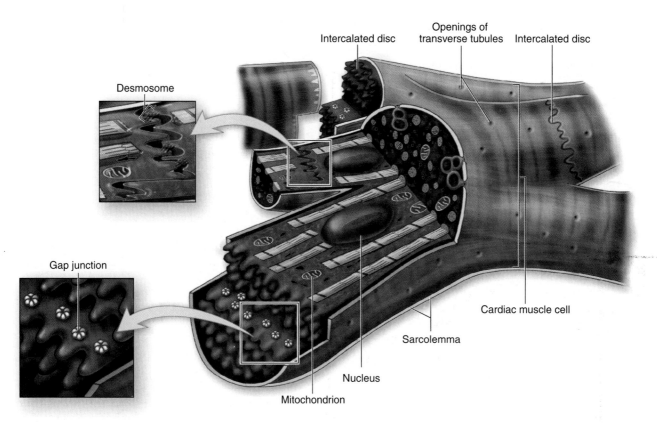

Figure 10–16. Cardiac muscle. Diagram of cardiac muscle cells indicates characteristic features of this muscle type. The fibers consist of separate cells with interdigitating processes where they are held together. These regions of contact are called the **intercalated discs**, which cross an entire fiber between two cells. The transverse regions of the steplike intercalated disc have abundant **desmosomes** and other adherent junctions which hold the cells firmly together. Longitudinal regions of these discs contain abundant **gap junctions**, which form "electrical synapses" allowing contraction signals to pass from cell to cell as a single wave. Cardiac muscle cells have central nuclei and myofibrils that are less dense and organized than those of skeletal muscle. Also the cells are often branched, allowing the muscle fibers to interweave in a more complicated arrangement within fascicles that produces an efficient contraction mechanism for emptying the heart.

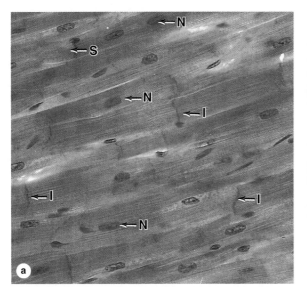
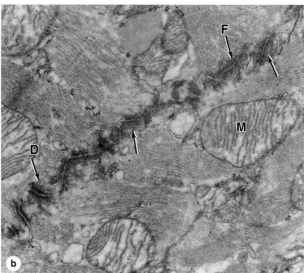

Figure 10–17. Cardiac muscle fibers. (a): Longitudinal sections of cardiac muscle at the light microscope level show nuclei (N) in the center of the muscle fibers and widely spaced intercalated discs (I) that cross the fibers. The occasional intercalated discs should not be confused with the repetitive, much more closely spaced striations (S), which are similar to those of skeletal muscle but less well-organized. Nuclei of fibroblasts in the endomysium are also present. X200. H&E. **(b):** TEM of an intercalated disc (arrows) shows a steplike structure representing the short interdigitating processes of the adjacent muscle cells. Transverse regions of the disc have many desmosomes (D) and adherent junctions called fascia adherentes (F), somewhat similar to the macula adherentes of epithelial cells. Fascia adherentes serve as anchoring sites for actin filaments of the terminal sarcomeres. Less electron-dense regions of the disc have abundant gap junctions. The sarcoplasm has numerous mitochondria (M) and myofibrillar structures similar to those of skeletal muscle but slightly less organized. X31,000.

Figure 10–18. Cardiac muscle ultrastructure. (a): TEM of cardiac muscle shows an abundance of mitochondria (M) and rather sparse sarcoplasmic reticulum (SR) in the areas between myofibrils. T tubules are less well-organized and are usually associated with one expanded terminal cisterna of SR, forming dyads (D) rather than the triads of skeletal muscle. Functionally these structures are similar in these two muscle types. X30,000. **(b):** Muscle cell from the cardiac atrium shows the presence of membrane-bound granules aggregated at the nuclear poles. These granules are most abundant in muscle cells of the right atrium (~600 per cell), but smaller quantities are also found in the left atrium and the ventricles. The atrial granules contain the precursor of a polypeptide hormone, **atrial natriuretic factor** (ANF). ANF targets cells of the kidneys to bring about sodium and water loss (natriuresis and diuresis). This hormone thus opposes the actions of aldosterone and antidiuretic hormone, whose effects on kidneys result in sodium and water conservation. X10,000. x, (Figure 10-18c, with permission, from Dr. J. C. Nogueira, Department of Morphology, Federal University of Minas Gerais, Belo Horizonte, Brazil.)

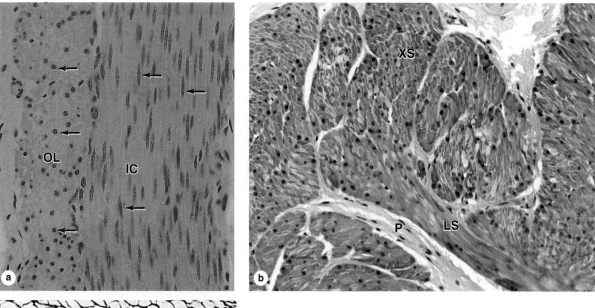

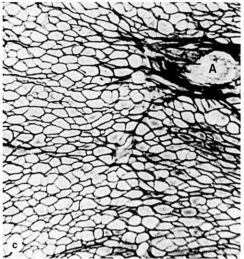

Figure 10–19. Smooth muscle. Cells or fibers of smooth muscle are long, tapering structures with elongated nuclei centrally located at the cell's widest part. **(a):** In a cross-section of smooth muscle in the wall of the small intestine, cells of the inner circular (IC) layer are cut lengthwise and cells of the outer longitudinal layer (OL) cross transversely. Only some nuclei (arrows) of the latter cells are in the plane of section, so that many cells appear to be devoid of nuclei. X140. H&E. **(b):** Section of smooth muscle in bladder, shows fibers in cross-section (XS) and longitudinal section (LS) with the same fascicle. There is much collagen in the branching perimysium (P), but very little evidence of endomysium is apparent. X140. Mallory trichrome. **(c):** Section stained only for reticulin reveals a thin endomysium around each fiber, with more reticulin in the connective tissue of small arteries (A). Reticulin fibers in the basal laminae of smooth muscle cells help hold the cells together as a functional unit during the slow, rhythmic contractions of this tissue. X200. Silver.

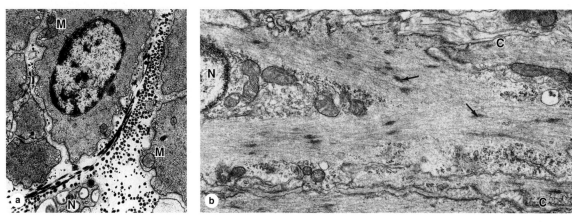

Figure 10–20. Smooth muscle ultrastructure. (a): TEM of a transverse section of smooth muscle showing six or seven cells sectioned at various points along their lengths, yielding profiles of various diameters with only the largest containing a nucleus. Thick and thin filaments are not organized into myofibril bundles and there are few mitochondria (M). There is evidence of a sparse external lamina around each cell and reticular fibers are abundant in the ECM. A small unmyelinated nerve (N) is also seen between the cells. X6650. **(b):** Longitudinal section showing several dense bodies in the cytoplasm (arrows) and at the cell membrane. Thin filaments and intermediate filaments both attach to the dense bodies. In the cytoplasm near the nucleus (N) are mitochondria, glycogen particles, and Golgi complexes. In the area shown at the lower right, the cell membrane shows invaginations called caveoli (C) (L. *caveoli*, little cavities), which in many cells are indicative of endocytosis, but in smooth muscle cells, where they are particularly numerous, may also function as the T tubules of skeletal muscle fibers and regulate release of Ca^{2+} from sarcoplasmic reticulum. X9,000.

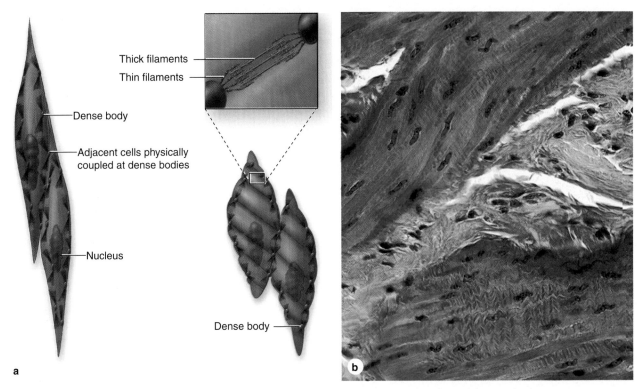

Thick filaments

Thin filaments

Dense body

Adjacent cells physically coupled at dense bodies

Nucleus

Dense body

a

b

Figure 10–21. Smooth muscle contraction. Most molecules that allow contraction are similar in the three types of muscle, but the filaments of smooth muscle are arranged differently and appear less organized. **(a):** The diagram shows thin filaments attach to **dense bodies** located in the cell membrane and deep in the cytoplasm. Dense bodies contain α-actinin for thin filament attachment. Dense bodies at the membrane are also attachment sites for intermediate filaments and for adhesive junctions between cells. This arrangement of both the cytoskeleton and contractile apparatus allows the multicellular tissue to contract as a unit, providing better efficiency and force. **(b):** Contraction decreases the length of the cell, deforming the nucleus and promoting contraction of the whole muscle. The micrograph shows a region of contracted tissue in the wall of a urinary bladder. The long nuclei of individual fibers assume a cork-screw shape when the fibers contract, reflecting the reduced cell length at this time. X240. Mallory trichrome.

each mature muscle fiber. Satellite cells are inactive, reserve myoblasts that persist after muscle differentiation. After injury or certain other stimuli, the normally quiescent satellite cells become activated, proliferating and fusing to form new skeletal muscle fibers. A similar activity of satellite cells has been implicated in muscle growth after extensive exercise, a process in which they fuse with their parent fibers to increase muscle mass beyond that occurring by cell hypertrophy. The regenerative capacity of skeletal muscle is limited, however, after major muscle trauma or degeneration.

Cardiac muscle lacks satellite cells and has virtually no regenerative capacity beyond early childhood. Defects or damage (eg, infarcts) in heart muscle are generally replaced by fibroblast proliferation and growth of connective tissue, forming myocardial scars. Smooth muscle, composed of simpler, mononucleated cells, is capable of a more active regenerative response. After injury, viable smooth muscle cells undergo mitosis and replace the damaged tissue. Contractile pericytes from the walls of small blood vessels (see Chapter 11) participate in the repair of vascular smooth muscle.

The Circulatory System

The circulatory system includes both the blood and lymphatic vascular systems. The **blood vascular system** (Figure 11–1) is composed of the following structures:

- The **heart,** an organ whose function is to pump the blood.
- The **arteries,** a series of efferent vessels that become smaller as they branch, and whose function is to carry the blood, with its nutrients and oxygen, to the tissues.
- The **capillaries,** the smallest blood vessels, constituting a complex network of thin tubules that branch profusely in almost every organ and through whose walls the interchange between blood and tissues takes place.
- The **veins,** which result from the convergence of capillaries into a system of larger channels that continue enlarging as they approach the heart, toward which they convey the blood to be pumped again.

The **lymphatic vascular system**, introduced with the discussion of interstitial fluid in Chapter 5, begins with the **lymphatic capillaries,** which are closed-ended tubules that merge to form vessels of steadily increasing size; these vessels terminate in the blood vascular system emptying into the large veins near the heart. One of the functions of the lymphatic system is to return the fluid of the tissue spaces to the blood. The internal surface of all components of the blood and lymphatic systems is lined by a single layer of a squamous epithelium, called endothelium.

The circulatory system is considered to consist of the macrovasculature, vessels that are more than 0.1 mm in diameter (large arterioles, muscular and elastic arteries, and muscular veins), and the **microvasculature** (arterioles, capillaries, and postcapillary venules) visible only with a microscope (Figure 11–2). The microvasculature is particularly important functionally, being the site of interchanges between blood and the surrounding tissues both under normal conditions and during inflammatory processes.

HEART

The heart is a muscular organ that contracts rhythmically, pumping the blood through the circulatory system (Figure 11–3). The right and left **ventricles** pump blood to the lungs and the rest of the body respectively; right and left **atria** receive blood from the body and the pulmonary veins respectively. The walls of all four heart chambers consist of three major layers or tunics: the internal endocardium; the middle myocardium; and the external epicardium.

The **endocardium** consists of a single layer of squamous endothelial cells on a thin layer of loose connective tissue containing elastic and collagen fibers as well as some smooth muscle cells. Connecting this subendothelial layer to the myocardium is additional connective tissue (often called the **subendocardial layer**) containing veins, nerves, and branches of the impulse-conducting system of the heart (Figure 11–4).

The **myocardium** is the thickest of the tunics and consists of cardiac muscle cells (see Chapter 10) arranged in layers that surround the heart chambers in a complex spiral. The myocardium is much thicker in the ventricles than in the atria. The arrangement of these muscle cells is extremely varied, so that in sections cells are seen to be oriented in many directions.

The heart is covered externally by simple squamous epithelium (mesothelium) supported by a thin layer of connective tissue that constitutes the **epicardium.** A subepicardial layer of loose connective tissue contains veins, nerves, and many adipocytes (Figure 11–5). The epicardium corresponds to the visceral layer of the **pericardium,** the serous membrane in which the heart lies. In the space between the pericardium's visceral layer (epicardium) and its parietal layer is a small amount of lubricant fluid that facilitates the heart's movements.

The cardiac valves consist of a central core of dense fibrous connective tissue (containing both collagen and elastic fibers), lined on both sides by endothelial layers. The bases of the valves are attached to strong fibrous rings that are part of the **fibrous skeleton**. This dense, fibrous region around the heart valves anchors the base of the valves and is the site of origin and insertion of the cardiac muscle fibers (Figure 11–6).

The heart has a specialized system to generate a rhythmic stimulus for contraction that is spread to the entire myocardium. This system (Figure 11–3) consists of two nodes located in the right atrium—the **sinoatrial (SA) node** (pacemaker) and the **atrioventricular (AV) node**—and the **atrioventricular bundle** (of His). The SA node is a small mass of modified cardiac muscle cells that are fusiform, smaller and with fewer myofibrils than neighboring muscle cells. The cells of the AV node are similar to those of the SA node but their cytoplasmic projections branch in various directions, forming a network. The AV bundle originates from the node of the same name, passes along the interventricular septum and splits into left and right bundles, and then branches further to both ventricles. The cells/fibers of the impulse-conducting system are modified cardiac muscle cells functionally integrated by gap junctions.

Distally fibers of the AV bundle become larger than ordinary cardiac muscle fibers and acquire a distinctive appearance. These **conducting myofibers** or **Purkinje fibers** have one or two

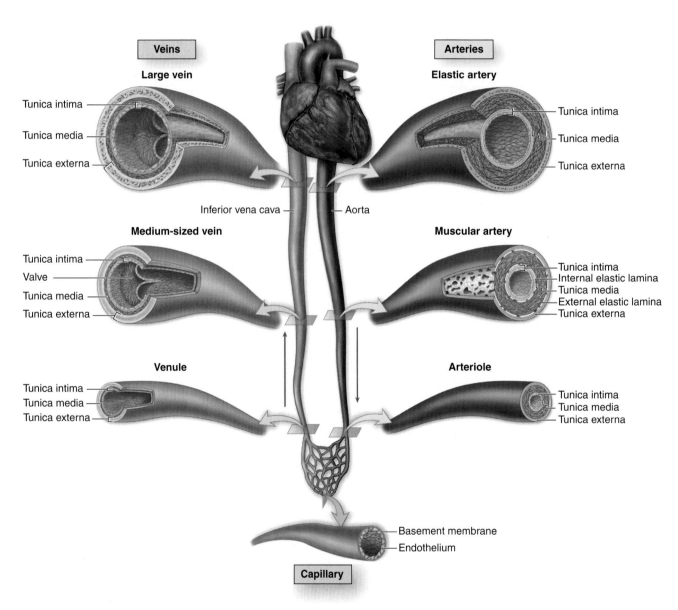

Figure 11–1. Vessels of the blood circulatory system. The heart is the principal organ of the blood circulatory system, pumping blood throughout the body and providing the force by which nutrients leave the capillaries and enter tissues. Large elastic arteries leave the heart and branch to form muscular arteries. These arteries branch further and enter organs, where they branch much further to form arterioles. These arterioles branch into the smallest vessels, the capillaries, the site of exchange between blood and surrounding tissue. Capillaries then merge to form venules, which merge further into small and then medium-sized veins. These veins leave organs, form larger veins which eventually bring blood back to the heart.

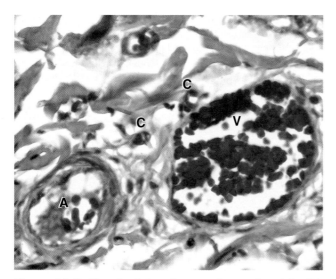

Figure 11-2. **Vessels of the microvasculature.** Arterioles (A), small capillaries (C) and venules (V) make up the microvasculature where, in almost every organ, exchange takes place between blood and the interstitial fluid of the tissues. X200. Masson trichrome.

central nuclei and their cytoplasm is rich in mitochondria and glycogen. Myofibrils are sparse and restricted to the periphery of the cytoplasm (Figure 11–4). After forming the subendocardial conducting network, these fibers penetrate the myocardial layer of both ventricles, an important arrangement that allows the stimulus for contraction to reach the innermost layers of the ventricular musculature.

Both parasympathetic and sympathetic neural components innervate the heart. Ganglionic nerve cells and nerve fibers are present in the regions close to the SA and AV nodes, where they affect heart rate and rhythm, such as during physical exercise and emotional stress. Stimulation of the parasympathetic division (vagus nerve) slows the heartbeat, whereas stimulation of the sympathetic nerve accelerates the rhythm of the pacemaker.

Between the muscular fibers of the myocardium are afferent free nerve endings related to sensibility and pain. Partial obstruction of the coronary arteries reduces the supply of oxygen to the myocardium and causes pain (angina pectoris).

TISSUES OF THE VASCULAR WALL

Walls of larger blood vessels contain three basic structural components: a simple squamous **endothelium**, **smooth muscle**, and **connective tissue** with elastic elements in addition to collagen.

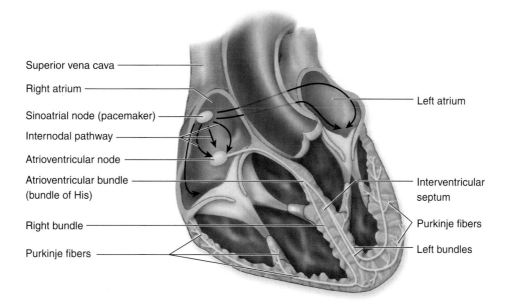

Figure 11-3. **Major histological features of the heart.** Longitudinal view of human heart showing the two atria and two ventricles. The ventricular walls are thicker than those of the atria, principally because of the much thicker myocardium. The **valves** are basically flaps of connective tissue anchored in the heart's dense **fibrous skeleton** region, shown in white. Other parts of the fibrous skeleton are the chordae tendinae, cords of dense connective tissue extending from the valves and attached to papillary muscles that help prevent valves from turning inside-out during ventricular contraction. All these parts of the fibrous skeleton are covered by endothelium. Shown in yellow are parts of the cardiac **conducting system**, which initiates the electrical impulse for heart's contraction (heartbeat) and spreads it through the ventricular myocardium. Both the sinoatrial (SA) node (pacemaker), in the posterior wall of the right atrium, and the atrioventricular (AV) node in the floor of the right atrium consist of myocardial tissue that is difficult to distinguish histologically from surrounding cardiac muscle. The AV node is continuous with specialized bundles of cardiac muscle fibers, the **AV bundle** (of His) which run along the interventricular septum to the apex of the heart, where they branch further as **conducting (Purkinje) fibers** which extend into myocardium of both ventricles.

The amount and arrangement of these tissues in vessels are influenced by **mechanical factors**, primarily blood pressure, and **metabolic factors** reflecting local needs of tissues.

The **endothelium** is a special type of epithelium that acts as a semipermeable barrier between two internal compartments: the blood plasma and the interstitial tissue fluid. Endothelium is highly differentiated to mediate and actively monitor the bidirectional exchange of small molecules and restrict the transport of some macromolecules.

Besides their role in the exchanges between blood and tissues, endothelial cells perform several other functions, including production of vasoactive factors that affect the vascular tone, such as nitric oxide, endothelins, and vasoconstrictive agents, and conversion of circulating angiotensin I to angiotensin II (see Chapter 19). Although morphologically similar, the endothelial cells of different blood vessels exert their various functional properties differently. Endothelial cells, especially those of arteries,

contain unique very small, elongated vesicles called Weibel-Palade bodies, which contain selectin and von Willebrand factor involved in blood coagulation.

Growth factors such as vascular endothelial growth factor (VEGF) help maintain the vasculature, regulate the formation of the vascular system from embryonic mesenchyme (vasculogenesis) and promote capillary outgrowth from existing vessels (angiogenesis) under normal and pathologic conditions in adults.

MEDICAL APPLICATION

The endothelium also has an antithrombogenic action, preventing blood coagulation. When endothelial cells are damaged by atherosclerotic

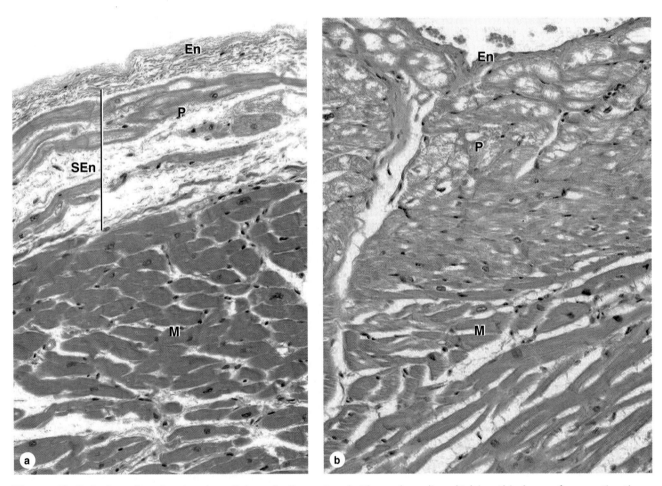

***Figure 11–4.* Endocardium & subendocardial conducting network.** The endocardium (En) is a thin layer of connective tissue lined by simple squamous endothelium. Between the endocardium and myocardium is a layer of variable thickness called the subendocardial layer (SEn) containing small nerves and in the ventricles the conducting (Purkinje) fibers (P) of the subendocardial conducting network. These fibers are cardiac muscle cells joined by intercalated disks but specialized for impulse conduction rather than contraction. Purkinje fibers are usually larger than contractile cardiac muscle fibers with large amounts of lightly stained glycogen filling most of the cytoplasm and displacing sparse myofibrils to the periphery. **(a):** Purkinje fibers running separately within the subendocardial layer. **(b):** Purkinje fibers intermingling with contractile fibers within the myocardium (M). Along with the nodes of specialized cardiac muscle in the right atrium which generate the electrical impulse, the network of conducting fibers comprises the conducting system of the heart. Both X200. H&E.

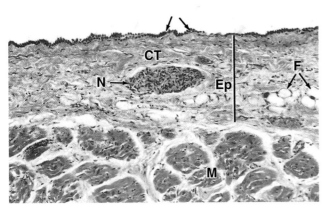

Figure 11–5. Epicardium or visceral pericardium. The external tunic of the heart, the epicardium, is the site of the coronary vessels and contains considerable adipose tissue. This section of atrium shows part of the myocardium (M) and epicardium (Ep). The epicardium consists of loose connective tissue (CT) containing both autonomic nerves (N) and fat (F). The epicardium is the visceral layer of the pericardium and is covered by the simple squamous-to-cuboidal epithelium (arrows) that also lines the pericardial space. These mesothelial cells secrete a lubricate fluid that prevents friction as the beating heart contacts the parietal pericardium on the other side of the pericardial cavity. X100. H&E.

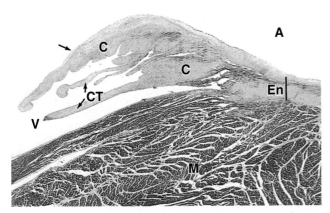

Figure 11–6. Valve leaflet and fibrous skeleton. The fibrous skeleton of the heart consists of masses of dense connective tissue in the endocardium which anchors the valves and surrounds the two atrioventricular canals, maintaining their proper shape. Section through a leaflet of the left atrioventricular valve (arrows) shows that valves are largely dense connective tissue (C) covered with a thin layer of endothelium. The collagen-rich connective tissue of the valves is stained pale green here and is continuous with the fibrous ring of connective tissue at the base of the valves, which fills the endocardium (En) of this area between the atrium (A) and ventricle (V). The **chordae tendinae** (CT), small strands of connective tissue which bind distal parts of valve leaflets, can also be seen here. The interwoven nature of the cardiac muscle fibers, with many small fascicles, in the myocardium (M) is also shown. X20. Masson trichrome.

lesions, for example, the uncovered subendothelial connective tissue induces the aggregation of blood platelets. This aggregation initiates a cascade of events that produce fibrin from circulating fibrinogen. An intravascular clot, or **thrombus** *(plural, thrombi), is formed that may grow until there is complete obstruction of the local vascular flow.*

From this thrombus, solid masses called **emboli** *(singular, embolus) may detach and be carried by the blood to obstruct distant blood vessels. In both cases the vascular flow may stop, a potentially life-threatening condition. Thus, the integrity of the endothelial layer preventing the contact between platelets and the subendothelial connective tissue is an important antithrombogenic mechanism.*

Smooth muscle cells or fibers occur in the walls of all vessels larger than capillaries and are arranged helically in layers. Each muscle cell is enclosed by an external lamina and by various amounts of other extracellular material, all of which these cells produce. In arterioles and small arteries the smooth muscle cells are frequently connected by communicating gap junctions.

Connective tissue components are present in vascular walls in amounts and proportions that vary based on local functional requirements. **Collagen fibers** are found throughout the wall: in the subendothelial layer, between muscle layers, and in the outer layers. **Elastic material** provides the resiliency for the vascular wall expanded under pressure. Elastin predominates in large arteries where it forms parallel lamellae regularly distributed between the muscle layers. **Ground substance** forms a heterogeneous gel in the extracellular spaces of the wall, contributing to the wall's physical properties and affecting permeability and diffusion of substances through the wall. Concentrations of glycosaminoglycans are higher in arterial than in venous tissues.

STRUCTURAL PLAN OF BLOOD VESSELS

All blood vessels greater than a certain diameter have many structural features in common and present a similar plan of construction. The distinction between different types of vessels often is not clear-cut because the transition from one type to another is gradual. Blood vessels are usually composed of the following layers, or tunics (L. *tunica*, coat), as shown in Figures 11–1 and 11–7.

- The **tunica intima** has one layer of endothelial cells supported by a thin subendothelial layer of loose connective tissue with occasional smooth muscle cells. In arteries, the intima is separated from the media by an **internal elastic lamina,** the most external component of the intima. This lamina, composed of elastin, has holes (fenestrae) that allow the diffusion of substances to nourish cells deep in the vessel wall. As a result of the loss of blood pressure and contraction of the vessel at death, the tunica intima of arteries may have a slightly folded appearance in tissue sections (Figure 11–8).

- The **tunica media,** the middle layer, consists chiefly of concentric layers of helically arranged smooth muscle cells (Figures 11–7 and 11–8). Interposed among the smooth muscle cells are variable amounts of elastic fibers and lamellae, reticular fibers of collagen type III, proteoglycans, and glycoproteins, all of which is produced by these cells. In arteries, the media has a thinner **external elastic lamina,** which separates it from the tunica adventitia.

- The **tunica adventitia** or tunica externa consists principally of type I collagen and elastic fibers (Figures 11–7 and 11–8). This adventitial layer is gradually continuous with the stromal connective tissue of the organ through which the blood vessel runs.

Large vessels usually have **vasa vasorum** ("vessels of the vessel"), which consist of arterioles, capillaries, and venules in the tunica adventitia and the outer part of the media (Figure 11–9). The vasa vasorum provide metabolites to cells of those layers, since in larger vessels the wall is too thick to be nourished solely by diffusion from the blood in the lumen. Luminal blood alone does provide nutrients and oxygen for cells of the tunica intima. Since they carry deoxygenated blood, large veins typically have more vasa vasorum than arteries.

Larger vessels are supplied with a network of unmyelinated sympathetic nerve fibers (**vasomotor nerves**) whose neurotransmitter is norepinephrine (Figure 11–9). Discharge of norepinephrine from these nerves produces vasoconstriction.

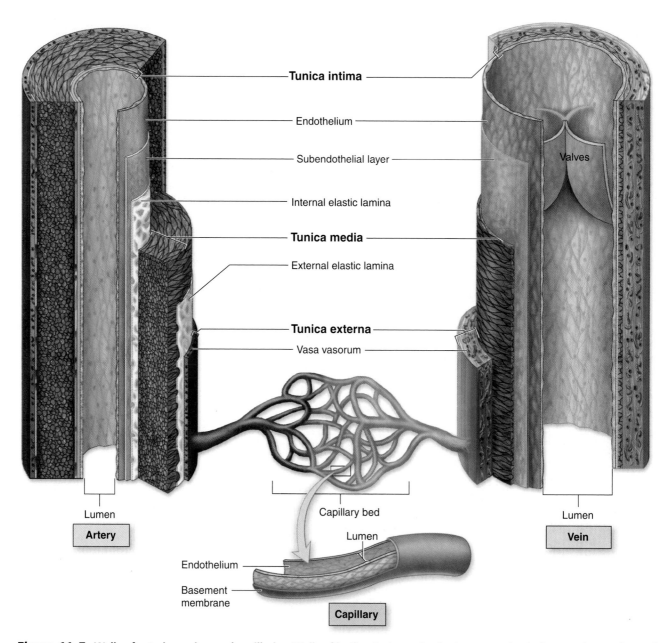

Figure 11–7. Walls of arteries, veins, and capillaries. Walls of both arteries and veins have a tunica intima, tunica media, and tunica externa (or adventitia), which correspond roughly to the heart's endocardium, myocardium and epicardium. An artery has a thicker tunica media and relatively narrow lumen. A vein has a larger lumen and its tunica externa is the thickest layer. The tunica intima of veins is often folded to form valves. Capillaries have only an endothelium, with no subendothelial layer or other tunics.

Because these efferent nerves generally do not enter the media of arteries, the neurotransmitter must diffuse for several micrometers to affect smooth muscle cells, where gap junctions propagate the response to the inner layers of muscle cells. In thinner-walled veins, nerve endings are found in both the adventitia and the media, but the overall density of innervation is less than that encountered in arteries. In skeletal muscle, arteries also receive a cholinergic vasodilator nerve supply. Acetylcholine released by these nerves acts on the endothelium to produce nitric oxide, which diffuses into the smooth muscle cells. The muscle cells then relax and the vessel lumen is dilated.

VASCULATURE

For didactic purposes vessels of the macrovasculature are classified arbitrarily as the types indicated in the following discussion.

Large Elastic Arteries

Large elastic arteries help to stabilize the blood flow. The elastic arteries include the aorta and its large branches. Freshly dissected, they have a yellowish color from the elastin in the media. The intima is thicker than the corresponding tunic of a muscular artery. An internal elastic lamina, although present, may not be easily discerned, since it is similar to the elastic laminae of the next layer (Figures 11–8 and 11–10). The media consists of elastic fibers and a series of concentrically arranged, perforated elastic laminae whose number increases with age (there are about 40 in the newborn, 70 in the adult). Between the elastic laminae are smooth muscle cells, reticular fibers, proteoglycans, and glycoproteins. The tunica adventitia is relatively underdeveloped.

The several elastic laminae contribute to the important function of making blood flow more uniform. During ventricular contraction (**systole**), the elastic laminae of large arteries are

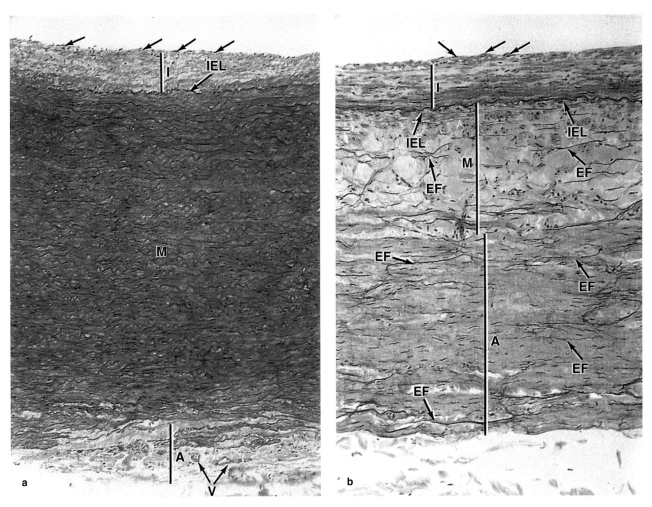

Figure 11–8. **Tunics of the vascular wall.** Comparison of the three major layers or tunics in the largest artery and vein. **(a):** aorta **(b):** vena cava. Simple squamous endothelial cells (arrows) line the tunica intima (I) which has subendothelial loose connective tissue and is separated from the tunica media by the internal elastic lamina (IEL), a prominent sheet of elastin. The media (M) contains elastic lamellae and fibers (EF) and multiple layers of smooth muscle not seen well here. The tunica media is much thicker in large arteries than veins, with relatively more elastin. Elastic fibers are also present in the outer tunica adventitia (A), which is relatively thicker in large veins. Vasa vasorum (V) are seen in the adventitia of the aorta. The connective tissue of the adventitia always merges with the less dense connective tissue around it. Both X122. Elastic.

stretched, reducing the force of the pressure somewhat. During ventricular relaxation (**diastole**), ventricular pressure drops to a low level, but the elastic rebound of large arteries helps to maintain arterial pressure. As a consequence, arterial pressure and blood velocity decrease and become less variable as the distance from the heart increases.

MEDICAL APPLICATION

Arterial Degenerative Alterations

Arteries undergo progressive and gradual changes from birth to death, and it is difficult to say where the normal growth processes end and the processes of involution begin. Each artery exhibits its own aging pattern.

*Atherosclerotic lesions are characterized by focal thickening of the intima, proliferation of smooth muscle cells and connective tissue elements, and the deposit of cholesterol in smooth muscle cells and macrophages. When heavily loaded with lipid, these cells may be referred to as **foam cells** and form the macroscopically visible fatty streaks and plaques that characterize **atherosclerosis**. These changes may extend to the inner part of the tunica media, and the thickening may become so great as to occlude the vessel. Coronary arteries are among those most predisposed to atherosclerosis. Uniform thickening of the intima is believed to be a normal phenomenon of aging.*

*Certain arteries irrigate only specific areas of specific organs and obstruction of this blood supply results in **necrosis** (death of tissues from a lack of metabolites). These **infarcts** occur*

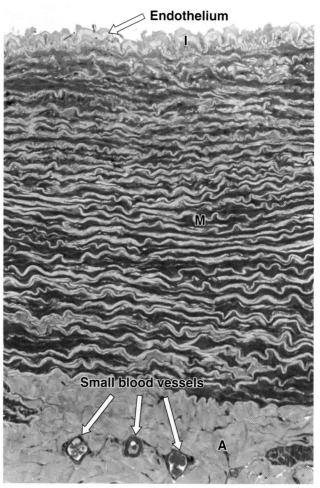

Figure 11–10. Elastic artery. The largest arteries contain considerable elastic material and expand with blood when the heart contracts. A transverse section through part of a large elastic artery shows a thick tunica media (M) consisting largely of many well-developed elastic lamellae. Strong pressure of blood pulsating into such arteries during systole expands the arterial wall, reducing the pressure and allowing strong blood flow to continue during diastole. The intima (I) of the empty aorta is typically folded and the adventitia (A) contains vasa vasorum. X200. PT.

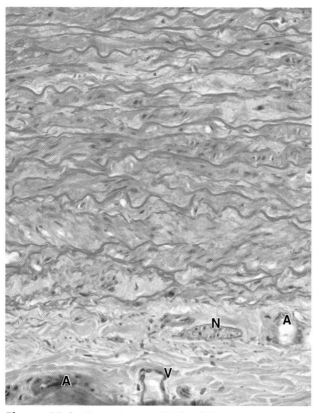

Figure 11–9. Vasa vasorum. Walls of the larger vessels, as the aorta, contain in the tunica adventitia a supply of microvasculature to bring O₂ and nutrients to local cells too far from the lumen to be nourished by blood there. These arterioles (A), capillaries and venules (V) constitute the vasa vasorum (vessels of vessels). The adventitia of large arteries is also supplied more sparsely with small sympathetic nerves (N) for control of vasoconstriction. X100. H&E.

commonly in the heart, kidneys, cerebrum, and certain other organs. In other regions (such as the skin), arteries anastomose frequently, and the obstruction of one artery does not lead to tissue necrosis, because the blood flow is maintained.

When the tunica media of an artery is weakened by an embryonic defect, disease, or lesion, the wall of the artery may dilate extensively. As this process of dilatation progresses, it becomes an aneurysm. Rupture of the aneurysm brings severe consequences and may cause death.

Muscular Arteries

The muscular arteries can control blood flow to organs by contracting or relaxing the smooth muscle cells of the tunica media. The intima has a very thin subendothelial layer and the internal elastic lamina, the most external component of the intima, is prominent (Figure 11–11). The tunica media may contain up to 40 layers of more prominent smooth muscle cells which are intermingled with a variable number of elastic lamellae (depending on the size of the vessel) as well as reticular fibers and proteoglycans. An external elastic lamina, the last component of the media, is present only in the larger muscular arteries. The adventitia consists of connective tissue. Lymphatic capillaries, vasa vasorum, and nerves are also found in the adventitia and these structures may penetrate to the outer part of the media.

Arterial Sensory Structures

Carotid sinuses are slight dilatations of the internal carotid arteries which contain **baroreceptors** detecting increases in blood pressure. The tunica media of each carotid sinus is thinner, allowing greater distension when blood pressure rises, and the intima and adventitia are rich in sensory nerve endings from cranial nerve IX, the glossopharyngeal nerve. The afferent nerve impulses are processed in the brain to trigger adjustments in vasoconstriction that return pressure to normal. Similar baroreceptors occur in aortic arches and other large arteries.

The **carotid bodies** are small, ganglia-like structures (paraganglia) near the bifurcation of the common carotid arteries that contain **chemoreceptors** sensitive to blood CO_2 and O_2 concentrations. A network of sinusoidal capillaries is intermixed with **glomus (type I) cells** containing numerous dense-core vesicles with dopamine, serotonin, and adrenaline (Figure 11–12). Dendritic fibers of cranial nerve IX, the glossopharyngeal nerve, synapse with the glomus cells. The sensory nerve is activated by neurotransmitter release from glomus cells in response to changes in the sinusoidal blood: increased CO_2, decreased O_2, or increased H^+ levels. **Aortic bodies** located on the arch of the aorta are similar in structure and function to carotid bodies.

Arterioles

Muscular arteries branch repeatedly into smaller and smaller arteries, until reaching a size with only two or three medial layers of muscle. The smallest arteries branch as **arterioles**, which have one or two smooth muscle layers and indicate the beginning of an organ's **microvasculature** (Figure 11–13) where exchanges between blood and tissue fluid occur. Arterioles are generally less than 0.5 mm in diameter, with lumens approximately

as wide as the wall is thick (Figures 11–2 and 11–14). The subendothelial layer is very thin, the elastic laminae are absent and the media is generally composed of circularly arranged smooth muscle cells. In both small arteries and arterioles, the tunica adventitia is very thin and inconspicuous.

In certain tissues and organs **arteriovenous shunts** or **anastomoses** regulate blood flow by allowing direct communication between arterioles and venules. Arterioles in such shunts have a relatively thick, capsule-like adventitia and a thick smooth muscle layer. Arteriovenous shunts are richly innervated by the sympathetic and parasympathetic nervous systems. These interconnections are abundant in skeletal muscle and in the skin of the hands and feet. When vessels of the arteriovenous anastomosis contract, all the blood must pass through the capillary network. When they relax, some blood flows directly to a venule instead of circulating in the capillaries. Their luminal diameters vary with the physiologic condition of the organ. Changes in

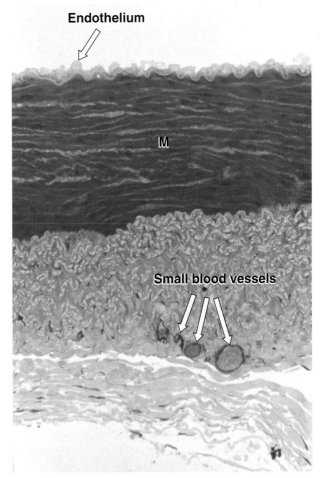

Figure 11–11. **Muscular artery.** With distance from the heart arteries gradually have relatively less elastin and more smooth muscle in their walls. Most arteries large enough to have names are of this muscular type. A transverse section through a muscular (medium caliber) artery shows multiple layers of smooth muscle in the media (M). The smooth muscle layers are more prominent than the elastic lamellae and fibers with which they intersperse. Vasa vasorum are seen in the tunica adventitia. X200. PT.

diameter of these vessels regulate blood pressure, blood flow, temperature and heat conservation in affected areas.

Capillaries

Capillaries permit different levels of metabolic exchange between blood and surrounding tissues. They are composed of a single layer of **endothelial cells** rolled up in the form of a tube. The average diameter of capillaries varies from 5 to 10 μm and their individual length is usually not more than 50 μm. Altogether capillaries comprise over 90% of all blood vessels in the body, with a total length of nearly 96,000 km (60,000 miles). The total diameter of the capillaries is approximately 800 times larger than that of the aorta. The velocity of blood in the aorta averages 320 mm/s, but in capillaries blood flows only about 0.3 mm/s. Because of their thin walls and slow blood flow, capillaries are a favorable place for the exchange of water, solutes, and macromolecules between blood and tissues.

Endothelial cells are functionally diverse according to the vessel they line. The capillaries are often referred to as exchange vessels, since it is at these sites that O₂, CO₂, substrates, and metabolites are transferred from blood to the tissues and from the tissues to blood. The mechanisms responsible for the interchange of materials between blood and tissue are not completely known. They depend on the kind of molecule and also on the structural characteristics and arrangement of endothelial cells in each type of capillary.

Small molecules, both hydrophobic and hydrophilic can diffuse or be actively transported across the plasmalemma of capillary endothelial cells. These substances are then transported by diffusion through the endothelial cytoplasm to the opposite cell surface, where they are discharged into the extracellular space. Water and some other hydrophilic molecules, less than 1.5 nm in diameter and below 10 kDa in molecular mass, can cross the capillary wall by diffusing through the intercellular junctions (paracellular pathway). The pores of fenestrated capillaries, the spaces between endothelial cells of sinusoidal capillaries, and the pinocytotic vesicles are other pathways for the passage of large molecules.

In general, endothelial cells are polygonal and elongated in the direction of blood flow (Figure 11–7). The nucleus causes that part of the cell to bulge into the capillary lumen. The cytoplasm contains a small Golgi appraratus, mitochondria, free ribosomes, and sparse cisternae of RER. Junctions of the tight

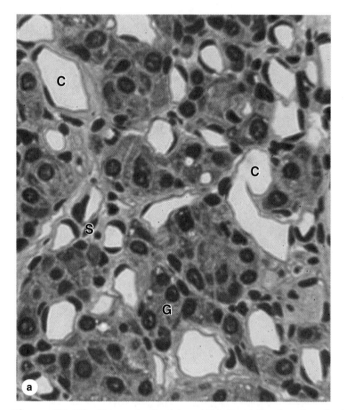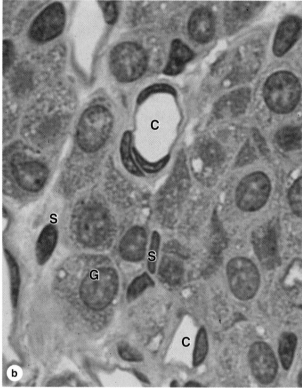

Figure 11–12. **Glomus body. (a)** and **(b)**: Specialized regions in the walls of specific arteries contain cells that act of chemoreceptors providing information to the brain regarding blood chemistry. The glomus bodies are two small (0.5–5 mm diameter) ganglion-like structures found near the bifurcations of the common carotid arteries. They contain many large sinusoidal capillaries (C) intermingled with clusters of large glomus cells (G) with round nuclei and cytoplasm filled with vesicles of various neurotransmitters that are clearly seen in (b). Supportive sheath cells (S) with elongated nuclei are associated with the groups of glomus cells. Glomus cells form synaptic connections with dendritic fibers of the glossopharyngeal nerve. Changes in the CO₂, O₂ and H⁺ concentrations in the sinusoidal blood are detected by the chemoreceptive glomus cells, which then release neurotransmitter that activates the sensory nerve to relay this information to the brain. a: X200; b: X400. Both PT.

zonula occludentes type are present between most endothelial cells, conferring the wall with variable permeability to macromolecules that plays significant roles in both normal and pathologic conditions.

MEDICAL APPLICATION

Junctions between endothelial cells of postcapillary venules are the loosest of the microvasculature. At these locations there is a characteristic loss of fluid from the circulatory system during the inflammatory response, leading to edema.

At various locations along capillaries and postcapillary venules are cells of mesenchymal origin with long cytoplasmic processes

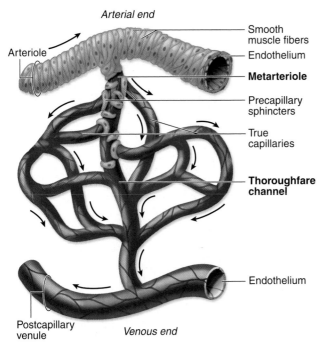

Figure 11–13. **Structure of microvasculature.** Microvasculature arises to meet nutritional needs of one organ or parts of one organ and consists of blood vessels of less than 0.5 mm diameter. Microvessels include **arterioles** and their smaller branches called **metarterioles** in which the layer of smooth muscle cells is dispersed as bands of cells that act as **precapillary sphincters**. The distal portion of the metarteriole, sometimes called a **thoroughfare channel**, lacks any smooth muscle cells. The wall of capillaries lacks smooth muscle cells altogether. The precapillary sphincters allow blood to enter the bed of capillaries in a pulsatile manner for maximally efficient exchange of nutrients, wastes, O_2, and CO_2 across the capillary wall. Capillaries and the metarteriole converge as **postcapillary venules**, the last component of the microvasculature. Blood enters microvasculature well-oxygenated and leaves poorly oxygenated.

partly surrounding the endothelial layer. These cells are called **pericytes** (Gr. *peri*, around, + *kytos*, cell). They are enclosed in their own basal lamina, which may fuse with that of the endothelial cells (Figure 11–15). Well-developed networks of myosin, actin, and tropomyosin in pericytes indicate these cells' primary contractile function. After tissue injuries, pericytes proliferate and differentiate to form both tunica media of new blood vessels and cells with various other functions in re-establishing the microvasculature and its ECM.

Capillaries have structural variations which permit different levels of metabolic exchange between blood and surrounding tissue. They can be grouped into three types, depending on the continuity of the endothelial cells and the external lamina (Figure 11–16).

1. The **continuous,** or tight, **capillary** (Figure 11–17) allows regulated exchange of material and is characterized by the distinct continuity of the endothelial cells in its wall. This is the most common type of capillary and is found in all kinds of muscle tissue, connective tissue, exocrine glands, and nervous tissue. In some places, but not in the nervous system, numerous pinocytotic vesicles are present on both endothelial cell surfaces. Vesicles also appear as isolated vesicles in the cytoplasm of these cells and are responsible for transcytosis of macromolecules in both directions across the endothelial cytoplasm.

2. The **fenestrated capillary** allows more extensive molecular exchange across the endothelium and is characterized by the presence of small circular fenestrae (L, *fenestra*, perforation) through the very thin squamous endothelial cells. Each fenestra is usually covered by a very thin diaphragm containing heparan proteoglycans but no lipid bilayer (Figures 11–18). The basal lamina of the fenestrated capillaries is continuous, covering the fenestrae. Fenestrated capillaries are found in tissues where rapid interchange of substances occurs between the tissues and the blood, as in the kidney, the intestine, the choroid plexus and the endocrine glands. Macromolecules experimentally injected into the bloodstream can cross the capillary wall through the fenestrae to enter tissue spaces.

3. The **sinusoid** or **discontinuous capillary** permits maximal exchange of macromolecules as well as cells between tissues and blood and has the following characteristics: endothelial cells have large fenestrae without diaphragms; the cells form a discontinuous layer and are separated from one another by wide spaces; the basal lamina is also discontinuous. Sinusoids are irregularly shaped and have diameters as large as 30–40 μm, much greater than those of other capillaries, properties which further slow blood flow at this site. Sinusoidal capillaries are found in the liver, spleen, some endocrine organs, and bone marrow (Figure 11–19).

Capillaries anastomose freely, forming a rich network or bed that interconnects the arterioles and venules (Figure 11–13). The arterioles may first branch into smaller vessels with a sparse layer of smooth muscle called **metarterioles**, which branch further into capillaries. Metarterioles often form a preferential channel for blood flow though a microcirculatory bed and help to regulate the circulation in capillaries. The richness of the capillary network is related to the metabolic activity of the tissues. Tissues with high metabolic rates, such as the kidney, liver, and cardiac and skeletal muscle, have an abundant capillary network; the opposite is true of tissues with low metabolic rates, such as smooth muscle and dense connective tissue.

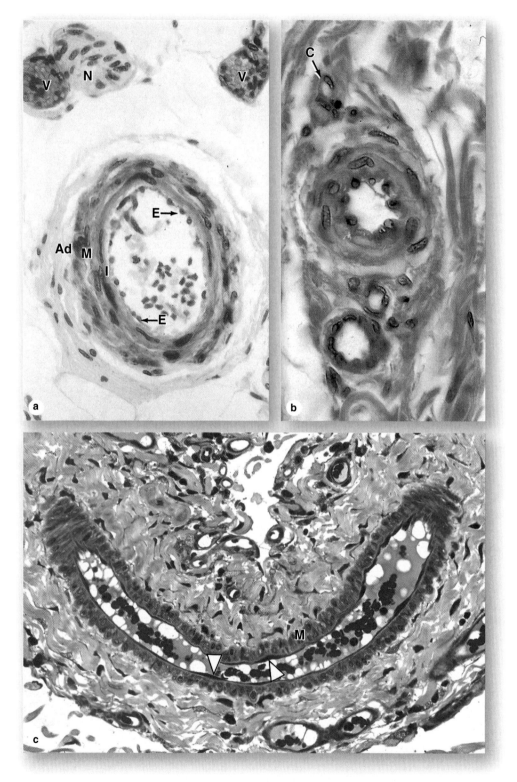

Figures 11–14. **Arterioles. (a):** Arterioles are microvessels with a tunica intima (I) that consists only of the endothelium (E), in which the cells may have rounded nuclei. They have tunica media (M) with only one or two layers of smooth muscle, and usually thin, inconspicuous adventitia (Ad). X350. Masson trichrome. **(b):** Three arterioles of various sizes are shown here and a capillary. X400. H&E. **(c):** A large mesenteric arteriole is cut obliquely and longitudinally and clearly shows the endothelial cells (arrow heads) and one or two layers of smooth muscle cells (M) cut transversely. Adventitia merges imperceptibly with neighboring connective tissue. X300. PT.

Venules

The transition from capillaries to venules occurs gradually. The immediate **postcapillary venules** are similar structurally to capillaries, with pericytes, but range in diameter from 15 to 20 μm. Postcapillary venules participate in the exchanges between the blood and the tissues and, as described in Chapter 12, are the primary site at which white blood cells leave the circulation at

sites of infection or tissue damage. These venules converge into larger **collecting venules** which have more contractile cells. With greater size the venules become surrounded by recognizable tunica media with two or three smooth muscle layers and are called **muscular venules**. A characteristic feature of all venules is the large diameter of the lumen compared to the overall thinness of the wall (Figure 11–20).

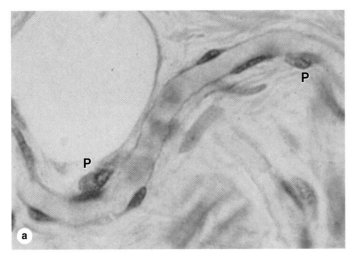
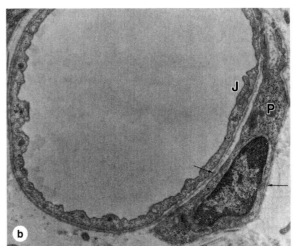

Figure 11–15. **Capillary with pericyte.** Capillaries consist only of an endothelium rolled as a tube, across which molecular exchange occurs between blood and tissue fluid. **(a):** Capillaries are normally associated with perivascular contractile cells called pericytes (P) which have a variety of functions. The more flattened nuclei belong to endothelial cells. X400. H&E. **(b):** TEM of a capillary cut transversely, showing the thin wall of one endothelial cell covered by an external lamina (arrows). Endothelial cells have numerous transcytotic vesicles and their edges overlap and are bound tightly together with occluding junctions (J). One pericyte (P) is shown, surrounded by its own external lamina. Pericytes can proliferate to form smooth muscle cells when a capillary is transformed into an arteriole or venule after tissue injury and repair. X13,000. (Figure 11–15b, reproduced, with permission, from Kelly DE, Wood RL, Enders AC (eds): *Bailey's Textbook of Microscopic Anatomy,* 18th ed. Williams & Wilkins, 1984. Reproduced, with permission, from Kelly D. E., Wood R.L., and Enders AC (eds): Bailey's Textbook of Microscopic Anatomy, 18th ed. Williams & Wilkins, 1984.)

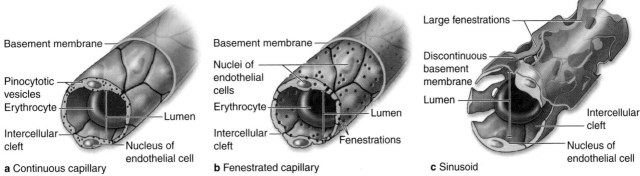

Figure 11–16. **Types of capillaries.** The vessels between arterioles and venules can be any of three types. **(a): Continuous capillaries,** the most common type, have tight, occluding junctions sealing the intercellular clefts between all the endothelial cells to produce minimal fluid leakage. All molecules exchanged across the endothelium must cross the cells by diffusion or transcytosis. **(b): Fenestrated capillaries** also have tight junctions, but perforations (fenestrae) through the endothelial cells allow greater exchange across the endothelium. The external lamina is continuous in both these capillary types. Fenestrated capillaries are found in organs where molecular exchange with the blood is important, such as endocrine organs, intestinal walls, and choroid plexus. **(c): Sinusoids** usually have a wider diameter than the other types of capillaries and have discontinuities between the endothelial cells, large fenestrae through the cells, and a partial, discontinuous basement membrane. Sinusoids are found in organs where exchange of macromolecules and cells occurs readily between tissue and blood, such as in bone marrow, liver, and spleen.

Veins

Blood entering veins is under very low pressure and moves toward the heart by contraction of the tunica media and external compressions from surrounding muscles and other organs. Valves project from the tunica intima to prevent back-flow of blood. Most veins are **small** or **medium veins** (Figure 11–21), with diameters less than one centimeter. Such veins are usually located in parallel with corresponding muscular arteries. The intima usually has a thin subendothelial layer and the media consists of small bundles of smooth muscle cells intermixed with reticular fibers and a delicate network of elastic fibers. The collagenous adventitial layer is well-developed.

The big venous trunks, paired with elastic arteries close to the heart, are **large veins** (Figure 11–8). Large veins have a well-developed tunica intima, but the tunica media is relatively thin, with few layers of smooth muscle and abundant connective tissue. The adventitial layer is thick in large veins and frequently contains longitudinal bundles of smooth muscle. Both the media and adventitia contain elastic fibers, but elastic laminae like those of arteries are not present.

Most veins have valves, but these are most prominent in large veins. Valves consist of paired semilunar folds of the tunica intima projecting across part of the lumen (Figures 11–21 and 11–22). They are rich in elastic fibers and are lined on both sides by endothelium. The valves, which are especially numerous in veins of the legs, help keep the flow of venous blood directed toward the heart.

LYMPHATIC VASCULAR SYSTEM

In addition to blood vessels, the body has a system of thin-walled endothelial channels that collect excess interstitial fluid from the tissue spaces and return it to the blood. This fluid is called lymph;

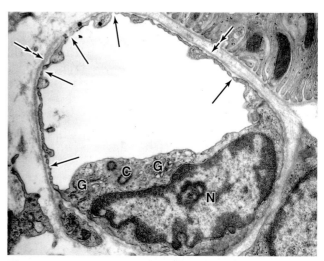

Figure 11–18. **Fenestrated capillary.** Fenestrated capillaries are specialized for uptake of molecules such as hormones in endocrine glands or for outflow of molecules such as in the kidney's filtration system. TEM of a transversely sectioned fenestrated capillary in the peritubular region of the kidney shows many typical fenestrae closed by diaphragms (arrows), with a continuous external lamina on the outer surface of the endothelial cell (double arrows). The diaphragms contain heparan sulfate proteoglycans, but their functional role is poorly understood at the molecular level. In this cell the Golgi apparatus (G), nucleus (N), and centrioles (C) can be seen. Fenestrated capillaries allow a freer exchange of molecules than continuous capillaries and are found in the intestinal wall, kidneys and endocrine glands. X10,000. (With permission, from Johannes Rhodin, Department of Cell Biology, New York University School of Medicine.)

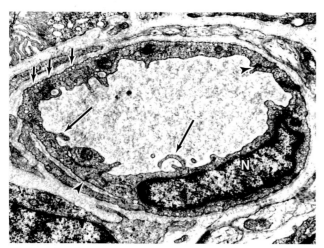

Figure 11–17. **Continuous capillary.** Continuous capillaries exert the tightest control over what molecules leave across their walls. TEM shows a continuous capillary in transverse section. A nucleus (N) is prominent, but tight or occluding junctions along overlapping folds between two cells can also be seen (arrowheads). Numerous transcytotic vesicles are evident (small arrows). The long arrows show extensions of broad cytoplasmic sheets suggesting phagocytosis, which is consistent with the presence of vacuoles and electron-dense lysosomes. All material that crosses continuous capillary endothelium must pass *through* the cells, usually by diffusion or transcytosis. X10,000.

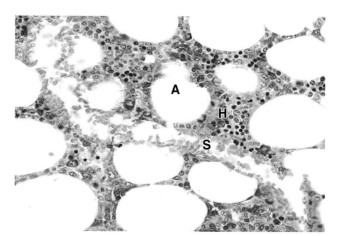

Figure 11–19. **Sinusoidal capillary.** Sinusoidal capillaries or sinusoids generally have much greater diameters than most capillaries and are specialized not only for maximal molecular exchange between blood and surrounding tissue, but also for easy movement of blood cells across the endothelium. The sinusoid (S) shown here is in bone marrow and is surrounded by tissue containing adipocytes (A) and masses of hematopoietic cells (H). The endothelium is very thin and cell nuclei are more difficult to find than in smaller capillaries. Ultrastructurally sinusoidal capillaries are seen to have large fenestrae through the cells and large discontinuities between the cells and through the basal lamina. X200. H&E.

unlike the blood, it flows in only one direction, toward the heart. The **lymphatic capillaries** originate in the various tissues as thin, closed-ended vessels that consist of a single layer of endothelium and an incomplete basal lamina. Lymphatic capillaries are held open by bundles of anchoring filaments of the elastic fiber system which also bind the vessels firmly to the surrounding connective tissue (Figure 11–23).

The thin lymphatic capillaries converge into larger lymphatic vessels. Interposed in the path of these lymphatics are lymph nodes, whose morphologic characteristics and functions are discussed in Chapter 14. With rare exceptions, such as the CNS and the bone marrow, lymphatic are found in almost all organs.

The larger lymphatics have a structure similar to that of veins except that they have thinner walls and lack a clear-cut separation between tunics. They also have more numerous internal valves (Figure 11–24). The lymphatic vessels are often dilated and assume a nodular, or beaded, appearance between the valves. As in veins, lymphatic circulation is aided by external forces (eg, contraction of surrounding skeletal muscle) and unidirectional lymph flow is mainly a result of the many valves. Contraction

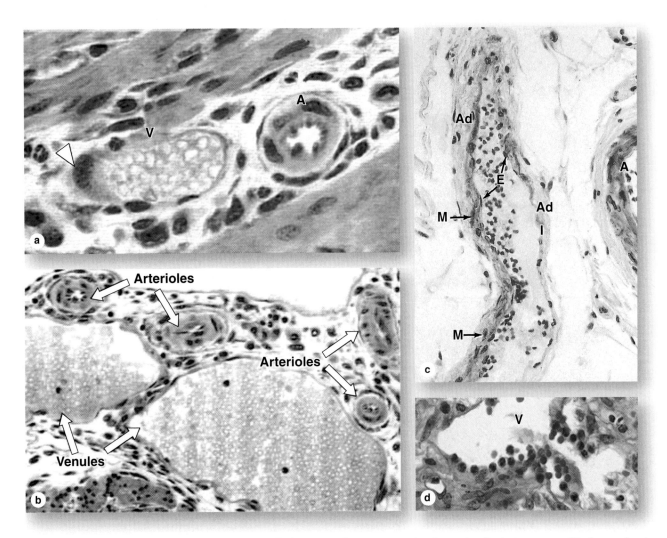

Figure 11–20. **Venules.** A series of increasingly larger and more organized venules lie between capillaries and veins. **(a): Postcapillary venules** resemble large capillaries, having only an endothelium with occasional pericytes (arrowhead). Their lumens and overall diameters are greater than those of nearby arterioles. X400. TB. **(b):** Large **collecting venules** have much greater diameters than arterioles but the wall is still very thin, consisting of an endothelium with more numerous pericytes or smooth muscle cells. X200. TB. **(c): Muscular venule** has a better defined tunica media, with as many as three layers of smooth muscle (M) in some areas, a very thin intima (I) of endothelial cells (E), and a more distinct tunica adventitia (Ad). Part of an arteriole (A) is included for comparison. Venules are the site in the vasculature where white blood cells leave the circulation to become functional in the interstitial space of surrounding tissues when such tissues are inflamed or infected. Such conditions cause endothelial cells of venules to loosen intercellular junctions and express new protein receptors on their luminal surface. Surface proteins on passing leukocytes bind these receptors, causing the cells to stick to the endothelial cells in a process termed margination. This adhesion is quickly followed by emigration from the venule between endothelial cells. X200. Masson trichrome. **(d):** Venule (V) from an infected small intestine shows several leukocytes adhering to and migrating across the endothelium. X200. H&E. (Figure 11–20a, with permission, from Telma M.T. Zorn, Department of Cell and Developmental Biology, University of São Paulo, Brazil.)

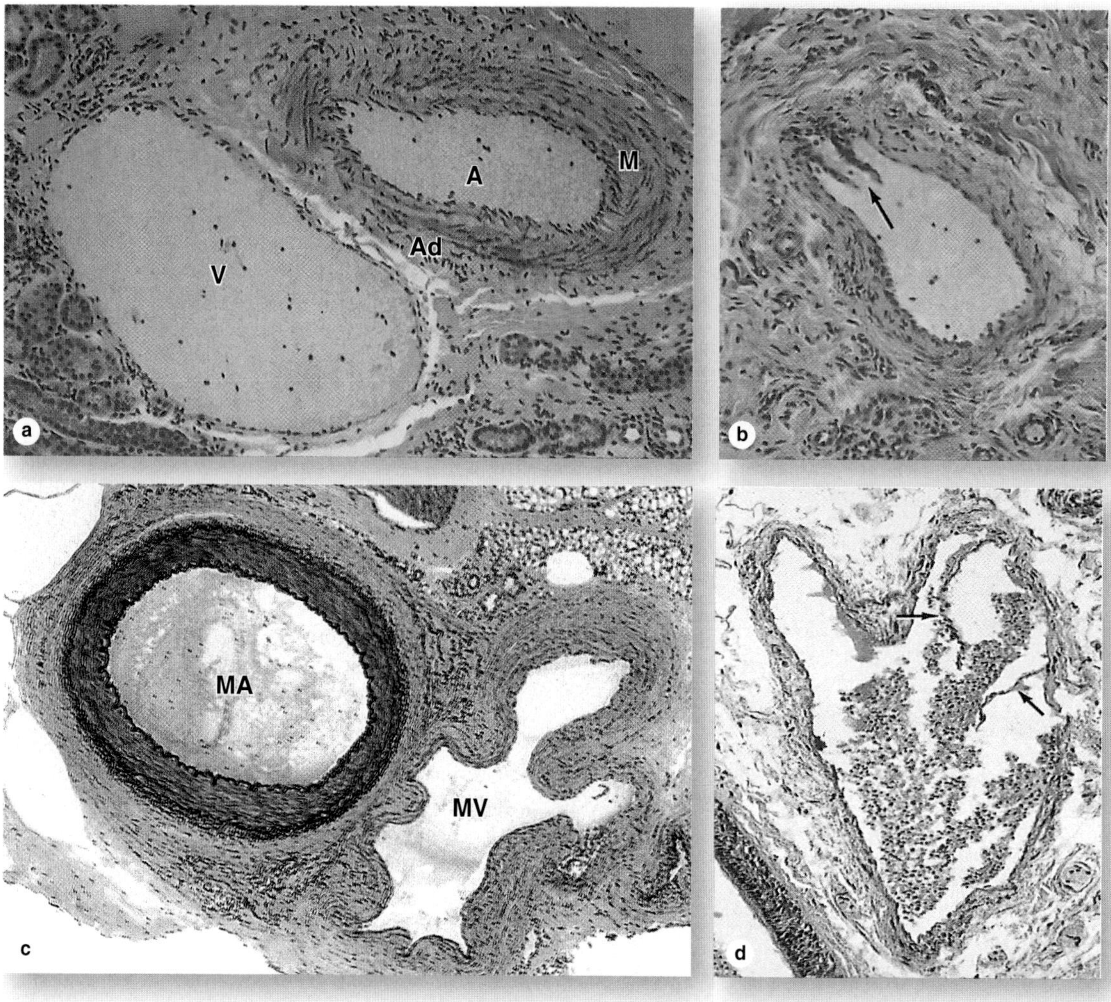

Figure 11–21. **Veins.** Veins usually travel near arteries and are classified as small, medium, or large based on size and development of the tunics. **(a):** Micrograph of small vein (V) shows a relatively large lumen compared to the small muscular artery (A) with its thick media (M) and adventitia (Ad). The wall of a small vein is very thin, containing only two or three layers of smooth muscle. X200. H&E. **(b):** Micrograph of a convergence between two small veins showing valves (arrow). Valves are thin folds of tunica intima projecting well into the lumen which act to prevent backflow of blood. X200. H&E. **(c):** Micrograph of a medium vein (MV) showing a thicker wall, but still less prominent than that of the accompanying muscular artery (MA). Both the media and adventitia are better developed, but the wall is often folded around the relatively large lumen. X100. H&E. **(d):** Micrograph of a medium vein containing blood and showing valve folds (arrows). X200. Masson trichrome.

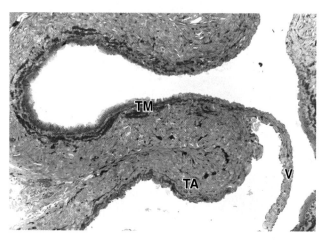

Figure 11–22. **Wall of large vein with valve.** Large veins have a muscular tunica media (TM) that is very thin compared to the tunica adventitia (TA) composed of dense irregular connective tissue. The wall is often folded as shown here. The tunica intima here projects into the lumen as a valve (V), composed of the subendothelial connective tissue with endothelium on both sides. X100, PT.

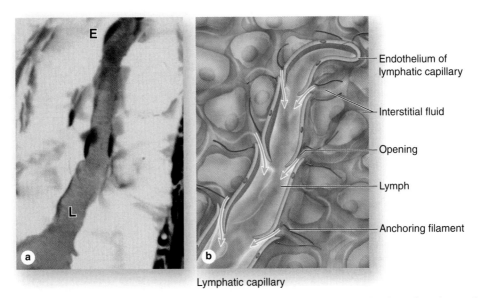

Lymphatic capillary

Figure 11–23. Lymphatic capillary. Lymphatic capillaries drain interstitial fluid produced when the plasma forced from the microvasculature by hydrostatic pressure does not all return to blood by the action of osmotic pressure. **(a):** Micrograph showing a lymphatic capillary filled with this fluid called lymph (L). Lymphatics are blind-ended vessels with a wall of very thin endothelial cells (E) and are quite variable in diameter (10-50 μm). Lymph is rich in proteins and other material and often stains somewhat better than the surrounding ground substance, as seen here. X200. Mallory trichrome. **(b):** Diagram indicating details of lymphatics, including the openings between the endothelial cells. The openings are held in place by anchoring filaments containing elastin and are covered by flaps of endothelium. Interstitial fluid enters primarily via these openings and the endothelial folds prevent backflow of lymph into tissue spaces. Lymphatic endothelial cells are typically larger than those of blood vessels.

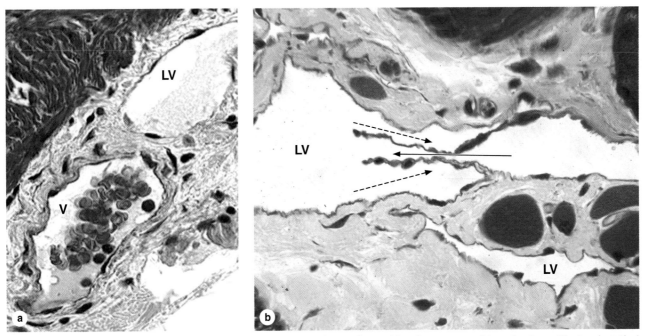

Figure 11–24. Lymphatic vessels and valve. Lymphatic vessels are formed by the merger of lymphatic capillaries, but their walls remain extremely thin. **(a):** Cross-section showing a lymphatic vessel (LV) near a venule (V), whose wall is thick by comparison. Lymphatic vessels normally do not contain red blood cells, which provides another characteristic distinguishing them from venules. X200. Mallory trichrome. **(b):** Lymphatic vessel (LV) in muscle is cut longitudinally showing a valve, the structure responsible for the unidirectional flow of lymph. The solid arrow shows the direction of the lymph flow, and the dotted arrows show how the valves prevent lymph backflow. The lower small lymphatic vessel is a lymphatic capillary with a wall consisting only of endothelium. X200. PT.

of smooth muscle in the walls of larger lymphatic vessels also helps to propel lymph toward the heart.

Lymphatic vessels ultimately end up as two large trunks: the **thoracic duct** and the **right lymphatic duct**, which respectively empty lymph into the junction of the left internal jugular vein with the left subclavian vein and into the confluence of the right subclavian vein and the right internal jugular vein. The structure of these lymphatic ducts is similar to that of large veins, with reinforced smooth muscle in the middle layer. In this layer, the muscle bundles are longitudinally and circularly arranged, with longitudinal fibers predominating. The adventitia is relatively underdeveloped, but contains vasa vasorum and a neural network.

Besides gathering interstitial fluid as lymph and returning it to the blood, the lymphatic system of vessels is a major distributor of lymphocytes, antibodies, and other immune components which it picks up at lymph nodes and other lymphoid tissues.

Blood

COMPOSITION OF PLASMA	Erythrocytes
BLOOD CELLS	Leukocytes
	Platelets

Blood is a specialized connective tissue in which cells are suspended in fluid extracellular material called **plasma**. Propelled mainly by rhythmic contractions of the heart, about five liters of blood in an average adult moves unidirectionally within the closed circulatory system. The so-called **formed elements** circulating in the plasma are **erythrocytes** (red blood cells), **leukocytes** (white blood cells) and **platelets**.

When blood leaves the circulatory system, either in a test tube or in the ECM surrounding blood vessels, plasma proteins react with one another to produce a clot, which includes formed elements and a yellowish liquid called **serum**. Serum contains growth factors and other proteins released from platelets during clot formation, which confer biological properties very different from those of plasma.

Collected blood in which clotting is prevented by the addition of anticoagulants (eg, heparin, citrate) can be separated by centrifugation into layers that reflect its heterogeneity (Figure 12–1). Erythrocytes make up the bottom layer and their volume, normally about 45% of the total blood volume in healthy adults, is called the **hematocrit**.

The yellowish translucent, slightly viscous supernatant comprising 55% at the top half of the centrifugation tube is the plasma. A thin layer between the plasma and the hematocrit, about 1% of the volume, is white or grayish in color and consists of leukocytes and platelets, both less dense than erythrocytes.

Blood is a distributing vehicle, transporting O_2 (Figure 12–2), CO_2, metabolites, hormones, and other substances to cells throughout the body. O_2 is bound mainly to hemoglobin in erythrocytes, while CO_2 is carried in solution as CO_2 or HCO_3^-, in addition to being hemoglobin bound. Nutrients are distributed from their sites of synthesis or absorption in the gut and metabolic residues are collected from all cells and removed from the blood by the excretory organs. Hormone distribution in blood permits the exchange of chemical messages between distant organs for normal cellular function. Blood further participates in heat distribution, the regulation of body temperature, and the maintenance of acid-base and osmotic balance.

Leukocytes have diversified functions and are one of the body's chief defenses against infection. These cells are generally spherical and inactive while suspended in circulating blood, but when called to sites of infection or inflammation they cross the wall of venules, migrate into the tissues, and display their defensive capabilities.

COMPOSITION OF PLASMA

Plasma is an aqueous solution, pH 7.4, containing substances of low or high molecular weight that make up 8–10% of its volume. Plasma proteins account for approximately 7% of the dissolved components, with the remainder including nutrients, nitrogenous waste products, hormones, and many inorganic ions collectively called electrolytes. Through the capillary walls, the low-molecular-weight components of plasma are in equilibrium with the interstitial fluid of the tissues. The composition of plasma is usually an indicator of the mean composition of the extracellular fluids in tissues.

The major plasma proteins include the following:

- **Albumin**, the most abundant plasma protein, is made in the liver and serves primarily in maintaining the osmotic pressure of the blood.

- **α- and β-globulins**, made by liver and other cells, include transferrin and other transport factors; fibronectin; prothrombin and other coagulation factors; lipoproteins and other proteins entering blood from tissues.

- **γ-globulins**, which are immunoglobulins (antibodies) secreted by lymphocytes in many locations.

- **Complement proteins**, a system of factors important in inflammation and destruction of microorganisms.

- **Fibrinogen**, the largest plasma protein (340 kD), also made in the liver, which during clotting polymerizes as insoluble, cross-linked fibers which block blood loss from small vessels.

BLOOD CELLS

Blood cells are generally studied in smears or films prepared by spreading a drop of blood in a thin layer on a microscope slide (Figure 12–3). In such films the cells are clearly visible and distinct from one another, facilitating observation of their nuclei and cytoplasmic characteristics. Blood smears are routinely stained

with special mixtures of acidic (eosin) and basic (methylene blue) dyes. These mixtures also contain dyes called **azures** that are more useful in staining cytoplasmic granules containing charged proteins and proteoglycans. Azurophilic granules produce metachromasia in stained leukocytes like that seen with mast cells. Some of these special stains, such as Giemsa and Wright stain, are named for hematologists who introduced their own modifications into the original mixture.

Erythrocytes

Erythrocytes (red blood cells) are terminally differentiated, lack nuclei, and are packed with the O_2-carrying protein hemoglobin. Under normal conditions, these corpuscles never leave the circulatory system.

Like most mammalian red blood cells, human erythrocytes suspended in an isotonic medium are flexible biconcave disks (Figure 12–4). They are approximately 7.5 μm in diameter, 2.6 μm thick at the rim, and only 0.75 μm thick in the center. This biconcave shape provides a large surface-to-volume ratio and facilitates gas exchange. The normal concentration of erythrocytes in blood is approximately 3.9–5.5 million per microliter in women and 4.1–6 million per microliter in men.

MEDICAL APPLICATION

*A decreased number of erythrocytes in the blood is usually associated with **anemia**. An increased number of erythrocytes (**erythrocytosis**, or*

polycythemia) may be a physiologic adaptation found, for example, in individuals who live at high altitudes, where O_2 tension is low. Polycythemia (Gr. polys, many + kytos, cell + haima, blood), usually an increase in hematocrit, is often associated with diseases of varying degrees of severity and increases blood viscosity; when severe, it can impair circulation through the capillaries.

*Abnormal erythrocytes with diameters greater than 9 μm are called **macrocytes**, and those with diameters less than 6 μm are called **microcytes**. The presence of a high percentage of erythrocytes with great variations in size is called **anisocytosis** (Gr. aniso, uneven, + kytos).*

Erythrocytes are normally quite flexible, which permits them to adapt to the irregular bends and small diameters of capillaries. Observations *in vivo* show that when traversing the angles of capillary bifurcations, erythrocytes with normal adult hemoglobin (HbA) are easily deformed and frequently assume a cuplike shape.

The plasmalemma of the erythrocyte, because of its ready availability, is the best-known membrane of any cell. It consists of about 40% lipid, 10% carbohydrate, and 50% protein. Most of the latter are integral membrane proteins (see Chapter 2), including ion channels, the anion transporter called **band 3 protein**, and **glycophorin A**. The glycosylated extracellular domains of these proteins include antigenic sites that form the basis for blood typing. Several peripheral proteins are associated with the inner surface of the membrane, including **spectrin**, which forms

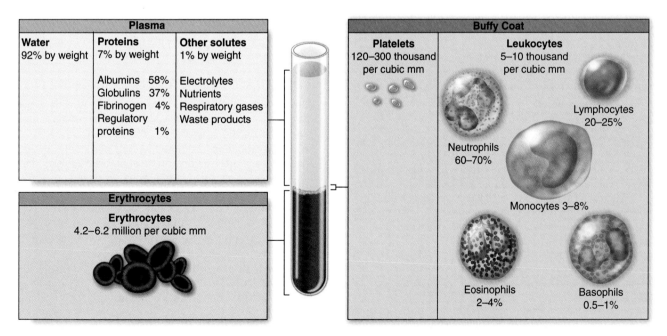

Figure 12–1. Composition of whole blood. A tube of blood after centrifugation (center) has about 43% of its volume represented by erythrocytes in the bottom half of the tube, a volume called the **hematocrit**. Between the sedimented erythrocytes and the supernatant light-colored plasma is a thin layer of leukocytes and platelets called the **buffy coat**. The average concentrations of erythrocytes, platelets and leukocytes in normal blood are included here, along with the percentage each type of leukocyte represent in the buffy coat. A cubic millimeter of blood is equivalent to a microliter (μL).

a lattice reinforcing the membrane, and **ankyrin**, which anchors the lattice to band 3 proteins. This meshwork permits the membrane and cell flexibility required for passage through capillaries and important for the normal low viscosity of blood.

Erythrocyte cytoplasm is densely filled with **hemoglobin**, the tetrameric O_2-carrying protein that accounts for the cells' uniform acidophilia. When combined with O_2 or CO_2, hemoglobin forms **oxyhemoglobin** or **carbaminohemoglobin,** respectively. The reversibility of these combinations is the basis for the gas-transporting capability of hemoglobin. The combination of hemoglobin with carbon monoxide (CO) is irreversible, however, reducing the cells' capacity to transport O_2.

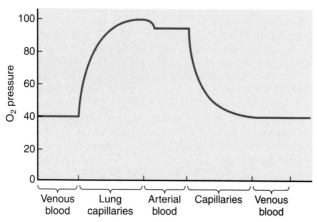

Figure 12–2. Blood O_2 content in each type of blood vessel. The amount of O_2 in blood (the O_2 pressure) is highest in arteries and lung capillaries and decreases in tissue capillaries, where exchange takes place between blood and tissues.

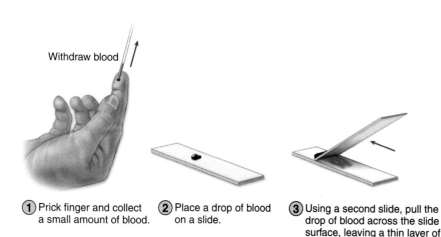

① Prick finger and collect a small amount of blood.
② Place a drop of blood on a slide.
③ Using a second slide, pull the drop of blood across the slide surface, leaving a thin layer of blood on the slide. After the blood dries, apply a stain for contrast. Place a coverslip on top.
④ When viewed under the microscope, blood smear reveals the components of the formed elements.

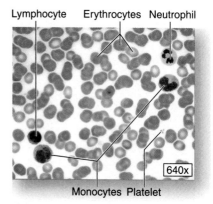

Figure 12–3. Preparing a blood smear.

erythrocytes with insufficient hemoglobin, usually related to iron deficiency in the diet; or accelerated destruction of blood cells.

Erythrocyte differentiation (presented in Chapter 13) includes loss of the nucleus and all organelles shortly before the cells are released by bone marrow into the circulation. Lacking mitochondria, mature erythrocytes rely on anaerobic glycolysis for their minimal energy needs. Lacking nuclei, they cannot replace defective proteins.

Human erythrocytes normally survive in the circulation for about 120 days. By this time defects in the membrane's cytoskeletal lattice or ion transport systems begin to produce swelling or other shape abnormalities, as well as changes in the cells' surface oligosaccharide complexes. Senescent or worn-out erythrocytes displaying such changes are removed from the circulation, mainly by macrophages of the spleen, liver, and bone marrow.

Leukocytes

Leukocytes (white blood cells) migrate to the tissues where they become functional and perform various activities (Figure 12–6). According to the type of cytoplasmic granules and the shape of their nuclei, leukocytes are divided into two groups: polymorphonuclear **granulocytes** and mononuclear **agranulocytes.** Both types are spherical while suspended in blood plasma, but become amoeboid and motile after leaving the blood vessels and invading the tissues. Their estimated sizes mentioned below refer to observations in blood smears in which the cells are spread and appear slightly larger than they are in the circulation.

Granulocytes possess two types of granules: the **specific granules** that bind neutral, basic, or acidic stains and have specific functions and the **azurophilic granules**, which are specialized lysosomes, stain darkly, and are present at some level in all leukocytes. When the cells phagocytose microorganisms, several azurophilic granule proteins act collectively to kill and then digest them. The bactericidal proteins include **myeloperoxidase**, which generates hypochlorite and other reactive agents toxic to

bacteria; cationic polypeptides called **defensins** which bind and produce holes in cell membranes of microorganisms; and **lysozyme**, which dissolves bacterial cell wall components. The major protein components of specific and azurophilic granules are listed in Table 12–1.

Granulocytes have **polymorphic nuclei** with two or more lobes and include the **neutrophils, eosinophils,** and **basophils** (Figures 12–1 and 12–6). All granulocytes are terminally differentiated cells with a life span of only a few days. Their Golgi complexes and rough ER are poorly developed. They have few mitochondria and depend largely on glycolysis for their low energy needs, containing glycogen that allows them to function in tissue with little O_2, such as inflamed areas. Granulocytes normally die by apoptosis in the connective tissue and billions of neutrophils alone die by apoptosis each day in the adult human. The resulting cellular debris is removed by macrophages and like all apoptotic cell death does not elicit an inflammatory response.

Agranulocytes do not have specific granules, but they do contain azurophilic granules (lysosomes). The nucleus is round or indented. This group includes **lymphocytes** and **monocytes** (Figures 12–1 and 12–6). The differential count of all types of leukocytes is presented in Table 12–2.

All leukocytes are key players in the defense against invading microorganisms, and in the repair of injured tissues. How they leave the circulation and become active at the specific sites where needed has been especially well-studied for neutrophils, the most abundant leukocyte specially adapted for bacteria removal. At sites of injury or infection, various substances termed cytokines are released that trigger loosening of intercellular junctions in the endothelial cells of local postcapillary venules and the rapid appearance of **P-selectin** on their luminal surfaces from Weibel-Palade bodies. Neutrophils and other leukocytes have on their surfaces ligands for P-selectin and interaction between these proteins causes cells flowing through the venules to slow down, like rolling tennis balls arriving at a patch of velcro. Other cytokines stimulate the now slowly rolling leukocytes to express integrins and other adhesion factors that produce firm attachment to the endothelium (Figure 11–20d). In a process called **diapedesis** (Gr. *dia,* through, + *pedesis,* to leap) the leukocytes quickly send

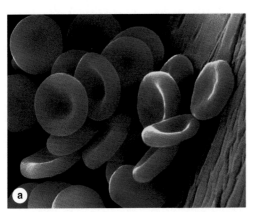
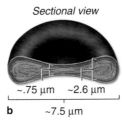
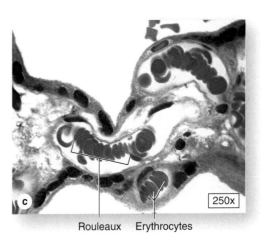

Sectional view

~.75 μm   ~2.6 μm

b   ~7.5 μm

250x

Rouleaux   Erythrocytes

Figure 12–4. Normal human erythrocytes. (a): Colorized SEM of normal erythrocytes with each side concave. X3000. **(b):** Diagram of an erythrocyte giving the cell's dimensions. The biconcave shape gives the cells a very high surface-to-volume ratio and places most hemoglobin within a short distance from the cell surface, both qualities which provide maximally efficient O_2 transport. Erythrocytes are also quite flexible and can easily bend to pass through small capillaries. **(c):** In small vessels red blood cells also often stack up in aggregates called rouleaux. X250. H&E.

extensions into the new intercellular openings, migrate out of the venules into surrounding tissue spaces, and head directly for the bacterial cells. The attraction of neutrophils to bacteria involves chemical mediators in a process of **chemotaxis**, which causes leukocytes to rapidly concentrate where their defensive actions are specifically needed.

The number of leukocytes in the blood varies according to age, sex, and physiologic conditions. In healthy adults, there are roughly 6000–10,000 leukocytes per microliter of blood (Table 12–2).

Neutrophils (Polymorphonuclear Leukocytes)

Neutrophils constitute 60–70% of circulating leukocytes. They are 12–15 μm in diameter in blood smears, with nuclei having two to five lobes linked by thin nuclear extensions (Figures 12–1, 12–6, and 12–7). In females, the inactive X chromosome may appear as a drumstick-like appendage on one of the lobes of the nucleus (Figure 12–7c) although this characteristic is not obvious in every neutrophil. Neutrophils are inactive and spherical while circulating but become actively amoeboid during diapedesis and upon adhering to solid substrates such as collagen in the ECM.

MEDICAL APPLICATION

Immature neutrophils that have recently entered the blood circulation have a nonsegmented nucleus in the shape of a horseshoe (band forms). An increased number of band neutrophils in the blood indicates a higher production of neutrophils, probably in response to a bacterial infection.

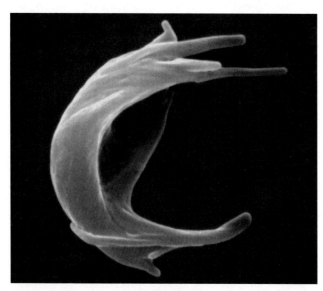

Figure 12–5. **Sickle cell erythrocyte.** A single nucleotide substitute in the hemoglobin gene produces a version of the protein that polymerizes to form rigid aggregates, leading to greatly misshapen cells with reduced flexibility. In individuals homozygous for the mutated HbS gene, this can lead to greater blood viscosity, and poor microvascular circulation, both features of sickle cell disease. X6500.

The cytoplasm of the neutrophil contains two main types of granules: the more abundant **specific granules,** which are very small and near the limit of light microscope resolution (Figure 12–7), and **azurophilic granules,** which are specialized lysosomes with components to kill ingested bacteria (Figure 12–8). Neutrophils are active phagocytes of bacteria and other small particles and are usually the first leukocytes to arrive at sites of infection, where they actively pursue bacterial cells using chemotaxis.

Neutrophils also contain glycogen, which is broken down into glucose to yield energy via the glycolytic pathway. The citric acid cycle is less important, as might be expected in view of the paucity of mitochondria in these cells. The ability of neutrophils to survive in an anaerobic environment is highly advantageous, since they can kill bacteria and help clean up debris in poorly oxygenated regions, eg, inflamed or necrotic tissue.

Neutrophils are short-lived cells with a half-life of 6–7 hours in blood and a life span of 1–4 days in connective tissues before dying by apoptosis.

MEDICAL APPLICATION

Neutrophils look for bacteria to engulf by pseudopodia and internalize them in vacuoles called phagosomes. Immediately thereafter, specific granules fuse with and discharge their contents into the phagosomes. By means of proton pumps in the phagosome membrane, the pH of the vacuole is lowered to about 5.0, a favorable pH for maximal activity of lysosomal enzymes. Azurophilic granules then discharge their enzymes into the acid environment, killing and digesting the engulfed microorganisms.

During phagocytosis, a burst of O_2 consumption leads to the formation of superoxide anions and hydrogen peroxide (H_2O_2). O_2^- is a short-lived free radical formed by the gain of one electron by O_2. It is a highly reactive radical that kills microorganisms ingested by neutrophils. Together with myeloperoxidase and halide ions, it forms a powerful killing system. Other strong oxidizing agents (eg, hypochlorite) can inactivate proteins. Lysozyme has the function of specifically cleaving a bond in the peptidoglycan that forms the cell wall of some gram-positive bacteria, thus causing their death. Lactoferrin avidly binds iron; because iron is a crucial element in bacterial nutrition, lack of its availability leads to bacterial death. The acid environment of phagocytic vacuoles can itself cause the death of certain microorganisms. A combination of these mechanisms will kill most microorganisms, which are then digested by lysosomal enzymes. Apoptotic neutrophils, bacteria, semidigested material, and tissue fluid form a viscous, usually yellow collection of fluid called pus.

Several neutrophil hereditary dysfunctions have been described. In one of them, actin does not polymerize normally, and the neutrophils are sluggish. In another, there is a failure to produce,

H₂O₂ and hypochlorite, and microbial killing power is reduced. This dysfunction results from a deficiency of NADPH oxidase, leading to a deficient respiratory burst. Children with these dysfunctions are subject to persistent bacterial infections. More severe infections result when neutrophil dysfunction and macrophage dysfunction occur simultaneously.

EOSINOPHILS

Eosinophils are far less numerous than neutrophils, constituting only 2–4% of leukocytes in normal blood. In blood smears, this cell is about the same size as a neutrophil, but with a characteristic bilobed nucleus (Figures 12–1, 12–6, and 12–9). The main identifying characteristic is the abundance of large, red specific granules (about 200 per cell) that are stained by eosin.

Ultrastructurally the eosinophilic specific granules are seen to be oval in shape, with many having a flattened crystalline core (Figure 12–9) containing **major basic protein**, an arginine-rich factor accounting for the granule's intense acidophilia. This protein constitutes 50% of the total granule protein. The major basic protein, along with eosinophilic peroxidase, other enzymes and toxins, have cytotoxic effects on parasites such as helminthic worms and protozoa. Eosinophils also phagocytose antigen-antibody complexes and modulate inflammatory responses in many ways. They are an important source of the factors mediating allergic reactions and asthma.

MEDICAL APPLICATION

*An increase in the number of eosinophils in blood (**eosinophilia**) is associated with allergic reactions and helminthic (parasitic) infections. In tissues, eosinophils are found in the connective tissues underlying epithelia of the bronchi, gastrointestinal tract, uterus, and vagina, and surrounding any*

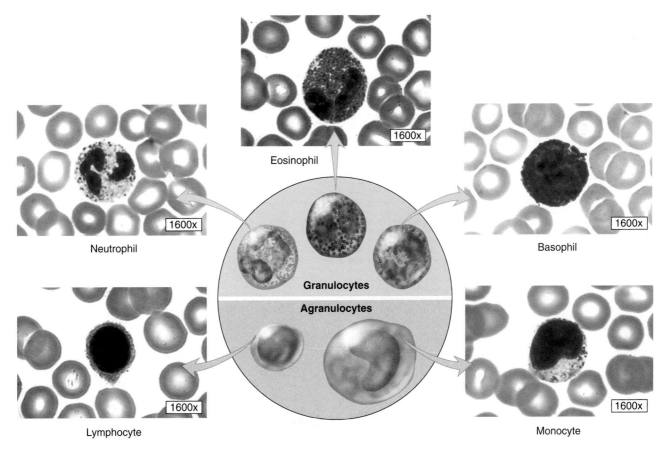

Figure 12–6. Five types of human leukocytes. Neutrophils, eosinophils, and basophils have granules that stain specifically with certain dyes and are called granulocytes. Lymphocytes and monocytes are considered agranulocytes, even though they may show azurophilic granules (lysosomes), which are also present in other leukocytes.

parasitic worms present. In addition, these cells produce substances that modulate inflammation by inactivating the leukotrienes and histamine produced by other cells. Corticosteroids (hormones from the adrenal cortex) produce a rapid

decrease in the number of blood eosinophils, probably by interfering with their release from the bone marrow into the bloodstream.

Table 12–1. Granule composition in human granulocytes.

Cell Type	Specific Granules	Azurophilic Granules
Neutrophil	Alkaline phosphatase Collagenase Lactoferrin Lysozyme Several nonenzymatic antibacterial basic proteins	Acid phosphatase α-Mannosidase Arylsulfatase β-Galactosidase β-Glucuronidase Cathepsin 5′-Nucleotidase Elastase Collagenase Myeloperoxidase Lysozyme Defensins
Eosinophil	Acid phosphatase Arylsulfatase β-Glucuronidase Cathepsin Phospholipase RNAase Eosinophilic peroxidase Major basic protein	
Basophil	Eosinophilic chemotactic factor Heparin Histamine Peroxidase	

Table 12–2. Number and percentage of blood corpuscles (blood count).

Corpuscle Type	Approximate Number per μL	Approximate Percentage
Erythrocyte	Female: $3.9–5.5 \times 10^6/\mu L$ Male: $4.1–6 \times 10^6/\mu L$	
Reticulocyte		1% of the erythrocyte count
Leukocyte	6000–10,000	
Neutrophil	5000	60-70%
Eosinophil	150	2-4%
Basophil	30	0.5%
Lymphocyte	2400	28%
Monocyte	350	5%
Platelet	300,000	

BASOPHILS

Basophils are also about 12–15 μm in diameter, but make up less than 1% of blood leukocytes and are therefore difficult to find in smears of normal blood. The nucleus is divided into two or more irregular lobes, but the large specific granules overlying the nucleus usually obscure its shape.

The azurophilic specific granules (0.5 μm in diameter) stain dark blue or metachromatically with the basic dye of blood smear stains and are fewer and more irregular in size and shape than the granules of the other granulocytes (Figures 12–1, 12–6, and 12–10). The metachromasia is due to the presence of heparin and other sulfated glycosaminoglycans (GAGs) in the granules. Basophilic specific granules also contain much histamine and various mediators of inflammation, including platelet activating factor, eosinophil chemotactic factor, and phospholipase A which produces low molecular weight factors called **leukotrienes**.

By migrating into connective tissues, basophils may supplement the functions of mast cells, with which they share a common progenitor cell origin. Both basophils and mast cells have

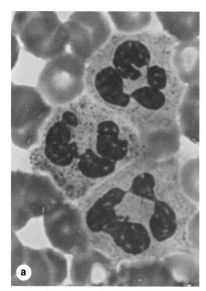
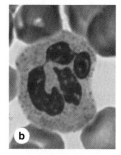
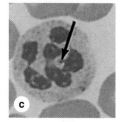

Figure 12–7. Neutrophils. (a): In blood smears neutrophils can be identified by their multilobulated nuclei, with lobules held together by thin strands. With this feature the cells are often called polymorphonuclear leukocytes, or just polymorphs. The cells are dynamic and the nuclear shape changes frequently. X1500. Giemsa. **(b):** Other identifying features of neutrophils include overall diameter of 12–15 μm, approximately twice that of the surrounding erythrocytes. The cytoplasmic granules are relatively sparse and heterogeneous in their staining properties, although generally pale and not obscuring the nucleus. X1500. Giemsa. **(c):** Micrograph shows a neutrophil from a female in which the condensed X chromosome appears as a drumstick appendage to a nuclear lobe (arrow). X1500. Wright.

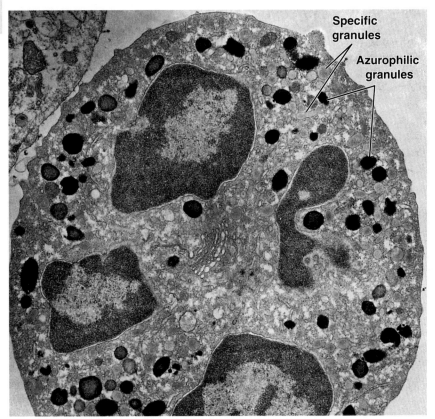

Specific granules

Azurophilic granules

Figure 12–8. **Neutrophil ultrastructure.** TEM of a sectioned human neutrophil immunostained for peroxidase reveals the two types of cytoplasmic granules: the small, pale, peroxidase-negative **specific granules** and the larger, dense, peroxidase-positive **azurophilic granules**. Specific granules undergo exocytosis during and after diapedesis, releasing many factors with various activities, including enzymes to digest ECM components and bacteriostatic factors. Azurophilic granules are modified lysosomes with components to kill engulfed bacteria. The nucleus is lobulated and the central Golgi apparatus is small. Rough ER and mitochondria are not abundant, because this cell utilizes glycolysis and is in the terminal stage of its differentiation. X27,000. (Reproduced, with permission, from Bainton D.F.: *Fed. Proc.* 1981;40:1443.)

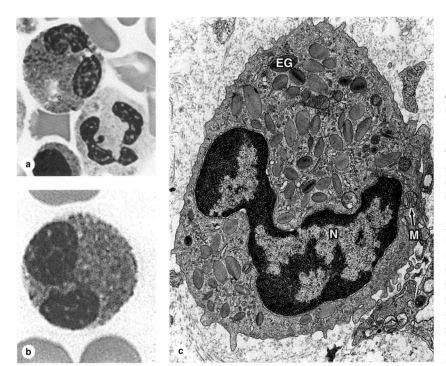

Figure 12–9. **Eosinophils.** Eosinophils are about the same size as neutrophils but have bilobed nuclei and abundant coarse cytoplasmic granules. The cytoplasm is often filled with brightly eosinophilic specific granules, but also includes some azurophilic granules. **(a):** Micrograph shows an eosinophil next to a neutrophil for comparison with its nucleus and granules. X1500. Wright. **(b):** Even with granules filling the cytoplasm, the two nuclear lobes of eosinophils are usually clear. X1500. Giemsa. **(c):** TEM of a sectioned eosinophil clearly shows the unique specific granules, as oval structures with disk-shaped electron-dense crystalline cores (EG). These along with lysosomes and a few mitochondria (M) fill the cytoplasm around the bilobed nucleus (N). X20,000.

metachromatic granules containing heparin and histamine, have IgE bound to surface receptors, and secrete their granular components in response to certain antigens (see Chapter 5).

MEDICAL APPLICATION

In some individuals a second exposure to a strong allergen, such as that delivered in a bee sting, may produce an intense, adverse systemic response. Basophils and mast cells may rapidly degranulate, producing vasodilation in many organs, a sudden drop in blood pressure, and other effects comprising a potentially lethal condition called anaphylaxis or **anaphylactic shock.**

In the dermatologic disease called **cutaneous basophil hypersensitivity,** *basophils are the major cell type at the site of inflammation.*

LYMPHOCYTES

Lymphocytes constitute a family of leukocytes with spherical nuclei (Figures 12–1 and 12–6). They can be subdivided into functional groups according to distinctive surface molecules (markers) that can best be distinguished immunocytochemically, notably **T lymphocytes**, **B lymphocytes**, and **natural killer (NK) cells**. Lymphocytes have diverse functional roles related to immune defense against invading microorganisms, foreign or abnormal antigens, and cancer cells. Additional information on the different types of lymphocytes and the functional characteristics in immune responses is presented in Chapter 14.

Most lymphocytes in the blood are small with diameters of 6–8 μm; medium and large lymphocytes range in size from 9 to 18 μm in diameter. Some larger lymphocytes may be cells that have been activated by specific antigens. The small lymphocytes that predominate in the blood are characterized by spherical nuclei, sometimes indented, and condensed, very basophilic chromatin, making them easily distinguishable from granulocytes.

The cytoplasm of the small lymphocyte is scanty, and in blood smears it appears as only a thin rim around the nucleus. It is slightly basophilic and may contain a few azurophilic granules,

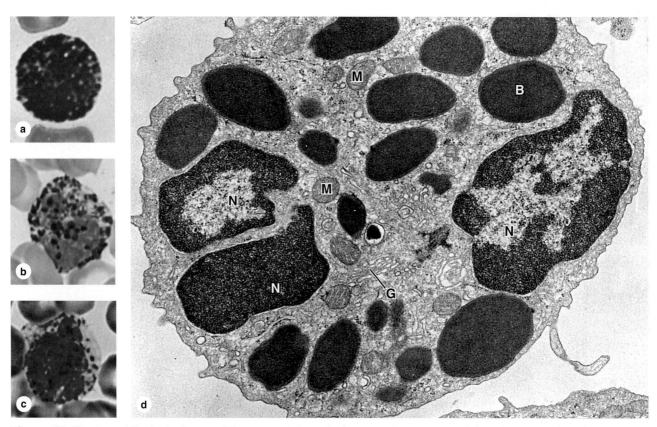

Figure 12–10. **Basophils. (a, b, c):** Basophils are approximately the same size as neutrophils and eosinophils, but have large, strongly basophilic specific granules which usually obstruct the appearance of the nucleus having two or three irregular lobes. a and b: X1500, Wright; c: X1500, Giemsa. **(d):** TEM of a sectioned basophil reveals the lobulated nucleus (N), appearing as three separated portions, the large specific basophilic granules (B), mitochondria (M), and Golgi complex (G). Basophils exert many activities modulating the immune response and inflammation and share many functions with mast cells, which are normal, longer term residents of connective tissue. X16,000. (Figure 12–10d reproduced, with permission, from Terry R.W. et al: *Lab. Invest* 1969;21:65.)

along with a few mitochondria and a small Golgi apparatus; it contains free polyribosomes (Figure 12–11).

Lymphocytes vary in life span according to their specific functions; some live only a few days and others survive in the circulating blood or other tissues for many years. They are the only type of leukocytes that, following diapedesis, can return from the tissues back to the blood.

MONOCYTES

Monocytes are bone marrow–derived agranulocytes with diameters varying from 12 to 20 μm. The nucleus is large, off-center, and may be oval, kidney-shaped, or distinctly U-shaped (Figure 12–12). The chromatin is less condensed than in lymphocytes and stains lighter than that of large lymphocytes.

The cytoplasm of the monocyte is basophilic and contains very small azurophilic granules (lysosomes), some of which are at the limit of the light microscope's resolution. These granules are distributed through the cytoplasm, giving it a bluish-gray color in stained smears. In the electron microscope, nucleoli may be seen in the nucleus, and a small quantity of rough ER, free polyribosomes, and many small mitochondria are observed. A Golgi apparatus involved in the formation of lysosomes is present and many microvilli and pinocytotic vesicles are found at the cell surface (Figure 12–12).

Circulating monocytes are precursor cells of the mononuclear phagocyte system (see Chapter 5). After crossing the walls of postcapillary venules, monocytes differentiate into **macrophages** in connective tissues, microglia in the CNS, osteoclasts in bone, etc.

Platelets

Blood platelets (**thrombocytes**) are nonnucleated, disklike cell fragments 2–4 μm in diameter. Platelets originate by fragmentation at the ends of cytoplasmic processes extending from giant polyploid cells called **megakaryocytes** in the bone marrow (Chapter 13). Platelets promote blood clotting and help repair minor tears or leaks in the walls of blood vessels, preventing loss

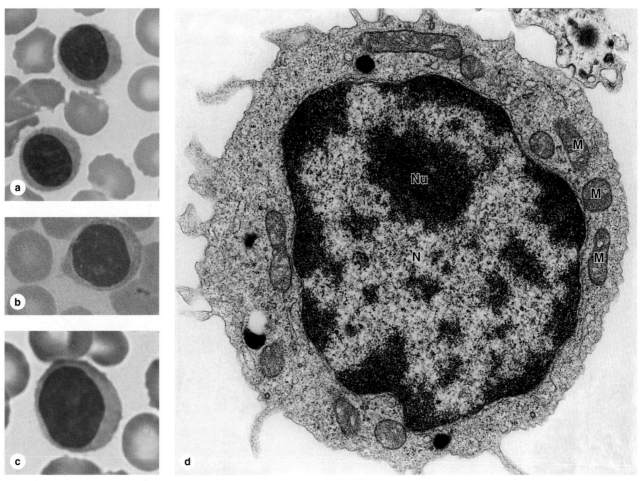

Figure 12–11. Lymphocytes. Lymphocytes are agranulocytes and lack the specific granules characteristic of granulocytes. Lymphocytes circulating in blood range in size from 6 to 15 μm in diameter and are sometimes classified arbitrarily as small, medium, and large. **(a):** The most numerous small lymphocytes shown here are slightly larger than the neighboring erythrocytes and often have only a thin rim of cytoplasm surrounding the spherical nucleus. X1500. Giemsa. **(b):** Medium lymphocytes are distinctly larger than erythrocytes. X1500. Wright. **(c):** Large lymphocytes, much larger than erythrocytes, may represent activated cells that have returned to the circulation. X1500. Giemsa. **(d):** Ultrastructurally a medium-sized lymphocytes is seen to be mostly filled with a euchromatic nucleus (N), with a nucleolus (Nu), surrounded by cytoplasm containing mitochondria (M), free polysomes, and a few lysosomes (azurophilic granules). X22,000.

of blood. Normal platelet counts range from 200,000 to 400,000 per microliter of blood. Platelets have a life span of about 10 days.

In stained blood smears, platelets often appear in clumps. Each platelet has a lightly stained peripheral zone, the **hyalomere,** and a central zone containing darker-staining granules, called the **granulomere** (Figure 12–13).

A coat rich in GAGs and glycoproteins, 15–20 nm thick, lies outside the plasmalemma and is involved in platelet adhesion. Ultrastructural analysis (Figure 12–13) reveals around the platelet periphery a **marginal bundle** of microtubules and microfilaments which helps to maintain the platelet's ovoid shape. Also in the hyalomere are two systems of membrane channels. An **open**

canalicular system of vesicles connected to invaginations of the plasma membrane, which may facilitate platelets' uptake of factors such as fibrinogen and serotonin from plasma. Another set of irregular tubular vesicles comprising the **dense tubular system** is derived from the ER and stores Ca^{2+} ions. Together these two membranous systems facilitate the extremely rapid exocytosis of proteins from platelets (degranulation) upon adhesion to collagen or other substrates outside the vascular endothelium.

The central granulomere possesses a variety of membrane-bound granules and a sparse population of mitochondria and glycogen particles (Figure 12–13). Electron-dense **delta granules,** 250–300 nm in diameter, contain adenosine diphosphate (ADP), adenosine triphosphate (ATP), and serotonin (5-hydroxytryptamine) taken up

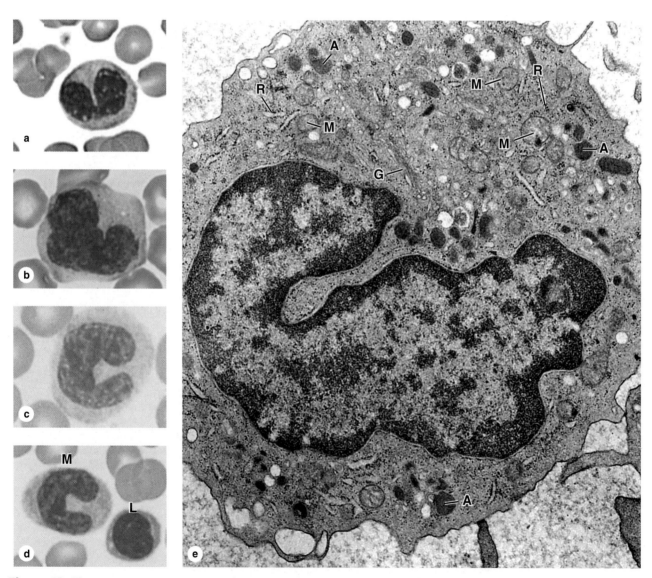

Figure 12–12. **Monocytes.** Monocytes are large agranulocytes with diameters from 12 to 20 μm that circulate as precursors to macrophages and other cells of the mononuclear phagocyte system. **(a, b, c, d):** Micrographs of monocytes that show their eccentric nuclei indented, kidney-shaped, or U-shaped. a: X1500, Giemsa; b–d: X1500, Wright. **(e):** TEM of the cytoplasm of a monocyte shows a Golgi apparatus (G), mitochondria (M), and lysosomes or azurophilic granules (A). Rough ER is poorly developed and there are some free ribosomes (R). X22,000. (Figure 12-12e, with permission, from D.F. Bainton and M.G. Farquhar, Department of Pathology, University of California at San Francisco.).

from plasma. **Alpha granules** are larger (300–500 nm in diameter) and contain platelet-derived growth factor, platelet factor 4, and several other platelet-specific proteins. Most of the stained granules seen with the light microscope in platelets are alpha granules. Small vesicles, 175–250 nm in diameter, have been shown to contain only lysosomal enzymes and have been termed **lambda granules.**

The role of platelets in controlling hemorrhage can be summarized as follows:

- **Primary aggregation.** Disruptions in the microvascular endothelium, which are common, allow platelet aggregation to collagen via collagen-binding protein in the platelet membrane. Thus, a **platelet plug** is formed as a first step to stop bleeding (Figure 12–13c).
- **Secondary aggregation.** Platelets in the plug release an adhesive glycoprotein and ADP, both of which are potent

inducers of platelet aggregation, increasing the size of the platelet plug.

- **Blood coagulation.** During platelet aggregation, **fibrinogen** from plasma, **von Willebrand factor** and others from damaged endothelium, and various factors from platelets promote the sequential interaction (cascade) of plasma proteins, giving rise to a **fibrin** polymer that forms a three-dimensional network of fibers trapping red blood cells and more platelets to form a **blood clot,** or **thrombus** (Figure 12–14).
- **Clot retraction.** The clot that initially bulges into the blood vessel lumen contracts slightly because of the interaction of platelet actin and myosin.
- **Clot removal.** Protected by the clot, the vessel wall is restored by new tissue, and the clot is then removed, mainly by the proteolytic enzyme **plasmin,** formed continuously

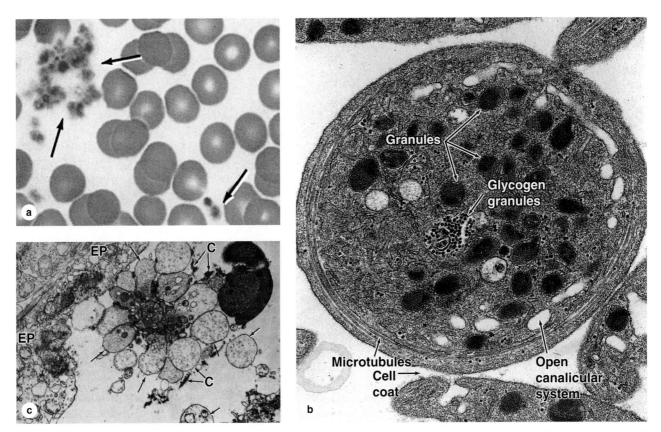

Figure 12–13. **Platelets.** Platelets are cell fragments 2–4 μm in diameter derived from megakaryocytes of bone marrow. Their primary function is to rapidly release the content of their granules upon contact with collagen (or other materials outside of the endothelium) to begin the process of clot formation and reduce blood loss from the vasculature. **(a):** In a blood smear, platelets (arrows) are often found as aggregates. Individually they show a lightly stained hyalomere region surrounding a more darkly stained central granulomere containing membrane-enclosed granules. X1500. Wright. **(b):** Ultrastructurally a platelet typically shows a system of microtubules and actin filaments near the periphery to help maintain its shape and an open canalicular system of vesicles continuous with the plasmalemma. The central granulomere region contains glycogen and secretory granules of different types. X40,000. **(c):** TEM section shows platelets adhering to collagen (C). Upon adhesion to collagen, platelets exocytose their granules into the canalicular system, which allows the very rapid secretion of factors involved in blood coagulation. Degranulating platelets (arrows) remain as an aggregate until their contents are exhausted. Other proteins involved in coagulation come from the plasma and from processes of adjacent endothelial cells (EP). The electron-dense structure on the right is part of an erythrocyte. X;7500. (Figure 12-13b, with permission, from Dr. M. J. G. Harrison, Middlesex Hospital and University College London.).

through the local action of **plasminogen activators** from the endothelium on **plasminogen** from plasma. Enzymes released from platelet lambda granules also contribute to clot removal.

Figure 12–14. Fibrin clot. Minor trauma to vessels of the microvasculature is a routine occurrence in active individuals and quickly results in a fibrin clot, shown here by SEM. A meshwork of polymeric proteins composed largely of fibrin traps erythrocytes and more degranulating platelets. Platelets in various states of degranulation are shown. Such a clot grows until blood loss from the vasculature stops. After repair of the vessel wall, fibrin clots are removed by proteolysis due primarily to locally generated plasmin, a nonspecific protease.

MEDICAL APPLICATION

Hemophilia A and B are clinically identical, differing only in the deficient factor. Both are due to sex-linked recessive inherited disorders. Blood from hemophiliac patients does not coagulate normally: the blood clotting time is prolonged. Persons with this disease bleed severely even after mild injuries, such as a skin cut, and may bleed to death after more severe injuries. The blood plasma of patients with hemophilia A is deficient in clotting factor VIII or contains a defective factor VIII, one of the plasma proteins involved in fibrin generation; in hemophilia B, the defect is in factor IX. In severe cases the blood is incoagulable. There are spontaneous hemorrhages in body cavities, such as major joints and the urinary tract. Generally, only males are affected by hemophilia A, because the recessive gene to factor VIII is on the X chromosome. Females may have one defective X chromosome, but the other one is usually normal. Females develop hemophilia only when they have the abnormal gene in both X chromosomes, a rare event. However, women with a defective X chromosome may transmit the disease to their male children.

Hemopoiesis

Mature blood cells have a relatively short life span and must be continuously replaced with stem cell progeny produced in the **hemopoietic** (Gr. *haima*, blood, + *poiesis*, a making) organs. In the earliest phase of human embryogenesis, blood cells arise from the yolk sac mesoderm. In the second trimester, hemopoiesis (also called hematopoiesis) occurs primarily in the developing liver, with the spleen also playing a role. Skeletal elements begin to ossify and bone marrow develops in their medullary cavities, so that, in the third trimester, bone marrow increasingly becomes the major hemopoietic organ.

After birth and on into childhood, erythrocytes, granulocytes, monocytes, and platelets are derived from stem cells located in bone marrow. The origin and maturation of these cells are termed, respectively, **erythropoiesis** (Gr. *erythros*, red, + *poiesis*), **granulopoiesis, monocytopoiesis,** and **thrombocytopoiesis.** Development of the major types of lymphocytes by **lymphopoiesis** occurs in the marrow and in the lymphoid organs to which precursor cells migrate, as discussed in Chapter 14.

Before reaching maturity and being released into the circulation, blood cells go through specific stages of differentiation and maturation. Because these processes are continuous, cells with characteristics that lie between the various stages are frequently encountered in smears of blood or bone marrow.

STEM CELLS, GROWTH FACTORS, & DIFFERENTIATION

Stem cells are **pluripotent** cells capable of asymmetric division and self-renewal. Some of their daughter cells form specific, irreversibly differentiated cell types, and other daughter cells remain stem cells. A constant number of pluripotent stem cells is maintained in a pool and cells recruited for differentiation are replaced with daughter cells from the pool.

Hemopoietic stem cells can be isolated by using fluorescence-labeled antibodies to mark specific cell-surface antigens and a fluorescence-activated cell-sorting (FACS) instrument. Stem cells are studied using experimental techniques that permit analysis of hemopoiesis *in vivo* and *in vitro.*

In vivo techniques include injecting the bone marrow of normal donor mice into lethally irradiated mice whose hematopoietic cells have been destroyed. In these animals, the transplanted bone marrow cells develop colonies of hematopoietic cells in the marrow cavities and spleen. This work led to the development of bone marrow transplants now used clinically to treat potentially lethal hemopoietic disorders.

In vitro techniques involve the use of semisolid tissue culture media made with a layer of cells derived from bone marrow stroma or purified protein growth factors produced by marrow stromal cells. Such methods create microenvironmental conditions sufficiently similar to the normal *in vivo* conditions to favor hemopoietic stem cell growth and differentiation.

Pluripotent Hemopoietic Stem Cells

It is believed that all blood cells arise from a single type of stem cell in the bone marrow called a **pluripotent stem cell** because it can produce all blood cell types (Figure 13–1). The pluripotent stem cells proliferate and form two major cell lineages: one for **lymphoid cells** (lymphocytes) and another for **myeloid cells** (Gr. *myelos*, marrow) that develop in bone marrow. Myeloid cells include granulocytes, monocytes, erythrocytes, and megakaryocytes. Early in their development, lymphoid cells migrate from the bone marrow to the thymus or to the lymph nodes, spleen, and other lymphoid structures, where they proliferate and differentiate (see Chapter 14).

Progenitor & Precursor Cells

The pluripotent stem cells give rise to daughter cells with restricted potentials called **progenitor cells** or **colony-forming units (CFUs)**, since they give rise to colonies of only one cell type when cultured or injected into a spleen. There are four types of progenitors/CFUs:

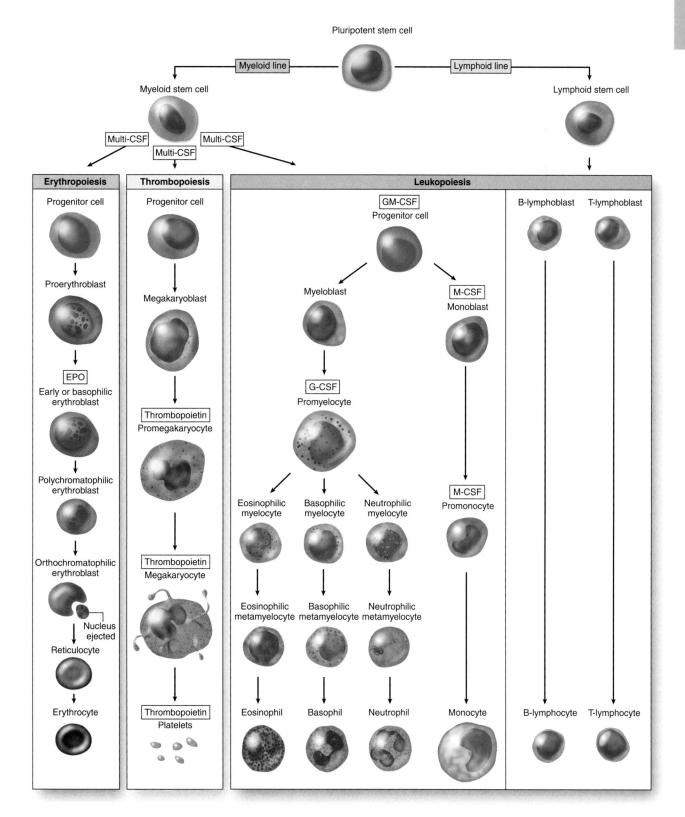

Figure 13–1. Origin and differentiative stages of blood cells. Rare pluripotent stem cells divide slowly, maintain their own population, and give rise to two major cell lineages of progenitor cells: the myeloid and lymphoid stem cells. The myeloid lineage includes precursor cells (blasts) for erythropoiesis, thrombopoiesis, granulopoiesis, and monocytopoiesis, all in the bone marrow. The lymphoid lineage forms lymphopoietic cells, partly in the bone marrow and partly in lymphoid organs.

- Erythroid lineage of CFU-erythrocytes (CFU-E)
- Thrombocytic lineage of CFU-megakaryocytes (CFU-Meg)
- Granulocyte-monocyte lineage of CFU-granulocytes-monocytes (CFU-GM)
- Lymphoid lineage of CFU-lymphocytes of all types (CFU-L).

All four progenitor/CFUs produce **precursor cells** or **blasts** in which the cells' morphologic characteristics begin to differentiate, suggesting the mature cell types they will become (Figure 13–1). In contrast, stem and progenitor cells cannot be morphologically distinguished and resemble large lymphocytes. Stem cells divide at a rate only sufficient to maintain their relatively small population. The rate of cell division is accelerated in progenitor and precursor cells, and large numbers of differentiated, mature cells are produced (3×10^9 erythrocytes and 0.85×10^9 granulocytes/kg/day in human bone marrow). Whereas progenitor cells divide asymmetrically to produce both progenitor and precursor cells, precursor cells produce only cells on the path to differentiation.

Hemopoiesis depends on favorable microenvironmental conditions and the presence of paracrine or endocrine growth factors. This microenvironment in hematopoietic organs is furnished largely by stromal cells and their extracellular matrix (ECM), which together create the niche in which stem cells are maintained. A general view of hemopoiesis shows that during this process, both the potential for various outcomes of differentiation and the self-renewing capacity of the cells gradually decrease. In contrast, the mitotic response to growth factors increases, attaining its maximum in the middle of the process. From that point on, mitotic activity decreases, morphologic characteristics and functional activity develop, and mature cells are formed (Table 13–1).

Hemopoietic growth factors, called **colony-stimulating factors** (**CSF**) or **hematopoietins** (**poietins**), are proteins with complex, overlapping functions in stimulating proliferation (mitogenic activity) of immature (mostly progenitor and precursor) cells, supporting differentiation of maturing cells, and enhancing the functions of mature cells. These three functions may all occur in the same growth factor or they may be expressed with different levels of intensity in different growth factors. The isolation and cloning of genes for several growth factors has permitted both the mass production of these proteins and tremendously advanced study of their effects *in vivo* and *in vitro*. The main characteristics of five well-characterized growth factors are presented in Table 13–2.

Table 13–1. Changes in properties of hematopoietic cells during differentiation.

Stem Cells	Progenitor Cells	Precursor Cells (Blasts)	Mature Cells
Potentiality			
		Mitotic activity	
			Typical morphologic characteristics
Self-renewing capacity			
	Influence of growth factors		
			Differentiated functional activity

Table 13–2. Main characteristics of five hemopoietic growth factors (colony-stimulating factors, CSF).

Name	Human Gene Location and Producing Cells	Main Biologic Activity
Granulocyte (G-CSF)	Chromosome 17 macrophages Endothelium fibroblasts	Stimulates formation (*in vitro* and *in vivo*) of granulocytes. Enhances metabolism of granulocytes. Stimulates malignant (leukemic) cells.
Granulocyte + macrophage (GM-CSF)	Chromosome 5 T lymphocytes Endothelium fibroblasts	Stimulates *in vitro* and *in vivo* production of granulocytes and macrophages.
Macrophage (M-CSF)	Chromosome 5 macrophages Endothelium fibroblasts	Stimulates formation of macrophages *in vitro*. Increases antitumor activity of macrophages.
Interleukin 3 (IL-3)	Chromosome 5 T lymphocytes	Stimulates *in vivo* and *in vitro* production of all myeloid cells.
Erythropoietin (EPO)	Chromosome 7 renal interstitial cells (outer cortex)	Stimulates red blood cell formation *in vivo* and *in vitro*.

Growth factors have been used clinically to increase marrow cellularity and blood cell counts. Potential therapeutic uses of growth factors include increasing the number of blood cells in diseases or induced conditions (eg, chemotherapy, irradiation) that result in low blood counts; increasing the efficiency of marrow transplants by enhancing cell proliferation; enhancing host defenses in patients with malignancies and infectious and immunodeficient diseases; and enhancing the treatment of parasitic diseases.

BONE MARROW

Under normal conditions, the production of blood cells by the bone marrow is adjusted to the body's needs, increasing its activity several-fold in a very short time. Bone marrow and adipocytes are found in the medullary canals of long bones and in the cavities of cancellous bones. There are two types of bone marrow based on their appearance at gross examination: blood-forming **red bone marrow,** whose color is produced by an abundance of blood and hemopoietic cells, and **yellow bone marrow,** which is filled with adipocytes and essentially excludes hemopoietic cells. In the newborn, all bone marrow is red and active in blood cell production, but as the child grows most of the marrow changes gradually to the yellow variety. Under certain conditions, such as severe bleeding or hypoxia, yellow marrow reverts to red.

Red bone marrow (Figure 13–2) is composed of a **stroma** (Gr: *stroma,* bed), **hemopoietic cords** or **islands** of cells, and **sinusoidal capillaries.** The stroma is a meshwork of specialized fibroblastic cells called reticular or adventitial cells and a delicate web of reticular fibers supporting hemopoietic cells and macrophages. The matrix of bone marrow also contains collagen type I, proteoglycans, fibronectin, and laminin, the latter glycoproteins interacting with integrins to bind cells to the matrix.

Sinusoids are formed by a thin layer of endothelial cells. Differentiated blood cells from the hemopoietic cords enter the circulation by passing through openings in the endothelium (Figure 13–3). Red bone marrow is also a site where macrophages phagocytose worn-out erythrocytes and store iron derived from hemoglobin breakdown.

Red bone marrow also contains stem cells that can produce other tissues in addition to blood cells. With their great potential for differentiation, these cells make it possible to generate specialized cells that are not rejected by the body because they are produced from stem cells from the marrow of the same person. The procedure is to collect bone marrow stem cells, cultivate them in appropriate medium to direct their differentiation to the cell type needed for transplantation, and then use the cells originating in tissue culture to replace certain defective cells. In this case the donor and the recipient are the same individual and histocompatibility is complete, excluding the possibility of rejection. These studies are at early stages, but results with animal models so far are promising.

MATURATION OF ERYTHROCYTES

A mature cell is one that has differentiated to the stage at which it can carry out all its specific functions. Erythrocyte maturation involves hemoglobin synthesis and formation of a small, enucleated,

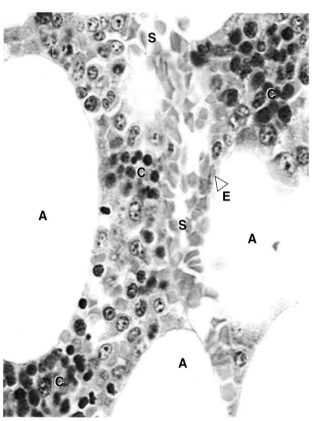

Figure 13–2. Red bone marrow (active in hemopoiesis). Red bone marrow contains adipocytes but is also active in hemopoiesis, with several cell lineages usually present. It can be examined histologically in sections of bones or in biopsies, but its cells can also be studied in smears. Marrow consists of capillary sinusoids running through a stroma of specialized, fibroblastic reticular cells and an ECM meshwork. Reticular cells secrete various colony-stimulating factors and the stroma forms the microenvironment for hemopoietic stem cell maintenance, proliferation, and differentiation. This section of red bone marrow shows some of its components. Sinusoid capillaries (S) containing erythrocytes are surrounded by stroma containing adipocytes (A) and islands or cords (C) of hemopoietic cells. Sinusoidal endothelial cells (one nucleus at E) are very thin. Most reticular cells and cells of the hemopoietic lineages are difficult to identify with certainty in routinely stained sections of marrow. X400. H&E.

biconcave corpuscle. Several major changes take place during erythrocyte maturation (Figure 13–4). Cell and nuclear volume decrease, and the nucleoli diminish in size and disappear. The chromatin becomes increasingly denser until the nucleus presents a pyknotic appearance and is finally extruded from the cell. There is a gradual decrease in the number of polyribosomes (basophilia decreases), with a simultaneous increase in the amount of hemoglobin (an acidophilic protein) within the cytoplasm. Mitochondria and other organelles gradually disappear.

There are three to five intervening cell divisions between the proerythroblast and the mature erythrocyte. The development of an erythrocyte from the first recognizable cell of the series to the release of reticulocytes into the blood takes approximately a week. The glycoprotein **erythropoietin (Epo)**, a growth factor produced in the kidneys, stimulates production of mRNA for **globin,** the protein component of hemoglobin and is essential for the production of erythrocytes.

The first recognizable cell in the erythroid series (Figure 13–5) is the **proerythroblast,** a large cell with loose, lacy chromatin, nucleoli, and basophilic cytoplasm. The next stage is represented by the **basophilic erythroblast,** with more strongly basophilic cytoplasm and a condensed nucleus with no visible nucleolus. The basophilia of these two cell types is caused by the large number of polyribosomes synthesizing hemoglobin. During the next stage cell volume is reduced, polyribosomes decrease and some cytoplasmic areas begin to be filled with hemoglobin, producing regions of both basophilia and acidophilia in the cell, now called a **polychromatophilic erythroblast.** In the next stage, the cell and nuclear volumes continue to condense and no basophilia is evident, resulting in a uniformly acidophilic cytoplasm—the **orthochromatophilic erythroblast.** Late in this stage, this cell ejects its nucleus which is phagocytosed by macrophages. The cell still has a small number of polyribosomes that, when treated with the dye brilliant cresyl blue, form a faintly stained network and the cell is called the **reticulocyte.** Reticulocytes pass to the circulation, where they may constitute 1% of the red blood cells, lose the polyribosomes and quickly mature as erythrocytes.

MATURATION OF GRANULOCYTES

Granulopoiesis involves cytoplasmic changes dominated by synthesis of proteins for the **azurophilic granules** and **specific granules.** These proteins are produced in the rough endoplasmic reticulum and the prominent Golgi apparatus in two successive stages (Figure 13–6). The azurophilic granules, which contain lysosomal hydrolases, stain with basic dyes, and are somewhat similar in all three types of granulocytes, are made first. Golgi activity then changes to produce proteins for the specific granules, whose contents differ in each of the three types of granulocytes and endow each type with certain different properties. In sections of bone marrow, cords of granulopoietic cells can be distinguished by their granule-filled cytoplasm from erythropoietic cords (Figure 13–7).

The **myeloblast** is the most immature recognizable cell in the myeloid series (Figure 13–8). It has a finely dispersed chromatin, and faint nucleoli. In the next stage, the **promyelocyte** is characterized by its basophilic cytoplasm and azurophilic granules containing lysosomal enzymes and myeloperoxidase. Different promyelocytes activate different sets of genes, resulting in lineages for the three types of granulocytes. The first visible sign of differentiation appears in the **myelocytes** (Figure 13–9), in which specific granules gradually increase in number and eventually occupy most of the cytoplasm at the metamyelocyte stage. These neutrophilic, basophilic, and eosinophilic metamyelocytes mature with further condensations of the nuclei. Before its complete maturation the neutrophilic granulocyte passes through an intermediate stage, the **stab** or **band cell,** in which its nucleus is elongated but not yet polymorphic.

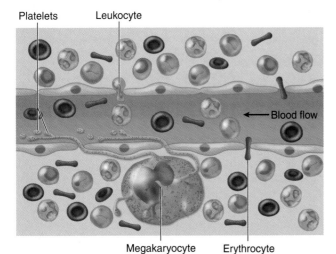

Platelets   Leukocyte

Blood flow →

Megakaryocyte   Erythrocyte

***Figure 13–3.* Sinusoidal endothelium in active marrow.** Diagram shows that mature, newly formed erythrocytes, leukocytes, and platelets in marrow enter the circulation by passing through the sinusoid capillary endothelium. Because erythrocytes (unlike leukocytes) cannot migrate through the wall of the sinusoid actively, they are believed to enter the sinusoid by a pressure gradient across its wall. Leukocytes cross the wall of the sinusoid by their own activity. All blood cells apparently penetrate through apertures and between the endothelial cells. Megakaryocytes form thin processes (proplatelets) that also pass through such apertures and liberate platelets at their tips.

MEDICAL APPLICATION

The appearance of large numbers of immature neutrophils (band cells) in the blood, sometimes called a "shift to the left", is clinically significant, usually indicating bacterial infection.

The vast majority of granulocytes are neutrophils. The total time taken for a myeloblast to emerge as a mature, circulating neutrophil is about 11 days. Under normal circumstances, five mitotic divisions occur in the myeloblast, promyelocyte, and neutrophilic myelocyte stages. Developing and mature neutrophils can be considered to exist in four functionally and anatomically defined compartments: the granulopoietic compartment in marrow; storage as mature cells in marrow until release; the circulating population; and a marginating population of cells adhering to endothelial cells of postcapillary venules and small veins (Figure 13–10). Margination of neutrophils in some organs can persist for several hours and is not always immediately followed by the cells' emigration from the microvasculature.

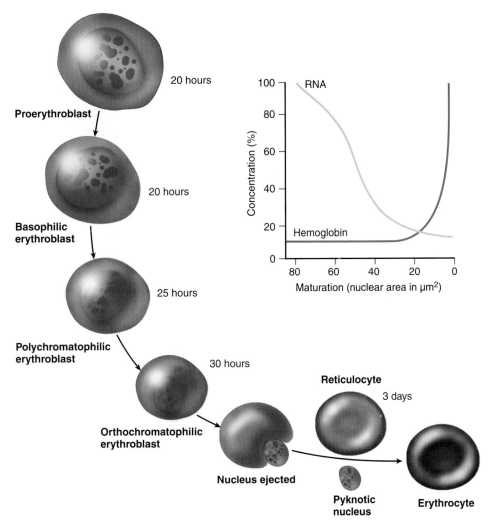

Figure 13–4. Summary of erythrocyte maturation. The color change in the cytoplasm shows the continuous decrease in basophilia and the increase in hemoglobin concentration from proerythroblast to erythrocyte. There is also a gradual decrease in nuclear volume and an increase in chromatin condensation, followed by extrusion of a pyknotic nucleus. The times are the average duration of each cell type. In the graph, 100% represents the highest recorded concentrations of hemoglobin and RNA.

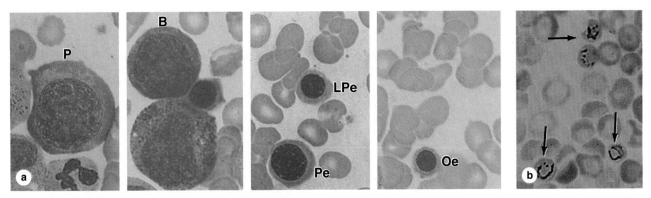

Figure 13–5. Erythropoiesis: Major erythrocyte precursors. (a): Micrographs showing a very large and scarce proerythroblast (P), a slightly smaller basophilic erythroblast (B) with very basophilic cytoplasm, typical and late polychromatophilic erythroblasts (Pe and LPe) with both basophilic and acidophilic cytoplasmic regions, and a small orthochromatophilic erythroblast (Oe) with cytoplasm nearly like that of the mature erythrocytes in the field. All X1400. Wright. **(b):** Micrograph containing reticulocytes (arrows) that have not yet completely lost the polyribosomes used to synthesize globin, as demonstrated by a stain for RNA. X1400. Brilliant cresyl blue.

Neutrophils and other granulocytes enter the connective tissues by migrating through intercellular junctions between endothelial cells of postcapillary venules in diapedesis. The connective tissues thus form a fifth terminal compartment for neutrophils, where the cells reside for a few days and then die by apoptosis, regardless of whether they have performed their major function of phagocytosis.

MATURATION OF AGRANULOCYTES

Study of the precursor cells of monocytes and lymphocytes is difficult, because these cells do not show specific cytoplasmic granules or nuclear lobulation, both of which facilitate the distinction between young and mature forms of granulocytes. Monocytes and lymphocytes in smear preparations are discriminated mainly on the basis of size, chromatin structure, and the presence of nucleoli.

Monocytes

The **monoblast** is a committed progenitor cell that is virtually identical to the myeloblast in its morphologic characteristics. Further differentiation leads to the **promonocyte,** a large cell (up to 18 μm in diameter) with basophilic cytoplasm and a large, slightly indented nucleus. The chromatin is lacy and nucleoli are evident. Promonocytes divide twice as they develop into **monocytes** (Figure 13–1). A large amount of rough ER is present, as is an extensive Golgi apparatus in which granule condensation occurs. These granules are primary lysosomes, which are observed as fine azurophilic granules in blood monocytes. Mature monocytes enter the bloodstream, circulate for about eight hours, and then enter tissues where they mature as **macrophages** (or other phagocytic cells) and function for several months.

Lymphocytes

As discussed in Chapter 14, circulating lymphocytes originate mainly in the thymus and the peripheral lymphoid organs

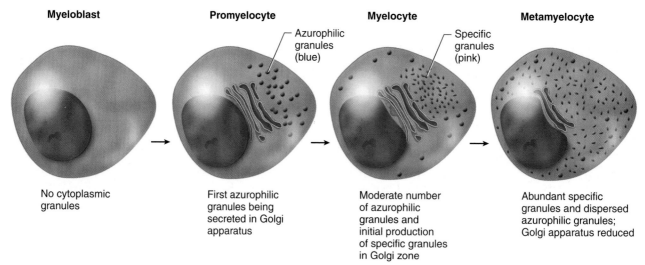

Figure 13–6. Granulopoiesis: Formation of granules. Diagram illustrating the sequence of cytoplasmic events in the maturation of granulocytes from myeloblasts. Modified lysosomes or azurophilic granules form first at the promyelocyte stage and are shown in blue; the specific granules of the particular cell type form at the myelocyte stage and are shown in pink. All granules are fully dispersed at the metamyelocyte stage, when indentation of the nucleus begins.

(eg, spleen, lymph nodes, tonsils etc.). However, all lymphocyte progenitor cells originate in the bone marrow. Some of these lymphocytes migrate to the thymus, where they acquire the full attributes of T lymphocytes. Subsequently, T lymphocytes populate specific regions of peripheral lymphoid organs. Other bone marrow lymphocytes differentiate into B lymphocytes in the bone marrow and then migrate to peripheral lymphoid organs, where they inhabit and multiply in their own special compartments.

As lymphocytes mature, their chromatin becomes more compact, nucleoli become less visible, and the cells decrease in size. In addition, subsets of the lymphocyte series acquire distinctive cell-surface receptors during differentiation that can be detected by immunocytochemical techniques. The first identifiable progenitor of lymphoid cells is the **lymphoblast,** a large cell capable of dividing two or three times to form **prolymphocytes.** Prolymphocytes are smaller and have relatively more condensed chromatin but none of the cell-surface antigens that mark T or B lymphocytes. In the bone marrow and in the thymus, these cells synthesize cell-surface receptors characteristic of the B or T lymphocyte lineages. In routine histological procedures B and T lymphocytes cannot be distinguished; immunocytochemical techniques using cell-specific markers are required to make the distinction.

MEDICAL APPLICATION

*Abnormal stem cells in bone marrow can produce diseases based on cells derived from that tissue. **Leukemias** are malignant clones of leukocyte precursors. They occur in lymphoid tissue (**lymphocytic leukemias**) and in bone marrow (**myelogenous***

*and **monocytic leukemias**). In these diseases, there is usually a release of large numbers of immature cells into the blood. Some symptoms of leukemias are a consequence of this shift in cell proliferation, with a lack of some cell types and excessive production of others. The patient is usually anemic and prone to infection.*

*A clinical technique that is helpful in the study of leukemias and other bone marrow disturbances is **bone marrow aspiration**. A needle is introduced through compact bone and a sample of marrow is withdrawn. The use of labeled monoclonal antibodies specific to membrane proteins of precursor blood cells aids in identifying cell types derived from these stem cells and contributes to a more precise diagnosis of the leukemia.*

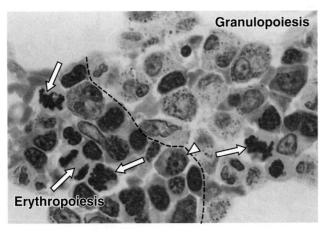

Figure 13–7. Developing erythrocytes and granulocytes in marrow. Precursor cells of different hemopoietic lineages develop side by side with some intermingling as various cell islands or cords in the bone marrow. Plastic section of red bone marrow showing mitotic figures (arrows), a plasma cell (arrowhead), and fairly distinct regions of erythropoiesis and granulopoiesis. Most immature granulocytes are in the myelocyte stage: their cytoplasm contains large, dark-stained azurophilic granules and small, less darkly stained specific granules. X400. Giemsa.

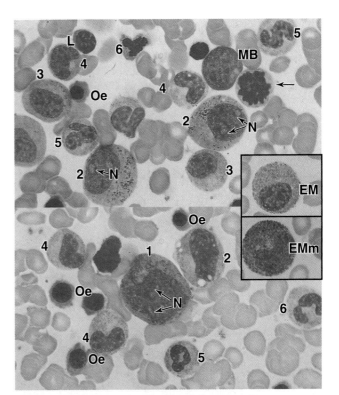

Figure 13–8. Granulopoiesis: Major granulocyte precursors. Two micrographs from smears of bone marrow showing the major cells of the neutrophilic granulocyte lineage. Typical precursor cells shown are labeled as follows: myeloblast (MB); promyelocyte (1); myelocytes (2); late myelocyte (3); metamyelocytes (4); stab or band cells (5); nearly mature segmented neutrophil (6). Some of the early stages show faint nucleoli (N). **Inset:** Eosinophilic myelocytes (EM) and metamyelocytes (EMm) with their specific granules having distinctly different staining. These and cells of the basophilic lineage are similar to developing neutrophils, except for their specific staining granules and lack of the stab cell form. Also seen among the erythrocytes of these marrow smears are some orthochromatophilic erythroblasts (Oe), a small lymphocyte (L), and a cell in mitosis (arrow). All X1400. Wright.

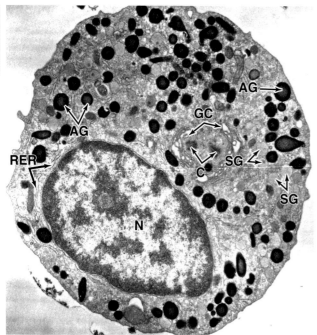

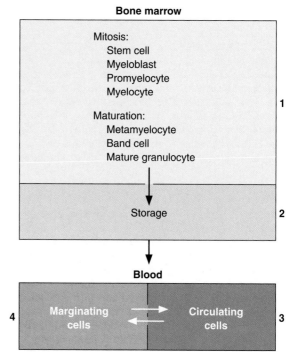

Figure 13–9. Neutrophilic myelocyte. At the myelocyte stage lysosomes (azurophilic granules) have formed and production of specific secretory granules is underway. This micrograph shows ultrastructurally a peroxidase-stained section of a neutrophilic myelocyte with cytoplasm containing both large, peroxidase-positive azurophilic granules (AG) and smaller specific granules (SG), which do not stain for peroxidase. The peroxidase reaction product is present only in mature azurophilic granules and is not seen in the rough ER (RER) or Golgi cisternae (GC), which are located around the centriole (C) near the nucleus (N). X15,000. (With permission, from Dr. Dorothy F. Bainton, Department of Pathology, University of California at San Francisco.)

Figure 13–10. Functional compartments of neutrophils. Neutrophils exist in at least four anatomically and functionally distinct compartments, the sizes of which are proportional to the number of cells. **(1)** Granulopoietic compartment subdivided into a mitotic part and a maturation part. **(2)** Storage (reserve) compartment, also in red marrow, acts as a buffer system, capable of releasing large numbers of mature neutrophils on demand. **(3)** Circulating compartment throughout the blood. **(4)** Marginating compartment, in which cells temporarily do not circulate, but rather adhere via selectins to vascular endothelial cells of postcapillary venules, particularly in the lungs. The marginating and circulating compartments are of about equal size, and there is a constant interchange of cells between them. The half-life of a neutrophil in these two compartments is less than 10 hours. The granulopoietic and storage compartments together include cells in the first 11 days of their existence and are about 10 times larger than the circulating and marginating compartments.

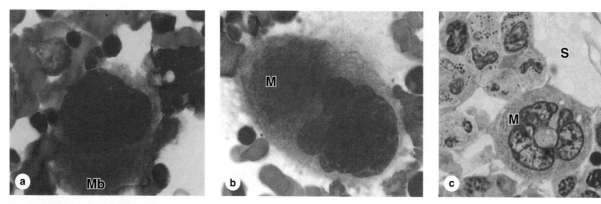

Figure 13–11. Megakaryoblast and megakaryocytes. (a): Megakaryoblasts (Mb) are very large, fairly rare cells in bone marrow, with very basophilic cytoplasm. X1400. Wright. **(b):** Megakaryoblasts undergo endomitosis (DNA replication without intervening cell divisions), becoming polyploid as they differentiate into megakaryocytes (M). These cells are even larger, but with cytoplasm that is less intensely basophilic. X1400. Wright. **(c):** Micrograph section of bone marrow megakaryocyte (M) shown near sinusoids (S). X400. Giemsa. Megakaryocytes produce all the characteristic components of platelets (membrane vesicles, specific granules, marginal microtubule bundles, etc) and in a complex process extend many long, branching pseudopodia-like projections called proplatelets, from the ends of which platelets are pinched off almost fully formed.

ORIGIN OF PLATELETS

In adults, the membrane-enclosed cell fragments called platelets originate in the red bone marrow by dissociating from mature **megakaryocytes** (Gr. *megas,* big, + *karyon,* nucleus, + *kytos*), which in turn differentiate from **megakaryoblasts** in a process driven by **thrombopotietin**. The megakaryoblast is 25–50 μm in diameter and has a large ovoid or kidney-shaped nucleus (Figure 13–11) with numerous small nucleoli. Before differentiating, these cells undergo endomitosis, with repeated rounds of DNA replication not separated by cell divisions, resulting in

a nucleus that is highly polyploid (ie, 64N or more than thirty times more DNA than a normal cell). The cytoplasm of this cell is homogeneous and intensely basophilic.

Megakaryocytes are giant cells, 35–150 μm in diameter, with irregularly lobulated polyploid nuclei, coarse chromatin, and no visible nucleoli. Their cytoplasm contains numerous mitochondria, a well-developed rough ER, and an extensive Golgi apparatus from which arise the conspicuous specific granules of platelets, or thrombocytes (Chapter 12). They are widely scattered in marrow, typically near sinusoidal capillaries.

To form platelets, megakaryocytes extend several long (>100 μm), wide (2–4 μm), branching processes called **proplatelets**. These extending proplatelets penetrate the sinusoidal endothelium and appear as long processes disposed lengthwise with the blood flow in these vessels (Figure 13–3). The proplatelet framework consists of actin filaments and a loose bundle of mixed polarity microtubules along which organelles, membrane vesicles, and specific granules are transported. A loop of microtubules forms a teardrop-shaped enlargement at the distal end of the proplatelet and cytoplasm within these loops is pinched off to form platelets with their characteristic marginal bundles of microtubules, vesicles and granules (Figure 12–13b).

During proplatelet growth microtubules polymerize in both directions. Proplatelet elongation does not depend on this polymerization, but on a dynein-based microtubule sliding mechanism similar to that of extension ladders. Mature megakaryocytes have numerous invaginations of plasma membrane ramifying throughout the cytoplasm, called **demarcation membranes** (Figure 13–12) which were formerly considered "fracture lines" or "perforations" for release of platelets, but are now thought to represent a membrane reservoir that facilitates continuous rapid proplatelet elongation. Each megakaryocyte produces a few thousand platelets, after which the remainder of the cell shows apoptotic changes and is removed by macrophages.

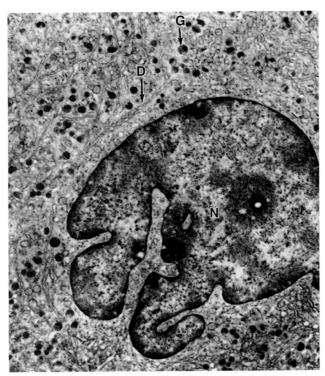

Figure 13–12. Megakaryocyte ultrastructure. TEM analysis of a megakaryocyte during platelet formation showing a lobulated nucleus (N), numerous cytoplasmic granules (G), and an extensive system of demarcation membranes (D) through the cytoplasm. Once believed to be perforations along which platelets were shed from the cells, this membrane system is now considered a reservoir of membrane used during elongation of the numerous proplatelets which extend from the megakaryocyte surface. X10,000.

MEDICAL APPLICATION

*In certain forms of **thrombocytopenic purpura**, a disease in which the number of blood platelets is reduced, the platelets appear to be bound to the cytoplasm of the megakaryocytes, indicating a defect in the liberation mechanism of these corpuscles. The life span of platelets is approximately 10 days.*

The Immune System & Lymphoid Organs

The body has a system of cells—the **immune system**—that has the ability to distinguish "self" (the organism's own molecules) from "non-self" (foreign substances). This system has the ability to neutralize or inactivate foreign molecules (such as soluble molecules as well as those present in viruses, bacteria, and parasites) and to destroy microorganisms or other cells (such as virus-infected cells, cells of transplanted organs, and cancer cells). On occasion, the immune system of an individual reacts against its own normal body tissues or molecules, causing **autoimmune diseases**.

The cells of the immune system (1) are distributed throughout the body in the blood, lymph, and epithelial and connective tissues; (2) are arranged in small spherical nodules called **lymphoid nodules** found in connective tissues and inside several organs; and (3) are organized in larger **lymphoid organs**—the lymph nodes, the spleen, the thymus, and the bone marrow. Lymphoid nodules are isolated cells of the immune system found in the mucosa of the digestive system (including the tonsils, Peyer's patches, and appendix), the respiratory system, the reproductive system, and the urinary system are collectively known as **mucosa-associated lymphoid tissue (MALT)** and may be considered a lymphoid organ. The wide distribution of immune system cells and the constant traffic of lymphocytes through the blood, lymph, connective tissues, and lymphoid organs provide the body with an elaborate and efficient system of surveillance and defense (Figure 14–1).

ANTIGENS

A molecule that is recognized by cells of the immune system is called an **antigen** and may elicit a response from these cells. Antigens may consist of soluble molecules (such as proteins, polysaccharides, and nucleoproteins) or molecules belonging to whole cells (bacteria, protozoa, tumor cells, or virus-infected cells). The cells of the immune system do not recognize and react to the whole antigen molecule but instead react to small molecular domains of the antigen known as **antigenic determinants** or **epitopes.** The response of the organism to antigens may be cellular (in which lymphocytes are primarily in charge of eliminating the antigen) or humoral (in which antibodies secreted by plasma cells are primarily responsible for the response). Some epitopes (eg, polysaccharides of bacterial walls or lipids) usually elicit a humoral response whereas proteins elicit both a cellular and humoral response. More details on cellular and humoral immune responses are provided below.

ANTIBODIES

An **antibody** is a glycoprotein that interacts specifically with an antigenic determinant. Antibodies belong to the **immunoglobulin** protein family. Free molecules of antibodies are secreted by plasma cells that arise by proliferation and terminal differentiation of clones of B lymphocytes whose receptors recognize and bind specific epitopes. These secreted antibodies either circulate in the plasma and may leave the blood vessels reaching the tissues or are present in the secretion of some epithelia (eg, of the mammary gland and salivary glands). Other antibodies are not free molecules, but are integral membrane proteins of the surface of lymphocytes. In any case, each antibody combines with the epitope that it specifically recognizes.

There are several classes of antibody molecules but all have a common design: they consist of two identical light chains and two identical heavy chains bound by disulfide bonds and noncovalent forces (Figure 14–2). The isolated carboxyl-terminal portion of the heavy chain molecules is called the **Fc region**. The Fc regions of some immunoglobulins are recognized by receptors on the membranes of several cell types and for this reason antibodies bind to the surface of these cells. The first 110 amino acids near the amino-terminal part of the light and heavy chains vary widely among different antibody molecules and this region is called the **variable region**. The **antigen-binding site** of an

antibody consists of the variable regions of one heavy and one light chain. Thus, each antibody has two antigen-binding sites, both for the same antigen. The molecules in some immunoglobulin classes may form dimers, trimers, or pentamers.

Classes of Antibodies

The main classes of immunoglobulins in humans are immunoglobulin G (IgG), IgA, IgM, IgE, and IgD (Table 14–1).

IgG is the most abundant class representing 75–80% of serum immunoglobulins. It is produced in large amounts during immune responses. IgG is the only immunoglobulin that crosses the placental barrier and is transported to the circulatory system of the fetus, protecting the newborn against infections for a certain period of time.

IgA is the main immunoglobulin found in secretions, such as nasal, bronchial, intestinal, and prostatic, as well as in tears, colostrum, saliva, and vaginal fluid. It is present in secretions as a dimer or trimer called **secretory IgA**, composed of two or three molecules of monomeric IgA united by a polypeptide chain called **protein J** and combined with another protein, the **secretory component**. Because it is resistant to several enzymes, secretory IgA persists in the secretions where it provides protection against the proliferation of microorganisms. IgA monomers and protein J are secreted by plasma cells in the lamina propria of the epithelium of the digestive, respiratory, and urinary passages; the secretory component is synthesized by the mucosal epithelial cells and is added to the IgA polymer as it is transported through the epithelial cells.

IgM constitutes about 10% of blood immunoglobulins and usually exits as a pentamer. Together with IgD, it is the major immunoglobulin found on the surface of B lymphocytes. These two classes of immunoglobulins have both membrane-bound and circulating forms. IgM bound to the membrane of a B lymphocyte functions as its specific receptor for antigens. The result of this interaction is the proliferation of B lymphocytes into antibody-secreting plasma cells. Secreted IgM, when bound to antigen, is very effective in activating the complement system.

IgE is much less abundant than the other classes and usually exists as a monomer. As its Fc region has a great affinity for receptors

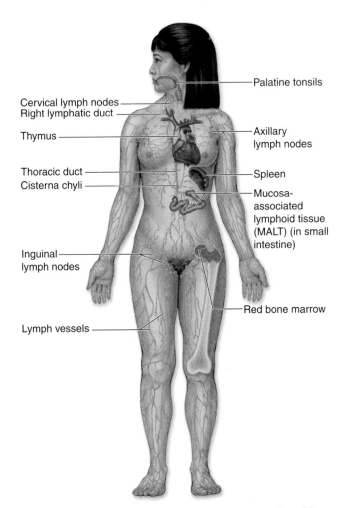

Figure 14–1. The lymphoid organs and main paths of lymphatic vessels. The lymphatic system is comprised of lymphatic vessels, which transport interstitial fluid (lymph) back to the blood circulation, and the lymphoid organs which house lymphocytes and other cells of the body's immune defense system. Primary lymphoid organs are the bone marrow and thymus, where B and T lymphocytes are formed respectively. The secondary lymphoid organs include the lymph nodes, mucosa-associated lymphoid tissue (MALT), and spleen.

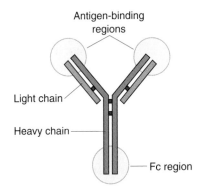

Figure 14–2. Basic structure of an immunoglobulin. Two light chains and two heavy chains form an antibody molecule ("monomer"). The chains are linked by disulfide bonds. The variable portions near the amino end of the light and heavy chains bind the antigen. The Fc region of the molecule may bind to surface receptors of several cell types.

Table 14–1. Summary of classes of antibodies.

	IgG	IgM	IgA	IgD	IgE
Structure	Monomer	Pentamer	Dimer or trimer with secretory component	Monomer	Monomer
Antibody percentage in the serum	75–80%	5–10%	10–15%	0.001%	0.002%
Presence in sites other than blood, connective tissue, and lymphoid organs	Fetal circulation in pregnant women	B lymphocyte surface (as a monomer)	Secretions (saliva, milk, tears, etc)	Surface of B lymphocytes	Bound to the surface of mast cells and basophils
Known functions	Activates phagocytosis, neutralizes antigens	First antibody produced in initial immune response; activates complement	Protects mucosae	Antigen receptor triggering initial B cell activation	Destroys parasitic worms and participates in allergies

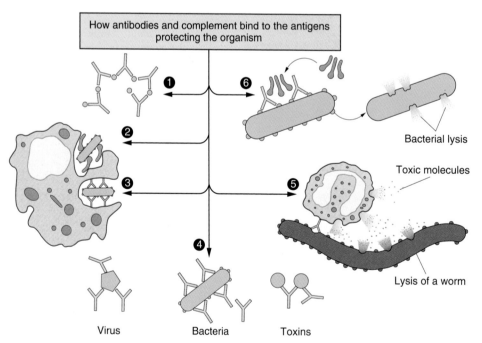

Figure 14–3. **Mechanisms of antigen inactivation.** Several mechanisms are used by components of the immune system to inactivate or remove potentially harmful material. **(1)** Agglutination, in which antibodies bind to antigens, forming aggregates and reducing the amount of free antigens; aggregates may be ingested by phagocytes; **(2)** opsonization of antigens by complement stimulates their phagocytosis; **(3)** opsonization of antigens by antibodies stimulates phagocytosis; **(4)** neutralization, in which the binding of antibody to microorganisms blocks their adhesion to cells and inactivates toxins; **(5)** cytotoxicity mediated by cells, which involves antibodies adhering to the surface of worms activating cells of the immune system (macrophages and eosinophils) and inducing them to liberate molecules that attack the surface of the animal; **(6)** complement activation, in which the binding of antibodies to the initial protein of the complement system triggers the complement cascade and causes cell lysis.

present on the surfaces of mast cells and basophils, it attaches to these cells after being secreted by plasma cells and only small amounts are found in the blood. When IgE molecules present on the surface of mast cells or basophils encounter the antigen that elicited the production of this specific IgE, the antigen–antibody complex triggers the liberation of several biologically active substances, such as histamine, heparin, leukotrienes, and eosinophil-chemotactic factor of anaphylaxis. This characterizes an **allergic reaction,** which is thus mediated by the binding of cell-bound IgE with the antigens **(allergens)** that stimulated its production (see Mast Cells in Chapter 5).

The properties and activities of **IgD** are not completely understood. It is monomeric and is even less abundant than IgE, constituting only 0.001% of the immunoglobulin in plasma. IgD is found on the cell membrane of B lymphocytes.

Actions of Antibodies

Some antibodies are able to agglutinate cells and to precipitate soluble antigens, thus neutralizing their harmful effects on the body (Figure 14–3). Although phagocytosis of microorganisms and other particles occurs spontaneously, this event is greatly stimulated when they are covered by antibodies produced against them, a phenomenon called **opsonization** (Figure 14–3). Because macrophages, neutrophils, and eosinophils have receptors for the Fc region of IgG on their surfaces, they phagocytose items that have been opsonized much more readily.

Antigen–antibody complexes and some antigens activate the **complement system,** a group of around 20 plasma proteins produced mainly in the liver and activated through a cascade of reactions. One of the most important proteins of this system is the component called **C3.** To defend the body against foreign molecules or cells, the complement system may (1) stimulate phagocytosis of microorganisms because of opsonization due to the binding of C3 fragments to specific C3 receptors on the surface of phagocytic cells (Figure 14–3) and (2) induce lysis of microorganisms by acting on their cell membranes (Figure 14–3).

CYTOKINES

The functions of cells in the immune system are regulated by a large number of molecules, mainly **cytokines,** which are peptides or glycoproteins usually with low molecular masses (between 8 and 80 kDa). They influence both the cellular and humoral immune responses (Table 14–2). Cytokines act on many cells that have receptors for them—not only cells of the immune system, but also cells of other systems, such as the nervous system and endocrine system. They are primarily produced by cells of the immune system, mainly lymphocytes, macrophages, and other leukocytes, but may also be synthesized by other cell types, such as endothelial cells and fibroblasts. **Chemotaxins,** or **chemokines,** are cytokines that induce diapedesis of leukocytes and migration to sites of inflammation (Chapter 12).

CELLS OF THE IMMUNE SYSTEM

The primary cells that participate in the immune response are lymphocytes, plasma cells, mast cells, neutrophils, eosinophils, and cells of the mononuclear phagocyte system. Antigen-presenting cells, a group of diverse cell types, assist other cells in the immune response. This group includes, among other cells, lymphocytes, macrophages, and dendritic cells.

Lymphocytes

Lymphocytes are classified as **B, T,** or **natural killer (NK) cells.** The B and T cells are the only cells that have the ability to selectively recognize a specific epitope among a vast number of different epitopes (on the order of 10^{18}). B and T cells differ based on their life history, surface receptors, and behavior during an immune response. Although B and T cells are morphologically indistinguishable in either the light or electron microscope, different surface proteins (markers) allow then to be distinguished by immunocytochemical methods. The precursors of all lymphocyte types originate in the bone marrow; some lymphocytes mature and become functional in the bone marrow, and after leaving the bone marrow enter the blood circulation to colonize connective tissues, epithelia, lymphoid nodules, and lymphoid organs. These are the **B lymphocytes** (Figure 14–4). **T lymphocyte precursors,** on the other hand, leave the bone marrow, and through the blood circulation reach the thymus where they undergo intense proliferation and differentiation or die by apoptosis. After their final maturation, T cells leave the thymus and are distributed throughout the body in connective tissues and lymphoid organs (Figure 14–4). Because of their function in lymphocyte production and maturation, the bone marrow and the thymus are called the **primary** or **central lymphoid organs.** The other lymphoid structures are all **secondary** or **peripheral lymphoid** organs (spleen, lymph nodes, solitary lymphoid nodules, tonsils, appendix, and Peyer's patches of the ileum). B and T cells are not anchored in the lymphoid organs; instead, they continuously move from one location to another, a process known as **lymphocyte recirculation** so that the cellular composition and microscopic anatomy of lymphoid tissues differ from one day to the next. B and T cells are not distributed uniformly throughout the lymphoid organs but occupy specific organs preferentially (Table 14–3).

Key features of B and T lymphocytes include the receptors they have on their surface. These receptors are fundamental for recognition of antigen epitopes and, thus, for triggering an immune response. T cells recognize a linear sequence of amino acids whereas

Table 14–2. Examples of cytokines, grouped by main function.

Cytokine[1]	Main Function
GM-CSF, M-CSF	Growth and differentiation factors for bone marrow cells
TNF-, IL-1, IL-6	Inflammation and fever
IL-12	Stimulation of innate and specific response
IL-2, IL-4, IL-3	Growth factors for T and B cells
IL-5	Eosinophil differentiation and activation
Interferon-γ	Activation of macrophages
IL-10, TGF-β	Regulation of the immune response
Interferon-α, Interferon-β	Antiviral activity

[1]GM-CSF, granulocyte-macrophage colony stimulating factor; M-CSF, macrophage colony-stimulating factor; TNF, tumor necrosis factor; IL, interleukin; TGF, transforming growth factor.

in B cells the spatial arrangement (ie, the molecular conformation) of proteins, nucleic acids, polysaccharides, or lipids is also important. Each B lymphocyte that leaves the bone marrow or each T lymphocyte that leaves the thymus has just one type of surface receptor that recognizes a specific epitope. As a result of gene rearrangement during the maturation of B and T cells, many millions of different cells, each carrying identical surface receptors able to recognize one specific epitope, are produced. Thus, each lymphocyte recognizes only one epitope.

In an organism not yet exposed to antigens, very few individual lymphocytes are able to recognize each of the millions of epitopes that exist, perhaps only from one to a few hundred such

cells for each. Soon after a lymphocyte is first exposed to the epitope it recognizes, a stimulus to cell proliferation occurs, leading to an amplification of that particular lymphocyte population, and thereby producing an expanded clone of lymphocytes able to recognize that epitope.

B LYMPHOCYTES

In B lymphocytes, the surface receptors able to recognize antigens are monomeric molecules of IgM; each B cell is covered by about 150,000 molecules of IgM. The encounter of a B lymphocyte with the epitope it recognizes leads to several cycles of cell proliferation, followed by a redifferentiation of most of these lymphocytes into **plasma cells.** This population of plasma cells secretes antibodies against the same epitope as that recognized by the B cell from which it arose. In most cases, the activation of B cells requires the assistance of a subclass of T lymphocytes known as **T helper cells.** Not all activated B cells, however, become plasma cells; some remain as long-lived **B memory cells,** which are able to react very rapidly to a second exposure to the same epitope.

T LYMPHOCYTES

T cells constitute 65–75% of blood lymphocytes. To recognize epitopes, all T cells have on their surfaces a molecule called a **T cell receptor (TCR).** In contrast to B cells, which recognize soluble antigens or antigens present on cell surfaces, T lymphocytes recognize only epitopes (mostly small peptides) that form complexes with special proteins on the cell surface of other cells (proteins of the major histocompatibility complex, see below).

Three important subpopulations of T cells are the following:

- **Helper cells,** which produce cytokines that promote differentiation of B cells into plasma cells, activate macrophages to become phagocytic, activate cytotoxic T lymphocytes, and induce many parts of an inflammatory reaction. Helper cells have a marker called CD4 on their surfaces and are, hence, called **CD4+ T cells.**
- **Cytotoxic T cells** are **CD8+** and act directly against foreign cells or virus-infected cells by two main mechanisms. In one, they attach to the cells to be killed and release proteins called **perforins** that create holes in the cell membrane of the target cell, with consequent cell lysis. In the other, they attach to a cell and kill it by triggering mechanisms that induce programmed cell death, or **apoptosis.**
- **Regulatory T cells** are CD4+CD25+ and play crucial roles in allowing immune tolerance, maintaining unresponsiveness to self-antigens and suppressing excessive immune responses.

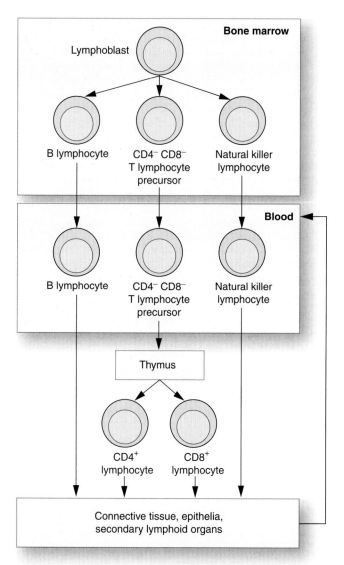

Figure 14–4. Origin of the main types of lymphocytes. B lymphocytes and natural killer (NK) lymphocytes are formed and become mature in the bone marrow and leave that compartment to seed the secondary lymphoid organs and transit through the blood to epithelia and connective tissues. Immature CD4− and CD8− T lymphocyte precursors are transported by the circulation from the bone marrow to the thymus, where they complete their maturation and leave as either CD4+ or CD8+ cells.

Table 14–3. Approximate percentages of B and T cells in lymphoid organs.

Lymphoid Organ	T lymphocytes (%)	B lymphocytes (%)
Thymus	100	0
Bone marrow	10	90
Spleen	45	55
Lymph nodes	60	40
Blood	70	30

These cells produce **peripheral tolerance**, which backs up the central tolerance emerging in the thymus.

The first encounter of a CD4 or CD8 T cell with its specific epitope is followed by amplification of that clone; some of the cells of this increased population become effector cells, doing the job for which they are specialized, and some remain memory helper or memory cytotoxic T cells, reacting rapidly to the next presentation of the same epitope.

NATURAL KILLER CELLS

The **natural killer** lymphocytes lack the marker molecules characteristic of B and T cells. They comprise about 10–15% of the lymphocytes of circulating blood. Their name derives from the fact that they attack virus-infected cells, transplanted cells, and cancer cells without previous stimulation; for this reason they are involved in what is called an **innate immune response**.

MEDICAL APPLICATION

*One of the primary causes of the immunodeficiency syndrome known as **AIDS** involves the killing of helper T cells by the infecting retrovirus. This cripples the patients' immune system rendering them susceptible to opportunistic infections by microorganisms that usually do not cause disease in immunocompetent individuals.*

Major Histocompatibility Complex (MHC) & Antigen Presentation

The **major histocompatibility complex (MHC)** is a complex of chromosomal loci encoding several proteins known as class I and class II MHC molecules. Because a great many alleles exist within each of the loci, there is great variation of these molecules among the general population. One individual, however, expresses only one set of class I proteins and one set of class II proteins; these proteins are unique to that person. All nucleated cells have class I proteins, but class II proteins exist on only a small group of cells operationally called **antigen-presenting cells (APCs).**

MHC molecules are integral membrane proteins present on the cell surface. They are synthesized by the rough ER like regular membrane proteins. However, on their way to the cell surface, they couple with small peptides of 10–30 amino acids whose origin differs depending on whether class I or class II molecules are involved.

In most cases, class I MHC molecules form complexes with peptides derived from cytosolic proteins synthesized in that cell. Proteins encoded by virus nucleic acid in an infected cell are an important example of this kind of cytosolic protein. The proteins are targeted by ubiquitin to be degraded by proteasomes, producing small peptides. These peptides are transported to membranes of the ER where they bind class I MHC molecules; the resulting complex is moved to the cell surface, exposing the peptides to the extracellular space (Figure 14–5).

The peptides that join class II MHC molecules result mostly from endocytosis and digestion in lysosomes. The vesicles that contain these peptides fuse with Golgi-derived vesicles that have

class II MHC embedded in their membranes. The peptides form complexes with the MHC proteins and, as in the case of class I molecules, the complex is transported to the cell surface exposing the peptides to the exterior of the cell (Figure 14–5).

Presentation of the MHC-I endogenous derived antigen complex at the cell surface

Presentation at the cell surface of exogenous derived antigens via MHC-II

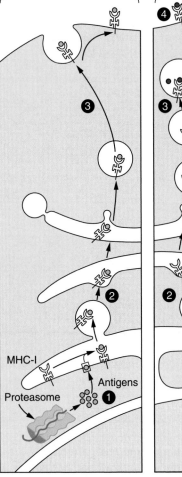
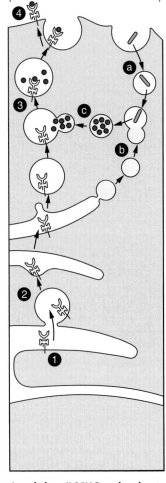

Figure 14–5. Binding of class I and class II MHC molecules to antigens. **Left:** The sequence of events by which antigens made in a cell (eg, in a virus-infected cell) are processed, bound to class I MHC proteins, and displayed at the cell surface. **(1)** Proteins in the cell are continuously digested by proteasomes and antigenic fragments are transferred to the rough ER where they associate with class I MHC proteins synthesized there. **(2)** The class I MHC–antigen complex is transferred to the Golgi region. **(3)** Golgi vesicles transport the complex to the cell membrane presenting the antigen at the outer surface. **Right:** Formation of complexes between class II MHC proteins and antigens internalized by the cell. **(1)** Synthesis of class II MHC molecules. **(2)** Transfer of class II MHC molecules to the Golgi region and formation of Golgi vesicles. The Golgi vesicles fuse with a lysosome containing antigens processed after endocytosis and digestion of microorganisms and debris by lysosomal enzymes **(a, b, c)**. **(3)** Antigens form complexes with class II MHC molecules. **(4)** The class II MHC–antigen complexes are exposed at the cell surface.

Contrary to B cells, which recognize soluble antigens or antigens present on cell surfaces, T lymphocytes recognize only small peptides displayed by MHC molecules. However, the T cell receptor does not interact only with the peptide; instead it interacts with the complex formed by the peptide and the portion of the MHC protein exposed on the surface of the antigen-presenting cell. T cells from an individual recognize this complex only if the MHC molecule belongs to the same individual (self MHC molecules). This happens because during T cell development in the thymus, the T cell precursors whose receptors did not recognize self MHC molecules were induced to die. The display of peptides on the surface of APCs is known as **antigen presentation.**

Because different individuals express different MHC molecules, cell or organ transplantation between genetically distinct individuals normally induces an intense immune reaction that leads to rejection of the transplant.

The cytosolic peptides displayed by class I MHC molecules may derive from (1) the cells' own proteins, in which case the T cell will recognize them as self-proteins, or (2) foreign proteins produced by virus-infected cells, tumor cells, or transplanted cells and organs. The peptides displayed by class II MHC molecules are mostly foreign proteins internalized by the cells through phagocytosis (Figure 14–5).

MEDICAL APPLICATION

Tissue grafts and organ transplants are classified as ***autografts*** *when the transplanted tissues are taken from the same individual receiving them,* ***isografts*** *when taken from an identical twin,* ***homografts*** *or* ***allografts*** *when taken from an individual (related or unrelated) of the same species, and* ***heterografts*** *or* ***xenografts*** *when taken from an animal of a different species.*

The body readily accepts autografts and isografts as long as an efficient blood supply is established for the organ. There is no rejection in such cases, because the transplanted cells are genetically identical to those of the host and present the same MHC on their surfaces. The organism recognizes the grafted cells as self (same MHC) and does not react with an immune response.

Homografts and heterografts, on the other hand, contain cells whose membranes have class I and class II MHC molecules that are foreign to the host; they are therefore recognized and treated as such. Transplant rejection is a complex process due to the activity of T lymphocytes and antibodies that react to and destroy the transplanted cells.

Antigen-Presenting Cells (APCs)

APCs are found in many tissues and constitute a heterogeneous population of cells that includes **dendritic cells**, macrophages, and B lymphocytes. Dendritic cells (not to be confused with cells of nervous tissue) occur not only in the lymphoid organs, but are also abundant in epidermis and many mucosae, where they

are called **Langerhans cells**. A common feature of APCs is the presence of class II MHC molecules on their surfaces. CD4$^+$ T (helper) cells interact with complexes formed by peptides and class II MHC molecules on APCs. However, CD8$^+$ T (cytotoxic) cells interact with peptides complexed with class I MHC molecules which can be presented by any nucleated cell. APCs, being recognized by helper lymphocytes, are essential for triggering and development of a complex immune response.

The Langerhans cells of the epidermis constitute a very efficient system for trapping antigens that enter the epidermis (Figure 14–6). Like APCs of other organs, these cells have many processes and upon capturing antigens, they retract the processes, move toward the dermis, and enter a lymphatic vessel.

TYPES OF IMMUNE RESPONSES

The two basic types of immune responses are the **innate response** and the **adaptive response.** The innate response, which can include the action of the complement system, defensins, and cells such as neutrophils, macrophages, mast cells, and natural killer cells, is fast, nonspecific, and older from an evolutionary point of view. It does not produce memory cells. The adaptive response, which depends on the initial recognition of antigens by B and T cells, is more complex, is slower and specific, produces memory cells, and is a more recent evolutionary development.

The adaptive mechanisms that lead to the elimination of antigens are classified as **humoral** or **cellular responses.** Humoral immunity is accomplished by antibodies produced by plasma cells derived from clones of activated B lymphocytes. Cellular immunity is mediated by T lymphocytes that (1) secrete cytokines that act on B lymphocytes, on other T cells, and on inflammatory cells such as macrophages and neutrophils, and (2) attack foreign cells or cells that exhibit foreign epitopes on their surfaces, such as cells infected by viruses or parasites, and some tumor cells. With few exceptions, both types of immune response are activated when foreign epitopes are recognized by lymphocytes.

Antigens such as molecules derived from microorganisms breaching the skin or a mucosa (or the connective tissue in the case of an injected vaccine) have two main fates. The antigen is phagocytosed either by a macrophage or by a dendritic cell and is then transported by these cells through the lymphatic vessels to the lymph node that drains that region of the body (**satellite lymph node**). Alternatively, the antigenic material is transported by lymph to the lymph nodes where macrophages or other APCs phagocytose them. Antigens that reach the lymph node are recognized by B lymphocytes. APCs that arrived from the skin or mucosa as well as APCs that processed antigens within the lymph node display the antigens to helper T lymphocytes as complexes with class II MHC molecules (Figure 14–7). B lymphocytes that recognize antigens are activated by helper cells to enter several cycles of cell division. Many of the daughter cells of B lymphocytes differentiate into plasma cells that secrete antibodies against the antigen recognized by the first B lymphocyte. The plasma cells secrete most of the antibodies into the lymph and the antibodies eventually reach the blood circulation and act on antigens in different ways (Figure 14–3). B cells that are not transformed into plasma cells remain as B memory cells.

Intracellular antigens, such as those synthesized in the cytosol of virus-infected cells, tumor cells, or transplanted cells, are presented to cytotoxic T lymphocytes bound to class I MHC molecules (Figure 14–8). Concurrently, APCs that phagocytose fragments of viruses, tumor cells, or transplanted cells display the antigens to helper T lymphocytes, bound to class II MHC molecules. Cytotoxic

T cells, activated by helper cells, enter several cycles of proliferation and some of these cells become effector cytotoxic T cells that will destroy the cells that hold the antigens (Figure 14–8). Some cells, instead of becoming effector cells, remain as memory cytotoxic T cells. A humoral response resulting from recognition of antigens by B lymphocytes generally occurs simultaneously.

MEDICAL APPLICATION

Diseases of immune system can be broadly grouped into three types:

1. *Some individuals develop abnormal and intense reactions in an attempt to neutralize the effects of some antigens. This exaggerated intolerance produces the events comprising* **allergic reactions**.

2. *The immune response can be impaired, a condition generally called* **immunodeficiency**; *this may have several causes, such as genetic or infectious factors (eg, by measles and human immunodeficiency virus). Immunodeficiencies of genetic origin may be caused by mutations or deletions in genes that code for molecules that participate in effector immune mechanisms or that are involved in the differentiation of T, B, and APC cell populations. As such, immunodeficiencies may affect the complement system, phagocytic activity, and the development and function of B and T lymphocytes*

3. *Autoimmune diseases are caused by T or B cell responses directed to self molecules. Tissues are affected or even destroyed by cytotoxic T lymphocytes or by autoantibodies.*

LYMPHOID TISSUE

Lymphoid tissue is connective tissue characterized by a rich supply of lymphocytes. It exists free within the regular connective tissue or is surrounded by capsules, forming the lymphoid organs. Because lymphocytes have very little cytoplasm, lymphoid tissue filled densely with such cells stains dark blue in H&E-stained sections. Lymphoid tissues are basically made up of free cells, typically with a rich network of reticular fibers of type III collagen supporting the cells (Figure 14–9). In most lymphoid organs, the reticular fibers are produced by a fibroblastic cell called a **reticular cell**, whose many processes rest on the fibers (Figures 5–23 and 14–9). The thymus is an exception in that its cells are supported by a reticulum of unusual epithelial cells.

The network of reticular fibers of the lymphoid tissue may be relatively dense and thus able to hold many free lymphocytes, macrophages, and plasma cells. Areas of more loosely organized lymphoid tissue, with fewer and larger spaces, allow easy movement of these cells (Figure 14–9).

In **nodular lymphoid tissue** groups of lymphocytes are arranged as spherical masses called **lymphoid nodules** or **lymphoid follicles**, containing primarily B lymphocytes. When lymphoid nodules become activated as a result of the arrival of antigen-carrying APCs and recognition of the antigens by B lymphocytes, these lymphocytes proliferate in the central portion of the nodule, which then stains lighter and is called a **germinal**

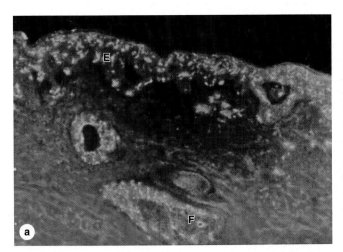
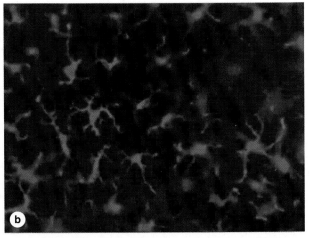

Figure 14–6. **Langerhans cells.** Langerhans cells are dendritic antigen-presenting cells of the epidermis and other epithelia of body surfaces, where they comprise an important defense against pathogens and environmental insults. Like other APCs they develop in the bone marrow, move into the blood circulation, and finally migrate into stratified squamous epithelia where they can be difficult to identify in routinely stained sections. **(a):** Section of immunostained skin showing Langerhans cells (yellow) abundant in hair follicles (F), where many microorganisms live, and throughout the epidermis (E). Keratin of the epidermis and follicles is stained green. X40. Antibody against langerin/CD207 and counterstained with anti- keratin. **(b):** Face-on view of an epidermal sheet stained using the same antibody showing that Langerhans cells form a network among the other epidermal cells which is difficult for invading microorganisms to avoid. Having sampled the invaders' antigens, Langerhans cells emerge from the epidermis and travel to the nearest lymph node to elicit lymphocytes that can mount a collective immune response. X200. Anti-langerin/CD207. (Reproduced, with permission, from N. Romani et al: *Acta Path. Micro. Immunol. Scandinavica.* 2003;111:725.)

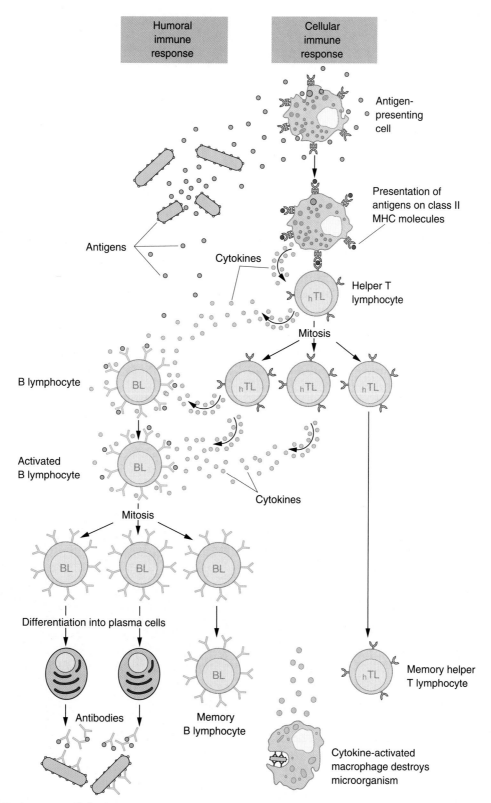

Figure 14–7. Basic events of the immune response to a microorganism. B lymphocytes recognize antigen and are activated by helper T lymphocytes which were presented that antigen by antigen-presenting cells. Using other cytokines, helper cells also stimulate B cells to enter several cycles of cell division followed by differentiation of many daughter cells into plasma cells that secrete antibodies to the antigen recognized by the first B lymphocyte. In practice, several different B cells recognize different epitopes, so that several different antibodies are produced. After the microorganism has been eliminated, some of the lymphocytes remain as long-lived memory cells.

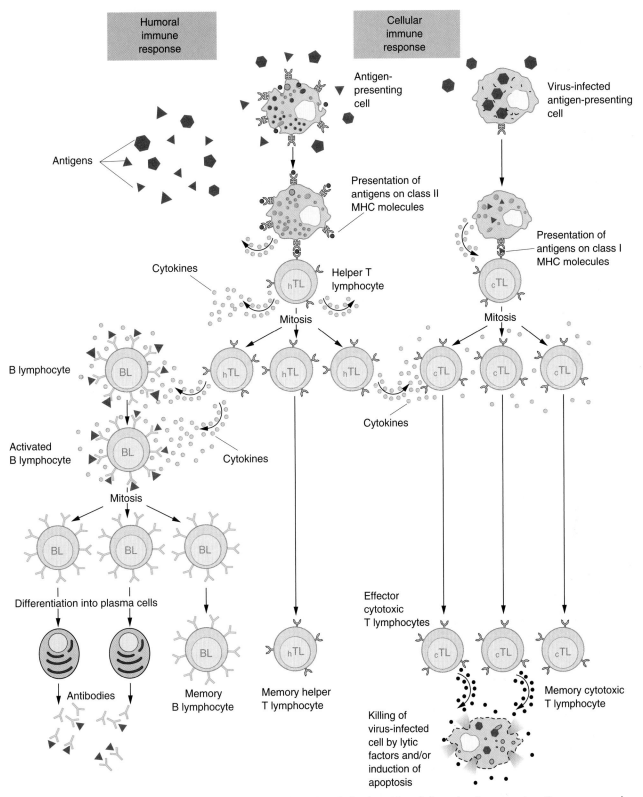

Humoral immune response

Cellular immune response

Antigens

Antigen-presenting cell

Virus-infected antigen-presenting cell

Presentation of antigens on class II MHC molecules

Presentation of antigens on class I MHC molecules

Cytokines

Helper T lymphocyte

Mitosis

Mitosis

B lymphocyte

Cytokines

Activated B lymphocyte

Cytokines

Mitosis

Differentiation into plasma cells

Effector cytotoxic T lymphocytes

Antibodies

Memory B lymphocyte

Memory helper T lymphocyte

Memory cytotoxic T lymphocyte

Killing of virus-infected cell by lytic factors and/or induction of apoptosis

Figure 14–8. **Basic events of the immune response to a virus infection.** Virus-infected cells present antigens as complexes with class I MHC cells. The complexes are recognized by cytotoxic T lymphocytes that are stimulated by helper T lymphocytes to enter several cycles of cell division. Many daughter cytotoxic cells turn into effector cells that kill the infected cells. A population of helper and cytotoxic T cells remains as memory cells. A simultaneous humoral response is usually launched by B lymphocytes that recognized the antigen.

center. After completion of the initial immune response, the germinal center may disappear. Germinal centers contain a special cell, the **follicular dendritic cell** (**FDC**, distinct from the dendritic APCs), with surfaces that have many extremely fine processes. Antigens bind surface proteins of FDCs in several ways (not including MHC proteins) but are not internalized or degraded by these cells. Rather antigen is retained by FDCs for extended periods (months to years) for interaction with B lymphocytes.

Lymphoid nodules vary widely in size, typically measuring a few hundred micrometers to one mm in diameter. They are found free in many connective tissues in the body and within lymph nodes, spleen, and tonsils, but not in the thymus which contains only T cells. Free lymphoid nodules are commonly present in the connective tissue of mucosal linings, where, together with free lymphocytes, they constitute the mucosa-associated lymphoid tissue (MALT). Individual lymphoid follicles are not encapsulated with connective tissue.

THYMUS

The thymus is a bilateral organ located in the mediastinum; it attains its peak development during youth. Like bone marrow and B cells, the thymus is considered a central or primary lymphoid organ because T lymphocytes form there. Whereas all other lymphoid organs originate exclusively from mesenchyme (mesoderm), the thymus has a dual embryonic origin. Its precursor lymphoblasts originate in the bone marrow, but then move to invade a unique epithelium that developed from the endoderm of the embryo's third and fourth pharyngeal pouches.

The thymus has a connective tissue capsule that penetrates the parenchyma and divides it into incomplete lobules, with continuity between the cortex and medulla of adjoining lobules (Figure 14–10). Each lobule has a peripheral darkly stained zone known as the **cortex** and a central light zone called the **medulla.** The cortex is richer in small lymphocytes than the medulla and therefore it stains more darkly.

The thymic cortex is composed of an extensive population of T lymphoblasts (also called **thymocytes**) and macrophages in a stroma of **epithelial reticular cells**. The epithelial reticular cells usually have large euchromatic nuclei and are diverse morphologically, but generally either squamous or stellate with long processes. They are typically joined to similar adjacent cells by desmosomes (Figure 14–11), forming an unusual **cytoreticulum.** Cytoplasmic bundles of intermediate keratin filaments (tonofilaments) give evidence of these cells' epithelial origin. Occluding junctions between flattened epithelial reticular cells at the boundary between cortex and medulla help to separate these two regions.

The thymic medulla also contains a cytoreticulum of epithelial reticular cells, many less densely packed differentiated T lymphocytes, and structures called **thymic (Hassall's) corpuscles,** which are characteristic of this region (Figure 14–12). Thymic corpuscles consist of epithelial reticular cells arranged concentrically, filled with keratin filaments, and sometimes calcified. They are absent in mice and a unique function for these structures in humans has not been established.

Arterioles and capillaries in the thymic cortex are sheathed by flattened epithelial reticular cells with tight junctions. The capillary endothelium is continuous and has a thick basal lamina. These features create a **blood-thymus barrier** and prevent most circulating antigens from leaving the microvasculature and entering the thymus cortex. No such barrier is present in the medulla and mature T lymphocytes exit the thymus via venules in this zone.

The thymus has no afferent lymphatic vessels and does not constitute a lymph filter, as do lymph nodes. The few lymphatic vessels of the thymus are in the connective tissue of the capsule, septa and blood vessel walls and are all efferent.

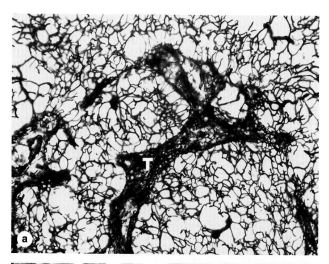
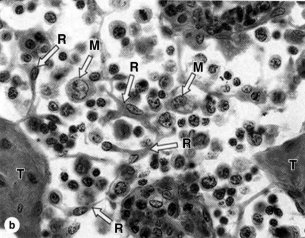

Figure 14–9. Reticular fibers and cells of lymphoid tissue. **(a):** A three-dimensional framework of reticular fibers (collagen type III) supports the cells of most lymphoid tissues and organs (except the thymus). Areas with larger spaces between the fibers offer more mobility to cells than areas in which the fiber mesh is tight, such as in trabeculae (T) and the cells here are generally more stationary. X10. Silver impregnation. **(b):** Cells of typical lymphoid tissue include the fibroblast-like reticular cells (R) which produce and maintain the trabeculae (T) and reticulin framework. Many cells are loosely attached to the reticulin fibers, including macrophages (M) and many lymphocytes. X240. H&E. (With permission, from Paulo A. Abrahamsohn, Institute of Biomedical Sciences, University of São Paulo, Brazil.)

Role of the Thymus in T Cell Maturation

The thymus is the site of T lymphocyte differentiation and removal of T lymphocytes reactive against self-antigens, an important part of **central self-tolerance induction**.

T lymphoblast surfaces do not yet exhibit the T cell receptor (TCR) or the CD4 and CD8 markers. Their progenitor cells

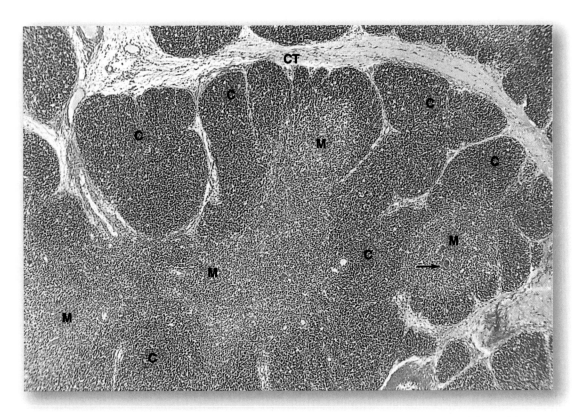

Figure 14–10. **Thymus.** The thymus, an encapsulated, bilateral organ in the mediastinum, is subdivided by connective tissue (CT) septa into connected lobes. Lobes of an active thymus shown have peripheral regions of cortex (C), where basophilic lymphocytes are fairly dense, and more central medulla (M) regions with fewer lymphocytes. Besides the differences in location and cell density, the medulla region is characterized by the scattered presence of distinct thymic corpuscles (arrow). X140. H&E.

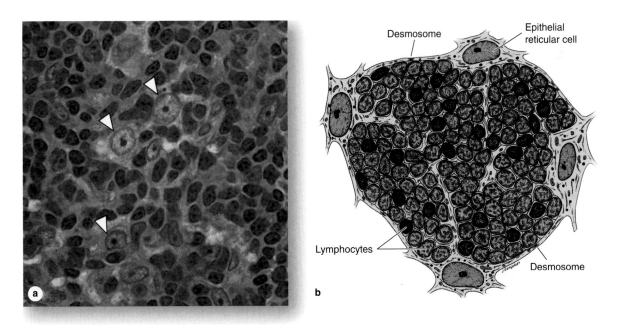

Figure 14–11. **Cortex of the thymus. (a):** The cortical zone of an active thymus is packed with lymphoblasts that proliferate as well as undergo positive and negative selection in that region. The lymphoblasts are supported on a meshwork of epithelial reticular cells (arrowheads). X400. PT. **(b):** The epithelial reticular cells extend long processes bound together by desmosomes to make the framework for the lymphocytes. The epithelial reticular cells also secrete polypeptide factors that promote T cell maturation.

arise in the fetal liver or bone marrow and migrate to the thymus during both fetal and postnatal life. After entering the thymus, T lymphoblasts populate the cortex where they proliferate extensively. As thymocytes mature and express T cell markers, they undergo **thymic selection**, a stringent quality control process, as they pass through a succession of microenvironments created by different mixes of the stromal epithelial reticular cells.

Differentiating thymocytes in the cortex are presented with antigens bound to class I and class II MHC proteins on the epithelial reticular cells, macrophages, and dendritic cells. Thymocytes whose TCRs cannot bind MHC molecules at all are nonfunctional and have no future as T cells; these cells (as many as 80% of the total) are induced to undergo apoptosis. Similarly, those thymocytes that strongly bind MHCs containing self-peptides are also deleted since such T cells could cause a damaging autoimmune response. Only 2–3% of the thymocytes pass both these positive and negative selection tests and survive to migrate into the thymic medulla. The others die by apoptosis and are removed by the numerous local macrophages. Movement into the medulla depends on the action of chemokines and on the interaction of thymocytes with the ECM and cytoreticulum. Mature, functional T cells enter the blood circulation by passing through the walls of venules in the medulla and are distributed throughout the body (Figure 14–4).

Besides their structural roles, the epithelial reticular cells produce a number of paracrine factors required for differentiation, selection and migration of mature T lymphocytes, notably thymopoietin and thymosins. Other polypeptides secreted by these cells, including thymulin and thymus humoral factor, also affect target cells outside the thymus.

The thymus reaches its maximum development in relation to body weight immediately after birth; it undergoes involution after attaining its greatest size in puberty, but continues to produce lymphocytes until old age (Figure 14–13).

MUCOSA-ASSOCIATED LYMPHOID TISSUE (MALT)

The digestive, respiratory, and genitourinary tracts are common sites of invasion by pathogens because their lumens are open to the external environment. To protect the organism, the mucosal connective tissue of these tracts contains large and diffuse collections of dendritic cells, lymphocytes, IgA-secreting plasma cells, APCs, and lymphoid nodules. Lymphocytes and dendritic cells are also present within the epithelia lining the lumens. Most of the lymphocytes are B cells; among T cells, CD4 helper cells predominate. In some places, these aggregates form large, conspicuous structures such as the **tonsils** and the **Peyer patches** in the ileum. Similar aggregates with lymphoid follicles are found in the **appendix**. Collectively the **mucosa-associated lymphoid tissue (MALT)** is one of the largest lymphoid organs, containing up to 70% of all the body's immune cells.

Tonsils are partially encapsulated lymphoid tissue lying beneath and in contact with the epithelium of the oral cavity and pharynx. According to their location they are called **palatine, pharyngeal,** or **lingual tonsils** (Figure 14–14).

Palatine tonsils, in the posterior parts of the soft palate, are covered by stratified squamous epithelium. Each has 10–20 epithelial invaginations that penetrate the tonsil deeply, forming **crypts** (Figure 14–15). The lymphoid tissue in these tonsils forms a band

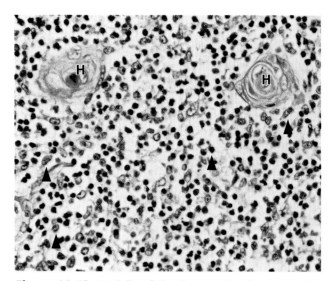

Figure 14–12. Medulla of the thymus. The thymic medulla contains fewer lymphocytes than the cortex and the epithelial reticular cells (arrowheads) have different morphology and function. The most characteristic feature of the medulla in humans is the presence of thymic (Hassall's) corpuscles (H). These are of variable size and contain layers of epithelial reticular cells undergoing keratinization and degeneration. X200. H&E.

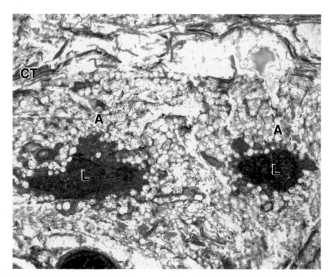

Figure 14–13. Adult thymus. The thymus is highly active at birth and remains so through early childhood, during which time most central tolerance is established. After puberty thymic activities decline and with many fewer lymphocytes present, its structure is reduced in a process of involution. In the adult thymus shown here, cortex and medulla regions are difficult to distinguish within the connective tissue (CT) capsule and only remnants of lymphoid tissue (L) remain, surrounded by much adipose tissue (A). Compare with Figure 14–10 to contrast the functional states of the active and involuted thymus. T lymphocytes continue to be produced in thymic tissue throughout adult life, but at a greatly reduced rate. X140. H&E.

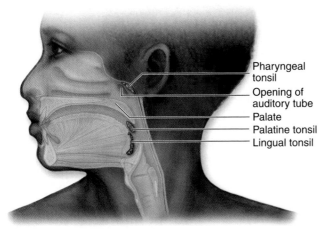

Pharyngeal
tonsil
Opening of
auditory tube
Palate
Palatine tonsil
Lingual tonsil

***Figure 14–14.* Tonsils.** Masses of lymphoid nodules comprising the tonsils are collected in three general locations in the wall of the pharynx. Palatine tonsils are located in the posterior lateral walls of the oral cavity and lingual tonsils are situated along the surface of the posterior third of the tongue. Both are covered with stratified squamous epithelium. The pharyngeal tonsil is a single tonsil situated in the posterior wall of the nasopharynx. It is usually covered by ciliated pseudostratified columnar epithelium typical of the upper respiratory tract, but areas of stratified epithelium can also be observed. Hypertrophied pharyngeal tonsils resulting from chronic inflammation are called **adenoids.**

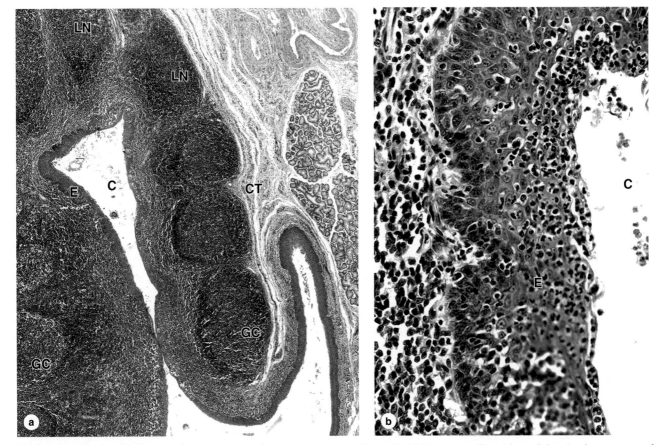

***Figure 14–15.* Detail of a tonsil.** Palatine tonsils are aggregates of lymphoid tissue, usually with nodules, in the mucosa of the posterior soft palate. **(a):** Micrograph showing several lymphoid nodules (LN), collectively covered by stratified squamous epithelium (E) on one side and a connective tissue capsule (CT) on the other. Some nodules show lighter staining germinal centers (GC). Infoldings of the mucosa in some tonsils form crypts (C), along which nodules are especially numerous. Lumens of crypts contain desquamated epithelial cells, live and dead lymphocytes, and bacteria. X140. H&E. **(b):** Epithelium surrounding crypts often becomes infiltrated with lymphocytes and neutrophils and can become difficult to recognize histologically. Lymphocytes abundant in the underlying connective tissue are seen on the left. X200. H&E. (Reproduced, with permission, from Paulo A. Abrahamsohn, Institute of Biomedical Sciences, University of São Paulo, Brazil.)

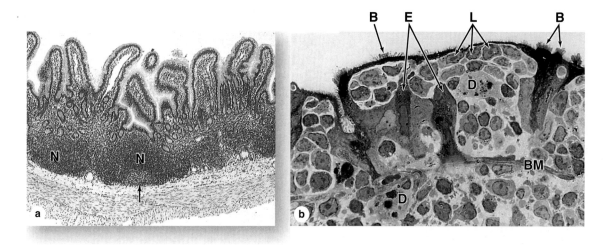

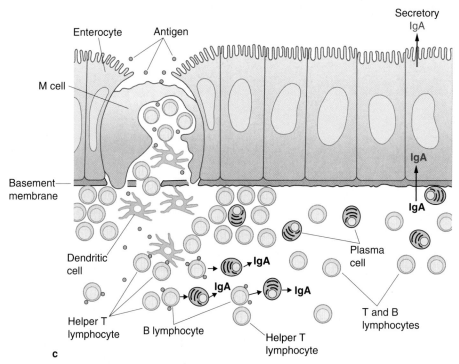

Figure 14–16. **Peyer patch and M cells.** The mucosa of the entire digestive tract is rich in diffuse lymphocytes and scattered follicles, which make up the gut-associated lymphoid tissue (GALT). Particularly large clusters of lymphoid follicles in the ileum of the small intestine are called Peyer patches, where microorganisms of the gut are continuously monitored by specialized sampling stations. **(a):** Section of Peyer patch shows a few of the typical lymphoid nodules (N), some with germinal centers (arrow), in these aggregates. The ileum is lined by an absorptive simple columnar epithelium and intraepithelial lymphocytes are frequently present between the columnar cells. X20. H&E. **(b):** Micrograph shows that the follicle-associated epithelium or FAE directly over each lymphoid follicle/nodule has other epithelial cells called M cells. M cells have distinctive short apical microfolds but lack the brush border and thick glycocalyx typical of enterocytes. The basal surface of M cells forms a unique large intraepithelial pocket harboring a transient population of T and B lymphocytes (L) and antigen-presenting dendritic cells (D) which pass through openings in the basement membrane (BM). Lymphocytes and phagocytic cells fill the lamina propria and the intraepithelial pockets of four large M cells shown in section here. Also seen are the brush border (B) and darker cytoplasm of enterocytes (E) within the FAE. These enterocytes release chemokines that attract dendritic cells and lymphocytes to the FAE and M cell intraepithelial pockets. X500.

(c): Diagram shows luminal antigens that are bound by M cells and transported by transcytosis directly to their intraepithelial pockets containing dendritic cells, helper T cells, and B cells. Dendritic cells then take up the antigen, process it, and present it to lymphocytes and induce adaptive immune responses. B lymphocytes are stimulated by T helper cells to differentiate into IgA-secreting plasma cells. These IgA molecules are coupled with secretory protein and transported into the intestinal lumen by enterocytes adjacent to the FAE. This IgA binds eliciting antigens, helping to neutralize potentially harmful microorganisms in the lumen, and binds the apical surfaces of M cells to promote antigen uptake and induction of immune responses.

(Figure 14–16b reproduced, with permission, from Marian R. Neutra, Children's Hospital, Harvard Medical School.)

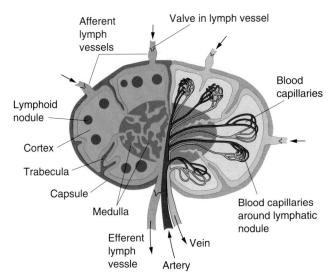

Afferent lymph vessels

Valve in lymph vessel

Blood capillaries

Lymphoid nodule

Cortex

Trabecula

Capsule

Medulla

Efferent lymph vessle

Vein

Artery

Blood capillaries around lymphatic nodule

Figure 14–17. **Schematic structure of a lymph node.** The left half of the figure shows the major regions and structural components of a lymph node and the flow of lymph within these organs, entering via afferent lymphatics on the convex side of the node, passing through unlined sinuses (shown in pink) in the lymphoid tissue, and leaving through an efferent lymphatic at the hilum. Valves in the lymphatic vessels assure the one-way flow of lymph. The right half depicts part of the blood circulation, with a small artery and vein both entering and leaving at the hilum.

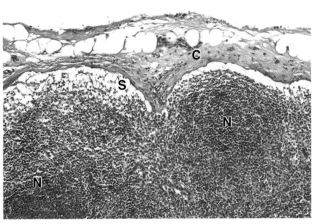

Figure 14–19. **Lymph node cortex.** The outer regions on the convex sides of a lymph node include the capsule (C), sub-capsular sinuses (S), and diffuse lymphoid tissue with lymphoid nodules (N). Afferent lymphatic vessels (which are only rarely shown well in sections) penetrate this capsule, dumping lymph into the sinus where its contents are processed by lymphocytes and APCs. X140. H&E. (With permission, from Paulo A. Abrahamsohn, Institute of Biomedical Sciences, University of São Paulo, Brazil.)

Figure 14–18. **Regions of a lymph node.** A low-magnification section of a lymph node showing the three functional regions: the cortex (C), the paracortex (P), and the medulla (M). Connective tissue of the capsule (CT) completely surrounds each lymph node and extends as several trabeculae (T) throughout the lymphoid tissue. Major spaces for lymph flow are present in this tissue under the capsule and along the trabeculae. A changing population of immune cells is suspended on reticular fibers throughout the cortex, paracortex, and medulla. Lymphoid nodules (LN) are normally restricted to the cortex and the medulla is characterized by sinuses (MS) and cords (MC) of lymphoid tissue. X40. H&E. (With permission, from Paulo A. Abrahamsohn, Institute of Biomedical Sciences, University of São Paulo, Brazil.)

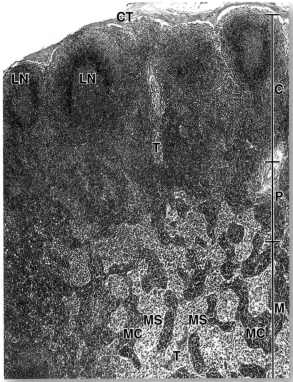

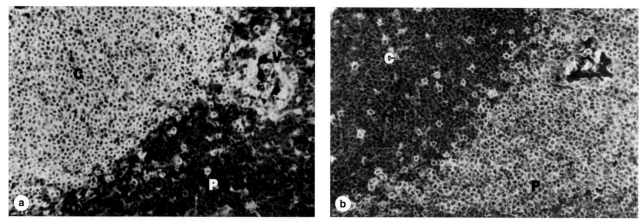

Figure 14–20. **Lymph node cortex and paracortex.** The region just inside the cortex is called the paracortex. Although most lymphocytes in the cortex are B cells, many located in nodules, the lymphocytes of the paracortex are largely T cells. This separation is indicated in the fluorescence micrographs here using immunohistochemistry on adjacent sections of lymph node. **(a):** Antibody against a B cell surface marker labels nearly all the lymphocytes in the cortex (C), as well as many cells around a high endothelial venule (V) in the paracortex, but few cells in the paracortex proper (P). **(b):** Stained with an antibody against a T cell marker, the paracortex is heavily labeled, but only a few cells in the cortex are stained, possibly T helper cells. X200. (Reproduced and modified, with permission, from IL Weissman: *Transplant Rev.* 1975;24:159.)

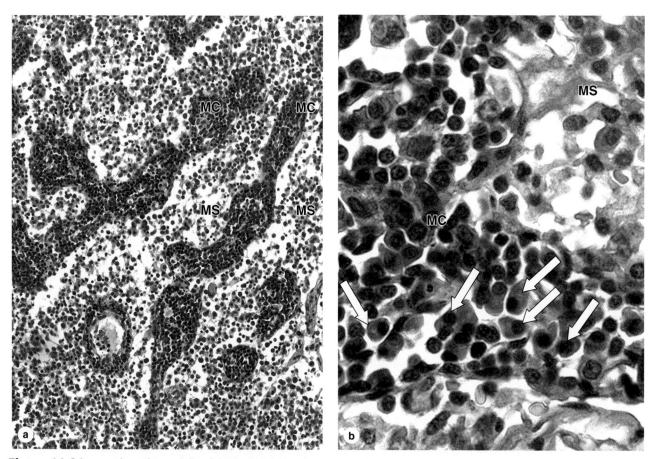

Figure 14–21. **Lymph node medulla. (a):** The medulla of a lymph node consists mainly of the medullary sinuses (MS) separated by intervening medullary cords (MC). Lymphocytes and plasma are abundant and predominate in number over other cell types. A blood vessel within a medullary cord is also seen. X200. H&E. **(b):** Higher magnification of a medullary cord (MC) shows plasma cells (arrows) with spherical, eccentric nuclei and much more cytoplasm than lymphocytes. Efferent lymph is rich in newly synthesized antibodies. A medullary sinus (MS) is also seen. X400. H&E. (With permission, from Paulo A. Abrahamsohn, Institute of Biomedical Sciences, University of São Paulo, Brazil.)

that contains free lymphocytes and lymphoid nodules, generally with germinal centers. The epithelium covering palatine tonsils can become so densely infiltrated by dendritic cells and lymphocytes that it may be difficult to recognize (Figure 14–15). Separating the lymphoid tissue from subjacent structures is a band of dense connective tissue that acts as a capsule or barrier against spreading tonsil infections.

The pharyngeal tonsil is situated in the posterior wall of the nasopharynx (Figure 14–14) and is usually covered by ciliated pseudostratified columnar epithelium, although areas of stratified epithelium can also be observed. The pharyngeal tonsil is composed of pleated mucosa containing diffuse lymphoid tissue and lymphoid nodules and a capsule thinner than that of the palatine tonsils. The lingual tonsils are situated along the posterior surface of the tongue (Figure 14–14) and are covered by stratified squamous epithelium with crypts. The lymphoid tissue of these tonsils has many of the same basic features as that of palatine tonsils (Figure 14–15). All these epithelia contain intraepithelial lymphocytes and dendritic cells.

MALT extends along the entire gastrointestinal tract, but in the wall of the ileum are particularly large aggregates of lymphoid follicles called Peyer patches, each consisting of 10–200 nodules and bulging into the gut lumen with no connective tissue capsule (Figure 14–16). The simple follicle-associated epithelium (FAE) covering these nodules contains specialized **M cells** with apical microfolds rather than the brush border and glycocalyx typical of enterocytes. M cells continuously sample antigens and microorganisms in the intestinal lumen. Each is characterized by a large basal intraepithelial pocket open to the underlying lymphoid tissue through a porous basement membrane (Figure 14–16), discussed further in Chapter 15. Antigenic material bound to the apical surface of M cells is rapidly translocated via transcytosis from the intestinal lumen to APCs and lymphocytes located in the pocket. T helper cells and B cells derived from these lymphocytes move away from the FAE and initiate adaptive responses to the antigens. These B cells give rise to plasma cells secreting

IgA, which is transported by enterocytes into the intestinal lumen to bind and neutralize potentially harmful antigens.

LYMPH NODES

Lymph nodes are bean-shaped, encapsulated structures, generally 2–10 mm in diameter, distributed throughout the body along the course of the lymphatic vessels (Figure 14–1). The nodes are found in the axillae (armpits) and groin, along the great vessels of the neck, and in large numbers in the thorax and abdomen, especially in mesenteries. Lymph nodes constitute a series of in-line filters that are important in the body's defense against microorganisms and the spread of tumor cells. All this lymph, derived from tissue fluid, is filtered by at least one node before returning to the circulation. These kidney-shaped organs have a convex surface that is the entrance site of lymphatic vessels and a concave depression, the **hilum**, through which arteries and nerves enter and veins and lymphatics leave the organ (Figure 14–17). A connective tissue **capsule** surrounds the lymph node, sending trabeculae into its interior.

The most common cells of lymph nodes are lymphocytes, macrophages and other APCs, plasma cells, and reticular cells; follicular dendritic cells are present within the lymphoid nodules. The different arrangement of the cells and of the reticular fiber stroma supporting the cells creates a **cortex**, a **medulla**, and an intervening **paracortex** (Figures 14–17, 4–18, 14–19).

The **cortex**, situated under the capsule, consists of the following components:

- Many reticular cells, macrophages, APCs, and lymphocytes (Figure 14–18).
- Lymphoid nodules, with or without germinal centers, formed mainly of B lymphocytes, embedded within the diffuse population of other cells (Figure 14–18).
- Areas immediately beneath the capsule, called the **subcapsular sinuses**, where the lymphoid tissue has wide

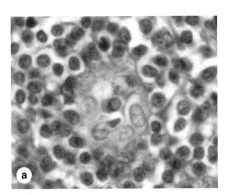
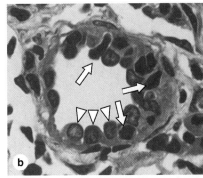
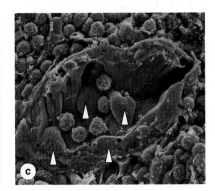

Figure 14–22. High endothelial venules (HEV). (a): High endothelial venules are found in the paracortex of lymph nodes, as shown, as well as in tonsils and Peyer patches. Their endothelial cells are unusually shaped but generally cuboidal and facilitate rapid translocation of lymphocytes into the lymphoid tissue. L-selectin on the lymphocytes recognizes sugar-rich ligands on the surfaces of these endothelial cells and as a consequence the lymphocytes stop there. Integrins promote adhesion between lymphocytes and the endothelial cells and the lymphocytes cross the vessel wall into the lymph node parenchyma. HEVs can be difficult to identify in H&E-stained paraffin sections. X400. H&E. **(b):** Plastic sections more clearly reveal the high endothelial cells (arrowheads) and the lymphocytes passing between them (arrows). X400. PT. **(c):** SEM of a sectioned HEV showing five typical lymphocytes adhering to endothelial cells (arrowheads) prior to migrating between them and joining other lymphocytes in the surrounding paracortex. X500. (Figure 14–22c reproduced, with permission from Fujita, T., 1989, Prog. *Clin. Biol. Res.*, 295:493.)

reticular fiber meshes (Figures 14–18 and 14–19). Lymph containing antigens, lymphocytes, and APCs circulates around the wide spaces of these sinuses after being delivered there by the afferent lymphatic vessels.

- **Cortical sinuses**, running between the lymphoid nodules, which arise from and share the structural features of the subcapsular sinuses. They communicate with the subcapsular sinuses through spaces similar to those present in the medulla (Figures 14–17, 14–18, 14–19, and 14–20).

The **paracortex** does not have precise boundaries with the cortex and medulla. It can be distinguished from the outer cortex by its lack of B cell lymphoid nodules (Figure 14–18) and its accumulation of T cells, which can be determined by immunohistochemistry (Figure 14–20). Venules in the paracortex comprise an important entry point for lymphocytes moving from blood into lymph nodes.

The lymph node **medulla** has two major components:

- **Medullary cords** (Figure 14–21) are branched cordlike extensions of lymphoid tissue arising from the paracortex. They contain primarily B lymphocytes and often plasma cells and macrophages (Figures 14–19 and 14–21).
- Medullary cords are separated by dilated spaces, frequently bridged by reticular cells and fibers, called **medullary sinuses** (Figure 14–21). They contain lymph, lymphocytes, often many macrophages, and sometimes even granulocytes if the lymph node is draining an infected region. These sinuses are continuous with the cortical sinuses and join at the hilum to deliver lymph to the efferent lymph vessel of the lymph node (Figure 14–17).

Afferent lymphatic vessels cross the capsule and pour lymph into the subcapsular sinus (Figure 14–17). From there, lymph passes through the cortical sinuses and then into the medullary sinuses. During this passage, the lymph infiltrates the cortex and

the medullary cords and is filtered and modified by immune cells. The lymph is collected by **efferent lymphatics** at the hilum and valves in both lymphatics assure the unidirectional flow of lymph.

Role of Lymph Nodes in the Immune Response

Lymph nodes are distributed throughout the body and lymph formed in tissues must pass through at least one node before entering the bloodstream. The lymph that arrives at a lymph node contains antigens as soluble molecules, portions of semi-destroyed microorganisms, or antigens already internalized and being transported by macrophages and other APCs. It may also contain microorganisms and cytokines, particularly if it is coming from a region with an infection or inflammation. Antigens that had not been phagocytosed before may be internalized by APCs in the lymph nodes. All antigens have the opportunity to be presented to B lymphocytes, to T helper cells, and to T cytotoxic lymphocytes for these cells to initiate an immune response.

The lymph node is an important site of lymphocyte proliferation (especially of B cells in the germinal centers) as well as of transformation of B lymphocytes into plasma cells. Because of this, the lymph that leaves a lymph node may be enriched in antibodies. When the lymph is returned to the blood circulation, these antibodies will be delivered to the entire body.

MEDICAL APPLICATION

As each satellite node receives lymph from a limited region of the body, malignant tumor cells often reach lymph nodes and are distributed to other parts of the body via the efferent lymph vessels and blood vessels, a process known as metastasis.

Infection and antigenic stimulation often cause lymph nodes to enlarge. These swollen nodules, which may be palpated under the skin as indicators of inflammation, have multiple germinal centers with active cell proliferation. Although plasma cells constitute only 1–3% of the cell population in resting nodes, their numbers increase greatly in stimulated lymph nodes.

Recirculation of Lymphocytes

Because all lymph formed in the body normally drains back into the blood, lymphocytes that leave the lymph nodes by efferent lymphatics eventually reach the bloodstream. These lymphocytes may then leave the blood vessels by entering the tissues and return with other lymph to another lymph node. However, most (90%) lymphocytes return to a lymph node by crossing the walls of specific postcapillary venules in the paracortex, the **high endothelial venules (HEVs)** (Figure 14–22). These vessels have an unusual endothelial lining of tall cuboidal cells, whose apical surface glycoproteins and integrins facilitate rapid diapedesis of lymphocytes out of the blood into the paracortex of the lymph

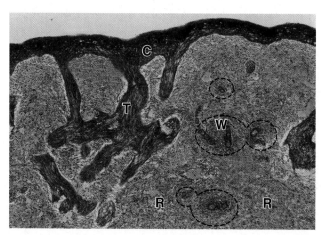

Figure 14–23. Spleen. The capsule (C) of the spleen connects to trabeculae (T) which partially subdivide the pulp-like interior of the organ. The red pulp (R) occupies most of the parenchyma, with white pulp (W) restricted to smaller areas, mainly around the central arterioles. Names of these splenic areas refer to their color in the fresh state: red pulp is filled with blood cells of all types, located both in cords and sinuses; white pulp is lymphoid tissue. Large blood vessels and lymphatics enter and leave the spleen at a hilum. X20. PSH.

node. High endothelial venules are also present in other lymphoid organs, such as the tonsils, Peyer patches, and appendix, but not in the spleen.

The continuous recirculation of lymphocytes enables most parts of the body to be constantly monitored, increasing the opportunity for lymphocytes to encounter activated APCs that have migrated to lymph nodes.

SPLEEN

The spleen is the largest single accumulation of lymphoid tissue in the body and the only one involved in filtration of blood, making it an important organ in defense against blood-borne antigens. It is also the main site of destruction of aged erythrocytes. As is true of other secondary lymphoid organs, the spleen is a

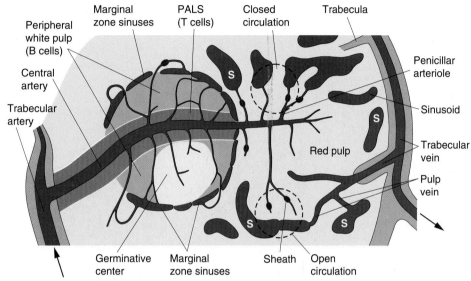

Figure 14–24. Blood flow in the spleen. Schematic view of the blood circulation and the structure of the spleen, from the trabecular artery to the trabecular vein. Small branches of these arteries are called central arteries and become enclosed within a sheath of lymphoid cells, the periarteriolar lymphoid sheath (PALS) in white pulp. B cells in these sheathes can form nodules as the largest masses of white pulp, and around these nodules are located the marginal sinuses. Emerging from the white pulp, the central arteriole branches as the penicillar arterioles, which lead to sheathed capillaries. From these, blood flows into either a closed circulation passing directly into splenic sinuses (S) or an open circulation, being dumped from the vasculature into the lymphoid tissue of the red pulp's splenic cords.

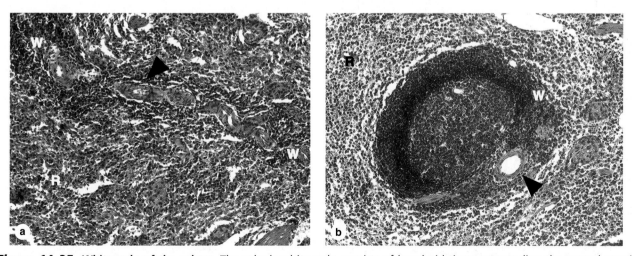

Figure 14–25. White pulp of the spleen. The splenic white pulp consists of lymphoid tissue surrounding the central arterioles as the periarteriolar lymphoid sheath (PALS) and the nodules of proliferating B cells in this sheath. **(a):** Longitudinal section of white pulp (W) in a PALS and the central arteriole (arrowhead) it surrounds. Surrounding the PALS is much red pulp (R). **(b):** A large nodule with a germinal center forms in the PALS. and the central arteriole (arrowhead) is displaced to the nodule's periphery. Small sinuses can be seen at the margin between white and red pulp. Both X20. H&E. (With permission, from Paulo A. Abrahamsohn, Institute of Biomedical Sciences, University of São Paulo, Brazil.)

production site of antibodies and activated lymphocytes, which are delivered to the blood. Any inert particles in blood are actively phagocytosed by spleen macrophages.

The spleen is surrounded by a **capsule** of dense connective tissue from which emerge **trabeculae**, which partially subdivide the parenchyma or **splenic pulp** (Figure 14–23). Large trabeculae originate at the hilum, on the medial surface of the spleen; these trabeculae carry nerves and arteries into the splenic pulp as well as veins that bring blood back into the circulation. Lymphatic vessels that arise in the splenic pulp also leave through the hilum via the trabeculae.

Splenic Pulp

The spleen is composed of reticular tissue containing reticular cells, many lymphocytes and other blood cells, macrophages, and APCs. The splenic pulp has two components, the **white pulp** and the **red pulp** (Figure 14–23). The small masses of white pulp consist of **lymphoid nodules** and the **periarteriolar lymphoid sheaths**, while the red pulp consists of blood-filled **sinusoids** and **splenic cords** (of Bilroth).

As expected of an organ specialized to process blood, the microvasculature of the spleen is important, although much about it remains to be learned. The splenic artery divides inside the hilum, branching into small **trabecular arteries** following this connective tissue. They leave the trabeculae and enter the parenchyma as arterioles enveloped by a sheath of T lymphocytes, the periarteriolar lymphoid sheath (**PALS**), which is part of the white pulp (Figure 14–24). Surrounded by the PALS, these vessels are known as **central arterioles** (Figure 14–25). After coursing through the parenchyma for variable stretches, the PALS receive large numbers of lymphocytes, mostly B cells, and may form lymphoid nodules (Figure 14–25). In these nodules the arteriole occupies an eccentric position but is still called the central arteriole. During its passage through the white pulp,

this arteriole sends off smaller branches that supply the surrounding lymphoid tissue (Figure 14–24).

Surrounding the lymphoid nodules is a **marginal zone** consisting of many blood sinuses and lymphoid tissue (Figures 14–24 and 14–25b). The marginal zone contains lymphocytes, many macrophages, and an abundance of blood antigens and thus plays an important role in the immunological activities of the spleen.

After leaving the white pulp, the sheath of lymphocytes slowly thins and the central arteriole subdivides to form straight **penicillar arterioles** (Figure 14–24). Some of the capillaries that emerge from these penicillar arterioles are sheathed with reticular cells, macrophages and lymphocytes, whose functional significance is not clear.

The red pulp is composed almost entirely of splenic cords and venous sinusoids (Figure 14–26). The splenic cords contain a network of reticular cells or reticular fibers that support T and B lymphocytes, macrophages, plasma cells, and many blood cells (erythrocytes, platelets, and granulocytes). The splenic cords are separated by wide, irregularly shaped sinusoids (Figures 14–26

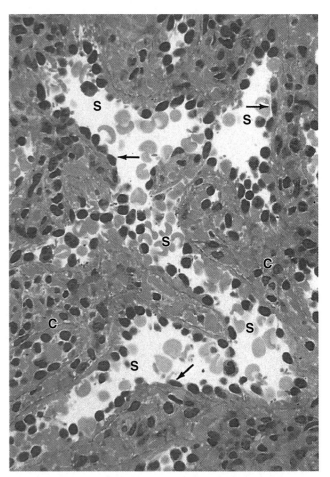

Figure 14–27. Splenic venous sinuses and stave cells. Higher magnification of splenic red pulp shows that the venous sinuses (S) are lined by endothelial cells (arrows) with large nuclei bulging into the sinusoidal lumens. The unusual endothelial cells are called stave cells and have special properties that allow selection of healthy red blood cells in the splenic cords (C). X100. H&E. (With permission, from Paulo A. Abrahamsohn, Institute of Biomedical Sciences, University of São Paulo, Brazil.)

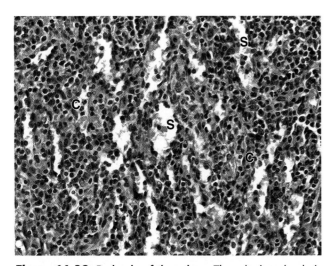

Figure 14–26. Red pulp of the spleen. The splenic red pulp is composed of splenic venous sinusoids (S) and splenic cords (C), both of which contain blood cells of all types. The cords, often called cords of Bilroth, are reticular tissue rich in lymphocytes. The sinuses are lined by unusual, nonsquamous endothelial cells. X40. H&E. (With permission, from Paulo A. Abrahamsohn, Institute of Biomedical Sciences, University of São Paulo, Brazil.)

and 14–27). Unusual elongated endothelial cells, called **stave cells**, line the splenic sinusoids, oriented in parallel with the sinusoid's blood flow. These cells are sparsely wrapped in reticular fibers set in a transverse direction, much like the hoops surrounding the staves of a wooden keg (Figure 14–28).

The highly permeable splenic sinusoids are surrounded by very incomplete basal laminae. The spaces between the endothelial cells of the splenic sinusoids are 2–3 μm or smaller and only flexible cells are able to pass easily from the red pulp cords into the lumen of the sinusoids. Because the lumen of splenic sinusoids is often blood-filled and very small and because the splenic cords are infiltrated with red blood cells, microscopic distinctions between splenic cords and sinusoids may be difficult.

Blood Flow in the Red Pulp

Blood flow through the splenic red pulp can take two routes (Figure 14–24). In the **closed circulation**, the penicillar arterioles

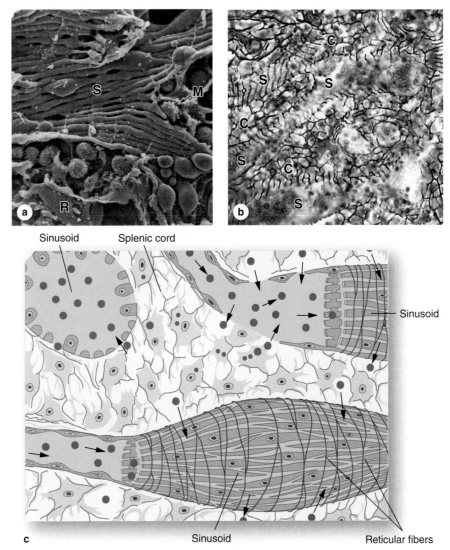

Figure 14–28. **Structure and function of splenic sinusoids.** The endothelial stave cells that line venous sinuses in red pulp are long cells oriented lengthwise along the sinuses. The elongated shape of the cells is difficult to appreciate from light micrographs (see Figure 14–26). **(a):** SEM clearly shows the parallel alignment of the stave cells (S), as well as many macrophages (M) in the surrounding red pulp (R). X500. **(b):** Silver-stained sections of spleen show black reticular fibers surrounding the sinuses, oriented 90 degrees to the long axis of the sinuses (S). These fibers appear similar to those in the surrounding splenic cords (C). The basement membrane of the stave cells is incomplete and open to the passage of cells. X400. Silver. **(c):** Diagram showing these components of splenic sinuses schematically, with the structures resembling a loosely organized wooden barrel. In the open circulation mode of blood flow, blood cells dumped into the cords of the red pulp move under pressure or by their own activity through the spaces between stave cells, reentering the vasculature and soon leaving the spleen via the splenic vein. Cells that cannot squeeze between the stave cells, mainly effete erythrocytes, are removed by macrophages. (Figure 14-28a reproduced, with permission, from Fujita T., 1989, *Prog. Clin. Biol. Res. 1989;295:493)*

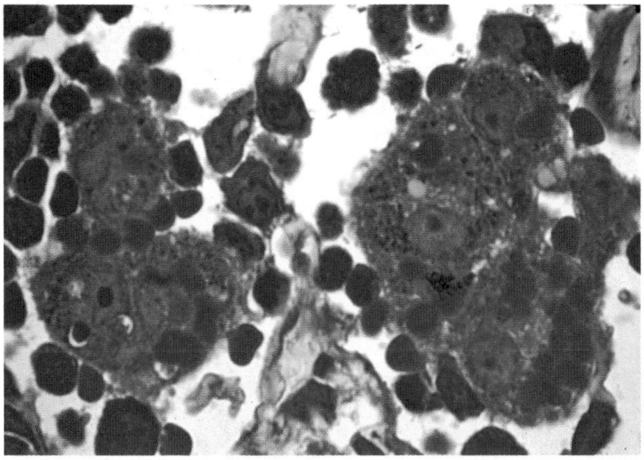

Figure 14–29. Erythrocyte removal by splenic macrophages. A micrograph of five macrophages in a splenic cord shows active phagocytosis of effete erythrocytes. After about 120 days of use erythrocytes undergo membrane changes and swell, signals for their engulfment by macrophages in the cords of reticular tissue between the venous sinuses. Healthy, flexible erythrocytes are pushed between stave cells and enter the sinuses. Phagocytosed erythrocytes are completely degraded within lysosomes. The iron released from hemoglobin binds its transport protein transferrin, returns to the circulation, and is reused primarily for erythropoiesis in bone marrow. Iron-free heme either binds its transport protein, hemopexin, or is metabolized to bilirubin and excreted in the bile by liver cells. After surgical removal of the spleen (splenectomy) the number of abnormal erythrocytes in the circulation increases, although many such cells are now removed in the bone marrow and liver. X400. PT.

or capillaries branching from them connect directly to the sinusoids, so that blood is always enclosed by the vascular endothelium. Alternatively, other penicillar arterioles are open-ended, dumping blood into the stroma of the splenic cords in a unique example of **open circulation**. In this route plasma and formed elements of blood reenter the vasculature by passing between the stave cells of sinusoids, which presents no problem for platelets, leukocytes, and healthy flexible erythrocytes. However, after their normal lifespan of 120 days effete erythrocytes undergo membrane changes, becoming swollen and less flexible, signalling their selective engulfment by macrophages in the splenic cords (Figure 14–29). From the sinusoids, blood proceeds to the red pulp veins that join together and enter the trabeculae, forming the **trabecular veins** (Figure 14–24). The splenic vein originates from these vessels and emerges from the hilum of the spleen. The trabecular veins do not have muscle in their wall and resemble channels hollowed out in the trabecular connective tissue and lined by endothelium.

MEDICAL APPLICATION

Although the spleen has numerous important functions in the body, it is not essential to life. In some situations the spleen must be removed (eg, an abdominal injury that results in rupture of the spleen capsule, certain anemias, and platelet disorders). In this case other organs such as the liver and bone marrow can take over many of the functions of the spleen, although the risk of infection may be higher in a splenectomized individual.

Digestive Tract

The digestive system consists of the digestive tract—oral cavity, esophagus, stomach, small and large intestines, rectum, and anus—and its associated glands—salivary glands, liver, and pancreas (Figure 15–1). Its function is to obtain from ingested food the molecules necessary for the maintenance, growth, and energy needs of the body. Macromolecules such as proteins, fats, complex carbohydrates, and nucleic acids are broken down into small molecules that are more easily absorbed through the lining of the digestive tract, mostly in the small intestine. Water, vitamins, and minerals from ingested food are also absorbed. In addition, the inner layer of the digestive tract is a protective barrier between the content of the tract's lumen and the internal milieu of the body.

The first step in digestion occurs in the mouth, where food is moistened by saliva and ground by the teeth into smaller pieces; saliva also initiates the breakdown of carbohydrates. Digestion continues in the stomach and small intestine, where the food's basic components (eg, amino acids, monosaccharides, free fatty acids) are absorbed. Water absorption occurs in the large intestine, causing undigested material to become semisolid.

GENERAL STRUCTURE OF THE DIGESTIVE TRACT

The entire gastrointestinal tract has certain common structural characteristics. It is a hollow tube with a lumen of variable diameter and a wall made up of four main layers: the **mucosa, submucosa, muscularis,** and **serosa.** The structure of these layers is summarized below and is illustrated for the small intestine in Figure 15–2.

The **mucosa** comprises an **epithelial lining;** an underlying **lamina propria** of loose connective tissue rich in blood vessels, lymphatics, lymphocytes and smooth muscle cells, sometimes also containing glands; and a thin layer of smooth muscle called the **muscularis mucosae** usually separating mucosa from submucosa. The mucosa is frequently called a **mucous membrane.**

The **submucosa** contains denser connective tissue with many blood and lymph vessels and the **submucosal plexus** of autonomic nerves. It may also contain glands and lymphoid tissue.

The thick **muscularis** is composed of smooth muscle cells that are spirally oriented and divided into two sublayers. In the internal sublayer (closer to the lumen), the orientation is generally circular; in the external sublayer, it is mostly longitudinal. In the connective tissue between the muscle sublayers are blood and lymph vessels, as well as another autonomic **myenteric nerve plexus**. This and the submucosal plexus together comprise the local **enteric nervous system** of the digestive tract, containing largely autonomic neurons functioning independently of the central nervous system (CNS).

The **serosa** is a thin layer of loose connective tissue, rich in blood vessels, lymphatics, and adipose tissue, with a simple squamous covering epithelium (**mesothelium**). In the abdominal cavity, the serosa is continuous with the mesenteries (thin membranes covered by mesothelium on both sides), which support the intestines, and with the peritoneum, a serous membrane that lines the cavity. In places where the digestive tract is not suspended in a cavity but bound to other structures, such as in the esophagus (Figure 15–1), the serosa is replaced by a thick **adventitia**, consisting of connective tissue containing vessels and nerves, lacking mesothelium.

The main functions of the digestive tract's epithelial lining are to:

- Provide a selectively permeable barrier between the contents of the tract and the tissues of the body,
- Facilitate the transport and digestion of food,

- Promote the absorption of the products of this digestion,
- Produce hormones that affect the activity of the digestive system,
- Produce mucus for lubrication and protection.

The abundant lymphoid nodules in the lamina propria and the submucosal layer protect the organism (in association with the epithelium) from bacterial invasion, as described in Chapter 14. The necessity for this immunologic support is obvious, because the entire digestive tract—with the exception of the oral cavity, esophagus, and anal canal—is lined by a simple thin, vulnerable epithelium. The lamina propria, located just below the epithelium, is a zone rich in macrophages and lymphocytes, some of which actively produce antibodies. These antibodies are mainly immunoglobulin A (IgA) and are secreted into the intestinal lumen bound to a secretory protein produced by the epithelial cells. This complex protects against viral and bacterial invasion. IgA is resistant to proteolytic enzymes and can therefore coexist with the proteases present in the lumen.

The muscularis mucosae allows local movements of the mucosa independent of other movements of the digestive tract, increasing contact of the lining with food. The contractions of the muscularis, generated and coordinated by autonomic nerve plexuses, propel and mix the food in the digestive tract. These plexuses are composed mainly of nerve cell aggregates (multipolar visceral neurons) that form small parasympathetic ganglia. A rich network of pre- and postganglionic fibers of the autonomic nervous system and some visceral sensory fibers in these ganglia permit communication between them. The number of these ganglia along the digestive tract is variable; they are most numerous in the regions of greatest motility.

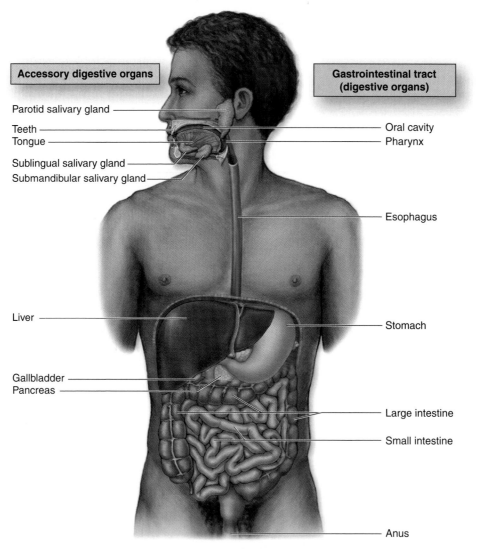

Figure 15–1. **The digestive system.** The digestive system consists of the tract from the mouth (oral cavity) to the anus, as well as the digestive glands emptying into this tract, primarily the salivary glands, liver, and pancreas. These accessory digestive glands are described in the next chapter.

MEDICAL APPLICATION

*In certain diseases, such as **Hirschsprung disease** (congenital megacolon) or **Chagas disease** (**Trypanosoma cruzi** infection), the plexuses in the digestive tract are severely injured and most of their neurons are destroyed. This results in disturbances of digestive tract motility, with frequent dilatations in some areas. The abundant innervation from the autonomic nervous system that the digestive tract receives provides an anatomic explanation of the widely observed action of emotional stress on this tract.*

ORAL CAVITY

The oral cavity (Figure 15–1) is lined with stratified squamous epithelium, keratinized or nonkeratinized, depending on the region. The keratin layer protects the oral mucosa from damage during masticatory function and is best developed on the gingiva (gum) and hard palate. The lamina propria in these regions has many papillae and rests directly on bony tissue. Nonkeratinized squamous epithelium covers the soft palate, lips, cheeks, and the floor of the mouth. Surface cells are shed continuously and replaced by progeny of stem cells in the basal epithelial layer. The lamina propria has papillae similar to those in the dermis of the skin and is continuous with a submucosa containing diffuse small salivary glands. The soft palate also has a core of skeletal muscle and lymphoid nodules. In the **lips**, there is also striated muscle and a transition from the oral nonkeratinized epithelium to the keratinized epithelium of the skin (Figure 15–3).

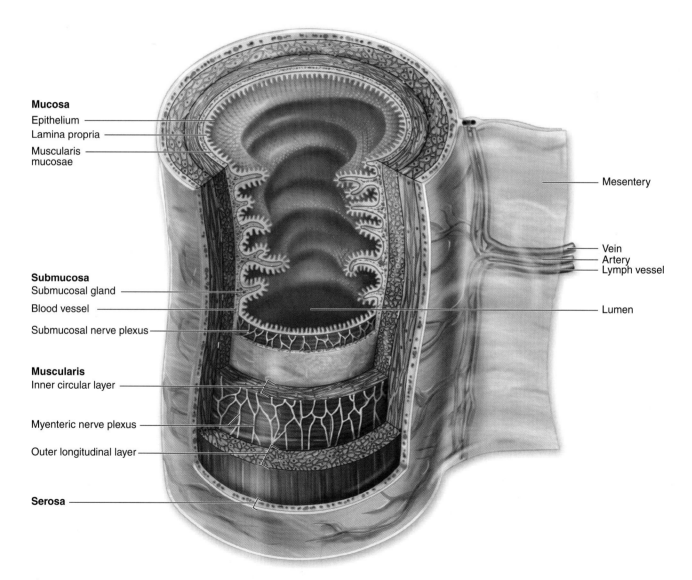

Figure 15–2. Major layers and organization of the digestive tract. Schematic diagram showing the structure of the small intestine portion of the digestive tract, with the four layers and their major components listed on the left. The intestines are suspended by mesenteries which are the sites of blood vessels and lymphatics from the intestines.

Tongue

The tongue is a mass of striated muscle covered by a mucous membrane whose structure varies according to the region. The muscle fibers cross one another in three planes and are grouped in bundles separated by connective tissue. Because the connective tissue of the lamina propria penetrates the spaces between the muscular bundles, the mucous membrane is strongly adherent to the muscle. The mucous membrane is smooth on the lower surface of the tongue. The tongue's dorsal surface is irregular, covered anteriorly by a great number of small eminences called **papillae.** The posterior third of the tongue's dorsal surface is separated from the anterior two thirds by a V-shaped groove, the **terminal sulcus.** Behind this boundary is the root of the tongue, whose surface shows the many bulges of the lingual tonsils and smaller collections of lymphoid nodules (Figure 15–4).

The numerous papillae on the anterior portion of the tongue are elevations of the mucous membrane that assume various forms and functions. Four types are recognized (Figure 15–4):

* **Filiform papillae** (Figure 15–5) are very numerous, have an elongated conical shape, and are heavily keratinized, which

gives their surface a gray or whitish appearance. Their epithelium lacks taste buds (described below) and their role is mechanical in providing a rough surface that facilitates food movement during chewing.
* **Fungiform papillae** (Figure 15–5) are less numerous, lightly keratinized, and mushroom-shaped with connective tissue cores and scattered taste buds on their upper surfaces. They are irregularly interspersed among the filiform papillae.
* **Foliate papillae** are poorly developed in adults, but consist of parallel ridges and furrows on the sides of the tongue, with taste buds.
* **Vallate** (or circumvallate) **papillae** (Figure 15–5) are the least numerous and largest lingual papillae, and have over half the taste buds on the human tongue. With diameters of one to three mm, seven to twelve circular vallate papillae normally form a V-shaped line just before the terminal sulcus. Ducts from several serous salivary (von Ebner) glands empty into the deep groove that surrounds each vallate papilla. This moatlike arrangement provides a continuous flow of fluid over the taste buds abundant on the sides of these papillae, which washes food particles from the vicinity so that the taste buds can receive and process new gustatory stimuli. These glands also secrete a lipase that prevents the formation of a hydrophobic film over the taste buds that would hinder their function.

Taste buds are also present in other parts of the oral cavity, such as the soft palate, and are continuously flushed by numerous small salivary glands dispersed throughout the oral mucosa.

Taste buds are ovoid structures, each containing 50–75 cells, within the stratified epithelium of the tongue and the oral mucosa (Figure 15–6). About half the cells are elongated **gustatory (taste) cells**, which turn over with a 7- to 10-day life span. Other cells present are slender **supportive cells**, immature cells, and basal **stem cells** which divide and give rise to the other two types. The base of each bud rests on the basal lamina and is entered by afferent sensory axons that form synapses on the gustatory cells. At the apical ends of the gustatory cells microvilli project through an opening called the taste pore. Molecules (tastants) dissolved in saliva contact the microvilli through the pore and interact with cell surface taste receptors (Figure 15–6).

Taste buds detect at least five broad categories of tastants: metal ions (salty); hydrogen ions from acids (sour); sugars and related organic compounds (sweet); alkaloids and certain toxins (bitter); and certain amino acids such as glutamate (umami; Jap. *umami*, savory). Salty and sour tastes are produced by ion channels; the other taste categories are mediated by G-protein-coupled receptors. Receptor binding produces depolarization of the gustatory cells, stimulating the sensory nerve fibers which send information to the brain for processing. Conscious perception of tastes in food requires olfactory and other sensations in addition to taste bud activity.

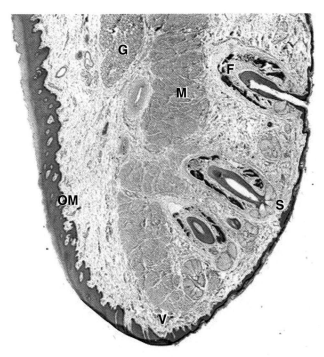

Figure 15–3. **Lip.** Low-magnification micrograph of a lip section showing one side covered by typical **oral mucosa (OM)**, the opposite side covered by skin (S) containing hair follicles (F) and associated glands. Between the oral portion of the lips and normal skin is the vermilion (V), or the vermilion zone, where epidermis is very thin, lightly keratinized, and transparent to blood in the rich microvasculature of the underlying connective tissue. Because this region lacks the glands for oil and sweat, it is prone to excessive dryness and chapping in cold, dry weather. Internally, the lips contain much-striated muscle (M) and many minor salivary glands (G). X10. H&E.

Pharynx

The pharynx, a transitional space between the oral cavity and the respiratory and digestive systems, forms an area of communication between the nasal region and the larynx (Figure 15–1). The pharynx is lined by stratified nonkeratinized squamous epithelium in the region continuous with the esophagus and by ciliated pseudostratified columnar epithelium containing goblet cells in the regions close to the nasal cavity.

The pharynx contains tonsils (described in Chapter 14) and the mucosa also has many small mucous salivary glands in its

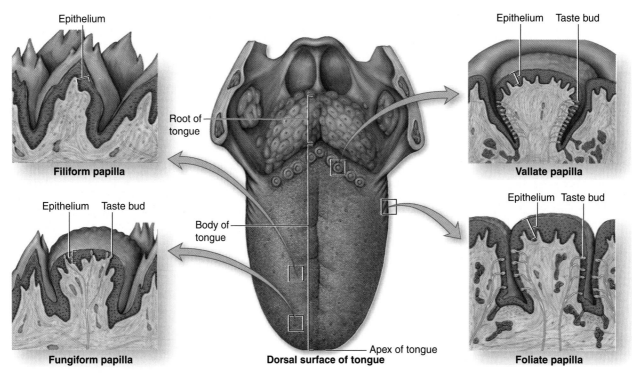

Figure 15–4. Tongue and lingual papillae. The posterior third of the tongue is the root and the anterior two-thirds the body of the tongue. The mucosa of the root is filled with masses of lymphoid nodules separated by crypts, all of which comprise the lingual tonsils. On the body of the tongue are papillae of four types, all containing cores of connective tissue covered by stratified squamous epithelium. Pointed filiform papillae provide friction to help move food during chewing. Ridge-like foliate papillae on the sides of the tongue are best-developed in young children. Fungiform papillae are scattered across the dorsal surface and 6–12 very large vallate papillae are present in a V-shaped line near the terminal sulcus. Taste buds are present on fungiform and foliate papillae but are much more abundant on vallate papillae.

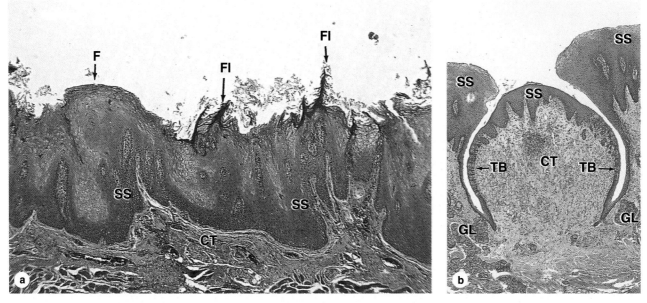

Figure 15–5. Lingual papillae. (a): Section of the dorsal surface of tongue shows both filiform (FI) and fungiform papillae (F). Both types are elevations of the connective tissue (CT) covered by stratified squamous epithelium (SS), but the filiform type is pointed and heavily keratinized while the fungiform type is mushroom-shaped, lightly keratinized, and has a few taste buds. **(b):** Micrograph shows a single very large vallate papilla with two distinctive features: many taste buds (TB) around the sides and several small salivary glands (GL) emptying into the cleft or moat formed by the elevated mucosa surrounding the papilla. These glands continuously flush the cleft, renewing the fluid in contact with the taste buds. The 7 to 12 vallate papillae on the tongue contain over half of the 10,000 or so taste buds in the human mouth and pharynx. Both X20. H&E.

lamina propria. The constrictor and longitudinal muscles of the pharynx are located outside this layer.

Teeth

In the adult human there are normally 32 **permanent teeth**, arranged in two bilaterally symmetric arches in the maxillary and mandibular bones (Figure 15–7). Each quadrant has eight teeth: two incisors, one canine, two premolars, and three permanent molars. Twenty of the permanent teeth are preceded by **deciduous (baby) teeth** which are shed; the others are permanent molars with no deciduous precursors. Each tooth has a **crown** exposed above the gingiva, a constricted **neck** at the gum, and one or more **roots** below the gingiva that hold the teeth in bony sockets called **alveoli,** one for each tooth (Figure 15–7).

The crown is covered by the extremely hard **enamel** and the roots by a bone-like tissue called **cementum.** These two coverings meet at the neck of the tooth. The bulk of a tooth is composed of another calcified material, **dentin,** which surrounds a soft connective tissue-filled space known as the **pulp cavity** (Figure 15–7). The pulp cavity narrows in the roots as the root canals, which extend to the tip of each root, where an opening (**apical foramen**) permits the entrance and exit of blood vessels,

lymphatics, and nerves of the pulp cavity. The **periodontal ligaments** are fibrous connective tissue bundles of collagen fibers inserted into both the cementum and alveolar bone, fixing the tooth firmly in its bony socket (alveolus).

DENTIN

Dentin is a calcified tissue consisting of 70% calcium hydroxyapatite, making it harder than bone. The organic matrix contains type I collagen fibers and glycosaminoglycans secreted by **odontoblasts,** tall polarized cells that line the tooth's internal pulp cavity (Figure 15–8). Mineralization of the **predentin** matrix involves matrix vesicles in a process similar to that in osteoid (Chapter 8). Long, slender apical **odontoblast processes** lie within **dentinal tubules** (Figure 15–9) which penetrate the full thickness of the dentin, gradually becoming longer as the dentin becomes thicker. Along their length the processes extend fine branches into smaller lateral branches of the tubules (Figure 15–8). Odontoblasts remain active in predentin secretion into adult life, gradually reducing the size of the pulp cavity.

Teeth are sensitive to stimuli such as cold, heat, and acidic pH, all of which can be perceived as pain. Pulp is highly innervated and some unmyelinated nerve fibers extend into the dental tubules near the pulp cavity (Figure 15–9). The different

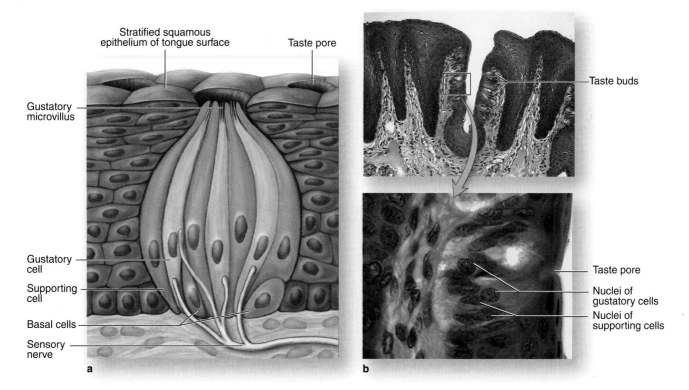

Figure 15–6. **Taste buds. (a):** Drawing of a single taste bud shows the gustatory (taste) cells, the supporting cells whose function is not well-understood, and the basal stem cells. Microvilli at the ends of the gustatory cells project through an opening in the epithelium, the taste pore. Afferent sensory axons enter the basal end of taste buds and synapse with the gustatory cells. **(b):** In the stratified squamous epithelium of the tongue surface or oral mucosa, taste buds form as distinct clusters of cells that recognizable histologically even at low magnification. At higher power the taste pore may be visible, as well as the elongated nuclei of gustatory and supporting cells and the fewer, round nuclei of basal stem cells. 140X and 500X. H&E.

stimuli can affect fluid inside dentinal tubules, stimulating these nerve fibers located near odontoblast processes.

MEDICAL APPLICATION

Unlike bone, dentin does not turn over or get remodeled, persisting as a mineralized tissue long after loss of the odontoblasts. It is therefore possible to maintain teeth whose pulp and odontoblasts have been destroyed by infection (canal treatment). In adult teeth, destruction of the covering enamel by erosion from use or dental caries (tooth decay) usually triggers a reaction in odontoblasts that causes them to resume the synthesis of dentin components.

ENAMEL

Enamel is the hardest component of the human body, consisting of nearly 98% hydroxyapatite and the rest organic material including at least two unique proteins, **amelogenin** and **enamelin**, but no collagen. Other ions, such as fluoride, can be incorporated or adsorbed by the hydroxyapatite crystals; enamel containing fluorapatite is more resistant to acidic dissolution caused by microorganisms, hence the addition of fluoride to toothpaste and water supplies.

Enamel consists of interlocking rods or columns, **enamel rods (prisms)**, bound together by other enamel. Each rod extends through the entire thickness of the enamel layer; the precise arrangement of rods in groups is very important for enamel's strength and mechanical properties.

In developing teeth enamel matrix is secreted by a layer of cells called **ameloblasts**, each of which produces one enamel prism (Figure 15–10). An ameloblast is a long, polarized cell with numerous mitochondria, well-developed RER and Golgi apparatus, and an apical extension, the **ameloblast process,** containing numerous secretory granules with proteins for the enamel matrix. After finishing the synthesis of enamel, ameloblasts form a protective epithelium that covers the crown until the eruption of the tooth, a function important in preventing several enamel defects.

Enamel is produced by cells of ectodermal origin, whereas most of the other structures of teeth derive from mesodermal and neural crest cells. Together these cells produce a series of structures around the developing oral cavity, the enamel organs, each of which forms one tooth (Figure 15–11).

PULP

Tooth pulp consists of connective tissue resembling mesenchyme. Its main components are the layer of odontoblasts, many fibroblasts, thin collagen fibrils, and ground substance (Figure 15–11). Pulp is a highly innervated and vascularized

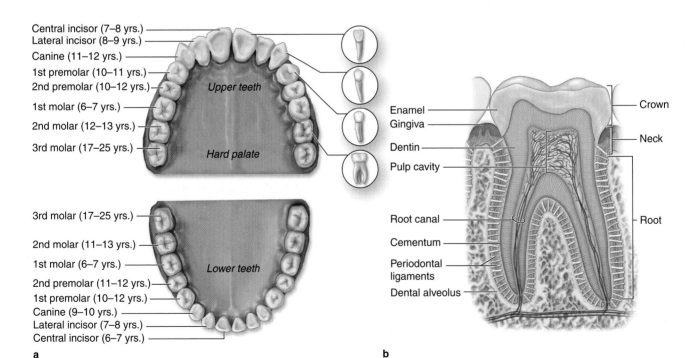

Figure 15–7. Teeth. All teeth are similar embryologically and histologically. **(a):** The dentition of the permanent teeth is shown, as well as the approximate age at eruption for each tooth. **(b):** Diagram of a molar's internal structure is similar to that of all teeth, with an enamel-covered **crown**, cementum-covered **roots** anchoring the tooth to alveolar bone of the jaw, and a slightly constricted **neck** where the enamel and cementum coverings meet at the gingiva. A pulp cavity extends into the neck and is filled with well-vascularized, well-innervated mesenchymal connective tissue. Blood vessels and nerves enter the tooth through apical foramina at the root tips.

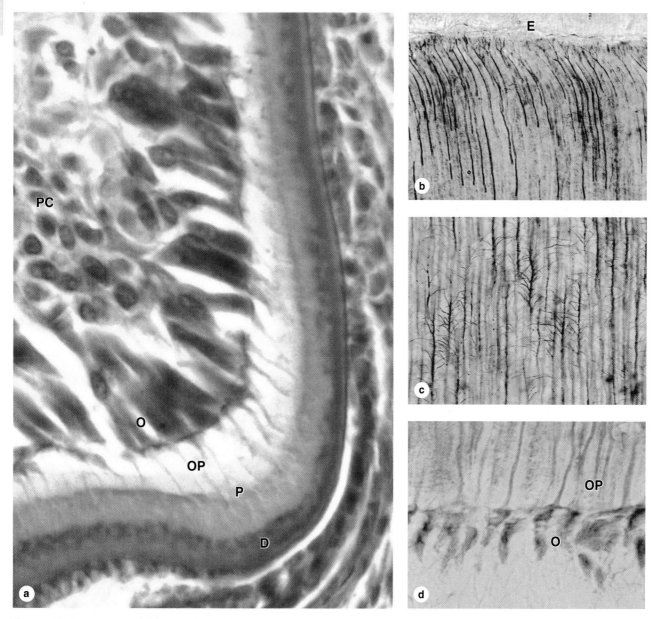

Figure 15–8. **Dentin and odontoblasts. (a):** Odontoblasts (O) are long polarized cells derived from mesenchyme of the developing pulp cavity (PC). Odontoblasts are specialized for collagen and GAG synthesis and are bound together by junctional complexes as a layer, with no basal lamina, so that a collagen-rich matrix called **predentin** (P) is secreted only from their apical ends at the dentinal surface. Within approximately one day of secretion predentin mineralizes to become dentin (D) as hydroxyapatite crystals form in a process similar to that occurring in osteoid of developing bones (Chapter 8). In this process the collagen is masked and calcified matrix becomes much more acidophilic and stains quite differently than that of predentin. When predentin secretion begins an apical extension from each cell, the odontoblast process (OP), forms and is surrounded by new matrix. As the dentin-predentin layer thickens, these processes lengthen. When tooth formation is complete odontoblasts persist and their processes are maintained in canals called dentinal tubules which run through the full thickness of the dentin. X400. Mallory trichrome. **(b, c):** Odontoblast processes can be silver-stained and shown to branch near the junction of dentin with enamel (E) and along their length closer to their source (c), with the lateral branches occupying smaller canaliculi within dentin. Both X400. Silver. **(d):** These odontoblast process (OP) connections to the odontoblasts (O), shown with stained nuclei here, are important for the maintenance of dentin in adult teeth. X400. Mallory trichrome. (Figure 15-8b, c and d used, with permission, from M.F. Santos, Department of Histology and Embryology, Institute of Biomedical Sciences, University of São Paulo, Brazil.)

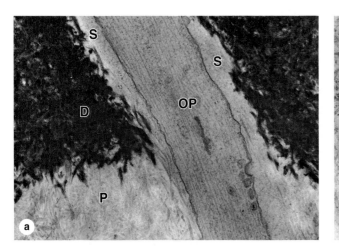

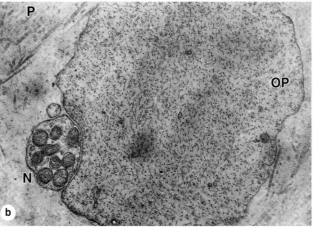

Figure 15–9. Ultrastructure of dentinal tubule. (a): TEM shows the calcification of dentin (D) at its border with not-yet-calcified predentin (P). An odontoblast process (OP) with microtubules and a few secretory vesicles occupies much of the space (S) in the dentinal tubule. A process extends from each odontoblast and the tubules continue completely across the dentin layer. X32,000. **(b):** Cross-section of an odontoblast process (OP) near predentin (P) shows its close association with an unmyelinated nerve fiber (N) extending there from fibers in the pulp cavity. These nerves respond to various stimuli, such as cold temperatures, reaching the nerve fibers through the dentinal tubules. X61,000.

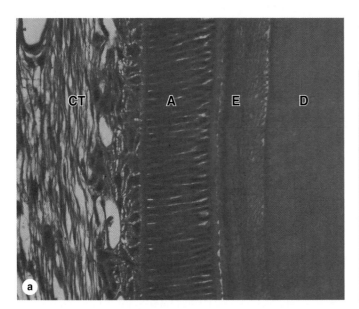

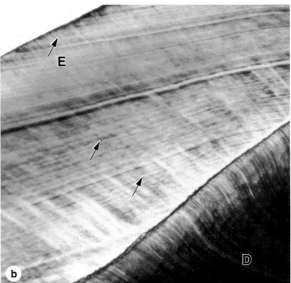

Figure 15–10. Ameloblasts and enamel. Ameloblasts (A) are tall polarized cells whose apical ends initially contact dentin (D). Ameloblasts are joined to form a cell layer surrounded basally by connective tissue (CT). As odontoblasts secrete predentin, ameloblasts secrete a matrix lacking collagens, but rich in a few glycoproteins which quickly initiate calcium hydroxyapatite formation to make enamel (E), the hardest material in the body. Enamel forms a layer, but consists of enamel rods or prisms, solidly fused together by more enamel. Each enamel rod represents the product of one ameloblast. No cellular processes occur in enamel and the layer of ameloblasts surrounding the developing crown is completely lost during tooth eruption. Teeth that have been decalcified for histological sectioning typically lose their enamel layer completely. X400. H&E.

(b): Micrograph of a thin preparation of a tooth prepared by grinding. Fine tubules can be observed in the dentin (D) and rods aligned the same way can be faintly observed (arrows) in the enamel (E). The more prominent lines that cross enamel diagonally represent incremental growth lines produced as the enamel matrix was secreted cyclically by the ameloblast layer. X400. Unstained.

tissue. Blood vessels and myelinated nerve fibers enter the apical foramen and divide into numerous branches. Some nerve fibers lose their myelin sheaths and extend into the dentinal tubules. Pulp fibers are sensitive to pain.

PERIODONTIUM

The periodontium comprises the structures responsible for maintaining the teeth in the maxillary and mandibular bones. It consists of the **cementum, periodontal ligament, alveolar bone, and gingiva.**

Cementum covers the dentin of the root and is similar in composition to bone, although osteons and blood vessels are absent. It is thicker in the apical region around the root, where there are **cementocytes,** cells resembling osteocytes, in lacunae. Unlike osteocytes, however, cementocytes do not communicate via canaliculi and their nourishment comes from external tissues. Like bone, cementum is labile and reacts to the stresses to which

it is subjected by resorbing old tissue or producing new tissue. Continuous production of cementum in the apex compensates for the physiologic wear of the teeth and maintains close contact between the roots of the teeth and their sockets.

MEDICAL APPLICATION

In comparison with bone, the cementum has lower metabolic activity because it is not irrigated by blood vessels. This feature allows the movement of teeth within alveolar bone by orthodontic appliances without significant root resorption.

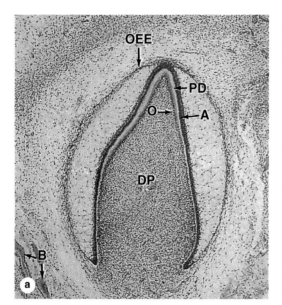
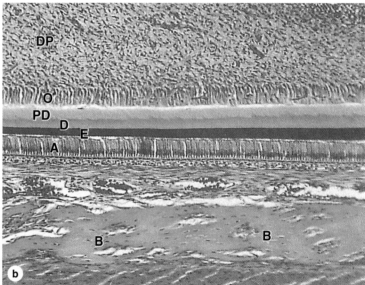

Figure 15–11. Tooth formation Tooth formation begins in the sixth week of human development when ectodermal epithelium lining the oral cavity begins to grow into the underlying mesenchyme of the developing jaws. At a series of sites corresponding to each future tooth, these epithelial cells proliferate extensively and become organized as **enamel organs**, each shaped rather like a wine glass with its stem initially still attached to the oral lining. Ameloblasts form from the innermost layer of cells in the enamel organ. Mesenchymal cells inside the concave portion of the enamel organ include neural crest cells which differentiate as the layer of odontoblasts with their apical ends in contact with the apical ends of the ameloblasts. **(a):** When production of dentin and enamel has begun, the enamel organ appears as shown in this micrograph. The ameloblast layer (A) is separated from the outer enamel epithelium (OEE) by a thick intervening region rich in GAGs but having fewer, widely separated cells. Surrounding the enamel organ is mesenchyme, some parts of which begin to undergo intramembranous bone formation (B) and form the jaws. Inside the cavity of each enamel organ, mesenchymal cells comprise the dental papilla (DP), in which the outermost cells are the layer of odontoblasts (O) facing the ameloblasts. These two cell layers begin to move apart as the odontoblasts begin to produce the layer of predentin (PD). Contact with dentin induces each ameloblast to begin secretion of a rod or prism of enamel matrix. More slowly calcifying interprismatic enamel fuses all the enamel rods into a very strong, solid mass. X20. H&E.

(b): Detail of an enamel organ at a later stage showing the layers of predentin (PD) and dentin (D) and a layer of enamel (E), along with the organized cell layers that produced this material. Odontoblasts (O) are in contact with the very cellular mesenchyme of the dental papilla (DP) which will become the pulp cavity. Ameloblasts (A) are prominent in the now much thinner enamel organ, which is very close to developing bone (B). Details of these cell layers are presented further in Figures 15-8, 15-9 and 15-10. Enamel formation continues until shortly before tooth eruption; formation of dentin continues after eruption until the tooth is fully formed. Odontoblasts persist around the pulp cavity, with processes penetrating the dental layer, producing factors to help maintain dentin. Mesenchymal cells immediately around the enamel organ differentiate into the cells of cementum and other periodontal tissues. X120. H&E.

The **periodontal ligament** is connective tissue 150 to 350 μm thick with collagen fiber bundles connecting the cementum and the alveolar bone of the tooth socket (Figure 15–12). It permits limited movement of the tooth within the socket and the fibers are organized to support the pressures exerted during mastication. This avoids transmission of pressure directly to the bone which would cause localized bone resorption. Unlike typical ligaments, it is highly cellular and has a rich supply of blood vessels and nerves, giving the periodontal ligament supportive, protective, sensory, and nutritive functions. Collagen of the periodontal ligament has an unusually high turnover rate (as demonstrated by autoradiography) and a high content of soluble collagens, with the space between its fibers filled with glycosaminoglycans (GAGs).

MEDICAL APPLICATION

The high rate of collagen renewal in the periodontal ligament allows processes affecting protein or collagen synthesis, eg, protein or vitamin C deficiency (scurvy), to cause atrophy of this ligament. As a consequence, teeth become loose in their sockets; in extreme cases they fall out.

The **alveolar bone** is in immediate contact with the periodontal ligament, which serves as its periosteum. It is primary (immature) bone, with the collagen fibers not arranged in the typical lamellar pattern of adult bone. Many of the collagen fiber bundles of the periodontal ligament penetrate this bone and bind it to the cementum (Figure 15–12). The bone closest to the roots of the teeth forms the socket. Vessels run through the alveolar bone and penetrate the periodontal ligament along the root, with some vessels and nerves entering the pulp at the apical foramen of each root.

The **gingiva** is a mucous membrane firmly bound to the periosteum of the maxillary and mandibular bones (Figure 15–13). It is composed of stratified squamous epithelium and lamina propria with numerous connective tissue papillae. A specialized part of this epithelium, named **junctional epithelium,** is bound to the tooth enamel by means of a cuticle resembling a thick basal lamina. The epithelial cells are attached to this cuticle by numerous hemidesmosomes. Between the enamel and the epithelium is the **gingival sulcus,** a groove up to 3 mm deep surrounding the neck (Figure 15–13a).

MEDICAL APPLICATION

The depth of the gingival sulcus, measured during clinical dental examinations, is an important indicator of potential periodontal disease.

ESOPHAGUS

The part of the gastrointestinal tract called the **esophagus** is a muscular tube whose function is to transport food from the mouth to the stomach. It is lined by nonkeratinized stratified squamous epithelium with stem cells scattered throughout the basal layer (Figure 15–14). In general, the esophagus has the same major layers as the rest of the digestive tract. In the submucosa are groups of small mucus-secreting glands, the **esophageal glands,** secretions of which facilitate the transport of foodstuffs and protect the mucosa. In the lamina propria of the region near the stomach are groups of glands, the **esophageal cardiac glands,** which also secrete mucus.

Swallowing begins with controllable motion, but finishes with involuntary peristalsis. In the proximal third of the esophagus the muscularis is exclusively skeletal muscle like that of the tongue. The middle third contains a combination of skeletal and smooth muscle fibers (Figure 15–14) and in the distal third the muscularis contains only smooth muscle. Also, only the most distal portion of the esophagus, in the peritoneal cavity, is covered by serosa. The rest is enclosed by a layer of loose connective tissue, the adventitia, which blends into the surrounding tissue.

STOMACH

The stomach, like the small intestine, is a mixed exocrine-endocrine organ that digests food and secretes hormones. It is a dilated segment of the digestive tract whose main functions are to continue the digestion of carbohydrates initiated in the mouth, add an acidic fluid to the ingested food, transform it by muscular activity into a viscous mass (**chyme**), and promote the initial digestion of proteins with the enzyme **pepsin.** It also produces a gastric lipase that digests triglycerides . Gross inspection reveals four regions: **cardia, fundus, body,** and **pylorus** (Figure 15–15). The fundus and body are identical in microscopic

Figure 15–12. Dental Pulp. The periphery of the dental pulp contains the organized odontoblasts (O) contacting the surrounding dentin (D). Centrally the pulp consists of delicate, highly cellular connective tissue resembling undifferentiated mesenchyme but with many thin-walled venules (V) and capillaries. Pulp has reticulin fibers and other fine collagen fibers, with much ground substance. Nerves fibers are also present. The blood and nerve supplies enter the pulp cavity via the apical foramen at the apex of the roots. X150. H&E.

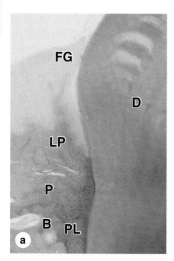
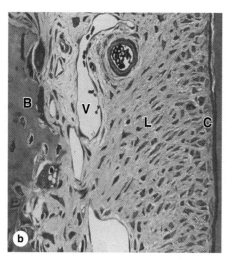
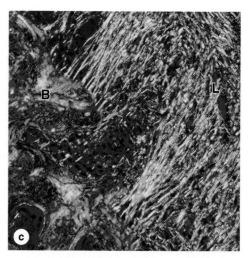

Figure 15–13. Periodontium. The periodontium of each tooth consists of the **cementum, periodontal ligament, alveolar bone, and gingiva. (a):** Micrograph of decalcified tooth showing the gingiva. The free gingiva (FG) is against the dentin (D), with little of the gingival sulcus apparent. Gingiva has many layers of stratified epithelial cells covering the connective tissue of the lamina propria (LP). The connective tissue is continuous with that of the periosteum (P) covering the alveolar bone (B) and with the periodontal ligament (PL). X10. H&E.

   (b): Micrograph shows the periodontal ligament (L) with its many blood vessels (V) and insertions into the alveolar bone (B). This ligament serves as the periosteum of the alveolar in tooth sockets and is also continuous with developing layers of cementum (C) that covers the dentin. Cementum forms a thin layer of bone-like material secreted by large, elongated cells called cementoblasts. X100. H&E. **(c):** Micrograph shows the continuity of collagen fibers in alveolar bone (B) with the bundles in the periodontal ligament (L). X200. Picrosirius in polarized light.

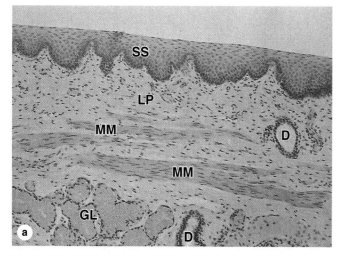
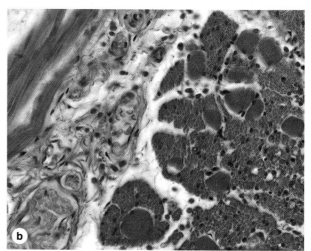

Figure 15–14. Esophagus. (a): Longitudinal section of esophagus shows mucosa consisting of nonkeratinized stratified squamous epithelium (SS), lamina propria (LP), and smooth muscles of the muscularis mucosae (MM). Beneath the mucosa is the submucosa containing esophageal mucous glands (GL) which empty via ducts (D) onto the luminal surface. X40. H&E. **(b):** Transverse section showing the muscularis halfway along the esophagus reveals a combination of skeletal muscle (right) and smooth muscle fibers (left) in the outer layer, which are cut both longitudinally and transversely here. This transition from muscles under voluntary control to the type controlled autonomically is important in the swallowing mechanism. X200. H&E.

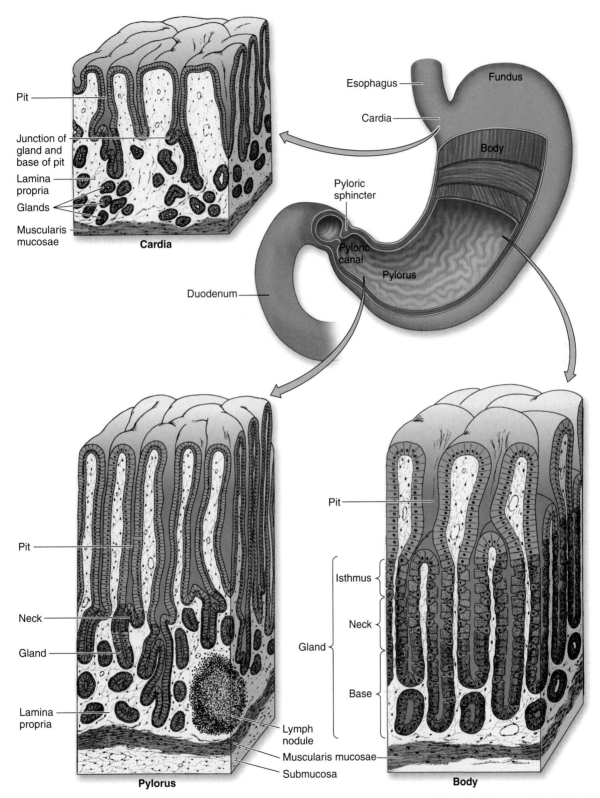

Figure 15–15. **Regions of the stomach.** The stomach is a muscular dilation of the digestive tract where mechanical and chemical digestion occurs. The muscularis consists of three layers for thorough mixing of the stomach contents as chyme: an outer longitudinal layer, a middle circular layer, and an inner oblique layer. The stomach mucosa shows distinct histological differences in the cardia, the fundus/body, and the pylorus. Cells that secrete HCl and pepsin are restricted mainly to the body and fundus regions. Glands of the cardia and pylorus produce primarily mucus.

structure so that only three histologically distinct regions are recognized. The mucosa and submucosa of the empty stomach have longitudinally directed folds known as **rugae**, which flatten when the stomach is filled with food. The wall in all regions of the stomach is made up of all four major layers (Figure 15–16).

Mucosa

Changing abruptly at the esophago-gastric junction, the mucosa of the stomach consists of a simple columnar **surface epithelium** that invaginates into the lamina propria, forming **gastric pits** (Figures 15–17 and 15–18). Emptying into the gastric pits are branched, tubular glands characteristic of the stomach region

(cardiac, gastric, and pyloric). Stem cells for the entire epithelial lining of the stomach are located in the upper regions of these glands near the gastric pits. The vascularized **lamina propria** that surrounds and supports these pits and glands contains smooth muscle fibers and lymphoid cells. Separating the mucosa from the underlying submucosa is a layer of smooth muscle, the **muscularis mucosae** (Figure 15–16).

When the luminal surface of the stomach is viewed under low magnification, numerous small circular or ovoid invaginations of the epithelial lining are observed. These are the openings of the gastric pits (Figures 15–17 and 15–18). The epithelium covering the surface and lining the pits is a simple columnar epithelium, the cells of which produce a protective mucus layer. Glycoproteins secreted by the epithelial cells are hydrated and mix with lipids and bicarbonate ions also released from the epithelium to form a thick, hydrophobic layer of gel with a pH gradient from almost 1 at the luminal surface to 7 at the epithelial cells. The mucus firmly adherent to the epithelial surface is very effective in protection, while the superficial luminal mucus layer is more soluble, partially digested by pepsin and mixed with the luminal contents. Hydrochloric acid, pepsin, lipases, and bile in the stomach lumen must all be considered as potential endogenous aggressors to the epithelial lining. Surface epithelial cells also form an important line of defense due to their mucus production, their tight intercellular junctions, and ion transporters to maintain intracellular pH and bicarbonate production. A third

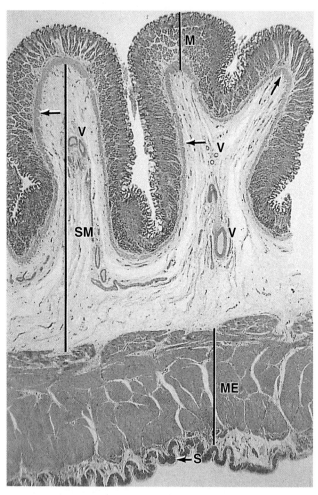

Figure 15–16. **Wall of the stomach with rugae.** Low magnification micrograph of the stomach wall at the fundus shows the relative thickness of the four major layers: the mucosa (M), the submucosa (SM), the muscularis externa (ME), and the serosa (S). Two rugae (folds) cut transversely and consisting of mucosa and submucosa are included. The mucosa is packed with branched tubular glands penetrating the full thickness of the lamina propria so that this sublayer cannot be distinguished at this magnification. The muscularis mucosae (arrows), immediately beneath the basal ends of the gastric glands, is shown. The submucosa is largely loose connective tissue, with blood vessels (V) and lymphatics. X12. H&E.

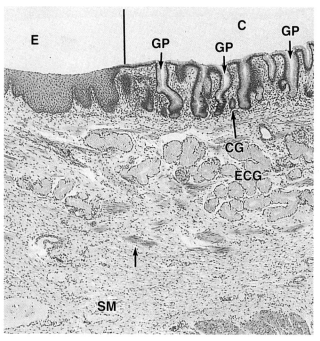

Figure 15–17. **Esophagogastric junction.** At the junction of the esophagus (E) and the cardiac region of the stomach (C) there is an abrupt change in the mucosa from stratified squamous epithelium to simple columnar epithelium invaginating as gastric pits (GP). The mucosa contains many mucus-secreting esophageal cardiac glands (ECG), whose function is supplemented by mucous cardiac glands (CG) opening into the superficial gastric pits. Strands of muscularis mucosae (arrow) separate the mucosa and submucosa (SM). X60. H&E.

line of defense is the underlying circulatory bed, which provides bicarbonate ions, nutrients, and oxygen to the mucosal cells, while removing toxic metabolic products. The rich vasculature also favors the rapid healing of superficial wounds to the mucosa.

MEDICAL APPLICATION

*Stress and other psychosomatic factors; ingested substances such as aspirin, nonsteroidal anti-inflammatory drugs or ethanol; the hyperosmolality of meals; and some microorganisms (eg, Helicobacter pylori) can disrupt this epithelial layer and lead to **ulceration**. The initial ulceration may heal, or it may be further aggravated by the local aggressive agents, leading to additional gastric and duodenal ulcers. Processes that enable the gastric mucosa to rapidly repair superficial damage incurred by several factors play a very important role in the defense mechanism, as does an adequate blood flow that supports gastric physiologic activity. Any imbalance between aggression and protection may lead to pathologic alterations. As an example, aspirin and ethanol irritate the mucosa partly by reducing mucosal blood flow. Several anti-inflammatory drugs inhibit the production of prostaglandins of the E type, which are very important substances for the alkalinization of the mucus layer and, consequently, important for protection.*

REGIONAL DIFFERENCES IN THE STOMACH MUCOSA

The **cardia** is a narrow circular region, only 1.5–3 cm in width, at the transition between the esophagus and the stomach (Figure 15–15). The **pylorus** is the funnel-shaped region opening into the small intestine. The mucosa of these two stomach regions contains tubular glands, usually branched, with coiled secretory portions called **cardial glands** and **pyloric glands** (Figure 15–19). The pits leading to these glands are longer in the pylorus. In both regions the glands secrete abundant **mucus**, as well as **lysozyme**, an enzyme that attacks bacterial walls.

In the **fundus** and **body**, the mucosa's lamina propria is filled with branched, tubular **gastric glands,** three to seven of which open into the bottom of each gastric pit. Each gastric gland has an isthmus, a neck, and a base; the distribution of epithelial cells in the glands is not uniform (Figures 15–15 and 15–20). The **isthmus**, near the gastric pit, contains differentiating mucous cells that migrate and replace surface mucous cells, a few undifferentiated stem cells, and a few parietal (oxyntic) cells; the **neck**

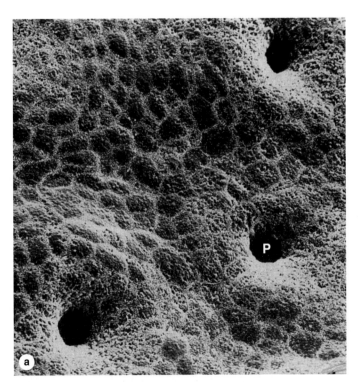 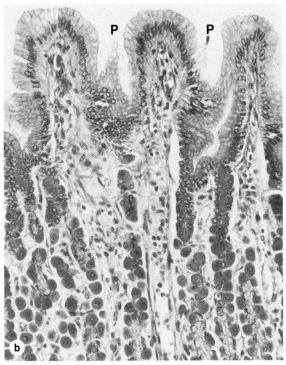

***Figure 15–18.* Gastric pits and glands. (a):** SEM of the stomach lining cleared of its mucus layer reveals closely placed gastric pits (P) surrounded by polygonal apical ends of surface mucous cells. X600. **(b):** Micrograph of the same lining shows that these surface mucous cells are part of a simple columnar epithelium continuous with the lining of the pits (P). Each pit extends into the lamina propria and then branches into several tubular glands. These glands branch further, coil slightly, and fill most of the volume of the mucosa. Around the glands, which contain other cells besides columnar cells, a small amount of connective tissue comprising the lamina propria is also seen. X200. H&E.

of the glands consists of stem cells, mucous neck cells (different from the isthmus mucous cells), and parietal cells (Figure 15–20); the **base** of the glands contains parietal cells and chief (zymogenic) cells. Various enteroendocrine cells are dispersed in the neck and the base of the glands.

These cells of the gastric glands provide key stomach functions. Important properties of each are as follows:

* **Mucous neck cells** are present in clusters or as single cells between parietal cells in the necks of gastric glands (Figure 15–20a). They are irregular in shape, with the nucleus at the base of the

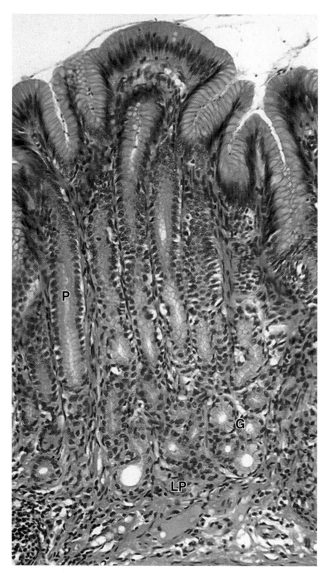

Figure 15–19. Pyloric glands. The pyloric region of the stomach has deep gastric pits (P) leading to short, coiled pyloric glands (G) in the lamina propria. Cardial glands are rather similar histologically and functionally. Cells of these glands secrete mucus and lysozyme primarily, with a few G cells also present. The glands and pits are surrounded by cells of the lamina propria (LP), connective tissue also containing lymphatics and MALT. Immediately beneath the glands is the smooth muscle layer of the muscularis mucosae. X140. H&E.

cell and the secretory granules near the apical surface. Their mucus secretion is less alkaline and quite different from that of the surface epithelial mucous cells.

* **Parietal cells** are present mainly in the upper half of gastric glands, with fewer in the base. They are large rounded or pyramidal cells, each with one central spherical nucleus and cytoplasm that is intensely eosinophilic due to the high density of mitochondria (Figures 15–20 and 15–21). A striking feature of the active secreting cell seen in the electron microscope is a deep, circular invagination of the apical plasma membrane, forming an **intracellular canaliculus** (Figure 15–22). Parietal cells secrete both **hydrochloric acid (HCl)** and **intrinsic factor**, a glycoprotein required for uptake of vitamin B_{12} in the small intestine. **Carbonic anhydrase** produces H_2CO_3 which dissociates in the cytoplasm into H^+ and HCO_3^+ (Figure 15–23). The active cell also releases K^+ and Cl^- and the Cl^- ions combine with H^+ to form HCl. The abundant mitochondria provide energy for the ion pumps located mainly in the extensive cell membrane of the microvilli projecting into the canaliculi. Secretory activity of parietal cells is stimulated both through cholinergic nerve endings (parasympathetic stimulation) and by histamine and a polypeptide called **gastrin**, both secreted by local enteroendocrine cells.

MEDICAL APPLICATION

*In cases of **atrophic gastritis**, both parietal and chief cells are much less numerous, and the gastric juice has little or no acid or pepsin activity. In humans, parietal cells are the site of production of intrinsic factor, a glycoprotein that binds avidly to vitamin B_{12}. In other species, however, the intrinsic factor may be produced by the chief cells.*

*The complex of vitamin B_{12} with intrinsic factor is absorbed by pinocytosis into the cells in the ileum; this explains why a lack of intrinsic factor can lead to vitamin B_{12} deficiency. This condition results in a disorder of the erythrocyte-forming mechanism known as **pernicious anemia**, usually caused by atrophic gastritis. In a certain percentage of cases, pernicious anemia seems to be an autoimmune disease, because antibodies against parietal cell proteins are often detected in the blood of patients with the disease.*

* **Chief (zymogenic) cells** predominate in the lower region of the tubular glands (Figure 15–24) and have all the characteristics of protein-synthesizing and -exporting cells. The cytoplasmic granules contain the inactive enzyme **pepsinogen**. This precursor is rapidly converted into the highly active proteolytic enzyme **pepsin** after being released into the acid environment of the stomach. Pepsins are aspartate endoproteinases of relatively broad specificity, active at pH <5. In humans chief cells also produce the enzyme lipase and the hormone leptin.

* **Enteroendocrine cells** are an epithelial cell type in the mucosa throughout the digestive tract, but are difficult to detect by routine H&E staining. Different enteroendocrine

cells secrete a variety of hormones, almost all short polypeptides (Table 15–1). They can be distinguished by TEM but are most easily identified by immunohistochemistry. In the fundus **enterochromaffin cells (EC cells)** are found on the basal lamina of gastric glands (Figure 15–24) and secrete principally **serotonin (5-hydroxytryptamine)**. In the pylorus and lower body of the stomach other enteroendocrine cells are located in contact with the glandular lumens, including **G cells** which produce the polypeptide **gastrin**. Gastrin stimulates the secretion of acid by parietal cells and has a trophic effect on gastric mucosa.

- **Stem cells** are few in number and found in the neck region of the glands. They are low columnar cells with basal nuclei and divide asymmetrically (Figure 3–20). Some of the daughter cells move upward to replace the pit and surface mucous cells, which have a turnover time of 4–7 days. Other daughter cells migrate more deeply into the glands and differentiate into mucous neck cells and parietal, chief, and

enteroendocrine cells. These cells are replaced much more slowly than are surface mucous cells.

MEDICAL APPLICATION

*Tumors called **carcinoids**, which arise from the EC cells, are responsible for the clinical symptoms caused by overproduction of serotonin. Serotonin increases gut motility, but high levels of this hormone/neurotransmitter have been related to mucosal vasoconstriction and damage.*

Other Layers of the Stomach

The **submucosa** is composed of connective tissue containing blood and lymph vessels; it is infiltrated by lymphoid cells,

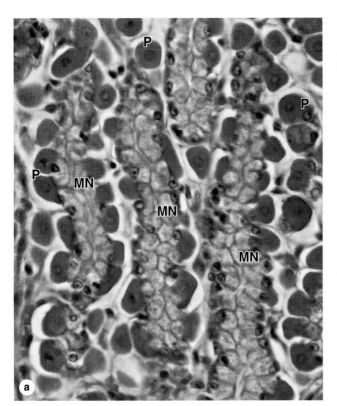

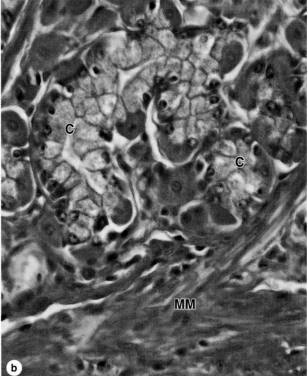

Figure 15–20. **Gastric glands.** Throughout the fundus and body regions of the stomach the gastric pits lead to glands with various cell types. **(a):** In the neck of the glands are mucous neck cells (MN), scattered or present as clusters of irregular, low columnar cells with basophilic, granular cytoplasm and basal nuclei. These cells produce mucus with a higher content of glycoproteins than that made by surface mucous cells. Among the neck mucous cells are stem cells that give rise to all epithelial cells of the glands. In the upper half of the glands are also numerous distinctive parietal cells (P), large rounded cells often bulging from the tubules, with large central nuclei surrounded by intensely eosinophilic cytoplasm with unusual ultrastructure. These cells produce HCl and the numerous mitochondria required for this process cause the eosinophilia. Around these tubular glands are various cells and microvasculature in connective tissue.

(b): Near the muscularis mucosae (MM), the bases of these glands contain fewer parietal cells, but another cell type, chief cells (C), is abundant here. Chief cells are also known as peptic or zymogenic cells. They are seen as clusters of cells with condensed, basal nuclei and basophilic cytoplasm. From their apical ends chief cells secrete pepsinogen, the zymogen precursor for pepsin, a major protease. Zymogen granules are often removed or stain poorly in routine preparations. Both X200. H&E.

macrophages, and mast cells. The **muscularis** is composed of smooth muscle fibers oriented in three main directions. The external layer is longitudinal, the middle layer is circular, and the internal layer is oblique. Rhythmic contractions of the muscularis serve to mix ingested food and chyme with the secretions from the gastric mucosa. At the pylorus, the middle layer is greatly thickened to form the **pyloric sphincter.** The stomach is covered by a thin **serosa.**

SMALL INTESTINE

The small intestine is the site of terminal food digestion, nutrient absorption, and endocrine secretion. The processes of digestion are completed in the small intestine, where the nutrients (products of digestion) are absorbed by cells of the epithelial lining. The small intestine is relatively long—approximately 5 m—and consists of three segments: **duodenum, jejunum,** and **ileum.** These segments have many characteristics in common and will be discussed together.

Mucous Membrane

Viewed with the naked eye, the lining of the small intestine shows a series of permanent circular or semilunar folds (**plicae circulares)**, consisting of mucosa and submucosa (Figures 15–25 and 15–26), which are best developed in the jejunum. Intestinal **villi** are 0.5- to 1.5-mm-long mucosal outgrowths (epithelium plus lamina propria) and project into the lumen (Figure 15–25). In the duodenum they are leaf-shaped, but gradually assume fingerlike shapes moving toward the ileum. Villi are covered by a simple columnar epithelium of **absorptive cells** and **goblet cells.**

Between the villi are small openings of short tubular glands called **intestinal crypts** or **crypts of Lieberkühn** (Figure 15–27). The epithelium of each villus is continuous with that of the intervening glands, which contain differentiating absorptive and goblet cells, Paneth cells, enteroendocrine cells, and stem cells that give rise to all these cell types.

Enterocytes, the absorptive cells, are tall columnar cells, each with an oval nucleus in the basal half of the cell (Figure 15–28). At the apex of each cell is a homogeneous layer called the **striated** (or **brush**) **border.** When viewed with the electron microscope, the striated border is seen to be a layer of densely packed **microvilli** (Figures 15–25e and 15–28c). As discussed in Chapter 4, each microvillus is a cylindrical protrusion of the apical cytoplasm approximately 1 μm tall and 0.1 μm in diameter containing actin filaments and enclosed by the cell membrane. Each absorptive cell is estimated to have an average of 3000 microvilli and 1 mm^2 of mucosa contains about 200 million of these

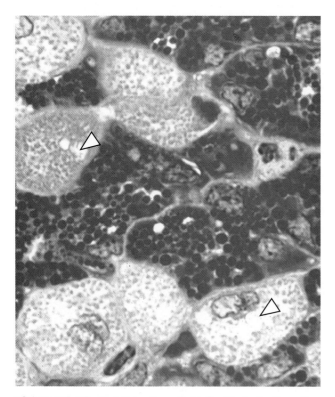

Figure 15–21. Parietal cells and chief cells. A plastic section of a gastric gland's basal portion shows better detail of parietal and chief cells than routine histological sections often allow. The large parietal cells' characteristic intracellular canaliculi (arrowheads) can be seen, along with the numerous acidophilic mitochondria. The smaller chief cells have cytoplasm containing numerous large red secretory (zymogen) granules. X400. PT.

Table 15–1. Principal enteroendocrine cells in the gastrointestinal tract.

Cell Type and Location	Hormone Produced	Major Action
X/A-like—stomach	Ghrelin	Increase sense of hunger
G—pylorus	Gastrin	Stimulation of gastric acid secretion
S—small intestine	Secretin	Pancreatic and biliary bicarbonate and water secretion
K—small intestine	Gastric inhibitory polypeptide	Inhibition of gastric acid secretion
L—small intestine	Glucagon-like peptide 1 (GLP-1)	Decrease sense of hunger
I—small intestine	Cholecystokinin (CCK)	Pancreatic enzyme secretion, gallbladder contraction
D—pylorus, duodenum	Somatostatin	Local inhibition of other endocrine cells
Mo—small intestine	Motilin	Increased gut motility
EC—digestive tract	Serotonin, substance P	Increased gut motility
D$_1$—digestive tract	Vasoactive intestinal polypeptide (VIP)	Ion and water secretion, increased gut motility

structures. Microvilli greatly increase the area of contact between the intestinal surface and the nutrients, a function also of the plicae and villi, which is an important feature in an organ specialized for absorption. It is estimated that plicae increase the intestinal surface three-fold, the villi increase it 10-fold, and the microvilli increase it 20-fold. Together, these processes are responsible for a 600-fold increase in the intestinal surface, resulting in a total absorptive area of 200 m^2!

Enterocytes absorb the nutrient molecules produced by digestion. Disaccharidases and peptidases secreted by these cells and bound to the microvilli hydrolyze the disaccharides and dipeptides into monosaccharides and amino acids that are easily absorbed through active transport. Digestion of fats results from the action of pancreatic lipase and bile. In humans, most of the lipid absorption takes place in the duodenum and upper jejunum and Figure 15–29 illustrates basic aspects of lipid absorption.

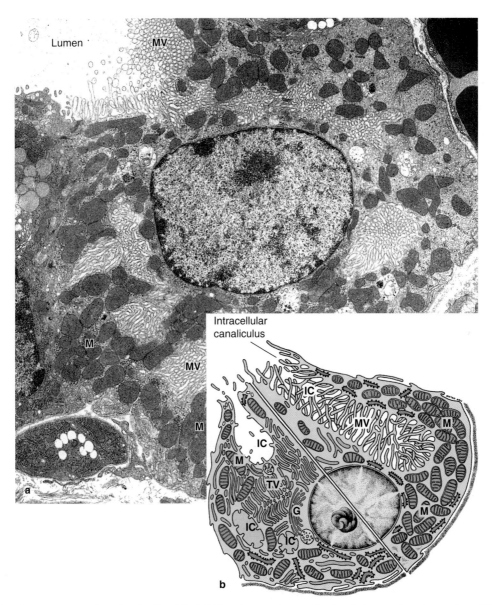

Figure 15–22. **Ultrastructure of parietal cells. (a):** A TEM of an active parietal cell shows abundant microvilli (MV) protruding into the intracellular canaliculi, near the lumen and deep in the cell. The remaining cytoplasm is filled with mitochondria (M). X10,200. **(b):** Composite diagram of a parietal cell, showing the ultrastructural differences between a resting cell (left) and an active cell (right). In the resting cell, a number of tubulovesicular structures can be seen in the apical region just below the plasmalemma (left), but the cell has few microvilli. When stimulated to produce hydrochloric acid (right), these vesicles fuse with the cell membrane to form the canaliculus and microvilli, thus providing a generous increase in the surface of the cell membrane for diffusion and ion pumps. (Figure 15-22a, with permission, from Dr. Susumu Ito, Department of Cell Biology, Harvard Medical School.)

Goblet cells are interspersed between the absorptive cells (Figures 15–25 and 15–28). They are less abundant in the duodenum and more numerous in the ileum. These cells produce glycoprotein mucins that are hydrated and cross-linked to form mucus, whose main function is to protect and lubricate the lining of the intestine.

Paneth cells, located in the basal portion of the intestinal crypts below the stem cells, are exocrine cells with large, eosinophilic secretory granules in their apical cytoplasm (Figures 15–27 and 15–30). Paneth cell granules undergo exocytosis to release lysozyme, phospholipase A2, and hydrophobic peptides called defensins, all of which bind and breakdown membranes of microorganisms and bacterial walls. Paneth cells have an important role in innate immunity and in regulating the microenvironment of the intestinal crypts.

Enteroendocrine cells are present in varying numbers throughout the length of the small intestine, secreting various peptides (Table 15–1) and representing part of the widely distributed **diffuse neuroendocrine system** (Chapter 20). Upon stimulation these cells release their secretory granules by

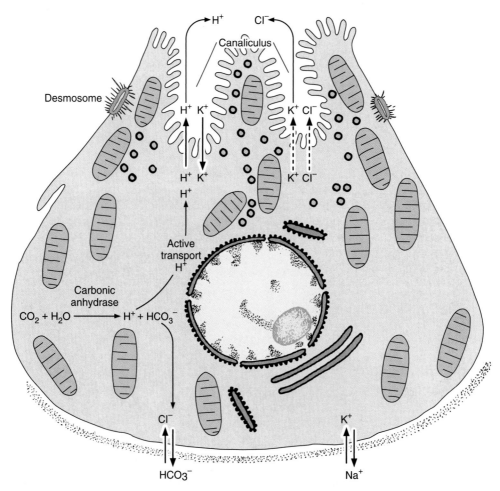

Figure 15–23. **Synthesis of HCl by parietal cells.** Diagram showing the main steps in the synthesis of hydrochloric acid. Active transport by ATPase is indicated by arrows and diffusion is indicated by dotted arrows. Under the action of carbonic anhydrase, carbonic acid is produced from CO_2. Carbonic acid dissociates into a bicarbonate ion and a proton (H^+), which is pumped into the stomach lumen in exchange for K^+. A high concentration of intracellular K^+ is maintained by the Na^+, K^+ ATPase, while HCO_3^- is exchanged for Cl^- by an antiport. The tubulovesicles of the cell apex are seen to be related to hydrochloric acid secretion, because their number decreases after parietal cell stimulation as microvilli increase. Most of the bicarbonate ion returns to the blood and is responsible for a measurable increase in blood pH during digestion, but some is taken up by surface mucous cells and used to raise the pH of mucus.

exocytosis and the hormones may then exert paracrine (local) or endocrine (blood-borne) effects. Polypeptide-secreting cells of the digestive tract fall into two classes: a "closed" type, in which the cellular apex is covered by neighboring epithelial cells (Figures 15–24 and 15–28) and an "open" type, in which the apex of the cell has microvilli and contacts the lumen (Figure 15–31). Peptides produced have both endocrine and paracrine effects, which include the control of peristalsis, regulation of secretions necessary for food digestion, and the sense of being satiated after eating.

MEDICAL APPLICATION

The hormone secretin, produced by enteroendocrine cells of the small intestine, was the very first hormone to be discovered. Two brothers-in-law, William Bayliss and Ernest Starling, working at University College London in 1900 observed that the factor caused the pancreas to secrete its alkaline digestive fluid and they named it "secretin." They decided further to call secretin a "hormone," from the Greek verb hormaein, "to excite or set in motion." Since that time hormones purified from tissue or made synthetically have had an enormous impact in treating innumerable medical disorders.

M (microfold) cells are specialized epithelial cells in the ileum overlying the lymphoid follicles of Peyer patches. As discussed in Chapter 14, these cells are characterized by the presence of basal membrane invaginations or pockets containing many intraepithelial lymphocytes and antigen-presenting cells (Figure 14–16). M cells selectively endocytose antigens and transport them to the underlying macrophages and lymphocytes, which then migrate to lymph nodes where immune responses to foreign antigens are initiated. M cells thus serve as sampling stations where material in the lumen of the gut is transferred to immune cells of the MALT in the lamina propria. The basement membrane under the M cells is porous, facilitating transit of cells between the lamina propria and the pockets of M cells (Figure 15–32).

Lamina Propria through Serosa

The lamina propria of the small intestine is composed of loose connective tissue with blood and lymph vessels, nerve fibers, and smooth muscle cells. The lamina propria penetrates the core of each intestinal villus, bringing with it microvasculature, lymphatics, and nerves (Figures 15–25 and 15–33). Smooth muscle fibers inside the villi are responsible for their rhythmic movements, which are important for efficient absorption. The muscularis mucosae also produces local movements of the villi and plicae circulares.

The proximal part of the duodenum has, primarily in its submucosa but extending into the mucosa, large clusters of branched tubular mucous glands, the **duodenal (or Brunner) glands**, with small excretory ducts opening among the intestinal crypts

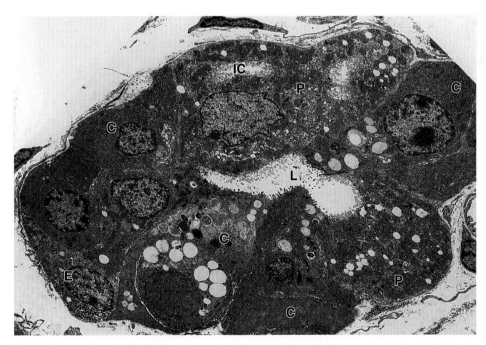

Figure 15–24. **Ultrastructure of parietal, chief, and enteroendocrine cells.** TEM of a transversely sectioned gastric gland shows the ultrastructure of three major cell types. Parietal cells (P) contain abundant mitochondria and intracellular canaliculi (IC). Most of the cells are chief cells (C), which have extensive rough ER and apical secretory granules near the lumen (L). An enteroendocrine cell (E) shows dense basal secretory granules. This example is an enterochromaffin cell (EC cell) secreting serotonin. It is a closed type enteroendocrine cell, ie, it has no contact with the gland's lumen, and secretes product in an endocrine/paracrine manner. X5300.

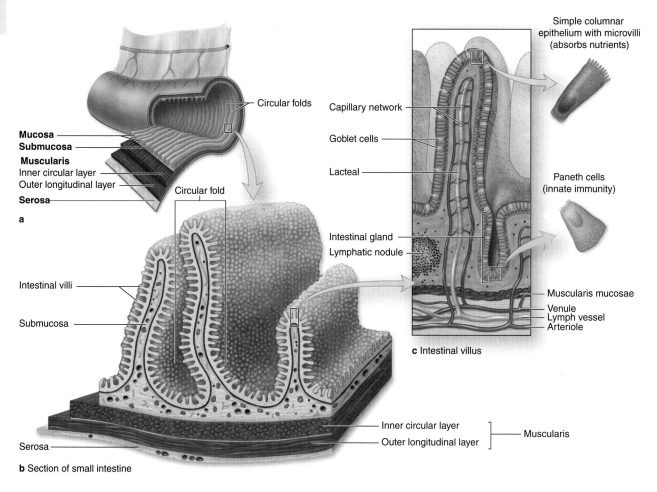

Simple columnar epithelium with microvilli (absorbs nutrients)

Capillary network

Goblet cells

Lacteal

Paneth cells (innate immunity)

Intestinal gland
Lymphatic nodule

Muscularis mucosae
Venule
Lymph vessel
Arteriole

c Intestinal villus

Circular folds

Mucosa
Submucosa
Muscularis
Inner circular layer
Outer longitudinal layer
Serosa

a

Circular fold

Intestinal villi

Submucosa

Inner circular layer
Outer longitudinal layer — Muscularis

Serosa

b Section of small intestine

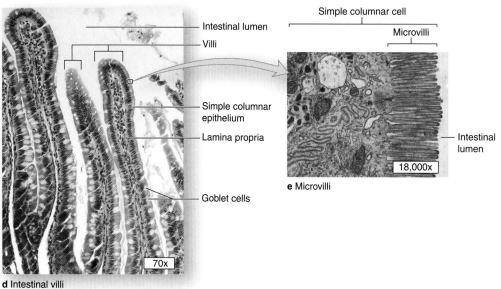

Intestinal lumen
Villi

Simple columnar cell
Microvilli

Simple columnar epithelium

Lamina propria

Intestinal lumen

Goblet cells

e Microvilli
18,000x

70x

d Intestinal villi

Figure 15–25. **Absorptive surface of the small intestine. (a):** The mucosa and submucosa are the inner two of the gut's four concentric layers. **(b):** They form circular folds or plicae circulares, which increase the absorptive area. **(c, d):** They are lined by a dense covering of finger-like projections called villi. Internally each villus contains lamina propria connective tissue with microvasculature and lymphatics called **lacteals**. Villi are covered with a simple columnar epithelium composed of absorptive enterocytes and goblet cells. **(e):** At the apical cell membrane of each enterocyte are located dense **microvilli,** which serve to increase greatly the absorptive surface of the cell. Between the villi the covering epithelium invaginates to form short tubular intestinal glands or crypts, which include stem cells for the epithelium and Paneth cells which prevent intestinal flora from becoming concentrated in these glands where damage to the stem cells could occur.

(Figure 15–34). The product of the glands is distinctly alkaline (pH 8.1–9.3), which neutralizes chyme entering the duodenum from the pylorus, protecting the mucous membrane and bringing the intestinal contents to the optimum pH for pancreatic enzyme action. In the ileum both the lamina propria and submucosa contain the lymphoid nodule aggregates known as **Peyer patches,** an important component of the MALT.

The muscularis is well developed in the small intestine, composed of an internal circular layer and an external longitudinal layer, and is covered by a thin serosa with mesothelium (Figures 15–25, 15–26 and 15–35).

Vessels & Nerves

The blood vessels that nourish the intestine and remove absorbed products of digestion penetrate the muscularis and form a large plexus in the submucosa (Figure 15–34). From the submucosa, branches extend through the muscularis mucosae and lamina propria and into the villi. Each villus receives, according to its size, one or more branches that form a capillary network just below its epithelium. At the tips of the villi, one or more venules arise from these capillaries and run in the opposite direction, reaching the veins of the submucosal plexus. The lymph vessels of the intestine begin as closed tubes in the cores of villi. These capillaries (**lacteals**), despite being larger than the blood capillaries, are often difficult to observe because their walls are so close together that they appear to be collapsed. Lacteals run to the region of lamina propria above the muscularis mucosae, where they form a plexus. From there they are directed to the submucosa, where they surround lymphoid nodules. Lacteals anastomose repeatedly and leave the intestine along with the blood vessels.

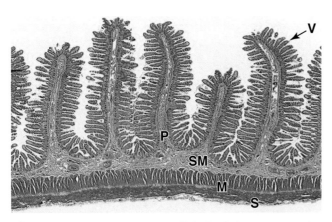

Figure 15–26. **Plicae ciculares (folds) of the jejunum.** The mucosa and submucosa (SM) of the small intestine form distinct projecting folds called plicae (P), which encircle or spiral around the inner circumference and are best developed in the jejunum. On each fold the mucosa forms a dense covering of projecting structures called villi (V). In this longitudinal section the two layers of the muscularis (M) are clearly distinguished. The inner layer has smooth muscle encircling the submucosa; the outer layer runs lengthwise just inside the serosa (S), the gut's outer layer. This arrangement of smooth muscle provides for strong peristaltic movement of the gut's contents. X12. H&E.

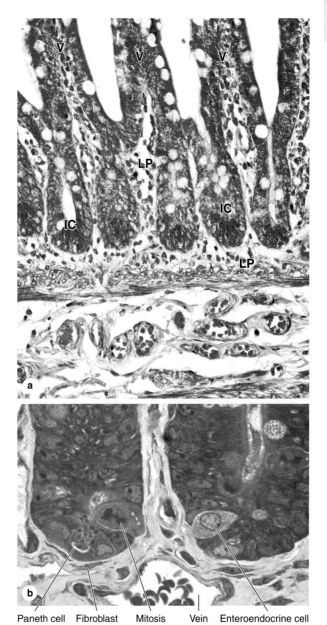

Paneth cell   Fibroblast   Mitosis   Vein   Enteroendocrine cell

Figure 15–27. **Intestinal crypts or glands. (a):** Between villi (V) throughout the small intestine, as seen in this micrograph, the covering epithelium invaginates into the lamina propria (LP) to form short tubular glands called intestinal glands or intestinal crypts (IC). The lining near the openings of the crypts contains a population of stem cells for the entire epithelial lining of the small intestine. Daughter cells slowly move with the growing epithelium out of the crypts, differentiating as goblet cells, enterocytes, and enteroendocrine cells. These cells continue to move up each villus and within a week are shed at the tip, with billions shed throughout the small intestine each day. At the base of the crypts are Paneth cells, derived from the same stem cell population, which contain eosinophilic secretory granules. X200. H&E. **(b):** Micrograph of a plastic section showing two crypts in which other cells can be distinguished, including a paler staining enteroendocrine cell, cells in mitosis, and cells differentiating as enterocytes and goblet cells. X400. PT.

They are especially important for lipid absorption; chylomicrons of lipoprotein are preferentially taken up by lacteals rather than blood capillaries.

Another process important for intestinal function is the rhythmic movement of the villi. This movement is the result of the contraction of smooth muscle fibers running vertically from the muscularis mucosae to the tip of the villi (Figure 15–33). These contractions occur at the rate of several strokes per minute and have a pumping action on the villi that propel the lymph to the mesenteric lymphatics.

The innervation of the intestines is formed by intrinsic and extrinsic components comprising the enteric nervous system. The intrinsic component comprises many small and diffuse groups of neurons that form the **myenteric** (Auerbach) **nerve plexus** (Figures 15–33 and 15–35) between the outer longitudinal and inner circular layers of the muscularis and the smaller **submucosal** (Meissner) **plexus** in the submucosa. The enteric nervous system contains some sensory neurons that receive information from nerve endings near the epithelial layer and in the muscularis

regarding the intestinal content (chemoreceptors) and the degree of intestinal wall expansion (mechanoreceptors). Other nerve cells are effectors innervating the muscle layers and hormone-secreting cells. The intrinsic innervation formed by these plexuses is responsible for the intestinal contractions that occur even in the absence of the extrinsic innervation that modulates the activity.

LARGE INTESTINE

The large intestine or bowel consists of a mucosal membrane with no folds except in its distal (rectal) portion and no villi (Figure 15–36). The mucosa is penetrated throughout its area by tubular intestinal glands lined by goblet and absorptive cells, with a small number of enteroendocrine cells (Figures 15–37 and 15–38). The absorptive cells or **colonocytes** are columnar and have short, irregular microvilli (Figure 15–38d). Stem cells for the epithelium of the large bowel are located in the bottom third of each gland. The large intestine is well suited to its

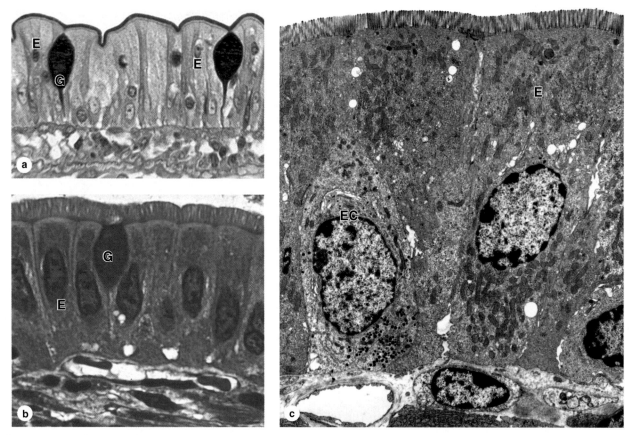

Figure 15–28. **Cells covering the villi. (a):** The columnar epithelium that covers intestinal villi consists mainly of the tall absorptive enterocytes (E). The apical ends of these cells are joined and covered by a brush border of microvilli. Covered by a coating of glycoproteins, the brush border, along with the mucus-secreting goblet cells (G), stains with carbohydrate staining methods. Other cells of the epithelium are scattered enteroendocrine cells, which are difficult to identify in routine preparations, and various immune cells such as intraepithelial lymphocytes. The small spherical nuclei of lymphocytes can be seen between the enterocytes. X200. PAS-hematoxylin. **(b):** At higher magnification individual microvilli of enterocytes are better seen and the striated appearance of the border is apparent. **(c):** TEM shows microvilli and densely packed mitochondria of enterocytes, and enteroendocrine cells (EC) with secretory granules can be distinguished along the basal lamina. X1850.

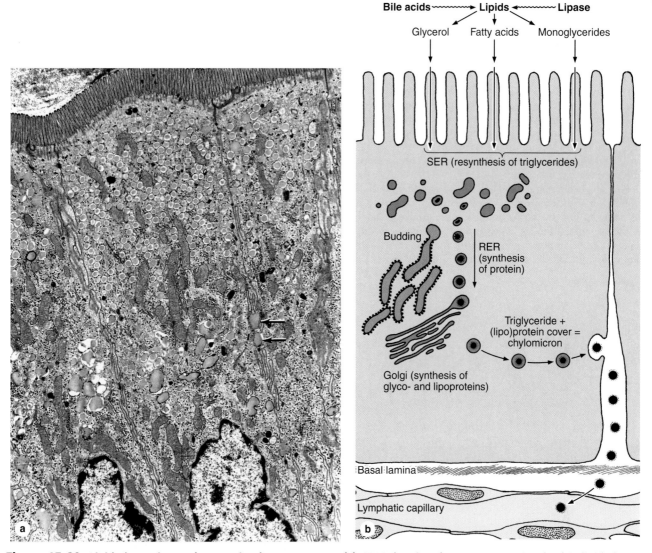

Figure 15–29. **Lipid absorption and processing by enterocytes. (a):** TEM showing that enterocytes involved in lipid absorption accumulate many small lipid droplets in vesicles of the smooth ER. These vesicles fuse near the nucleus, forming larger globules that are moved laterally and cross the cell membrane to the extracellular space (arrows) for eventual uptake by lymphatic capillaries (lacteals) in the lamina propria. X5000.

(b): Diagram explaining how lipids are processed by enterocytes. Bile components in the lumen emulsify fats into lipid droplets, which are broken down further by lipases to monoglycerides and fatty acids. These compounds are stabilized in an emulsion by the action of bile acids. The products of hydrolysis diffuse passively across the microvilli membranes and are collected in the cisternae of the smooth ER, where they are resynthesized as triglycerides. Processed through the RER and Golgi, these triglycerides are surrounded by a thin layer of proteins and packaged in vesicles containing chylomicrons (0.2–1 μm in diameter) of lipid complexed with protein. Chylomicrons are transferred to the lateral cell membrane, secreted by exocytosis, and flow into the extracellular space in the direction of the lamina propria, where most enter the lymph in lacteals. (Figure 15–29a, with permission, from Robert R. Cardell, Jr, Department of Cancer and Cell Biology, University of Cincinnati College of Medicine.)

main functions: absorption of water, formation of the fecal mass from undigestible material, and production of mucus that lubricates the intestinal surface.

The lamina propria is rich in lymphoid cells and in lymphoid nodules that frequently extend into the submucosa (Figure 15–37). The richness in MALT is related to the large bacterial population of the large intestine. The muscularis comprises longitudinal and circular strands, but differs from that of the small intestine, with fibers of the outer layer gathered in three longitudinal bands called **taeniae coli** (Figure 15–37). Intraperitoneal portions of the colon are covered by serosa, which is characterized by small, pendulous protuberances of adipose tissue.

Near the beginning of the large intestine, the **appendix** is an evagination of the cecum. It is characterized by a relatively small and irregular lumen, shorter and less dense tubular glands, and no taeniae coli. Although it has no function in digestion, the appendix is a significant component of the MALT, with abundant lymphoid follicles in its wall (Figure 15–39).

MEDICAL APPLICATION

*Because the appendix is a closed sac and its contents are relatively static, it can easily become a site of inflammation (**appendicitis**). With the small lumen and relatively thin wall of the appendix, inflammation and the growth of lymphoid follicles in the wall can produce swelling that can lead to bursting of the appendix. Severe appendicitis is a medical emergency since a burst appendix will produce infection of the peritoneal cavity.*

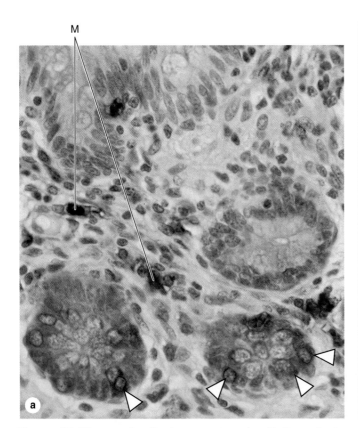
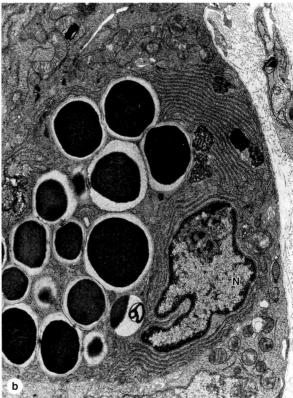

Figure 15–30. **Paneth cells.** Secretory Paneth cells lie at the base of each intestinal crypt, below the level of the stem cells from which they and cells covering the villi all arise. **(a):** Micrograph of Paneth cell granules are shown immunohistochemically to contain lysozyme (arrowheads), which is also present in macrophages (M). X100.

(b): TEM shows that a Paneth cell has a prominent nucleolus in the basal nucleus (N), abundant rough ER, and large secretory granules in which the protein cores are surrounded by halos of polysaccharide-rich material. Paneth cells are key components of the gut's innate immunity, secreting into the crypt lumens enzymes and peptides called defensins that prevent microorganisms from permanently lodging in the crypts and affecting stem cell and differentiative activities. X3000.

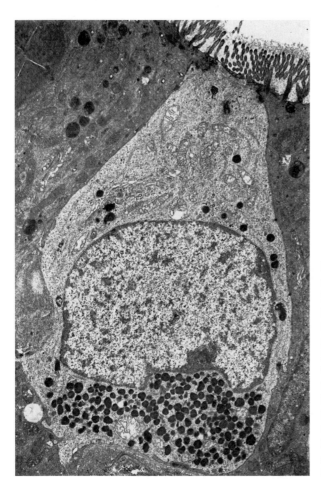

Figure 15–31. Enteroendocrine cell. TEM of an open type enteroendocrine cell in the epithelium of the duodenum showing microvilli at its apical end in contact with the lumen. The microvilli have components of nutrient-sensing and signal transduction systems similar in some components to those of taste bud gustatory cells. Activation of these cells by nutrients triggers the release at the basolateral membranes of peptide factors, including satiation peptides, which diffuse through extracellular fluid to enter capillaries (endocrine) or to bind receptors on nearby nerve terminals, smooth muscle fibers, or other cells (paracrine). Hormones from the various enteroendocrine cells act in a coordinated manner to control gut motility, regulate secretion of enzymes, HCl, bile and other components for digestion, and produce the sense of satiety in the brain. X6900. (With permission, from A.G.E. Pearse, Department of Histochemistry, Royal Postgraduate Medical School, London.)

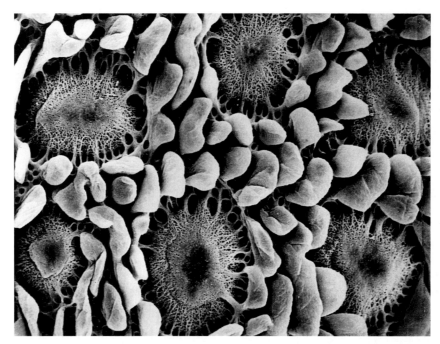

Figure 15–32. Ultrastructure of basement membrane at Peyer patches. In the ileum of the small intestine are specialized regions of mucosa called Peyer patches, which are important parts of the mucosa-associated lymphoid tissue (MALT) in the wall of the gut. Between the villi in these regions are groups of M cells which selectively sample the lumen's contents and transfer antigens to underlying cells of the MALT. SEM of a Peyer patch after removal of the epithelial cells reveals the basement membrane. Over the broad villi the basement membrane is continuous, but over lymphoid follicles it is sieve-like, which facilitates movement of immune cells to and from the intraepithelial pockets of M cells (see Chapter 14). X1000. (With permission, from Samuel G. McClugage, Department of Cell Biology and Anatomy, Louisiana State University Health Sciences Center.)

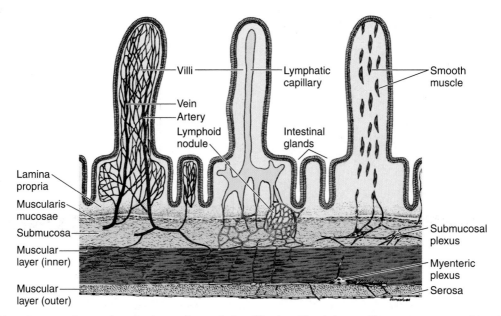

Villi
Lymphatic capillary
Smooth muscle
Vein
Artery
Lymphoid nodule
Intestinal glands
Lamina propria
Muscularis mucosae
Submucosa
Submucosal plexus
Muscular layer (inner)
Myenteric plexus
Muscular layer (outer)
Serosa

***Figure 15–33.* Microvasculature, lymphatics, and muscle in villi.** The villi of the small intestine contain blood microvasculature (left), lymphatic capillaries called lacteals (center), and both innervation and smooth muscle fibers (right).

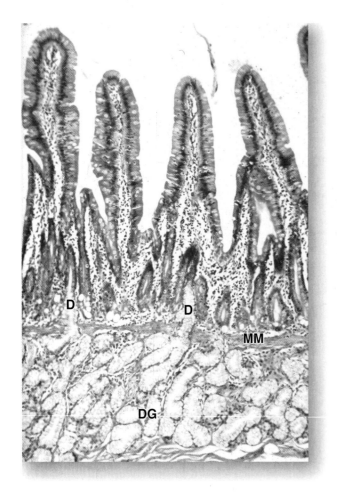

***Figure 15–34.* Duodenal (Brunner) glands.** Concentrated mainly in the upper duodenum are large masses of compound branched mucous glands, the duodenal glands (DG), with many lobules that occupy much of the submucosa and may extend above the muscularis mucosae (MM) into the mucosa. Many small excretory ducts (D) extend from these lobules through the lamina propria and empty into the lumen among the small intestinal crypts (IC). Alkaline mucus from duodenal glands neutralizes the pH of material entering the duodenum and supplements the mucus from goblet cells in lubricating and protecting the lining of the small intestine. X100. H&E.

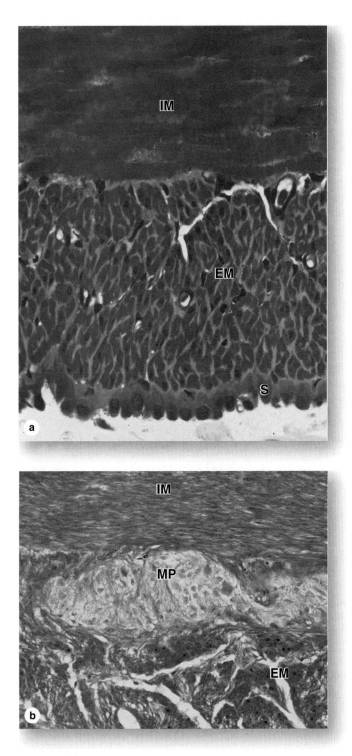

Figure 15–35. **Small intestine muscularis and myenteric plexus. (a):** Transverse sections of the small intestinal wall show the orientation of the internal (IM) and external (EM) smooth muscle layers. The inner layer is predominantly circular while the outer layer is longitudinal. The serosa (S) is a thin connective tissue covered here by a mesothelium of cuboidal cells. X200. PT. **(b):** Pale-staining neurons and other cells in one myenteric plexus (MP) are seen between the two muscle layers. X100. H&E. Along the entire digestive tract autonomic neurons from the numerous myenteric ganglia and smaller submucosal ganglia innervate the wall and comprise the enteric nervous system, which is of key importance for gut function and involved in many digestive problems. Local activity of these neurons is controlled by sensory neurons and effector neurons supplying the musculature, both of which are intrinsic to the enteric nervous system. Extrinsic innervation in this system includes parasympathetic cholinergic nerve fibers that stimulate the activity of smooth muscle and sympathetic adrenergic nerve fibers that depress muscle activity.

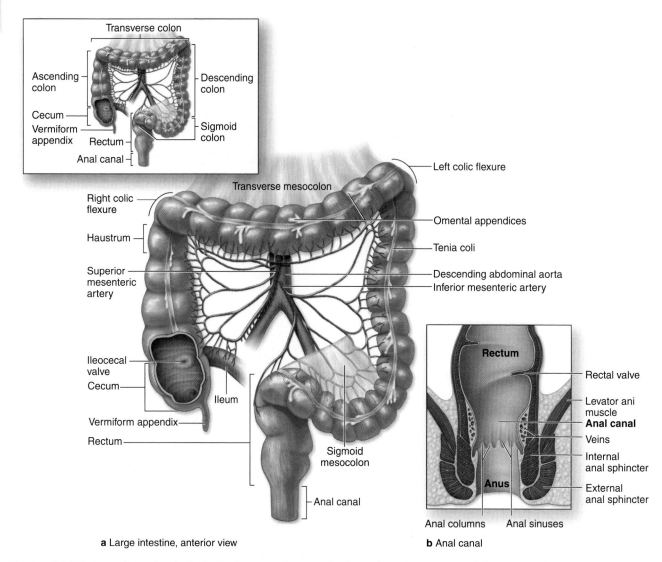

a Large intestine, anterior view

b Anal canal

Figure 15–36. **Large intestine (colon).** As shown at the top, the large intestine consists of the cecum; the ascending, transverse, descending, and sigmoid regions of the colon; and the rectum. **(a):** Anterior view of the large intestine with the proximal end exposed shows the ileocecal valve at its attachment to the ileum, along with the blind sac called the cecum and its extension, the appendix. The mucosa has shallow plicae but no villi. The muscularis has two layers, but the outer longitudinal layer consists only of three distinct bundles of muscle fibers called **taeniae coli** (ribbons of the colon). These bands cause the colon wall to form a series of sacs called **haustra**. The serosa of the colon is continuous with that of the supporting mesenteries and displays a series of suspended masses of adipose tissue called **omental appendages**.

(b): At the distal end of the rectum, the anal canal, the mucosa and submucosa are highly vascularized, with venous sinuses, and are folded as a series of longitudinal anal folds with intervening anal sinuses. Fecal material accumulates in the rectum is eliminated by muscular contraction, including action of an internal anal sphincter of smooth (involuntary) muscle and an external sphincter of striated (voluntary) muscle.

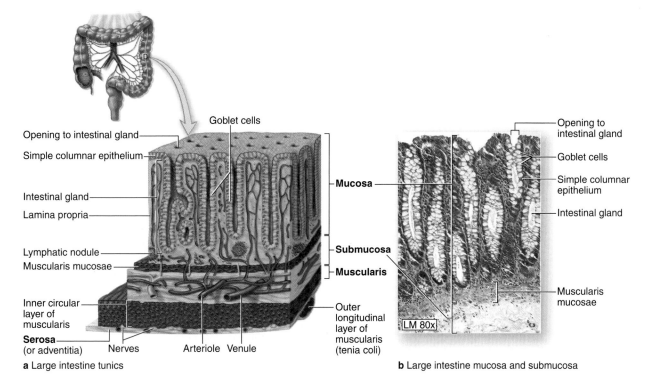

Figure 15–37. Wall of the large intestine. (a): Diagram showing the wall of the large intestine composed of the four typical layers. **(b):** The mucosa is occupied mostly by tubular intestinal glands extending as deep as the muscularis mucosae and by lamina propria rich in MALT. The submucosa is well vascularized. The muscularis has a typical inner circular layer, but the outer longitudinal muscle is only present in three equally spaced bands, the taeniae coli (a). b: X80, H&E

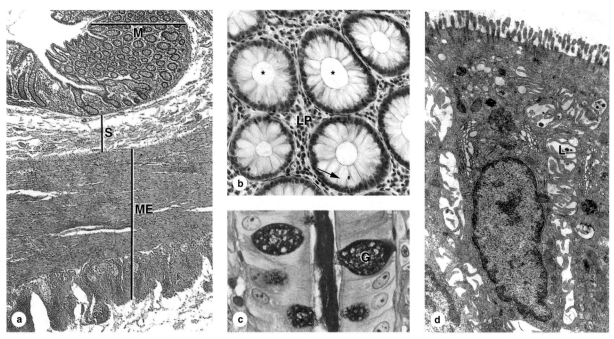

Figure 15–38. Mucosa of the large intestine (colon). (a): Transverse section of the colon shows the muscularis externa (ME), including a taenia coli cut transversely in the lower part of the figure, the submucosa (S), the mucosa (M) filled with tubular intestinal glands. Some of these glands are cut longitudinally, but most seen here are cut transversely. X14. H&E. **(b):** Transversely cut glands are seen to consist of simple columnar epithelium surrounded a tubular lumen (asterisk) and embedded in lamina propria (LP) with many free lymphocytes. Lymphocytes can also be seen penetrating the epithelium (arrow). X200. H&E. **(c):** Longitudinal section of one gland stained for glycoproteins shows mucus in the lumen and two major cell types in the epithelium: goblet cells (G) and other columnar cells specialized for water absorption. X400. PAS. **(d):** TEM micrograph of the absorptive cells, called **colonocytes**, reveals short microvilli at their apical ends, prominent Golgi complexes above the nuclei, and dilated intercellular spaces with interdigitating leaflets of cell membrane (L), a sign of active water transport. The absorption of water is passive, following the active transport of sodium from the basolateral surfaces of the epithelial cells. X3900.

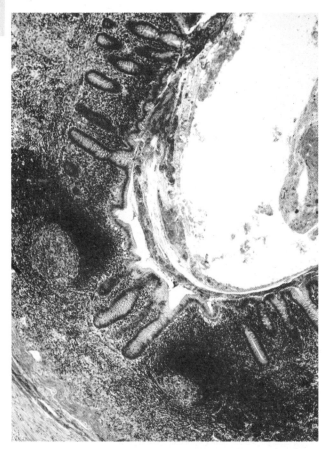

Figure 15–39. Appendix. A blind evagination off the cecum, the appendix, has a very small lumen, fewer glands in its mucosa, and no taeniae coli. The laminar propria and submucosa are generally filled with lymphocytes and lymphoid follicles, making the appendix a significant part of the MALT. X40. H&E.

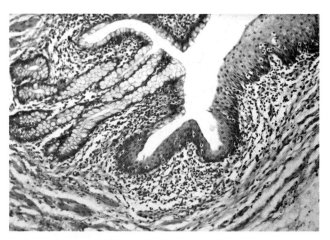

Figure 15–40. Mucosa of the recto-anal junction. The simple columnar epithelium with tubular glands that lines the rectum (left side) changes abruptly to stratified squamous epithelium in the anal canal (right side), as seen in this longitudinal section. The connective tissue of the lamina propria is seen to contain many free lymphocytes. X40. H&E.

In the anal region, the mucous membrane forms a series of longitudinal folds, the **anal columns** (Figure 15–36). About two cm above the anal opening, at the recto-anal junction, the lining of the mucosa is replaced by stratified squamous epithelium (Figure 15–40). In this region, the lamina propria contains a plexus of large veins that, when excessively dilated and varicose, can produce hemorrhoids.

MEDICAL APPLICATION

*Approximately 90–95% of malignant tumors of the digestive system are derived from gastric or intestinal epithelial cells, usually in the large intestine. Malignant tumors of the colon are derived almost exclusively from its glandular epithelium (**adenocarcinomas**) and are the second most common cause of cancer deaths in the United States.*

Organs Associated with the Digestive Tract

16

SALIVARY GLANDS
PANCREAS
LIVER
    Stroma
    Hepatic Lobules

Blood Supply
The Hepatocyte
HEPATIC LOBULE STRUCTURE AND FUNCTION
    Liver Regeneration
BILIARY TRACT & GALLBLADDER

The organs associated with the digestive tract include the salivary glands, the pancreas, the liver, and the gallbladder. Products of these organs facilitate transport and digestion of food within the gastrointestinal tract. The main functions of the salivary glands are to wet and lubricate ingested food and the oral mucosa, to initiate the digestion of carbohydrates and lipids with amylase and lipase, and to secrete protective bacteriostatic substances such as the immunoglobulin IgA, lysozyme, and lactoferrin.

The pancreas produces digestive enzymes that act in the small intestine and hormones important for the metabolism of the absorbed nutrients. The liver produces bile, an important fluid in the digestion of fats. The gallbladder absorbs water from the bile and stores it in a concentrated form. The liver also plays a major role in carbohydrate and protein metabolism and inactivates and metabolizes many toxic substances and drugs. It also synthesizes most blood plasma proteins and the factors necessary for blood coagulation.

SALIVARY GLANDS

Exocrine glands in the mouth produce saliva, which has digestive, lubricating, and protective functions. With a usual pH of 6.5–6.9, saliva also has an important buffering function and in many nonhuman species is also very important for evaporative cooling. There are three pairs of large salivary glands: the **parotid, submandibular**, and **sublingual glands** (Figure 16–1), in addition to minor glands in mucosa and submucosa throughout the oral cavity which secrete 10% of the total volume of saliva.

MEDICAL APPLICATION

Reduced function of the major salivary glands due to diseases or radiotherapy is associated with caries, atrophy of the oral mucosa and speech difficulties.

A capsule of connective tissue surrounds each major salivary gland. The parenchyma of each consists of secretory endpieces and a branching duct system arranged in **lobules**, separated by connective tissue septa originating from the capsule. The secretion of each gland is either serous, seromucous, or mucous, depending on its glycoprotein mucin content. Saliva from the parotids is serous and watery. The submandibular and sublingual glands produce a seromucous secretion, with mostly mucus from the minor glands. Saliva is modified by the cells of the duct system draining the secretory units, with much Na^+ and Cl^- reabsorbed while certain growth factors and digestive enzymes are added.

Two major kinds of secretory cells occur, arranged in separate units. **Serous cells** are polarized protein-secreting cells, usually pyramidal in shape, with a broad base resting on the basal lamina and a narrow apical surface facing the lumen (Figures 16–2 and 16-3). Adjacent cells are joined together by junctional complexes and usually form a spherical mass of cells called an **acinus,** with a very small lumen in the center (Figure 16–2). Acini and their duct system resemble grapes attached to a stem. Serous acinar cells largely produce digestive enzymes and other proteins.

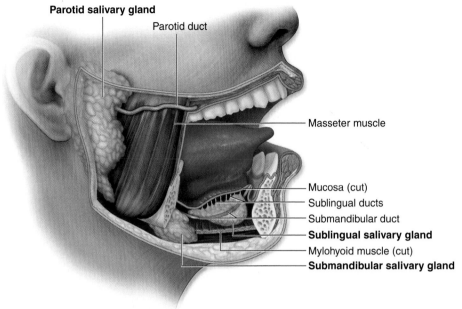

Parotid salivary gland
Parotid duct

Masseter muscle

Mucosa (cut)
Sublingual ducts
Submandibular duct
Sublingual salivary gland
Mylohyoid muscle (cut)
Submandibular salivary gland

Figure 16–1. **Major salivary glands.** About 90% of saliva is produced by three bilateral pairs of salivary glands: the parotid, submandibular, and sublingual glands. Locations and relative sizes of these glands are shown here diagrammatically. These glands plus microscopic salivary glands throughout the oral mucosa produce 0.75 to 1.50 L of saliva daily.

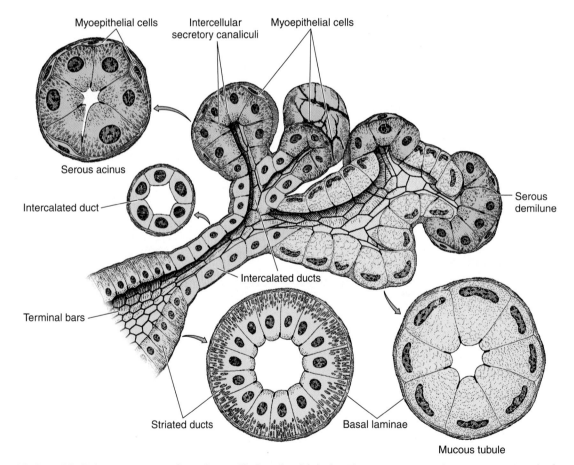

Myoepithelial cells
Intercellular secretory canaliculi
Myoepithelial cells

Serous acinus

Intercalated duct

Serous demilune

Intercalated ducts

Terminal bars

Striated ducts

Basal laminae

Mucous tubule

Figure 16–2. **Epithelial components of a submandibular gland lobule.** The secretory portions are composed of pyramidal serous (light blue) and mucous (light brown) cells. Serous cells are typical protein-secreting cells, with rounded nuclei, accumulation of rough ER in the basal third, and an apex filled with protein-rich secretory granules. The nuclei of mucous cells, flattened with condensed chromatin, are located near the bases of the cells. The short intercalated ducts are lined with cuboidal epithelium. The striated ducts are composed of columnar cells with characteristics of ion-transporting cells: basal membrane invaginations with mitochondrial accumulations. Myoepithelial cells are shown in the serous acini.

Mucous cells are somewhat more cuboidal or columnar in shape, with nuclei pressed toward the bases of the cells. They exhibit the characteristics of mucus-secreting cells (Figures 16–3 and 16–4), containing hydrophilic glycoprotein mucins which are important for the moistening and lubricating functions of the saliva. Mucous cells are most often organized as **tubules** rather than acini and produce mostly mucins.

Myoepithelial cells, described in Chapter 4, are found inside the basal lamina of the secretory units and (to a lesser extent) the initial part of the duct system (Figure 16–2). Surrounding the secretory portion myoepithelial cells are well developed and branched (and are sometimes called basket cells), whereas those associated with the initial ducts are spindle-shaped and lie parallel to the duct's length. Myoepithelial cells prevent distention of the endpiece when the lumen fills with saliva and their contraction accelerates secretion of the product.

In the **intralobular duct system,** secretory endpieces empty into **intercalated ducts,** lined by cuboidal epithelial cells, and several of these short ducts join to form **striated ducts** (Figure 16–2). The columnar cells of striated ducts often show radial striations extending from the cell bases to the level of the nuclei. Ultrastructurally the striations consist of infoldings of the basal plasma membrane (Figure 16–5). Numerous mitochondria are aligned parallel to the infolded membranes which contain ion transporters. Such folds greatly increase the cell surface area, facilitating ion absorption, and are characteristic of cells specialized for ion transport.

In the large salivary glands, the connective tissue contains many lymphocytes and plasma cells. The plasma cells release **IgA**, which forms a complex with a secretory component synthesized by the epithelial cells of serous acini and intralobular ducts. The IgA-secretory complex released into the saliva resists enzymatic digestion and constitutes an immunologic defense mechanism against pathogens in the oral cavity.

The striated ducts of each lobule converge and drain into ducts located in the connective tissue septa separating lobules, where they become **interlobular,** or **excretory, ducts.** They are initially lined with pseudostratified or stratified cuboidal epithelium, but more distal parts of the excretory ducts are lined with stratified columnar epithelium containing a few mucous cells.

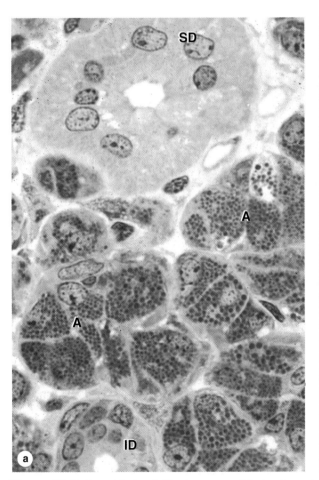
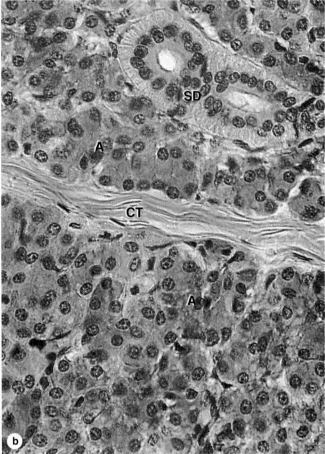

Figure 16–3. **Parotid gland.** The large parotid gland consists entirely of serous acini with cells producing amylase and other proteins for storage in secretory granules. **(a):** Micrograph of a parotid gland shows densely packed serous acini (A) with ducts. Secretory granules of serous cells are clearly shown in this plastic section, as well as both an intercalated duct (ID) and striated duct (SD), both cut transversely. X400. PT. **(b):** Striations of a duct (SD) are better seen here, along with a septum (CT) and numerous serous acini (A). The connective tissue often includes adipocytes. X200. H&E.

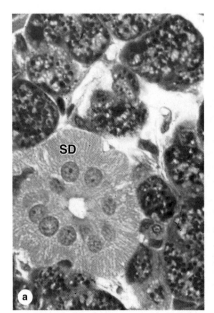

Figure 16–4. Ultrastructure of serous and mucous cells. A micrograph of a mixed acinus from a submandibular gland shows both serous and mucous cells. Mucous cells (upper area shown here) have large, hydrophilic granules like those of goblet cells. Serous cells (lower area) have small, dense granules that stain more intensely with most stains. X2500. (With permission, from John D. Harrison, Department of Oral Pathology, King's College, London.)

The main duct of each large salivary gland ultimately empties into the oral cavity and is lined with nonkeratinized-stratified squamous epithelium.

Vessels and nerves enter the large salivary glands at a hilum and gradually branch into the lobules. A rich vascular and nerve plexus surrounds the secretory and ductal components of each lobule. The capillaries surrounding the secretory endpieces are very important for the secretion of saliva, which is stimulated by the autonomic nervous system. Parasympathetic stimulation, usually elicited through the smell or taste of food, provokes a copious watery secretion with relatively little organic content. Sympathetic stimulation inhibits such secretion, and produces the potential for dry mouth often associated with anxiety.

Features specific to each group of major salivary glands include the following:

- **Parotid gland**, located in each cheek near the ear, is a branched acinar gland with secretory portions composed exclusively of serous cells surrounding very small lumens (Figure 16–3). Serous cells contain secretory granules with abundant α-**amylase** and **proline-rich proteins**. Amylase activity is responsible for most of the hydrolysis of ingested carbohydrates which begins in the mouth. Proline-rich proteins, the most abundant factors in parotid saliva, have antimicrobial properties and Ca^{2+} binding properties that may help maintain the surface of enamel.
- **Submandibular gland** is a branched tubuloacinar gland (Figures 16–4 and 16–6), with secretory portions containing both mucous and serous cells. The serous cells are the main component of this gland and are easily distinguished from mucous cells by their rounded nuclei and basophilic cytoplasm. Most of the secretory units in this gland are serous

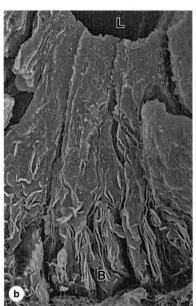
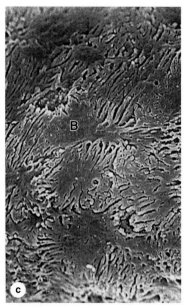

Figure 16–5. Striated ducts. (a): This light micrograph of a striated duct (SD) shows very faint pink striations in the basal half of the columnar cells. The striations are produced by mitochondria located in the folds of the lateral cell membrane. X200. H&E. **(b):** SEM indicates that the apical ends of the cells are joined together near the small lumen (L), with interdigitating folds of cell membrane best developed at the basal end (B). X4000. **(c):** This SEM shows the bases (B) of several such cells with the basal lamina removed, revealing the extensive interlocking of folded membrane between neighboring cells. Mitochondria between the folds supply energy for membrane ion pumps and ion uptake from saliva is rapid and efficient. X4000

acinar, with about 10% consisting of mucous tubules capped with serous cells. Such caps are called **serous demilunes** (Figure 16–6a). Lateral and basal membrane infoldings of the serous cells increase the ion-transporting surface area and facilitate electrolyte and water transport. In addition to α-amylase and proline-rich proteins, serous cells of the submandibular gland secrete other enzymes, including **lysozyme,** which hydrolyzes the walls in many types of bacteria.

- **Sublingual gland**, like the submandibular gland, is a branched tubuloacinar gland formed of serous and mucous cells. Here mucous cells predominate (Figure 16–6b), with serous cells only present in demilunes on mucous tubules. The major salivary product is mucus, but cells of the serous demilunes in this gland secrete amylase and lysozyme.

Small, nonencapsulated salivary glands are distributed throughout the oral mucosa and submucosa with short ducts to the oral cavity. Minor salivary glands are usually mucous, an exception being the small serous glands at the bases of circumvallate papillae on the back of the tongue (Chapter 15). B lymphocytes releasing IgA are common within the minor salivary glands.

PANCREAS

The pancreas is a mixed exocrine-endocrine gland that produces both digestive enzymes and hormones (Figures 16–7 and 16–8). A thin capsule of connective tissue covers the pancreas and sends septa into it, separating the pancreatic lobules. The secretory acini are surrounded by a basal lamina that is supported by a delicate sheath of reticular fibers and a rich capillary network.

The digestive enzymes are produced by cells of the larger exocrine portion and the hormones are synthesized in clusters of endocrine epithelial cells known as **pancreatic islets** (islets of Langerhans) (see Chapter 20). The exocrine portion of the pancreas is a compound acinar gland, similar in structure to the parotid gland. The two glands can be distinguished histologically by the absence of striated ducts and the presence of the

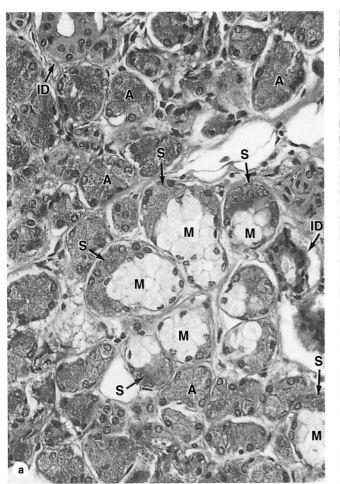
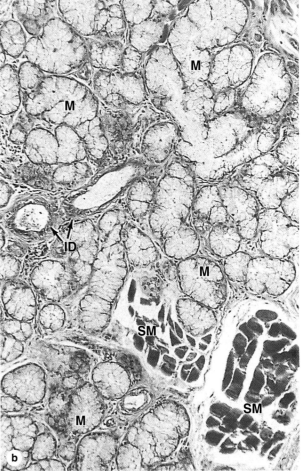

Figure 16–6. **Submandibular gland and sublingual gland. (a):** Submandibular gland is a mixed serous and mucous gland (serous cells predominate) and shows well-stained cells in serous acini (A) and in serous demilunes (S) and pale-staining mucous cells (M) grouped as tubules in this tubuloacinar gland. Small intralobular ducts (ID) drain each lobule, but these are not composed of columnar cells with well-developed striations. X340. H&E. **(b):** Sublingual gland is a mixed but largely mucous gland with a tubuloacinar arrangement of poorly stained mucous cells (M). Small intralobular ducts (ID) are seen in connective tissue, as well as small fascicles of lingual striated muscle (SM). X140. H&E.

islets in the pancreas. Another characteristic detail is that in the pancreas the initial portions of intercalated ducts penetrate the lumens of the acini (Figure 16–9). Small pale-staining **centroacinar cells** constitute the intraacinar portion of the intercalated duct and are found only in pancreatic acini. **Intercalated ducts** merge to form larger interlobular ducts lined by columnar epithelium. No ducts in the pancreas are striated.

Each exocrine acinus of the pancreas is composed of several serous cells surrounding a very small lumen (Figure 16–9). The acinar cells are highly polarized, with a spherical nucleus, and are typical protein-secreting cells (Figure 16–10). The number of zymogen granules present in each cell varies and is maximal in animals that have fasted.

The exocrine pancreas secretes 1.5 to 2 L of fluid per day. Pancreatic juice is rich in bicarbonate ions (HCO_3^-) and digestive enzymes, including several **proteases (trypsinogens, chymotrypsinogen, proelastases, protease E, kallikreinogen, procarboxipeptidases)**, **α-amylase**, **lipases**, and **nucleases (DNAase** and **RNAase)**. The proteases are stored as inactive zymogens in the secretory granules of acinar cells. After secretion trypsinogens are cleaved and activated by enterokinase only in the lumen of the small intestine, generating trypsins which activate the other proteases in a cascade. This, along with production of protease inhibitors by the acinar cells, prevents the pancreas from digesting itself.

MEDICAL APPLICATION

*In acute necrotizing **pancreatitis**, the proenzymes may be activated and digest pancreatic tissues, leading to very serious complications. Possible causes include infection, gallstones, alcoholism, drugs, and trauma.*

Pancreatic secretion is controlled mainly through two polypeptide hormones—**secretin** and **cholecystokinin (CCK)**—produced by enteroendocrine cells of the intestinal mucosa (duodenum and jejunum). The vagus (parasympathetic) nerve also stimulates pancreatic secretion and the autonomic system works in concert with the hormones to control pancreatic secretion.

Acid and partially digested food in the gastric **chyme** enters the duodenum and stimulates local release of CCK and secretin. CCK promotes the exocytosis of zymogens and enzymes from the pancreatic acinar cells. Secretin causes acinar and duct cells to add water and bicarbonate ions to the secreted proteins, producing an abundant alkaline fluid with its enzymes diluted but rich in electrolytic ions. This fluid neutralizes the acidic chyme and allows the pancreatic enzymes to function at their optimal

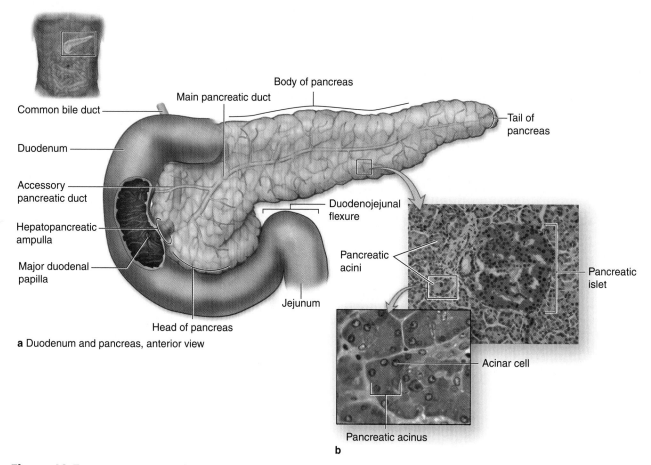

a Duodenum and pancreas, anterior view

b

Figure 16–7. Pancreas and duodenum. (a): The main regions of the pancreas are shown in relation to the two pancreatic ducts and the duodenum. **(b):** Micrographs show a pancreatic islet and several pancreatic acini. X75 and X200. H&E.

pH range. The coordinated action of both hormones provides for proper secretion of enzyme-rich, alkaline pancreatic juice.

MEDICAL APPLICATION

*In conditions of extreme malnutrition such as **kwashiorkor**, pancreatic acinar cells and other active protein-secreting cells atrophy and lose much of their rough ER, hindering production of digestive enzymes.*

LIVER

Except for the skin the liver is the body's biggest organ, weighing about 1.5 kg or about 2% of an adult's body weight. With a large right lobe and smaller left lobe, it is the largest gland and is situated in the abdominal cavity beneath the diaphragm (Figure 16–11). The liver is an interface between the digestive system and the blood: the organ in which nutrients absorbed in the digestive tract are processed for use by other parts of the body. Most blood in the liver (70–80%) comes from the portal vein arising from the stomach, intestines, and spleen; the rest (20–30%) is supplied by the hepatic artery. All the materials absorbed via the intestines reach the liver through the portal vein, except the complex lipids (chylomicrons), which are transported mainly by lymph vessels. The position of the liver in the circulatory system is optimal for gathering, transforming, and accumulating metabolites from blood and for neutralizing and eliminating toxic substances in blood. The elimination occurs in the **bile**, an exocrine secretion of the liver that is important for lipid digestion in the gut. The liver also produces plasma proteins such as albumin, fibrinogen, and various carrier proteins.

Stroma

The liver is covered by a thin fibrous capsule of connective tissue that becomes thicker at the hilum, where the portal vein and the hepatic artery enter the organ and where the right and left hepatic ducts and lymphatics exit. These vessels and ducts are

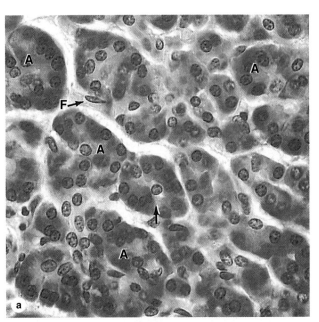

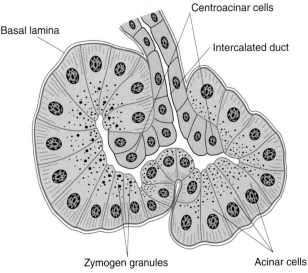

Figure 16–9. Pancreatic acini. (a): Micrograph of exocrine pancreas shows the serous, enzyme-producing cells arranged in small acini (A) with very small lumens. Acini are surrounded by small amounts of connective tissue with fibroblasts (F). Each acinus is drained by an intercalated duct with its initial cells, the centroacinar cells (arrow), inserted into the acinar lumen. X200. H&E. **(b):** The diagram shows the arrangement of cells more clearly. Under the influence of secretin, the centroacinar and other cells of these small ducts secrete a copious HCO_3^- - rich fluid that hydrates and alkalinizes the enzymatic secretions of the acinar cells. Pancreatic acini lack myoepithelial cells and their intercalated ducts lack striations.

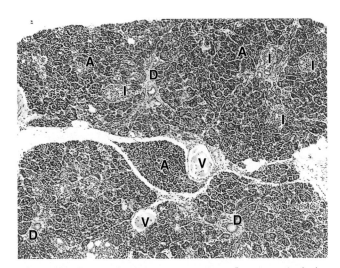

Figure 16–8. Pancreas. Low power view of pancreas includes several islets (I) surrounded by many serous acini (A). The larger interlobular ducts (D) are lined by simple columnar epithelium. The ducts and blood vessels (V) are located in connective tissue, which also provides a thin capsule to the entire gland and thin septa separating the lobules of secretory acini. X20. H&E.

surrounded by connective tissue all the way to their termination (or origin) in the portal spaces between the liver lobules. At this point,, a delicate reticular fiber network surrounds and supports the liver cells and the sinusoidal endothelial cells of the liver lobules (Figures 16–12 and 16–13).

Hepatic Lobules

Liver cells or **hepatocytes** (Gr. *hepar,* liver, + *kytos,* cell) are epithelial cells grouped in interconnected plates. Hepatocytes are arranged into thousands of small (~0.7 × 2 mm), polyhedral **hepatic lobules** which are the classic structural and functional units of the liver (Figure 16–11). Each lobule has three to six **portal areas** at its periphery and a venule called a **central vein** in its center (Figures 16–11, 16–12, and 16–13). The portal zones at the corners of the lobules consist of connective tissue

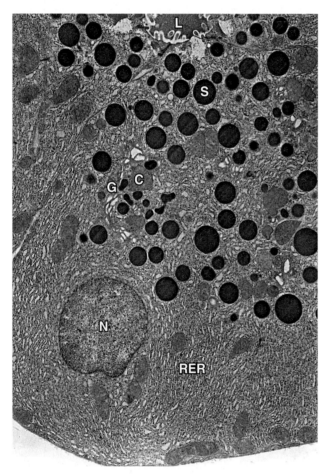

***Figure 16–10.* Pancreatic acinar cells.** TEM of a pancreatic acinar cell shows its pyramidal shape and the round, basal nucleus (N) surrounded by cytoplasm packed with cisternae of rough ER (RER). The Golgi apparatus (G) is situated at the apical side of the nucleus and is associated with condensing vacuoles (C) and numerous secretory (zymogen) granules (S). The small lumen (L) of the acinus contains proteins recently released from the cell by exocytosis. Exocytosis of digestive enzymes from secretory granules is promoted by CCK, released from the duodenum when food enters that region from the stomach. X8000.

in which are embedded a venule (a branch of the portal vein), an arteriole (a branch of the hepatic artery), and a duct of cuboidal epithelium (a branch of the bile duct system) – three structures called the **portal triad** (Figure 16–12). The venule contains blood from the superior and inferior mesenteric and splenic veins. The arteriole contains blood from the celiac trunk of the abdominal aorta. The duct carries bile synthesized by the parenchymal cells (hepatocytes) and eventually empties into the hepatic duct. Portal areas also have nerve fibers and lymphatics. In some animals (eg, pigs), the lobules are separated from each other by a layer of connective tissue, making them easy to distinguish. In humans the lobules are in close contact along most of their length and it is more difficult to establish the exact limits between different lobules (Figure 16–12).

Hepatocytes make up each of the interconnected plates like the bricks of a wall and the plates are arranged radially around the central vein (Figure 16–11). From the periphery of the lobule to its center, the plates of hepatocytes branch and anastomose freely, forming a rather sponge-like structure (Figure 16–12). The spaces between these plates contain important microvascular components, the **liver sinusoids** (Figures 16–11, 16–12, and 16–13). These irregularly dilated sinusoids consist only of a discontinuous layer of fenestrated endothelial cells (Figures 16–14 and 16–15). The endothelial cells are separated from the underlying hepatocytes by a thin, discontinuous basal lamina and a very narrow **perisinusoidal space** (the space of Disse), into which project microvilli of the hepatocytes for exchanges between these cells and plasma (Figure 16–14). This exchange is the key to liver function, not only because of the large number of macromolecules (eg, lipoproteins, albumin, and fibrinogen) secreted into the blood by hepatocytes but also because the liver takes up and catabolizes many of these large molecules.

Liver sinusoids are surrounded and supported by delicate sheaths of reticular fibers (Figure 16–13). Two noteworthy cells are associated with these sinusoids in addition to the endothelial cells:

- Abundant specialized **stellate macrophages**, also known as **Kupffer cells**, are found between sinusoidal endothelial cells and on the luminal surface within the sinusoids, mainly near the portal areas (Figure 16–14). Their main functions are to break down aged erythrocytes and free heme for re-use, remove bacteria or debris that may enter the portal blood from the gut, and act as antigen-presenting cells in adaptive immunity.
- In the perisinusoidal space (not the lumen) are stellate **fat storing cells** (or **Ito cells**) with small lipid droplets containing vitamin A (Figure 16–14). These cells, which make up about 8% of the cells in a liver but are difficult to see in routine preparations, store much of the body's vitamin A, produce ECM components, and have a regulatory role in local immunity.

MEDICAL APPLICATION

*In chronically diseased liver, Ito cells proliferate and acquire the features of myofibroblasts, with or without the lipid droplets. Under these conditions, these cells are found close to the damaged hepatocytes and play a major role in the development of fibrosis, including the fibrosis secondary to **alcoholic liver disease**.*

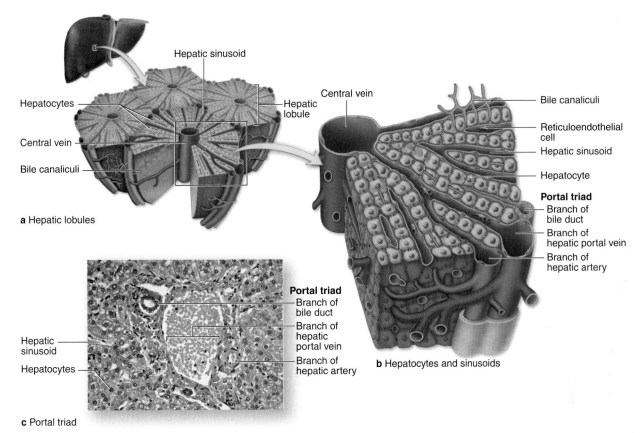

a Hepatic lobules

c Portal triad

b Hepatocytes and sinusoids

Figure 16–11. Liver. A large organ in the upper right quadrant of the abdomen, immediately below the diaphragm, the liver is composed of thousands of polygonal structures called hepatic lobules, which are the basic functional units of the organ. **(a):** Diagram shows a small central vein projecting through the center of each hepatic lobule and several sets of blood vessels defining the periphery. The peripheral vessels are grouped primarily in connective tissue comprising the portal tracts, which usually include a branch of the portal vein and a branch of the hepatic artery, as well as a branch of the bile duct. These comprise the portal triad. **(b):** Both blood vessels to each lobule give off sinusoids, which run between plates of hepatocytes and drain into the central vein. **(c):** Micrograph showing components of the portal triad. X220. H&E.

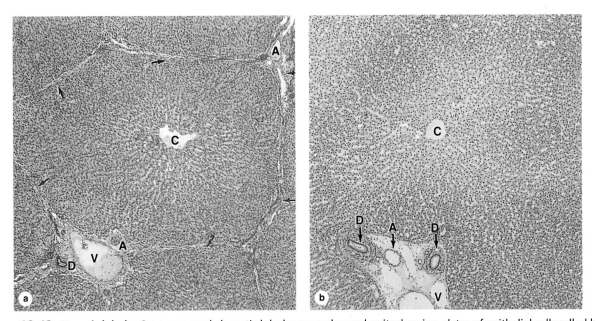

Figure 16–12. Hepatic lobule. Cut transversely hepatic lobules are polygonal units showing plates of epithelial cells called hepatocytes radiating from a central venule (C). **(a):** Hepatic lobule of some mammals, such as the pig, are delimited on all sides by connective tissue. **(b):** Hepatic units of humans have much less connective tissue and their boundaries are more difficult to distinguish. In all cases peripheral connective tissue of portal areas with microvasculature and small bile duct (D) branches can be seen and in humans as in other mammals these are present at the boundaries between two or more hepatic lobules. The vessels near the bile ducts branches are a venule (V) off the portal vein and an arteriole (A) off the hepatic artery. Both X150. H&E.

Blood Supply

As an important interface for processing blood from the digestive system, the liver gets most of its blood from the **portal vein**, which carries nutrient-rich but oxygen-poor blood from the abdominal viscera. Oxygenated blood is brought in with the smaller portion derived from the **hepatic artery** (Figure 16–11).

The portal system conveys blood from the pancreas, spleen, and the intestines. Nutrients are accumulated and transformed in the liver and toxic substances are neutralized and eliminated there. In the liver the portal vein branches repeatedly and sends small **portal venules** to the portal spaces. The portal venules branch into smaller distributing venules that run around the periphery of each lobule and lead into the sinusoids. The sinusoids run radially, converging in the center of the lobule to form the **central** or **centrolobular vein** (Figures 16–11, 16-12, and 16–13). This vessel, like the sinusoids, has very thin walls

consisting only of endothelial cells supported by a sparse population of collagen fibers (Figure 16–13). Central venules from each lobule converge into veins, which eventually form two or more large **hepatic veins** that empty into the inferior vena cava.

The hepatic artery branches repeatedly and forms arterioles in the portal areas, some of which lead directly into the sinusoids at various distances from the portal spaces, thus adding oxygen-rich arterial blood to the portal venous blood in the sinusoids.

Blood always flows from the periphery to the center of each hepatic lobule. Consequently, oxygen and metabolites, as well as all other toxic or nontoxic substances absorbed in the intestines, reach the lobule's peripheral cells first and then the more central cells. This direction of blood flow partly explains why the properties and function of the periportal hepatocytes differ from that of the centrolobular cells. Hepatocytes near the portal areas can rely on aerobic metabolism and are often more active in protein synthesis, while the more central cells are exposed to lower

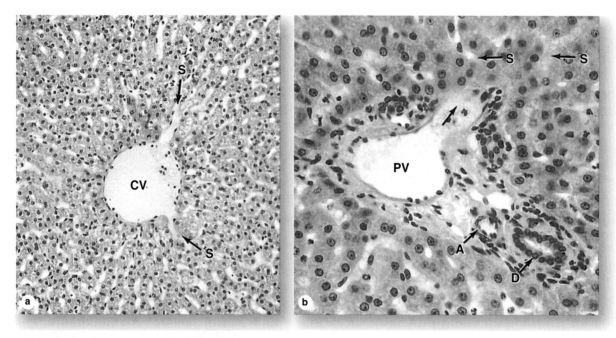

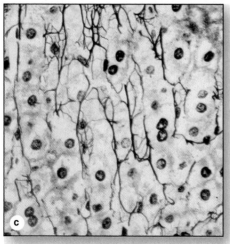

Figure 16–13. **Hepatic lobule microvasculature. (a):** The hepatic lobule's central vein (CV) is actually a venule consisting of little more than an endothelial tube with smaller sinusoids (S) coming in from all directions. X200. H&E. **(b):** Peripheral portal areas contain more connective tissue and are the sites of the portal triad: a portal venule (PV), an arteriole (A) branching off the hepatic artery, and one or two branches of the bile duct (D). Blood flows (arrow) from these arterioles and venules into the sinusoids between the plates of hepatocytes that run to the central venule. In the portal area the bile ductules are lined by simple cuboidal epithelium. X400. H&E. **(c):** Reticulin (collagen type III) fibers running along the plates of hepatocytes are the major support for the sinusoids and central venules. Most connective tissue in the liver is found in the septa and portal tracts. X400. Silver.

concentrations of nutrients and oxygen and are more involved with detoxification and glycogen metabolism.

The Hepatocyte

Hepatocytes are large polyhedral cells, with six or more surfaces, and typical diameters of 20–30 μm. In H&E-stained sections their cytoplasm is usually eosinophilic because of the large number of mitochondria, up to 2000 per cell. Hepatocytes have large spherical nuclei with nucleoli. The cells frequently have two or more nuclei and about 50% of them are polyploid, with two, four, eight or more times the normal diploid chromosome number. Polyploid nuclei are characterized by greater size, which is proportional to their ploidy.

The surface of each hepatocyte is in contact with the wall of a sinusoid, through the perisinusoidal space, and with the surfaces of other hepatocytes (Figure 16–16). Where two hepatocytes abut, they delimit a tubular space between them known as the **bile canaliculus** (Figure 16–16).

The canaliculi, the first portions of the bile duct system, are long spaces 1–2 μm in diameter. They are limited only by the plasma membranes of two hepatocytes, which extend a small number of microvilli into their interiors (Figure 16–16). The cell membranes near these canaliculi are firmly joined by tight junctions. Gap junctions also occur between hepatocytes, allowing intercellular communication and coordination of the cells'

activities. The bile canaliculi form a complex anastomosing network progressing along the plates of the hepatic lobule and terminating in the region of the portal spaces (Figure 16–11). The bile flow therefore progresses in a direction opposite to that of the blood, ie, from the center of the lobule to its periphery. Near the peripheral portal areas, bile canaliculi empty into **bile ductules** (Figure 16–17) composed of cuboidal epithelial cells called **cholangiocytes** (Figures 16–18 and 16–19). After a short distance, these ductules cross the limiting hepatocytes of the lobule and end in the **bile ducts** in the portal spaces. Bile ducts are lined by cuboidal or columnar epithelium and have a distinct connective tissue sheath. They gradually enlarge and fuse, forming right and left **hepatic ducts,** which subsequently leave the liver.

Functionally the hepatocyte may be the most versatile cell in the body. The hepatocyte has an abundant endoplasmic reticulum—both smooth and rough (Figure 16–18). Rough ER for synthesis of plasma proteins causes cytoplasmic basophilia, which is often more pronounced in hepatocytes near the portal areas (Figure 16–12). Various important processes take place in the smooth ER, which is distributed diffusely throughout the cytoplasm. This organelle is responsible for the processes of oxidation, methylation, and conjugation required for inactivation or detoxification of various substances before their excretion. The smooth ER is a labile system that reacts promptly to the molecules received by the hepatocyte.

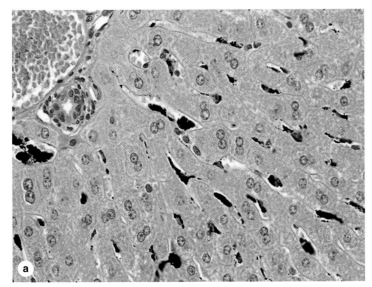
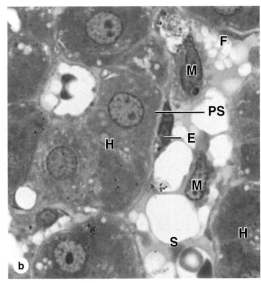

Figure 16–14. **Hepatic sinusoids.** Nutrient-rich blood from the portal vein and oxygen-rich blood from the arteriole mix in the sinusoids running between the plates of hepatocytes from the portal areas to the central venules. Molecules in the blood are processed mainly by the hepatocytes, but other cells in or near the sinusoids are also important. Specialized stellate macrophages, often called Kupffer cells, are bound to the endothelial lumen of the sinusoids, where they detect and phagocytose effete erythrocytes. **(a):** Stellate macrophages are seen as black cells in a liver lobule from a rat injected with particulate India ink. X200. H&E.

(b): In this plastic section, stellate macrophages (M) are seen in the sinusoid (S) between two groups of hepatocytes (H). They are larger than the flattened endothelial cells (E). Between the endothelium and the hepatocytes is a thin space called the perisinusoidal space (PS), in which are located fibroblastic fat-storing cells (F), or Ito cells, that maintain the very sparse ECM of this compartment and are also specialized for vitamin A storage in small lipid droplets. These cells are numerous but are difficult to demonstrate in routine histological preparations. If their lipid droplets become very large or abundant, the fat-storing cells resemble adipocytes. X750. PT.

One of the main processes that occurs in the smooth ER is the conjugation of hydrophobic (water-insoluble) toxic bilirubin by glucuronyl-transferase to form a water-soluble nontoxic bilirubin glucuronide. This conjugate is excreted by hepatocytes into the bile. When bilirubin or bilirubin glucuronide is not excreted properly, various diseases characterized by **jaundice** *can result.*

*One of the frequent causes of jaundice in newborns is the often underdeveloped state of the smooth ER in their hepatocytes (**neonatal hyperbilirubinemia**). The current treatment for these cases is exposure to blue light from ordinary fluorescent tubes, which transforms unconjugated bilirubin into a water-soluble photoisomer that can be excreted by the kidneys.*

The hepatocyte frequently contains deposits of glycogen, which appear ultrastructurally as coarse, electron-dense granules in the cytosol, often close to the smooth ER (Figure 16–16). Liver glycogen is a depot for glucose and is mobilized if the blood glucose level falls below normal. In this way, hepatocytes maintain

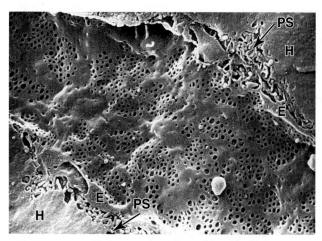

Figure 16–15. Ultrastructure of the sinusoid wall. SEM of the luminal surface of the endothelium lining a sinusoid in liver shows grouped fenestrations. At the border are seen cut edges of the endothelial cell (E) in this discontinuous sinusoid and hepatocytes (H). Between these two cells is the thin perisinusoidal space (PS), into which project microvilli from the hepatocytes surface. Blood plasma passes freely through the fenestrations into the perisinusoidal space, where the voluminous membrane of hepatocytes acts to remove many high and low molecular weight blood components and nutrients for storage and processing. Proteins synthesized and secreted from hepatocytes, such as albumin, fibrinogen, and other blood proteins, are released into the perisinusoidal space. X6500. (With permission, from Eddie Wisse, Electron Microscopy Unit, Department of Pathology, University of Maastricht, The Netherlands.)

a steady level of blood glucose, one of the main sources of energy for the body. Hepatocytes also normally store triglycerides in small lipid droplets. This capacity to store metabolites is important, because it supplies the body with energy between meals.

Hepatocytes do not usually store proteins in secretory granules but continuously release them into the bloodstream. About 5% of the protein exported by the liver is produced by the sinusoidal stellate macrophages.

The hepatocyte is responsible for converting lipids and amino acids into glucose by means of a complex enzymatic process called **gluconeogenesis** (Gr. *glykys*, sweet, + *neos*, new, + *genesis*, production). It is also the main site of amino acid deamination, resulting in the production of urea which is transported in blood to the kidney and is excreted there.

Hepatocyte lysosomes are important in the turnover and degradation of intracellular organelles. Peroxisomes are also abundant and are important for oxidation of excess fatty acids, breakdown of the hydrogen peroxide generated by this oxidation (by means of catalase activity), breakdown of excess purines to uric acid, and participation in the synthesis of cholesterol, bile acids, and some lipids used by neurons to make myelin. Each hepatocyte may contain up to 50 Golgi complexes involved in the formation of lysosomes and the secretion of proteins, glycoproteins, and lipoproteins into plasma.

A variety of rare inherited disorders of peroxisome function occur in humans, most involving mutations of the enzymes found within peroxisomes. For example, X-linked adrenoleukodystrophy (X-ALD) results from a failure to metabolize fatty acids properly, resulting in the deterioration of the myelin sheaths of neurons.

Bile secretion is an exocrine function in the sense that hepatocytes are involved in the uptake, transformation, and excretion of blood components into the bile canaliculi. Bile has several essential components in addition to water and electrolytes: bile acids (low molecular weight organic acids such as cholic acid, and their deprotonated forms called bile salts), phospholipids, cholesterol, and heme-containing bile pigments such as yellowish-green bilirubin. The secretion of bile acids is illustrated in Figure 16–18. Bile acids have an important function in emulsifying the lipids in the digestive tract, promoting easier digestion by lipases and subsequent absorption.

*Abnormal proportions of bile acids may lead to the formation of **gallstones** (cholelithiasis). Gallstones can block bile flow and cause jaundice— the presence of bile pigments such as bilirubin in blood—from the rupture of tight junctions around the bile canaliculi. The presence of bilirubin in* *blood leads to the temporary yellowing of the skin and sclera of the eyes characteristic of jaundice.*

Most of the bile pigment bilirubin is derived from the degradation of the hemoglobin in senescent erythrocytes, which occurs mainly in macrophages of the spleen, but also in other macrophages including those in the liver sinusoids. Released from macrophages bilirubin binds albumen, circulates, and is taken up from plasma by hepatocytes. In the smooth ER hydrophobic bilirubin is conjugated to glucuronate, forming water-soluble bilirubin glucuronide, which is then secreted into the bile canaliculi. Released into the gut with bile, some bilirubin is metabolized by bacteria to other pigments which give feces its characteristic color. Other bilirubin is absorbed in the gut and removed from blood in the kidneys, giving urine its yellow color.

Various drugs and potentially toxic substances can be inactivated by oxidation, methylation, or conjugation. The enzymes participating in these processes are located mainly in the smooth ER of hepatocytes. Glucuronyltransferase, an enzyme that conjugates glucuronate to bilirubin, also causes conjugation of several other compounds such as steroids, barbiturates, antihistamines, and anticonvulsants. Under certain conditions, drugs that are inactivated in the liver can induce an increase in the hepatocyte's smooth ER, thus improving the detoxification capacity of the organ.

MEDICAL APPLICATION

The administration of barbiturates to laboratory animals results in rapid development of smooth ER in hepatocytes. Barbiturates can also increase synthesis of glucuronyltransferase. This finding has led to the use of barbiturates in the treatment of glucuronyltransferase deficiencies.

HEPATIC LOBULE STRUCTURE & FUNCTION

The different categories of hepatocyte functions—including secretion of protein factors into blood, the secretion of bile components, and the removal of oxygen and small compounds of all kinds from blood—has led to three ways to think of liver lobule structure, which are summarized in Figure 16–19. The **classic hepatic lobule**, with blood flowing past hepatocytes from up to six portal triad areas to a central venule, emphasizes the endocrine function of the structure producing factors for uptake by plasma. The concept of **portal lobules** of hepatocytes is more useful

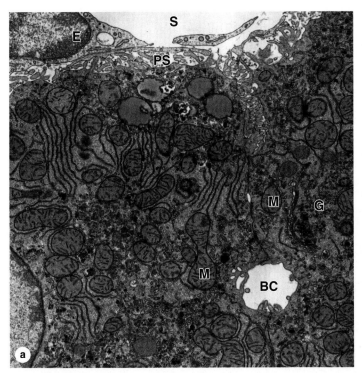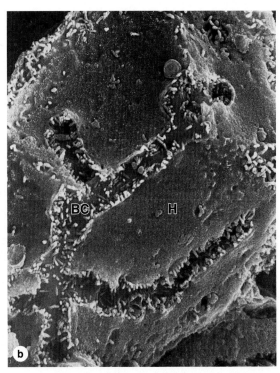

Figure 16–16. Ultrastructure of hepatocytes and bile canaliculi. (a): TEM of hepatocytes show small bile canaliculi (BC) between two cells, with junctional complexes binding the cells firmly and tightly at these sites. The bile canaliculus is the site of exocrine secretion by hepatocytes. The two adjoining hepatocytes extend short microvilli and secrete bile components into this space. The hepatocytes have many mitochondria (M), small electron-dense glycogen granules, and Golgi complexes (G) and extend more numerous microvilli into the perisinusoidal space (PS), which is the site where hepatocytes remove and add components in plasma. The endothelial cell (E) lining the sinusoid (S) is also seen. X9500. **(b):** SEM of hepatocytes (H) broken apart from one another reveals the length of a bile canaliculus (BC) along the cell's surface. Such canaliculi run between the cells of the hepatocyte plates in the hepatic lobules and carry bile toward the portal areas where the canaliculi join cuboidal bile ductules. (Figure 16–16a, with permission, from Douglas L. Schmucker, Department of Anatomy, University of California, San Francisco.)

when considering the exocrine function of these cells, ie, bile secretion. The portal area has the bile ductule at the center and bile, moving in the opposite direction as the blood, flows toward it from all the surrounding hepatocytes. The tissue draining bile into each portal area duct is roughly triangular in shape, with the central veins of three classic lobules at its angles.

The **liver acinus**, a third way of viewing liver cells, emphasizes the nature of the blood supply to the hepatocytes and the oxygen gradient from the hepatic artery branch to the central vein. The acinus contains the hepatocytes in an irregular oval or diamond-shaped area extending from two portal triads to the two closest central veins (Figure 16–19). Hepatocytes nearest the hepatic arteriole, zone I in this concept, get the most oxygen and nutrients and can most readily carry out functions requiring oxidative metabolism such as protein synthesis. Hepatocytes in zone III, near the central vein, get the least oxygen and nutrients. They are the preferential sites of glycolysis, lipid formation, and drug biotransformations and are the first hepatocytes to undergo fatty accumulation and ischemic necrosis. In the intervening zone II, hepatocytes have an intermediate range of metabolic functions between those in zones I and III. The major activities in any given hepatocyte result from the cell adapting to the microenvironment produced by the contents of the blood to which it is exposed.

Liver Regeneration

Unlike the salivary glands and pancreas, the liver has a strong capacity for regeneration despite its slow rate of cell renewal. The loss of hepatic tissue from the action of toxic substances triggers a mechanism by which the remaining healthy hepatocytes begin to divide, in a process of **compensatory hyperplasia**, continuing until the original mass of tissue is restored. Surgical removal of a liver portion produces a similar response in the hepatocytes of the remaining lobe(s). The regenerated liver tissue is usually well organized, with the typical lobular arrangement, and replaces the functions of the destroyed tissue. In humans this capacity is important, because

one liver lobe can often be donated by a living relative for surgical transplantation and full liver function restored in both donor and recipient.

Besides proliferation of existing hepatocytes, a role for liver stem cells in regeneration has been shown in some experimental models. Stem cells called oval cells are present in the initial epithelium of bile ductules near the portal areas and these can give rise to both hepatocytes and cholangiocytes.

MEDICAL APPLICATION

When there is continuous or repeated damage to hepatocytes over a long period of time, the multiplication of liver cells is followed by a pronounced increase in the amount of connective tissue. Instead of normal liver tissue there is the formation of

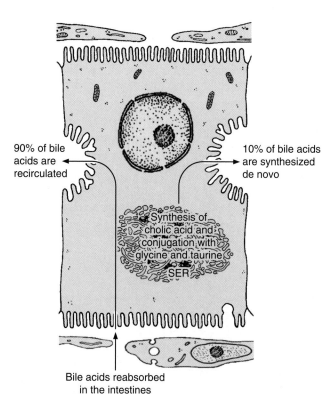

90% of bile acids are recirculated

10% of bile acids are synthesized de novo

Synthesis of cholic acid and conjugation with glycine and taurine SER

Bile acids reabsorbed in the intestines

Figure 16–18. **Secretion of bile acids.** Bile acids are organic molecules, primarily cholic acid and its derivatives, found in bile excreted from hepatocytes into the system of bile canaliculi and ducts. These compounds are often deprotonated, becoming the bile salts. Stored in the gall bladder and released into the duodenum after a meal, bile acids emulsify fats, facilitating lipid breakdown and absorption. About 90% of bile acids are absorbed from the intestinal epithelium and are transported by the hepatocyte from the blood to bile canaliculi (enterohepatic recirculation). About 10% of bile acids are synthesized in the smooth ER of hepatocytes by conjugation of cholic acid (synthesized in hepatocytes from cholesterol) with the amino acid glycine or taurine, producing glycocholic and taurocholic acids.

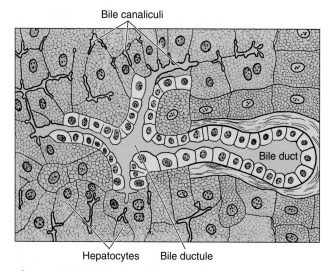

Bile canaliculi

Bile duct

Hepatocytes   Bile ductule

Figure 16–17. **Bile ductules.** Near the periphery of hepatic lobules, the bile canaliculi join with bile ductules lined by cuboidal epithelial cells called cholangiocytes. The ductules soon merge with branches of the bile ducts in the portal spaces. The branches merge from all over the liver and give rise to the left and right hepatic ducts which leave the organ at the hilum.

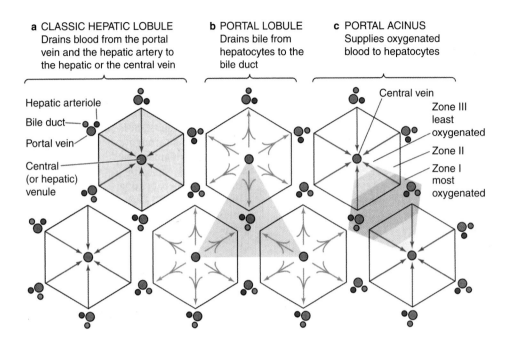

a CLASSIC HEPATIC LOBULE
Drains blood from the portal vein and the hepatic artery to the hepatic or the central vein

b PORTAL LOBULE
Drains bile from hepatocytes to the bile duct

c PORTAL ACINUS
Supplies oxygenated blood to hepatocytes

Hepatic arteriole
Bile duct
Portal vein
Central (or hepatic) venule

Central vein
Zone III least oxygenated
Zone II
Zone I most oxygenated

Figure 16–19. Concepts of structure-function relationships in liver. Studies of liver microanatomy, physiology, and pathology have given rise to three related ways to view the liver's organization which emphasize different aspects of hepatocyte activity. **(a):** The **classic lobule** concept offers a basic understanding of the structure-function relationship in liver organization and emphasizes the endocrine function of hepatocytes as blood flows past them toward the central vein. **(b):** The **portal lobule** emphasizes the hepatocytes' exocrine function and the flow of bile from regions of three classic lobules toward the bile duct in the portal triad at the center here. The area drained by each bile duct is roughly triangular.

(c): The **liver acinus** concept emphasizes the different oxygen and nutrient contents of blood at different distances along the sinusoids, with blood from each portal area supplying cells in two or more classic lobules. Each hepatocyte's major activity is determined by its location along the oxygen/nutrient gradient: periportal cells of zone I get the most oxygen and nutrients and show metabolic activity generally different from the pericentral hepatocytes of zone III, exposed to the lowest oxygen and nutrient concentrations. Many pathological changes in liver are best understood from the point-of-view of liver acini. (Reproduced, with permission from Boron WF, Boulpaep EL: *Medical Physiology: A Cellular and Molecular Approach*, Saunders. Copyright Elsevier, 2005.)

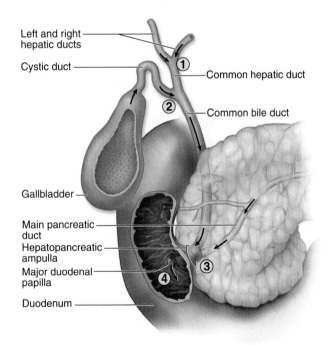

Left and right hepatic ducts
Cystic duct
① Common hepatic duct
② Common bile duct
Gallbladder
Main pancreatic duct
Hepatopancreatic ampulla
Major duodenal papilla
Duodenum
③
④

Figure 16–20. Biliary tract and gall bladder. Bile leaves the liver in the left and right hepatic ducts, which merge (1) to form the common hepatic duct, which connects to the cystic duct serving the gall bladder. The latter two ducts merge (2) to form a common bile duct. The main pancreatic duct merges with the common bile duct at the hepatopancreatic ampulla (3) which enters the wall of the duodenum. Bile and pancreatic juices together are secreted from the major duodenal papilla (of Vater) into the duodenal lumen (4). All these ducts carrying bile are lined by cuboidal or low columnar cells called cholangiocytes, similar to those of the small bile ductules in the liver.

*various sized nodules composed of a central mass of disorganized hepatocytes surrounded by a great amount of connective tissue. This disorder, called **cirrhosis**, is a progressive and irreversible process, causes liver failure, and is usually fatal. The fibrosis is diffuse, affecting the entire liver.*

Cirrhosis is a consequence of any sustained injury to hepatocytes produced by various agents, such as ethanol, drugs or other chemicals, hepatitis virus (mainly types B, C, or D), parasites, and autoimmune liver disease.

Alcohol-induced liver damage is responsible for most cases of cirrhosis, because ethanol is metabolized primarily in the liver. Ethanol also alters hepatic regeneration through an unknown mechanism, favoring the development of cirrhosis.

BILIARY TRACT & GALLBLADDER

The bile produced by the hepatocytes flows through the **bile canaliculi, bile ductules,** and **bile ducts.** These structures gradually merge, forming a network that converges to form the **hepatic duct.** The hepatic duct, after receiving the **cystic duct** from the gallbladder, continues to the duodenum as the **common bile duct** (Figure 16–20).

The hepatic, cystic, and common bile ducts are lined with a mucous membrane having a simple columnar epithelium of cholangiocytes. The lamina propria and submucosa are relatively thin, with mucous glands in some areas of the cystic duct, and surrounded by a thin muscularis. This muscle layer becomes thicker near the duodenum and finally, in the portion within the duodenal wall, forms a sphincter that regulates bile flow.

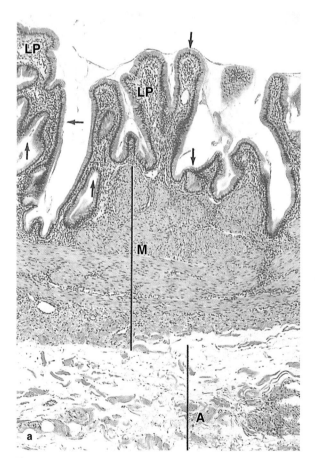
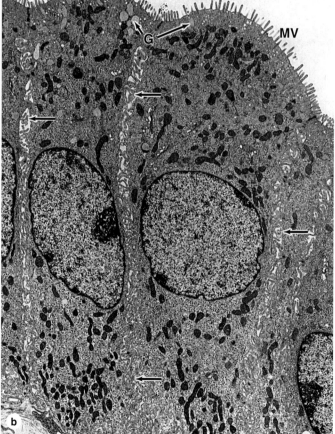

***Figure 16–21.* Gallbladder.** The gallbladder is a saclike structure that stores and concentrates bile, and releases it into the duodenum after a meal. **(a):** Its wall consists largely of a highly folded mucosa, with a simple columnar epithelium (arrows) overlying a typical lamina propria (LP); a muscularis (M) with bundles of muscle fibers oriented in all directions to facilitate emptying of the organ; an external adventitia (A) where it is against the liver and a serosa where it is exposed. X60. H&E. **(b):** TEM of the epithelium shows cells specialized for water uptake across apical microvilli (MV) and release into the intercellular spaces (arrows) along the folded basolateral cell membranes. Abundant mitochondria provide the energy for this pumping process. Scattered apical secretory granules (G) contain mucus. X5600.

The gallbladder is a hollow, pear-shaped organ (Figure 16-20) attached to the lower surface of the liver, capable of storing 30–50 mL of bile. The wall of the gallbladder consists of a mucosa composed of simple columnar epithelium and lamina propria, a thin muscularis with bundles of muscle fibers oriented in several directions, and an external adventitia or serosa (Figure 16–21). The mucosa has abundant folds that are particularly evident when the gallbladder is empty.

The lining epithelial cells have prominent mitochondria, microvilli, and intercellular spaces, all indicative of active absorptive cells (Figure 16–21). The main function of the gallbladder is to store bile, concentrate it by absorbing its water, and release it when necessary into the digestive tract. This process depends on an active sodium-transporting mechanism in the gallbladder's epithelium, with water absorption from bile an osmotic consequence of the sodium pump. Contraction of the smooth muscle of the gallbladder is induced by cholecystokinin (CCK) released from enteroendocrine cells of the small intestine. Release of CCK is, in turn, stimulated by the presence of dietary fats in the small intestine. Removal of the gallbladder due to obstruction or chronic inflammation leads to direct flow of bile from liver to gut, with few major consequences on digestion.

The Respiratory System

RESPIRATORY EPITHELIUM
NASAL CAVITIES
   Smell (Olfaction)
SINUSES & NASOPHARYNX
LARYNX
TRACHEA
BRONCHIAL TREE & LUNG
   Bronchi

Bronchioles
Respiratory Bronchioles
Alveolar Ducts
Alveoli
Regeneration in the Alveolar Lining
PULMONARY VASCULATURE & NERVES
PLEURA
RESPIRATORY MOVEMENTS

The respiratory system includes the **lungs** and a branching system of tubes that link the sites of gas exchange with the external environment. Air is moved through the lungs by a ventilating mechanism, consisting of the thoracic cage, intercostal muscles, diaphragm, and elastic components of the lung tissue. The respiratory system is divided anatomically into structures of the upper and lower respiratory tracts (Figure 17–1). Functionally, these structures make up the system's **conducting portion**, which consists of the nasal cavities, nasopharynx, larynx, trachea, bronchi (Gr. *bronchos*, windpipe), bronchioles, and terminal bronchioles; and a **respiratory portion** (where gas exchange takes place), consisting of respiratory bronchioles, alveolar ducts, and alveoli. **Alveoli** are saclike structures that make up the greater part of the lungs. They are the main sites for the principal function of the lungs—the exchange of O_2 and CO_2 between inspired air and blood.

The conducting portion serves two main functions: to provide a conduit through which air moves to and from the lungs and to condition the inspired air. To ensure an uninterrupted supply of air, a combination of cartilage, elastic and collagen fibers, and smooth muscle provides the conducting portion with rigid structural support and the necessary flexibility and extensibility.

RESPIRATORY EPITHELIUM

Most of the conducting portion is lined with ciliated pseudostratified columnar epithelium known as **respiratory epithelium** (Figure 17–2). This epithelium has at least five cell types, all of which touch the thick basement membrane:

- **Ciliated columnar cells** (described in Chapter 2) are the most abundant, each with about 300 cilia on its apical surface (Figure 17–2).

MEDICAL APPLICATION

*Immotile cilia syndrome, a disorder that causes infertility in men and chronic respiratory tract infections in both sexes, is caused by immobility of cilia and flagella induced, in some cases, by deficiency of **dynein**, a protein normally present in the cilia. Dynein participates in the ciliary movement (see Chapter 2).*

- **Goblet cells** are also abundant in some areas of the respiratory epithelium (Figure 17–2), filled in their apical portions with granules of mucin glycoproteins.
- **Brush cells** are a much more sparsely scattered and less easily found, columnar cell type, which has a small apical surface bearing a tuft of many short, blunt microvilli (Figure 17–2c). Brush cells express some signal transduction components like those of gustatory cells and have afferent nerve endings on their basal surfaces and are considered to be chemosensory receptors.
- **Small granule cells** are also difficult to distinguish in routine preparations, but possess numerous dense core granules 100–300 nm in diameter. Like brush cells, they represent about 3% of the total cells and are part of the diffuse neuroendocrine system (Chapter 20).
- **Basal cells**, small rounded cells on the basement membrane and not extending to the luminal surface, are stem cells that give rise to the other cell types.

MEDICAL APPLICATION

From the nasal cavities through the larynx, portions of the epithelial lining are stratified squamous. This type of epithelium is evident in regions exposed to direct airflow or physical abrasion (eg, oropharynx, epiglottis, vocal folds); it provides more protection from wear and abrasion than does respiratory epithelium. In smokers the proportion of ciliated cells to goblet cells is altered to aid in clearing the increased particulate and gaseous pollutants (eg, CO, SO$_2$). Although the greater numbers of goblet cells in a smoker's epithelium provide for a more rapid clearance of pollutants, the reduction in ciliated cells caused by excessive intake of CO results in decreased movement of the mucus layer and frequently leads to congestion of the smaller airways.

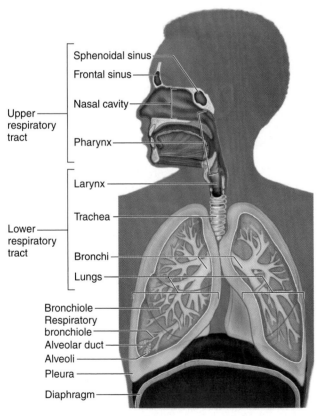

Figure 17–1. Anatomy of the respiratory system. Anatomically, the respiratory tract has upper and lower parts. Histologically and functionally, the respiratory system has a **conducting portion**, which consists of all the components that condition air and bring it into the lungs, and a **respiratory portion**, where gas exchange actually occurs, consisting of respiratory bronchioles, alveolar ducts, and alveoli in the lungs. Portions of two sets of paranasal sinuses are shown here.

NASAL CAVITIES

The left and right nasal cavity each has two components: the external **vestibule** and the internal **nasal cavities (or fossae).** The vestibule is the most anterior and dilated portion of each nasal cavity. Skin of the nose enters the **nares** (nostrils) partway up the vestibule and has sweat glands, sebaceous glands, and short coarse **vibrissae** (hairs) that filter out particulate material from the inspired air. Within the vestibule, the epithelium loses its keratinized nature and undergoes a transition into typical respiratory epithelium before entering the nasal fossae.

The nasal cavities lie within the skull as two cavernous chambers separated by the osseous **nasal septum.** Extending from each lateral wall are three bony shelflike projections (Figure 17–1) called **conchae.** The middle and inferior conchae are covered with respiratory epithelium; the superior conchae are covered with a specialized **olfactory epithelium.** The narrow passages between the conchae improve the conditioning of the inspired air by increasing the surface area of moist, warm respiratory epithelium and by slowing and increasing turbulence in the airflow. The result is increased contact between air streams and the mucous layer. Within the lamina propria of the conchae are large venous plexuses known as **swell bodies.** Every 20–30 minutes, the swell bodies on one side become temporarily engorged with blood, resulting in distension of the conchal mucosa and a concomitant decrease in the flow of air. During this time, most of the air is directed through the other nasal fossa, allowing the engorged respiratory mucosa to recover from dehydration.

MEDICAL APPLICATION

Allergic reactions and inflammation can cause abnormal engorgement of swell bodies in both fossae, severely restricting the air flow. The abundance of thin-walled venules in the lining of the nasal cavities and their proximity to the epithelial surface explains why nosebleeds occur so commonly.

In addition to swell bodies, the nasal cavities' mucosa has a rich vascular system with a complex organization. Large vessels form a close-meshed latticework next to the underlying periosteum, from which arcading branches lead toward the surface. Blood in these vessels flows from the rear of the cavities in a direction opposite to that of the inspired air, transferring heat to warm it quickly.

A major function of the entire conducting portion is to condition inspired air by cleaning, moistening, and warming it before it enters the lungs. In addition to the moist vibrissae, the rich vasculature in the lamina propria, and the ciliated and mucus-secreting cells of respiratory epithelium, conditioning also involves numerous mucous and serous glands in the mucosa. Once the air reaches the nasal fossae, particulate and gaseous impurities are trapped in a layer of mucus. This mucus, in conjunction with serous secretions, also serves to moisten the incoming air, protecting the delicate alveoli of the lungs from desiccation.

Smell (Olfaction)

The olfactory chemoreceptors are located in the **olfactory epithelium**, a specialized region of the mucous membrane covering the superior conchae at the roof of the nasal cavity. In humans, it is about 10 cm^2 in area and up to 100 μm in thickness. It is a pseudostratified columnar epithelium composed of three types of cells (Figure 17–3):

- **Basal cells** are small, spherical or cone-shaped and form a layer at the basal lamina. They are the stem cells for the other two types.
- **Supporting cells** are columnar, with broad, cylindrical apexes and narrower bases. On their free surface are microvilli submerged in a fluid layer. Well-developed junctional complexes bind the supporting cells to the adjacent olfactory cells. The supportive role of these cells is not well-understood, but they express abundant ion channels whose function appears to be required to maintain a microenvironment conducive to olfactory function and survival.

- **Olfactory neurons** are bipolar neurons present throughout this epithelium. They are distinguished from supporting cells by the position of their nuclei, which lie between those of the supporting cells and the basal cells. The dendrite end of each olfactory neuron is the apical (luminal) pole of the cell and has a knoblike swelling with about a dozen basal bodies. From the basal bodies emerge long nonmotile cilia with defective axonemes but a considerable surface area for membrane chemoreceptors. These receptors respond to odoriferous substances by generating an action potential along the (basal) axons of these neurons, which leave the epithelium and unite in the lamina propria as very small nerves which then pass through foramina in the cribriform plate of the ethmoid bone to the brain (Figure 17–3). There they form cranial nerve I, the olfactory nerve, and eventually synapse with other neurons in the olfactory bulb.

The lamina propria of the olfactory epithelium possesses large serous glands (glands of Bowman), which produce a flow of fluid surrounding the olfactory cilia and facilitating the access of new odoriferous substances.

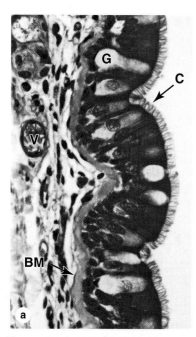  

Figure 17–2. Respiratory epithelium. Respiratory epithelium is the classic example of pseudostratified ciliated columnar epithelium. **(a):** Details of its structure vary in different regions of the respiratory tract, but it usually rests on a very thick basement membrane (BM) and has several cell types, some columnar, some basal and all contacting the basement membrane. Ciliated columnar cells are the most abundant, with hundreds of long robust cilia (C) on each of their bulging apical ends which provide a lush cover of cilia on the luminal surface. Most of the small rounded cells at the basement membrane are stem cells and their differentiating progeny, which together make up about 30% of the epithelium. Intraepithelial lymphocytes and dendritic cells are also present in respiratory epithelium. Mucus-secreting goblet cells (G) are also present. The lamina propria is well-vascularized (V). X400. Mallory trichrome. **(b):** SEM shows the luminal surface of goblet cells (G) among the numerous ciliated cells. X2500. **(c):** As shown by SEM of another region, goblet cells (G) predominate in some areas, with subsurface accumulations of mucus evident in some (arrows). The film of mucus traps most airborne dust particles and microorganisms and the ciliary movements continuously propel the sheet of mucus toward the esophagus for elimination. Other columnar cells, representing only about 3% of the cells in respiratory epithelium, are brush cells (B) with small apical surfaces bearing a tuft of short, blunt microvilli. Brush cells have features of chemosensory receptors but their physiological significance is highly uncertain. X3000. (Figure 17–2b and 17–2c reprinted, with permission, from John Wiley & Sons, Inc., *Am. J. Anat.* 1974;139:421. Copyright © 1974.)

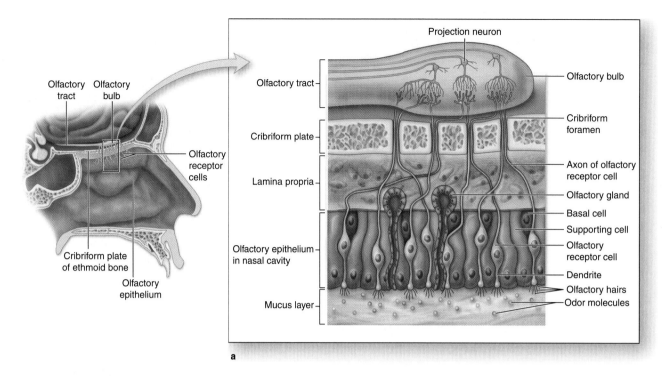

Projection neuron

Olfactory tract

Cribriform plate

Lamina propria

Olfactory epithelium in nasal cavity

Mucus layer

Olfactory bulb

Cribriform foramen

Axon of olfactory receptor cell

Olfactory gland

Basal cell

Supporting cell

Olfactory receptor cell

Dendrite

Olfactory hairs

Odor molecules

Olfactory tract

Olfactory bulb

Olfactory receptor cells

Cribriform plate of ethmoid bone

Olfactory epithelium

a

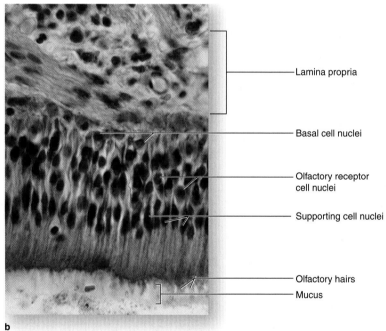

Lamina propria

Basal cell nuclei

Olfactory receptor cell nuclei

Supporting cell nuclei

Olfactory hairs

Mucus

b

Figure 17–3. Olfactory epithelium. (a, b): The olfactory epithelium covers the superior conchae bilaterally and sends axons from throughout its entire 10 cm² area to the brain via small openings in the cribriform plate of the ethmoid bone. It is a pseudostratified epithelium, containing basal stem cells and columnar support cells in addition to the bipolar olfactory neurons. The dendrites of these neurons are at the luminal ends and have cilia specialized with many membrane receptors for odor molecules. Binding such ligands causes depolarization which passes along basal axons to the olfactory bulb of the brain X200. H&E.

The olfactory neurons are some of the only neurons to be replaced regularly and constantly due to regenerative activity of the epithelial stem cells from which they arise. For this reason loss of the sense of smell due to toxic fumes or physical injury to the epithelium is usually temporary. However damage to the ethmoid bone at the base of the skull can shear the olfactory axons and lead to more permanent loss of olfaction if axonal regeneration through the cribriform plate is also blocked.

SINUSES & NASOPHARYNX

The **paranasal sinuses** are bilateral cavities in the frontal, maxillary, ethmoid, and sphenoid bones of the skull (Figure 17–1). They are lined with a thinner respiratory epithelium with fewer goblet cells. The lamina propria contains only a few small glands and is continuous with the underlying periosteum. The paranasal sinuses communicate with the nasal cavities through small openings and mucus produced in the sinuses is moved into the nasal passages by the activity of the ciliated epithelial cells.

MEDICAL APPLICATION

Sinusitis is an inflammatory process of the sinuses that may persist for long periods of time, mainly because of obstruction of drainage orifices. Chronic sinusitis and bronchitis are components of immotile cilia syndrome, which is characterized by defective ciliary action.

Posterior to the nasal cavities, the **nasopharynx** is the first part of the pharynx, continuing caudally with the oropharynx, the posterior part of the oral cavity (Figure 17–1). It is lined

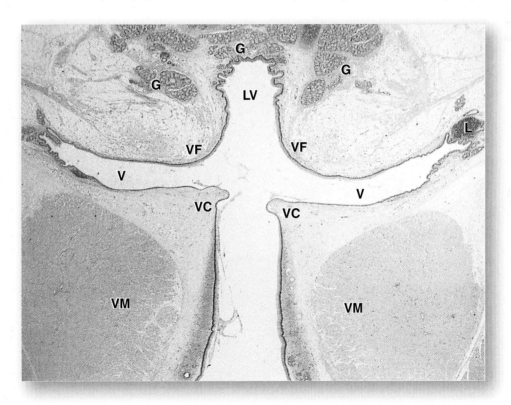

Figure 17–4. Larynx. The larynx is a short passageway for air between the pharynx and trachea. Its wall contains skeletal muscles and pieces of cartilage, all of which make the larynx specialized for sound production. The low-power micrograph shows the upper laryngeal vestibule (LV), which is surrounded by seromucous glands (G). The lateral walls of this region bulge as a pair of broad folds, the vestibular folds (VF). These also contain seromucous glands and areolar tissue with MALT, often with lymphoid nodules (L) and are largely covered by respiratory epithelium, with regions near the epiglottis having stratified squamous epithelium. Below each large vestibular fold is a narrow space or ventricle (V), below which is another pair of lateral folds, the vocal folds or cords (VC). These are covered by stratified squamous epithelium and project more sharply into the lumen, defining the rim of the opening into the larynx itself. Each contains a large striated vocalis muscle (VM) and nearer the surface a small ligament, which is cut transversely and therefore difficult to see here. Variable tension of these ligaments caused by the muscles produces different sounds as air is expelled across the vocal cords. All the structures and spaces above these folds add resonance to the sounds and assist in phonation. X15. H&E.

with respiratory epithelium and contains the medial pharyngeal tonsil and the bilateral openings of the auditory tubes to each middle ear.

LARYNX

The **larynx** is a rigid, short (4 cm × 4 cm) passage for air between the pharynx and the trachea (Figure 7–1). Its wall is reinforced by hyaline cartilage (in the thyroid, cricoid, and the inferior arytenoid cartilages) and smaller elastic cartilages (in the epiglottis, cuneiform, corniculate, and the superior arytenoid cartilages), all connected by ligaments. In addition to maintaining an open airway, movements of these cartilages by skeletal muscles participate in sound production during phonation and the epiglottis serves as a valve to prevent swallowed food or fluid from entering the trachea.

The **epiglottis,** which projects from the upper rim of the larynx, extends into the pharynx and has lingual and laryngeal surfaces. The entire lingual surface and the apical portion of the laryngeal surface are covered with stratified squamous epithelium. At variable points on the laryngeal surface of the epiglottis the epithelium undergoes a transition to ciliated pseudostratified columnar epithelium. Mixed mucous and serous glands are found in the lamina propria beneath the epithelium.

Below the epiglottis, the mucosa of the larynx extends two pairs of folds bilaterally into the lumen (Figure 17–4). The upper pair, the **vestibular folds** or **false vocal cords**, is partly covered with typical respiratory epithelium beneath which lie numerous seromucous glands. The lower pair of folds constitutes the **vocal folds** or vocal cords. These are covered with stratified squamous epithelium and contain bundles of parallel elastic fibers (vocal ligament) and large bundles of striated **vocalis muscles**. The muscles regulate the tension of each vocal fold and its ligaments. As expelled air is forced between the folds, variable tension in these vocal cords produces different sounds. All structures and spaces in the respiratory tract above the vocal folds are involved in modifying the resonance of the sounds.

TRACHEA

The **trachea** is 12-14 cm long and lined with a typical respiratory mucosa (Figure 17–5). In the lamina propria numerous seromucous glands produce watery mucus and in the submucosa 16–20 C-shaped rings of hyaline cartilage keep the tracheal lumen open (Figure 17–6). The open ends of the cartilage rings are on the posterior surface, against the esophagus, and are bridged by a bundle of smooth muscle (**trachealis muscle**) and a sheet of fibroelastic tissue attached to the perichondrium. The entire organ is surrounded by adventitia.

The trachealis relaxes during swallowing to facilitate the passage of food by allowing the esophagus to bulge into the lumen of the trachea, with the elastic layer preventing excessive distention of the lumen. In the cough reflex the muscle contracts to narrow the tracheal lumen and provide for increased velocity of the expelled air and better loosening of material in the air passage.

BRONCHIAL TREE & LUNG

The trachea divides into two **primary bronchi** that enter the lungs at the hilum, along with arteries, veins, and lymphatic vessels. After entering the lungs, the primary bronchi course downward and outward, giving rise to three **secondary (lobar) bronchi** in the right lung and two in the left lung (Figure 17–6),

each of which supplies a pulmonary lobe. These lobar bronchi again divide, forming **tertiary (segmental) bronchi**. Each of these tertiary bronchi, together with the smaller branches it supplies, constitutes a **bronchopulmonary segment**—approximately 10–12% of each lung with its own connective tissue capsule and blood supply. The existence of such lung segments facilitates the specific surgical resection of diseased lung tissue without affecting nearby healthy tissue.

The tertiary bronchi give rise to smaller and smaller bronchi, whose terminal branches are called **bronchioles**. Each bronchiole enters a pulmonary lobule, where it branches to form five to seven **terminal bronchioles**. The pulmonary lobules are pyramid-shaped, with the apex directed toward the pulmonary hilum. Each lobule is delineated by a thin connective tissue septum, best seen in the fetus. In adults these septa are frequently

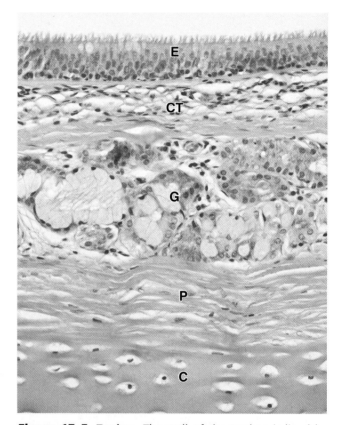

Figure 17–5. Trachea. The wall of the trachea is lined by typical respiratory epithelium (E) underlain by connective tissue (CT) and seromucous glands (G) in the lamina propria. The submucosa contains C-shaped rings of hyaline cartilage (C) covered by perichondrium (P). The watery mucous fluid produced by goblet cells and by the glands forms a layer that permits the ciliary movement to propel foreign particles continuously out of the respiratory system in the mucociliary escalator. The openings in the cartilage rings are on the posterior surface, against the esophagus, and contain smooth muscle and elastic tissue. These allow distention of the tracheal lumen when large pieces of food pass through the esophagus. The trachealis muscle in the opening of the C also contracts during the cough reflex to narrow the tracheal lumen and produce stronger expulsion of air and dislodged mucus in the air passages. X50. H&E.

incomplete, resulting in a poor delineation of the lobules. Moving through the smaller bronchi and bronchioles toward the respiratory portion, the histologic organization of both the epithelium and the underlying lamina propria gradually becomes more simplified.

Bronchi

Each primary bronchus branches repeatedly, with each branch becoming progressively smaller until it reaches a diameter of about 5 mm. The mucosa of the larger bronchi is structurally similar to the tracheal mucosa except for the organization of cartilage and smooth muscle (Figure 17–7). In the primary bronchi most cartilage rings completely encircle the lumen, but as the bronchial diameter decreases, cartilage rings are gradually replaced with isolated plates of hyaline cartilage. Abundant mucous and serous glands are also present, with ducts opening into the bronchial lumen. In the bronchial lamina propria is a layer of crisscrossing bundles of spirally arranged smooth muscle (Figures 17–7 and 17–8), which become more prominent in the smaller bronchial branches. Contraction of this muscle layer is responsible for the folded appearance of the bronchial mucosa observed in histologic section.

The lamina propria also contains elastic fibers and abundant mucous and serous glands (Figure 17–8) whose ducts open into the bronchial lumen. Numerous lymphocytes are found both

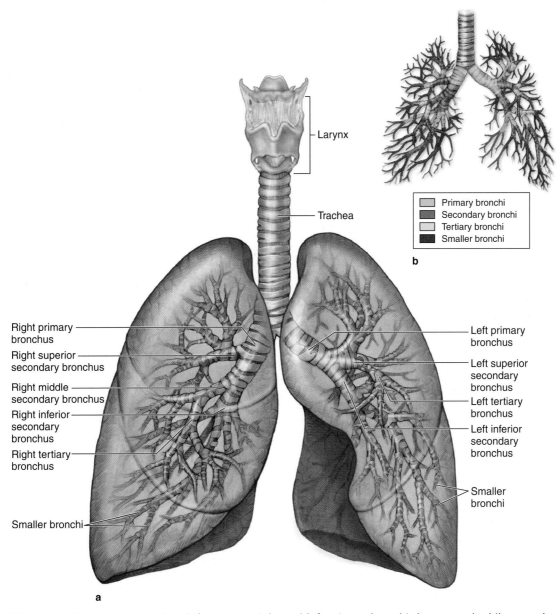

Figure 17–6. Bronchial tree. The trachea bifurcates as right and left primary bronchi that enter the hilum on the posterior side of each lung along with the pulmonary vessels, lymphatics, and nerves. **(a):** Within each lung bronchi subdivide further to form the bronchial tree, the last component of the air conducting system. **(b):** Diagram shows color-coding of the major branches of the bronchial tree.

within the lamina propria and among the epithelial cells. Lymphatic nodules are present and are particularly numerous at the branching points of the bronchial tree. Elastic fibers, smooth muscle, and MALT become relatively more abundant as bronchi become smaller and cartilage and other connective tissue are reduced.

Bronchioles

Bronchioles are the intralobular airways with diameters of 5 mm or less, formed after about the tenth generation of branching, and have neither cartilage nor glands in their mucosa (Figure 17–9). In the larger bronchioles, the epithelium is still ciliated pseudostratified columnar, but this decreases in height and complexity to become ciliated simple columnar or cuboidal epithelium in the smaller terminal bronchioles. Goblet cells disappear during this transition, but the epithelium of terminal bronchioles instead contains other numerous columnar cells: the **exocrine bronchiolar cells,** commonly called **Clara cells** (Figure 17–10). These mitotically active cells secrete surfactant components and have various important defensive roles. Scattered neuroendocrine cells (Chapter 20) are also present, producing serotonin and other peptides that help control the tone of the local smooth

muscle. Groups of similar cells, called **neuroepithelial bodies**, occur in some bronchioles and at higher levels in the bronchial tree. These are innervated by autonomic and sensory fibers and some of the cells appear to function as chemosensory receptors in monitoring air O_2 levels. Epithelial stem cells are also present in these groups of cells.

The bronchiolar lamina propria is composed largely of smooth muscle and elastic fibers. The musculature of both the bronchi and the bronchioles is under the control of the vagus nerve and the sympathetic nervous system, in addition to the influence of neuroendocrine peptides. Stimulation of the vagus nerve decreases the diameter of these structures; sympathetic stimulation produces the opposite effect.

MEDICAL APPLICATION

The increase in bronchiole diameter in response to stimulation of the sympathetic nervous system explains why epinephrine and other sympathomimetic drugs are frequently used to relax smooth muscle during asthma attacks. When the thickness of the bronchial walls is compared with that of the bronchiolar walls, the bronchiolar muscle layer is seen to be proportionately greater. Increased airway resistance in asthma is believed to be due mainly to contraction of bronchiolar smooth muscle.

Respiratory Bronchioles

Each terminal bronchiole subdivides into two or more respiratory bronchioles that serve as regions of transition between the conducting and respiratory portions of the respiratory system (Figure 17–11). The respiratory bronchiolar mucosa is structurally identical to that of the terminal bronchioles, except that their walls are interrupted by the openings to saclike alveoli where gas exchange occurs. Portions of the respiratory bronchioles are lined with ciliated cuboidal epithelial cells and Clara cells, but at the rim of the alveolar openings the bronchiolar epithelium becomes continuous with the squamous alveolar lining cells (type I alveolar cells; see below). Proceeding distally along these bronchioles, the alveoli increase in number, and the distance between them is reduced. Between alveoli the bronchiolar epithelium consists of ciliated cuboidal epithelium, although cilia may be absent in more distal portions. Smooth muscle and elastic connective tissue lie beneath the epithelium of respiratory bronchioles.

Alveolar Ducts

Proceeding distally along the respiratory bronchioles, the number of alveolar openings in the bronchiolar wall slowly increases. Respiratory bronchioles branch into tubes called **alveolar ducts** that are completely lined by the openings of alveoli (Figure 17–12). Both the alveolar ducts and the alveoli are lined with extremely attenuated squamous alveolar cells. In the lamina propria surrounding the rim of the alveoli is a thin network of smooth muscle cells, which disappears at the distal ends of alveolar ducts. A rich

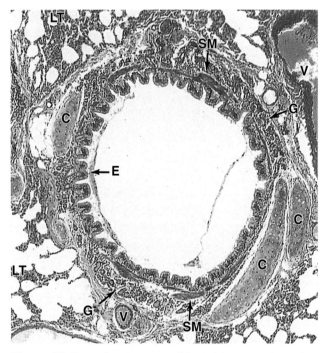

Figure 17–7. **Tertiary (segmental) bronchus.** In a cross-section of a large bronchus the lining of respiratory epithelium (E) and the mucosa are folded due to contraction of its smooth muscle (SM). At this stage in the bronchial tree, the wall is also surrounded by many pieces of hyaline cartilage (C) and contains many seromucous glands (G) in the submucosa which drain into the lumen. In the connective tissue surrounding the bronchi can be seen arteries and veins (V), which are also branching as smaller and smaller vessels in the approach to the respiratory bronchioles. All bronchi are surrounded by distinctive lung tissue (LT) showing the many empty spaces of pulmonary alveoli. X56. H&E.

matrix of elastic and collagen fibers provides the only support of the duct and its alveoli.

Alveolar ducts open into atria of two or more **alveolar sacs** (Figure 17–12). Elastic and reticular fibers form a network encircling the openings of atria, alveolar sacs, and alveoli. The elastic fibers enable the alveoli to expand with inspiration and to contract passively with expiration. The reticular fibers serve as a support that prevents overdistention and damage to the delicate capillaries and thin alveolar septa. Both fibers contribute to the connective tissue housing the network of capillaries around each alveolus.

Alveoli

Alveoli are saclike evaginations (about 200 μm in diameter) of the respiratory bronchioles, alveolar ducts, and alveolar sacs. Alveoli are responsible for the spongy structure of the lungs (Figures 17–11 and 17–12). Structurally, alveoli resemble small pockets that are

open on one side, similar to the honeycombs of a beehive. Within these cuplike structures, O_2 and CO_2 are exchanged between the air and the blood. The structure of alveolar walls is specialized to enhance diffusion between the external and internal environments. Generally, each wall lies between two neighboring alveoli and is therefore called an **interalveolar septum**. These septa contain the cells and ECM of connective tissue, notably the elastic and collagen fibers, which is vascularized with the richest capillary network in the body (Figure 17–11).

Air in the alveoli is separated from capillary blood by three components referred to collectively as the **respiratory membrane** or blood-air barrier:

- Surface lining and cytoplasm of the alveolar cells,
- Fused basal laminae of the closely apposed alveolar cells and capillary endothelial cells, and
- Cytoplasm of the endothelial cells (Figures 17–13, 17–14, and 17–15).

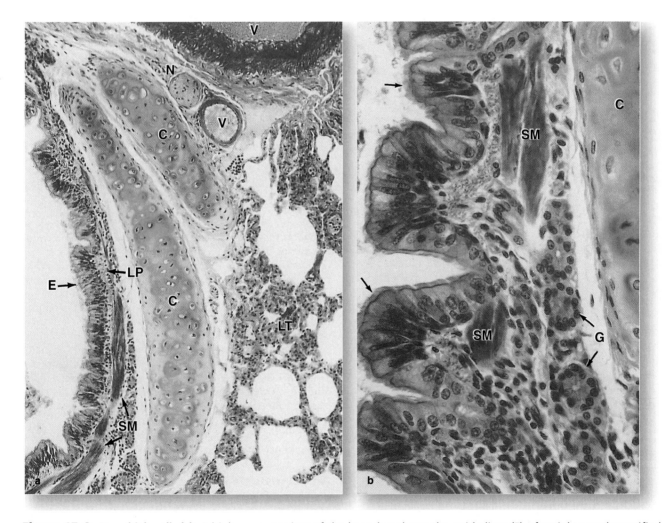

Figure 17–8. Bronchial wall. (a): A higher power view of the bronchus shows the epithelium (E) of mainly pseudostratified ciliated columnar cells with a few goblet cells. The lamina propria (LP) contains the distinct layer of smooth muscle (SM) surrounding the entire bronchus. The submucosa is the site of the supporting cartilage (C) and the adventitia includes blood vessels (V) and nerves (N). Lung tissue (LT) directly surrounds the adventitia of bronchi. X140. H&E. (b): This micrograph shows the epithelium of a smaller bronchus, in which the epithelium is primarily of columnar cells with cilia (arrows), with fewer goblet cells. The lamina propria has both smooth muscle (SM) and small serous glands (G) near cartilage (C). X400. H&E.

The total thickness of these layers varies from 0.1 to 1.5 μm. Within the interalveolar septum, densely anastomosing pulmonary capillaries are supported by the meshwork of reticular and elastic fibers, which are the primary structural support of the alveoli. Macrophages and other leukocytes can also be found within the interstitium of the septum (Figures 17–13 and 17-14). The basal laminae of the capillary endothelial cells and the epithelial (alveolar) cells fuse as a single membranous structure (Figures 17–13 and 17–15).

Pores 10–15 μm in diameter occur in the interalveolar septum (Figure 17–13) and connect neighboring alveoli opening to different bronchioles. These pores equalize air pressure in the alveoli and promote collateral circulation of air when a bronchiole is obstructed.

O_2 from the alveolar air passes into the capillary blood through the blood-air barrier; CO_2 diffuses in the opposite direction. Liberation of CO_2 from H_2CO_3 is catalyzed by the enzyme **carbonic anhydrase** present in erythrocytes. The approximately 300 million alveoli in the lungs provide a vast internal surface for gas exchange, which has been calculated to be approximately 140 m^2.

Capillary endothelial cells are extremely thin and can be easily confused with type I alveolar epithelial cells. The endothelial lining of the capillaries is continuous and not fenestrated (Figure 17–15). Clustering of the nuclei and other organelles allows the remaining areas of the cell to become extremely thin, increasing the efficiency of gas exchange. The most prominent feature of the cytoplasm in the flattened portions of the cell are numerous pinocytotic vesicles.

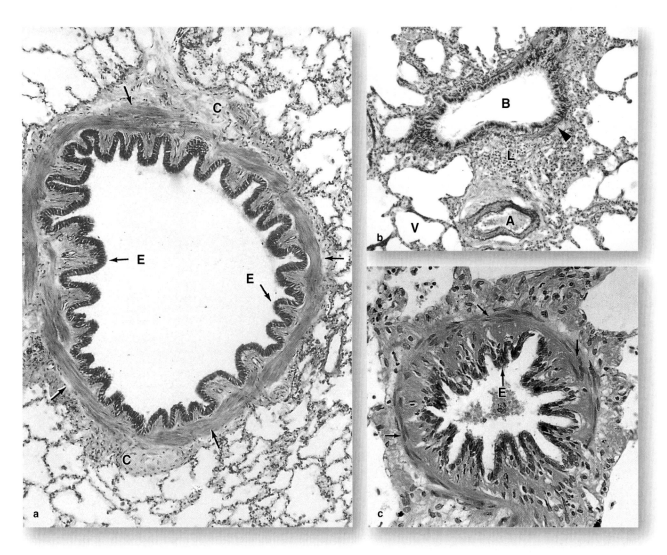

Figure 17–9. **Bronchioles.** Bronchial branches less than about 5 mm in diameter lack supporting cartilage and are called bronchioles. **(a):** A large bronchiole has the characteristically folded respiratory epithelium (E) and prominent smooth muscle (arrows), but is supported only by fibrous connective tissue (C) with no glands. X140. H&E. **(b):** Staining for elastic fibers reveals the high elastic content of the smooth muscle (arrowhead) associated with the muscle of a smaller bronchiole (B) in which the epithelium is simple columnar. Darkly stained elastic fibers are also present in the tunica media of a large arteriole (A) nearby and to a lesser extent in the accompanying venule (V). The connective tissue includes many lymphocytes (L) of MALT and lymphoid nodules are also common at this level. X180. Elastic stain. **(c):** In very small bronchioles the epithelium (E) is reduced to simple low columnar and the several layers of smooth muscle cells (arrows) comprise a high proportion of the wall. X300. H&E.

Type I alveolar cells (also called type I pneumocytes or squamous alveolar cells) are extremely attenuated cells that line the alveolar surfaces. Type I cells cover 97% of the alveolar surface (type II cells covering the remainder). These cells are so thin (sometimes only 25 nm) that the electron microscope was needed to prove that all alveoli are covered with an epithelial lining (Figure 17–15). Organelles such as the ER, Golgi apparatus, and mitochondria are grouped around the nucleus, leaving large areas of cytoplasm virtually free of organelles and reducing the thickness of the blood-air barrier. The cytoplasm in the thin portion

contains pinocytotic vesicles, which may play a role in the turnover of surfactant and the removal of small particulate contaminants from the outer surface. In addition to desmosomes, all type I epithelial cells have occluding junctions that prevent the leakage of tissue fluid into the alveolar air space (Figure 17–16). The main role of these cells is to provide a barrier of minimal thickness that is readily permeable to gases.

Type II alveolar cells (type II pneumocytes) are interspersed among the type I alveolar cells with which they have occluding and desmosomal junctions (Figure 17–16). Type II cells are rounded

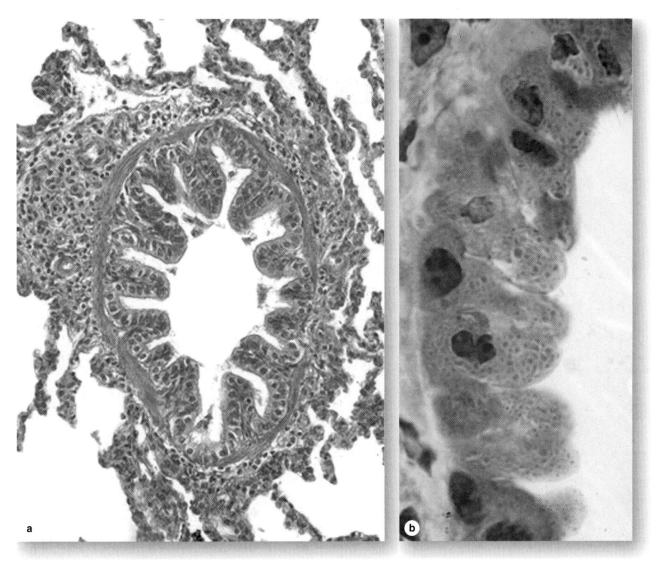

Figure 17-10. **Terminal bronchiole and Clara cells.** The last parts of the air conducting system before the sites of gas exchange appear are called the terminal bronchioles, which generally have diameters of one to two mm. **(a):** Cross-section shows that a terminal bronchiole has only one or two layers of smooth muscle cells. The epithelium contains ciliated cuboidal cells and many low columnar nonciliated cells. X300. PT. **(b):** The nonciliated Clara cells with bulging domes of apical cytoplasm contain granules, as seen better in a plastic section. Named for Dr. Max Clara, the histologist who first described them in 1937, these cells have several important functions. They secrete components of surfactant which reduces surface tension and helps prevent collapse of the bronchioles. In addition, Clara cells produce enzymes that help break down mucus locally. The P450 enzyme system of their smooth ER detoxifies potentially harmful compounds in air. In other defensive functions, Clara cells also produce the secretory component for the transfer of IgA into the bronchiolar lumen; lysozyme and other enzymes active against bacteria and viruses; and several cytokines that regulate local inflammatory responses. Mitotically active cells are also present and include the stem cells for the bronchiolar epithelium. X500. PT.

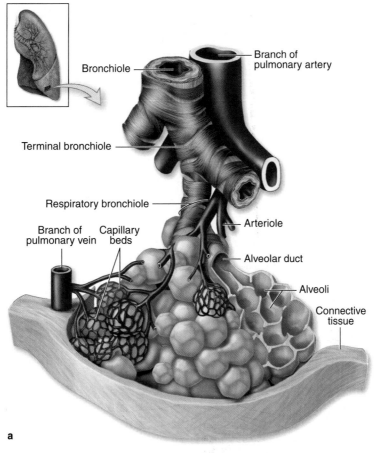

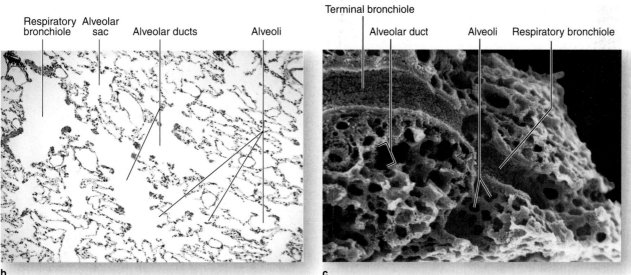

Figure 17–11. **Terminal bronchioles, respiratory bronchioles, and alveoli.** Terminal bronchioles branch into respiratory bronchioles, which then branch further into alveolar ducts and individual alveoli. Respiratory bronchioles are similar in most respects to terminal bronchioles except for the presence of scattered alveoli along their length.

(a): Diagram shows the branching relationship, as well as the pulmonary blood vessels that travel with the bronchioles and the dense layer of branching capillaries that surrounds each alveolus for gas exchange between blood and air. **(b):** The micrograph shows the branching nature of the bronchioles in two dimensions. X60. H&E. **(c):** SEM shows in three dimensions the relationship of alveoli to terminal and respiratory bronchioles. X180.

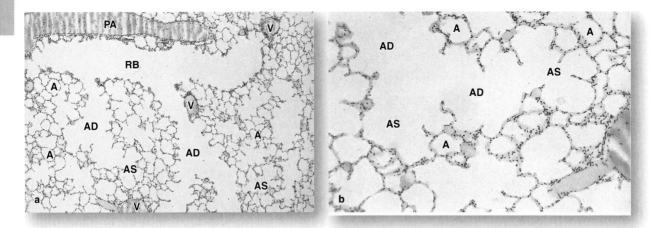

***Figure 17–12.* Respiratory bronchioles, alveolar ducts, and alveoli.** Lung tissue has a spongy structure because of the abundant air passages and pockets called alveoli. **(a):** Typical section of lung tissue including many bronchioles, some of which are respiratory bronchioles (RB) cut lengthwise, and showing the branching continuity with alveolar ducts (AD) and sacs (AS). Respiratory bronchioles still have a layer of smooth muscle and some regions of cuboidal epithelium, but alveolar ducts have only sparse strands of smooth muscle and an epithelium consisting of only a series of neighboring alveoli. The smooth muscle fibers are sphincter-like and appear as knobs between adjacent alveoli. Individual alveoli (A) all open to the sacs or ducts. The respiratory bronchiole runs along a thin-walled branch of the pulmonary artery (PA), which has a relatively thin wall, while branches of the pulmonary vein (V) course elsewhere in the parenchyma. X14. H&E.

(b): Higher magnification shows the relationship of the many rounded, thin-walled alveoli (A) to alveolar ducts (AD). Alveolar ducts end in two or more clusters of alveoli called alveolar sacs (AS). Those alveoli shown here that do not show openings to the ducts or the sacs have their connections in adjacent planes of other sections. X140. H&E.

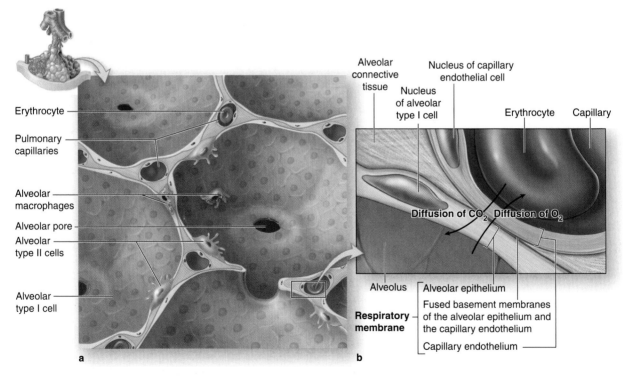

***Figure 17–13.* Alveoli and the blood-air barrier.** Gas exchange between air and blood occurs at a membranous barrier between each alveolus and the capillaries surrounding it. The total area of this air-blood barrier in each lung has been calculated at approximately 70 m². **(a):** Diagram shows the relationship between capillaries and two or more saclike alveoli. **(b):** The air-blood barrier consists of an alveolar type I cell, a capillary endothelial cell, and their fused basement membranes. Oxygen diffuses from alveolar air into capillary blood and carbon dioxide moves in the opposite direction. The inner lining of alveoli is covered by a layer of surfactant, not depicted here, which lowers fluid surface tension and helps prevent collapse of alveoli.

cells that often occur in groups of two or three along the alveolar surface at points where the alveolar walls unite. These cells rest on the basement membrane and are part of the epithelium, with the same origin as the type I cells that line the alveolar walls. They divide by mitosis to replace their own population and also the type I population. In histologic sections, they exhibit a characteristic vesicular or foamy cytoplasm. These vesicles are caused by the presence of **lamellar bodies** (Figures 17–16 and 17–17) that are best preserved and evident in tissue prepared for electron microscopy. Lamellar bodies, which average 1–2 μm in diameter, contain concentric or parallel lamellae limited by a unit membrane. Histochemical studies show that these bodies, which contain phospholipids, glycosaminoglycans, and proteins, are continuously synthesized and released at the apical surface of the cells. The lamellar bodies give rise to a material that spreads over the alveolar surfaces as the **pulmonary** surfactant, providing an extracellular coating that lowers surface tension.

Surfactant consists of an aqueous hypophase covered with a monomolecular phospholipid film composed mainly of **dipalmitoyl phosphatidylcholine** and **phosphatidylglycerol** (Figure 17–17). Surfactant also contains several specific proteins. Pulmonary surfactant serves several major functions in the economy of the lung, but acts primarily to reduce the surface tension on the alveoli. The reduction of surface tension means that less inspiratory force is needed to inflate the alveoli, easing the work of breathing. Without surfactant, alveoli would tend to collapse during expiration. In fetal development, surfactant appears in the last weeks of gestation as lamellar bodies develop in the type II cells.

MEDICAL APPLICATION

Respiratory distress syndrome (RDS) *of the newborn is a life-threatening disorder of the lungs caused by a deficiency of surfactant. It is principally associated with prematurity and is the leading cause of mortality among premature infants. The incidence of respiratory distress syndrome varies inversely with gestation age. The immature lung is deficient in both the amount and composition of surfactant. In the normal newborn, the onset of breathing is associated with a massive release of stored surfactant, which reduces the surface tension of the alveolar cells. This means that less inspiratory force is needed to inflate the alveoli, and thus the work of breathing is reduced. Microscopically, the alveoli are collapsed and the respiratory bronchioles and alveolar ducts are dilated and contain edema fluid. RDS is treated by introducing synthetic or animal-derived surfactant into the lungs through a tube. Surfactant also has bactericidal effects, aiding in the removal of potentially dangerous bacteria that reach the alveoli.*

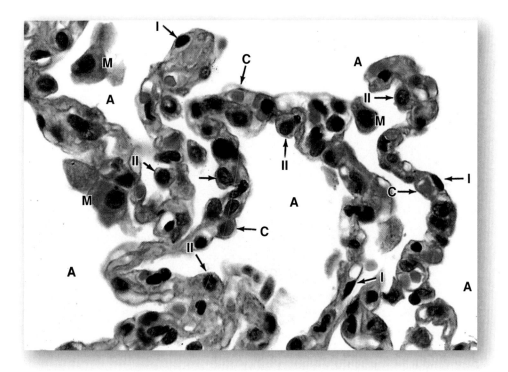

Figure 17–14. **Alveolar walls.** The wall between alveoli (A) contains several cell types. As seen here the capillaries (C) contain erythrocytes and leukocytes. The alveoli are lined mainly by squamous type I alveolar cells (I), which line almost the entire alveolus surface and across which gas exchange occurs. Type II alveolar cells line a bit of each alveolus and are large rounded cells, often bulging into the alveolus (II). These type II cells have many functions of Clara cells, including production of surfactant. Also present are alveolar macrophages (M), sometimes called dust cells, which may be in the alveoli or in the interalveolar septa.

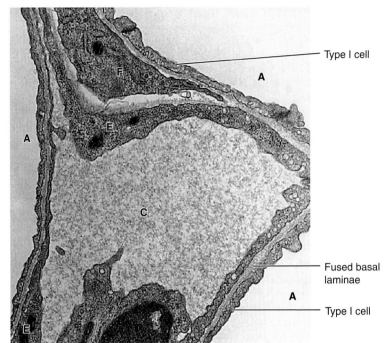

Figure 17–15. **Ultrastructure of the blood-air barrier.** TEM of a transversely sectioned capillary (C) in an interalveolar septum shows areas for gas exchange between blood and air in three alveoli (A). The endothelium is extremely thin but not fenestrated and its basal lamina fuses with that of the alveolar cells. A fibroblast (F) can be seen in the septum and the thickened nuclear regions of two endothelial cells (E) are also included. The nucleus at the bottom belongs to an endothelial cell or a circulating leukocyte. X30,000.

Type I cell

A

Fused basal laminae

Type I cell

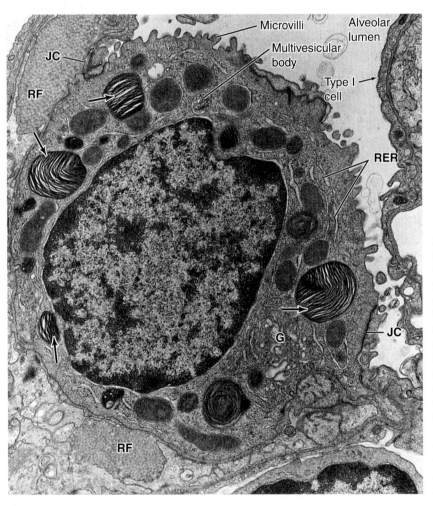

Microvilli

Alveolar lumen

Multivesicular body

JC

Type I cell

RF

RER

RER

G

JC

RF

Figure 17–16. **Ultrastructure of type II alveolar cells.** TEM of a type II alveolar cell protruding into the alveolar lumen shows unusual cytoplasmic features. Arrows indicate lamellar bodies which store newly synthesized pulmonary surfactant after processing of its components in rough ER (RER) and the Golgi apparatus (G). Smaller multivesicular bodies with intralumenal vesicles are also often present. Short microvilli are present and the type II cell is attached via junctional complexes (JC) with the adjacent, very thin type I epithelial cell. The ECM contains prominent reticular fibers (RF). X17,000. (Reproduced, with permission from Dr. Mary C. Williams, Pulmonary Center and Department of Anatomy, Boston University School of Medicine.)

The surfactant layer is not static but is constantly being turned over. The lipoproteins are gradually removed by pinocytosis in both types of alveolar cells and by macrophages.

Alveolar macrophages, also called dust cells, are found in alveoli and in the interalveolar septum (Figures 17–13 and 17–14). Tens of millions of monocytes migrate daily from the microvasculature into the lung tissue, where they phagocytose erythrocytes lost from damaged capillaries and air-borne particulate matter that has entered alveoli. Some debris within these cells was most likely passed from the alveolar lumen into the interstitium following the pinocytotic activity of type I alveolar cells. Active macrophages in lung are often slightly darker due to their content of dust and carbon from air and complexed iron (hemosiderin) from erythrocytes (Figure 17–14). Filled macrophages have various fates: most migrate into bronchioles where they move up the mucociliary escalator for removal in the pharynx; others exit the lungs in the lymphatic drainage, while some remain in the interalveolar septa connective tissue for years.

Alveolar lining fluids are also removed via the conducting passages as a result of ciliary activity. As the secretions pass up through the airways, they combine with bronchial mucus, forming **bronchoalveolar fluid,** which aids in the removal of particulate components brought in with inspired air. The bronchoalveolar fluid contains several lytic enzymes (eg, lysozyme, collagenase, β-glucuronidase) that are derived from Clara cells, type II cells, and alveolar macrophages.

MEDICAL APPLICATION

*In congestive heart failure, the lungs become congested with blood, and erythrocytes pass into the alveoli, where they are phagocytized by alveolar macrophages. In such cases, these macrophages are called **heart failure cells** when present in the lung and sputum; they are identified by a positive histochemical reaction for iron pigment (hemosiderin).*

Increased production of type I collagen is common, and many diseases that lead to respiratory distress are known to be associated with lung fibrosis.

Regeneration in the Alveolar Lining

Inhalation of toxic gases or similar materials can kill type I and type II cells lining pulmonary alveoli. Death of the first cells results in increased mitotic activity in the remaining type II cells, the progeny of which become both cell types. The normal turnover rate of type II cells is estimated to be 1% per day and results in a continuous renewal of both alveolar cells. With

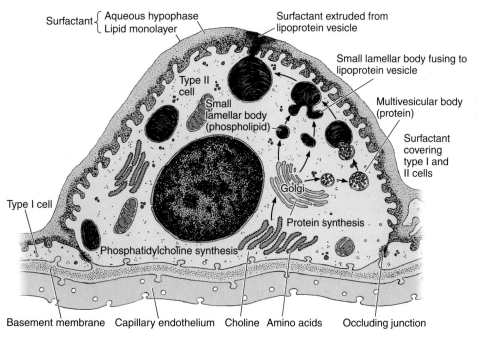

Figure 17–17. Type II alveolar cell function. Diagram illustrates secretion of surfactant by a type II cell. Surfactant contains protein-lipid complexes synthesized initially in the ER and Golgi, with further processing and storage in large organelles called lamellar bodies. Multivesicular bodies are organelles smaller than most lamellar bodies that are frequently seen in type II alveolar cells. Such bodies form when the membrane components of an early endosome are sorted, invaginate, and pinch off into smaller vesicles inside the endosome's lumen. Multivesicular bodies interact with Golgi complexes, with some or most intralumenal vesicle components being ubiquinated for degradation while other components and the surrounding membrane are recycled to the cell membrane, or in the case of type II alveolar cells, first added to the content of lamellar bodies. Surfactant is secreted continuously by exocytosis and forms an oily monomolecular film of lipid over an aqueous hypophase containing proteins. Occluding junctions around the margins of the epithelial cells prevent leakage of tissue fluid into the alveolar lumen.

increased toxic stress Clara cells can also divide and give rise to alveolar cells.

PULMONARY VASCULATURE & NERVES

Circulation in the lungs includes both nutrient (systemic) and functional (pulmonary) vessels. Pulmonary arteries and veins represent the functional circulation and the arteries are relatively thin-walled as a result of the low pressures (25 mm Hg systolic, 5 mm Hg diastolic) encountered in the pulmonary circuit. Within the lung the pulmonary artery branches and accompa-nies the bronchial tree (Figures 17–11 and 17–12), with its branches surrounded by adventitia of the bronchi and bronchi-oles. At the level of the alveolar duct, the branches of this artery form the capillary network in the interalveolar septum and in close contact with the alveoli. The lung has the best-developed capillary network of any organ, with capillaries between all alve-oli, including those in the respiratory bronchioles.

Venules that originate in the capillary network are found singly in the parenchyma, somewhat removed from the airways (Figures 17–11 and 17–12), supported by a thin covering of con-nective tissue. After veins leave a lobule, they follow the bronchial tree toward the hilum.

Nutrient vessels follow the bronchial tree and distribute blood to most of the lung up to the respiratory bronchioles, at which point they anastomose with small branches of the pulmonary artery.

The lymphatic vessels originate in the connective tissue of bron-chioles. They follow the bronchioles, bronchi, and pulmonary ves-sels and all drain into lymph nodes in the region of the hilum. This lymphatic network is called the **deep network** to distinguish it from the **superficial network** of lymphatic vessels in the visceral pleura. Both networks drain toward the hilum, either following the entire length of the pleura or after entering lung tissue via the inter-lobular septa. Lymphatic vessels are not found in the terminal por-tions of the bronchial tree beyond the alveolar ducts.

Both parasympathetic and sympathetic efferent fibers inner-vate the lungs and general visceral afferent fibers, carrying poorly localized pain sensations, are also present. Most of the nerves are found in the connective tissue surrounding the larger airways.

PLEURA

The lung's outer surface and the internal wall of the thoracic cavity are covered by a serous membrane called the **pleura**

(Figure 17–18). The membrane attached to lung tissue is called the **visceral pleura** and the membrane lining the thoracic walls is the **parietal pleura**. The two layers are continuous at the hilum and are both composed of simple squamous mesothelial cells on a thin connective tissue layer containing collagen and elastic fibers. The elastic fibers of the visceral pleura are continuous with those of the pulmonary parenchyma.

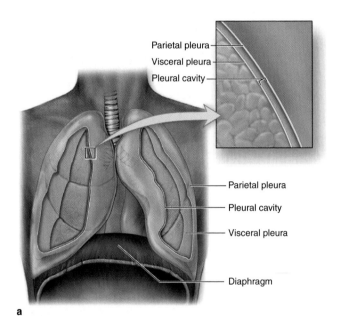

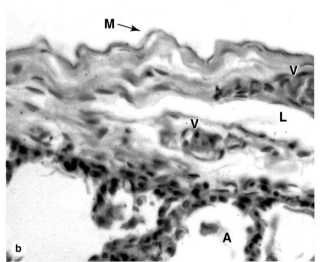

Figure 17–18. Pleura. The pleura are serous membranes (serosa) associated with each lung and thoracic cavity. **(a):** Dia-gram illustrates the parietal pleura lining the inner surface of the thoracic cavity and the visceral pleura covering the outer surface of the lung. Between these layers is the narrow space of the pleural cavity. **(b):** Both layers are similar histologically and consist of a simple squamous mesothelium (M) on a thin layer of connective tissue, as shown here for visceral pleura covering alveoli (A). The connective tissue is rich in both col-lagen and elastic fibers and contains both blood vessels (V) and lymphatics (L). X140.

The narrow **pleural cavity** (Figure 17–18) between the parietal and visceral layers is entirely lined with mesothelial cells that normally produce a thin film of serous fluid that acts as a lubricant, facilitating the smooth sliding of one surface over the other during respiratory movements.

In certain pathologic states, the pleural cavity may contain liquid or air. Like the walls of the peritoneal and pericardial cavities, the serosa of the pleural cavity is quite permeable to water and fluid exuded from blood plasma commonly accumulates (as a pleural effusion) in this cavity under abnormal conditions.

RESPIRATORY MOVEMENTS

During inhalation, contraction of the intercostal muscles elevates the ribs and contraction of the diaphragm lowers the bottom of the thoracic cavity, increasing its diameter and resulting in pulmonary expansion. The bronchi and bronchioles increase in diameter and length during inhalation. The respiratory portion also enlarges, mainly as a result of expansion of the alveolar ducts.

Individual alveoli enlarge only slightly. The elastic fibers of the pulmonary parenchyma are stretched by this expansion. During exhalation, the lungs retract passively due to muscle relaxation and the elastic fibers' return to the unstretched condition.

MEDICAL APPLICATION

Squamous cell carcinoma, the principal lung tumor type, usually results from the effects of cigarette smoking on the bronchial and bronchiolar epithelial lining. Chronic smoking induces the transformation of the respiratory epithelium into a stratified squamous epithelium, an initial step in its eventual differentiation into a tumor.

Skin

The skin is the largest single organ of the body, typically accounting for 15–20% of total body weight and, in adults, presenting 1.5–2 m^2 of surface to the external environment. Also known as the **integument** (L. *integumentum*, covering) or **cutaneous layer**, the skin is composed of the **epidermis,** an epithelial layer of ectodermal origin, and the **dermis,** a layer of mesodermal connective tissue (Figure 18–1). The junction of dermis and epidermis is irregular, and projections of the dermis called **papillae** interdigitate with evaginations of the epidermis known as **epidermal ridges**. Epidermal derivatives include hairs, nails, and sebaceous and sweat glands. Beneath the dermis lies the **subcutaneous tissue** or **hypodermis** (Gr. *hypo*, under, + *derma*, skin), a loose connective tissue that may contain pads of adipocytes. The subcutaneous tissue binds skin loosely to the underlying tissues and corresponds to the superficial fascia of gross anatomy.

The specific functions of the skin fall into several broad categories.

- **Protective**. It provides a physical barrier against thermal and mechanical insults such as frictional forces and against most potential pathogens and other material. Microorganisms that do penetrate skin alert resident lymphocytes and antigen-presenting cells of skin (Figure 14–6) and an immune response is mounted. The dark pigment melanin in epidermis protects cells against ultraviolet radiation. Skin is also a permeability barrier against excessive loss or uptake of water, which has allowed for terrestrial life. Skin's selective permeability allows some lipophilic drugs such as certain steroid hormones and medications to be administered via skin patches.

- **Sensory**. Many types of sensory receptors allow skin to constantly monitor the environment and various mechanoreceptors with specific locations in skin are important for the body's interactions with physical objects.

- **Thermoregulatory**. A constant body temperature is normally more easily maintained thanks to the skin's insulating components (eg, the fatty layer and hair on the head) and its mechanisms for accelerating heat loss (sweat production and a dense superficial microvasculature).

- **Metabolic**. Cells of skin synthesize vitamin D$_3$, needed in calcium metabolism and proper bone formation, through the local action of UV light on the vitamin's precursor. Excess electrolytes can be removed in sweat and the subcutaneous layer stores a significant amount of energy in the form of fat.

- **Sexual signaling**. Many features of skin, such as pigmentation and hair, are visual indicators of health involved in attraction between the sexes in all vertebrate species, including humans. The effects of sex pheromones produced by the apocrine sweat glands and other glands of skin are also important for this attraction.

The dermal-epidermal interdigitations are of the peg-and-socket variety in most skin, but occur as well-formed ridges and grooves in the thick skin of the palms and soles which is more subject to friction. These ridges and the intervening sulci form distinctive patterns unique for each individual, appearing as combinations of loops, arches, and whorls, called dermatoglyphs, also known as fingerprints and footprints. Skin is elastic and can expand rapidly to cover swollen areas and like the gut lining is self-renewing throughout life. In healthy individuals injured skin is repaired rapidly. The molecular basis of skin healing is increasingly well-understood and provides a basis for better understanding of repair and regeneration in other organs.

EPIDERMIS

The epidermis consists mainly of a stratified squamous keratinized epithelium composed of cells called **keratinocytes**. Three less abundant epidermal cell types are also present: pigment-producing **melanocytes**, antigen-presenting **Langerhans cells,** and tactile epithelial cells or **Merkel cells** (Figure 18–2).

The epidermis forms the major distinction between **thick skin** (Figure 18–2), found on the palms and soles, and **thin skin** (Figure 18–3) found elsewhere on the body. The designations

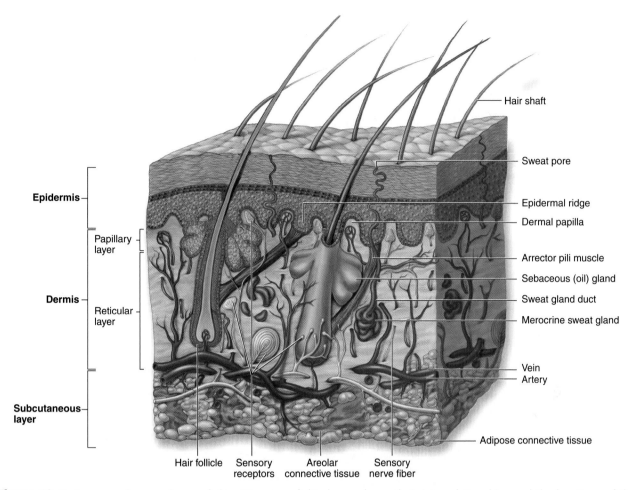

Figure 18–1. **Layers and appendages of skin.** Diagram of skin layers shows their interrelationships and the locations of the epidermal appendages (hair follicles, sweat and sebaceous glands), the vasculature, and the major sensory receptors.

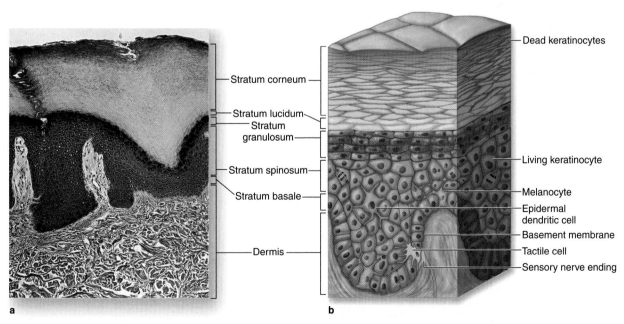

Figure 18–2. **Layers (strata) of epidermis in thick skin. (a):** Micrograph shows the sequence of the epidermal layers in thick skin and the approximate sizes and shape of keratinocytes in these layers. Also shown are the coarse bundles of collagen in the dermis and on the far left, the duct from a sweat gland entering the epidermis from a dermal papilla and coiling to a surface pore through all the strata. X100. H&E. **(b):** Diagram illustrating the sequence of the epidermal layers also indicates the normal locations of three important nonkeratinocyte cells in the epidermis: melanocytes, a dendritic (Langerhans) cell, and a tactile cell.

"thick" and "thin" refer to the thickness of the epidermal layer, which varies from 75 to 150 μm for thin skin and from 400 to 1400 μm (1.4 mm) for thick skin. Total skin thickness (epidermis plus dermis) also varies according to site. For example, skin on the back is about 4 mm thick, whereas that of the scalp is about 1.5 mm thick.

From the dermis outward, the epidermis consists of four layers of keratinocytes, five layers in thick skin (Figure 18–2):

- The **basal layer (stratum basale)** is a single layer of basophilic columnar or cuboidal cells on the basement membrane at the dermal-epidermal junction (Figures 18–2 and 18–3). Hemidesmosomes in the basal plasmalemma help bind these cells to the basal lamina and desmosomes bind the cells of this layer together in their lateral and upper surfaces. The stratum basale is characterized by intense mitotic activity and is responsible, in conjunction with the initial portion of the next layer, for constant production of epidermal cells. Although stem cells for keratinocytes are found in the basal layer, a niche for such cells also occurs in a specialized bulge in the hair follicle sheath continuous with the epidermis. The human epidermis is renewed about every 15–30 days, depending on age, the region of the body, and other factors. All keratinocytes in the stratum basale contain intermediate filaments about 10 nm in diameter composed of **keratins**. As the cells progress upward, the amount and types of keratin filaments increase until they represent half the total protein in the outermost layer.

- The **spinous layer (stratum spinosum)**, normally the thickest epidermal layer (Figures 18–2 and 18–3), consists of polyhedral or slightly flattened cells having central nuclei with nucleoli and cytoplasm actively synthesizing keratin filaments.

Just above the basal layer some cells may still divide and this combined zone is sometimes called the stratum germinativum. The keratin filaments form microscopically visible bundles called **tonofibrils** which converge and terminate at the numerous desmosomes, by which the cells are joined together strongly to resist friction. Cytoplasm is drawn into short cellular extensions around the tonofibrils on both sides of each desmosome (and these are elongated if the cells shrink slightly when processed histologically), leading to the appearance of many short spines or prickles at the cell surfaces (Figure 18–4). The epidermis of areas subjected to continuous friction and pressure (such as the soles of the feet) has a thicker stratum spinosum with more abundant tonofibrils and desmosomes.

MEDICAL APPLICATION

In adults, one third of all cancers are of the skin. Most of these derive from cells of the basal or spinous layers, producing, respectively, basal cell carcinomas and squamous cell carcinomas. Fortunately both types of tumors can be diagnosed and excised early and consequently are rarely lethal. Skin cancer shows an increased incidence in fair-skinned individuals residing in regions with high amounts of solar radiation.

- The **granular layer (stratum granulosum)** consists of 3–5 layers of flattened polygonal cells undergoing terminal differentiation. Their cytoplasm is filled with intensely basophilic masses (Figures 18–2, 18–3, and 18–5) called **keratohyaline granules.** These structures are not membrane-bound and consist of dense masses of **filaggrin** and other proteins that associate with the keratins of tonofibrils, linking them into large cytoplasmic structures in the important process of **keratinization.** Other characteristic features visible only with the TEM in cells of the granular layer are the membrane-coated **lamellar granules,** small (0.1–0.3 μm) ovoid structures containing many lamellae composed of various lipids. Lamellar granules undergo exocytosis, discharging their contents into the intercellular spaces of the stratum granulosum. There this lipid-rich material produces sheets that envelop the cells, which are now little more than flattened sacs filled with keratins and associated proteins. The layer of lipid envelopes is a major component of the epidermal barrier against the loss of water from skin. Formation of this barrier, which appeared first in reptiles, was one of the important evolutionary events that permitted animals to develop on land. Together, keratinization and production of the lipid-rich layer also have a crucial sealing effect in skin, forming the barrier to penetration by most foreign materials.

- The **stratum lucidum** is only seen in thick skin, where it consists of a thin, translucent layer of extremely flattened eosinophilic cells (Figures 18–1 and 18–5). The nuclei and organelles have been lost and the cytoplasm consists almost only of densely packed keratin filaments embedded in an electron-dense matrix. Desmosomes are still evident between adjacent cells.

- The **stratum corneum** (Figures 18–2 and 18–3) consists of 15–20 layers of flattened, nonnucleated keratinized cells whose

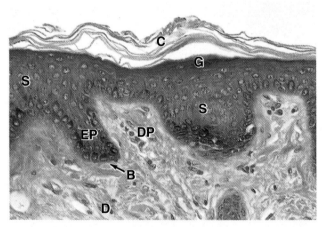

Figure 18–3. **Layers of epidermis in thin skin.** The interface between dermis and epidermis in thin skin is held together firmly by interlocking epidermal pegs (EP) and dermal papillae (DP). The dermis (D) is more cellular and well vascularized than that of thick skin, with elastin and less coarse bundles of collagen. The epidermis usually shows only four layers in thin skin: the one-cell thick stratum basale (B) containing most mitotic cells; the stratum spinosum (S) where synthesis of much keratin and other proteins takes places; the stratum granulosum (G); and the stratum corneum (C), consisting of dead squames composed mostly of keratin. X240. H&E.

cytoplasm is filled with birefringent filamentous keratins. Keratin filaments contain at least six different polypeptides, with molecular mass ranging from 40 to 70 kDa, their composition changing as the epidermal cells differentiate and when the tonofibrils become heavily massed with other proteins from the keratohyaline granules. After keratinization, the cells contain only fibrillar and amorphous proteins with thickened plasma membranes and are called **squames** or horny, cornified cells. These cells are continuously shed at the surface of the stratum corneum.

MEDICAL APPLICATION

*In **psoriasis**, a common skin disease, there is an increase in the number of proliferating cells in the stratum basale and the stratum spinosum as well as a decrease in the cycle time of these cells. This results in greater epidermal thickness and more rapid renewal of epidermis, but also can produce abnormal keratinization with a defective skin barrier.*

Melanocytes

The color of the skin is the result of several factors, the most important of which are the keratinocytes' content of **melanin** and **carotene** and the number of blood vessels in the dermis.

Eumelanin is a brownish black pigment produced by the **melanocyte** (Figures 18–6 and 18–7), a specialized cell of the epidermis found among the cells of the basal layer and in the hair follicles. The similar pigment found in red hair is called **pheomelanin** (Gr. *phaios*, dusky, + *melas*, black). Melanocytes are neural crest derivatives which migrate into the developing epidermis' stratum basale, where eventually one melanocyte accumulates for every five or six basal keratinocytes (600–1200/mm² of skin).

They have rounded cell bodies and form hemidesmosomes with the basal lamina, but no desmosomes with adjacent keratinocytes. Long irregular dendritic extensions from each melanocyte branch into the epidermis, running between the cells of the basal and spinous layers and terminating in invaginations of the neighboring five to ten keratinocytes. Ultrastructurally, a melanocyte is a pale-staining cell with numerous small mitochondria, short cisternae of rough endoplasmic reticulum (RER), and a well-developed Golgi apparatus (Figure 18–6).

MEDICAL APPLICATION

***Malignant melanoma** is an invasive tumor of melanocytes. Dividing rapidly, malignantly transformed melanocytes penetrate the basal lamina, enter the dermis, and invade the blood and lymphatic vessels to gain wide distribution throughout the body.*

Melanin is synthesized in the melanocyte, with **tyrosinase** playing an important role in the process. Tyrosinase and tyrosinase-related proteins are transmembrane proteins synthesized in the rough ER, which accumulate in vesicles formed in the Golgi complex (Figure 18–7). Tyrosinase activity converts tyrosine first into **3,4-dihydroxyphenylalanine** (**DOPA**), which is then further transformed and polymerized into melanin. This pigment is then linked to a matrix of structural proteins in the vesicles. Melanin accumulates in these vesicles until they form mature granules called **melanosomes**, which are elliptical structures about 1 μm in length.

Once formed, melanin granules are transported via kinesin along microtubules to the actin-rich tips of the melanocyte's dendrites. The associated keratinocytes in both the basal and spinous layers phagocytose the tips of these dendrites and the

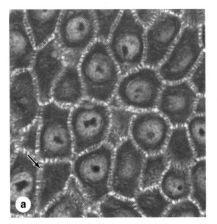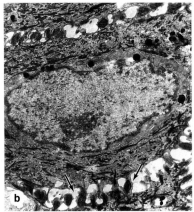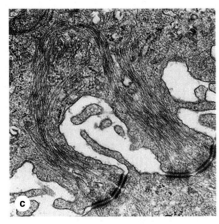

Figure 18–4. Keratinocytes of the stratum spinosum. **(a):** Light micrograph of a section of skin from the sole of the foot (thick skin), showing only the stratum spinosum, highlights cells with numerous, short cytoplasmic projections (arrow). X400. PT. **(b):** TEMs show a single spinous keratinocyte with arrows marking some desmosomes at the projections. X8400. **(c):** Detail of the desmosomes joining two cells showing intermediate filaments associated with desmosomes. Keratinocytes of the stratum spinosum undergo considerable protein synthesis, primarily making keratins which form large bundles called tonofibrils. These terminate at the desmosomes linking the cells (arrows) and form the short cellular extensions characteristic of this layer when the cells shrink slightly. The black granules next to the nucleus are melanin granules. X40,000.

ingested material fuses with lysosomes. These are transported along keratinocyte microtubules via dynein to the region near the nucleus, where the melanosomes are released. Within each keratinocyte they accumulate as a supranuclear cap which absorbs and scatters sunlight, protecting nuclear DNA from the deleterious effects of UV radiation.

Although melanocytes synthesize melanin, the keratinocytes act as a depot and contain more of this pigment than the cells that make it. One melanocyte plus the keratinocytes into which it transfers melanosomes make up an **epidermal-melanin unit**. The density of such units is similar in all individuals. Melanocytes of people with ancestral origins near the equator, where the need for protection against the sun is greatest, produce melanin granules more rapidly and accumulate them in keratinocytes more abundantly. UV radiation causes keratinocytes to secrete various paracrine factors that stimulate melanocyte activity.

Darkening of the skin (tanning) after exposure to solar radiation (wavelength of 290–320 nm) is the result of a two-step process. First, a physicochemical reaction darkens preexisting melanin. Next, the rates of melanin synthesis in the melanocytes and transfer to keratinocytes accelerate, increasing the amount of this pigment.

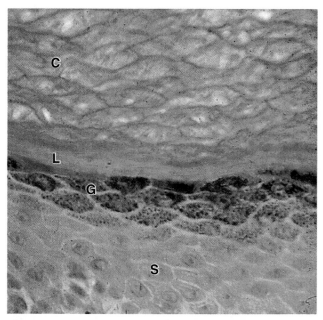

Figure 18–5. Stratum granulosum and stratum lucidum: thick skin. In keratinocytes moving upward from the stratum spinosum (S) differentiation proceeds with the cells becoming filled with numerous large, amorphous masses of protein called keratohyaline granules which are highly basophilic. Cells that contain such granules form a stratum granulosum (G) only three to five cells thick, where keratin filaments are cross-linked with filaggrin and other proteins from these granules to produce tight bundles filling the cytoplasm and flattening the cells. Smaller organelles called lamellar granules undergo exocytosis in this layer, secreting a lipid-rich layer around the cells which makes the epidermis impermeable to water. Together the lipid envelope and the keratin-filled cells determine most of the physical properties of the epidermis. The cells leaving the stratum granulosum, still bound together by desmosomes, undergo terminal differentiation and in thick skin appear as a dense, thin layer called the stratum lucidum (L). The acidic proteins in the granular, basophilic masses are dispersed through the tonofibril bundles, giving the cells of this new layer an eosinophilic, clear appearance. In the most superficial layers, the stratum corneum (C), the cells are fully differentiated and have lost nuclei and cytoplasm. They consist only of flattened, keratinized structures called squames bound by hydrophobic, lipid-rich intercellular cement and at the surface they are worn away (thick skin) or flake off (thin skin). X560. H&E.

MEDICAL APPLICATION

*In humans, lack of cortisol from the adrenal cortex causes overproduction of adrenocorticotropic hormone (ACTH), which can increase the pigmentation of the skin. An example of this is **Addison disease**, which is caused by dysfunction of the adrenal glands.*

* **Albinism**, a hereditary inability of the melanocytes to synthesize melanin, is caused by the absence of tyrosinase activity or the inability of cells to take up tyrosine. As a result, the skin is not protected from solar radiation by melanin, and there is a greater incidence of basal and squamous cell carcinomas (skin cancers).*

* The degeneration and disappearance of entire melanocytes causes a patchy loss of pigment in the skin disorder called **vitiligo**.*

Dendritic (Langerhans) Cells

Antigen-presenting **dendritic cells** (**Langerhans cells**), which are usually most clearly seen in the spinous layer, represent 2–8% of the epidermal cells. Cytoplasmic processes extend from these dendritic cells between keratinocytes of all the layers, forming a fairly dense network in the epidermis (Figure 14–6). They are bone marrow–derived, blood-borne cells, capable of binding, processing, and presenting antigens to T lymphocytes in the same manner as immune dendritic cells in other organs. Microorganisms cannot penetrate the epidermis without alerting its dendritic cells and triggering an immune response. Langerhans cells, along with more scattered epidermal lymphocytes and similar immune cells in the dermis, make up a major component of the skin's adaptive immunity.

Because of its location the skin is continuously in close contact with many antigenic molecules. Various epidermal features participate in both innate and adaptive immunity (Chapter 14), providing an immunological component to the skin's overall protective function.

Tactile (Merkel) Cells

Epithelial tactile cells (commonly called **Merkel cells**) are mechanoreceptors that resemble pale-staining keratinocytes with keratin filaments in their cytoplasm but few if any melanosomes. Small, Golgi-derived dense-core neurosecretory granules containing peptides like those of neuroendocrine cells are a characteristic

feature. Derived from neural crest cells, Merkel cells are located in the basal epidermal layer (Figure 18–2) in areas of high tactile sensitivity and at the bases of hair follicles. The basolateral surfaces of the cells contact expanded terminal discs of unmyelinated sensory fibers that penetrate the basal lamina (Figure 18–8). Tactile cells have functions related to the diffuse neuroendocrine system (Chapter 20) in addition to their contributions as mechanoreceptors in the sense of touch.

MEDICAL APPLICATION

Merkel cells are of clinical importance because an uncommon carcinoma derived from them is very aggressive and difficult to treat. Merkel cell carcinoma is 40 times less common than malignant melanoma, but has twice the mortality of that disease.

DERMIS

The **dermis** is the connective tissue (Figures 18–1 and 18–2) that supports the epidermis and binds it to the subcutaneous tissue (hypodermis). The thickness of the dermis varies according to the region of the body, and reaches its maximum of 4 mm on the back. The surface of the dermis is very irregular and has many projections (dermal papillae) that interdigitate with

projections (epidermal pegs or ridges) of the epidermis (Figure 18–1). Dermal papillae are more numerous in skin that is subjected to frequent pressure, where they reinforce the dermal-epidermal junction. During embryonic development, the dermal mesenchyme determines the differentiative fate of the overlying epidermis. For example, in mouse experiments, dermis obtained from the fetal foot sole always induces the formation of a thick, heavily keratinized epidermis irrespective of the site of origin of the epidermal cells.

A **basement membrane** is always found between the stratum basale and the papillary layer of the dermis and follows the contour of the interdigitations between these layers. The basement membrane is a composite structure consisting of the **basal lamina** and the **reticular lamina** and can usually be seen with the light microscope. Nutrients for keratinocytes must diffuse into the avascular epidermis from the dermis vasculature through this basement membrane.

MEDICAL APPLICATION

*Abnormalities of the dermal-epidermal junction can lead to one type of blistering disorder (**bullous pemphigoid**). Another type of blistering disorder (**pemphigus**) is caused by autoimmune damage to intercellular junctions between keratinocytes.*

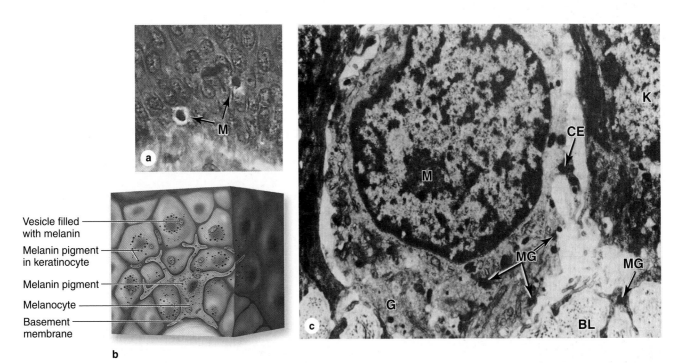

Figure 18–6. Melanocytes. (a): Micrograph showing melanocytes (M) in the epidermal basal layer which synthesize melanin granules and transfer them into neighboring keratinocytes of the basal and spinous layers. Typically melanocytes are pale-staining cells on the basement membrane, with lower total melanin content than the keratinocytes. X400. H&E. **(b):** Diagram of a melanocyte. It sends irregular dendritic processes between neighboring keratinocytes for transfer of melanin to those cells. **(c):** Ultrastructurally, a melanocyte is located on the basal lamina (BL) and has well-developed Golgi complexes (G) producing the vesicles in which melanin is synthesized. As they fill, these vesicles become melanin granules (MG), which accumulate at the tips of the dendritic cytoplasmic extensions (CE) before transfer to keratinocytes (K). X14,000.

322 / CHAPTER 18

The dermis contains two layers with rather indistinct boundaries—the outermost papillary layer and the deeper reticular layer (Figure 18–1). The thin **papillary layer**, which constitutes the major part of the dermal papillae, is composed of loose connective tissue, with fibroblasts and other connective tissue cells, such as mast cells and macrophages. Extravasated leukocytes are also seen. From this layer, **anchoring fibrils** of type VII collagen insert into the basal lamina and bind the dermis to the epidermis. The **reticular layer** is thicker, composed of irregular dense connective tissue (mainly bundles of type I collagen), and has more fibers and fewer cells than the papillary layer. A network of elastic fibers is also present (Figure 18–9), providing

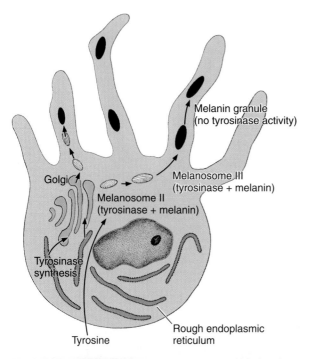

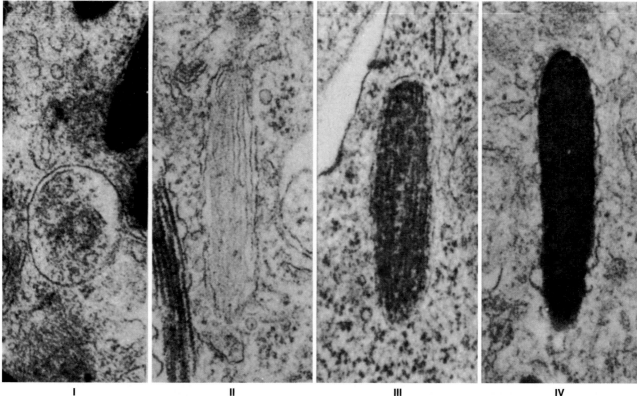

Figure 18–7. Melanosome formation. **(Upper part):** Diagram of a melanocyte, illustrating the main features of melanin formation. **(Lower part):** Maturation of the granules shown ultrastructurally. Tyrosinase is synthesized in the rough ER, is processed through the Golgi apparatus, and accumulates in vesicles that also have a fine granular matrix of other proteins (stage I melanosomes). Melanin synthesis begins in the ovoid stage II melanosomes, in which the matrix has been organized into parallel filaments on which polymerized melanin is deposited. Melanin accumulates on the matrix, forming a stage III melanosome and finally a mature melanin granule (stage IV) in which melanin completely fills the vesicle. This structure loses its tyrosinase activity and the internal matrix appears completely filled with melanin. The mature granules are ellipsoid, approximately 0.5 by 1 μm in size, and visible by light microscopy. Melanin granules are transported to the tips of the melanocyte's processes and are then transferred to the neighboring keratinocytes of the basal and spinous layers. In keratinocytes the melanin granules are transported to a region near the nucleus, where they accumulate as a supranuclear cap shading the DNA against the harmful effects of UV radiation.

elasticity to the skin. Spaces between the collagen and elastic fibers are filled with proteoglycans rich in dermatan sulfate.

The dermis is the site of such epidermal derivatives as the hair follicles and glands. There is also a rich supply of nerves in the dermis. The effector nerves to dermal structures are postganglionic fibers of sympathetic ganglia; no parasympathetic innervation is present. Sensory afferent nerve fibers form a network in the papillary dermis and around hair follicles, ending at epithelial tactile cells, at the encapsulated sensory receptors in dermis, and as free (uncovered) nerve endings among cells of the epidermis.

SUBCUTANEOUS TISSUE

The **subcutaneous layer** (Figure 18–1) consists of loose connective tissue that binds the skin loosely to the subjacent organs, making it possible for the skin to slide over them. This layer, also called the hypodermis or superficial fascia, often contains fat cells that vary in number in different regions of the body and vary in size according to nutritional state. An extensive vascular supply in the subcutaneous layer promotes rapid uptake of insulin or drugs injected into this tissue.

VESSELS & SENSORY RECEPTORS

The connective tissue of the skin contains a rich network of blood and lymphatic vessels. Blood vessels that nourish the cells of skin form two major plexuses (Figure 18–1): deep at the interface

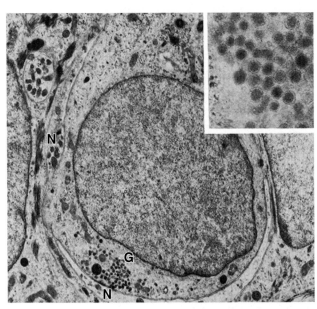

Figure 18–8. Tactile (Merkel) cell. Epithelial tactile cells in the basal epidermal layer of skin with high tactile sensitivity are neural crest-derived cells that function as mechanoreceptors. TEM of a tactile cell from the finger epidermis of a 21-week fetus shows a mass of dense-core cytoplasmic granules (G) near the basolateral cell membrane, which is in direct contact with the expanded ending of a nerve (N). X14,000. **Inset:** Granules are similar in morphology and content to the granules of many neuroendocrine cells. X61,500. (Reproduced, with permission from Fitzpatrick TB et al: *Dermatology in General Medicine.* The McGraw-Hill Companies, 2008.)

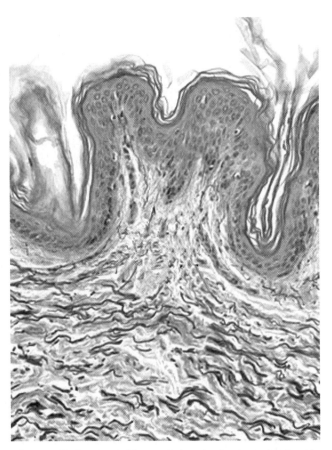

Figure 18–9. Elastic fibers of dermis. Section of thin skin stained for elastic fibers shows the extensive distribution of darkly stained elastic fibers among the eosinophilic collagen bundles. In the dermal papillary layer, the diameter of fibers decreases as they approach the epidermis and insert into the basement membrane. X100. Weigert elastic stain.

between hypodermis and dermis, and superficially between the papillary and reticular dermal layers. The latter **subpapillary plexus** sends branches into the dermal papillae and supplies a rich, nutritive capillary network just below the epidermis.

In addition to the nutritive function, dermal vasculature has a thermoregulatory function which involves numerous **arteriovenous anastomoses** or shunts (Chapter 11) located between the two major plexuses. The shunts decrease blood flow in the papillary layer to minimize heat loss in cold conditions and increase this flow to facilitate heat loss when it is hot, thus helping maintain a constant body temperature. Lymphatic vessels begin as closed sacs in the dermal papillae and converge to form two plexuses located with the blood vessels.

With its large surface and external location, the skin functions as an extensive receiver for certain stimuli from the environment.

A variety of sensory receptors are present in skin, including both simple nerve endings with no glial or collagenous covering and more complex structures with sensory fibers enclosed by glia and delicate connective tissue capsules (Figure 18–10). The *unencapsulated* receptors include the following:

- **Tactile discs** associated with the epidermal tactile cells (Figure 18–8), which function as receptors for light touch.
- **Free nerve endings** in the papillary dermis and extending into lower epidermal layers, which respond primarily to high and low temperatures, pain, and itching, but also function as tactile receptors.
- **Root hair plexuses**, a web of sensory fibers surrounding the bases of hair follicles in the reticular dermis that detects movements of the hairs.

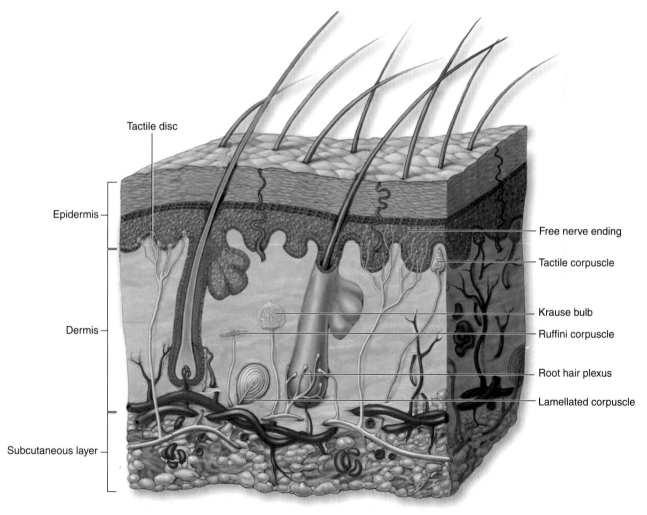

Figure 18–10. Tactile receptors. Skin contains several types of sensory receptors, primarily involved in the sense of touch. In the epidermis are free nerve endings and tactile discs on nerve fibers associated with tactile (Merkel) cells of the basal layer. Both have unencapsulated nerve fibers, as does the root hair plexus around the bases of hair follicles in the dermis. They detect light touch or movement of hair, although epidermal free nerve endings also detect pain and temperature extremes. More complex tactile receptors encapsulated with connective tissue layers are all in the dermis and include tactile (Meissner) receptors (light touch), lamellated (Pacinian) corpuscles (pressure and high-frequency vibration), Krause bulbs (pressure and low-frequency vibrations), and Ruffini corpuscles (continuous pressure and tissue distortion). The latter two receptors are less well characterized structurally and functionally.

The following *encapsulated* receptors are tactile mechanoreceptors:

- **Tactile corpuscles** (also called **Meissner corpuscles**) are elliptical structures, about 30–75 μm by 150 μm, perpendicular to the epidermis in the dermal papillae (Figure 18–11a) and papillary layer of the fingertips, palms and soles. They detect light touch.
- **Lamellated (Pacinian) corpuscles** are large oval structures, approximately 0.5 mm by 1 mm, found deep in the reticular dermis or hypodermis, with an outer capsule and 15 to 50 thin, concentric lamellae of flat Schwann-type cells and collagen surrounding a highly branched, unmyelinated axon (Figure 18–11b). Lamellated corpuscles are specialized for sensing coarse touch, pressure (sustained touch), and vibrations, with distortion of the capsule amplifying a mechanical stimulus to the axonal core where an impulse is initiated.

- Krause corpuscles and Ruffini corpuscles are other encapsulated, pressure-sensing mechanoreceptors in dermis, but are more poorly characterized structurally (Figure 18–10).

The encapsulated, lamellated mechanoreceptors are also found in the connective tissue of organs located deep in the body, including the wall of the rectum and urinary bladder, where they also produce the sensation of pressure when the surrounding tissue is distorted.

HAIR

Hairs are elongated keratinized structures derived from invaginations of the epidermal epithelium called **hair follicles** (Figure 18–12). The color, size, shape and texture of hairs vary according to age, genetic background, and region of the body. All skin has at least minimal hair except that of the palms, soles, lips, glans penis,

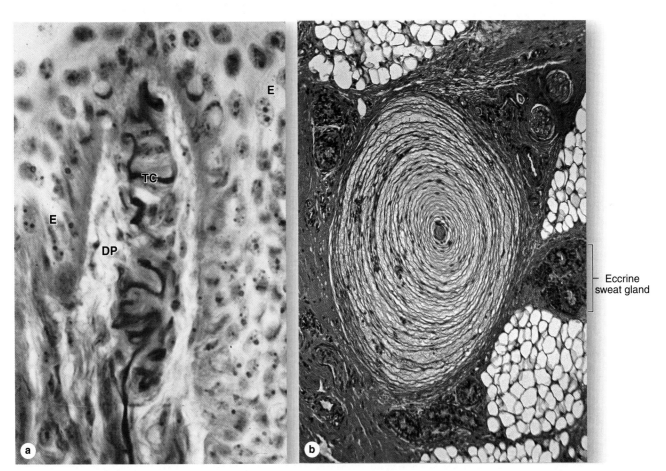

Figure 18–11. **Tactile and lamellated corpuscles.** Micrographs showing the two most commonly seen sensory receptors of skin. **(a):** Tactile (Meissner) corpuscle. X400. H&E. **(b):** Lamellated (Pacinian) corpuscle. X40. H&E. Tactile corpuscles (TC) are specialized to detect light touch and are frequently located in dermal papillae (DP), very close to the epidermis (E). They are elliptical in shape, approximately 150 μm long, with an outer capsule (from the perineurium) and thin, stacked inner layers of modified Schwann cells, around which course several nerve fibers.

Lamellated corpuscles detect coarse touch or pressure and are much larger oval structures, frequently 1 mm in length, found deep in the reticular dermis near the subcutaneous tissue. Here the outer connective tissue capsule surrounds 15 to 50 thin, concentric layers of modified Schwann cells, each separated by slightly viscous interstitial fluid. Several axons enter one end of the corpuscle and lie in the cylindrical, inner core of the structure. Movement or pressure of this corpuscle from any direction displaces the inner core, leading to a nerve impulse.

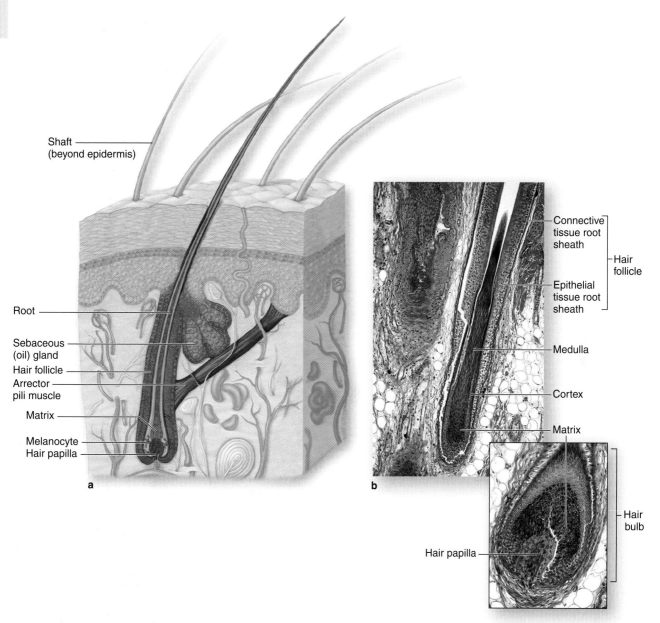

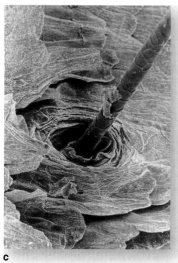

Figure 18–12. Hair. All types of body hair have a similar composition and form in hair follicles derived from the epidermis but extending deep into the dermis. One or more sebaceous glands form from the same epidermal down-growth and the entire structure is referred to as a pilosebaceous unit, which contains a specialized niche with stem cells for keratinocytes of the unit and neighboring epidermis. **(a):** Schematic diagram shows major parts of a hair follicle, including the arrector pili muscle which pulls the hair erect and sebaceous glands which empty into the follicle near the epidermis.

(b): Micrograph shows the medulla and cortex in the root of a hair cut longitudinally and the epithelial and connective tissue sheaths surrounding the growing hair. Both figures show the dermal hair papilla with microvasculature entering the base of the follicle. This nutritive papilla is surrounded by a matrix of epithelial cells similar to those of the stratum germinativum. Cells of the matrix proliferate, take up melanin granules, and undergo keratinization to differentiate as the three concentric layers of the hair. The outermost layer of the hair is the thin cuticle, composed of shingle-like cells. X70 and X180. H&E. **(c):** SEM shows the cuticle on a shaft of hair emerging at the stratum corneum from its follicle. X260.

clitoris, and labia minora. The face has about 600 hairs/cm² and the remainder of the body has about 60/cm². Hairs grow discontinuously, with periods of growth followed by periods of rest, and this growth does not occur synchronously in all regions of the body or even in the same area. The duration of the growth and rest periods also varies according to the region of the body. In the scalp, growth periods (**anagen**) may last for several years, whereas the periods of follicle regression (**catagen**) and inactivity (**telogen**) may together last only 3 to 4 months. Hair growth on the face and pubis is strongly influenced by sex hormones, especially androgens.

During anagen the hair follicle has a terminal dilatation called a **hair bulb** (Figure 18–12). A **dermal papilla** inserts into the base of the hair bulb and contains a capillary network required to sustain the hair follicle. Loss of this blood flow results in death of the follicle. The epidermal cells covering this dermal papilla form the **hair root** that produces and is continuous with the **hair shaft** protruding beyond the skin surface.

The epithelial cells (keratinocytes) that make up the hair bulb are similar to those in the basal and spinous layers of epidermis. They divide constantly and then undergo keratinization, differentiating into specific cell types. In certain types of thick hairs, the cells of the central region of the root at the apex of the dermal papilla produce large, vacuolated, and moderately keratinized cells that form the **medulla** of the hair (Figures 18–12b and 18–13).

Other cells differentiate into heavily keratinized, compactly grouped fusiform cells that form the hair **cortex**. The most peripheral cells produce the hair **cuticle,** a thin layer of heavily keratinized, shingle-like cells covering the cortex (Figures 18–12c and 18–13). Melanocytes in the hair bulb transfer melanin granules into the epithelial cells that will later differentiate to form the hair.

The outermost cells of the hair bulb are continuous with the epithelial root sheath, in which two layers can be recognized. The **internal root sheath** completely surrounds the initial part of the hair shaft but degenerates above the level of the attached sebaceous glands. The **external root sheath** covers the internal sheath and extends all the way to the epidermis, where it is continuous with the basal and spinous layers.

Separating the hair follicle from the dermis is an acellular hyaline layer, the thickened basement membrane called the **glassy membrane** (Figure 18–13). The surrounding dermis forms a connective tissue sheath. Running from a midpoint on this sheath and to the dermal papillary layer is a small bundle of smooth muscle cells, the **arrector pili muscle** (Figures 18–1 and 18–12). Contraction of these muscles pulls the hair shafts to a more erect position, usually when it is cold in an effort to trap a layer of warm air near the skin. In regions where hair is fine, contraction of arrector pili muscles is seen to produce tiny bumps on the skin surface ("goose bumps") where each contracting muscle distorts the attached dermis.

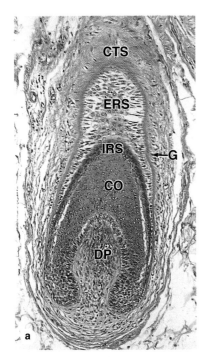 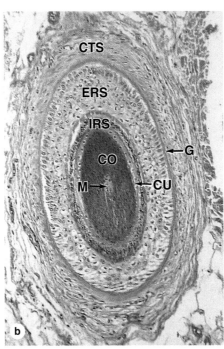 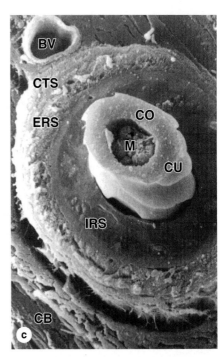

Figure 18–13. Layers of a hair and its follicle. **(a):** The base of a hair follicle sectioned obliquely shows the vascularized dermal papilla (DP) continuous with the surrounding connective tissue sheath (CTS). The papilla is surrounded by the deepest part of the epithelial sheath, which is continuous with both the internal root sheath (IRS) and external root sheath (ERS). Both of these layers are in turn continuous with the stratified epidermis. Just outside the ERS is the glassy membrane (G) which is continuous with the basement membrane of the epidermis. The epithelial cells (keratinocytes) around the papilla proliferate and differentiate as the root of the hair itself. Above the papilla only the cortex (CO) of the hair is clearly seen in this section. X140. H&E. **(b):** A hair root sectioned more transversely shows the same layers of the follicular sheath, but the layers of the hair root are now seen to include the medulla (M), cortex (CO), and cuticle (CU). X140. H&E. **(c):** SEM of a similar specimen gives a different perspective on these layers, including the shingle-like nature of the thin cuticle surface, and the small blood vessel (BV) and collagen bundles (CB) in the surrounding dermis. X2600. (Figure 18–13c, with permission, from W.H. Freeman & Co., Kessel, R.G. and Kardon, R.H., 1979, *Tissues and Organs: A Text-Atlas of Scanning Electron Microscopy.*)

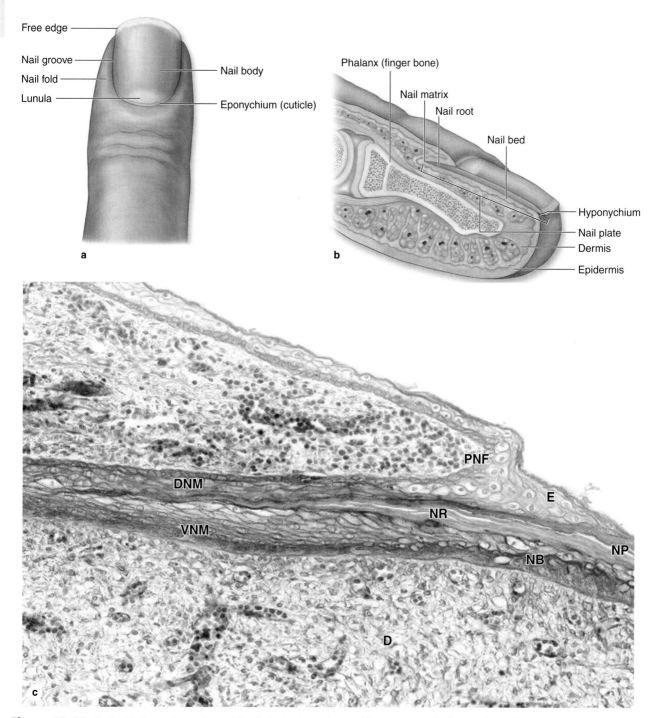

Free edge
Nail groove
Nail fold
Lunula
Nail body
Eponychium (cuticle)

a

Phalanx (finger bone)
Nail matrix
Nail root
Nail bed
Hyponychium
Nail plate
Dermis
Epidermis

b

PNF
DNM
E
NR
VNM
NP
NB
D

c

Figure 18–14. **Nails.** Nails are hard, keratinized derivatives formed in a process similar to that of the stratum corneum and hair. **(a):** Surface view of a finger shows the nail's major parts, including the crescent-shaped white area called the lunula, which derives its color from the opaque nail matrix and immature nail plate below it. **(b):** Diagram of a sagittal section includes major internal details and shows the hyponychium at which the free end of the nail plate is bound to epidermis.

(c): A micrograph of a sagittal section from a fetal finger shows of the proximal nail fold (PNF) and its epidermal extension, the eponychium (E) or cuticle. The nail root (NR), the most proximal region of the nail plate (NP), is formed like the hair root by a matrix of proliferating, differentiating keratinocytes. These cells make up the dorsal nail matrix (DNM) and ventral nail matrix (VNM), which contribute keratinized cells to the nail root. The mature nail plate remains attached to the nail bed (NB), which consists of basal and spinous epidermal layers over dermis (D), but is pushed forward on this bed by continuous growth in the nail matrix. X100. Mallory trichrome.

Hair color is produced by the activity of melanocytes located between the papilla and the epithelial cells of the hair root. The melanocytes produce and transfer melanin granules to these keratinocytes by a mechanism that is generally similar to that described for the epidermis. However, keratinization to produce hair does differ in some respects. Unlike epidermal keratinization where terminal differentiation of all cells gives rise to the stratum corneum, cells in the hair root differentiate into the cell types of the hair medulla, cortex, and cuticle which differ somewhat in ultrastructure, histochemical characteristics, and function. Keratin of hair has a harder and more compact nature that that of stratum corneum, maintaining its structure much longer. Finally, although keratinization in the epidermis occurs continuously and over the entire surface, it is intermittent in the hair and occurs only in the hair root.

NAILS

A similar process of keratinization produces the **nails**, which are hard, flexible plates of keratin on the dorsal surface of each distal phalanx (Figure 18–14). The proximal part of the nail is the **nail root** and is covered by the proximal skin fold which is thin and lacks both hair and glands. The epidermal stratum corneum extending from the proximal nail fold forms the **cuticle**, or **eponychium**. The keratinized **nail plate** is bound to a bed of

epidermis called the **nail bed**, which contains only the basal and spinous layers (Figure 18–14). The nail plate arises from the **nail matrix,** which extends from the nail root. Cells of the matrix divide, move distally, and become keratinized, forming the nail root. This matures as the nail plate, which continuous growth in the matrix pushes forward over the nail bed (which makes no contribution to the plate) at about 3 mm/month for fingernails and 1 mm/month for toenails. The distal end of the plate becomes free of the nail bed at the epidermal fold called the **hyponychium** and is worn away or cut off. The nearly transparent nail plate and the thin epithelium of the nail bed provide a useful window on the amount of oxygen in the blood by showing the color of blood in the dermal vessels.

GLANDS OF THE SKIN

Sebaceous Glands

Sebaceous glands are embedded in the dermis over most of the body surface, except the thick, hairless (glabrous) skin of the palms and soles. There is an average of about 100 such glands per square centimeter of skin, but the frequency increases to 400–900/cm^2 in the face and scalp. Sebaceous glands are branched acinar glands with several acini converging at a short duct which usually empties into the upper portion of a hair

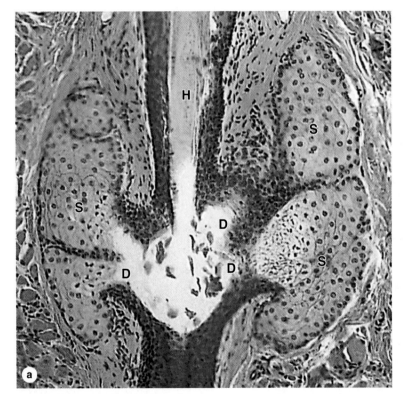
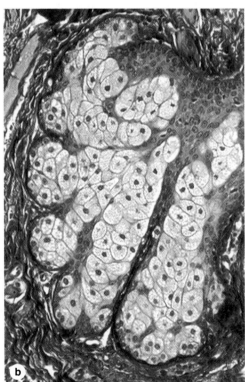

Figure 18–15. **Sebaceous glands.** Sebaceous glands secrete a complex mixture of lipids called sebum into short ducts which generally open into hair follicles **(a):** Micrograph shows small cells near the connective tissue capsule which proliferate and give rise to an acinus composed of large sebocytes (S), which undergo terminal differentiation by filling with small lipid droplets and then disintegrating at the ducts (D) near the hair (H) shaft, with the loss of nuclei and other organelles. X122. H&E. **(b):** Micrograph showing the gland's capsule and differentiating sebocytes at higher magnification X400. H&E. Sebum production is the classic example of holocrine secretion, in which the entire cell dies and contributes to the secretory product. Steady proliferation of the peripheral cells inside the capsule pushes sebum slowly and continuously into the ducts. Myoepithelial cells are not present.

follicle (Figure 18–12). The bulge region of the follicle is a stem cell niche generating cells of the hair follicle and matrix, the neighboring epidermis, and associated sebaceous glands. In certain hairless regions, such as the genital glands, eyelids, and nipples, sebaceous ducts open directly onto the epidermal surface. The acini consist of a basal layer of undifferentiated flattened epithelial cells on the basal lamina. These cells proliferate and are displaced toward the middle of the acinus, undergoing terminal differentiation as distinctly large, lipid-producing **sebocytes** which have their cytoplasm filled with small fat droplets (Figure 18–15). Their nuclei shrink and undergo autophagy along with other organelles and near the duct the cells disintegrate and release the lipids via holocrine secretion. The product of this process is **sebum,** which is gradually moved to the surface of the skin along the hair follicle or duct.

Sebum is a complex mixture of lipids that includes wax esters, squalene, cholesterol and triglycerides which are hydrolyzed by bacterial enzymes after secretion. Secretion from sebaceous glands greatly increases at puberty, stimulated primarily by testosterone in men and by ovarian and adrenal androgens in women. Specific functions of sebum appear to include helping maintain the stratum corneum and hair, as well as exerting weak antibacterial and antifungal properties on the skin surface.

MEDICAL APPLICATION

*The flow of sebum is continuous, and a disturbance in the normal secretion and flow of sebum is one of the reasons for the development of **acne**, a chronic inflammation of obstructed sebaceous glands common during and after puberty.*

Sweat Glands

Sweat glands are epithelial derivatives embedded in the dermis which open to the skin surface (Figure 18–1) or into hair follicles. Eccrine sweat glands and apocrine sweat glands have different distributions, functions, and structural details.

Eccrine sweat glands (Figures 18–16 and 18–17) are widely distributed in the skin and are most numerous on the soles of the feet ($620/cm^2$). Collectively the 3 million eccrine sweat glands of the average person roughly equal the mass of a kidney and can produce as much as 10 L/day, a secretory rate far exceeding

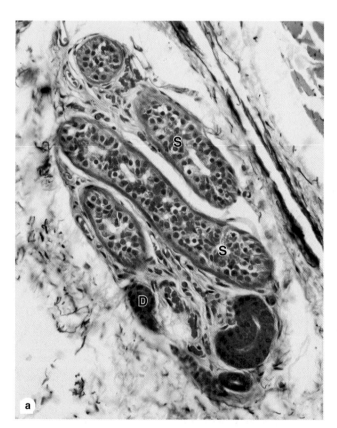
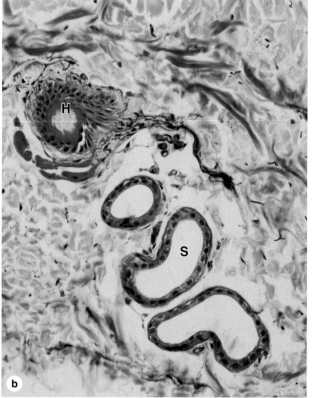

Figure 18–16. **Eccrine and apocrine sweat glands. (a):** Micrograph of an eccrine sweat gland which plays an important thermoregulatory function in production of fluid which evaporates on the body surface, thereby cooling that surface. Histologically eccrine glands have small lumens in the secretory portions (S) and the ducts (D), both of which have an irregular stratified cuboidal appearance. **(b):** Apocrine sweat glands are restricted mainly to the axillae and perineum and produce a more protein-rich secretion with pheromonal properties. The lumens of apocrine gland secretory portion (S) are much larger than those of eccrine glands and their ducts open into hair follicles (H) rather than to the epidermal surface. Both X200. Mallory trichrome.

that of other exocrine glands. Sweating is the physiological response to increased body temperature during physical exercise or thermal stress and in humans the most effective means of temperature regulation.

Both the secretory portions and ducts of eccrine sweat glands are coiled and have small lumens. The secretory part is generally more pale-staining than the ducts and has stratified cuboidal epithelium consisting of three cell types (Figure 18–17). Pale pyramidal or columnar **clear cells** produce the sweat, having abundant mitochondria and microvilli to provide large surface areas. Interstitial fluid from the capillary-rich dermis around the gland is transported through the clear cells, either directly into the lumen or into intercellular canaliculi that open to the lumen. As numerous as the clear cells are pyramidal **dark cells** which line most of the luminal surface and do not touch the basal lamina (Figure 18–17). Dark cells are mucoid and filled with glycoprotein-containing granules whose functions are not well-understood but include components of innate immunity with bactericidal activity. **Myoepithelial cells** on the basal lamina (Figure 4–27) produce contractions that help discharge secretion into the duct.

The ducts of eccrine sweat glands consist of two layers of more acidophilic epithelial cells filled with mitochondria and having membranes rich in Na^+, K^+-ATPase. These duct cells

absorb Na^+ ions to prevent excessive loss of this electrolyte. After its release on the surface of the skin sweat evaporates, cooling the skin. Besides its important cooling role, sweat glands also function as an auxiliary excretory organ, eliminating small amounts of nitrogenous waste and excess salts.

Apocrine sweat glands are largely confined to skin of the axillary and perineal regions. Their development (but not functional activity) depends on sex hormones and is not complete until puberty. The most obvious histological difference between the two kinds of sweat glands is the much larger lumen of apocrine glands (Figure 18–16). The secretory portions of apocrine sweat glands consist of simple cuboidal, eosinophilic cells with numerous apical secretory granules that undergo exocytosis. Thus the glands are misnamed: their cells show merocrine, not apocrine, secretion. Lumens of apocrine glands often show stored, protein-rich product, which myoepithelial cells help move into ducts opening into hair follicles. The wall of the ducts is similar to that of the eccrine glands. The slightly viscous secretion is initially odorless but may acquire a distinctive odor as a result of bacterial activity. The production of pheromones by apocrine glands is well-established in many mammals and likely in humans, although in a reduced or vestigial capacity. Apocrine sweat glands are innervated by adrenergic nerve endings, whereas eccrine sweat glands receive cholinergic fibers.

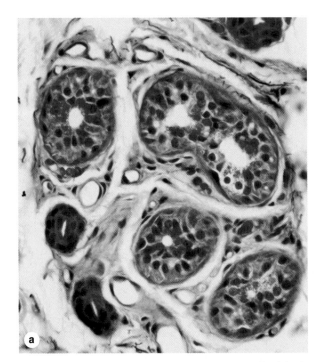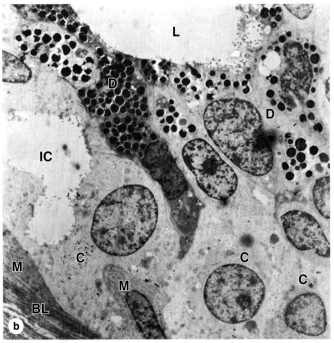

Figure 18–17. **Eccrine sweat gland secretory cells. (a):** The secretory portions of eccrine sweat glands have a stratified cuboidal epithelium, containing cell types with different staining properties. Cells closest to the lumen contain eosinophilic granules. X400. Mallory trichrome.

(b): TEM reveals three cell types. Myoepithelial cells (M) are thin cells present at the basal lamina (BL). Irregular pyramidal cells called dark cells (D) border the lumen (L) and are filled with the eosinophilic secretory granules which are electron-dense. Among the products released from these granules are bactericidal peptides and other components of innate immunity. Clear cells (C) are columnar and have their basal ends on the basal lamina and function in the rapid transport of water from interstitial fluid in the capillary-rich dermis directly into the lumen or into intercellular canaliculi (IC) which are continuous with the lumen. Na^+ ions are recovered from this fluid through the action of cells in the ducts, which are seen in the lower left corner of (a). X6500.

The Urinary System

The urinary system consists of the paired kidneys and ureters, the bladder, and the urethra. This system helps maintain homeostasis by a complex combination of processes that involves the following:

- **Filtration of cellular wastes from blood**
- **Selective reabsorption of water and solutes**
- **Excretion of the wastes and excess water as urine.**

Urine produced in the kidneys passes through the ureters to the bladder for temporary storage and is then released to the exterior through the urethra. The two kidneys produce about 125 mL of filtrate per minute, of which 124 mL is reabsorbed in these organs and 1 mL is released into the ureters as urine. About 1500 mL of urine is formed every 24 hours. The kidneys also regulate the fluid and electrolyte balance of the body and are the site of production of **renin**, a protease that participates in the regulation of blood pressure by cleaving circulating angiotensinogen to angiotensin I. **Erythropoietin**, a glycoprotein that stimulates the production of erythrocytes, is also produced in the kidneys. The steroid prohormone vitamin D, initially produced in skin keratinocytes, is hydroxylated in kidneys to an active form (**1,25-dihydroxyvitamin D$_3$** or **calcitriol**) involved in regulating calcium balance.

KIDNEYS

Each kidney has a concave medial border, the **hilum**—where nerves enter, the ureter exits, and blood and lymph vessels enter and exit—and a convex lateral surface, both covered by a thin fibrous capsule (Figure 19–1). The expanded upper end of the ureter, called the **renal pelvis,** divides into two or three **major calyces.** Smaller branches, the **minor calyces,** arise from each major calyx. The area surrounding the calyces, called the **renal sinus**, usually contains considerable adipose tissue.

The kidney has an outer **cortex** and an inner **medulla** (Figures 19–1 and 19–2). In humans, the renal medulla consists of 8–15 conical structures called **renal pyramids**, which are separated by cortical extensions called **renal columns**. Each

medullary pyramid plus the cortical tissue at its base and along its sides constitutes a **renal lobe** (Figure 19–1).

Each kidney contains 1–1.4 million functional units called **nephrons** (Figure 19–2). The major divisions of each nephron are:

- **Renal corpuscle**, an initial dilated portion in the cortex
- **Proximal convoluted tubule**, located primarily in the cortex
- **Thin** and **thick limbs** of the **nephron loop** (loop of Henle), which descend into the medulla, then ascend back to the cortex
- **Distal convoluted tubule**
- **Collecting tubule**.

Collecting tubules from several nephrons converge into **collecting ducts** which carry urine to the calyces and the ureter. **Cortical nephrons** are located almost completely in the cortex while **juxtamedullary nephrons** close to the medulla have long loops in the medulla (Figure 19–2).

BLOOD CIRCULATION

As expected for an organ specialized to process the blood, the anatomical organization of the kidney vasculature and its associations with nephron components are very important. Blood vessels of the kidney are named according to their precise locations or shapes (Figure 19–3).

Each kidney receives blood from a **renal artery,** which divides into two or more segmental arteries at the hilum. In the renal sinus these branch further to form the **interlobar arteries** extending between the renal pyramids toward the corticomedullary junction (Figure 19–3). Here the interlobar arteries branch further to form the **arcuate arteries** which travel in an arc along this junction at the base of each renal pyramid. Smaller **interlobular arteries** branch off at right angles from the arcuate arteries and enter the cortex.

From the interlobular arteries arise the microvascular **afferent arterioles,** which supply blood to a tuft of capillaries called the **glomerulus,** each of which is associated with a renal corpuscle (Figures 19–3 and 19–4). Blood leaves the glomerular capillaries, not via venules, but via the **efferent arterioles,** which at once branch again to form another capillary network, the **peritubular capillaries** that nourish cells of the proximal and distal tubules and carry away reabsorbed substances. The efferent arterioles associated with glomeruli near the medulla continue as long, straight vessels directly into the medulla providing nutrients and oxygen there, and then loop back into the cortex as venules. These small medullary vessels and their intervening capillary plexuses comprise the **vasa recta** (L. *recta,* straight).

Blood leaves the kidney in veins that follow the same courses as arteries and have the same names (Figure 19–3). The outermost peritubular capillaries and capillaries in the kidney capsule converge into small stellate veins which empty into the interlobular veins.

Renal Corpuscles & Blood Filtration

At the beginning of each nephron is a renal corpuscle, about 200 μm in diameter and containing a loose knot of capillaries, the glomerulus, surrounded by a double-walled epithelial capsule called the **glomerular (Bowman's) capsule** (Figures 19–2 and 19–5). The internal layer (**visceral layer**) of the capsule closely envelops the glomerular capillaries . The external **parietal layer** forms the outer surface of the capsule. Between the two

capsular layers is the urinary or **capsular space,** which receives the fluid filtered through the capillary wall and the visceral layer. Each renal corpuscle has a **vascular pole,** where the afferent arteriole enters and the efferent arteriole leaves, and a urinary or **tubular pole,** where the proximal convoluted tubule begins (Figure 19–5). After entering the renal corpuscle, the afferent arteriole usually divides and subdivides into the two to five capillaries of the renal glomerulus.

The parietal layer of a glomerular capsule consists of a simple squamous epithelium supported externally by a basal lamina and a thin layer of reticular fibers. At the tubular pole, this epithelium changes to the simple cuboidal epithelium characteristic of the proximal tubule (Figure 19–5).

During embryonic development, the simple epithelium of the parietal layer remains relatively unchanged, whereas the internal or visceral layer is greatly modified. The cells of this layer, the **podocytes** (Figures 19–5d and 19–6), have a cell body from which arise several **primary processes.** Each primary process gives rise to numerous **secondary (foot) processes** or **pedicels** (L. *pedicellus,* little foot) that embrace a portion of one glomerular capillary (Figures 19–5d and 19–6). The cell bodies of podocytes do not contact the basement membrane of the capillary, but each pedicel is in direct contact with this structure (Figure 19–6).

The pedicels interdigitate, defining elongated spaces 30–40 nm wide—the **filtration slits** (Figure 19–6). Spanning adjacent processes (and thus bridging the filtration slits) is a thin semipermeable diaphragm of uniform thickness (Figure 19–6).

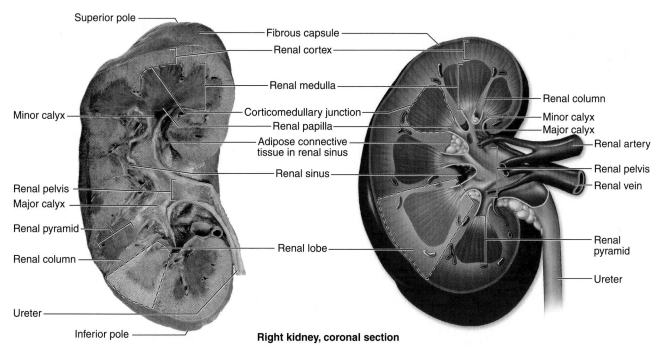

Figure 19–1. Kidney. Each kidney is bean-shaped, with a concave hilum where the ureter and the renal artery and veins enter. The ureter divides and subdivides into several major and minor calyces, around which is located the renal sinus containing adipose tissue. Division of the parenchyma into cortex and medulla can also be seen grossly. Attached to each minor calyx is a renal pyramid, a conical region of medulla delimited by extensions of cortex. A renal pyramid with associated cortex constitutes a renal lobe. The cortex and hilum are covered with a fibrous capsule.

These **slit diaphragms** are a highly specialized type of intercellular junction in which the large transmembrane protein **nephrin** is important both structurally and functionally. Projecting from the cell membrane on each side of the slit, nephrin molecules interact to form a porous structure within the diaphragm.

Between the highly fenestrated endothelial cells of the capillaries and the covering podocytes is the thick (~0.1 μm) glomerular basement membrane (Figure 19–6). This membrane is the most substantial part of the filtration barrier separating the blood in the capillaries from the capsular space. It is formed by the fusion of capillary- and podocyte-produced basal laminae and is maintained by the podocytes. Laminin and fibronectin in this fused basement membrane bind integrins of both the podocyte and endothelial cell membranes. The meshwork of type IV collagen cross-linked in a matrix of negatively charged proteoglycans may help restrict the passage of cationic molecules. Thus, the glomerular basement membrane (GBM) is a selective macromolecular barrier which acts as a physical filter and as a barrier against negatively charged molecules.

The initial glomerular filtrate has a chemical composition similar to that of the blood plasma except that it contains very little protein because macromolecules do not readily cross the glomerular filter. Proteins and other particles greater than 10 nm in diameter or exceeding 70 kDa, the approximate molecular mass of albumin, do not readily cross the glomerular barrier.

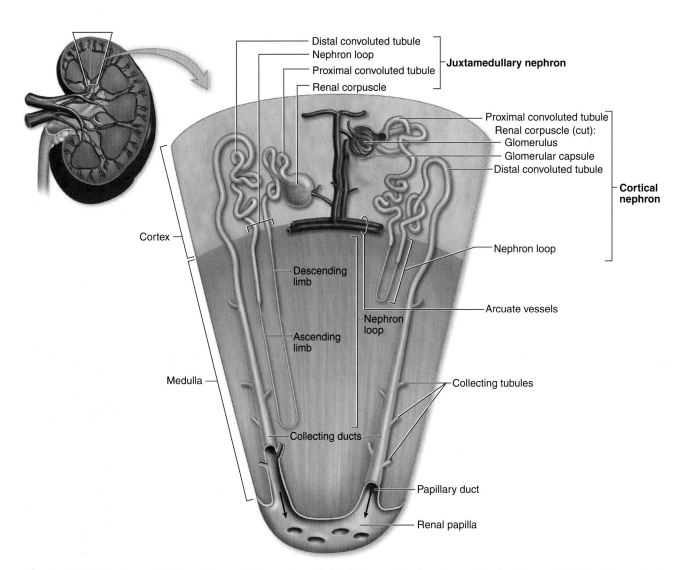

Figure 19–2. Nephrons. Within each renal lobe are hundreds of thousands of nephrons, the function unit of the kidney. Each nephron originates in the cortex, at the renal corpuscle associated with glomerular capillaries. Extending from the corpuscle is the proximal convoluted tubule, then the nephron loop (of Henle) into the medulla and back to the cortex, then the distal convoluted tubule and collecting tubule which merges into a collecting duct for urine transport to the calyx. All nephrons are located completely within the cortex except for their medullary loops. Juxtamedullary nephrons usually have much longer loops than cortical nephrons.

Glomerular capillaries are uniquely situated between two arterioles—afferent and efferent—the muscle of which allows increased hydrostatic pressure in these vessels, favoring movement of plasma across the glomerular filter. The glomerular filtration rate (GFR) is constantly regulated by neural and hormonal inputs affecting the degree of constriction in each of these arterioles. The total glomerular filtration area of an average adult has been estimated at 500 cm^2 and the average GFR at 125 mL per minute or 180 liters per day. Because the total amount of circulating plasma averages 3 L, it follows that the kidneys typically filter the entire blood volume 60 times every day.

In addition to capillary endothelial cells and podocytes, renal corpuscles also contain **mesangial cells** (Gr. *mesos*, in the midst, + *angeion*, vessel) (Figure 19–7), which resemble pericytes in producing components of an enveloping external lamina. Mesangial cells are difficult to distinguish in routine sections form podocytes, but often stain more darkly. They and their surrounding matrix comprise the mesangium (Figure 19-7), which fills interstices between capillaries that lack podocytes. Functions of the mesangium are many and varied, and include the following:

* Physical support and contraction—the mesangium provides internal structural support to the glomerulus and like pericytes, its cells respond to vasoactive substances to help maintain hydrostatic pressure for the optimal rate of filtration.
* Phagocytosis—mesangial cells phagocytose protein aggregates that adhere to the glomerular filter, including antibody-antigen complexes abundant in many pathological conditions.
* Secretion—the cells synthesize and secrete several cytokines, prostaglandins, and other factors important for immune defense and repair in the glomerulus.

Proximal Convoluted Tubule

At the tubular pole of the renal corpuscle, the squamous epithelium of the capsule's parietal layer is continuous with the cuboidal epithelium of the **proximal convoluted tubule** (Figures 19–8 and 19–9). This very tortuous tubule is longer than the distal convoluted tubule and is therefore more frequently seen in sections of renal cortex. Cells of the proximal tubule reabsorb 60–65% of the

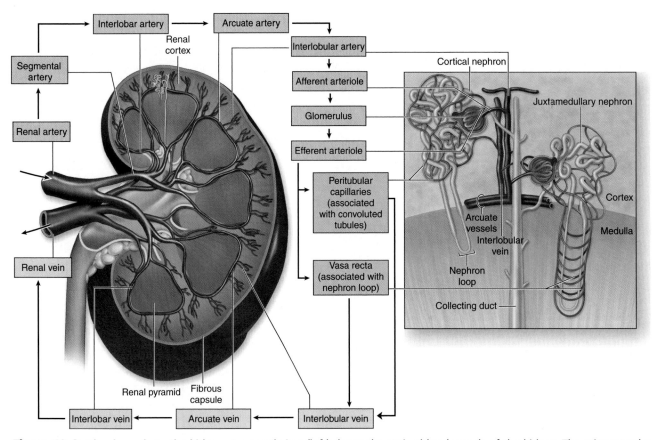

Figure 19–3. Blood supply to the kidney. A coronal view (left) shows the major blood vessels of the kidney. The microvascular components extending into the cortex and medulla from the interlobular vessels are shown on the right. Pink boxes indicate vessels with arterial blood and blue indicate the venous return. The intervening lavender boxes and vessels are intermediate sites where most reabsorbed material re-enters the blood.

water filtered in the renal corpuscle, along with almost all of the nutrients, ions, vitamins, and small plasma proteins. The water and its solutes are transferred directly across the tubular wall and immediately taken up by the **peritubular capillaries**.

The cells of the proximal tubules have acidophilic cytoplasm (Figures 19–8 and 19–9) because of the presence of numerous mitochondria. The cell apex has abundant long microvilli which form a prominent **brush border** for reabsorption (Figures 19–8,

19–9, and 19–10). Because the cells are large, each transverse section of a proximal tubule typically contains only three to five rounded nuclei. In routine histologic preparations, the brush border may be disorganized and give the lumens a fuzz-filled appearance. Capillaries and other microvascular components are abundant in the sparse surrounding connective tissue (Figure 19–8).

Ultrastructurally the apical cytoplasm of these cells has numerous pits and vesicles near the bases of the microvilli,

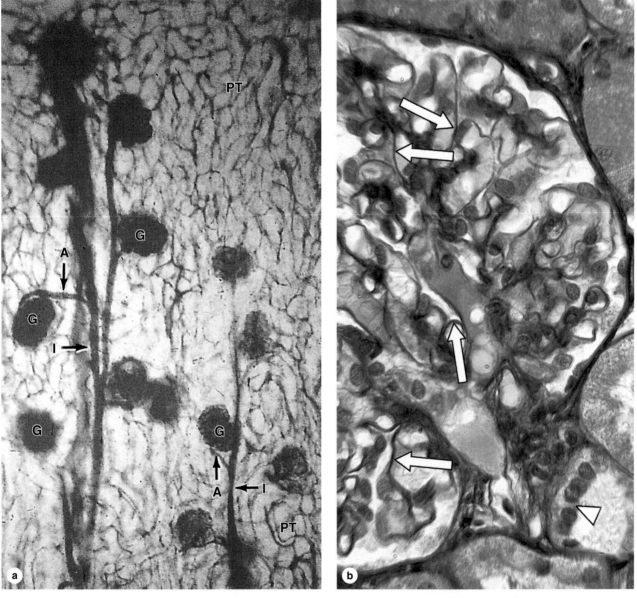

Figure 19–4. Microvasculature of the renal cortex. **(a):** Cortical vasculature is revealed in a section of kidney with the renal artery injected with carmine dye before fixation. Having branched at right angles off the arcuate arteries, small interlobular arteries (I) run straight out through the cortex and give off the afferent arterioles (A) which bring blood to the glomerular capillaries. Each glomerulus (G) contains a loose mass of capillaries with nearly a centimeter in total length. These drain into an efferent arteriole which then branches as a large, diffuse network of peritubular capillaries (PT) throughout the cortex. X125. **(b):** A section of one glomerulus shows many capillaries and the closely associated cells of the renal corpuscle's internal visceral layer. The thick basement membrane of these glomerular capillaries contains much type IV collagen and is visible around the cut capillaries (arrows). Also shown are the simple squamous external parietal layer of the capsule and the vascular pole where the arterioles enter the corpuscle and the macula densa (arrowhead) is located. X400. PSH.

indicating active pinocytosis (Figure 19–10). The **pinocytotic vesicles** contain small plasma proteins (with a molecular mass less than 70 kDa) that passed through the glomerular filter. The vesicles fuse with lysosomes for proteolysis and amino acids are released to the circulation. The cells also have many long basal **membrane invaginations** and lateral interdigitations with neighboring cells (Figure 19–10). The Na$^+$/K$^+$-ATPase (sodium pump) responsible for actively transporting sodium ions out of the cells is localized in these basolateral membranes. Long

mitochondria are concentrated along the basal invaginations (Figure 19–9), characteristically for cells engaged in active ion transport. Because of the extensive interdigitations of the lateral membranes, discrete limits between cells of the proximal tubule are difficult to see in the light microscope. The proximal convoluted tubules actively reabsorb all the glucose and amino acids in the filtrate and about 85% of the sodium chloride and other ions. This absorption involves the membrane sodium pumps. Water diffuses passively, following the osmotic gradient. When

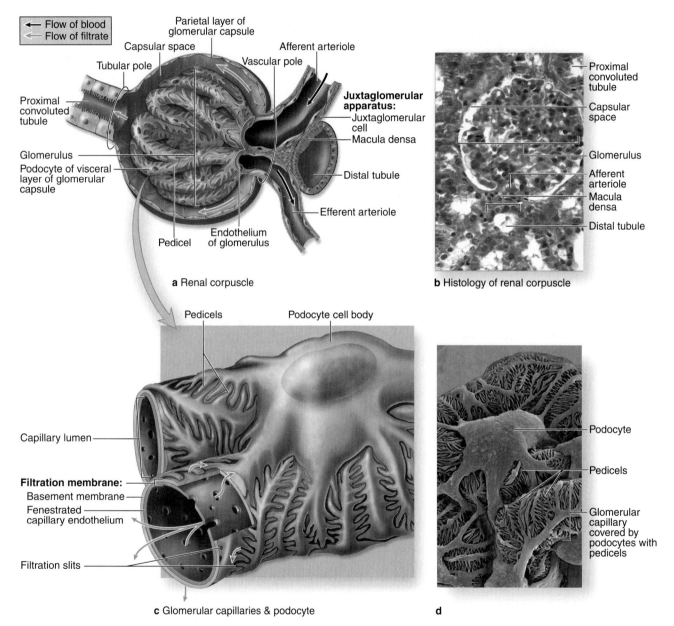

Figure 19–5. Renal corpuscles. **(a):** The renal corpuscle is a small mass of capillaries called the glomerulus housed within a bulbous glomerular capsule. The internal lining of the capsule is composed of complex epithelial cells called podocytes, which cover each capillary, forming filtration slits between interdigitating processes called pedicels. Blood enters and leaves the glomerulus through the afferent and efferent arterioles respectively. **(b):** The micrograph shows the major histological features of a renal corpuscle. H&E. X300. **(c):** Filtrate is produced in the corpuscle when blood plasma is forced under pressure across the filtration membrane of the glomerular capillary wall and through the filtration slits between the pedicels of podocyte processes. **(d):** The SEM shows the distinctive appearance of podocytes and their processes covering glomerular capillaries. X800

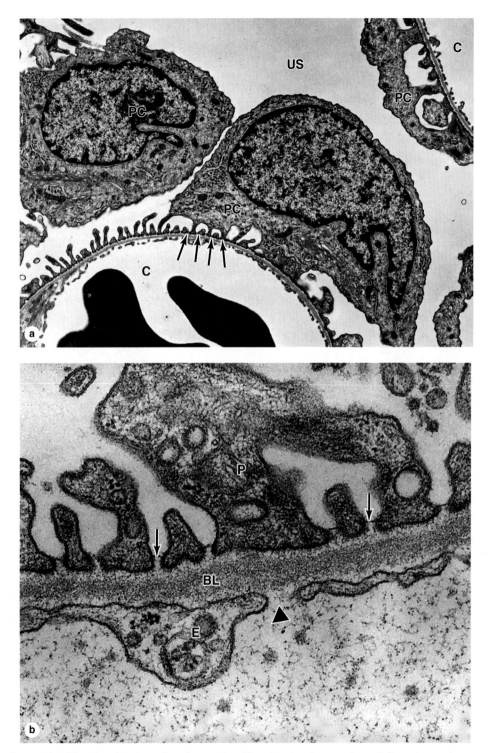

Figure 19–6. Glomerular filtration barrier. The glomerular filtration barrier consists of three layered components: the fenestrated capillary endothelium, the glomerular basement membrane, and filtration slits between podocyte processes. The major component of the filter is formed by fusion of the basal laminae of a podocyte and a capillary endothelial cell. **(a):** TEM showing cell bodies of two podocytes (PC) and the series of pedicels on the glomerular basement membrane separated by the filtration slits (arrows). On the other side of the membrane is the thin lining of a capillary (C) endothelial cell, with fenestrations. Together these openings allow filtration of liquid from plasma into the urinary space (US) of Bowman's capsule. X10,000. **(b):** At higher magnification, both the fenestrations (arrowhead) in the capillary endothelium (E) and the filtration slits (arrows) separating the pedicels (P) are better seen on the two sides of the fused basal laminae (BL). The endothelial fenestrations in glomeruli lack diaphragms, but very thin slit diaphragms cross the space between pedicels and play an important role in filtration. X45,750.

the amount of glucose in the filtrate exceeds the absorbing capacity of the proximal tubule, as in diabetes, urine becomes more abundant and contains glucose.

In addition to these activities, cells of the proximal convoluted tubules can also move substances from the peritubular capillaries into the tubular lumen, an active process referred to as tubular secretion. Organic anions such as choline and creatinine and many foreign compounds such as penicillin are excreted in this manner, which allows kidneys to dispose of such substances at a higher rate than by glomerular filtration alone. The cells of the proximal tubule are also involved in vitamin D hydroxylation.

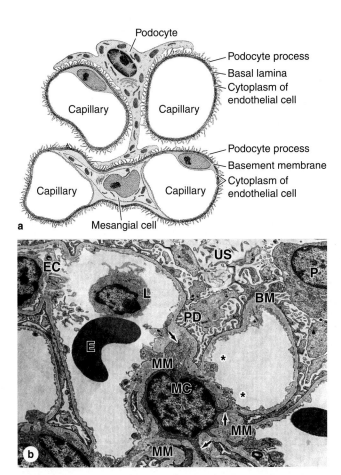

Figure 19–7. Mesangium. (a): Diagram showing that mesangial cells in renal corpuscles are located between capillaries and are enveloped by a dense extracellular matrix like that of the basement membrane around the capillaries. **(b):** The TEM shows one mesangial cell (MC) and the amorphous mesangial matrix (MM) surrounding it. This matrix appears similar and in many places continuous with basement membrane (BM). The matrix helps support capillary loops where podocytes are lacking. Mesangial cells extend processes (arrows) around capillaries which may affect their state of contraction. Some mesangial processes appear to pass between endothelial cells (EC) into the capillary lumen (asterisks) where they may help remove or endocytose adherent protein aggregates. The capillary at the left contains an erythrocyte (E) and a leukocyte (L). Podocytes (P) and their pedicels (PD) open to the urinary space (US) and associate with the capillary surfaces not covered by mesangial cells.

Nephron Loop (of Henle)

The proximal convoluted tubule continues as a much shorter proximal straight tubule which enters the medulla and becomes the **nephron loop**. This is a U-shaped structure with a **descending limb** and an **ascending limb**, both composed of simple epithelia, cuboidal near the cortex, but squamous deeper in the medulla (Figure 19–2). In the outer medulla, the straight portion of the proximal tubule, with an outer diameter of about 60 µm, narrows abruptly to about 12 µm and continues as the nephron loop's thin descending limb. The lumen of this segment of the nephron is wide and the wall consists of squamous epithelial cells whose nuclei protrude slightly into the lumen (Figures 19–9 and 19–11).

Approximately one seventh of all nephrons are located near the corticomedullary junction and are therefore called **juxtamedullary nephrons**, which are of prime importance in the mechanism that allows the kidneys to produce concentrated, hypertonic urine. Juxtamedullary nephrons usually have long loops, extending deep into the medulla, with short thick proximal straight segments, long thin descending and ascending limbs, and long thick ascending limbs (Figure 19–2).

The nephron loop and surrounding tissue are involved in making urine hypertonic and conserving water; only animals with such loops are capable of concentrating urine and thus conserving body water. Cuboidal cells of the loops' thick ascending limbs actively transport sodium chloride out of the tubule against a concentration gradient into the hyaluronate-rich interstitial connective tissue, making that compartment hyperosmotic. Squamous cells of the loops' thin descending limbs are freely permeable to water but not salts, while the thin ascending limbs are permeable to NaCl but impermeable to water. Flow of the filtrate in opposite directions (countercurrent flow) in the two parallel limbs of nephron loops establishes a gradient of osmolarity in the interstitium of the medullary pyramids and countercurrent blood flow in the loops of the **vasa recta** help maintain this gradient. The interstitial osmolarity at the pyramid tips is about four times that of the blood. The high interstitial osmolarity draws water passively from the collecting ducts in the medullary pyramids (Figures 19–2 and 19–11), concentrating the urine. Water permeability of these ducts is increased by antidiuretic hormone (ADH), which is released from the pituitary when body water is low. The water thus saved immediately enters the blood in the adjacent capillaries of the vasa recta. The role of the nephron loop and vasa recta in establishing the conditions for urine concentration is called the **countercurrent multiplier effect**.

Distal Convoluted Tubule & Juxtaglomerular Apparatus

The thick ascending limb of the nephron loop is straight as it enters the cortex, and then becomes tortuous as the **distal convoluted tubule** (Figure 19–2). The simple cuboidal cells of these tubules differ from those of the proximal convoluted tubules in being smaller and having no brush border (Figure 19–9). Because distal tubule cells are flatter and smaller than those of the proximal tubule, more nuclei are typically seen in sections of distal tubules than in those of proximal tubules (Figure 19–8). Cells of the distal convoluted tubule do have basal membrane invaginations and associated mitochondria similar to those of proximal tubules, indicating their similar ion-transporting function (Figure 19–9). The rate of Na^+ absorption and K^+ secretion by the sodium pumps is regulated by **aldosterone** from the adrenal glands and is important for the body's water-salt balance.

The distal tubule also secretes H^+ and NH_4^+ into tubular urine, an activity essential for maintenance of the acid-base balance in the blood.

The initial, straight part of the distal tubule makes contact with the vascular pole of the renal corpuscle of its parent nephron and forms part of a specialized structure, the **juxtaglomerular apparatus (JGA)** (Figures 19–9, 19–5, and 19-12). Cells of this structure establish a feedback mechanism that allows autoregulation of renal blood flow and keeps the rate of glomerular filtration relatively constant. At the point of contact with the arterioles, the cells of the distal tubule become columnar and more closely packed, with apical nuclei, basal Golgi complexes, and a more elaborate and varied system of ion channels and transporters. This thickened spot of the distal tubule wall is called

the **macula densa** (Figures 19–5 and 19–12). Adjacent to the macula densa, the tunica media of the afferent arteriole is also modified. The smooth muscle cells develop a secretory phenotype with more rounded nuclei, rough ER, Golgi complexes and zymogen granules and are called **juxtaglomerular granular (JG) cells** (Figures 19–5 and 19–12). Also at the vascular pole are **lacis cells** (Fr. *lacis*, meshwork), which are extraglomerular mesangial cells that probably have many of the same supportive functions as these cells inside the glomerulus. Lacis cells may also transmit signals from the macula densa into the glomerulus, affecting vasoconstriction there.

Basic functions of the JGA in the autoregulation of the glomerular filtration rate (GFR) and in controlling blood pressure are thought to include the following. Elevated arterial pressure

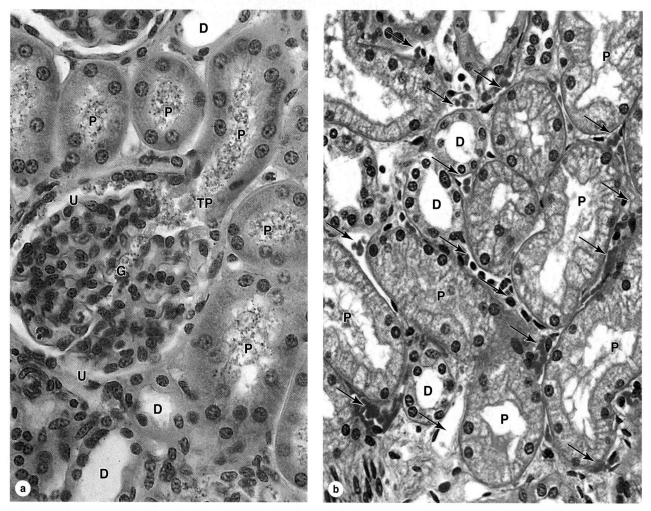

Figure 19–8. **Renal cortex: Proximal and distal convoluted tubules. (a):** The micrograph shows the continuity at a renal corpuscle's tubular pole (TP) between the simple cuboidal epithelium of a proximal convoluted tubule (P) and the simple squamous epithelium of the capsule's parietal layer. The urinary space (U) between the parietal layer and the glomerulus (G) drains into the lumen of the proximal tubule. The lumen of the proximal tubules appear filled, due to the long microvilli of the brush border and aggregates of small plasma proteins bound to this structure. By contrast, the lumens of distal convoluted tubules (D) appear empty, lacking a brush border and protein.

(b): In another section the abundant peritubular capillaries and draining venules surrounding the proximal and distal convoluted tubules are indicated (arrows). Fibroblastic interstitial cells of the cortex are the source of erythropoietin, the growth factor secreted in response to a prolonged decrease in the local oxygen concentration. Both X400. H&E.

increases glomerular-capillary pressure, which increases the GFR. Higher GFR leads to higher luminal concentrations of Na$^+$ and Cl$^-$ in nephrons which are monitored by cells of the macula densa. Increased ion levels cause these cells to release ATP, adenosine, and other vasoactive compounds that trigger contraction of the afferent arteriole, which lowers glomerular pressure and decreases the GFR. This lowers tubular ion concentrations, which turns off the release of vasoconstrictors from the macula densa.

Decreased arterial pressure leads to increased autonomic stimulation to the JGA as a result of baroreceptor function, including local baroreceptors in the afferent arteriole, possibly the JG cells themselves. This causes the JG cells to release their major secretory product **renin**, an aspartyl protease, into the blood. There renin cleaves the plasma protein **angiotensinogen** into the inactive decapeptide **angiotensin I**. Angiotensin converting enzyme (ACE) on lung capillaries clips this further to **angiotensin II**, a potent vasoconstrictor which directly raises systemic blood pressure and stimulates the adrenals to secrete **aldosterone**. Aldosterone promotes Na$^+$ and water reabsorption in the distal convoluted tubules, which raises blood volume to help increase blood pressure. The return of normal blood pressure turns off secretion of renin by JG cells.

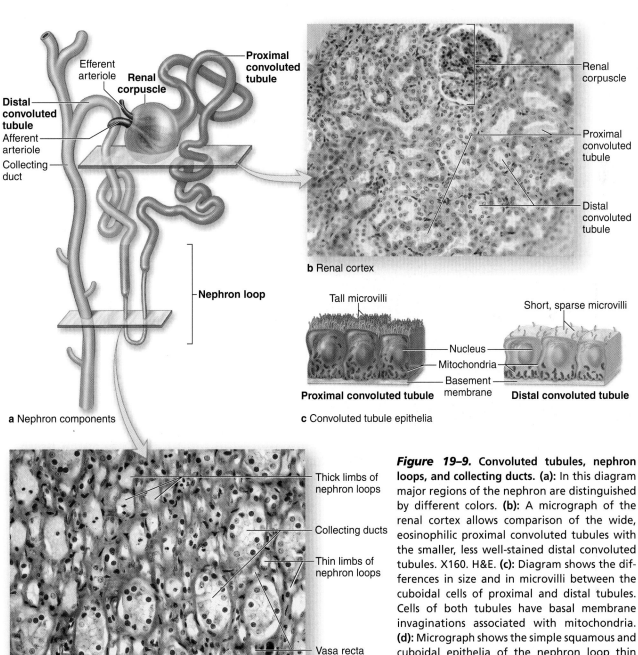

a Nephron components

b Renal cortex

c Convoluted tubule epithelia

d Cross-section of renal medulla

Figure 19–9. **Convoluted tubules, nephron loops, and collecting ducts. (a):** In this diagram major regions of the nephron are distinguished by different colors. **(b):** A micrograph of the renal cortex allows comparison of the wide, eosinophilic proximal convoluted tubules with the smaller, less well-stained distal convoluted tubules. X160. H&E. **(c):** Diagram shows the differences in size and in microvilli between the cuboidal cells of proximal and distal tubules. Cells of both tubules have basal membrane invaginations associated with mitochondria. **(d):** Micrograph shows the simple squamous and cuboidal epithelia of the nephron loop thin limbs and thick limbs respectively, as well as the pale columnar cells of the collecting ducts. X160. Mallory trichrome.

MEDICAL APPLICATION

A significant hemorrhage decreases blood volume, which promotes decreased blood pressure and increased renin secretion. Angiotensin II and aldosterone act in concert to increase blood pressure and to help restore blood volume. Other factors (eg, sodium depletion, dehydration) that lower blood pressure by decreasing blood volume also activate the renin–angiotensin II–aldosterone mechanism that contributes to the maintenance of normal blood pressure.

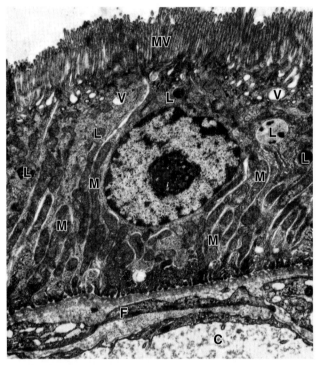

Figure 19–10. Ultrastructure of proximal convoluted tubule cells. TEM reveals important features of the cuboidal cells of the proximal convoluted epithelium: the long, dense apical microvilli (MV), the abundant pinocytotic pits and vesicles (V) in the apical regions near lysosomes (L). Small proteins brought into the cells nonspecifically by pinocytosis are degraded in lysosomes and the amino acids released basally. Apical ends of adjacent cells are sealed with zonula occludens, but the basolateral sides are characterized by long invaginating folds of membrane along which many long mitochondria (M) are situated. These folds provide a greatly increased surface area for pumping of ions across the membrane. Water and the small molecules released from the proximal convoluted tubules are taken up immediately by the adjacent peritubular capillaries (C). Between the basement membranes of the tubule and the capillary shown here is an extension of a fibroblast (F). X10,500.

Collecting Tubules & Ducts

Urine passes from the distal convoluted tubules to **collecting tubules**, the last part of each nephron, which join each other to form larger, straight **collecting ducts** that run to the tips of the medullary pyramids and empty into the minor calyces (Figure 19–2). The collecting tubules are lined with cuboidal epithelium and have a diameter of approximately 40 μm. Cells of the converging collecting ducts are more columnar and the duct diameters reach 200 μm near the tips of the medullary pyramids (Figures 19–13 and 19–14).

Along their entire extent, collecting tubules and ducts are composed mainly of weakly staining **principal cells** with few organelles and scanty microvilli (Figure 19–13). The intercellular limits of the cells are clearly visible in the light microscope. Ultrastructurally the principal cells can be seen to have basal membrane infoldings, consistent with their role in ion transport. Scattered among the principal cells are variably darker **intercalated cells** with more abundant mitochondria which help regulate the acid-base balance by secreting H^+ and absorbing HCO_3^-.

In the medulla, collecting ducts are a major component of the urine-concentrating mechanism. Cells of collecting ducts are particularly rich in **aquaporins**, integral proteins found in most cell membranes that function as selective pores for passage of water molecules. Here aquaporins are sequestered in membranous cytoplasmic vesicles. As mentioned earlier, the pituitary hormone ADH (also known as arginine vasopressin) makes collecting ducts more permeable to water, increasing the rate at which water molecules are pulled osmotically from their lumens and transferred to the vasa recta, and thus retained in the body. This effect is produced when activated ADH receptors on the basolateral cell membrane stimulate the movement of vesicles with specific aquaporins and their insertion into either the apical or basolateral membranes, increasing the number of membrane channels for water movement through the cells.

URETERS, BLADDER, & URETHRA

Urine is transported by the **ureters** to the **bladder** where it is stored until emptied during micturition via the **urethra**. The calyces, renal pelvis, ureter, and bladder have the same basic histologic structure, with the walls becoming gradually thicker closer to the bladder. The mucosa of these organs is lined by unique stratified **transitional epithelium** or urothelium (Figures 19–14). This is surrounded by a folded lamina propria and submucosa, followed by a dense sheath of interwoven smooth muscle layers and adventitia (Figure 19–15). Urine moves from the renal pelvises to the bladder by peristaltic contractions.

The urothelium is composed of the following three layers:

- a single layer of small basal cells resting on a very thin basement membrane,
- an intermediate region containing from one to several layers of more columnar cells,
- a superficial layer of very large, polyhedral or bulbous cells called **umbrella cells** which are occasionally bi- or multinucleated and are highly differentiated to protect underlying cells against the cytotoxic effects of hypertonic urine.

Umbrella cells are especially well developed in the bladder (Figures 19–16 and 19–17) where contact with urine is the greatest. These cells, up to 100 μm in diameter, have extensive intercellular junctional complexes surrounding unique apical membranes. Most of the apical surface consists of **asymmetric unit**

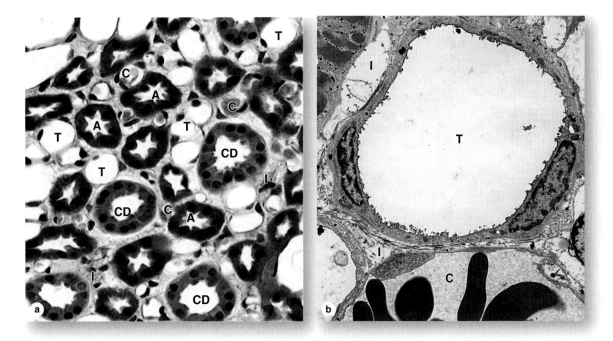

Figure 19–11. Renal medulla: Nephron loops and collecting ducts. (a): A micrograph of a medullary pyramid cut transversely shows closely packed cross sections of the many nephron loops' thin descending limbs (T) and thick ascending limbs (A), intermingled with parallel vasa recta capillaries (C). All these structures are embedded in the interstitium (I) which contains sparse myofibroblast-like cells in a matrix very rich in hydrophilic hyaluronate. The specialized nature of the interstitial tissue helps maintain the osmolarity gradient established by differential salt and water transport across the wall of the nephron loop which is required to concentrate urine and conserve body water. X400. Mallory trichrome. **(b):** The TEM reveals the slightly fibrous nature of the interstitium (I) and shows that the simple squamous epithelium of the thin limbs (T) is slightly thicker than that of the nearby vasa recta capillaries (C). X3300. (Figure 19–11b, with permission, from Johannes Rhodin, Department of Cell Biology and Anatomy, University of South Florida College of Medicine.)

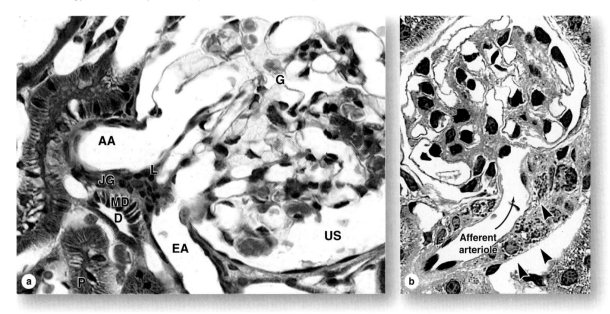

Figure 19–12. Juxtaglomerular apparatus or JGA. (a): Micrograph shows that the juxtaglomerular apparatus forms at the point of contact between a nephron's distal tubule (D) and the vascular pole of its glomerulus (G). At that point cells of the distal tubule become columnar as a thickened region called the macula densa (MD). Smooth muscle cells of the afferent arteriole's (AA) tunica media are converted from a contractile to a secretory morphology as juxtaglomerular granule cells (JG). Also present are lacis cells (L), which are extraglomerular mesangial cells adjacent to the macula densa, the afferent arteriole, and the efferent arteriole (EA). In this specimen the lumens of proximal tubules (P) appear filled and the urinary space (US) is somewhat swollen. X400. Mallory trichrome.

(b): A plastic section through an afferent arteriole of a JGA shows the JG cells (arrowheads) with secretory granules containing renin. Activities of the macula densa and renin released by JG cells together produce an incompletely understood autoregulatory tubuloglomerular feedback loop that not only helps control arterial blood pressure, but also helps maintain a relatively constant rate of glomerular filtration despite changes in blood pressure. X500.

344 / CHAPTER 19

membranes, in which regions of the outer lipid layer appear twice as thick as the inner leaflet. These regions are lipid rafts containing mostly integral membrane proteins called **uroplakins** which assemble into paracrystalline arrays of stiffened plaques 16 nm in diameter. Urine contacts primarily these membranous plaques, which are impermeable and protect cytoplasm and underlying cells from its hyperosmotic effects. Plaques are hinged together by more narrow regions of typical membrane. When the bladder is emptied, not only does the mucosa fold extensively, but individual umbrella cells decrease their apical surface area by folding the membrane at its hinge domains and internalizing the folded plaques in discoidal vesicles. As the bladder fills again the discoidal vesicles rejoin the apical membrane, increasing its surface area as the cell shape changes from round to flat. The urothelium becomes thinner, apparently the result of the intermediate cells being pushed and pulled laterally to accommodate the increased volume of urine.

The bladder's lamina propria and dense irregular connective tissue of the submucosa are highly vascularized. The muscularis consists of three poorly delineated layers, collectively called the detrusor muscle, which contract to empty the bladder (Figure 19–16). Three muscular layers are seen most distinctly at the neck of the bladder near the urethra (Figure 19–17). The ureters pass through the wall of the bladder obliquely, forming a valve that prevents the backflow of urine into the ureters. All the urinary passages are covered externally by an adventitial layer, except for the upper part of the bladder which is covered by serous peritoneum.

The **urethra** is a tube that carries the urine from the bladder to the exterior. The urethral mucosa has large longitudinal folds, giving it a distinctive appearance in cross section. In men, the two ducts for sperm transport during ejaculation join the urethra at the prostate gland (Chapter 21). The male urethra is longer and consists of three segments:

- The **prostatic urethra**, 3–4 cm long, extends through the prostate gland and is lined by urothelium
- The **membranous urethra**, a short segment, passes through an **external sphincter** of striated muscle and is lined by stratified columnar and pseudostratified epithelium

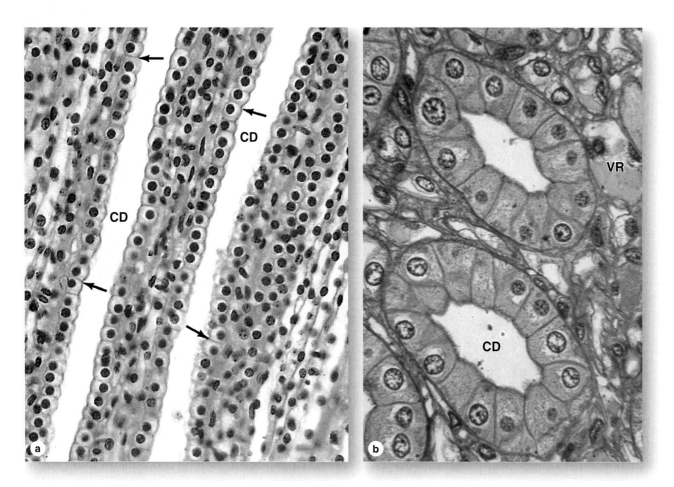

Figure 19–13. **Collecting tubules and ducts. (a):** Longitudinally sectioned renal pyramid showing two collecting ducts (CD) and their distinct lateral cell boundaries (arrows), with interstitial connective tissue. X400. H&E. **(b):** Similar features are seen in transversely sectioned collecting ducts, with vasa recta (VR) present in the interstitium. X600. PT. Weakly stained duct principal cells are initially cuboidal and become increasingly columnar along the ducts. Collecting ducts adjust the ionic composition of urine in their lumens and allow increased water reabsorption from this urine when fluid levels in the body are low. This occurs under the influence of the posterior pituitary hormone ADH, which causes a greatly increased number of aquaporin water channels to be temporarily inserted into the apical cell membranes of these cells.

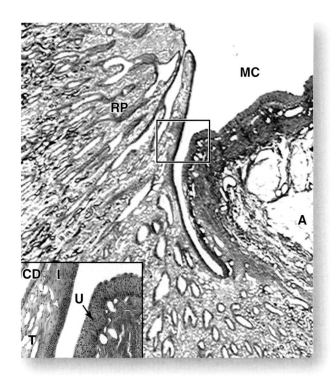

***Figure 19–14.* Renal papilla, collecting ducts, and minor calyx.** A sagittal section of a renal papilla shows numerous collecting ducts (sometimes called ducts of Bellini at this level of the renal pyramid) converging at the end of the renal papilla (RP) where they empty into the minor calyx (MC). The mucosa of the calyx contains dense connective tissue stained blue here and adipose tissue (A). The ducts are embedded in interstitial tissue that also contains thin limbs of the nephron loops. X50. Mallory trichrome. **Inset:** An enlarged area shows the columnar epithelium of the collecting ducts (CD), the interstitium (I) and thin limbs (T), and the protective urothelium (U) that lines the minor calyx. X200.

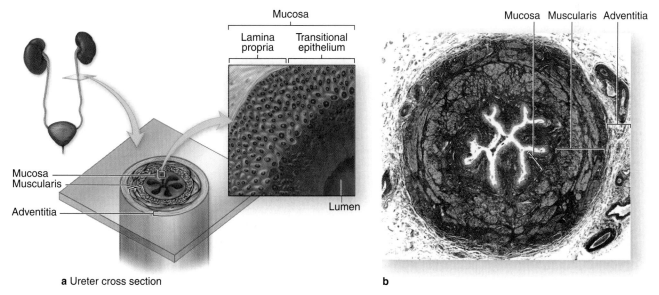

a Ureter cross section   b

***Figure 19–15.* Ureters.** Each ureter carries urine from the renal pelvis of the kidney to the urinary bladder for storage prior to voiding via the urethra. **(a):** Diagram of a ureter in cross section shows a characteristic pattern of longitudinally folded mucosa, surrounded by a thick muscularis that moves urine by regular waves of peristalsis. The lamina propria is lined by a unique stratified epithelium called transitional epithelium or urothelium resistant to the potentially deleterious effects of contact with hypertonic urine. **(b:)** Histologically the muscularis is much thicker than the mucosa and an adventitia is also present. X18. H&E.

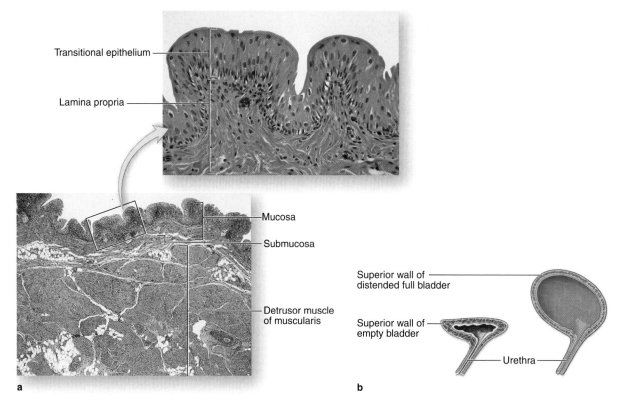

Transitional epithelium

Lamina propria

Mucosa

Submucosa

Detrusor muscle
of muscularis

Superior wall of
distended full bladder

Superior wall of
empty bladder

Urethra

a

b

Figure 19–16. Urinary bladder. The bladder is a muscular sac that is expandable as it fills with urine. **(a):** The micrograph shows the folded mucosa, submucosa, and muscularis of the empty bladder wall, with an enlargement showing the stratified transitional epithelium (urothelium) and lamina propria. X18 and X80. H&E. **(b):** The diagram depicts a sagittal view showing that the bladder expands primarily upward and becomes more oval in shape as it fills with urine. The bladder in an average adult can hold 400–600 mL of urine, with the urge to empty appearing at a volume of 150–200 mL.

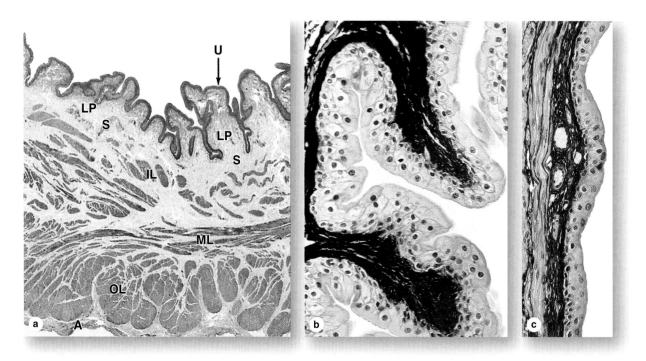

U

LP

S

LP

S

IL

ML

OL

a

A

b

c

Figure 19–17. Bladder wall and urothelium. (a): In the neck of the bladder, near the urethra, the wall shows four layers: the mucosa with urothelium (U) and lamina propria (LP); the thin submucosa (S); inner, middle, and outer layers of smooth muscle (IL, ML, and OL), and the adventitia (A). X15. H&E. **(b):** When the bladder is empty, the mucosa is highly folded and the urothelium has bulbous umbrella cells. X250. PSH. **(c):** when the bladder is full, the mucosa is pulled smooth, the urothelium is thinner, and the umbrella cells are flatter. X250. H&E.

- The **spongy urethra**, 15 cm in length, is enclosed within erectile tissue of the penis (Chapter 21) and is lined by stratified columnar and pseudostratified columnar epithelium (Figure 19–18), with stratified squamous distally.

In women, the urethra is exclusively a urinary organ. The female urethra is a tube 4 to 5 cm long, lined initially with transitional epithelium, then by stratified squamous epithelium and some areas of pseudostratified columnar epithelium. The middle part of the female urethra is surrounded by the external striated muscle sphincter.

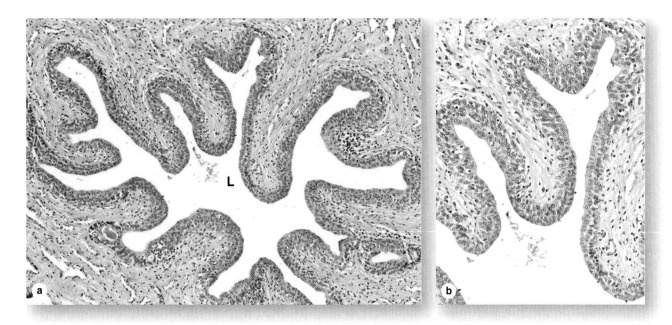

Figure 19–18. Urethra. The urethra is a fibromuscular tube that carries urine from the bladder to the exterior of the body. **(a):** A transverse section shows that the mucosa has large longitudinal folds around the lumen (L). X50. H&E. **(b):** A higher magnification of the urethral epithelium is shown in this micrograph. The thick epithelial lining is stratified columnar in some areas and pseudostratified columnar elsewhere, but becomes stratified squamous at the distal end of the urethra. X250. H&E.

Endocrine Glands

Secretory cells of endocrine glands release their products, signaling molecules called **hormones**, into a neighboring vascularized compartment for uptake by capillaries and distribution throughout the body, rather than directly into an epithelial duct like the cells of exocrine glands. Endocrine cells typically aggregate as cords, or as follicles in the case of the thyroid gland. Besides the specialized endocrine glands discussed in this chapter, many other organs specialized for other functions, such as the heart, thymus, gut, kidneys, testis, and ovaries contain various endocrine cells (Figure 20–1).

Distribution by the circulation allows hormones to act on target cells with receptor proteins for those hormones at a distance from the site of their secretion. As discussed briefly in Chapter 2, other endocrine cells produce hormones that act very quickly at only a short distance. This may involve **paracrine** secretion, with localized dispersal in interstitial fluid or through short loops of blood vessels, as when gastrin made by pyloric G cells reaches target cells in the fundic glands, or **juxtacrine** secretion, in which a signaling molecule remains on the secreting cell's surface or adjacent extracellular matrix and effects target cells when the cells make contact. Juxtacrine signaling is particularly important in developmental tissue interactions. In **autocrine** secretion, cells may produce molecules that act on themselves or on cells of the same type. Insulin-like growth factor (IGF) produced by several cell types may act on the same cells that produced it. Frequently, endocrine glands are also target organs for other hormones. This helps allow the body to control hormone secretion through a feedback mechanism and to keep blood hormonal levels within strict limits.

Hormones, like neurotransmitters, are frequently hydrophilic molecules such proteins, glycoproteins, peptides, or modified amino acids. Their receptor proteins are on the surface of target cells. Alternatively, hydrophobic steroid and thyroid hormones must circulate on transport proteins but can diffuse through the membranes of cells and activate cytoplasmic protein receptors in target cells (Chapter 2).

PITUITARY GLAND (HYPOPHYSIS)

The **pituitary gland**, or **hypophysis** (Gr. *hypo*, under, + *physis*, growth), weighs about 0.5 g in adults and has dimensions of about $10 \times 13 \times 6$ mm. It lies below the brain in a cavity of the sphenoid bone—the sella turcica (Figure 20–2). The pituitary develops in the embryo partly from oral ectoderm and partly from the developing brain (Figure 20–3). The neural component arises as a bud growing down from the floor of the diencephalon and caudally as a stalk or infundibulum still attached to the brain. The oral component arises as an outpocketing of ectoderm from the roof of the primitive mouth and grows cranially, forming a structure called the **hypophyseal (Rathke's) pouch**. Later, the base of this pouch constricts and separates it from the pharynx. The anterior wall then thickens greatly, reducing the pouch lumen to a small fissure (Figure 20–3).

Because of its dual origin, the pituitary actually consists of two glands—the posterior **neurohypophysis** and the anterior **adenohypophysis**—united anatomically but with different functions. The neurohypophysis retains features of CNS tissue from which it developed and consists of a large part, the **pars nervosa**, and the smaller **infundibulum** with its stalk attached to the hypothalamus at the median eminence (Figures 20–2 and 20–4). The **adenohypophysis**, derived from the oral ectoderm, has three parts: a large **pars distalis** (or the **anterior lobe**); the **pars tuberalis**,

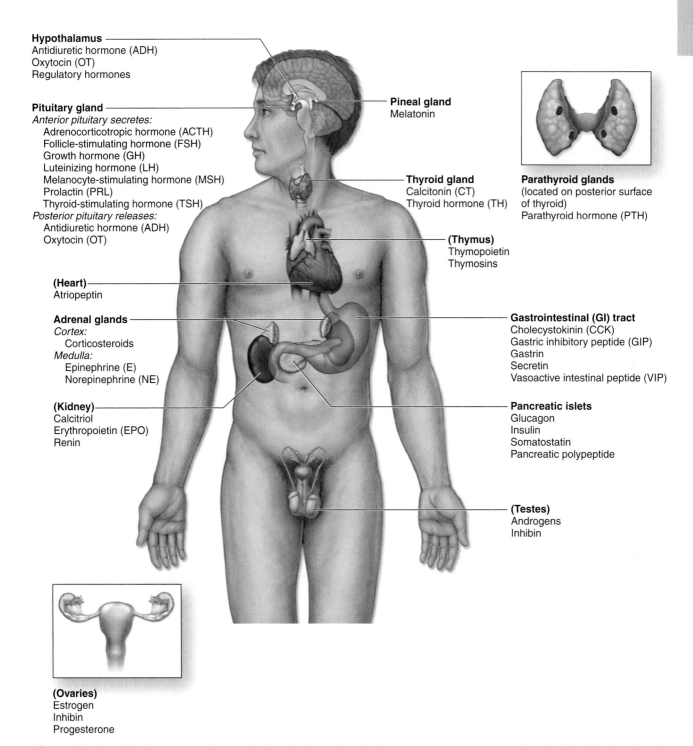

Hypothalamus
Antidiuretic hormone (ADH)
Oxytocin (OT)
Regulatory hormones

Pituitary gland
Anterior pituitary secretes:
  Adrenocorticotropic hormone (ACTH)
  Follicle-stimulating hormone (FSH)
  Growth hormone (GH)
  Luteinizing hormone (LH)
  Melanocyte-stimulating hormone (MSH)
  Prolactin (PRL)
  Thyroid-stimulating hormone (TSH)
Posterior pituitary releases:
  Antidiuretic hormone (ADH)
  Oxytocin (OT)

(Heart)
Atriopeptin

Adrenal glands
Cortex:
  Corticosteroids
Medulla:
  Epinephrine (E)
  Norepinephrine (NE)

(Kidney)
Calcitriol
Erythropoietin (EPO)
Renin

Pineal gland
Melatonin

Thyroid gland
Calcitonin (CT)
Thyroid hormone (TH)

Parathyroid glands
(located on posterior surface
of thyroid)
Parathyroid hormone (PTH)

(Thymus)
Thymopoietin
Thymosins

Gastrointestinal (GI) tract
Cholecystokinin (CCK)
Gastric inhibitory peptide (GIP)
Gastrin
Secretin
Vasoactive intestinal peptide (VIP)

Pancreatic islets
Glucagon
Insulin
Somatostatin
Pancreatic polypeptide

(Testes)
Androgens
Inhibin

(Ovaries)
Estrogen
Inhibin
Progesterone

Figure 20–1. **Endocrine system.** The endocrine glands and major hormones they secrete are listed with their locations. In parentheses are shown other organs, including the heart, kidney, thymus, gut, and gonads, that contain endocrine cells and have important endocrine functions. In addition many widely distributed tissues and cells throughout the body have endocrine functions but are not shown here. These include adipose cells which secrete the hormone leptin and vascular endothelial cells which produce polypeptides called endothelins which promote vasoconstriction.

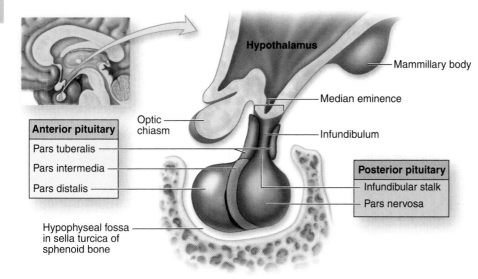

Anterior pituitary
- Pars tuberalis
- Pars intermedia
- Pars distalis

Hypophyseal fossa in sella turcica of sphenoid bone

Hypothalamus

Mammillary body

Median eminence

Optic chiasm

Infundibulum

Posterior pituitary
- Infundibular stalk
- Pars nervosa

Figure 20–2. **Pituitary gland.** The pituitary gland is composed of an anterior part and a posterior part, which is directly attached to the hypothalamus region of the brain by a stalk called the infundibulum. The gland occupies a fossa of the sphenoid bone called the sella turcica (L. Turkish saddle), which is also used as a radiological landmark.

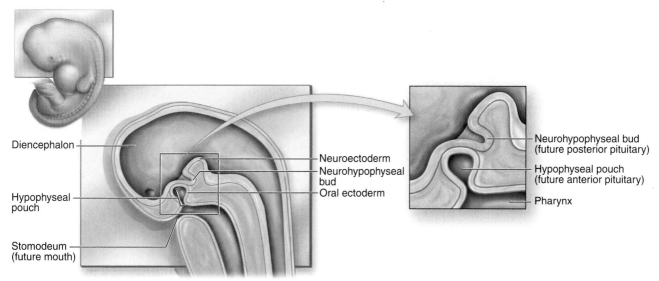

Diencephalon

Hypophyseal pouch

Stomodeum (future mouth)

Neuroectoderm
Neurohypophyseal bud
Oral ectoderm

Neurohypophyseal bud (future posterior pituitary)

Hypophyseal pouch (future anterior pituitary)

Pharynx

a Week 3: Hypophyseal pouch and neurohypophyseal bud form.

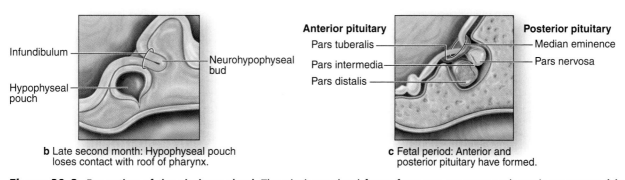

Infundibulum

Hypophyseal pouch

Neurohypophyseal bud

b Late second month: Hypophyseal pouch loses contact with roof of pharynx.

Anterior pituitary
- Pars tuberalis
- Pars intermedia
- Pars distalis

Posterior pituitary
- Median eminence
- Pars nervosa

c Fetal period: Anterior and posterior pituitary have formed.

Figure 20–3. **Formation of the pituitary gland.** The pituitary gland forms from two separate embryonic structures. **(a):** During the third week of development, a hypophyseal pouch (or Rathke's pouch, the future anterior pituitary) grows from the roof of the pharynx, while a neurohypophyseal bud (future posterior pituitary) forms from the diencephalon. **(b):** By late in the second month, the hypophyseal pouch detaches from the roof of the pharynx and merges with the neurohypophyseal bud. **(c):** During the fetal period, the anterior and posterior parts of the pituitary complete development.

which wraps around the infundibulum; and the thin **pars intermedia** (Figures 20–2 and 20–4).

Blood Supply & the Hypothalamo-Hypophyseal Portal System

Study of the pituitary gland's blood supply is important to understand its function. The blood supply derives from two groups of vessels coming off the internal carotid artery. The **superior hypophyseal arteries** supply the median eminence and the infundibular stalk; the **inferior hypophyseal arteries** provide blood mainly for the neurohypophysis, with a small supply to the stalk. The superior hypophyseal arteries form a **primary capillary network** irrigating the stalk and median eminence. The capillaries then rejoin to form venules that branch again as a larger **secondary capillary network** in the adenohypophysis (Figure 20–5). Capillaries of both networks are fenestrated and this **hypothalamo-hypophyseal portal system** is of utmost importance because it carries neuropeptides from the median eminence a short distance to the adenohypophysis where they either stimulate or inhibit hormone release by the endocrine cells there.

Embryologically, anatomically, and functionally, the pituitary gland is connected to the hypothalamus at the base of the brain. In the hypothalamo-hypophyseal system there are three groups of hormones released at three sites (Figure 20–5):

1. Peptide hormones synthesized by neurons in specific aggregates (nuclei) in the hypothalamus, the **supraoptic** and the **paraventricular nuclei**, undergo axonal transport and accumulate distally in these axons, which are situated in the neurohypophysis' pars nervosa (Figure 20–5).

2. Another group of peptides is produced by neurons in other hypothalamic nuclei and carried in axons for temporary axonal storage and secretion in the median eminence. There these peptides enter the capillaries of the primary plexus and

are transported to the adenohypophysis where they diffuse among endocrine cells and control hormone release from their target cells (Figure 20–5).

3. The third group of hormones consists of proteins and glycoproteins released from the endocrine cells of the adenohypophysis' pars distalis (under the control of the neuropeptides just mentioned) and picked up by capillaries of the second portal system plexus, from which they enter the general circulation (Figure 20–5).

Adenohypophysis (Anterior Pituitary)

The three parts of the adenohypophysis are derived from the hypophyseal pouch off the embryonic pharynx.

PARS DISTALIS

The **pars distalis** accounts for 75% of the adenohypophysis and is covered by a thin fibrous capsule. The main components are cords of epithelial cells interspersed with fenestrated capillaries (Figures 20–4 and 20–6). Fibroblasts are present and produce reticular fibers supporting the cords of hormone-secreting cells. Common stains suggest two broad groups of cells in the pars distalis based on staining affinity: **chromophils** and **chromophobes**. Chromophils are secretory cells in which hormone is stored in cytoplasmic granules. They are also called basophils and acidophils according to their affinity for basic and acidic dyes, respectively (Figures 20–6 and 20–7). Subtypes of basophilic and acidophilic cells are identified by TEM or more easily by immunohistochemistry and are named for their specific hormones or target cells (Table 20–1). Acidophils include the **somatotropic** and **mammotropic** cells, while the basophilic cells are the **gonadotropic**, **corticotropic**, and **thyrotropic** cells. Somatotropic cells typically constitute about half the cells of the pars distalis in humans, with thyrotropic cells the least abundant. Chromophobes stain weakly, with few or no secretory granules, and also represent a heterogeneous group, including stem and undifferentiated progenitor cells as well as any degranulated cells present. Each granular cell makes one kind of hormone, except gonadotropic cells which produce two proteins and corticotropic cells, in which the major gene product, proopiomelanocortin (POMC), is cleaved posttranslationally into the smaller polypeptide hormones adrenocortical trophic hormone (ACTH) and β-lipotropin (β-LPH). Hormones produced by the pars distalis have widespread functional activities; they regulate almost all other endocrine glands, milk secretion, melanocyte activity, and the metabolism of muscle, bone, and adipose tissue (Figure 20–8 and Table 20–1).

PARS TUBERALIS

The **pars tuberalis** is a funnel-shaped region surrounding the infundibulum of the neurohypophysis (Figure 20–2). Most of the cells of the pars tuberalis are basophilic gonadotropic cells that secrete follicle-stimulating hormone (FSH) and luteinizing hormone (LH).

PARS INTERMEDIA

The **pars intermedia** is a thin zone of basophilic cells between the pars distalis and the pars nervosa of the neurohypophysis, which is often invaded by these basophils (Figure 20–9). The pars intermedia develops from the dorsal wall of the hypophyseal pouch and usually contains colloid-filled cysts that represent remnants of that structure's lumen (Figures 20–3 and 20–9). During fetal life parenchymal cells of this region, like the

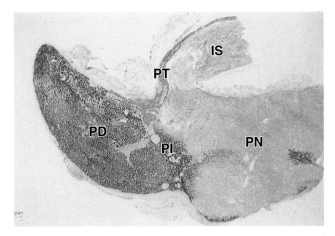

Figure 20–4. Pituitary gland. Histologically the two parts of the pituitary gland reflect their origins, as seen in this low magnification section of an entire gland. The infundibular stalk (IS) and pars nervosa (PN) of the neurohypophysis resemble CNS tissue, while the adenohypophysis' pars distalis (PD), pars intermediate (PI), and pars tuberalis (PT) are typically glandular in their level of staining. X15. H&E.

corticotropic cells of the pars distalis, express POMC. However in these cells POMC is cleaved by a different protease to produce smaller peptide hormones, including two forms of melanocyte-stimulating hormone (MSH), γ-LPH, and β-endorphin. MSH increases melanocyte activity and cells of the pars intermedia are often referred to as melanotropic cells, but the overall physiological significance of this region remains uncertain, especially in adults.

CONTROL OF SECRETION IN THE ADENOHYPOPHYSIS

The activities of the cells of the anterior pituitary are controlled primarily by peptide hormones produced by specialized neurons in certain hypothalamic nuclei and stored in their axons that run to the median eminence (Table 20–2). Most of these hormones are **hypothalamic-releasing hormones**; liberated from the axons, they are transported by capillaries to the pars distalis (Figure 20–5) where they stimulate hormone synthesis and/or release. Two of the hypothalamic factors, however, act to inhibit hormone release by specific cells of the pars distalis (**hypothalamic-inhibiting hormones**; Table 20–2). Because of the strategic position of the hypothalamic neurons and the control they exert on the hypophysis and therefore on many bodily functions, many sensory stimuli coming to the brain, as well as stimuli arising within the CNS, can affect the function of the pituitary gland and then quickly also affect the function of many other organs and tissues.

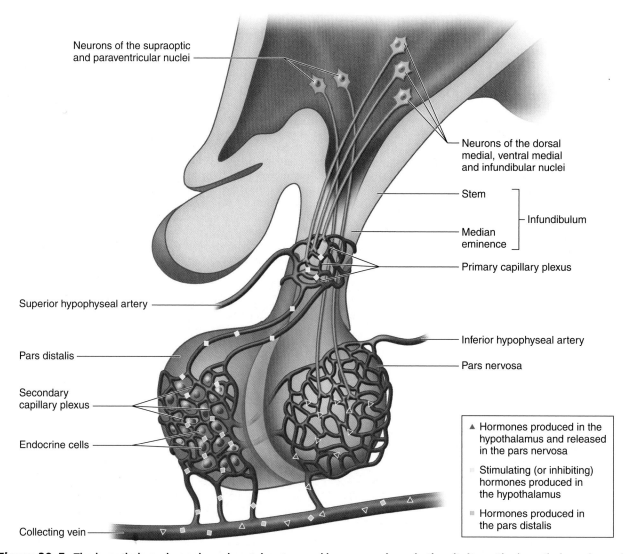

Neurons of the supraoptic and paraventricular nuclei

Neurons of the dorsal medial, ventral medial and infundibular nuclei

Stem

Median eminence

Infundibulum

Primary capillary plexus

Superior hypophyseal artery

Inferior hypophyseal artery

Pars distalis

Pars nervosa

Secondary capillary plexus

Endocrine cells

Collecting vein

▲ Hormones produced in the hypothalamus and released in the pars nervosa

Stimulating (or inhibiting) hormones produced in the hypothalamus

Hormones produced in the pars distalis

Figure 20–5. **The hypothalamo-hypophyseal portal system and hormone release in the pituitary.** The hypothalamo-hypophyseal portal system, with blood from the superior and inferior hypophyseal arteries, consists of two successive capillary networks: one in the pars nervosa around the infundibulum and median eminence, and the second throughout the pars distalis which drains into the collecting hypophyseal veins. Also shown in the diagram are (yellow) neurons extending axons to the median eminence, where they secrete peptides that are then carried in the capillaries to the pars distalis to regulate hormone release from cells there, and (green) neurons from the supraoptic and paraventricular nuclei in the hypothalamus extending axons to the pars nervosa, where they secrete peptides that are picked up by capillaries and carried to distal target cells.

Another mechanism controlling cells of the anterior pituitary is **negative feedback** by hormones from the target organs on secretion of the relevant hypothalamic factors and on hormone secretion by the relevant pituitary cells. Figure 20–10 illustrates this mechanism, using the thyroid as an example, and shows the complex chain of events that begins with the action of neural stimuli on the neurosecretory cells of the hypothalamic nuclei and ends with the effects of hormones from the pituitary's target organs.

Finally, hormone secretion in the pars distalis is affected by other hormones from outside the feedback loop or even outside the major target tissues. Examples include: the proteins inhibin and activin, members of the transforming growth factor-β family produced in the gonads, which control release of FSH and LH; the 28-amino acid polypeptide ghrelin produced mainly in the stomach mucosa, which stimulates GH secretion; and

oxytocin, liberated in the posterior pituitary in the course of breast-feeding, which increases secretion of prolactin.

All these mechanisms allow the fine-tuning of hormone secretion by cells of the anterior pituitary.

Neurohypophysis (Posterior Pituitary)

The neurohypophysis consists of the pars nervosa and the infundibular stalk (Figure 20–2). The pars nervosa, unlike the adenohypophysis, does not contain secretory cells. It is composed of neural tissue, containing some 100,000 unmyelinated axons of secretory neurons situated in the supraoptic and paraventricular nuclei of the hypothalamus (Figure 20–5). Also present are highly branched glial cells called **pituicytes** that resemble astrocytes and are the most abundant cell type in the posterior pituitary (Figure 20–11).

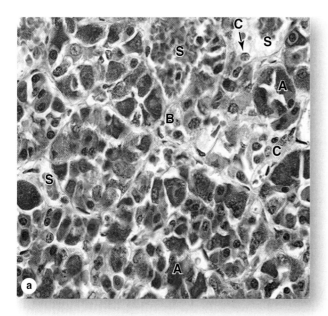

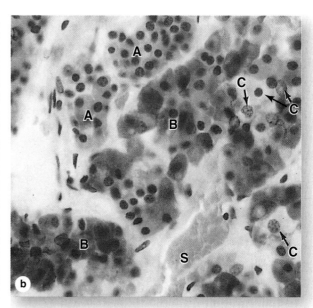

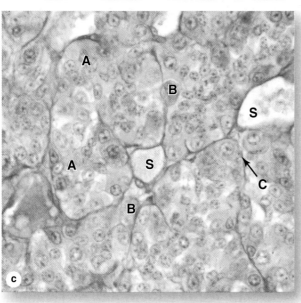

Figure 20–6. **Pars distalis: Acidophils, basophils, and chromophobes. (a,b):** Most general staining methods simply allow the parenchymal cells of the pars distalis to be subdivided into acidophil cells (A), basophils (B), and chromophobes (C) in which the cytoplasm is poorly stained. X400. H&E. **(c):** Gomori trichrome provides similar information. X400. Cords of acidophils and basophils vary in distribution and number in different regions of the pars distalis, but are always closely associated with capillaries and sinusoids (S) in the second capillary plexus of the portal system. The vascular plexus carries off secreted hormones into the general circulation. Specific acidophil or basophil cells can be identified immunohistologically with antibodies against their hormone products. Chromophobes are less numerous and represent various undifferentiated parenchymal cells. Their number and density also vary in different regions.

The secretory neurons have all the characteristics of typical neurons, including the ability to conduct an action potential, but have larger diameter axons and well-developed synthetic components related to the production of the 9-amino acid peptide hormones **vasopressin**—also called **antidiuretic hormone (ADH)**—and **oxytocin**. These hormones are transported axonally into the pars nervosa and accumulate in axonal dilations called **neurosecretory bodies** or **Herring bodies**, visible in the light microscope as faintly eosinophilic structures (Figure 20–11). The neurosecretory bodies contain numerous membrane-enclosed granules with either oxytocin or vasopressin bound to 10 kDa carrier proteins called **neurophysin I and II** respectively. The hormone-neurophysin complex is synthesized as a single polypeptide and then cleaved to produce the peptide hormone and its specific binding protein. Nerve impulses along the axons trigger the release of the peptides from the neurosecretory bodies for uptake by the fenestrated capillaries of the pars nervosa and the hormones are then distributed to the general circulation. Axons from the supraoptic nuclei are mainly concerned with vasopressin/ADH secretion, whereas most of the fibers from the paraventricular nuclei are concerned with oxytocin secretion.

ADH is released in response to increased tonicity of the blood, recognized by osmoreceptor cells in the hypothalamus, which then stimulate synthesis of the hormone in supraoptic neurons. The main effect of ADH is to increase the permeability of the collecting ducts of the kidney to water (Chapter 19). As a result, more water is absorbed by these tubules and replaced into blood instead of being eliminated in the urine (Table 20–3). Thus, ADH helps to regulate the osmotic balance of body fluids.

Oxytocin stimulates contraction of the myoepithelial cells around the alveoli and ducts of the mammary glands during nursing and of uterine smooth muscle during childbirth (Table 20–3). The secretion of oxytocin is stimulated by breast-feeding via sensory tracts that act on the hypothalamus in a neurohormonal reflex called the milk-ejection reflex.

ADRENAL GLANDS

The **adrenal (suprarenal) glands** are paired organs that lie near the superior poles of the kidneys, embedded in the perirenal adipose tissue (Figures 20–1 and 20–12). They are flattened structures with a half-moon shape, about 4–6 cm long, 1–2 cm wide, and 4–6 mm thick in adults. Together they weigh about 8 g, but their weight and size vary with the age and physiologic condition

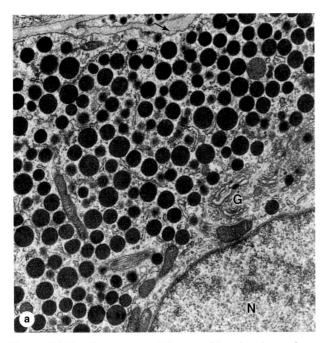
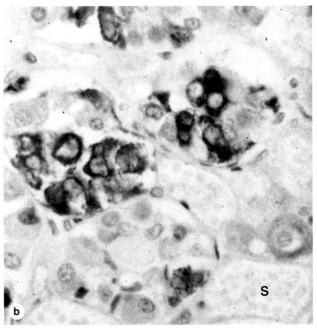

Figure 20–7. **Ultrastructure & immunohistochemistry of somatotropic cells. (a):** Ultrastructurally, cytoplasm of all chromophil cells is shown to have well-developed Golgi complexes (G), euchromatic nuclei (N), and cytoplasm filled with secretory granules, as shown here for a somatotropic cell, the most common acidophil. The arrow indicates the cell membrane. Granules of somatotropic cells can be distinguished from those of other chromophils by TEM, but all of the various chromophils are more easily identified using immunohistochemistry and antibodies against the hormone products. X10,000. **(b):** The micrograph shows somatotropic cells (S) stained using an antibody against somatotropin. X400. Hematoxylin counterstain.

of the individual. Adrenal glands are each covered by a dense connective tissue capsule that sends thin septa to the interior of the gland as trabeculae. The stroma consists mainly of a rich network of reticular fibers that support the secretory cells. The gland consists of two concentric layers: a yellowish peripheral layer, the **adrenal cortex**, and a reddish-brown central layer, the **adrenal medulla**.

The adrenal cortex and the adrenal medulla can be considered two organs with distinct origins, functions, and morphologic characteristics that become united during embryonic development. They arise from different embryonic germ layers. The cortex arises from mesoderm and the medulla consists of cells derived from the neural crest, from which sympathetic ganglion cells also originate.

The general histological appearance of the adrenal gland is typical of an endocrine gland in which cells of both cortex and medulla are grouped in cords along capillaries.

Blood Supply

The adrenal glands are supplied by several arteries that enter at various points around their periphery (Figure 20–12). The branches of these arteries form three groups: those supplying the capsule; the **cortical arterioles**, which quickly form capillaries and sinusoids that irrigate all cells of the cortex and eventually join the medullary capillaries; and **medullary arterioles**, which pass directly through the cortex and form an extensive capillary network in the medulla.

The cells of the adrenal medulla, thus, receive both arterial blood from the medullary arteries and venous blood originating from capillaries of the cortex. The capillary and sinusoidal endothelium is highly attenuated and fenestrated. Capillaries of both the cortex and the medulla form the central **medullary veins**, which join to leave the gland as the **adrenal** or **suprarenal vein** (Figure 20–12).

Adrenal Cortex

Cells of the adrenal cortex have characteristic features of steroid-secreting cells. These include central nuclei and acidophilic cytoplasm, usually rich in lipid droplets. As shown in Figure 20–13, their cytoplasm can be seen ultrastructurally to have an exceptionally profuse smooth ER of interconnected tubules, which contain the enzymes for cholesterol synthesis and conversion of the steroid prohormone pregnenolone into specific active steroid hormones. The mitochondria are often spherical, with tubular rather than shelflike cristae (Figure 20–13). Besides being the sites of ATP production, these mitochondria contain the enzymatic equipment for converting cholesterol to pregnenolone and for some steps in steroid hormone synthesis. The differentiated function of steroid-producing cells results, therefore, from close collaboration between smooth ER and mitochondria.

Steroid hormone-secreting cells do not store their product in granules. As low-molecular-weight, lipid-soluble molecules, steroids diffuse freely through the plasma membrane and do not require exocytosis to be released from the cells.

The adrenal cortex has three concentric zones in which the cords of epithelial cells are arranged somewhat differently and are specialized to produce different classes of steroid hormones (Figure 20–14):

- Immediately inside the connective tissue capsule is the **zona glomerulosa**, consisting of closely packed, rounded or arched cords of columnar or pyramidal cells surrounded by many capillaries and comprising about 15% of the cortex (Figure 20–15). The steroids made by these cells are called **mineralcorticoids** because they affect uptake of Na+, K+, and water by epithelial cells. The principal product is **aldosterone**, the major regulator of salt balance, which acts to stimulate Na+ reabsorption in the distal convoluted tubules of the kidneys (Chapter 19). Aldosterone secretion in the zona glomerulosa is stimulated primarily by angiotensin II and also by an increase in plasma K+ concentration, but only weakly by ACTH.

Table 20–1. Secretory cells of the pars distalis.

Cell Type	Stain Affinity	% of Total Cells	Hormone Produced	Main Physiologic Activity
Somatotropic cell	Acidophilic	50	Somatotropin (growth hormone, GH)	Acts on growth of long bones via insulin-like growth factors synthesized in liver
Mammotropic cell (or actotropic cell)	Acidophilic	15–20	Prolactin (PRL)	Promotes milk secretion
Gonadotropic cell	Basophilic	10	Follicle-stimulating hormone (FSH) and luteinizing hormone (LH) in the same cell type	FSH promotes ovarian follicle development and estrogen secretion in women and spermatogenesis in men. LH promotes ovarian follicle maturation and progesterone secretion in women and interstitial cell androgen secretion in men.
Thyrotropic cell	Basophilic	5	Thyrotropin (TSH)	Stimulates thyroid hormone synthesis, storage, and liberation
Corticotropic cell	Basophilic	15–20	Adrenal corticotropin (ACTH) Lipotrophins	Stimulates secretion of adrenal cortex hormones Lipid metabolism regulation

- The middle zone, the **zona fasciculata**, occupies 65–80% of the cortex and consists of long cords of large polyhedral cells, one or two cells thick, separated by fenestrated sinusoidal capillaries (Figure 20–15). The cells are most densely filled with cytoplasmic lipid droplets and, as a result of lipid dissolution during tissue preparation, often appear vacuolated or spongy in common histological preparations. Cells of this zone secrete **glucocorticoids**, especially **cortisol**, which primarily affect carbohydrate metabolism by stimulating production of glucose from amino acids or fatty acids (gluconeogenesis) in many cells and glucose conversion into glycogen in the liver. Cortisol induces fat mobilization in subcutaneous adipose tissue and protein breakdown in muscle. Cortisol also suppresses many aspects of the immune response, including cytokine release and lymphopoiesis, and has other effects in other tissues. Secretion of glucocorticoids in the zona fasciculata is controlled by ACTH from the anterior pituitary and negative feedback proportional to the concentration of circulating glucocorticoids is exerted at both the pituitary and hypothalamic levels (Figure 20–10). Cells of the zona fasciculata also secrete small amounts of androgens.

- The innermost **zona reticularis** comprises about 10% of the cortex and contacts the adrenal medulla. It consists of smaller cells disposed in a network of irregular cords interwoven with wide capillaries (Figure 20–15). The cells are usually more heavily stained than those of the other zones because they contain fewer lipid droplets and more lipofuscin pigment. Cells of the zona reticularis also produce cortisol, but primarily secrete the weak androgen **dehydroepiandrosterone (DHEA)** which is converted to testosterone in several other tissues. Secretion by these cells is also stimulated by ACTH and is under feedback regulation with the pituitary and hypothalamus.

MEDICAL APPLICATION

Because of the feedback mechanism controlling the adrenal cortex, patients who are treated with corticoids for long periods should never stop taking these hormones suddenly: secretion of ACTH

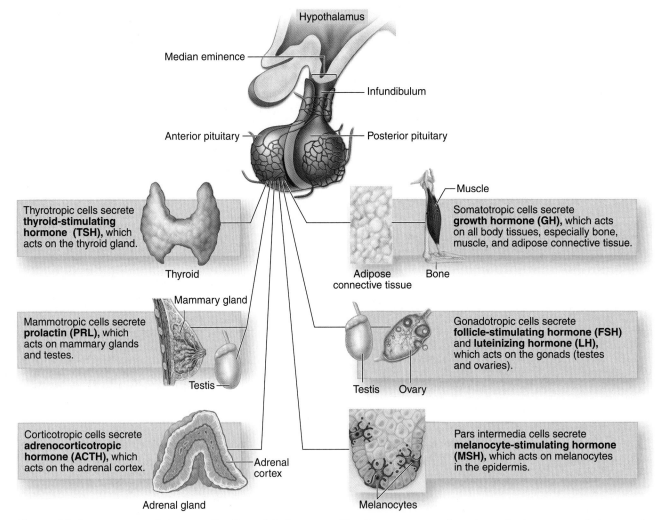

Figure 20–8. Hormones of the pars distalis and their targets. The diagram summarizes the major hormones of the anterior pituitary and indicates their most important targets.

in these patients is inhibited, and thus the cortex will not be induced to produce corticoids, causing severe drops in the levels of sodium and potassium.

Fetal Adrenal Cortex

At birth in humans (but not most other mammals) the adrenal gland is larger than that of the adult and produces up to 200 mg of corticosteroids per day, twice that of an adult. At this age, a layer known as the fetal or **provisional cortex**, comprising 80% of the total gland, is present between the thin permanent cortex and an under-developed medulla. The fetal cortex is thick and contains mostly cords of large, steroid-secreting cells under the control of the fetal pituitary. The principal function of the cells is secretion of sulfated DHEA which is converted in the placenta

to active estrogens (and androgens), which mostly enter the maternal circulation. The fetal adrenal cortex is an important part of a **fetoplacental unit** which affects both endocrine systems during pregnancy but whose physiological significance remains largely unclear. After birth, the provisional cortex undergoes involution while the permanent cortex organizes the three layers (zones) described above.

Adrenal Medulla

The adrenal medulla is composed of large, pale-staining polyhedral cells arranged in cords or clumps and supported by a reticular fiber network (Figure 20–16). A profuse supply of sinusoidal capillaries intervenes between adjacent cords and a few parasympathetic ganglion cells are present. Medullary parenchymal cells, known as **chromaffin cells**, arise from neural crest cells, as do the postganglionic neurons of sympathetic and parasympathetic ganglia. Chromaffin cells can be considered modified sympathetic postganglionic neurons, lacking axons and dendrites and specialized as secretory cells.

Unlike cells of the cortex, medullary chromaffin cells contain many electron-dense granules, 150–350 nm in diameter, for hormone storage and secretion. These granules contain one or the other of the catecholamines, **epinephrine** or **norepinephrine**. Ultrastructurally the granules of epinephrine-secreting cells are less electron-dense and generally smaller than those of norepinephrine-secreting cells. Catecholamine, together with Ca^{2+} and ATP, are bound in a granular storage complex with 49 kDa proteins called **chromogranins**.

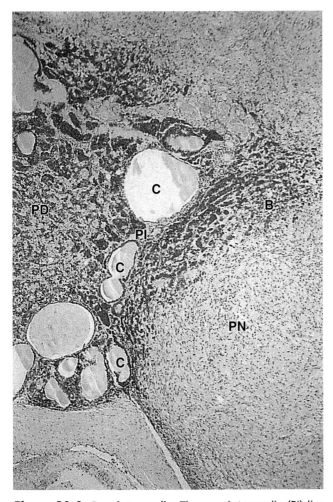

Figure 20–9. **Pars intermedia.** The pars intermedia (PI) lies between the pars distalis (PD) and the pars nervosa (PN), with many of its basophilic cells (B) usually invading the latter. Remnants of the embryonic hypophyseal pouch's lumen are usually present in this region as colloid-filled cysts (C) of various sizes. Function of this region in adults is not clear, but in the fetus the basophils produce melanocyte-stimulating hormones important for melanocyte activity. X56. H&E.

Table 20–2. Hypothalamic hormones regulating the anterior pituitary.

Hormone	Chemical Form	Functions
Thyrotropin-releasing hormone (TRH)	3-amino acid peptide	Stimulates synthesis and release of both thyrotropin (TSH) and prolactin
Gonadotropin-releasing hormone (GnRH)	10-amino acid peptide	Stimulates the release of both FSH and LH
Somatostatin	14-amino acid	Inhibits release of both somatotropin (GH) and thyrotropin (TSH)
Growth hormone–releasing hormone (GHRH)	40- or 44-amino acid polypeptides (2 forms)	Stimulates synthesis and release of somatotropin (GH)
Dopamine (prolactin-inhibiting hormone)	Modified amino acid	Inhibits release of prolactin
Corticotropin releasing hormone (CRH)	41-amino acid polypeptide	Stimulates synthesis of POMC and release of both β-lipotropin (β-LPH) and corticotropin (ACTH)

Norepinephrine-secreting cells are also found in paraganglia (collections of catecholamine-secreting cells adjacent to the autonomic ganglia) and in various viscera. The conversion of norepinephrine to epinephrine (adrenalin) occurs only in chromaffin cells of the adrenal medulla. About 80% of the catecholamine secreted from the adrenal is epinephrine.

Medullary chromaffin cells are innervated by cholinergic endings of preganglionic sympathetic neurons, from which impulses trigger hormone release by exocytosis. Epinephrine and norepinephrine are released to the blood in large quantities during intense emotional reactions, such as fright, and produce vasoconstriction, increased blood pressure, changes in heart rate, and metabolic effects such as elevated blood glucose. These effects facilitate various defensive reactions to the stressor (the fight-or-flight response). During normal activity, the adrenal medulla continuously secretes small quantities of the hormones.

MEDICAL APPLICATION

*One disorder of the adrenal medulla is **pheochromocytoma**, a tumor of its cells that causes hyperglycemia and transient elevations of blood pressure.*

*Disorders of the adrenal cortex can be classified as hyperfunctional or hypofunctional. Tumors of the adrenal cortex can result in excessive production of glucocorticoids (**Cushing syndrome**) or aldosterone (**Conn syndrome**). Cushing syndrome is most often (90%) due to a pituitary adenoma that results in excessive production of ACTH; it is rarely caused by adrenal hyperplasia or an adrenal tumor. Excessive production of adrenal androgens has little effect in men, but precocious puberty (in boys) and hirsutism (abnormal hair growth) and virilization (in girls) are encountered in prepubertal children.*

*Adrenocortical insufficiency (**Addison disease**) is caused by destruction of the adrenal cortex in some diseases. The signs and symptoms suggest failure of secretion of both glucocorticoids and mineralocorticoids by the adrenal cortex.*

Carcinomas of the adrenal cortex are rare, but most are highly malignant. About 90% of these

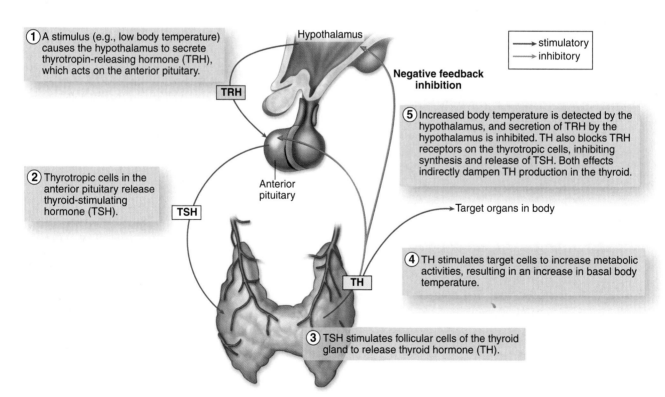

Figure 20–10. Negative feedback loops affecting anterior pituitary secretion. Relationship between the hypothalamus, the anterior pituitary, and its target organs is shown, using the thyroid as an example. Hypothalamic thyrotropin-releasing hormone (TRH) stimulates secretion of thyrotropin (TSH), which stimulates synthesis and secretion of thyroid hormone (TH). In addition to their effects on target organs, TH inhibits TSH secretion from the pars distalis and TRH secretion from the hypothalamus by negative-feedback.

tumors produce steroids associated with endocrine glands.

PANCREATIC ISLETS

The **pancreatic islets** (islets of Langerhans) are compact spherical or egg-shaped masses of endocrine tissue embedded within the acinar exocrine tissue of the pancreas (Figure 20–17). Most islets are 100–200 μm in diameter and contain several hundred cells, but some are much smaller with only a few cells. There are more than 1 million islets in the human pancreas, with the gland's narrow tail region most enriched for islets, but they only constitute 1–2% of the organ's volume. A very thin capsule of reticular fibers surrounds each islet, separating it from the adjacent acinar tissue. Pancreatic islets have the same embryonic origin as the pancreatic acinar tissue: masses of cells in epithelial outgrowths from the intestinal lining (endoderm) near the common bile duct.

Each islet consists of polygonal or rounded cells, smaller and more lightly stained than the surrounding acinar cells, arranged

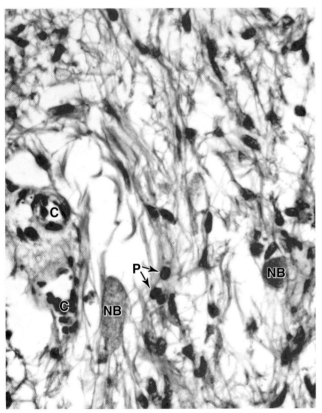

Figure 20–11. Pars nervosa: Neurosecretory bodies and pituicytes. The pars nervosa of the posterior pituitary consists of modified neural tissues containing unmyelinated axons supported and ensheathed by glia cells called pituicytes (P), the most numerous cell type present. The axons run from the supraoptic and paraventricular hypothalamic nuclei and have swellings called neurosecretory (Herring) bodies (NB) from which either oxytocin or vasopressin is released upon neural stimulation. The released hormones are picked up by capillaries (C) for distribution throughout the body. X400. H&E.

in cords that are separated by a network of fenestrated capillaries (Figure 20–17). Autonomic nerve fibers contact some of the endocrine cells and the blood vessels.

Routine stains or trichrome stains show that most islet cells are acidophilic or basophilic with fine cytoplasmic granules (Figure 20–17). Ultrastructural analysis reveals features of active polypeptide-secreting cells and secretory granules that vary in size, morphology, and electron density from cell to cell. The major hormone-producing islet cells are most easily identified and studied by immunohistochemistry:

- **α or A cells** secrete primarily **glucagon** and are usually located near the periphery of islets.
- **β or B cells** produce **insulin** (L. *insula*, island), are located centrally in islets and are the most numerous cell type.
- **δ or D cells**, secreting **somatostatin**, are scattered and much less abundant.

Insulin is a heterodimeric protein and the other two hormones are smaller single-chain polypeptides. A minor fourth cell type, more common in islets located within the head of the pancreas, are **F or PP cells**, which secrete **pancreatic polypeptide**. Table 20–4 summarizes the types, quantities, and main functions of the major hormones produced by islet cells. Pancreatic islets also normally contain a few enterochromaffin cells, like those of the digestive tract, which secrete other polypeptide hormones having other effects within the digestive system and which are also scattered in the pancreatic acini and ducts.

Activity of the two major islet cells, α and β cells, is regulated largely by blood glucose levels above or below the normal level of 70 mg/dL. Increased glucose levels stimulate β cells to release insulin and inhibit α cells from releasing glucagon; decreased glucose levels stimulate α cells to release glucagon. Opposing actions of these hormones (Table 20–4) help to precisely control blood glucose concentration, an important factor in body homeostasis. Increased secretion of these hormones or somatostatin also acts in a paracrine manner to affect hormone release within an islet as well as activity of the neighboring acinar cells.

Sympathetic and parasympathetic nerve endings are closely associated with about 10% of the α, β, and δ cells and can also function as part of the control system for insulin and glucagon secretion. Gap junctions transfer the autonomic neural stimulus to the other cells. Sympathetic fibers increase glucagon release and inhibit insulin release; parasympathetic fibers increase secretion of both glucagon and insulin.

MEDICAL APPLICATION

Insulin-dependent or *type 1 diabetes (juvenile diabetes) results from partial or total autoimmune destruction of β cells and subsequent lack of insulin. Insulin-independent diabetes or type 2 diabetes occurs later in life, results from a failure of cells to respond to insulin, and is frequently associated with obesity.*

Tumors arising from islet cells may produce insulin, glucagon, somatostatin, and pancreatic polypeptide. Some pancreatic tumors produce two or more of these hormones simultaneously, generating complex clinical symptoms.

Table 20–3. Hormones of the posterior pituitary.

Hormone	Function
Vasopressin/antidiuretic hormone (ADH)	Increases water permeability of renal collecting ducts
Oxytocin	Stimulates contraction of mammary gland myoepithelial cells and uterine smooth muscle

DIFFUSE NEUROENDOCRINE SYSTEM

The enterochromaffin cells scattered in both the islets and small ducts of the pancreas are similar to those of the digestive tract (Chapter 15). Collectively these dispersed cells, as well as similar cells in the respiratory mucosa, make up the **diffuse neuroendocrine system** (DNES). Like the pancreatic islets, most of these cells are derived from endodermal cells of the embryonic gut. Such secretory cells are considered neuroendocrine because they produce many of the same polypeptides and neurotransmitter-like molecules such as serotonin (5-hydroxytryptamine) also released by neurosecretory cells in the CNS. Cells of the DNES are also referred to as **gastroenteropancreatic (GEP)** endocrine cells. Several such cells, along with their hormones and major functions, are summarized in Table 15–1. Most of these hormones are polypeptides and many act in a paracrine manner, affecting primarily the activities of neighboring contractile cells and secretory cells (both exocrine and endocrine). GEP endocrine or enteroendocrine cells of the stomach and small bowel are shown in Figures 15–24, 15–28, and 15–31.

Many cells of the DNES are stained by solutions of chromium salts and have therefore been called **enterochromaffin cells**. Those cells that stain with silver nitrate are sometimes called **argentaffin cells**. Those DNES cells secreting serotonin or certain other amine derivatives demonstrate amine precursor uptake and decarboxylation and are often referred to acronymically as **APUD cells**. Such names are still widely used but have been largely replaced by letter designations like those used for pancreatic islet cells (Table 15–1). Whatever name is used, cells of the DNES are highly important due to their role in regulating motility and secretions of all types within the digestive system.

THYROID GLAND

The thyroid gland, located in the cervical region anterior to the larynx, consists of two lobes united by an isthmus (Figure 20–18). It originates in early embryonic life from the foregut endoderm

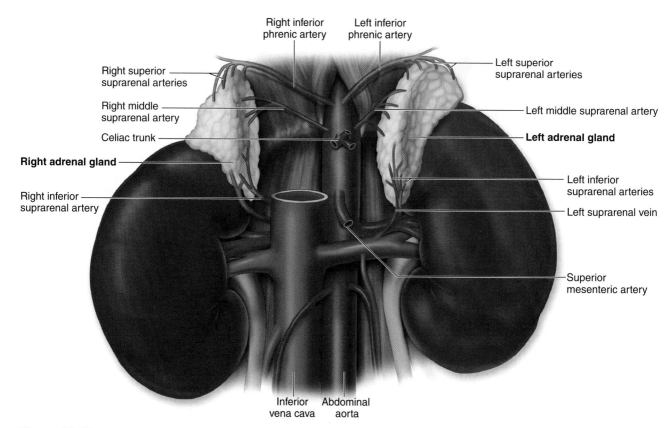

Figure 20–12. Location and blood supply of the adrenal glands. The paired adrenal glands are located at the superior pole of each kidney and each consists of an outer cortex that produces a variety of steroid hormones and an inner medulla that produces epinephrine and norepinephrine. This anterior view shows the relationship of the adrenal to the kidneys and the vasculature supplying these glands.

near the base of the future tongue. Its function is to synthesize the thyroid hormones: **thyroxine** (tetra-iodothyronine or T_4) and **tri-iodothyronine** (T_3), which are important for growth, for cell differentiation, and for the control of the basal metabolic rate and

oxygen consumption in cells throughout the body. Thyroid hormones affect protein, lipid, and carbohydrate metabolism.

The parenchyma of the thyroid is composed of millions of rounded epithelial structures called **thyroid follicles**. Each follicle consists of a simple epithelium and a central lumen filled with a gelatinous substance called **colloid** (Figure 20–19). The thyroid is the only endocrine gland in which a large quantity of secretory product is stored. Moreover, the accumulation is outside the cells, in the colloid of the follicles, which is also unusual. In humans there is sufficient hormone in follicles to supply the body for up to three months with no additional synthesis. Thyroid colloid contains the large glycoprotein **thyroglobulin** (660 kDa), the precursor for the active thyroid hormones.

The thyroid gland is covered by a fibrous capsule from which septa extend into the parenchyma, dividing it into lobules and carrying blood vessels, nerves, and lymphatics. Follicles are densely packed together, separated from one another only by sparse reticular connective tissue (Figure 20–19). This stroma is very well vascularized with an extensive network of fenestrated capillaries closely surrounding the follicles, which facilitates molecular transfer between the follicular cells and blood.

Follicular cells range in shape from squamous to low columnar and the follicles are quite variable in diameter (Figure 20–19). The size and cellular features of thyroid follicles vary with their functional activity. Active glands have more follicles of low columnar epithelium; glands with mostly squamous follicular cells are considered hypoactive.

The follicular epithelial cells have typical junctional complexes apically and rest on a basal lamina. The cells exhibit organelles indicating active protein synthesis and secretion, as well as phagocytosis and digestion. The nucleus is generally round and in the center of the cell. Basally the cells are rich in rough ER and apically, facing the follicular lumen, are Golgi complexes, secretory granules filled with colloidal material, large phagosomes and abundant lysosomes. The cell membrane of the apical pole has a moderate number

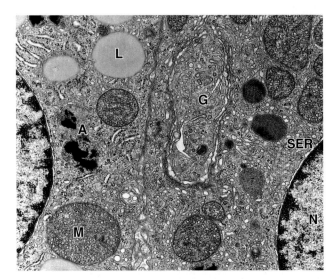

Figure 20–13. Ultrastructure of cortical adrenalocytes. TEM of two adjacent steroid-secreting cells from the zona fasciculate shows features typical of steroid-producing cells: lipid droplets (L) containing cholesterol esters, mitochondria (M) with tubular and vesicular cristae, abundant smooth endoplasmic reticulum (SER), and autophagosomes (A), which remove mitochondria and SER between periods of active steroid synthesis. Also seen are the euchromatic nuclei (N), a Golgi apparatus (G), RER and lysosomes. X25,700.

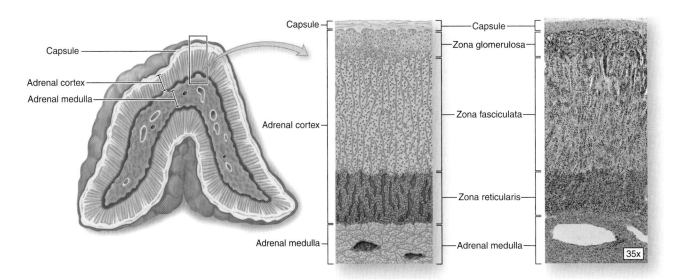

Figure 20–14. Adrenal gland. Inside the capsule of each adrenal gland is an adrenal cortex, formed from embryonic mesodermal cells, which completely surrounds an innermost adrenal medulla derived embryologically from neural crest cells. Both regions are very well vascularized with fenestrated sinusoidal capillaries. Cortical cells are arranged as three layers: the zona glomerulosa near the capsule, the zona fasciculata (the thickest layer), and the zona reticularis.

of microvilli. Mitochondria and other cisternae of rough ER are dispersed throughout the cytoplasm.

Another endocrine cell type, the **parafollicular,** or **C, cell,** is also found inside the basal lamina of the follicular epithelium or as isolated clusters between follicles (Figure 20–20). Parafollicular cells, derived from neural crest cells migrating into the area of the embryonic foregut, are usually somewhat larger than follicular cells and stain less intensely. They have a smaller amount of rough ER, large Golgi complexes, and numerous small (100–180 nm in diameter) granules containing polypeptide hormone (Figure 20–20). These cells synthesize and secrete **calcitonin,** one function of which is to suppress bone resorption

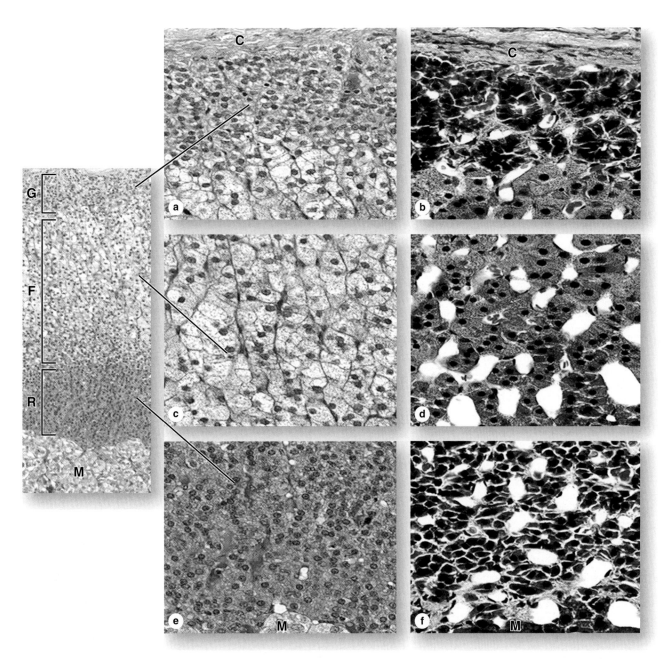

Figure 20–15. Adrenal cortex. The steroid-secreting cells of the adrenal cortex are arranged differently to form three fairly distinct concentric layers surrounding the medulla (M). Shown here are sections from two adrenal glands, stained with H&E (left) and Mallory trichrome, in which the sparse collagen appears blue (right). **(a, b):** Immediately beneath the capsule (C) the zona glomerulosa consists of rounded clusters of columnar cells principally secreting the mineralcorticoid aldosterone. **(c, d):** The thick middle layer, the zona fasciculata, consists of long cords of large, spongy-looking cells mainly secreting glucocorticoids such as cortisol. **(e, f):** Cells of the innermost layer are small, better stained, arranged in a close network and secrete mainly sex steroids. Cells of all the layers are closely associated with sinusoidal capillaries. Left: X20 H&E.; a, c, e: X200. H&E. b, d, f: X200. Mallory trichrome.

by osteoclasts. Calcitonin secretion is triggered by elevated blood Ca^{2+} levels.

Control of Thyroid Function

The major regulator of the anatomic and functional state of thyroid follicles is thyroid-stimulating hormone (TSH; thyrotropin), which is secreted by the anterior pituitary (Figure 20–8). TSH increases the height of the follicular epithelium and stimulates all stages of thyroid hormone production and release. Thyroid hormones inhibit the release of TSH, maintaining an adequate quantity of T_4 and T_3 in the organism (Figure 20–10). TSH receptors are abundant on the basal cell membrane of follicular cells. Secretion of TSH is also increased by exposure to cold and decreased by heat and stressful stimuli.

Storage & Release of Thyroid Hormone

Production, storage, and release of thyroid hormones involve an unusual, multistage process with both an exocrine phase and an endocrine phase in the follicular cells. Both phases are promoted by TSH and can be occurring in the same cell, as summarized in Figure 20–21. The major activities of this process include the following:

1. The **production of thyroglobulin**, which is similar to that in other glycoprotein-exporting cells, with synthesis in the rough ER and the addition of carbohydrate there and in the Golgi apparatus. This large glycoprotein has no hormonal activity itself but contains 140 tyrosyl residues used to make

thyroid hormone. As part of the exocrine phase of cellular activity, thyroglobulin is released from large vesicles at the apical surface of the cell into the lumen of the follicle.

2. The **uptake of circulating iodide** is accomplished in follicular cells by the Na/I symporter or cotransporter in the basolateral cell membrane, which allows for 30-fold concentration of dietary iodide in the normal thyroid relative to the plasma. Low levels of circulating iodide trigger synthesis of the Na/I symporter, increasing the uptake of iodide and compensating for the lower serum concentration.

3. At the apical surface of the cells, iodide is transferred to the follicular lumen by the anion transport protein pendrin and there it undergoes oxidation **to active iodine** by membrane-bound thyroid peroxidase at the cell surface.

4. In the lumen **tyrosine residues of thyroglobulin are iodinated** covalently with either one or two iodine atoms. Following this, two iodinated tyrosines, still part of thyroglobulin, are conjugated by oxidative coupling reactions to form T_3 or T_4.

5. Immediately or after a delay, in the endocrine phase of the process the follicular cells take up the iodinated thyroglobulin in colloid by endocytosis or pinocytosis. The large endocytic vesicles fuse with lysosomes and move to the basolateral cell membrane while the thyroglobulin inside is thoroughly degraded by lysosomal proteases. T_4 and T_3, freed in this manner from thyroglobulin, then cross the cell membrane and basement membrane, and are taken up into the capillaries.

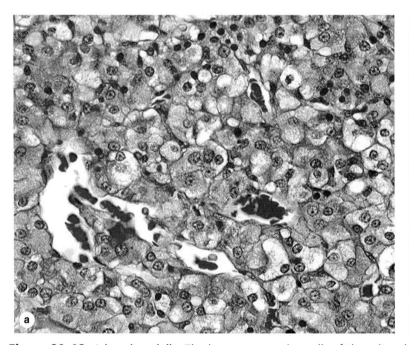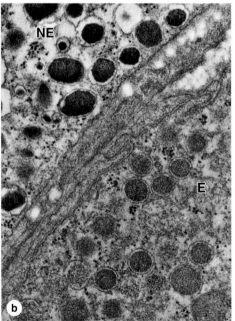

***Figure 20–16.* Adrenal medulla.** The hormone-secreting cells of the adrenal medulla are chromaffin cells, which resemble sympathetic neurons. **(a):** The micrograph shows they are large pale-staining cells, arranged in cords interspersed with wide capillaries. Faintly stained cytoplasmic granules can be seen in most chromaffin cells. X200. H&E. **(b):** TEM reveals that the granules of norepinephrine-secreting cells (NE) are more electron-dense than those of cells secreting epinephrine (E), which is a function of the chromogranins to which the catecholamines are bound in the granules. Most of the hormone produced is epinephrine, which is only made in the adrenal medulla. X33,000.

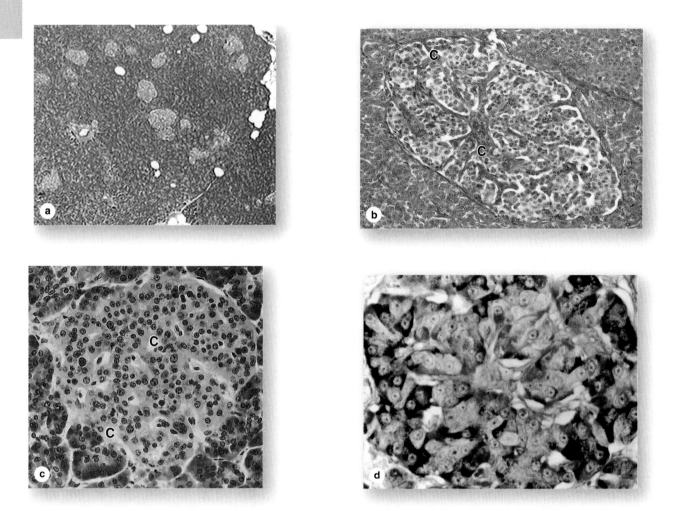

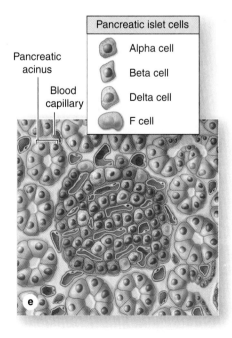

Pancreatic islet cells

Alpha cell

Beta cell

Delta cell

F cell

Pancreatic acinus

Blood capillary

Figure 20–17. **Pancreatic islets.** Pancreatic islets are clumped masses of pale-staining endocrine cells embedded in the exocrine acinar tissue of the pancreas. **(a):** A low-magnification micrograph through the tail of the pancreas reveals many islets stained lighter than the surrounding tissue. The white spots are adipocytes in fine pancreatic septa, like those of the adipose tissue outside the gland. X12.5. H&E. **(b):** Micrograph of an enlarged islet showing its capillary system. Several arterioles enter each islet, branch into fenestrated capillaries (C) among the peripheral islet cells, then converge centrally before leaving the islet as efferent capillaries carrying blood to the acini surrounding the islet. This local vascular system allows specific islet hormones to help control secretion of other islet cells and the neighboring acini. X40. H&E. **(c):** With H&E staining all cells of an islet appear similar, although slight differences in cell size and basophilia may be apparent. Capillaries (C) are also apparent. X55. H&E. **(d):** An islet prepared with a modified aldehyde fuchsin stain shows granules in the peripheral α cells are a deep brownish purple and the central β cells granules are brownish orange. Reticulin connective tissue of the islet capsule and along the capillaries stains green in this preparation. Immunohistochemistry with antibodies against the various islet polypeptide hormones allows definitive identification of each islet cell type. X300. Modified aldehyde fuchsin and light green. **(e):** Diagram shows the four major islet hormones and the cells secreting them: α cells making glucagon, β cells making insulin, δ cells making somatostatin, and F (PP) cells making pancreatic polypeptide. (Figure 20–17d, with permission, from Arthur A. Like, Department of Pathology, University of Massachusetts Medical School, Worcester.)

Nearly all of both thyroid hormones are carried in blood tightly bound to plasma proteins. T_4 is the more abundant compound, constituting 90% of the circulating thyroid hormone. Both molecules bind the same intracellular receptors of target cells, but T_3 is two- to tenfold more active than T_4. The half-life of T_3 is 1.5 days in comparison with a week for T_4. Both thyroid hormones increase the number of mitochondria and their cristae and stimulate mitochondrial protein synthesis.

MEDICAL APPLICATION

*A diet low in iodide hinders the synthesis of thyroid hormones, causing increased secretion of TSH and compensatory growth of the thyroid gland, a condition known as **iodine deficiency goiter**. Goiters are endemic in some regions of the world, where dietary iodide is scarce and addition of iodide to table salt is not required. Hypothyroidism in the fetus may present at birth as **cretinism**, characterized by arrested or retarded physical and mental development.*

*Adult hypothyroidism may result from diseases of the thyroid gland (eg, due to defects in hormone synthesis or release) or may be secondary to pituitary or hypothalamic failure. Autoimmune diseases of the thyroid, such as **Hashimoto disease**, may impair its function, with consequent hypothyroidism.*

*Hyperthyroidism may be caused by a variety of thyroid diseases, of which the most common form is **Graves' disease**, characterized by inflammation and growth of the extraocular adipose tissue, which leads to the bulging of the eyes (exophthalmos). In this thyroid disorder hyperfunction is due to an autoimmune response involving antibodies to TSH receptors. These antibodies can bind the receptors on follicular cells and act as long-lasting thyroid stimulators, continuously stimulating thyroid hormone secretion and producing many effects of hyperthyroidism such as decreased body weight and accelerated heart rate.*

PARATHYROID GLANDS

The **parathyroid glands** are four small oval masses—each 3×6 mm—with a total weight of about 0.4 g. They are located on the back of the thyroid gland, one at each end of the upper and lower poles, usually embedded in the larger gland's capsule (Figure 20–22). The parathyroid glands are derived from the pharyngeal pouches—the superior glands from the fourth pouch and the inferior glands from the third pouch. Their embryonic migration to the developing thyroid gland is sometimes misdirected so that the number and locations of the glands are somewhat variable. Up to 10% of individuals may have parathyroid tissue attached to the thymus, which originates from the same pharyngeal pouches.

Each parathyroid gland is contained within a capsule which sends septa into the gland, where they merge with reticular fibers that support elongated cordlike clusters of secretory cells. With increasing age many secretory cells are replaced with adipocytes, which may constitute more than 50% of the gland in older people.

Two types of cells are present in parathyroid glands: chief (or principal) cells and oxyphil cells (Figure 20–23). The **chief cells** are small polygonal cells with round nuclei and pale-staining, slightly acidophilic cytoplasm. Ultrastructurally the cytoplasm is

Table 20–4. Major cell types and hormones of pancreatic islets.

Cell Type	Quantity	Hormone Produced	Hormone Structure and Size	Hormone Function
α	~20%	Glucagon	Polypeptide; 3500 Da	Acts on several tissues to make energy stored in glycogen and fat available through glycogenolysis and lipolysis; increases blood glucose content
β	~70%	Insulin	Dimer of α and β chains with S-S bridges; 5700-6000 Da	Acts on several tissues to cause entry of glucose into cells and promotes decrease of blood glucose content
δ or D	5–10%	Somatostatin	Polypeptide; 1650 Da	Inhibits release of other islet cell hormones through local paracrine action; inhibits release of GH and TSH in anterior pituitary and HCl secretion by gastric parietal cells
F or PP	Rare	Pancreatic polypeptide	Polypeptide; 4200 Da	Stimulates activity of gastric chief cells; inhibits bile secretion, pancreatic enzyme and bicarbonate secretion, and intestinal motility

seen to be filled with irregularly shaped granules 200–400 nm in diameter. These are secretory granules containing the polypeptide **parathyroid hormone** (PTH), a major regulator of blood calcium levels. Much smaller, often clustered, populations of **oxyphil cells** are sometimes present, more commonly in older individuals. These are much larger than the principal cells and are characterized by acidophilic cytoplasm filled with abnormally shaped mitochondria. Some oxyphil cells show low levels of PTH synthesis, suggesting these cells are transitional derivatives from chief cells. Parathyroid hormone targets osteoblasts, which respond by producing an osteoclast-stimulating factor to increase the number and activity of osteoclasts. This promotes resorption of the calcified bone matrix and the release of Ca^{2+}, increasing the concentration of Ca^{2+} in the blood, which suppresses

parathyroid hormone production. Calcitonin from the thyroid gland inhibits osteoclast activity, lowering the blood Ca^{2+} concentration and promoting osteogenesis. Parathyroid hormone and calcitonin thus have opposing effects and constitute a dual mechanism to regulate blood levels of Ca^{2+}, an important factor in homeostasis. Parathyroid hormone also indirectly increases the absorption of Ca^{2+} from the gastrointestinal tract by stimulating the synthesis of vitamin D, which is necessary for this absorption.

In addition to increasing the concentration of Ca^{2+}, parathyroid hormone reduces blood phosphate levels. This effect results from another target cell of parathyroid hormone, renal tubule cells, which reduce their absorption of phosphate and allow more phosphate excretion in urine.

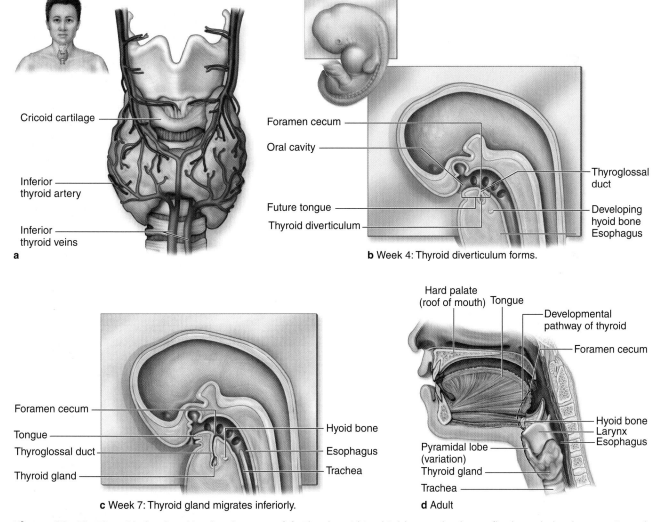

a

b Week 4: Thyroid diverticulum forms.

c Week 7: Thyroid gland migrates inferiorly.

d Adult

Figure 20–18. **Thyroid gland and its development. (a):** The thyroid is a highly vascular, butterfly-shaped gland, approximately 5 cm × 5 cm and weighing 20–30 g in adults, surrounding the anterior surface of the trachea just below the larynx. Immigrating neural crest cells infiltrate the epithelium as precursors to the thyroid's parafollicular C cells. **(b):** Thyroid development begins in the fourth week as an epithelial diverticulum growing down from the endodermal lining of the foregut. **(c):** The thyroid diverticulum continues to grow in an inferior direction and its connection to the developing pharynx, the thyroglossal duct, later regresses. **(d):** By fetal stages, the thyroid has attained its normal adult position.

In **hyperparathyroidism**, concentrations of blood phosphate are decreased and concentrations of blood Ca²⁺ are increased. This condition frequently produces pathologic deposits of calcium in several organs, such as the kidneys and arteries. The bone disease caused by hyperparathyroidism, which is characterized by an increased number of osteoclasts and multiple bone cavities, is known as **osteitis fibrosa cystica**. Bones from patients with this disorder are less stress-resistant and prone to fractures.

In **hypoparathyroidism**, concentrations of blood phosphate are increased and concentrations of blood Ca²⁺ are decreased. The bones become denser and more mineralized. This condition causes spastic contractions of the skeletal muscles and generalized convulsions called **tetany**. These symptoms are caused by the exaggerated excitability of the nervous system, due to the lack of Ca²⁺ in the blood. Patients with hypoparathyroidism are treated with calcium salts and vitamin D to promote Ca²⁺ uptake in the gut.

PINEAL GLAND

The **pineal gland**, also known as the pineal body or epiphysis cerebri, regulates the daily rhythms of bodily activities. It is a very small, pine cone-shaped organ in the brain measuring approximately 5–8 mm in length and 3–5 mm at its greatest width and weighing about 150 mg. The pineal develops with the brain from neuroectoderm in the roof of the diencephalon

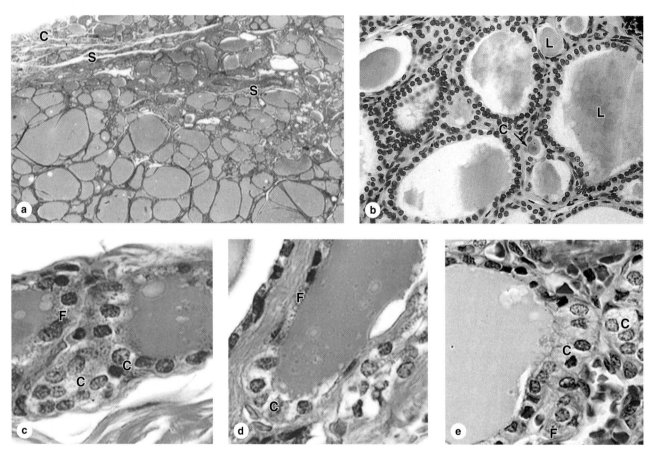

Figure 20–19. **Thyroid follicular cells and parafollicular cells. (a):** A low-power micrograph of thyroid gland shows the thin capsule (C), from which septa (S) with the larger blood vessels, lymphatics, and nerves enter the gland. The parenchyma of the organ is distinctive, consisting of colloid-filled epithelial follicles of many sizes. The lumen of each follicle is filled with a lightly staining colloid of a large gelatinous protein called thyroglobulin. X12. H&E. **(b):** The lumen (L) of each follicle is surrounded by a simple epithelium in which the cell height ranges from squamous to low columnar. Also present are large pale-staining parafollicular or C cells (C) which secrete calcitonin, a polypeptide involved with calcium metabolism. X200. H&E. **(c,d,e):** C cells may be part of the follicular epithelium or present singly or in groups outside of follicles. Follicular cells (F) can usually be distinguished from C cells (C) by the smaller size and darker staining properties. Unlike follicular cells, C cells seldom vary in their size or pale staining characteristics. C cells are somewhat easier to locate in or between small follicles. c and d: X400. H&E; e: X400. Mallory trichrome.

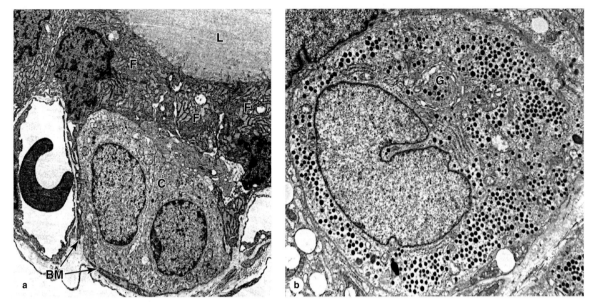

Figure 20–20. Ultrastructure of thyroid follicular and parafollicular cells. **(a):** TEM of the follicular epithelium shows pseudopodia and microvilli extending from the follicular cells (F) into the colloid of the lumen (L). The cells have apical junctional complexes, much RER, well-developed Golgi complexes, and many lysosomes. Inside the basement membrane (BM) of the follicle, but often not contacting the colloid in the lumen, are occasional C or parafollicular cells (C). To the left and right of the two C cells seen here are capillaries intimately associated with the follicular cells, but outside the basement membrane. X2000. **(b):** A TEM of a C cell, with its large Golgi apparatus (G), extensive RER, and cytoplasm filled with small secretory granules containing calcitonin. X5000.

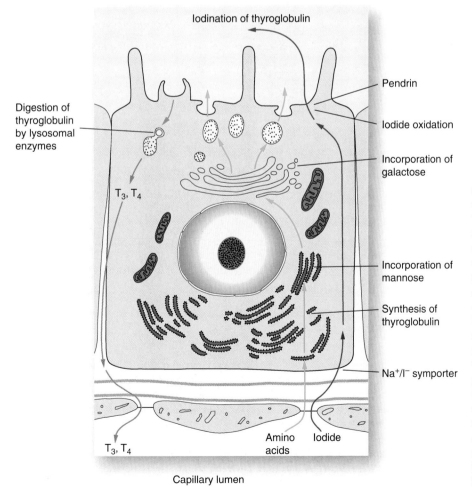

Figure 20–21. Thyroid follicular cell functions. The diagram shows the multistep process by which thyroid hormones are produced via the stored thyroglobulin intermediate. In an exocrine phase of the process, the glycoprotein thyroglobulin is made and secreted into the follicular lumen and iodide is pumped across the cells into the lumen. In the lumen tyrosine residues of thyroglobulin are iodinated and then covalently coupled to form T_3 and T_4 still within the glycoprotein. The iodinated thyroglobulin is then endocytosed by the follicular cells and degraded by lysosomes, releasing free active T_3 and T_4 to the adjacent capillaries in an endocrine manner. Both phases are promoted by TSH and may occur simultaneously in the same cell.

and is found in the posterior of the third ventricle, attached to the brain by a short stalk.

The pineal gland is covered by connective tissue of the pia mater, from which emerge septa containing small blood vessels and subdividing various sized groups of secretory cells as lobules. The prominent and abundant secretory cells are the **pinealocytes**, which have slightly basophilic cytoplasm and large, irregular euchromatic nuclei and nucleoli (Figure 20–24). Ultrastructurally pinealocytes are seen to have secretory vesicles, many mitochondria, and long cytoplasmic processes extending to the vascularized

septa, where they end in dilatations near capillaries, indicating an endocrine function. These cells produce **melatonin**, a low molecular-weight tryptophan derivative. Unmyelinated sympathetic nerve fibers enter the pineal gland and end among pinealocytes, with some forming synapses.

Interstitial glial cells of the pineal gland stain positively for glial fibrillary acidic protein and thus most closely resemble astrocytes. They have elongated nuclei more heavily stained than those of pinealocytes, long cytoplasmic processes, and are usually found in perivascular areas and between the groups of pinealocytes. Pineal astrocytes represent only about 5% of the cells in the gland.

A characteristic feature of the pineal gland is the presence of variously sized concretions of calcium and magnesium salts called

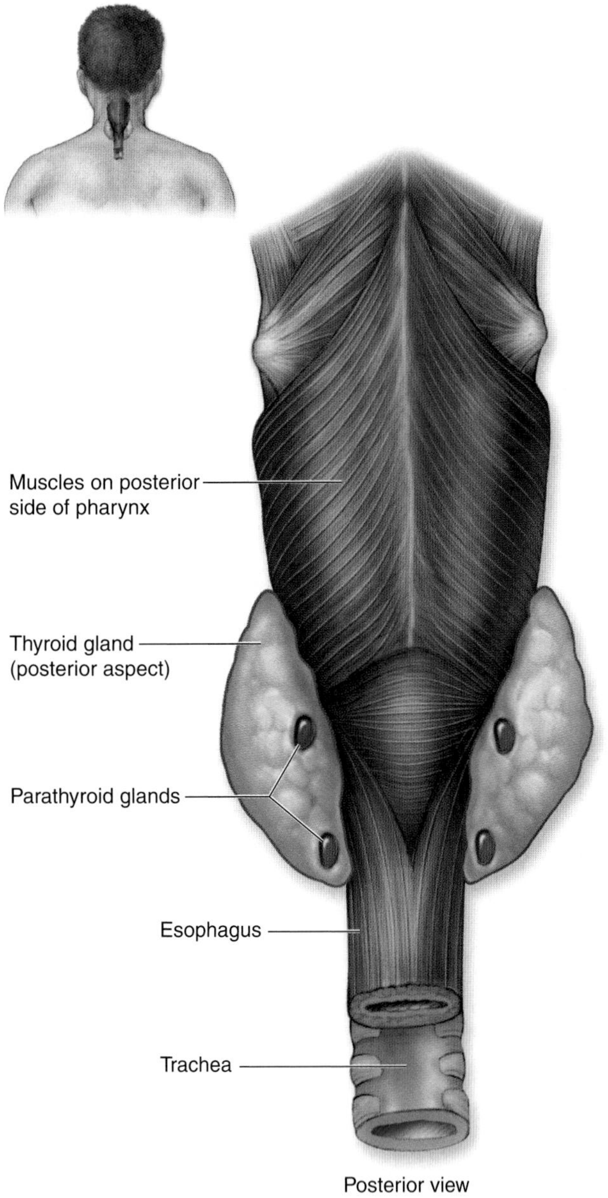

Figure 20–22. Parathyroid glands. The parathyroid glands are four small nodules normally embedded in the capsule on the posterior surface of the thyroid gland. They arise embryologically from the third and fourth pharyngeal pouches and migrate to the developing thyroid, a process that frequently leads to ectopic or additional parathyroid glands, often associated with the thymus.

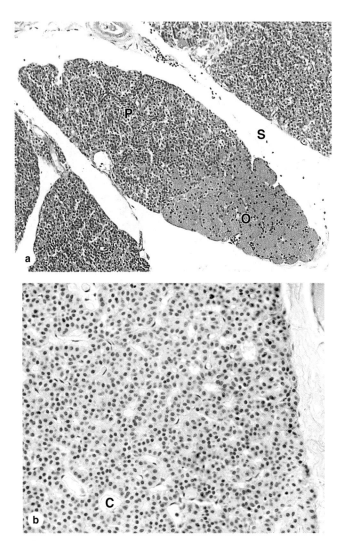

Figure 20–23. **Parathyroid principal cells. (a):** A small lobe of parathyroid gland, surrounded by connective tissue septa (S), shows mainly densely packed cords of small principal cells (P), also called chief cells. Older parathyroid glands show increasing numbers of much larger and acidophilic nonfunctional oxyphil cells (O) which may occur singly or in clumps of varying sizes. X60. H&E. **(b):** The micrograph shows that principal cells are slightly eosinophilic, with round central nuclei, present in cords separated by capillaries (C). These cells secrete the polypeptide parathyroid hormone (PTH). X300. H&E.

corpora arenacea or brain sand, which form by precipitation around extracellular protein deposits. Such concretions appear during childhood and gradually increase in number and size with age, with no apparent effect on the gland's function. Accumulations of brain sand are opaque to x-rays and allow the pineal to serve as a good midline marker in radiological and computer-assisted tomography studies of the brain.

Melatonin release from pinealocytes is promoted by darkness and inhibited by daylight and the resulting diurnal fluctuation in blood melatonin levels induces rhythmic changes in the activity of the hypothalamus, pituitary gland, and other endocrine tissues that characterize the circadian (24 hours, day/night) rhythm of physiological functions and behaviors. In humans and other mammals the cycle of light and darkness is detected within the retinas and transmitted to the pinealocytes via the retinohypothalamic tract, the suprachiasmatic nucleus, and the tracts of sympathetic fibers entering the pineal. The pineal gland acts therefore as a neuroendocrine transducer, converting nerve input regarding light and darkness into variations in many hormonal functions.

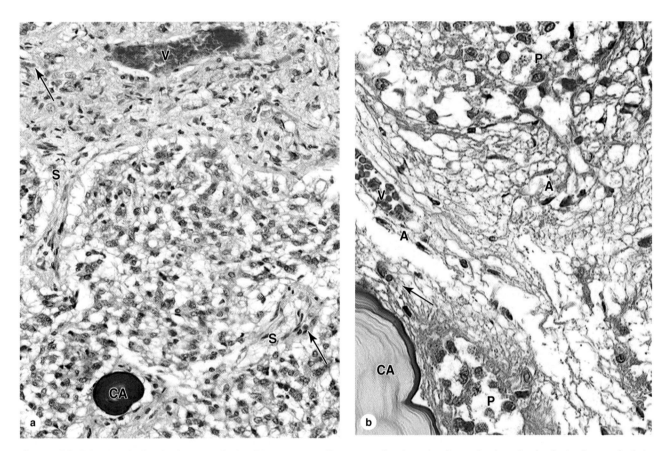

Figure 20–24. Pineal gland. The pineal gland is a very small neuroendocrine gland attached to the brain in the roof of the third ventricle. It is covered by the pia mater, which sends septa of connective tissue into the glands, subdividing groups of parenchymal cells called pinealocytes. **(a):** The micrograph shows a group of pinealocytes surrounded by septa (S) containing venules (V) and capillaries (arrows). Also seen is an extracellular mineral deposit called a corpus arenaceum (CA) of unknown physiological significance but an excellent marker for the pineal. X40. H&E.

(b): At higher magnification the numerous large pinealocytes (P) with euchromatic nuclei can be compared to the relatively few astrocytes (A) which have darker, more elongated nuclei. Astrocytes are located mainly within septa and near small blood vessels (V). Capillaries (arrow) are not nearly as numerous as in other endocrine glands. At the lower left is a port of a very large corpus arenaceum (CA), the calcified structures also known as brain sand. Along the septa run unmyelinated tracts of sympathetic fibers, associated indirectly with photoreceptive neurons in the retinas and running to the pinealocytes to stimulate melatonin release in periods of darkness. Levels of circulating melatonin are one factor determining the diurnal rhythms of hormone release and physiological activities throughout the body. X100. H&E.

The Male Reproductive System

TESTES
    Seminiferous Tubules
    Spermatogenesis
    The Clonal Nature of Male Germ Cells
    Spermiogenesis
    Sertoli Cells

    Interstitial Tissue
INTRATESTICULAR DUCTS
EXCRETORY GENITAL DUCTS
ACCESSORY GLANDS
PENIS

The male reproductive system is composed of the testes, genital ducts, accessory glands, and penis (Figure 21–1). Testes function in the production of hormones and spermatozoa. Although testosterone is the main hormone produced in the testes, both testosterone and its metabolite, dihydrotestosterone, are necessary for male reproductive physiology. Testosterone is important for spermatogenesis, sexual differentiation during embryonic and fetal development, and control of gonadotropin secretion. Dihydrotestosterone acts on many organs and tissues of the body during puberty and adulthood (eg, muscle, hair pattern, and hair growth).

The genital ducts and accessory glands produce secretions and aided by smooth muscle contractions, propel spermatozoa toward the exterior. These secretions also provide nutrients for spermatozoa while they are confined to the male reproductive tract. Spermatozoa and the secretions of the genital ducts and accessory glands make up the **semen** (L, *semen*, seed), which is introduced into the female reproductive tract through the penis.

TESTES

Each **testis** (testicle) is surrounded by a capsule of dense connective tissue, the **tunica albuginea.** The tunica albuginea is thickened on the posterior side of the testis to form the **mediastinum testis,** from which fibrous septa penetrate the organ and divide it into about 250 pyramidal compartments or **testicular lobules** (Figure 21–2). The septa are incomplete, and there is frequently intercommunication between lobules. Each lobule is occupied by one to four **seminiferous tubules** that are surrounded by interstitial loose connective tissue rich in blood and lymphatic vessels, nerves, and endocrine **interstitial cells (Leydig cells)** which secrete testosterone. Seminiferous tubules produce male reproductive cells, the spermatozoa, whereas interstitial cells secrete testicular androgens.

The testes develop retroperitoneally in the dorsal wall of the embryonic abdominal cavity. They are moved during fetal development and eventually are suspended within the two sides of the scrotum at the ends of the spermatic cords. Because of the migration from the abdominal cavity, each testis carries with it a serous sac, the **tunica vaginalis** (Figure 21–2), derived from the peritoneum. The tunic consists of an outer parietal layer lining the scrotum and an inner visceral layer, covering the tunica albuginea on the anterior and lateral sides of the testis.

Temperature is important in the regulation of spermatogenesis, which occurs only below the core body temperature of 37°C. A testicular temperature of about 34°C is maintained in the scrotal sac by various mechanisms. Each testicular artery is surrounded by a rich **pampiniform venous plexus** with cooler blood from the testis which can draw heat from the arterial blood by a countercurrent heat-exchange system. Evaporation of sweat from the scrotum also contributes to heat loss. Relaxation or contraction of the dartos muscle of the scrotum and the cremaster muscles of the spermatic cords move the testes away from, or closer to, the body respectively, allowing further control on testicular temperature.

MEDICAL APPLICATION

*Failure of descent of the testes into the scrotum (**cryptorchidism** [Gr. kryptos, hidden, + orchis, testis]) maintains the testes at the core body temperature which inhibits spermatogenesis. In some cases, spermatogenesis can occur normally if the testes are moved surgically to the scrotum. For this reason, it is important to examine male*

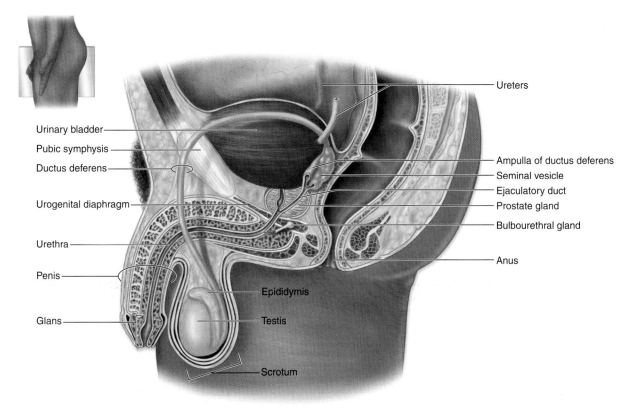

Figure 21–1. The male reproductive system. The diagram shows the locations and relationships of the testes, epididymis, glands, and the ductus deferens running from the scrotum to the urethra. The ductus deferens is located along the anterior and superior sides of the bladder as a result of the testes descending into the scrotum from the abdominal cavity during fetal development.

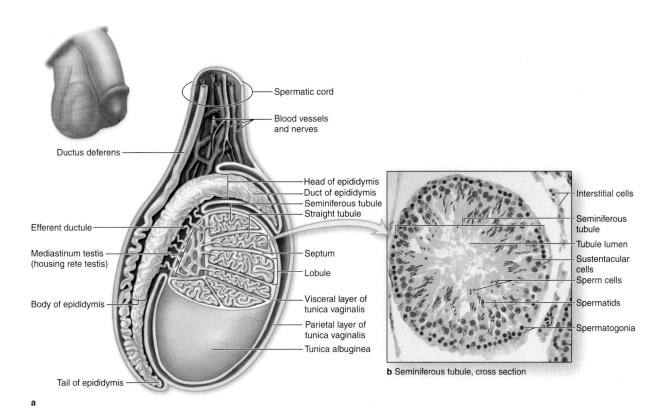

b Seminiferous tubule, cross section

a

Figure 21–2. Testes and seminiferous tubules. The anatomy of a testis is shown. **(a):** The diagram shows a partially cut-away sagittal section. **(b):** The micrograph shows a cross section of one seminiferous tubule. X250. H & light green.

newborns to check if the testicles are present in the scrotum. Although germ cell proliferation is inhibited by abdominal temperature, testosterone synthesis by interstitial cells is not affected. Therefore men with cryptorchidism can be sterile but still develop secondary male characteristics.

Malnutrition, alcoholism, and the action of certain classes of drugs lead to alterations in spermatogonia, with a resulting decrease in production of spermatozoa. X-irradiation is quite toxic to cells of the spermatogenic lineage, causing the death of those cells and sterility in animals.

Androgen-producing interstitial cell tumors can cause precocious puberty in males.

Seminiferous Tubules

Sperm are produced in the seminiferous tubules at a rate of about 2×10^8 per day in the adult male. Each testicle has 250–1000 seminiferous tubules in its lobules, with each tubule measuring about 150–250 μm in diameter and 30–70 cm in length. The combined length of the tubules of one testis is about 250 m. Each tubule is a convoluted loop linked via a short, narrower segment, the **straight tubule**, to the **rete testis**, a labyrinth of epithelium-lined channels embedded in the mediastinum testis (Figures 21–2 and 21–3). Ten to twenty **efferent ductules** connect the rete testis to the head of the **epididymis** (Figure 21–2).

Each seminiferous tubule is lined with a complex, specialized stratified epithelium called germinal or **seminiferous epithelium** (Figure 21–2). The basement membrane of this epithelium is covered by fibrous connective tissue, with an innermost layer containing flattened, smooth muscle-like **myoid cells** (Figure 21–4), which allow weak contractions of the tubule. Interstitial cells occur in the connective tissue between the seminiferous tubules (Figures 21–4).

The seminiferous epithelium consists of two types of cells: nondividing **supporting** or **sustentacular cells (Sertoli cells)** and proliferative cells of the **spermatogenic lineage** (Figure 21–5). The cells of the spermatogenic lineage comprise four to eight concentric cell layers and their function is to produce the cells that become sperm. The part of sperm production that includes cell division through mitosis and meiosis is called **spermatogenesis**. The final differentiation of the haploid male germ cells is called **spermiogenesis**.

Spermatogenesis

Spermatogenesis begins at puberty with a primitive germ cell, the **spermatogonium** (Gr. *sperma* + *gone*, generation), a relatively small round cell, about 12 μm in diameter. These cells are located basally in the epithelium next to the basement membrane (Figures 21–5 and 21–6) and different stages of their development are recognized mainly by the shape and staining properties of their nuclei. Spermatogonia with dark, ovoid nuclei act as stem cells, dividing infrequently and giving rise both to new stem cells and to cells with more pale-staining, ovoid nuclei that divide more rapidly as transit amplifying (progenitor) cells (Figure 21–7). These **type A spermatogonia** each undergo several unique clonal divisions, remaining interconnected as a syncytium (see below), and form **type B spermatogonia**, which have more spherical pale nuclei.

Each type B spermatogonium then undergoes a final mitotic division to produce two cells that grow in size and become **primary spermatocytes**, which are spherical cells with euchromatic nuclei (Figures 21–6 and 21–7). Primary spermatocytes replicate their DNA, so each chromosome consists of duplicate chromatids, and enter meiosis, during which homologous chromosomes come together in synapsis, DNA recombination occurs, and two rapid cell divisions produce haploid cells (Chapter 3). The primary spermatocyte has 46 (44 + XY) chromosomes, the diploid number, and a DNA content of 4N. (The letter N denotes either the haploid number of chromosomes 23 in humans, or the amount of DNA in this set.) Soon after their formation, these cells enter the first meiotic prophase which lasts about 22 days. Most spermatocytes seen in sections are in this phase of meiosis. The primary spermatocytes are the largest cells of the spermatogenic lineage and are characterized by the presence

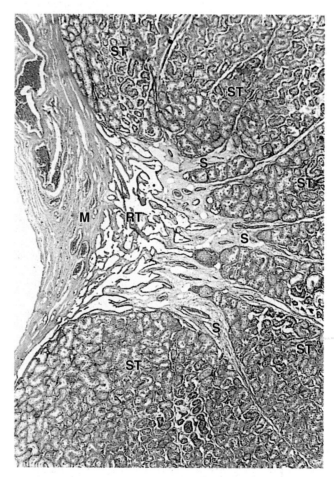

Figure 21–3. Lobules converging at rete testis. The capsule of the testis, the tunica albuginea, thickens on the posterior surface as the mediastinum (M) testis, from which many thin septa (S) subdivide the organ into about 250 lobules. Each lobule contains from one to four convoluted seminiferous tubules (ST) in a sparse connective tissue interstitium. Each tubule is a loop attached by means of a short straight tubule to the rete testis (RT), a maze of channels embedded in the mediastinum testis. From the rete testis sperm move into the epididymis. X60. H&E

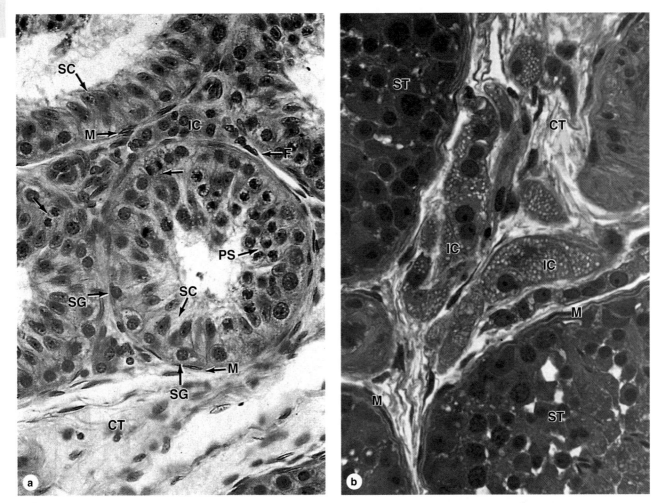

Figure 21–4. Seminiferous tubule and interstitial cells. (a): The micrograph shows seminiferous tubules surrounded by connective tissue (CT), containing many large rounded or polygonal interstitial cells (IC) secreting androgens. Immediately surrounding each tubule are flattened myoid cells (M), which contract to help move sperm out of the tubule, and layers of fibroblasts (F). Inside the tubule itself is a unique seminiferous epithelium composed of columnar supporting cells called Sertoli cells (SC), which usually have oval nuclei and distinct nucleoli, and germ cells of the spermatogenic lineage. Prominent among the latter are spermatogonia (SG), diploid cells always located near the basement membrane, and primary spermatocytes (PS) which are undergoing meiosis closer to the lumen of the tubule. At the upper left corner is a portion of a straight tubule, which lacks germ cells and consists solely of Sertoli cells. X300. H&E. **(b):** A higher magnification of a plastic section shows lipid droplets filling the cytoplasm of the clumped interstitial cells (IC) in the connective tissue (CT) between tubules, which is typical of steroid-secreting endocrine cells. The epithelium of the seminiferous tubules (ST) is immediately surrounded by myoid cells (M). X400. PT.

Figure 21–5. Spermatogenesis and spermiogenesis. (a): The diagram shows a small part of a seminiferous tubule with its surrounding tissue. The seminiferous epithelium is formed by two cell populations: the supporting or Sertoli cells and the cells of the spermatogenic lineage. The latter are intimately associated with and almost completely surrounded by the surface of the columnar supporting Sertoli cells. Nearest the basement membrane are spermatogonia, which divide by mitosis to produce more spermatogonia and cells committed to meiosis. The latter cells grow to become primary spermatocytes while arrested in the first meiotic division. They then divide to become secondary spermatocytes, which divide quickly again to become haploid spermatids, which differentiate as sperm cells. These last events take place near the apical ends of the Sertoli cells at the lumen of the tubules. The small dots at the junction of the Sertoli cells represent occluding junctions which form part of the blood-testis barrier and which are opened transiently as the germ cells move across those areas, forming new junctions between themselves and the support cells to maintain the barrier function. Usually all stages of this process, called spermatogenesis, do not take place within one small area of tubule as shown schematically here.

(b): The diagram depicts the major morphological changes that occur within spermatids as they undergo the differentiation process, called spermiogenesis, and become highly specialized sperm cells. These changes involve flattening of the nucleus, formation of an acrosome which resembles a large lysosome, growth of a flagellum (tail) from the basal body, reorganization of the mitochondria in the midpiece region, and shedding of unneeded cytoplasm as a residual body.

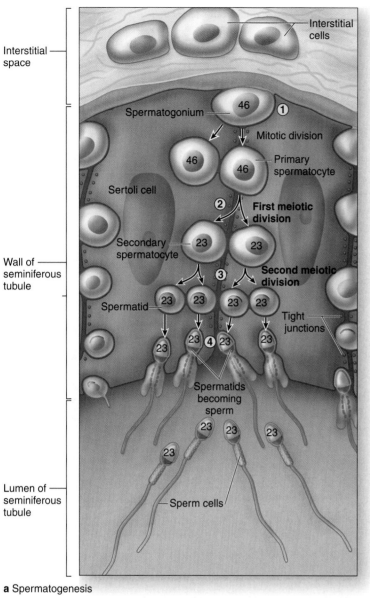

a Spermatogenesis

1 Germ cells that are the origin of sperm cells are *diploid cells* (containing 46 chromosomes, or 23 pairs) called spermatogonia. Mitotic divisions of these cells produce a new germ cell and a committed cell. The committed cell is a primary spermatocyte.

2 The first meiotic division begins in the *diploid* primary spermatocytes. The *haploid cells* (containing 23 chromosomes only) produced by the first meiotic division are called secondary spermatocytes.

3 The second meiotic division originates with the secondary spermatocytes and produces spermatids.

4 The process of spermiogenesis begins with spermatids and results in morphological changes needed to form sperm that will be motile.

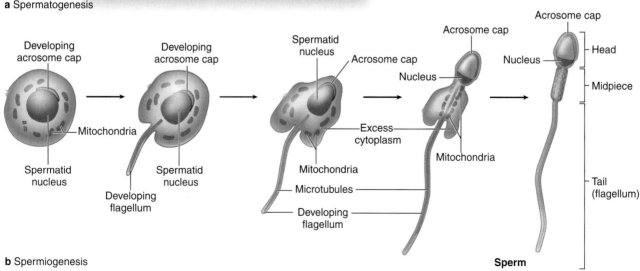

b Spermiogenesis

of partially condensed chromosomes in various stages of synapsis and recombination (Figure 21–6).

Homologous chromosomes separate in the first meiotic division, which produces smaller cells called **secondary spermatocytes** (Figures 21–5 and 21–7) with only 23 chromosomes (22 + X or 22 + Y), but each still consists of two chromatids so the amount of DNA is 2N (Chapter 3). Secondary spermatocytes are rare in testis sections because they are short-lived cells that remain in interphase only very briefly and quickly undergo the second meiotic division. Division of each secondary spermatocyte separates the chromatids of each chromosome and produces two haploid cells called **spermatids** that each contain 23 chromosomes (Figures 21–5, 21–6, and 21–7). Because no S phase (DNA replication) occurs between the first and second meiotic

divisions, the amount of DNA per cell is reduced by half when the chromatids separate and the cells formed are haploid (1N). With fertilization, a haploid ovum and sperm produced by meiosis unite and the normal diploid number for the species is restored.

The Clonal Nature of Male Germ Cells

The stem cells produced by divisions of type A spermatogonia remain as separate cells. However all subsequent divisions of the daughter cells that become transit amplifying progenitor cells have incomplete cytokinesis after telophase and the cells remain attached to one another by **intercellular bridges** of cytoplasm (Figure 21–7). These allow free communication among the cells

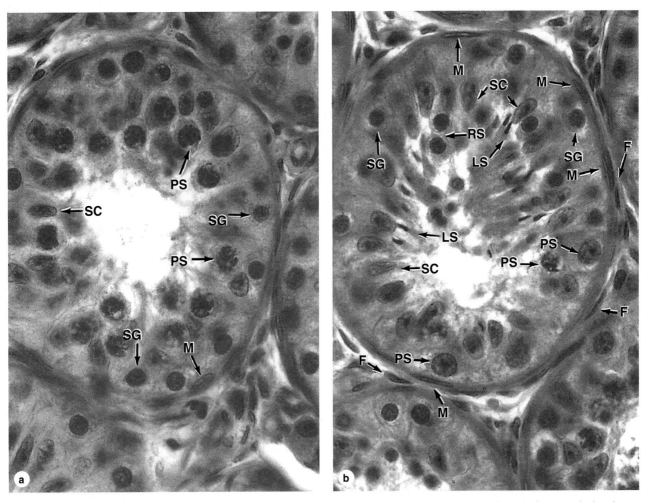

Figure 21–6. **Seminiferous tubules: Sertoli cells and spermatogenesis.** In the two cross-sections of seminiferous tubules shown, most of the associated cell types can be seen. Outside the tubules are myoid cells (M) and fibroblasts (F). Inside near the basement membrane are many prominent spermatogonia (SG), small cells which divide mitotically but give rise to a population that enters meiosis. The meiotic cells grow and undergo chromosomal synapsis to become primary spermatocytes (PS), arrested for 3 weeks in prophase of the first meiotic division during which recombination occurs. Primary spermatocytes are the largest spermatogenic cells and are usually abundant at all levels between the basement membrane and the lumen. Each divides to form two secondary spermatocytes, which are seldom seen in sections because they undergo the second meiotic division almost immediately to form two haploid spermatids. Newly formed round spermatids (RS) differentiate and lose volume in becoming late spermatids (LS) and finally motile, highly specialized sperm cells. All stages of spermatogenesis and spermiogenesis occur with the cells intimately associated with the surfaces of adjacent Sertoli cells (SC) which perform several supportive functions. Both X750. H&E.

derived from a single type A spermatogonium through their remaining mitotic and meiotic divisions. Although some cells degenerate without completing spermatogenesis and some cells may separate, approximately a hundred cells may remain linked through the period of meiosis. The complete significance of this **spermatogenic syncytium** is not clear, but the cytoplasmic bridges allow each developing spermatid to share the cytoplasm of neighboring cells. The haploid cells can thus be supplied with products of the complete diploid genome, including proteins and RNA encoded by genes on the X or Y chromosome missing in

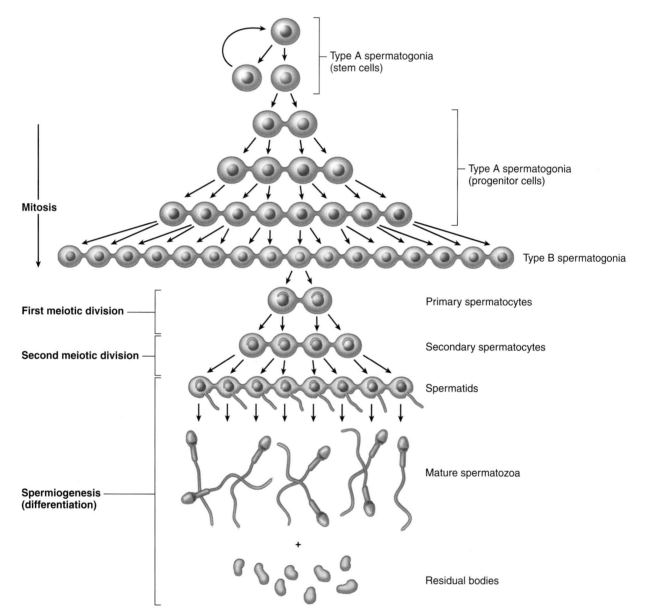

Figure 21–7. Clonal nature of spermatogenesis. The diagram shows the clonal nature of the germ cells during spermatogenesis. A subpopulation of type A spermatogonia act as stem cells, dividing to produce new stem cells and other type A spermatogonia that undergo transit amplification as progenitor cells for spermatocytes. Mitosis in these cells occurs with incomplete cytokinesis, leaving the cytoplasm of most or all of these cells connected by intercellular bridges. Type A spermatogonia divide mitotically two or three more times, then differentiate as type B spermatogonia which undergo a final round of mitosis to form the cells that then enter meiosis and become a primary spermatocytes (two are shown), with their cytoplasm still interconnected. The intercellular bridges persist during the first and second meiotic divisions and are finally lost as the haploid spermatids complete their differentiation into sperm (spermiogenesis). During differentiation each spermatid sheds excess cytoplasm as a residual body which is phagocytosed by Sertoli cells, and any germ cells that cannot complete this process and degenerate. The interconnected state of these spermatogonia and the sperm to which they give rise allows free intercellular communication and facilitates their coordinated progress through meiosis and spermiogenesis.

their haploid nuclei. The germ cells finally become separated from one another during the differentiation process (Figure 21–7).

The cellular events and changes between the final mitoses of spermatogonia and the formation of spermatids take about 2 months. The spermatogenic cells are not randomly distributed in the seminiferous epithelium; cells at different stages of development are grouped together in specific associations. The intercellular bridges between cells in each group may help to coordinate their divisions and differentiation.

Spermiogenesis

Spermiogenesis is the final stage in sperm production and is the process by which spermatids transform into spermatozoa, cells that are highly specialized to deliver male DNA to the ovum. No cell division occurs during this process.

The spermatids can be distinguished by their small size (7–8 μm in diameter), haploid nuclei with highly condensed chromatin, and position near the lumen of the seminiferous tubules (Figures 21–5 and 21–6). Spermiogenesis includes formation of the acrosome (Gr. *akron*, extremity, + *soma*, body), condensation and elongation of the nucleus, development of the flagellum, and the loss of much of the cytoplasm. The end result is the mature spermatozoon, which is then released into the lumen of the seminiferous tubule. Spermiogenesis can be divided into three phases:

- During the *early Golgi phase*, the cytoplasm of spermatids contains a prominent Golgi apparatus near the nucleus,

mitochondria, a pair of centrioles, free ribosomes, and tubules of smooth ER (Figure 21–5). Small **proacrosomal vesicles** accumulate in the Golgi apparatus and subsequently coalesce to form a single membrane-limited **acrosomal cap** close to one end of the nucleus (Figures 21–5 and 21–8). The centrioles migrate to a position near the cell surface and opposite the forming acrosome. One centriole acts as a basal body, serving to organize the axoneme of the flagellum which is similar in structure to that of a cilium.

- During the *acrosome phase*, the acrosomal cap, or **acrosome**, spreads to cover about half of the condensing nucleus (Figures 21–5 and 21–8). The acrosome is a specialized type of lysosome containing several hydrolytic enzymes, including hyaluronidase, neuraminidases, acid phosphatase, and a trypsin-like protease called acrosin. These enzymes are released when spermatozoa encounter an oocyte and the outer membrane of the acrosome fuses with the sperm's plasma membrane. They dissociate cells of the corona radiata and digest the zona pellucida, both structures that surround the egg (Chapter 22). This process, the **acrosomal reaction,** is one of the first steps in fertilization.

Also during this phase of spermiogenesis, the spermatids become oriented toward the base of the Sertoli cells and the axonemes project toward the lumen of the tubule (Figure 21–6b). In addition, the nuclei become more elongated and the chromatin very highly condensed, with the histones of nucleosomes replaced by small basic peptides called protamines. Flagella growth continues and mitochondria aggregate around

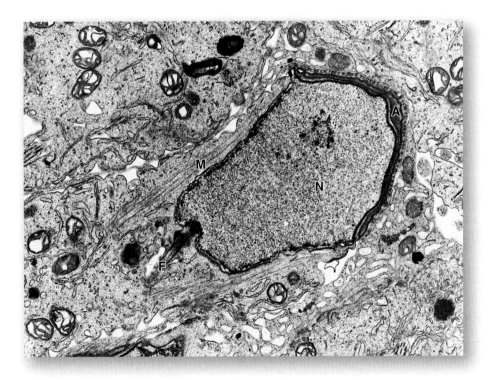

Figure 21–8. Spermatid in acrosome phase of differentiation. A TEM of a spermatid during the acrosome phase of spermiogenesis shows the nucleus (N) in the center of the cell, half covered by the thin Golgi-derived acrosome (A). The flagellum (F) can be seen emerging from a basal body near the nucleus on the side opposite the acrosome. A cylindrical bundle of microtubules and actin filaments called the manchette (M), surrounds the nucleus behind the acrosome. The manchette is a temporary structure in which vesicles, mitochondria and keratins are shuttled into position as the spermatid elongates in preparation for its final maturation. The spermatid is almost completely surrounded by a Sertoli cell. X7500.

the proximal part of each flagellum, forming a thickened region known as the **middle piece,** the region where the ATP for flagellar movements of the spermatozoa is generated (Figure 21–5). As in cilia, movement of the flagellum results from the interactions of microtubules, ATP, and **dynein**, a protein with ATPase activity (Chapter 2).

MEDICAL APPLICATION

Immotile cilia syndrome in men is characterized by immotile spermatozoa and consequent infertility. It is caused by a lack of dynein or other proteins required for ciliar and flagellar motility in the cells of the diseased person. This disorder usually coincides with chronic respiratory infections because of impaired motility of the ciliary axonemes of respiratory epithelial cells.

- During the final *maturation phase* of spermiogenesis, unneeded cytoplasm is shed as a **residual body** from each spermatozoon and is phagocytosed by Sertoli cells. Mature sperm (Figure 21–5) are then released into the lumen of the tubule.

Sertoli Cells

The **Sertoli cells**, named for Enrico Sertoli (1842–1910) who first demonstrated their physiological significance, are highly important for the function of the testes. They are columnar or pyramidal cells that largely envelop cells of the spermatogenic lineage and function as supporting cells or nurse cells. The bases of the Sertoli cells adhere to the basal lamina and their apical ends frequently extend into the lumen of the seminiferous tubule. In the light microscope, the outlines of Sertoli cells are very poorly defined because of the numerous lateral processes that surround spermatogenic cells (Figure 21–5). Each Sertoli cell supports 30–50 germ cells at various stages of development. Studies with the TEM reveal that Sertoli cells contain abundant smooth ER, some rough ER, well-developed Golgi complexes, as well as numerous mitochondria and lysosomes. The elongated nucleus, which can be triangular in outline, has infoldings and a prominent nucleolus, often exhibiting very little heterochromatin (Figure 21–6).

Elaborate tight occluding junctions between the basolateral membranes of adjacent Sertoli cells form a **blood-testis barrier** in the seminiferous epithelium, the tightest blood-tissue barrier in mammals. This physical barrier is one part of a system that prevents autoimmune attacks against the unique spermatogenic cells, which first appear long after the immune system is mature and central self-tolerance established. Spermatogonia lie in a **basal compartment**, situated below the junctions and open to the vascularized interstitial tissue which contains lymphocytes and antigen-presenting cells. Early in meiosis, newly formed spermatocytes temporarily disrupt the cell adhesion molecules of the most basal junctions, transiently establishing new junctions between adhesion factors in their own membranes and those of Sertoli cells, and move into the **adluminal compartment** without compromising the blood-testis barrier. Spermatocytes and spermatids adhere closely to the Sertoli cells, lying within deep invaginations of these cells' lateral and apical membranes, above

the barrier. Movement of the spermatogenic cells between the support cells while maintaining effective occluding junctions between all the cells is all the more impressive when one remembers that the germ cells remain linked by intercellular bridges. As the flagellar tails of the spermatids develop, they appear as tufts extending from the apical ends of the Sertoli cells. Sertoli cells are also connected and coupled ionically by numerous gap junctions, which may help regulate the transient changes in the occluding junctions and in coordinating the cycle of the seminiferous epithelium described above.

Sertoli cells have several specific functions within the seminiferous epithelium, mostly involving the blood-testis barrier:

- **Support, protection, and nutrition of the developing spermatogenic cells.** Because spermatocytes, spermatids, and sperm are isolated from plasma proteins and nutrients by the blood-testis barrier, these spermatogenic cells depend on the Sertoli cells for production or transport into the lumen of metabolites and nutritive factors such as the iron-transport protein transferrin. Thus, while protecting spermatogenic cells from immune components in plasma, Sertoli cells must supply the plasma factors needed for growth and differentiation.

- **Exocrine and endocrine secretion.** Sertoli cells continuously secrete into the seminiferous tubules a fluid used for sperm transport in the direction of the genital ducts. Secretion of nutrients and **androgen-binding protein** (ABP), which concentrates testosterone to a level required for spermiogenesis, is promoted by follicle-stimulating hormone (FSH). In an endocrine manner, Sertoli cells release the steroid estradiol derived from testosterone and secrete the 39 kDa glycoprotein **inhibin,** which in a feedback loop with the anterior pituitary gland suppresses synthesis and release of FSH. In the fetus Sertoli cells also secrete a 140 kDa glycoprotein called **müllerian-inhibiting substance (MIS)** that causes regression of the embryonic müllerian (paramesonephric) ducts; without MIS these ducts persist and become parts of the female reproductive tract.

- **Phagocytosis.** During spermiogenesis, excess cytoplasm shed as residual bodies is phagocytosed and digested by Sertoli cell lysosomes. No proteins from sperm normally pass back across the blood-testis barrier.

MEDICAL APPLICATION

Spermiogenesis produces sperm-specific proteins. Because sexual maturity occurs long after the development of immunocompetence, differentiating sperm cells could be recognized as foreign and provoke an autoimmune response that would damage germ cells. The blood-testis barrier, together with various mechanisms to ensure a local immunotolerant environment, eliminates any interaction between developing sperm and the immune system. This barrier prevents the passage of immunoglobulins into the seminiferous tubule and accounts for the lack of impaired fertility even in men whose serum contains high levels of antibodies against sperm antigens.

Interstitial Tissue

The interstitial tissue of the testis is the site of androgen production. The spaces between the seminiferous tubules are filled with connective tissue that contains mast cells, macrophages, nerves, lymphatics, and blood vessels including fenestrated capillaries. During puberty, **interstitial,** or **Leydig, cells** become apparent as either rounded or polygonal cells with central nuclei and eosinophilic cytoplasm rich in small lipid droplets (Figures 21–4). These cells produce the male hormone **testosterone,** which is responsible for the development of the secondary male sex characteristics. Testosterone is synthesized by enzymes present in mitochondria and the smooth ER in a system similar to that of adrenal cortical cells.

Just as Sertoli cells are stimulated by FSH, testosterone secretion by interstitial cells is promoted by the other gonadotropic hormone of the pituitary, **luteinizing hormone (LH),** which is also called **interstitial cell stimulating hormone (ICSH).** Testosterone synthesis thus begins at puberty, when the hypothalamus begins producing gonadotropin-releasing hormone.

In the late embryonic testes gonadotropic hormone from the placenta stimulates interstitial cells to synthesize the testosterone needed for development of the ducts and other parts of the male reproductive system. These fetal interstitial cells are very active during the third and fourth months of pregnancy, then regress and become quiescent cells resembling fibroblasts until puberty when they resume testosterone synthesis in response to the pituitary gonadotropin.

INTRATESTICULAR DUCTS

The intratesticular genital ducts are the **straight tubules (tubuli recti),** the **rete testis,** and the **efferent ductules** (Figure 21–2). These ducts carry spermatozoa and liquid from the seminiferous tubules to the duct of the epididymis.

Most seminiferous tubules are in the form of loops, both ends of which join the rete testis by the short straight tubules. These tubules are recognized by the gradual loss of spermatogenic cells, with an initial segment in which the walls are lined only by Sertoli cells (Figures 21–4a and 21–9), followed by a main segment

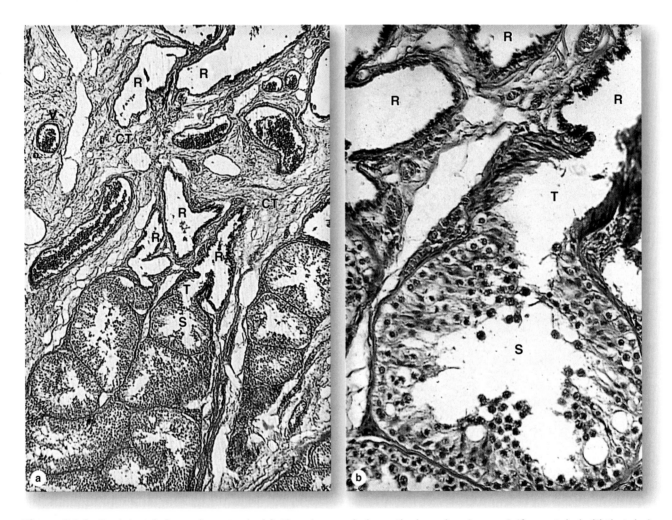

Figure 21–9. **Straight tubules and rete testis. (a):** The micrograph shows the long, looping seminiferous tubule (S) that drains into a short straight tubule (T), called a tubulus rectus. X120. H&E. **(b):** A higher magnification of one such junction shows that the transition to the straight tubule (T) is characterized by many tall Sertoli cells devoid of germ cells. The straight tubules all empty into the rete testis (R), a network of interconnected channels embedded along with blood vessels (V) in the connective tissue (CT) of the mediastinum. Channels of the rete testis are lined with simple cuboidal epithelium. X300. H&E.

consisting of cuboidal epithelium supported by a dense connective tissue sheath. All the straight tubules empty into the **rete testis,** an interconnected network of channels lined with cuboidal epithelium. The channels of the rete testis are embedded within the connective tissue of the mediastinum (Figure 21–9).

The rete testis drains into about 20 **efferent ductules** (Figure 21–10). They are lined by an unusual epithelium with groups of nonciliated cuboidal cells alternating with groups of taller ciliated cells. This gives the epithelium a characteristic scalloped appearance (Figure 21–10c). The nonciliated cells absorb most of the fluid secreted by the seminiferous tubules. This absorption and the ciliary activity create a fluid flow that sweeps sperm toward the epididymis. A thin layer of circularly oriented smooth muscle cells is seen outside the basal lamina of the epithelium which aids movement of the sperm. The efferent ductules empty into the duct of the epididymis.

EXCRETORY GENITAL DUCTS

Excretory genital ducts are those of the **epididymis,** the **ductus (vas) deferens,** and the **urethra**. They transport sperm from the epididymis to the penis during ejaculation.

The duct of the **epididymis** is a single highly coiled tube about 4–5 m in length. Together with a connective tissue capsule and blood vessels, this long duct forms the head, body, and tail of the epididymis, which lies along the superior and posterior sides of each testis (Figure 21–2). Sperm are stored in the epididymis and attain there various final characteristics including motility, membrane receptors for zona pellucida proteins, maturation of the acrosome, and ability to fertilize. The efferent ductules join the duct in the head of the epididymis and it opens into the **ductus (vas) deferens** at the tail. The epididymal duct is lined with pseudostratified columnar epithelium composed of rounded basal cells and columnar cells with long, branched, irregular microvilli called

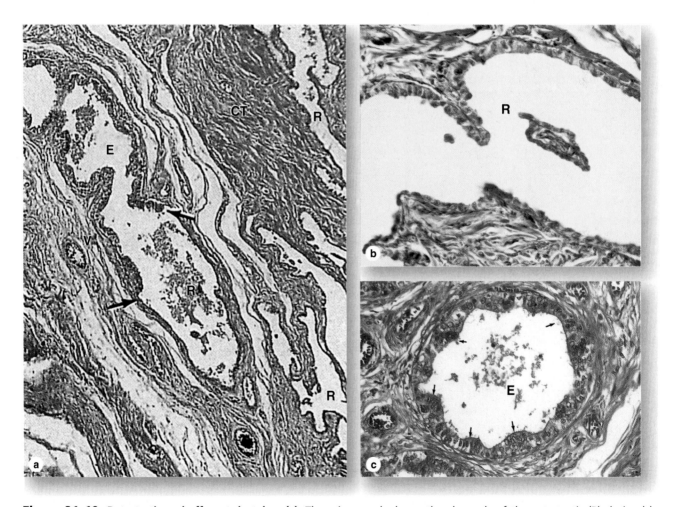

Figure 21–10. **Rete testis and efferent ductules. (a):** The micrograph shows the channels of the rete testis (R) drained by efferent ductules (E). A transition from rete testis to an efferent ductule is seen (arrows) and blood vessels (V) are seen in the mediastinum connective tissue (CT). X150. H&E. **(b):** The micrograph shows simple cuboidal epithelium that lines the rete testis (R). X350. Mallory trichrome. **(c):** The efferent ductules (E) are lined by a simple epithelium with a characteristic scalloped appearance in section, consisting of patches of cuboidal cells with water-absorbing microvilli alternating with patches of taller cells with cilia. This epithelium creates a fluid flow that, together with contractile activity of the thin muscularis around the efferent ductules, carries sperm toward the epididymis. X350. H&E.

stereocilia. (Figure 21–11). The epithelial cells of the epididymal duct absorb water and participate in the uptake and digestion of residual bodies produced during spermiogenesis. These cells are supported on a basal lamina surrounded by smooth muscle cells, whose peristaltic contractions move the sperm along the duct, and by loose connective tissue rich in capillaries.

From the epididymis the **ductus (vas) deferens**, a long straight tube with a thick, muscular wall, continues toward the prostatic urethra and empties into it (Figure 21–1). It is characterized by a narrow lumen and a thick layer of smooth muscle (Figure 21–12). Its mucosa is folded longitudinally and is lined along most of its length by pseudostratified columnar epithelium with sparse stereocilia. The lamina propria is rich in elastic fibers and the very thick muscularis consists of longitudinal inner and outer layers and a middle circular layer. The muscles produce strong peristaltic contractions during ejaculation which rapidly move sperm along this duct from the epididymis.

The ductus deferens forms part of the spermatic cord, which includes the testicular artery, the pampiniform plexus, and nerves (Figure 21–2). After passing over the urinary bladder, the ductus deferens dilates to form an **ampulla** (Figure 21–1), in which the epithelium is thicker and more extensively folded. At the final portion of the ampulla, the seminal vesicles join the duct. From there on, the ductus deferens enters the prostate gland and opens into the prostatic **urethra.** The segment entering the prostate is called the **ejaculatory duct.** The mucosa of the ductus deferens continues through the ejaculatory duct, but the muscle layers disappear beyond the ampulla.

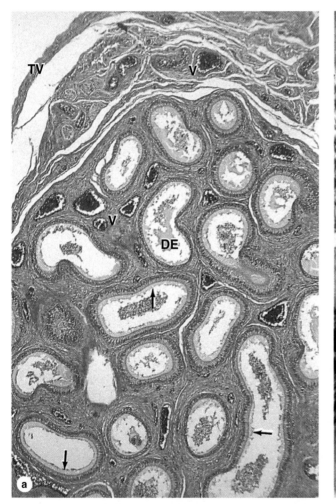

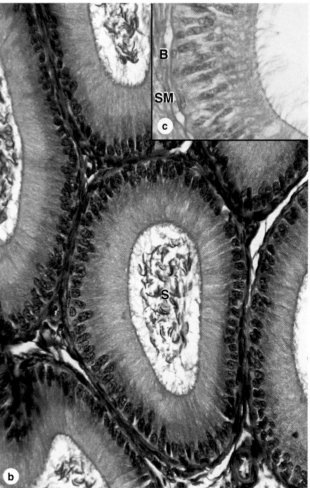

***Figure 21–11.* Epididymis. (a):** The epididymis contains a long, highly coiled duct in which sperm are temporarily stored and undergo the final maturation steps required for their ability to fertilize an egg. The duct of the epididymis (DE) is enclosed by connective tissue that contain many blood vessels (V) and the connective tissue is covered by a capsule and the tunica vaginalis (TV). The duct is lined by a pseudostratified columnar epithelium with long stereocilia (arrows). The columnar cells are very tall and the stereocilia very long in the head of the epididymis and both gradually shorten toward the tail. X140. H&E.

(b): The higher power micrograph shows the duct is surrounded by a thin circular layer of smooth muscle cells and the lumen is seen to contain sperm (S). The smooth muscle becomes thicker and a longitudinal layer develops in the body and tail of the epididymis. X400. H&E. **(c):** Another section of the smooth muscle (SM) and wall of the duct shows two abundant cell types in the epithelium: the tall principal cells with stereocilia and small basal cells (B) on the basal lamina. Macrophages and intraepithelial lymphocytes are also commonly seen in the epididymal duct. X500. H&E.

ACCESSORY GLANDS

The accessory glands of the male reproductive tract produce secretions that are added to sperm during ejaculation to produce semen and which are essential for reproduction. The accessory genital glands are the **seminal vesicles,** the **prostate gland,** and the **bulbourethral glands** (Figure 21–13).

The two **seminal vesicles** consist of highly tortuous tubes about 15 cm in length. The unusual mucosa displays a great number of thin, complex folds that fill most of the lumen (Figure 21–14). The folds are lined with simple or pseudostratified columnar epithelial cells rich in secretory granules. The lamina propria contains elastic fibers and is surrounded by smooth muscle with inner circular and outer longitudinal layers. The seminal vesicles are exocrine glands that produce a viscid, yellowish secretion containing **fructose,** citrate, inositol, **prostaglandins, fibrinogen,**

as well as enzymes and other proteins. These semen components, which typically make up about 70% of the ejaculate, provide nutrient energy sources for the sperm, coagulate semen after ejaculation, and affect activity of the female reproductive tract. The height of the seminal vesicle epithelial cells and their degree of secretory activity are dependent on adequate levels of testosterone.

The **prostate gland** is a dense organ surrounding the urethra below the bladder. It is approximately 2 cm × 3 cm × 4 cm in size and weighs about 20 g. The prostate is a collection of 30–50 branched tubuloalveolar glands, all surrounded by a dense fibromuscular stroma covered by a capsule. The glands are arranged in concentric layers around the urethra: the inner layer of **mucosal glands**, an intermediate layer of **submucosal glands**, and a peripheral layer with the prostate's **main glands** (Figure 21–15). Ducts from individual glands may converge but all empty directly into the prostatic urethra, which runs through the center of the

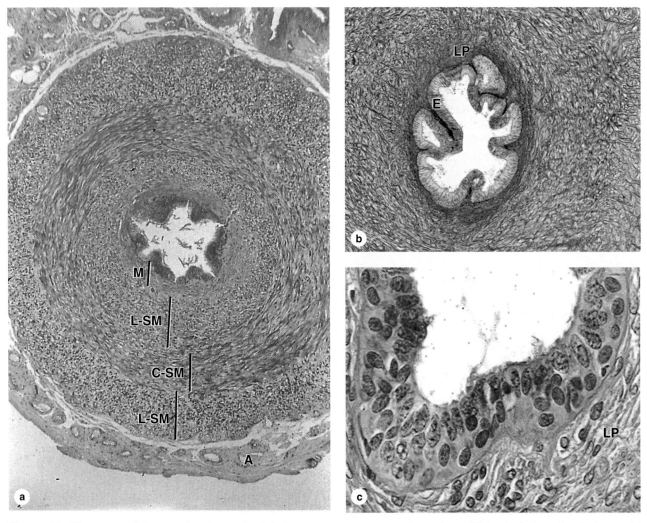

Figure 21–12. **Ductus deferens.** The ductus (vas) deferens transports sperm from the epididymis during ejaculation and is contained within the spermatic cord. **(a):** The micrograph of a vas deferens in cross-section shows that it consists of a mucosa (M), a thick muscularis with inner and outer layers of longitudinal smooth muscle (L-SM) and an intervening layer of circular smooth muscle (C-SM), and an external adventitia (A). The muscularis is specialized for powerful peristaltic movement of sperm at ejaculation. X60. H&E. **(b):** The lamina propria (LP) is rich in elastic fibers and the thick epithelial lining (E) shows longitudinal folds. X150. Mallory trichrome. **(c):** Higher magnification of the mucosa shows that the epithelium is pseudostratified with basal cells and many columnar cells with some stereocilia. X400. H&E.

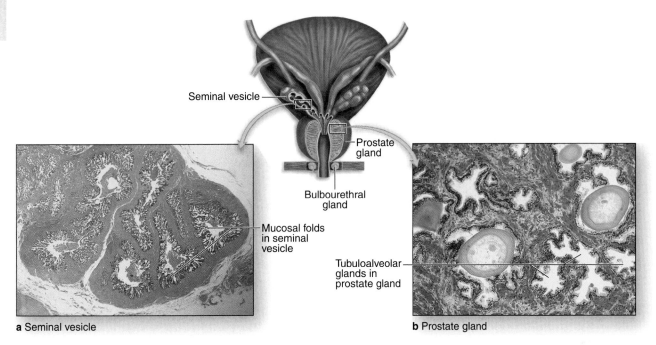

a Seminal vesicle

b Prostate gland

Figure 21–13. **Accessory glands of the male reproductive tract.** Three sets of glands connect to the ductus deferens or urethra: the paired seminal vesicles, the prostate, and the paired bulbourethral glands. The first two types of glands contribute the major volume to semen and the latter produces a secretion that lubricates the urethra before ejaculation. **(a):** Micrograph shows the characteristically folded mucosa of the seminal vesicles. X25. H&E. **(b):** Micrograph shows the characteristic individual tubuloalveolar glands of the prostate. X80. H&E.

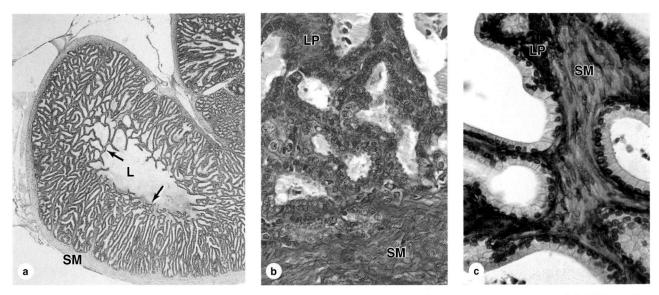

Figure 21–14. **Seminal vesicles.** The seminal vesicles are paired exocrine glands that secrete most seminal fluid, including sperm nutrients. **(a):** A low-power micrograph shows that each consists of a highly coiled duct surrounded by two layers of smooth muscle (SM) that expel the luminal contents during ejaculation. The mucosa characteristically displays thin primary, secondary, and tertiary folds (arrows) that give the lumen (L) a distinctive appearance. X20. H&E. **(b, c):** Both micrographs show that the folds include smooth muscle (SM) covered by a thin lamina propria (LP) and an epithelium. The epithelial cells are simple or pseudostratified columnar, varying with activity and location in the gland, and contain lipid droplets, secretory granules, and also commonly lipofuscin (L). Both: X300. b: H&E; c: PSH.

prostate (Figures 21–1). The prostate has three zones, corresponding to the glandular layers:

- The **transition zone** occupies about 5% of the prostate volume, surrounds the prostatic urethra, and contains the mucosal glands emptying directly into the urethra.

- The **central zone** occupies 25% of the gland's volume and contains the submucosal glands with longer ducts.

- The **peripheral zone** occupies about 70% of the prostate and contains the main glands with still longer ducts. Glands of this area are the most common location of both inflammation and cancer.

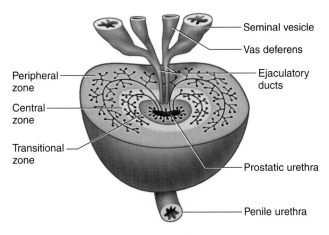

The tuboalveolar glands of the prostate are lined by a simple or pseudostratified columnar epithelium. The glands produce prostatic fluid containing various glycoproteins and enzymes and store this fluid for expulsion during ejaculation. An exceptionally rich fibromuscular stroma surrounds the glands (Figure 21–16). The prostate is surrounded by a fibroelastic capsule. Septa from this capsule penetrate the gland and divide it into indistinct lobes. As with the seminal vesicle, the structure and function of the prostate depend on the level of testosterone.

Figure 21–15. Organization of the prostate gland. The prostate secretes and stores a significant contribution to the seminal fluid that is released at ejaculation. It consists of 30–50 branched tubuloalveolar glands organized into three layers, shown here schematically. Around the prostatic urethra is the transition zone which contains the mucosal glands. Surrounding most of that zone is the intermediate central zone, which contains the submucosal glands. The outermost and largest layer is the peripheral zone, which contains the most numerous main glands. Glands of all the layers contribute to the prostatic secretion.

MEDICAL APPLICATION

Benign prostatic hypertrophy is present in 50% of men older than 50 years and in 95% of men older than 70 years. It generally occurs in the transition zone surrounding the urethra and can lead to compression of the urethra with clinical symptoms.

Prostatic cancer is one of the most common forms of cancer in men. A secretory product of the prostate, a serine protease known clinically as prostate-specific antigen (PSA), leaks into the blood in the glandular disruption caused by prostate cancer. Low levels of circulating PSA are also produced in the liver, but an increased PSA

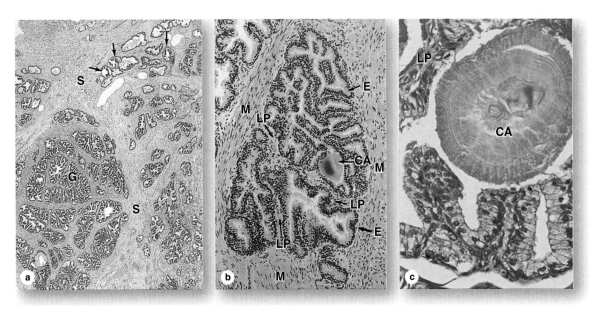

Figure 21–16. Prostate gland. (a): The prostate has a dense fibromuscular stroma (S) in which are embedded a large number of small tubuloalveolar glands (G). The arrows indicate sites of calcified concretions that have been lost during sectioning. X20. H&E. **(b):** A micrograph of one gland, including a corpus amylaceum (CA) concretion, shows a secretory simple or pseudostratified columnar epithelium (E) surrounded by lamina propria (LP), which is in turn surrounded by smooth muscle (M). X122. H&E. **(c):** Higher magnification shows the lamellar nature of a corpus amylaceum (CA) and the columnar epithelium underlain by sparse lamina propria (LP). X300. Mallory trichrome.

concentration in the blood is routinely used to diagnose and monitor prostate cancer.

Small spherical concretions, 0.2–2 mm in diameter and often calcified, are frequently observed in the lumens of the prostatic glands (Figure 21–16). They are called **corpora amylacea,** or **prostatic concretions** and contain primarily deposited glycoproteins and sulfated glycosaminoglycans (GAGs), particularly keratan sulfate. Their number often increases with age, but they seem to have no physiological or clinical significance.

The paired round **bulbourethral glands** (Cowper's glands), 3–5 mm in diameter, are located in the urogenital diaphragm (Figure 21–1) and empty into the proximal part of the penile urethra. Each gland has several lobules with tubuloalveolar secretory units lined by a mucus-secreting simple columnar epithelium dependent on testosterone. The septa between lobules contain smooth muscle cells. During erection the bulbourethral glands, as well as numerous, small, and histologically similar urethral glands along the urethra, release a clear mucus-like secretion containing various small carbohydrates, which coats and lubricates the urethral lining in preparation for the imminent passage of sperm.

PENIS

The main components of the penis are three cylindrical masses of erectile tissue, plus the penile urethra, surrounded by skin (Figure 21–1). Two of these cylinders—the **corpora cavernosa**—are placed dorsally. The other—the **corpus spongiosum**—is ventral and surrounds the urethra (Figures 21–1 and 21–17). At its end the corpus spongiosum expands, forming the **glans** (Figure 21–1). Most of the penile urethra is lined with pseudostratified columnar epithelium. In the glans, it becomes stratified squamous epithelium continuous with that of the thin epidermis covering the glans. Small mucus-secreting **urethra glands** (glands of Littre) are found along the length of the penile urethra. In uncircumcised men the surface of the glans is covered by the **prepuce** or foreskin, a retractile fold of thin skin with sebaceous glands in the internal fold.

The corpora cavernosa are covered by a resistant layer of dense connective tissue, the **tunica albuginea** (Figures 21–17 and 21–18). The corpora cavernosa and the corpus spongiosum are both composed of erectile tissue, which contains a large number of venous **cavernous spaces** lined with endothelial cells and separated by trabeculae of connective tissue fibers and smooth muscle cells.

The arterial supply of the penis derives from the internal pudendal arteries, which give rise to the deep arteries and the dorsal arteries of the penis. Deep arteries branch to form nutritive arteries to the trabeculae and coiling **helicine arteries**, which empty directly into the cavernous spaces of erectile tissue. There are arteriovenous shunts between the helicine arteries and the deep dorsal vein.

Penile erection involves filling the cavernous spaces of the corpora cavernosa and corpus spongiosum with blood. This is initiated by external stimuli to the CNS and is controlled by autonomic nerve input to smooth muscle in vascular walls of the penis. Parasympathetic stimulation relaxes muscle in the trabeculae and dilates the helicine arteries, leading to increased blood flow into the cavernous spaces. The filled spaces compress the venules and veins against the dense tunica albuginea, blocking outflow of the blood and producing tumescence and rigidity in the cylinders

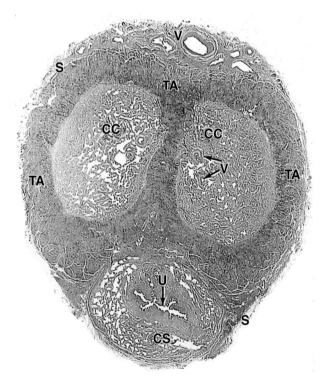

Figure 21–18. Penis. The corpus spongiosum (CS) is on the ventral side of the penis and surrounds the urethra (U). Two corpora cavernosa (CC) fill the dorsal side and all three bodies of cavernous or erectile tissue are surrounded by dense, fibrous tunica albuginea (TA). Along the dorsal side run the major blood vessels (V) and deep in each mass of erectile tissue are smaller blood vessels (V), including the central arteries. Externally the penis is covered by skin (S) attached to the tunica albuginea or neighboring connective tissue. X15. H&E.

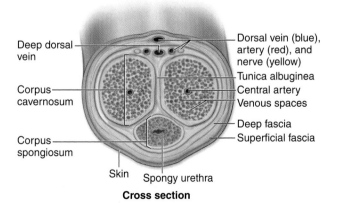

Cross section

Figure 21–17. Structure of the penis. A diagram of a transverse section of the penis shows the relationships of the three erectile bodies, tunica albuginea, and major blood vessels. Compare this section to the diagram in Figure 21–1.

THE MALE REPRODUCTIVE SYSTEM / 387

of erectile tissue. Beginning with ejaculation, the firing of sympathetic nerves stimulates constriction of the helicine arteries, decreasing blood flow into the spaces, lowering the pressure there and allowing the veins to open and drain most blood from the erectile tissue.

MEDICAL APPLICATION

In producing an erection parasympathetic nerves in the vasculature release nitric oxide (NO) which leads to increased NO synthesis by endothelial cells. NO stimulates cGMP (cyclic guanosine monophosphate) production in adjacent smooth muscle cells and this intracellular messenger triggers sequestration of Ca^{2+} ions in smooth ER, causing the muscle cells to relax. Diffusion of NO and cGMP throughout the smooth muscle via gap junctions produces relaxation and vasodilatation of the entire local vasculature and rapid in-flow of blood.

Arterial or neural insufficiency to tissue of the penis can lead to erectile dysfunction which can be treated effectively in many cases with compounds like sildenafil, which inhibit the phosphodiesterase that degrades cGMP and thereby enhance and prolong the effect of NO on the smooth muscle.

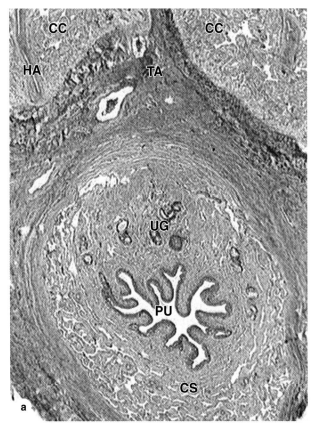

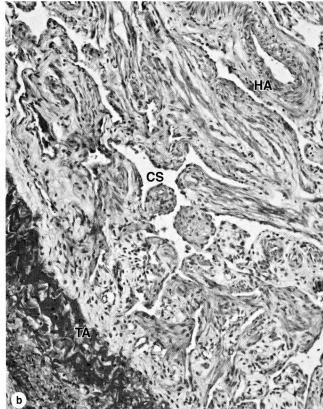

Figure 21–19. **Penile urethra and erectile tissue. (a):** The micrograph shows the corpus spongiosum (CS) surrounding the penile urethra (PU) with its longitudinally folded wall. Near the penile urethra are small urethral glands (UG) with short ducts for the release of a mucus-like secretion into the urethra during erection. These glands supplement the similar function of the larger, paired bulbourethral glands. In one of the two dorsal corpora cavernosa (CC) small helicine arteries (HA) are visible. The bodies of erectile tissue are ensheathed by dense, nonelastic tunica albuginea (TA) connective tissue. X100. H&E.

(b): A higher magnification of erectile tissue is shown with a small portion of tunica albuginea (TA). The erectile bodies are composed of fibrous, elastic connective tissue completely infiltrated with hundreds of small, flaccid vascular cavernous spaces (CS) lined by endothelium. Very little blood normally passes through this vasculature due to constriction of the helicine arteries (HA) serving them. During the erection process, smooth muscle of the helicine arteries relaxes, allowing rapid blood flow into the cavernous spaces, filling them and causing compression of their venous drainage in the tunica albuginea and fibroelastic trabeculae. This leads to full capacity filling of the cavernous spaces, swelling of the tissue volume, and turgidity of the erectile tissue bodies. X200. H&E.

The Female Reproductive System

The female reproductive system consists of two ovaries and oviducts (or uterine tubes), the uterus, the vagina, and the external genitalia (Figure 22–1). The functions of this system are to produce female gametes (**oocytes**), provide the environment for fertilization, and to hold the embryo during its complete development through the fetal stage until birth. As in the male, the female reproductive system produces steroidal sex hormones that control organs of the reproductive system and influence other organs of the body. Beginning at **menarche**, when the first menses occurs, the reproductive system undergoes cyclic changes in structure and functional activity. These modifications are controlled by neurohumoral mechanisms. **Menopause** is a variable period during which the cyclic changes become irregular and eventually disappear. In the postmenopausal period there is a slow involution of the reproductive organs. Although the mammary glands do not belong to the genital system, they are included here because they undergo changes directly connected to the functional state of the reproductive system.

OVARIES

Ovaries are almond-shaped bodies approximately 3 cm long, 1.5 cm wide, and 1 cm thick. Each ovary is covered by a simple cuboidal epithelium, the **germinal epithelium**, continuous with the mesothelium and overlying a layer of dense connective tissue capsule, the **tunica albuginea**, like that of the testis and responsible for the whitish color of the ovary. Most of the ovary consists of the **cortex**, a region filled with a highly cellular connective tissue stroma and many ovarian follicles, which in the adult ovary vary greatly in size (Figure 22–1). The most internal part of the ovary is the **medulla**, which contains loose connective tissue and blood vessels entering the organ through the hilum from mesenteries suspending the ovary (Figures 22–1 and 22–2).

There are no sharp limits between the ovarian cortical and medullary regions.

Early Development of the Ovary

In the first month of embryonic life, a small population of **primordial germ cells** migrates from the yolk sac to the gonadal primordia. In the gonads these cells divide extensively and differentiate as **oogonia**. In developing ovaries of a two-month embryo there are about 600,000 oogonia which produce more than 7 million by the fifth month. Beginning in the third month, oogonia begin to enter the prophase of the first meiotic division but arrest after completing synapsis and recombination, without progressing to later stages of meiosis (see Chapter 3). These cells arrested in meiosis are called **primary oocytes** (Gr. *oon*, egg, + *kytos*, cell). Each primary oocyte becomes surrounded by flattened supportive cells called **follicular cells** within an **ovarian follicle**. By the seventh month of development, most oogonia have transformed into primary oocytes within follicles. Many primary oocytes, however, are lost through a slow, continuous degenerative process called **atresia,** which continues through a woman's reproductive life. At puberty the ovaries contain about 300,000 oocytes. Because generally only one oocyte resumes meiosis with ovulation during each menstrual cycle (average duration, 28 days) and the reproductive life of a woman lasts about 30–40 years, only about 450 oocytes are liberated from ovaries by ovulation. All others degenerate through atresia.

Ovarian Follicles

An ovarian follicle consists of an oocyte surrounded by one or more layers of epithelial cells. The follicles that are formed

during fetal life—**primordial follicles**—consist of a primary oocyte enveloped by a single layer of the flattened follicular cells (Figures 22–2, 22–3, and 22–4). These follicles are found in the superficial areas of the cortex. The oocyte in the primordial follicle is a spherical cell about 25 μm in diameter, with a large nucleus and mostly uncoiled chromosomes in the first meiotic prophase. The organelles tend to be concentrated near the nucleus and include numerous mitochondria, several Golgi complexes, and extensive ER cisternae. A basal lamina surrounds the follicular cells, marking a clear boundary between the follicle and the vascularized stroma.

Follicular Growth

Beginning in puberty with the release of follicle-stimulating hormone (FSH) from the pituitary, a small group of primordial follicles each month begins a process of follicular growth. This involves growth of the oocyte, proliferation and changes in the follicular cells, and proliferation and differentiation of the stromal fibroblasts around each follicle. Selection of the primordial follicles that undergo growth and recruitment early in each cycle and of the dominant follicle destined to ovulate that month both involve complex hormonal balances and subtle differences among

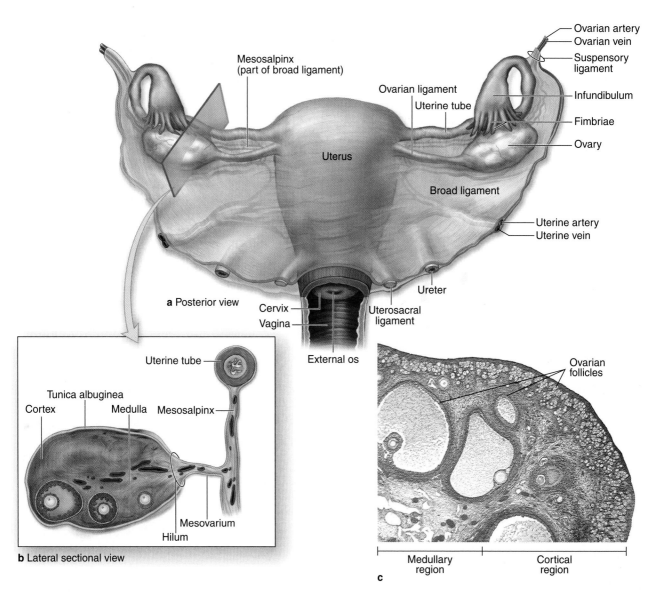

Figure 22–1. **The female reproductive system and overview of ovary. (a):** The diagram shows the internal organs of the female reproductive system, which includes as the principal organs the ovaries, uterine tubes, uterus, and vagina. **(b):** A lateral sectional view of an ovary shows the ovary and the relationship of its main supporting mesenteries, the mesovarium and the mesosalpinx of the broad ligament. **(c):** Micrograph of a sectioned ovary, indicating the medullary and cortical regions, with follicles of several different sizes in the cortex. X15. H&E.

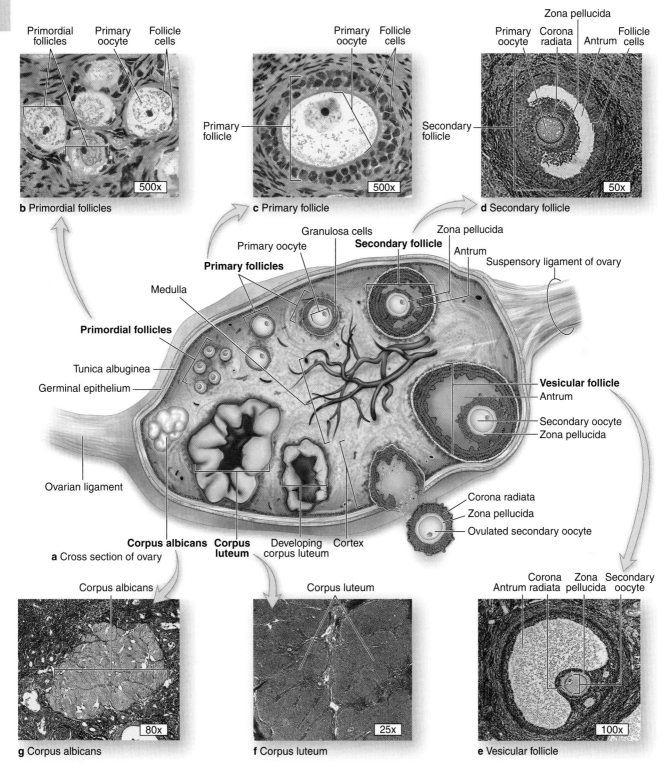

Primordial follicles | Primary oocyte | Follicle cells

b Primordial follicles

500x

Primary oocyte | Follicle cells

Primary follicle

c Primary follicle

500x

Zona pellucida

Primary oocyte | Corona radiata | Antrum | Follicle cells

Secondary follicle

d Secondary follicle

50x

Granulosa cells

Primary oocyte

Primary follicles

Medulla

Primordial follicles

Tunica albuginea

Germinal epithelium

Ovarian ligament

Zona pellucida

Secondary follicle

Antrum

Suspensory ligament of ovary

Vesicular follicle

Antrum

Secondary oocyte

Zona pellucida

Corona radiata

Zona pellucida

Ovulated secondary oocyte

Corpus albicans | **Corpus luteum** | Developing corpus luteum | Cortex

a Cross section of ovary

Corpus albicans

g Corpus albicans

80x

Corpus luteum

f Corpus luteum

25x

Corona

Antrum radiata | Zona pellucida | Secondary oocyte

e Vesicular follicle

100x

Figure 22–2. **Follicle development and changes within the ovary.** The ovary produces both oocytes and sex hormones. A diagram of a sectioned ovary **(a)**, shows the different stages of follicle maturation, ovulation, and corpus luteum formation and degeneration. All of the stages and structures shown in this diagram actually would appear at different times during the ovarian cycle and do not occur simultaneously. Follicles are arranged here for easy comparisons. The primordial follicles shown are greatly enlarged. The histological sections identify primordial follicles **(b)**, a primary follicle **(c)**, a secondary follicle **(d)**, and a large vesicular follicle **(e)**. After ovulation, the portion of the follicle left behind forms the corpus luteum **(f)**, which then degenerates into the corpus albicans **(g)**. All H&E.

follicles in FSH receptor numbers, aromatase activity and estrogen synthesis, and other variables.

Prompted by FSH, an oocyte grows most rapidly during the first part of follicular development, reaching a maximum diameter of about 120 μm. The nucleus enlarges; the mitochondria become more numerous and uniformly distributed; the ER becomes much more extensive; and the Golgi complexes enlarge

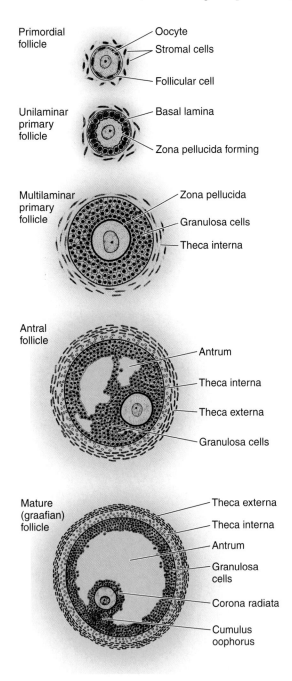

Figure 22–3. Stages of ovarian follicles, from primordial to mature. Diagrams of sectioned ovarian follicles show the changing size and morphology of follicular/granulosa cells at each stage and the disposition of the surrounding thecal cells. However, the relative proportions of the follicles are not maintained in the series of drawings: mature follicles are much larger relative to the early follicles.

and move peripherally. Follicular cells undergo mitosis and form a simple cuboidal epithelium around the growing oocyte. The follicle is now called a **unilaminar primary follicle** (Figures 22–3 and 22–5a). The follicular cells continue to proliferate, forming a stratified follicular epithelium, the **granulosa**, in which the cells communicate through gap junctions. Follicle cells are now termed **granulosa cells** and the follicle is a **multilayered primary follicle** (Figures 22–3 and 22–5b) still surrounded by basement membrane. Between the oocyte and granulosa cells, a layer of extracellular material called the **zona pellucida** develops, 5 to 10 μm thick and consisting of four glycoproteins secreted by the oocyte (Figures 22–5 and 22–6). Zona pellucida components, ZP1 through 4, bind proteins on the surfaces of sperm and induce acrosomal activation. Filopodia of follicular cells and microvilli of the oocyte penetrate the zona pellucida, allowing communication between these cells via gap junctions.

As the follicles grow with increasing oocyte size and numbers of granulosa cells, they move deeper in the ovarian cortex. Small spaces develop within the granulosa layer as the cells secrete **follicular fluid** (or **liquor folliculi**). This fluid accumulates, the spaces gradually coalesce, and the granulosa cells reorganize themselves around a larger cavity, the **antrum** (Figures 22–3 and 22–7a), producing follicles now called secondary or **antral follicles.** Follicular fluid contains hyaluronate, growth factors, plasminogen, fibrinogen, the anticoagulant heparan sulfate proteoglycan, and high concentrations of steroids (progesterone, androstenedione, and estrogens) with binding proteins.

During the reorganization of the granulosa layer to form the antrum, some cells form a small hillock, the **cumulus oophorus,** surrounding the oocyte and protruding into the antrum (Figures 22–3 and 22–7b). The granulosa cells immediately around and linked to the oocyte make up the **corona radiata** and accompany the oocyte when it leaves the ovary.

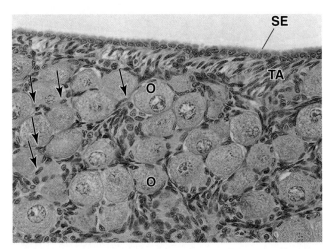

Figure 22–4. Primordial ovarian follicles. The cortical region of an ovary is surrounded by the surface epithelium (SE), a mesothelium with usually cuboidal cells. This layer is sometimes called the germinal epithelium because of an early erroneous view that it was the source of oogonia precursor cells. Underlying the epithelium is a connective tissue layer, the tunica albuginea (TA). Groups of primordial follicles, each formed by an oocyte (O) surrounded by a layer of flat epithelial follicular cells (arrows), are present in the ovarian connective tissue (stroma). X200. H&E.

While the follicle develops, the stromal cells immediately around the follicle differentiate to form the **follicular theca** (Gr. *theca*, outer covering). This layer subsequently differentiates further as two tissues around the follicle: a well-vascularized endocrine tissue, the **theca interna**, and a more fibrous outer **theca externa** containing smooth muscle and fibroblasts (Figures 22–3, 22–7, and 22–8). The cells of the theca interna differentiate as steroid-producing cells, with abundant smooth ER, mitochondria with tubular cristae, and numerous lipid droplets. These cells secrete a steroid hormone—androstenedione—that is transported to the granulosa where, under the influence of FSH, the cells synthesize an enzyme, aromatase, that transforms the steroid into estradiol. This estrogen returns to the thecae and stroma around the follicle, enters capillaries, and is distributed throughout the body. The boundary between the two thecal layers is not clear and neither is there a sharp boundary between the theca externa and the rest of the stroma. On the other hand, the boundary between the theca interna and the granulosa layer is well-defined, since their cells are quite different and there is a thick basement membrane between them (Figure 22–8).

During each menstrual cycle, usually one follicle grows much more than the others and becomes the dominant follicle, while many of the other follicles eventually enter atresia. The dominant follicle usually reaches the most developed stage of follicular growth and may undergo ovulation. This **mature** or **preovulatory follicle** (sometimes called the **graafian follicle** after the 17th century Dutch reproductive biologist Regnier De Graaf) reaches a diameter of 20-30 mm or more prior to ovulation, large enough to protrude from the surface of the ovary and be detected by ultrasound imaging. The antrum increases greatly in size by accumulating follicular fluid and the oocyte adheres to the wall of the follicle through the cumulus oophorus of granulosa cells (Figure 22–3). Since the granulosa cells of the follicle wall do not multiply in proportion to the growth of the antrum, the granulosa layer becomes thinner. A mature follicle has very thick thecal layers and normally develops from a primordial follicle over a period of about 90 days.

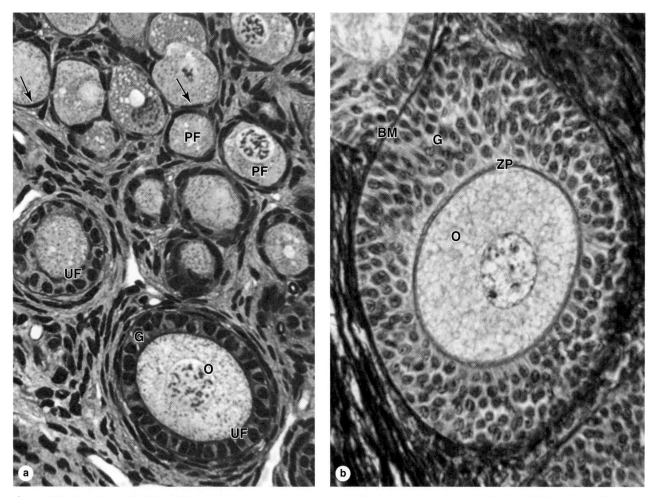

Figure 22–5. **Primary follicles. (a):** A micrograph of ovarian cortex shows several primordial follicles (PF) and their flattened follicles cells (arrows), and two unilaminar primary follicles (UF) in which the follicles cells or granulosa cells (G) form a single cuboidal layer around the large primary oocyte (O). X200. PT. **(b):** This micrograph was taken at the same magnification and shows a larger multilayered primary follicle. Granulosa cells (G) have now proliferated to form several layers. Between them and the oocyte (O) is the 5–10 μm thick zona pellucida (ZP), a glycoprotein layer produced by the oocyte that is required for sperm binding and fertilization. The primary oocyte is now a very large cell. With this stain the basement membrane (BM) that separates the follicle from the surrounding stroma can also be seen. X200. PSH.

Follicular Atresia

Most ovarian follicles undergo the degenerative process called **atresia**, in which follicular cells and oocytes die and are disposed of by phagocytic cells. Follicles at any stage of development, including nearly mature follicles, may become atretic (Figure 22–9). Atresia involves apoptosis and detachment of the granulosa cells, autolysis of the oocyte and collapse of the zona pellucida. Early in this process, macrophages invade the degenerating follicle and phagocytose the debris. Later fibroblasts occupy the area of the follicle and produce a collagen scar that may persist for a long time. Although follicular atresia takes place from before birth until a few years after menopause, it is most prominent just after birth, when levels of maternal hormones decline rapidly, and during both puberty and pregnancy, when qualitative and quantitative hormonal changes occur again.

Ovulation

At **ovulation** the large mature primary oocyte escapes from the ovary and is caught by the dilated end of the uterine tube which is closely applied to the ovarian surface at that time. Ovulation normally occurs midway through the menstrual cycle, ie, around the fourteenth day of a typical 28-day cycle. In humans usually only one oocyte is liberated during each cycle, but sometimes either no oocyte or two or more simultaneous oocytes may be expelled.

In the hours before ovulation the large mature follicle bulging against the tunica albuginea develops a whitish or translucent ischemic area, the **stigma**, in which the compaction of the tissue has blocked blood flow. Concurrently the granulosa cells and theca interna begin to secret progesterone as well as estrogen. The stimulus for ovulation is a surge of LH secreted by the anterior pituitary gland in response to the rapidly rising level of

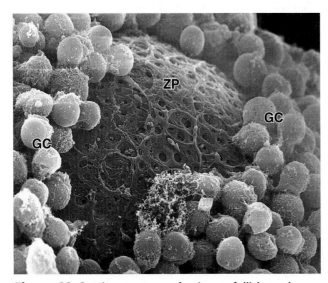

Figure 22–6. Ultrastructure of primary follicle and zona pellucida. An SEM of a fractured primary follicle shows the oocyte surrounded by granulosa cells (GC). Between the very large oocyte surface and the granulosa cells is a fibrous layer of extracellular material, the zona pellucida (ZP), which contains four related glycoproteins that bind sperm and form an irregular meshwork. X3000.

estrogen produced by the mature dominant follicle. LH stimulates hyaluronate and prostaglandin synthesis and overall fluid production within the preovulatory follicle. Progesterone, LH and FSH activate several proteolytic enzymes, including **plasmin** and **collagenases**, within and around the mature follicle which rapidly weaken the granulosa layer (and the cumulus oophorus) as well as the overlying tunica albuginea. The increasing pressure of the follicular fluid and weakening of the follicular wall lead to ballooning and then rupture of the ovarian surface at the stigma. The oocyte and corona radiata, along with follicular fluid and cells from the cumulus, are expelled through this opening by contraction of theca externa smooth muscle triggered by prostaglandins from the follicular fluid.

Just before ovulation the oocyte completes the first meiotic division, which it began and arrested in prophase during fetal life. The chromosomes are equally divided between the two daughter cells, but one of these retains almost all of the cytoplasm. That cell is now the **secondary oocyte** and the other becomes the **first polar body**, a very small nonviable cell containing a nucleus and a minimal amount of cytoplasm. Immediately after expulsion of the first polar body, the nucleus of the oocyte begins the second meiotic division, which arrests this time in metaphase.

The ovulated secondary oocyte adheres loosely to the ovary surface because of the hyaluronate-rich, coagulating follicular fluid released with it and, as described later, is drawn into the opening of the uterine tube where fertilization may occur. If not fertilized within about 24 hours, the secondary oocyte begins to degenerate.

Corpus Luteum

After ovulation, the granulosa cells and theca interna of the ovulated follicle reorganize to form a larger temporary endocrine gland, the **corpus luteum** (L., yellowish body), in the ovarian cortex. Ovulation causes the collapse and folding of the granulosa and thecal layers of the follicle's wall and blood from disrupted capillaries typically accumulates as a clot in the former antrum (Figure 22–10). The former granulosa is now invaded by capillaries and cells of both the granulosa and theca interna change histologically and functionally under the influence of LH, becoming specialized for more extensive production of the steroid progesterone in addition to estrogens.

Granulosa cells increase greatly in size (20–35 μm in diameter), without dividing, and eventually comprise about 80% of the parenchyma of the corpus luteum. They are called **granulosa lutein cells** (Figure 22–10) and now have lost many features of protein-secreting cells to expand their role in aromatase conversion of androstenedione into estradiol. The former theca interna contributes the other component of the corpus luteum, giving rise to **theca lutein cells** (Figure 22–10). These cells are less than half the size of the granulosa lutein cells and often stain more darkly, with cytoplasmic ultrastructural features of steroid-synthesizing cells. LH causes these cells to produce large amounts of progesterone as well as androstenedione. Theca lutein cells are typically aggregated in the folds of the wall of the corpus luteum, which like all endocrine glands becomes well vascularized.

The fate of the corpus luteum depends on whether a pregnancy occurs. Following the LH surge, the corpus luteum is programmed to secrete progesterone for 10–12 days. Without further LH stimulation and in the absence of pregnancy, both major cell types of the corpus luteum cease steroid production and undergo apoptosis while the tissue regresses. A consequence of

the decreased secretion of progesterone is menstruation, the shedding of part of the uterine mucosa. Estrogen produced by the active corpus luteum inhibits FSH release from the pituitary. However, after the corpus luteum degenerates, the concentration of blood steroids decreases and FSH secretion increases again, stimulating the growth of another group of follicles and beginning the next menstrual cycle. The corpus luteum that persists for part of only one menstrual cycle is called a **corpus luteum of menstruation.** Remnants from its degeneration and regression are phagocytosed by macrophages, after which fibroblasts invade the area and produce a scar of dense connective tissue called **corpus albicans** (L., white body) (Figure 22–11).

If pregnancy occurs, the uterine mucosa cannot be allowed to undergo menstruation since the embryo would be lost. To prevent the drop in circulating progesterone, trophoblast cells of the implanting embryo produce a glycoprotein hormone called **human chorionic gonadotropin** (**HCG**) whose action is similar to that of LH. HCG targets the corpus luteum, maintaining it and promoting further growth of this endocrine gland and stimulating secretion of progesterone to maintain the uterine mucosa. Progesterone also stimulates secretion in the uterine mucosal glands, which is important for embryonic nutrition before the placenta is functional. This **corpus luteum of pregnancy** becomes very large and is maintained by HCG for 4–5 months, by which time the placenta itself produces progesterone (and estrogens) at levels adequate to maintain the uterine mucosa. It then degenerates and is replaced by a large corpus albicans.

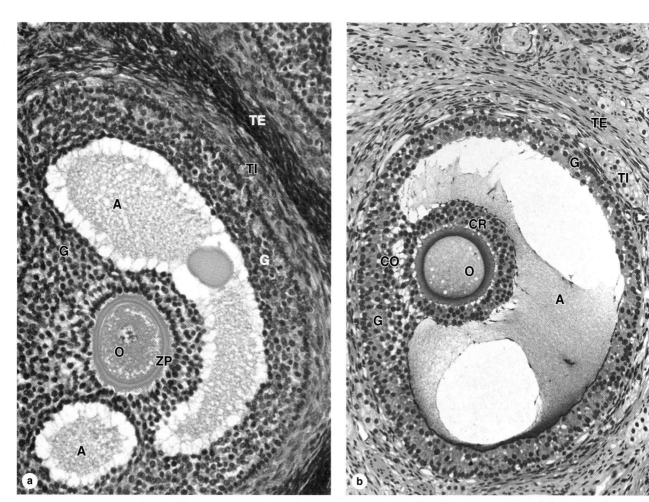

Figure 22–7. **Antral follicles. (a):** A micrograph with part of an antral follicle shows the large, fluid-filled antral cavities or vesicles (A) that appear in the granulosa layer as the cells produce follicular fluid. The oocyte (O) is surrounded by the zona pellucida (ZP) and granulosa cells (G), which also line the wall of the follicle. Fibroblastic cells immediately outside the growing follicles have developed as a steroid-secreting theca interna (TI) and a covering theca externa (TE). X100. H&E.

(b): A slightly more developed antral follicle shows a verge large single antrum (A) filled with follicular fluid in which the proteins formed a thin film during fixation. The oocyte (O) now projects into this fluid-filled cavity, still surrounded by granulosa cells which now make up the corona radiata (CR). Cells of the corona radiata are linked to the oocyte via gap junctions between processes that cross the zona pellucida. The corona radiata and oocyte are attached to the side of the follicle via a less dense mass of granulosa cells called the cumulus oophorus (CO) which is continuous with the remaining granulosa cells that form the wall of the follicle and surround the antrum. Thecae interna (TI) and externa (TE) surround the whole follicle. X100. PT.

UTERINE TUBES

The **uterine tubes,** or **oviducts** are two muscular tubes (Figure 22–12) with considerable mobility, each measuring about 12 cm in length. Each has a funnel-shaped end, the **infundibulum,** opening into the peritoneal cavity next to the ovary and with a fringe of fingerlike extensions called **fimbriae** (L., fringes). The sequence of regions along each tube is as follows:

- the infundibulum,
- the **ampulla,** the longest and expanded area where fertilization normally occurs,
- the **isthmus,** a more narrow region nearer the uterus, and
- the **uterine** or **intramural part,** which passes through the wall of the uterus and opens into the interior of this organ.

The wall of the oviduct consists of a folded mucosa, a thick muscularis with somewhat interwoven circular (or spiral) and longitudinal layers of smooth muscle (Figure 22–13), and a thin serosa covered by visceral peritoneum with mesothelium.

MEDICAL APPLICATION

Uterine tubes are commonly called fallopian tubes, after the 16th century anatomist Fallopius, and in medical terminology are frequently denoted by the prefix "salping-" (Gr., salpinx, trumpet), as in "salpingitis" for inflammation of the tubal lining and "salpingectomy" for surgical removal of these structures.

The mucosa has numerous branching, longitudinal folds that are most prominent in the ampulla, which in cross section resembles a labyrinth (Figure 22–12). These mucosal folds become smaller in the segments of the tube closer to the uterus and are not present in the intramural portion of the tube.

The mucosa is composed of a simple columnar epithelium on a lamina propria of loose connective tissue (Figures 22–12 and 22–13). The epithelium contains two interspersed, functionally important cell types: **ciliated cells** and darker staining **secretory cells,** or **peg cells,** whose apical ends typically bulge into the lumen (Figure 22–13). The cilia beat toward the uterus, causing movement of the viscous liquid film that covers the epithelial surface and contains glycoproteins and nutritive components produced by the secretory cells. Triggered primarily by estrogens, the cilia elongate and both cell types undergo hypertrophy during the follicular growth phase of the ovarian cycle and undergo atrophy with loss of cilia during the late luteal phase.

At the time of ovulation, the uterine tube shows active movement. The fringed infundibulum moves very close to the ovary and partially covers its surface. This favors the transport of the ovulated secondary oocyte into the tube. Promoted by sweeping muscular contractions of the fimbriae and ciliary activity of the epithelium, the oocyte enters the infundibulum and moves to the ampulla. The secretion covering the mucosa has nutrient and protective functions for the oocyte and the sperm, including factors that promote sperm activation (**capacitation**). The oocyte typically remains viable for a maximum of about 24 hours if it is not fertilized.

Usually occurring in the ampulla, fertilization culminates with fusion of the haploid sperm and egg nuclei and reconstitutes the diploid number of chromosomes typical of the species. First however, fertilization triggers the oocyte to complete the

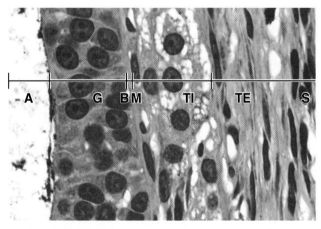

Figure 22–8. Wall of antral follicle. At higher magnification, a small part of the wall of an antral follicle shows the cell layers of the granulosa (G) next to the antrum (A), in which proteins have aggregated on cells in contact with the follicular fluid. The theca interna (TI) surrounds the follicle, its cells appearing vacuolated and lightly stained because of their cytoplasmic lipid droplets, a characteristic of steroid-producing cells. The overlying theca externa (TE) contains fibroblasts and smooth muscle cells and merges with the stroma (S). A basement membrane (BM) separates the theca interna from the granulosa, blocking vascularization of the latter. X400. PT.

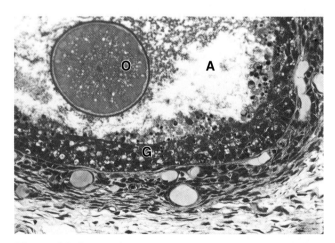

Figure 22–9. Atresia. Atresia or degeneration of a follicle can occur at any stage of its development and is shown here in a follicle that had developed a large antrum. Atresia is characterized by apoptosis of granulosa cells (G) and autolysis of the oocyte, with macrophages entering the degenerating structure to clean up debris. Many apoptotic cells are seen loose in the antrum (A) here and the cells of the corona radiata have already disappeared, leaving the degenerative oocyte (O) free within the antrum. X200. PT.

second meiotic division, with formation of the **ovum** and release of a second polar body. The corona radiata is generally still present when the sperm fertilizes the oocyte and its cells gradually detach over the next several hours.

The diploid cell formed during fertilization, the **zygote** (Gr. *zygotos,* yoked together), begins cell division and is transported to the uterus, which takes about 5 days. Contractions of the oviduct muscle layers, together with ciliary movement of the film covering the mucosa, transport the early embryo toward the uterus. The ciliary activity is apparently not essential, since the transport process occurs normally in women with immotile cilia syndrome.

MEDICAL APPLICATION

In a woman whose uterine tube is blocked by postinflammation scar tissue, the embryo cannot reach the uterus and may implant itself in the oviduct wall (**ectopic** or **tubal pregnancy**). In this case, the lamina propria may react like the uterine endometrium and form decidual cells. Because of its small diameter and inability to expand, the tube cannot contain the growing embryo and will rupture, causing extensive hemorrhage that can be fatal if not treated immediately.

UTERUS

As shown in Figure 22–12, the uterus is a pear-shaped organ with thick, muscular walls. Its largest part, the **body**, is entered by the left and right uterine tubes and the curved, superior area between the tubes is called the **fundus**. The uterus narrows in the **isthmus** and ends in a lower cylindrical structure, the **cervix**, with the lumen in these regions termed the **internal os** (L. *os*, mouth) and **cervical canal,** respectively.

Supported by the set of ligaments and mesenteries also associated with the ovaries and uterine tubes (Figure 22–1), the uterine wall has three major layers (Figure 22–12):

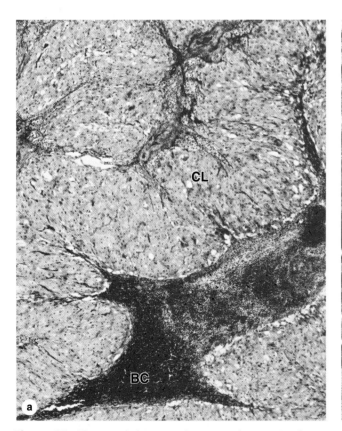
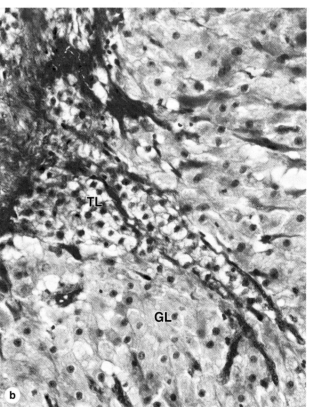

Figure 22–10. **Corpus luteum.** The corpus luteum is a large endocrine structure formed from the remains of the large dominant follicle after it undergoes ovulation. **(a):** The low-power micrograph shows the corpus luteum (CL), characterized by folds of the former granulosa which collapses as the theca externa contracts at ovulation. The former antrum often contains a blood clot (BC) from vessels in the thecal layers disrupted during ovulation. Cells of the granulosa and theca interna become reorganized under the influence of pituitary LH and their names are changed. X15. H&E. **(b):** Granulosa lutein cells (GL), seen at higher magnification here, undergo significant hypertrophy, producing most of the corpus luteum's increased size, and begin producing progesterone. The theca lutein cells (TL) increase only slightly in size, are somewhat darker-staining than the granulosa lutein cells, and continue to produce estrogens. Theca lutein cells, derived from the theca interna, are typically located within the folds that comprise the bulk of this tissue. X100. H&E.

- an outer connective tissue layer, the **perimetrium**, continuous with the ligaments, which is adventitial in some areas, but largely a serosa covered by mesothelium,
- a thick tunic of highly vascular smooth muscle, the **myometrium** (Figure 22–14), and
- a mucosa, the **endometrium**, lined by simple columnar epithelium.

The three layers are continuous with their counterparts in the uterine tubes. The thickness and structure of the endometrium, even more than that of the tubal mucosa, are influenced cyclically by the shifting levels of ovarian hormones (Figure 22–15).

Myometrium

The myometrium (Gr. *myo,* muscle, + *metra,* uterus), the thickest tunic of the uterus (Figure 22–14), is composed of bundles of smooth muscle fibers separated by connective tissue containing many blood vessels. The bundles of smooth muscle form four interwoven, poorly defined layers. The first and fourth layers are composed mainly of fibers disposed generally parallel to the long axis of the organ, with the middle layers circularly disposed and containing the larger blood vessels.

During pregnancy, the myometrium goes through a period of extensive growth involving both **hyperplasia** (increasing the number of smooth muscle cells) and **hypertrophy** (increasing cell size). During this growth, many of the smooth muscle cells also actively synthesize collagen, strengthening the uterine wall. After pregnancy, uterine smooth muscle cells shrink and many undergo apoptosis, with removal of unneeded collagen, and the uterus returns almost to its prepregnancy size.

Endometrium

The lamina propria or stromal connective tissue of the endometrium contains primarily type III collagen fibers with abundant fibroblasts and ground substance. Its covering simple columnar epithelium has both ciliated and secretory cells, the latter forming the lining of the numerous tubular **uterine**

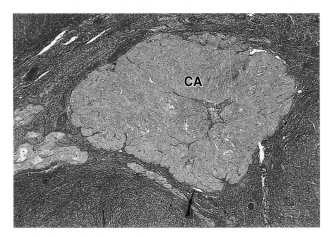

Figure 22–11. **Corpus albicans.** A corpus albicans (CA) is the scar of connective tissue that forms at the site of a corpus luteum after its involution. It contains mostly collagen, with few fibroblasts or other cells, and gradually becomes very small and lost in the ovarian stroma. Involution of the corpus luteum does not involve atresia. X60. H&E.

glands, which penetrate the full thickness of the endometrium (Figure 22–14).

The endometrium can be subdivided into two zones. (1) The **basal layer** adjacent to the myometrium, contains highly cellular lamina propria and the deep basal ends of the uterine glands. (2) The superficial **functional layer** (or functionalis) contains more spongy and less cellular lamina propria, richer in ground substance, most of the length of the glands, as well as the surface epithelium (Figure 22–14). The functional layer undergoes profound changes during the menstrual cycles, but the basal layer remains relatively unchanged (Figure 22–15).

The blood vessels supplying the endometrium are of special significance in the periodic sloughing of the functional layer during menses. Arcuate arteries in the middle layers of the myometrium send two sets of smaller arteries to the endometrium (Figures 22–14 and 22–16): **straight arteries,** which supply only the basal layer, and long, progesterone-sensitive **spiral arteries,** which extend farther and bring blood throughout the functional layer. Spiral arteries branch with numerous arterioles supplying a rich capillary bed that includes many dilated, thin-walled vessels called **vascular lacunae.**

Menstrual Cycle

Throughout the female reproductive system, estrogens and progesterone control growth and differentiation of epithelial cells and the associated connective tissue. Even before birth, these cells are influenced by circulating maternal estrogen and progesterone that reach the fetus through the placenta. After menopause, diminished synthesis of these hormones results in a general involution of tissues in the reproductive organs.

From the time of puberty until menopause at about age 45–50, the pituitary gonadotropins produce cyclic changes in levels og the ovarian hormones, which cause the endometrium to undergo cyclic structural modifications during the menstrual cycle (Figures 22–15 and 22–17). The duration of the menstrual cycle is variable but averages 28 days. Because menstrual cycles are a consequence of ovarian follicle changes related to oocyte production, a woman is fertile only during the years when she is having menstrual cycles.

Day one of the menstrual cycle is usually taken as the day when menstrual bleeding appears. The menstrual discharge consists of degenerating endometrium mixed with blood from its ruptured blood vessels. The **menstrual period** of the cycle lasts 3–4 days on average. The next phase, the **proliferative phase,** is of variable length, 8–10 days on average, and the **secretory phase** begins at ovulation and lasts about 14 days (Figure 22–15). The structural changes that occur during the cycle are gradual, and the activities characterizing these phases overlap to some extent.

Proliferative Phase

After the menstrual phase, the uterine mucosa is relatively thin (about 0.5 mm). The beginning of the **proliferative phase,** also called the **follicular** or **estrogenic phase,** coincides with the rapid growth of a small group of ovarian follicles undergoing the transition from preantral to antral follicles. With development of their thecae interna, these follicles actively secrete estrogen, increasing its plasma concentrations.

Estrogens act on the endometrium, inducing cell proliferation and reconstituting the functional layer lost during menstruation. Cells in the basal ends of glands proliferate, migrate, and form the new epithelial covering over the endometrial surface exposed

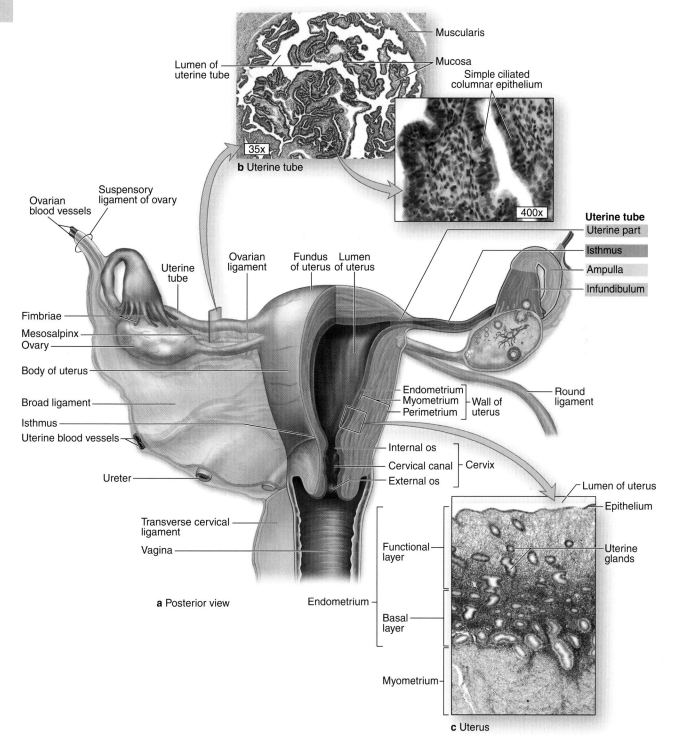

Figure 22–12. Uterine tubes and uterus. The uterine tubes or oviducts are paired ducts that catch the ovulated secondary oocyte, nourish both the oocyte and sperm, provide the microenvironment for fertilization, and transport the embryo undergoing cleavage to the uterus. **(a):** The diagram shows the relationship between the uterine tubes and the uterus in an intact posterior view (left) and in a cut-away view (right). **(b):** The micrograph shows a uterine tube in cross-section. H&E. **(c):** Micrograph shows the layers of the endometrium and myometrium in the wall of the uterus. X45. H&E.

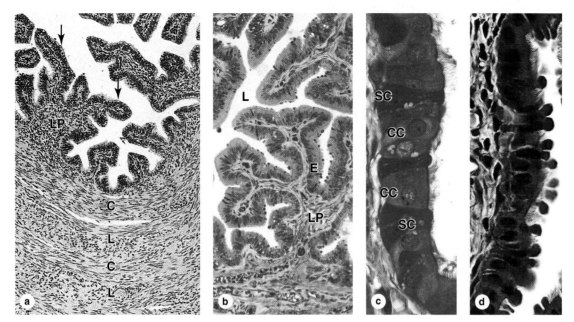

Figure 22–13. Mucosa of the uterine tube wall. The muscularis of the uterine tube contracts to move the embryo toward the uterus and its mucosa facilitates sperm and oocyte movement and provides a nutritive and protective environment for fertilization and early embryonic development. **(a):** A cross-section of the uterine tube at the antrum shows the interwoven circular (C) and longitudinal (L) layers of smooth muscle in the muscularis and in the complex of folded mucosa, the lamina propria (LP) underlying a simple columnar epithelium (arrows). X140. H&E. **(b):** The micrograph shows the epithelium (E) contains primarily two columnar cell types, ciliated and nonciliated, with the latter showing darker staining apical pegs bulging into the lumen (L). X200. PT. **(c, d):** Higher magnification of the epithelium shows the ciliated cells (CC) interspersed with the secretory cells (SC), which produce the nutritive fluid covering the epithelium. All aspects of the epithelial cells are controlled by hormones, mainly estrogens, and the cells' histological and functional features vary during the ovarian cycle due to fluctuations in these hormones. In (d) the secretory cells shown are at their most developed and most active state in the period shortly after ovulation when an embryo might be present. c: X400, PT; d: X400, Mallory trichrome.

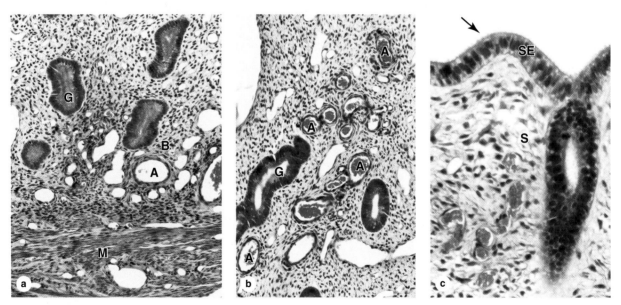

Figure 22–14. Uterus. Most of the uterine wall is composed of the myometrium, consisting of multiple interwoven layers of well-vascularized smooth muscle. The inner layer of the uterus, corresponding to a mucosa, is the endometrium. **(a):** The micrograph shows the basal layer (B) of the endometrium, bordering the myometrium (M). The basal layer contains the basal ends of the uterine glands (G) and many small arteries (A) embedded in a distinctive connective tissue stroma with many fibroblasts, ground substance and primarily fine type III collagen, but no adipocytes. X100. Mallory trichrome. **(b):** Superficial to the basal layer of the endometrium is its functional layer, the part that changes histologically and functionally depending on estrogen levels. This micrograph shows only functional layer and includes parts of the long uterine glands (G) as well as one spiral artery (A). X100. Mallory trichrome. **(c):** The surface epithelium (SE) lining the endometrium is simple columnar, with many cells having cilia (arrow). The underlying stroma (S) has an extensive microvasculature, much ground substance, and fibroblastic cells with large, active nuclei. X400. Mallory trichrome.

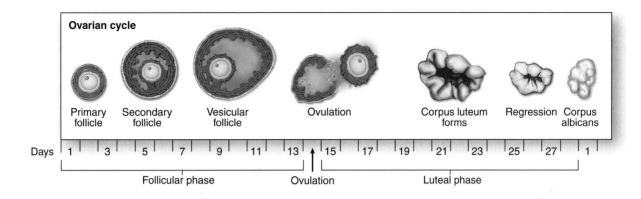

Days 1 3 5 7 9 11 13 15 17 19 21 23 25 27 1

Follicular phase — Ovulation — Luteal phase

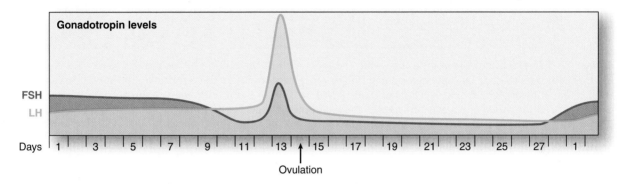

Days 1 3 5 7 9 11 13 15 17 19 21 23 25 27 1

Ovulation

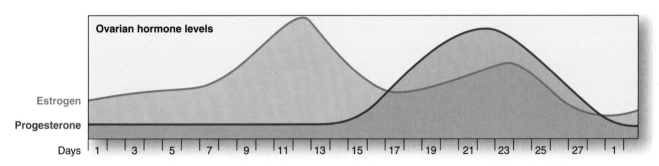

Days 1 3 5 7 9 11 13 15 17 19 21 23 25 27 1

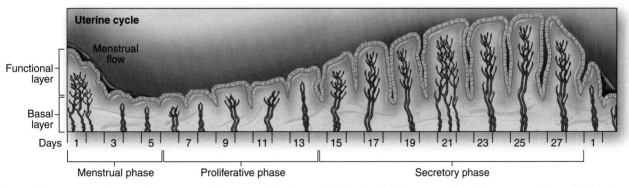

Days 1 3 5 7 9 11 13 15 17 19 21 23 25 27 1

Menstrual phase — Proliferative phase — Secretory phase

Figure 22–15. Correlation of ovarian and menstrual cycles with levels of their controlling hormones. The cyclic development of ovarian follicles and the corpus luteum, controlled by the pituitary gonadotropins FSH and LH, lead to cyclic shifts in the levels of the major ovarian hormones: steroidal estrogens and progesterone. Estrogen stimulates the proliferative phase of the uterine cycle and its level peaks near the day of ovulation, which marks the midpoint of the ovarian cycle. After ovulation the corpus luteum forms and produces both progesterone and estrogens, which together promote growth and development of the endometrial functional layer. Unless fertilization and implantation of an embryo occur, regression of the corpus luteum

leads to declining levels of the steroid hormones and failure of the new endometrial tissue to be maintained. This tissue sloughs off as the menstrual flow, the first day of which is taken to mark day 1 of both the ovarian cycle and the uterine cycle. The basal layer of endometrium is not sensitive to the loss of progesterone and is retained during menstruation, serving to regenerate the functional layer during the ensuing proliferative phase.

during the preceding menstruation. During the proliferative phase, the endometrial lining is a simple columnar surface epithelium and the uterine glands are relatively straight tubules with narrow, nearly empty lumens (Figure 22–17). Mitotic figures can be found among both the epithelial cells and fibroblasts.

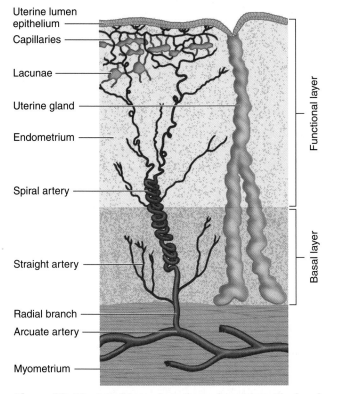

Figure 22–16. Arterial supply to the endometrium. The basal and functional layers of the endometrium are supplied by different sets of small arteries emerging from the uterine arcuate arteries in the myometrium: the straight arteries and spiral arteries respectively. The spiral arteries are uniquely sensitive to progesterone, growing rapidly in a spiral fashion as the functional layer thickens under the influence of that luteal steroid and providing blood to a microvasculature that includes many lacunae lined by thin endothelium. This blood supply brings oxygen and nutrients to cells of the functionalis and to an embryo implanting itself into that tissue. If no embryo is present to produce the gonadotropin replacing LH, the corpus luteum undergoes regression 8–10 days after ovulation. The rapid decline in the level of progesterone causes constriction of the spiral arteries and other changes that quickly lead to local ischemia in the functional layer and its separation from the basal layer during menstruation.

Cells of the glands gradually accumulate cisternae of rough ER and larger Golgi complexes in preparation for their secretory activity. Spiral arteries lengthen as the functional layer is reestablished and grows (Figure 22–14) and extensive microvasculature forms near the surface of the functional layer. At the end of the proliferative phase, the endometrium is 2–3 mm thick.

SECRETORY PHASE

After ovulation, the **secretory** or **luteal phase** starts as a result of the progesterone secreted by the corpus luteum. Progesterone stimulates epithelial cells of the uterine glands that formed during the proliferative phase and these cells begin to accumulate glycogen, which then undergoes apocrine secretion. Soon a thick secretion rich in glycogen and glycoproteins dilates the glandular lumens. Histologically the glands become highly coiled during this period. The superficial microvasculature now includes thin-walled, blood-filled lacunae (Figures 22–16 and 22–17). The endometrium reaches its maximum thickness (5 mm) during the secretory phase as a result of the accumulation of secretions and edema in the stroma.

If fertilization has taken place in the day following ovulation, the embryo has been transported to the uterus about 5 days later and now attaches to the uterine epithelium when endometrial thickness and secretory activity are optimal for embryo implantation and nutrition. The uterine gland secretion is the major source of embryonic nutrition before and during implantation. In addition to promoting secretion, progesterone inhibits strong contractions of the myometrium that might interfere with embryo implantation.

MENSTRUAL PHASE

When fertilization of the oocyte and embryonic implantation do not occur, the corpus luteum regresses and circulating levels of progesterone and estrogens begin to decrease 8–10 days after ovulation, which leads to the onset of menstruation (Figure 22–15). The drop-off in progesterone produces (1) spasms of muscle contraction in the small spiral arteries of the functional layer, interrupting normal blood flow, and (2) increased synthesis by arterial cells of prostaglandins, which produce strong vasoconstriction and local hypoxia. Cells undergoing hypoxic injury release cytokines that increase vascular permeability and immigration of leukocytes. The leukocytes release collagenase and several other matrix metalloproteinases (MMPs) which degrade basement membranes and other ECM components (Figure 22–17). The basal layer of the endometrium, not dependent on the progesterone-sensitive spiral arteries, is relatively unaffected by these activities. However, major portions of the functional layer, including the surface epithelium, most of each gland, the stroma and blood-filled lacunae, detach from the endometrium and slough away as the menstrual flow or **menses**. Arterial constriction normally limits blood loss during menstruation, but some blood does emerge from the open ends of venules. The amount of endometrium and blood lost in menstruation varies between women and in the same woman at different times.

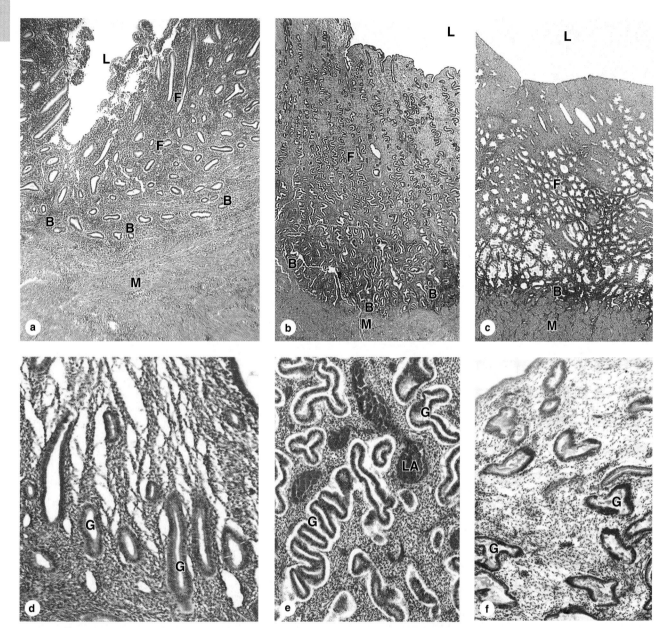

Figure 22–17. **Proliferative, secretory, and premenstrual phases in the uterus.** The major phases of the uterine cycle overlap, but produce distinctly different and characteristic changes in the functional layer (F) closest to the lumen (L) with little effect on the basal layer (B) and myometrium (M). Characteristic features of each phase include the following. During most of the proliferative phase **(a and d)** the functional layer is still relatively thin, the stroma is more cellular and the glands (G) are relatively straight, narrow, and empty. In the secretory phase **(b and e)** the functional layer is less heavily cellular and perhaps four times thicker than the basal layer. The tubular glands have wider lumens containing secretory product and coil tightly up through the stroma, giving a zig-zag or folded appearance histologically. Superficially in the functional layer, lacunae (La) are widespread and filled with blood. The short premenstrual phase **(c and f)** begins with constriction of the spiral arteries, which produces hypoxia that causes swelling and dissolution of the glands (G). The stroma of the peripheral functionalis is more compact and that near the basal layer typically appears more sponge-like during this time of blood stasis, apoptosis and breakdown of the stromal matrix. a: X20; b and c: X12; d, e, and f: X50. All H&E.

At the end of the menstrual phase, the endometrium is usually reduced to a thin layer and is ready to begin a new cycle as its cells begin dividing to reconstitute the mucosa. Table 22–1 summarizes of the main events of the menstrual cycle.

MEDICAL APPLICATION

Endometriosis is a fairly common disorder in which viable cells of the endometrium, displaced during menstruation, reflux into one or both uterine tubes, move upward and grow outside the uterus rather than undergoing vaginal discharge. Common sites of such growth are the tubes themselves, the ovarian surface, or peritoneal lining. Under the influence of estrogen and progesterone, the ectopic tissue grows and degenerates monthly without effective removal, leading to pain, inflammation, cysts, adhesions, and scar tissue that can result in infertility.

Embryonic Implantation, Decidua, & the Placenta

The oocyte is fertilized in the ampulla of the uterine tube and the resulting zygote undergoes mitotic cleavages as it is moved passively toward the uterus. During this time the embryo remains within the covering zona pellucida and is about the same size as the oocyte at fertilization. The cells that result from segmentation of the zygote are called **blastomeres** (Gr. *blastos,* germ, + *meros,* part) and the compact aggregate of blastomeres is the **morula** (L. *morum,* mulberry). Because the zygote is not growing in size, the blastomeres become smaller with each division.

The embryo reaches the uterine cavity 4–5 days after fertilization and loses the zona pellucida. At this time a cavity develops in the center of the morula and the embryo enters the **blastocyst** stage of development. The blastomeres arrange themselves as a peripheral layer called the **trophoblast** around the cavity, while a few cells collect just inside this layer, forming the **embryoblast** or **inner cell mass** (Figure 22–18). The blastocyst remains in the lumen of the uterus for two or three days, immersed in the endometrial glands' secretion.

Implantation, or nidation, involves the blastocyst attaching to the surface epithelial cells of the late secretory phase endometrium and its proteolytic penetration through this epithelium into the underlying stroma (Figure 22–18), a process that lasts about three days. Cells of the trophoblast drive the events of implantation, during which time cells of the embryoblast rearrange as two new cavities, the **amnion** and the **yolk sac.** Where the cells lining these cavities make contact, the **bilaminar embryonic disc** develops with its **epiblast** layer continuous with the amnion and its **hypoblast** layer continuous with the yolk sac (Figure 22–18).

All parts of the embryo develop from this early embryonic disc. The yolk sac and amnion form extraembryonic structures, but only the latter persists throughout pregnancy. The trophoblast differentiates during implantation into a **cytotrophoblast** and a more superficial **syncytial trophoblast** (an invasive, multinucleated mass), both of which contribute to the embryonic portion of the placenta. By about the ninth day after ovulation, the embryo is totally implanted in the endometrium and derives nutrients from blood and secretions there. Trophoblast cells release anti-inflammatory cytokines to prevent an adverse reaction of the uterus to the implanted embryo and these are supplemented later by various embryonic factors that produce local immune tolerance for the embryo throughout the pregnancy.

The endometrial stroma goes through profound changes in the days following implantation. The fibroblasts become enlarged and polygonal, more active in protein synthesis, and are now called **decidual cells.** The whole endometrium is now referred to as the **decidua** (L. *deciduus,* falling off, shedding). The decidua include the **decidua basalis,** situated between the embryo itself and the myometrium; **decidua capsularis,** between the embryo and the lumen of the uterus; and **decidua parietalis,** the remainder of the decidua (Figure 22–19).

The **placenta** is the site of nutrient, waste, O_2, and CO_2 exchanges between the mother and the fetus and contains tissues from both individuals. The embryonic part is the **chorion,** derived from the former trophoblast, and the maternal part is

Table 22–1. Summary of events of the menstrual cycle.

	Stage of Cycle			
	Proliferative	**Secretory or Luteal**		**Menstrual**
Main actions of pituitary hormones	Follicle-stimulating hormone stimulates rapid growth of ovarian follicles.	Peak of luteinizing hormone at the beginning of secretory stage, secreted by stimulation of estrogen, induces ovulation and development of the corpus luteum.		
Main events in the ovary	Growth of ovarian follicles; dominant follicle reaches preovulatory stage.	Ovulation.	Development of the corpus luteum.	Degeneration of the corpus luteum.
Dominant ovarian hormone	Estrogens, produced by the growing follicles, act on vagina, tubes, and uterus.	Progesterone, produced by the corpus luteum, acts mainly on the uterus.		Progesterone production ceases.
Main events in the endometrium	Growth of the mucosa after menstruation.	Further growth of the mucosa, coiling of glands, secretion.		Shedding of part of the mucosa about 14 days after ovulation.

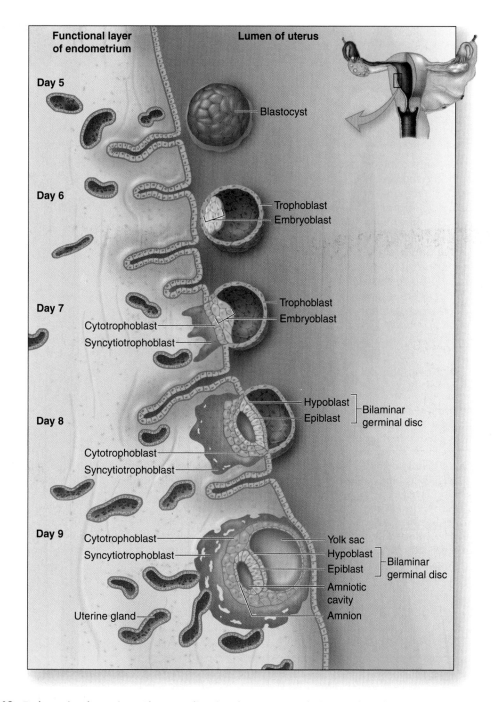

Figure 22–18. Embryo implantation. The coordination between ovulation and endometrial development results in the embryo arriving as a blastocyst about 5 days after ovulation or fertilization, when the uterus is in the late secretory phase and best prepared for implantation. After the zona pellucida is shed, receptor proteins on embryonic trophoblast cells bind ligands and proteoglycans on the endometrial epithelial cells. The trophoblast sends processes between the latter cells and promotes their apoptotic destruction. The trophoblast now also forms an invasive, outer syncytial layer called the syncytiotrophoblast. MMPs are activated and/or released locally to digest the basal lamina and other stroma components, allowing the developing embryo to become enclosed within the stroma. Until chorionic villi of the early placenta are formed, the implanted embryo absorbs nutrients and oxygen from the local endometrial tissue and lacunae of blood.

from the decidua basalis. The two trophoblast layers together form **chorionic villi** that project into the blood-filled lacunae of the decidua and provide a larger surface for nutrient and O$_2$ absorption. The villi are invaded by embryonic mesenchyme and by the end of the third week of pregnancy wide capillaries continuous with the embryonic vasculature develop in the villi. Eventually the placenta contains thousands of chorionic villi, each branching many times and each branch containing one or more capillary loops. Suspended in the pools of maternal blood, the chorionic villi have an enormous area for metabolite exchange (Figure 22–20). Exchange of gases, nutrients, and wastes occurs between fetal blood in the capillaries and maternal blood bathing the villi, with diffusion occurring across the trophoblast layer, the connective tissue of the villus, and the capillary endothelium.

The placenta is also an endocrine organ, producing HCG, chorionic thyrotropin, chorionic corticotropin, estrogens, progesterone, and other hormones. More detailed information on the developing embryo and on the formation and structure of the placenta should be sought in embryology textbooks.

MEDICAL APPLICATION

*The initial attachment of the embryo usually occurs on the ventral or dorsal walls of the body of the uterus. Sometimes the embryo attaches close to the internal os. In this case the placenta will be interposed between the fetus and the vagina, obstructing the passage of the fetus at parturition. This situation, called **placenta previa**, must be recognized by the physician, and the fetus must be delivered by cesarean section; otherwise, the fetus may die. Sometimes, as already mentioned, the embryo attaches to the epithelium of the uterine tube. Very rarely the zygote may enter the abdominal cavity, attach to the peritoneum, and develop there.*

Uterine Cervix

The **cervix** is the lower, more cylindrical part of the uterus (Figure 22–1) and it differs in histologic structure from the rest of the uterus. The mucosal lining of the endocervix is a mucus-secreting simple columnar epithelium on a thick lamina propria. The region of the cervix where the endocervical canal opens into the vagina is called the **external os**, which bulges into the upper vagina and is covered by the exocervical mucosa which has stratified squamous epithelium. A distinct junction, or **transformation zone**, occurs where the simple columnar epithelium undergoes an abrupt transition to stratified squamous epithelium (Figure 22–21). The deeper, middle layer of the cervix has little smooth muscle and consists mainly of dense connective tissue. From this stroma many lymphocytes and other leukocytes penetrate the stratified epithelium to reinforce the local immune defense against microorganisms. Before parturition the cervix dilates greatly and softens due to intense collagenolytic activity in the stroma.

The endocervical mucosa contains many branched **cervical glands**, which produce mucus and are often dilated. Hormonal shifts during the uterine cycle cause the mucosa to swell periodically and influence activity of the cervical glands. However, this mucosa is affected histologically much less than the endometrium and it does not desquamate during menstruation.

Cervical secretions change cyclically, however, and play a significant role in fertilization and early pregnancy. At the time of ovulation, the mucous secretions are maximal, watery, and facilitate movement through the uterus by sperm. In the luteal phase high progesterone levels cause the mucous secretions to become viscous and hinder the passage of both sperm and microorganisms into the body of the uterus. During pregnancy, the cervical glands proliferate and secrete abundant, highly viscous mucus which forms a plug in the endocervical canal.

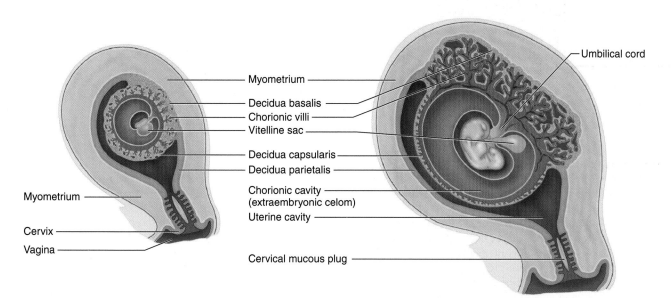

Figure 22–19. Decidua, early placenta, and extraembryonic membranes. After implantation and throughout pregnancy the endometrial connective tissue cells exist as large, synthetically active decidual cells. The endometrium is called the decidua, with three regions recognized in the locations indicated in the diagrams: decidua basalis, capsularis, and parietalis.

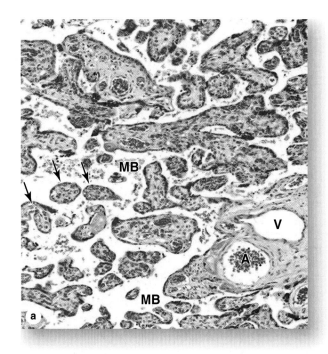

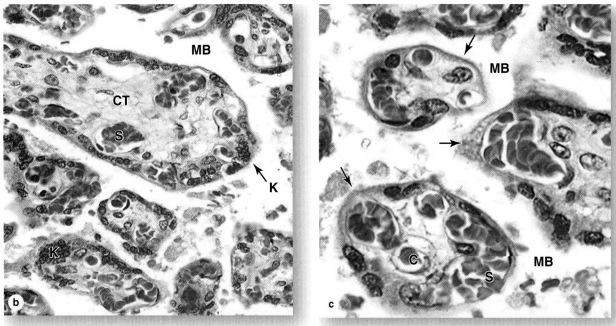

Figure 22–20. **Term placenta.** The placenta contains chorionic villi of the fetus and maternal blood pooled in spaces of the decidua. **(a):** At low magnification, a full-term placenta includes sections of many villus stems, containing arteries (A) and (V) of the extraembryonic vasculature, and hundreds of smaller villus branches (arrows) which contain connective tissue and microvasculature. Maternal blood (MB) normally fills the space around all the villi. X50. H&E. **(b):** At higher magnification, the villus connective tissue (CT) can be seen to still resemble mesenchyme and to be surrounded by epithelial cells of the trophoblast, including both the inner cytotrophoblast epithelium and the overlying syncytial trophoblast. In many areas nuclei of the syncytiotrophoblast layer have formed clusters or knots (K) on the surfaces of villi. The trophoblast separates the sinusoids (S) and other vessels containing fetal blood from the maternal blood (MB) in the intervillus space. X200. H&E. **(c):** Still higher magnification of the same section shows that the villus branches each contain several capillaries (C) and wide sinusoids (S) filled with fetal blood. By the end of pregnancy cells of the cytotrophoblast have greatly decreased in number in many areas of the villi and only a thin syncytiotrophoblast underlain by basement membrane surrounds the villus in these regions (arrows). The external syncytiotrophoblast surface is densely covered with microvilli which increase the absorptive surface and have many receptors and transporters for uptake of material from maternal blood. The extraembryonic blood vessels become closely associated with these areas of thin trophoblast for maximal diffusion of material between the two pools of blood. X400. H&E.

Cancer of the cervix (**cervical carcinoma**) is derived from its stratified squamous epithelium. Although it is observed fairly frequently, the mortality rate is decreasing rapidly worldwide because this carcinoma is now usually detected in its early stages by routine screening programs. For such tests cervical cells are studied by **exfoliative cytology**, using cells that have been scraped with a brush or wooden spatula from the cervix, especially from the region of the junction between the simple columnar and stratified epithelia, stained on a slide and examined microscopically. Abnormal cells suggestive of precancerous changes in the epithelium are detected in such "cervical

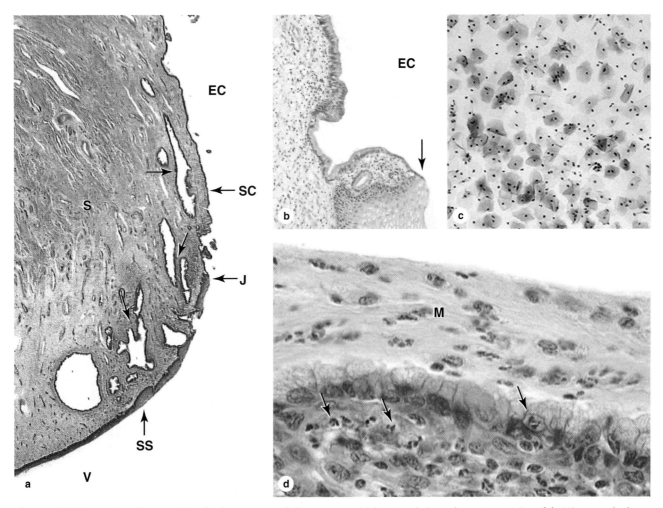

Figure 22–21. **Cervix.** The cervix is the lower part of the uterus, which extends into the upper vagina. **(a):** Micrograph shows that the mucosa of the endocervical canal (EC) is continuous with the endometrium and like that tissue is lined by simple columnar epithelium (SC). The endocervical mucosa has folds and many large branched cervical glands (arrows) secreting mucus under the influence of the ovarian hormones and often becoming quite dilated. At the external os, the point at which this canal opens into the vagina (V), there is an abrupt junction (J) between this simple epithelium and the stratified squamous epithelium (SS) covering the exocervix and vagina. X15. H&E.

   (b): Higher magnification shows the junction (arrow) and the lining of the endocervical canal (EC) more clearly. X50. H&E. **(c):** Micrograph shows exfoliative cytology of cells scraped from a normal exocervix in a routine cervical smear. The squamous cells are stained on a slide by the Papanicolaou procedure using hematoxylin, orange G, and eosin and stain differently according to their content of keratins. Surface cells have denser cytoplasmic keratin and stain pinkish orange, while less fully differentiated subsurface cells have blue-green cytoplasm and larger nuclei. Unusually high numbers of blue-green cells, cells with atypical nuclei, or other cytological abnormalities detected by this method prompt further tests for the possibility of cervical carcinoma, which is not uncommon. X200. Papanicolaou stain. **(d):** Micrograph shows a higher magnification of the endocervical lining. Exposed to a relatively high population of microorganisms, this mucosa normally has a large number of neutrophils and other leukocytes which form an important part of the innate immune defense in this region. Such cells are seen here in the lamina propria and epithelium (arrows), but are also numerous and readily apparent in the layer of mucus (M) that was fixed in place on this specimen. X400. H&E.

smears" or "Pap smears," named after their developer, George Papanicolaou, who introduced the diagnostic technique in the 1920s.

Abnormal cell growth (dysplasia) with the possibility of progressing to carcinoma (neoplasia) occurs most commonly at the transformation zone between the columnar and stratified epithelia, especially when the cells also contain the human papillomavirus (HPV), which is very common.

VAGINA

The wall of the **vagina** (L., *vagina*, sheath, scabbard) lacks glands and consists of three layers: a **mucosa,** a **muscular layer,** and an **adventitia.** Mucus covering the lumen of the vagina is produced by the glands of the uterine cervix. During intercourse additional, lubricating mucus is provided by a pair of large and many small **vestibular glands** opening into the vestibule, a space enclosed within the labia minora that also contains the vaginal and urethral orifices and the anterior erectile tissue of the clitoris. The stratified squamous epithelium covering these various components of the vestibule, which together make up the **external genitalia**, merges with epidermis of the surrounding skin. The mucosa of these structures is abundantly supplied with sensory nerves and the range of tactile receptors normally found in skin (Chapter 18), which are important in the physiology of sexual arousal.

The epithelium of the vaginal mucosa is stratified squamous, with a thickness of 150–200 μm in adults (Figure 22–22). Its cells contain a small amount of keratohyaline, but do undergo keratinization to form keratin plates as in the epidermis. Stimulated by

estrogens, the epithelial cells synthesize and accumulate glycogen. When the cells desquamate, bacteria metabolize glycogen to lactic acid, causing a relatively low pH within the vagina which helps provide protection against pathogenic microorganisms.

The lamina propria of the mucosa is rich in elastic fibers and has numerous narrow papillae projecting into the epithelial layer (Figure 22–22). The vaginal connective tissue normally contains lymphocytes and neutrophils in relatively large quantities. During the premenstrual and menstrual phases of the cycle, leukocytes are particularly numerous throughout the mucosa and in the lumen of the vagina. The vaginal mucosa itself has few sensory nerve endings.

The muscular layer of the vagina is composed mainly of two indistinct layers of smooth muscle, disposed as circular bundles next to the mucosa and as thicker longitudinal bundles near the adventitial layer. The dense connective tissue of the adventitia is rich in elastic fibers, making the vaginal wall strong and elastic while binding it the surrounding tissues. This outer layer also contains an extensive venous plexus, lymphatics, and nerves.

MAMMARY GLANDS

The **mammary glands** of the breasts develop embryologically as invaginations of surface ectoderm along two ventral lines, the milk lines, from the axillae to the groin. In humans one set of glands resembling highly modified apocrine sweat glands persists on each side of the chest. Each mammary gland consists of 15–25 **lobes** of the compound tubuloalveolar type whose function is to secrete milk to nourish newborns. Each lobe, separated from the others by dense connective tissue with much adipose tissue, is a separate gland with its own excretory **lactiferous duct** (Figure 22–23). These ducts, each 2–4.5 cm long, emerge

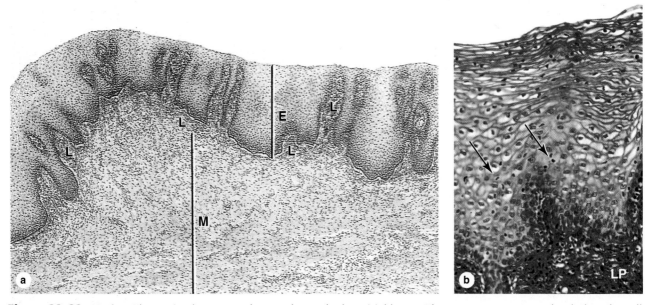

Figure 22–22. Vagina. The vagina has mucosal, muscular, and adventitial layers. There are no secretory glands, but the cells of the thick, nonkeratinized stratified squamous epithelium become filled with glycogen before desquamation and thin-walled veins of the mucosa and muscular layers exude fluid into the epithelium. **(a):** The micrograph shows the lamina propria (L) is highly cellular and extends narrow papillae into the epithelium (E). The papillae and entire lamina propria are very rich in protective lymphocytes and neutrophils. The muscular layer (M) has bundles of smooth muscle arranged in a circular manner near the mucosa and longitudinally near the adventitia. X60. H&E. **(b):** Higher magnification of the epithelium and lamina propria (LP) shows invasion of leukocytes (arrows) between epithelial cells from the connective tissue. X200. PSH.

independently in the **nipple,** which has 15–25 pore-like openings, each about 0.5 mm in diameter. The histologic structure of the mammary glands varies according to sex, age, and physiologic status.

Breast Development during Puberty

Before puberty, the mammary glands in both sexes are composed only of **lactiferous sinuses** near the nipple, with small, branching ducts emerging from these sinuses. In girls undergoing puberty and having higher levels of circulating estrogens, the breasts increase in size as a result of adipocyte accumulation in the connective tissue and increased growth and branching of the duct system. The nipple enlarges with growth of the lactiferous sinuses.

In nonpregnant adult women the characteristic parenchymal structure of the gland, the lobe, consists of many **lobules,** sometimes called **terminal duct lobular units (TDLU).** Each lobule has several small, branching ducts, but the attached secretory units are small and rudimentary (Figure 22–23). The duct system is embedded in loose, vascular connective tissue and a denser, less cellular connective tissue separates the lobes.

The lactiferous sinuses are lined with stratified cuboidal epithelium and the lining of the lactiferous ducts and terminal ducts is simple cuboidal epithelium covered by closely packed myoepithelial cells. Sparse fibers of smooth muscle also encircle

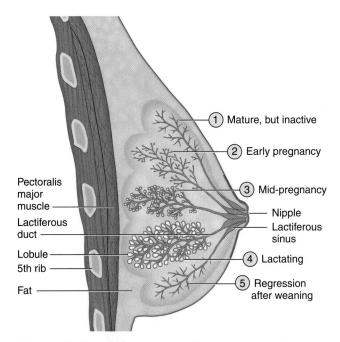

Figure 22–23. Mammary gland. Shown here are the major tissues and structures in a breast, along with the sequence of changes that occur in the duct system and secretory units before, during, and after pregnancy and lactation. **(1)** Before pregnancy, the gland is inactive, with small ducts and only a few small secretory alveoli. **(2)** Alveoli develop and begin to grow early in a pregnancy. **(3)** By mid-pregnancy, the alveoli and ducts have become large and have dilated lumens. **(4)** At parturition and during the time of lactation, the alveoli are greatly dilated and maximally active in production of milk components. **(5)** After weaning, the alveoli and ducts regress with apoptotic cell death.

(Figure labels: Mature, but inactive (1); Early pregnancy (2); Mid-pregnancy (3); Nipple; Lactiferous sinus; Lactating (4); Regression after weaning (5); Pectoralis major muscle; Lactiferous duct; Lobule; 5th rib; Fat)

the larger ducts. Epithelial cells of the ducts become slightly more columnar at the time of peak estrogen levels around ovulation and in the premenstrual phase of the cycle connective tissue of the breast becomes somewhat edematous, making the breasts slightly larger.

Skin covering the nipple constituted the **areola** and is fairly typical thin skin with sebaceous glands (Chapter 18). The epidermis is continuous with the lining of the lactiferous sinuses. The areola contains more melanin than skin elsewhere on the breast and darkens further during pregnancy. Skin of the nipple is abundantly supplied with sensory nerve endings. Connective tissue of the nipple is rich in smooth muscle fibers that run parallel to the lactiferous sinuses and produce nipple erection when they contract.

Breasts during Pregnancy & Lactation

The mammary glands undergo growth during pregnancy as a result of the synergistic action of several hormones, mainly estrogen, progesterone, prolactin, and human placental lactogen. One result of these hormones is the proliferation of secretory **alveoli** at the ends of the intralobular ducts (Figures 22–24 and 22–25). The spherical alveoli are composed of cuboidal epithelium, with stellate myoepithelial cells between the secretory cells and the basal lamina. The degree of glandular development varies among lobules and even within a single lobule.

While alveoli and the system of ducts grow and develop during pregnancy in preparation for lactation, the stroma becomes less prominent. The loose connective tissue within lobules is infiltrated in lymphocytes and plasma cells, the latter becoming more numerous late in pregnancy when they begin to produce immunoglobulins (secretory IgA).

Late in pregnancy the glandular alveoli and ducts are dilated by an accumulation of **colostrum,** a fluid rich in proteins, vitamin A, and certain electrolytes that is produced under the influence of prolactin. Antibodies are synthesized abundantly by plasma cells and transferred into colostrum, from which passive acquired immunity is conferred on the breast-fed newborn.

Following parturition levels of circulating estrogens and progesterone decline and the glandular alveoli of the breasts become very active in milk production, primarily influenced by prolactin from the anterior pituitary (Chapter 20). Epithelial cells of the alveoli enlarge and engage actively in synthesis of proteins and lipids for secretion. Large amounts of protein are made on rough ER, processed through the Golgi apparatus and packaged into secretory vesicles, which undergo exocytosis during **merocrine secretion** into the lumen (Figure 22–26). Spherical lipid droplets, containing primarily neutral triglycerides and cholesterol, form in the cytoplasm of the alveolar cells, grow greatly in size by accretion of more lipids, and eventually pass out of the cells into the lumen by the process of **apocrine secretion,** during which the droplets become enveloped with a portion of the apical cell membrane (Figure 22–26).

Throughout lactation secretion of proteins, membrane-bound lipid droplets, and other components is on-going, with the products accumulating as milk in the lumens of the duct system (Figure 22–25). Proteins normally constitute approximately 1.5% of human milk and include mainly various caseins, which aggregate as micelles, as well as soluble β-lactoglobulin and α-lactoalbumin, all of which are digested as a source of amino acids by the infant. Less abundant proteins in milk include many which assist digestion and use of other milk nutrients, immunoglobulins and several proteins with antimicrobial activity, and various

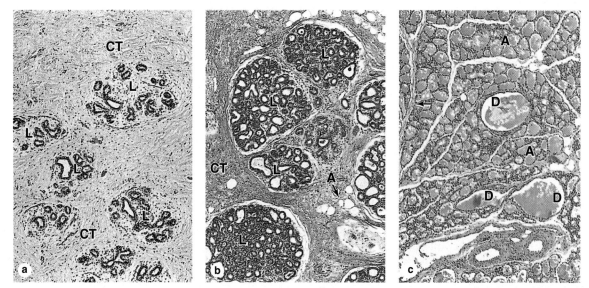

Figure 22–24. Alveolar development in the breast during pregnancy. The micrographs show the major changes in the mammary glands during pregnancy and lactation that involve mainly the secretory alveoli, lobules and ducts. **(a):** The mammary glands of adult, nonpregnant women are inactive, with small ducts and few lobules (L) having secretory alveoli which are not well-developed. The structure with the large lumen in each lobule is part of the duct; the smaller structures are the small, undeveloped alveoli. The breasts are composed largely of connective tissue (CT), having considerable fat. **(b):** The glands become active during pregnancy, with the duct system growing rapidly and the secretory units of each lobule becoming much larger and more extensively branched. In this micrograph adipocytes (A) are included but these are only a small fraction of those present. **(c):** During lactation, the lobules are greatly enlarged and the lumens of both the numerous glandular alveoli (A) and the excretory ducts (D) are filled with milk. The protein content of milk makes it eosinophilic in histological sections. At this time the intralobular connective tissue is more sparse and difficult to see, except for small septa (arrows). All X60, H&E.

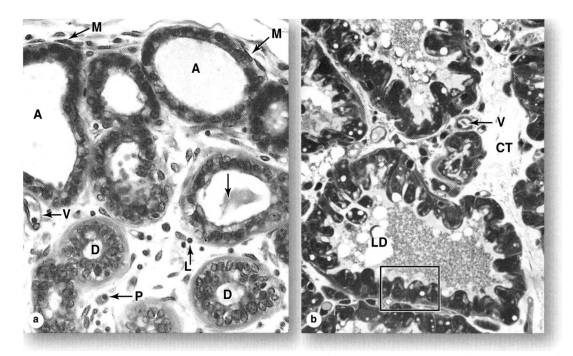

Figure 22–25. Active versus lactating alveoli. Glandular alveoli develop completely only during pregnancy and begin milk production near the end of pregnancy. **(a):** Micrograph shows alveoli (A) develop as spherical structures composed of cuboidal epithelial cells surrounded by the contractile processes of myoepithelial cells (M). Development occurs at different rates throughout the breast. Late in pregnancy lymphocytes (L) leave venules (V), accumulate in the intralobular connective tissue and differentiate as plasma cells (P). IgA secreted by these cells is transferred into milk and helps confer passive immunity from the mother to the nursing infant. A small amount of milk is beginning to accumulate in the lumen of the duct (arrow). X400. H&E. **(b):** Micrograph from a plastic section shows secretory cells of the lactating gland, to be more columnar and contain variously sized lipid droplets, which are also visible in the milk (LD). Connective tissue appears less cellular, although blood vessels (V) and lymphocytes remain present. Secretory cells in the enclosed area are shown diagrammatically in Figure 22–26. X400. PT.

mitogenic growth factors. Lipids normally constitute about 4% of milk in humans, while the major sugar, lactose, makes up as much as 7–8% and is a major source of energy. Lactose is synthesized in the Golgi apparatus and also serves to help draw water osmotically into the protein secretory vesicles, which adds greatly to the volume of milk.

MEDICAL APPLICATION

When a woman is breast-feeding, the nursing action of the child stimulates tactile receptors in the nipple, resulting in liberation of the posterior pituitary hormone **oxytocin**. *This hormone causes contraction of the smooth muscle of the lactiferous sinuses and ducts, as well as the myoepithelial cells of alveoli, resulting in the* **milk-ejection reflex**. *Negative emotional stimuli, such as frustration, anxiety, or anger, can inhibit the liberation of oxytocin and thus prevent the reflex.*

Postlactational Regression in the Mammary Glands

When breast-feeding is stopped (weaning), most alveoli that developed secretory properties during pregnancy degenerate. There is apoptosis and sloughing of whole cells (Figure 22–27), with dead cells and debris removed by macrophages, as well as autophagy in most other epithelial cells. The duct system of the gland returns to its general appearance in the inactive state before pregnancy. After menopause, alveoli and ducts of the mammary glands are reduced further in size and there is some loss of fibroblasts, collagen, and elastic fibers in the stroma.

MEDICAL APPLICATION

*Most instances of breast cancer (**mammary carcinomas**) arise from epithelial cells of the lactiferous ducts. Cell spreading (or metastasizing) from the carcinoma via the circulatory or lymphatic vessels to critical organs such as the lungs or brain are responsible for the mortality associated with breast cancer. Axillary lymph nodes are removed surgically and examined histologically for the presence of metastatic mammary carcinoma cells. Early detection (eg, through self-examination, mammography, ultrasound, and other techniques) and consequent early treatment have significantly reduced the mortality rate.*

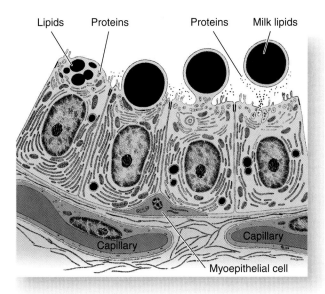

Figure 22–26. Secretion in the mammary gland. Alveolar cells of the lactating mammary gland are highly active in protein synthesis on rough ER and lipid synthesis. Most proteins are packaged into secretory vesicles in the Golgi apparatus and secreted at the apical end of the cells by typical exocytosis or merocrine secretion. Lipids coalesce as free cytoplasmic droplets. These grow in size and eventually undergo aprcrine secretion, in which they are extruded from the cell along a portion of the apical cell membrane (and often a small amount of attached cytoplasm.) Both types of secretion are shown here in a sequence moving from left to right. Similar cells are seen in the enclosed area in the light micrograph shown in Figure 22–25.

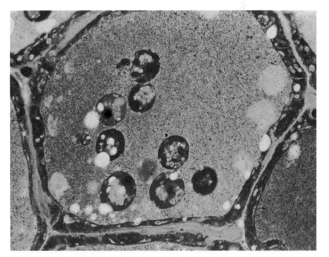

Figure 22–27. Apoptosis during postlactational mammary gland regression. After weaning, all glandular alveoli of the breast regress, as shown in this plastic section of a single alveolus. The secretory cells now have a low cuboidal structure and many cells are undergoing apoptosis and have sloughed into the lumen. Milk with lipid droplets is also still present there. The dead cells and other tissue debris are removed by macrophages. X400. PT.

Information about the external world is conveyed to the central nervous system from sensory **receptors**. Chemoreceptor units for the senses of taste and smell were discussed in Chapters 15 and 17, respectively, and the range of mechanoreceptors that mediate the sense of touch in its various components was presented in Chapter 18. This chapter will examine the systems responsible for vision via photoreceptors of the eye and for the senses of equilibrium and hearing that involve mechanoreceptors in the vestibulocochlear apparatus of the ear.

EYES: THE PHOTORECEPTOR SYSTEM

The **eye** (Figure 23–1) is a complex and highly developed photosensitive organ that analyses the form, intensity, and color of light reflected from objects, providing the sense of sight. The eyes are located in protective areas of the skull, the **orbits**, which also contain cushions of adipose tissue. Each eyeball includes a tough, fibrous globe to maintain its shape, a system of transparent tissues that refract light to focus the image, a layer of photosensitive cells, and a system of neurons whose function it is to collect, process, and transmit visual information to the brain. Each eye is composed of three concentric tunics or layers (Figure 23–2): a tough external layer consisting of the **sclera** and the **cornea;** a more vascular middle layer consisting of the **choroid, ciliary body,** and **iris;** and an inner sensory layer, the **retina,** which consists of an outer pigmented epithelium and an inner retina proper. The photosensitive inner layer of the retina communicates with the cerebrum through the **optic nerve** on the eye's posterior side; its anterior edge is called the **ora serrata** (Figure 23–1).

The **lens** of the eye is a biconvex transparent structure held in place by a circular system of **zonular fibers,** which extend from the lens into a thickening of the middle layer, the **ciliary body,** and by close apposition to the vitreous body on its posterior side (Figure 23–1). Partly covering the anterior surface of the lens is an opaque pigmented expansion of the middle layer

called the **iris.** The round hole in the middle of the iris is the **pupil** (Figure 23–1).

The eye contains two fluid-filled cavities: the **anterior chamber,** which occupies the space between the cornea and the iris and the **posterior chamber,** between the iris, ciliary processes, zonular attachments, and lens (Figure 23–1). Interconnected at the pupil, these contain a clear fluid called **aqueous humor.** The **vitreous chamber** lies behind the lens and its zonular attachments and is surrounded by the retina. This chamber is filled with a transparent, gelatinous mass of connective tissue called the **vitreous body**.

Eye formation begins in the early embryo with epithelial optic vesicles bulging bilaterally from the developing forebrain. These elongate and form optic stalks bearing **optic cups** (Figure 23–3). Interaction between the optic cups and the overlying surface ectoderm causes the latter to invaginate and detach bilaterally, forming the **lens vesicles.** In the ensuing weeks, head mesenchyme differentiates to form most of the tissue in the eye's three layers and the vitreous, with the ectoderm of the optic cup and optic stalk giving rise to the retina and optic nerve, respectively, and with surface ectoderm contributing to the cornea (Figure 23–3).

Fibrous Layer

SCLERA

The fibrous, external layer of the eyeball protects the more delicate internal structures and provides sites for muscle insertion. (In reference to the eye, the terms "external/outer" and "internal/inner" refer to structures closer to the eyeball's surface or in its interior, respectively.) The opaque white posterior five-sixths of the external layer is the **sclera** (Figure 23–1); this forms a segment of a sphere with a diameter of approximately 22 mm in adults. The sclera averages 0.5 mm in thickness, is relatively avascular, and consists of tough, dense connective tissue containing flat type I collagen bundles which intersect in various directions while remaining parallel to the surface of the organ, with a moderate amount of ground substance and scattered fibroblasts. Tendons of the extraocular muscles that move the eyes insert into anterior areas of the sclera. Posteriorly the sclera thickens to approximately 1 mm and joins with the epineurium covering the optic nerve. A thin inner region of the sclera, adjacent to the choroid, is slightly less dense, with thinner collagen fibers, more fibroblasts, elastic fibers, and melanocytes.

CORNEA

In contrast to the sclera, the anterior one-sixth of the eye—the **cornea**—is colorless, transparent, and completely avascular (Figure 23–1). A section of the cornea shows that it consists of five layers:

- an **external stratified squamous epithelium**,
- an **anterior limiting membrane** (**Bowman's membrane**, the basement membrane of the stratified epithelium),
- the **stroma**,
- a **posterior limiting membrane** (**Descemet's membrane**, the basement membrane of the endothelium), and
- an inner simple squamous **endothelium**.

The stratified surface epithelium is nonkeratinized, with five or six cell layers of cells comprising about 10% of the corneal thickness (Figure 23–4). Numerous mitotic figures are present in the basal layers, particularly near the periphery of the cornea, reflecting the epithelium's high capacity for cell renewal and repair. The flattened surface cells have microvilli and folds

protruding into a protective layer or tear film of lipid, glycoprotein, and water about 7 μm thick. As another protective adaptation, the corneal epithelium also has one of the richest sensory nerve supplies of any tissue. The basement membrane of this epithelium is very thick (8–12 μm) and contributes to the stability and strength of the cornea, helping to protect against infection of the underlying stroma.

The thick **stroma**, or substantia propria, is formed of approximately 60 layers of parallel collagen bundles that align at approximately right angles to each other and may cross the complete corneal diameter. The uniform orthogonal array of collagen fibrils contributes to the transparency of this avascular tissue. Between the collagen lamellae are cytoplasmic extensions of flattened fibroblast-like cells called **keratocytes** (Figure 23–4). The ground substance surrounding these cells is rich in proteoglycans such as lumican, containing keratan sulfate and chondroitin sulfate, which help maintain the precise organization and spacing of the collagen fibrils.

The posterior surface of the stroma is bounded by another thick (~10 μm) structure (Descemet's membrane) composed of fine, interwoven collagen fibers, upon which lies the corneal endothelium (Figure 23–4). Cells of this simple squamous epithelium are active in protein synthesis to maintain this basement membrane and in pumping sodium ions into the adjacent anterior chamber. Chloride ions and water follow passively from the corneal stroma. In this way, the endothelium is largely responsible for maintaining a state of hydration within the cornea that helps provide maximum transparency and optimal light refraction.

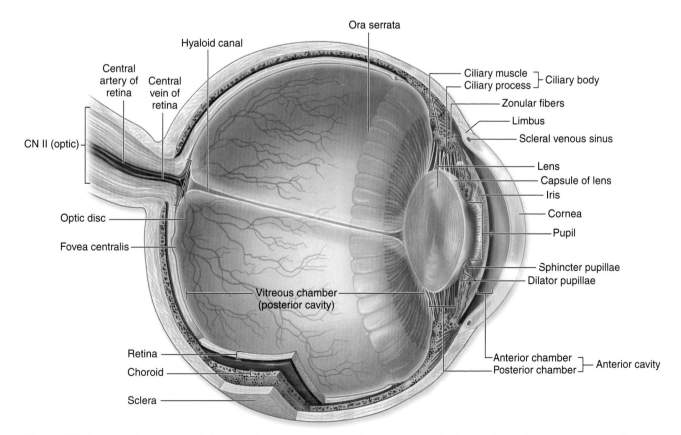

Figure 23–1. Internal anatomy of the eye. The sagittal section of an eye shows the inter-relationships among the major ocular structures, the three major layers or tunics of the wall, important regions within those layers and the refractive elements (cornea, lens, and vitreous).

MEDICAL APPLICATION

*The shape or curvature of the cornea can be changed surgically to improve certain visual abnormalities involving the ability to focus. In the common ophthalmological procedure, **laser-assisted in situ keratomileusis (LASIK) surgery),** the corneal epithelium is displaced as a flap and the stroma reshaped by an excimer laser which vaporizes collagen and keratocytes in a highly controlled manner with no damage to adjacent cells or ECM. After reshaping the stroma, the epithelial flap is repositioned and a relatively rapid regenerative response reestablishes normal corneal physiology.*

LIMBUS

Encircling the cornea is the **corneoscleral junction,** or **limbus,** a transitional area where the transparent stroma merges with the opaque sclera (Figures 23–1 and 23–5). This region does have microvasculature which, along with aqueous humor in the anterior chamber, provides metabolites for the corneal cells by diffusion. Stem cells for the stratified epithelium are concentrated at the limbus, from which rapidly dividing transit amplifying cells move centripetally into the corneal epithelium.

At the corneoscleral junction Descemet's membrane and its simple endothelium are replaced with a system of irregular endothelium-lined channels called the **trabecular meshwork,** which penetrate the stroma and allow slow, continuous drainage of aqueous humor from the anterior cavity (Figure 23–6). The fluid is pumped from these channels into the adjacent larger space of the **scleral venous sinus,** or **canal of Schlemm** (Figures 23–1, 23–5, and 23–6), from which it drains into the aqueous and episcleral veins of the sclera.

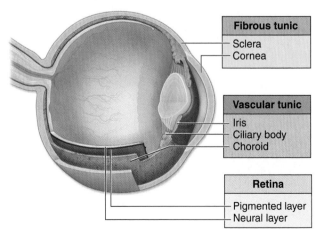

Figure 23–2. Layers of the eye. The sagittal view of an eye shows its three major layers or tunics, with the lens in the anterior opening of the vascular layer and retina.

Vascular Layer

The eye's more vascular middle layer, also known as the **uvea,** consists of three parts, from posterior to anterior: the **choroid,** the **ciliary body,** and the **iris** (Figure 23–2).

CHOROID

The choroid (Figure 23–7) is a highly vascular tunic in the posterior two-thirds of the eye, with loose, well-vascularized connective tissue rich in collagen and elastic fibers, fibroblasts, melanocytes, macrophages, lymphocytes, mast cells, and plasma cells. The abundant melanocytes give the layer its characteristic black color and block light from entering the eye except through the pupil.

The outer part of the choroid bound to the sclera is the **suprachoroidal lamina.** The inner region is richer than the outer layer in microvasculature and is called the **choriocapillary lamina.** Its microvasculature is important for nutrition and normal maintenance of the retina. A thin (2–4 μm) amorphous hyaline sheet known as **Bruch's membrane** separates the choriocapillary layer from the retina (Figure 23–7). This membrane extends from the ora serrata back to the optic nerve (Figure 23–1).

Ultrastructurally, Bruch's membrane is seen to consist of several layers. In the middle is a network of elastic fibers, sandwiched on both sides by collagen fibers, which are in turn covered on one side by the basal laminae of the choriocapillary endothelial cells and on the other side by the basal lamina of the pigmented retina epithelium.

CILIARY BODY

The ciliary body, an anterior expansion of the choroid at the level of the lens (Figures 23–1, 23–2, and 23–5), is a thickened ring of tissue lying just inside the anterior portion of the sclera. In transverse section the ciliary body is roughly a triangle, with its long base contacting the sclera, another side in contact with the vitreous body, and the third facing the posterior chamber (Figure 23–5). The ciliary body has a stroma of loose connective tissue, rich in microvasculature, elastic fibers, and melanocytes, surrounding much smooth muscle (Figure 23–5). The **ciliary muscle** has small fascicles of muscle that insert on the sclera and are arranged in such a way that their contraction in response to parasympathetic nerves decreases the internal diameter of the ciliary body ring, reducing tension on the fibers that run from this body to the lens. This allows the lens to become more rounded and better focus light from nearby objects onto the retina. The ciliary muscles are therefore important in visual accommodation (see Lens, below).

The surfaces of the ciliary body that face the vitreous body, posterior chamber, and lens are covered by a double layer of low columnar epithelial cells, the **ciliary epithelium,** formed from the rim of the embryonic optic cup (Figure 23–3). The epithelial cells directly covering the ciliary stroma are rich in melanin (Figure 23–8) and correspond to the anterior projection of the pigmented retina epithelium. The surface layer of cells lacks melanin and is contiguous with the sensory layer of the retina.

This stratified columnar epithelium covers the **ciliary processes,** a series of about 75 radial ridges extending from the surface of the ciliary body. Cells of the nonpigmented layer have tight junctions and extensive basal infoldings characteristic of ion-transporting cells, with Na⁺/K⁺-ATPase in their lateral plasma membranes. These cells actively transport fluid from the vascular stroma into the posterior chamber, thus forming the **aqueous humor.** This fluid has an inorganic ion composition similar to that of plasma but contains less than 0.1% protein (plasma has about 7% protein). Secreted into the posterior chamber, aqueous humor flows toward the lens, passing between it and the iris to reach the anterior chamber through the pupil (Figure 23–9). The aqueous then flows into the angle formed by the cornea with the basal part of the iris and penetrates the channels of the trabecular meshwork at the corneoscleral junction (limbus), from which it is pumped into the scleral venous sinus.

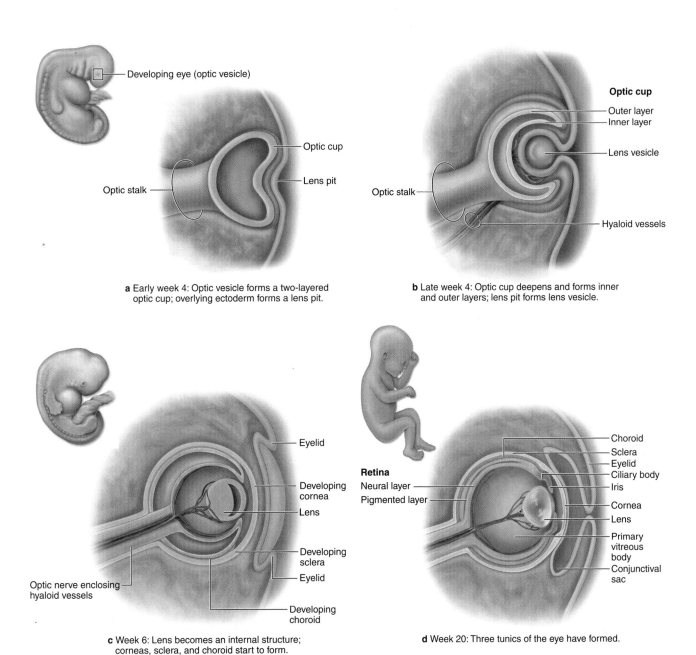

a Early week 4: Optic vesicle forms a two-layered optic cup; overlying ectoderm forms a lens pit.

b Late week 4: Optic cup deepens and forms inner and outer layers; lens pit forms lens vesicle.

c Week 6: Lens becomes an internal structure; corneas, sclera, and choroid start to form.

d Week 20: Three tunics of the eye have formed.

Figure 23–3. **Development of eye.** Eyes begin to form early in development when the optic vesicles bulge bilaterally from the diencephalic region of the forebrain (prosencephalon). These grow, remaining connected to the developing brain by the optic stalks, and approach the surface ectoderm. At this point each vesicle folds in on itself to form the inner and outer layers of the optic cup and inducing surface ectoderm to invaginate into the cup as the lens vesicle, which soon detaches from the surface and lies in the opening of the optic cup. Blood vessels, called the hyaloid vessels, grow along the optic stalk, enter the optic cup and grow toward the developing lens. In the following weeks, head mesenchyme associates with the developing optic cup, which is forming the two major layers of the retina. The mesenchymal cells differentiate around the pigmented layer of the developing retina as the iris, ciliary body and choroid of the vascular layer. Other mesenchymal cells give rise to the more external fibrous layer. The hyaloid vessels regress, leaving a space called the hyaloid canal, in the vitreous body. Folds of skin develop features of the eyelids and conjunctiva. The epidermis lining the latter structures develops in continuity with the surface epithelium of the cornea.

Aqueous humor is produced continuously. If its drainage from the anterior chamber is impeded, typically by obstruction of the trabecular meshwork or scleral venous sinus, intraocular pressure can increase, causing the condition called glaucoma. Untreated glaucoma can cause pressing of the vitreous body against the retina, affecting visual function and possibly leading to neuropathy in that tissue.

The surface epithelial cells in the grooves between the ciliary processes secrete elastin, fibrillin, and proteoglycans, which assemble as thin fibers attached to the surface of the lens capsule (Figure 23–10). The lens is thus anchored within the lumen of the ciliary body by this circular system of zonular fibers, which together comprise the ciliary zonule (also called the suspensory ligament of the lens.)

Iris

The **iris** is the most anterior extension of the uvea (middle layer) that partially covers the lens, leaving a round opening in the center called the **pupil** (Figures 23–1 and 23–2). The anterior surface of the iris, exposed to the anterior chamber, is not covered by epithelium, but consists of an irregular, discontinuous layer of fibroblasts and melanocytes, densely packed and with interdigitating processes. Deeper in the iris the stroma is more typical loose connective tissue with microvasculature (Figure 23–11). The posterior surface of the iris is smooth, with a two-layered epithelium continuous with that covering the ciliary body and its processes. Here however, the epithelial cells in direct contact with the posterior chamber are filled with melanin granules, which obscure most cellular features. The heavy pigmented epithelium of the iris prevents light from entering the interior

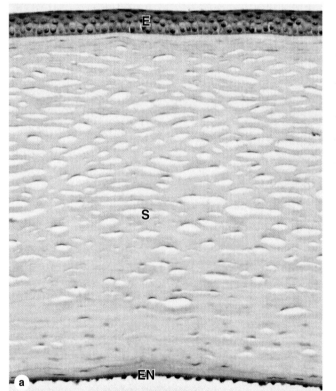
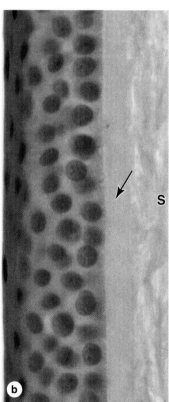

Figure 23–4. **Cornea.** The anterior structure of the eye, the cornea has five layers. **(a):** The micrograph shows the external stratified squamous epithelium (E), which is nonkeratinized and five or six cells thick. It is densely supplied with sensory free nerve endings that trigger the blinking reflex and its surface is covered with a tear film produced by glands in the eyelids and superior orbit. The stroma (S) comprises approximately 90% of the cornea's thickness, consisting of some 60 layers of long type I collagen fibers arranged in a precise orthogonal array and alternating with flattened cells called keratocytes. The stroma is lined internally by endothelium (EN). X100. H&E. **(b):** The corneal epithelium rests firmly on the thick homogeneous Bowman's membrane (arrow). The stroma is completely avascular and nutrients reach the keratocytes and epithelial cells by diffusion from the surrounding limbus and aqueous humor behind the cornea. X400. H&E. **(c):** The posterior surface of the cornea is covered by simple squamous epithelium (endothelium) that rests on another thick, strong layer of collagen and other extracellular material called Descemet's membrane (arrow). Na/K ATPase of the endothelial cells is responsible for pumping Na+ and drawing water out of the cornea, maintaining its proper state of hydration. In this state the cornea is perfectly transparent and with its curvature is a major refractive structure of the eye. X400. H&E.

of the eye except through the pupil. The underlying epithelial layer is composed of **myoepithelial cells** which are also at least partially pigmented. Radially extended processes from these myoepithelial cells make up the very thin **dilator pupillae muscle** along the posterior side of the iris (Figure 23–11).

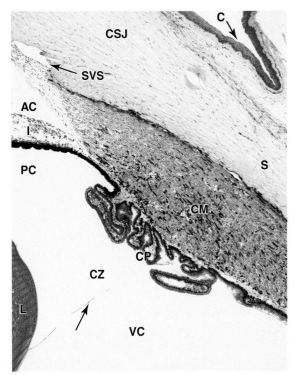

Figure 23–5. Corneoscleral junction (limbus) and ciliary body. At the circumference of the cornea is the limbus or corneoscleral junction (CSJ), where the transparent corneal stroma merges with the opaque, vascular sclera (S). The epithelium of the limbus is slightly thicker than the corneal epithelium, containing stem cells for the latter, and is continuous with the conjunctive (C) covering the anterior part of the sclera and lining the eyelids. The stroma of the limbus contains the scleral venous sinus (SVS), or canal of Schlemm, which receives aqueous humor from an adjacent trabecular meshwork at the surface of the anterior chamber (AC). Internal to the limbus, the middle layer of the eye consists of the ciliary body and its anterior extension, the iris (I). The thick ring of the ciliary body includes loose connective tissue containing melanocytes, smooth ciliary muscle (CM), numerous extensions covered by epithelium called the ciliary processes (CP), and the ciliary zonule (CZ), a system of fibrillin-rich fibers that attach to the capsule of the lens (L) in the center of the ciliary body. Pieces of one zonular fiber can be seen (arrow). Projecting into the posterior chamber (PC), the ciliary processes produce aqueous humor which then flows into the anterior chamber through the pupil. Changes in tension on the zonular fibers produced by contraction and relaxation of the ciliary muscles change the shape of the lens and allow visual accommodation. Behind the ciliary zonule and lens a thin, transparent membrane (not shown) surrounds the vitreous body and separates the posterior chamber from the vitreous chamber (VC). X12.5. H&E.

The abundant melanocytes in the vascular layer of the eye act collectively to keep stray light rays from interfering with image formation. Melanocytes of the iris stroma also provide the color of the eyes. In individuals with very few lightly pigmented cells in the stroma, light with a blue color is reflected back from the black pigmented epithelium on the posterior iris surface. As the number of melanocytes and amount of collagen increases in the stroma, the iris color changes through various shades of green, gray, and brown. Individuals with albinism have almost no pigment and the pink color of their irises is due to the reflection of incident light from the blood vessels of the stroma.

The iris contains smooth muscle bundles disposed in a circular array near the pupillary margin as the **sphincter pupillae muscle**. The dilator and sphincter muscles of the iris have sympathetic and parasympathetic innervation, respectively, for enlarging and constricting the pupil.

Lens

The **lens** is a transparent biconvex structure immediately behind the iris, used to focus light on the retina (Figure 23–1). Derived from an invagination of the embryonic surface epithelium (ectoderm), the lens is a unique avascular tissue (Figure 23–3). It is highly elastic, a feature that is lost with age as lens tissue hardens. The lens has three principal components.

Lens Capsule

The lens is covered by a thick (10–20 μm), homogeneous capsule rich in proteoglycans and type IV collagen (Figure 23–12). Originally the basement membrane of embryonic surface ectoderm, the lens capsule protects the underlying cells and provides the place of attachment for zonular fibers (Figure 23–10).

Lens Epithelium

Subcapsular **lens epithelium** consists of a single layer of cuboidal epithelial cells and is present only on the anterior surface of the lens. The basal ends of the epithelial cells attach to the lens capsule and their apical surfaces have interdigitations that bind the epithelium to the internal lens fibers (Figure 23–12). At the posterior edge of this epithelium, near the equator of the lens, the cells divide to provide new cells that differentiate as lens fibers. This process allows for growth of the lens and continues at a slow, decreasing rate near the equator of the lens throughout adult life.

Lens Fibers

Lens fibers are highly elongated and appear as thin, flattened structures (Figure 23–12). Developing from stem cells in the lens epithelium, the differentiating lens fibers eventually lose their nuclei and other organelles, fill the cytoplasm with a group of proteins called **crystallins**, and become very long. Mature lens fibers are typically 7–10 mm long, 8–10 μm wide, and 2 μm thick. The fibers are densely packed together forming a perfectly transparent tissue highly specialized for light refraction.

The lens is held in place by a radially oriented group of fibers, the elastic **ciliary zonule**, which inserts on both the lens capsule and on the ciliary body (Figures 23–1 and 23–10). This system is important in the process known as **accommodation,** which permits focusing on near and far objects by changing the curvature of the lens. When the eye is at rest or gazing at distant objects, the lens is kept stretched by the zonule in a plane perpendicular to the optical axis. To focus on a near object, the ciliary muscles contract, causing forward displacement of the choroid and ciliary body. This relieves some of the tension exerted by the zonule on the lens, allowing the latter to round up and become thicker, keeping the object in focus.

Vitreous Body

The **vitreous body** occupies the vitreous chamber behind the lens (Figure 23–1). It is composed of transparent connective tissue containing mostly (99%) water (vitreous humor), bound to hyaluronate, and a small amount of collagen. This gel-like connective tissue is contained within the vitreous membrane composed of type IV collagen and other proteins of external laminae. The only cells in the vitreous body are a few macrophages and a small population of cells near the membrane called hyalocytes, which synthesize the hyaluronate and collagen.

Retina

The **retina**, the inner layer of the eye, is derived from the embryonic optic cup (Figure 23–3). Like the optic cup, the retina consists of two major layers (Figure 23–2). The inner one, the **neural retina**, contains the neurons and photoreceptors. This layer's visual region extends anterior only as far as the **ora serrata** (Figure 23–1), but it continues as a cuboidal epithelium lining the surface of the ciliary body and posterior iris. The outer **pigmented layer** is an epithelium resting on Bruch's membrane just inside the choroid (Figure 23–7). This pigmented, cuboidal epithelium also lines the ciliary body and posterior iris, contributing to the double epithelium described with those structures.

The pigmented epithelium consists of low columnar cells with basal nuclei. The cells have well-developed junctional complexes, gap junctions, and numerous invaginations of the basal membranes associated with mitochondria. The apical ends of the cells extend processes and sheath-like projections that surround the tips of the photoreceptors. Melanin granules are numerous in the extensions and apical cytoplasm (Figure 23–13). This

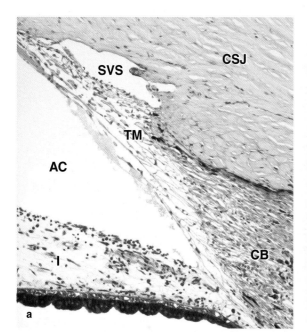
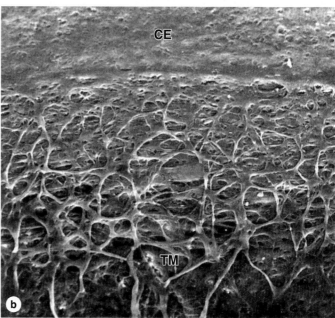

Figure 23–6. **Trabecular meshwork and scleral venous sinus. (a):** At the corneoscleral junction (CSJ), or limbus, encircling the cornea, the posterior endothelium and its thick underlying (Descemet's) membrane are replaced by a meshwork of irregular channels lined by endothelium and supported by trabeculae of connective tissue. Aqueous humor from the anterior cavity (AC) fills the channels of this trabecular meshwork (TM) and is pumped by endothelial cells into an adjacent space, the scleral venous sinus (SVS). X50. H&E. **(b):** The SEM surface view shows that the transition from corneal endothelium (CE) to trabecular meshwork (TM) is gradual and the channels formed are large. Movement of aqueous humor into this corner formed by the iris (I) and the trabecular meshwork—the iridocorneal angle—for removal via the scleral venous sinus is of major importance in regulating intraocular pressure. Factors causing impaired aqueous removal lead to glaucoma, a condition in which elevated pressure affects proper function of the retina and vision. X300.

cellular region also contains numerous phagocytic vacuoles and secondary lysosomes, peroxisomes, and abundant smooth ER, with specialized regions in these cells for isomerization of *all-trans*-retinal (derived from vitamin A) and its transport to the photoreceptors. The diverse functions of the cells in the retinal pigmented epithelium include the following:

- serve as an important part of the blood-retina barrier,
- absorb light passing through the retina to prevent its reflection,
- phagocytose shed components from the adjacent rods and cones,
- remove free radicals, and
- isomerize and regenerate the retinoids used as chromophores by the rods and cones.

MEDICAL APPLICATION

*The pigmented epithelium and the photoreceptor layer of the retina, derived from the two layers of the optic cup, are not firmly joined to one another. Head trauma or other conditions can cause the two layers to separate with an intervening space. In such regions of **detached retina** the photoreceptor cells no longer have access to metabolic support from the pigmented layer and choroid and will eventually die. Prompt*

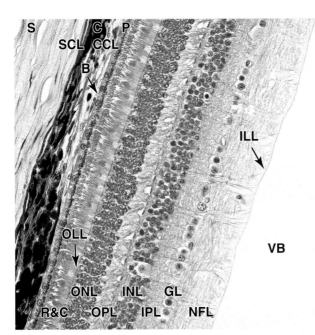

Figure 23–7. Sclera, choroid, and retina. This section of the wall of an eye shows the dense connective tissue of the sclera (S) and the loose, vascular connective tissue of the choroid (C). Melanocytes are prominent in the choroid, especially in its outer region, the suprachoroidal lamina (SCL). The choroid's inner region, the choroidocapillary lamina (CCL), has a rich microvasculature which helps provide O_2 and nutrients to the adjacent retina. Between the choroid and the retina is a thin layer of extracellular material known as Bruch's layer (B). The outer layer of the retina is the pigmented layer (P) of cuboidal epithelium containing melanin. Adjacent to this are the packed photoreceptor components of the rods and cones (R&C), the cell bodies of which make up the outer nuclear layer (ONL). Junctional complexes between these cells are aligned and can be seen as a thin line called the outer limiting layer (OLL). Axons of the rods and cones extend into the outer plexiform layer (OPL) forming synapses there with dendrites of the neurons in the inner nuclear layer (INL). These neurons send axons into the inner plexiform layer (IPL), where they synapse with dendrites of cells in the ganglionic layer (GL). Axons from these cells fill most of the nerve fiber layer (NFL) which is separated by the inner limiting layer (ILL) from the gelatin-like connective tissue of the vitreous body (VB). X200. H&E.

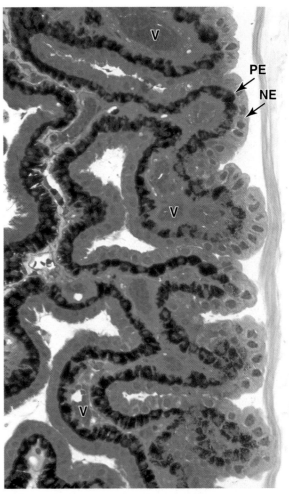

Figure 23–8. Epithelium of ciliary processes. This section of ciliary processes shows that their surface epithelium is a double layer of pigmented (PE) and nonpigmented epithelial (NE) low columnar or cuboidal cells. The two layers are derived developmentally from the folded rim of the embryonic optic cup, so that the exposed surface of the nonpigmented layer is actually the basal surface of the cells. No true basal lamina is present, but instead these cells produce the components that give rise to the fibers of the ciliary zonule in the embryo. Beneath the double epithelium is a core of connective tissue with many small blood vessels (V). Fluid from these vessels is pumped by the epithelial cells out of the ciliary processes as aqueous humor. X200. PT.

repositioning of the retina and reattaching it with laser surgery is an effective treatment.

The posterior, photosensitive part of the retina is a complex structure containing more than 30 subtypes of neurons interconnected via synapses. The neural retina has three major layers of neurons (Figures 23–7 and 23–14): an outer layer of photosensitive cells, the **rod and cone cells**; an intermediate layer of **bipolar neurons**, which connects to the rod and cone cells; and an internal layer of **ganglion cells**, which synapse with the bipolar cells through their dendrites and send out axons that converge to form the **optic nerve** which leaves the eye and passes to the brain.

Between the rod and cone cell layer and the bipolar cells is a region called the **outer plexiform layer** that contains fibers and synapses connecting the neurons in these two cellular layers. The similar region of synapses between the bipolar and ganglion cells is called the **inner plexiform layer** (Figures 23–7 and 23–15). The retina has an inverted structure, with the light first passing through the ganglion layer and then the bipolar layer to reach the rod and cone cells (Figure 23–14).

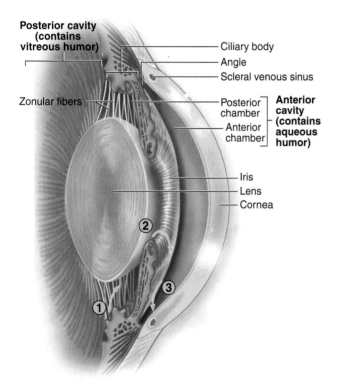

Figure 23–9. Production and removal of aqueous humor. Aqueous humor is a continuously flowing liquid that carries metabolites to and from cells and helps maintain an optimal microenvironment within the anterior cavity of the eye. Epithelial cells covering the ciliary body secrete the aqueous into the posterior chamber of the anterior cavity (1), from which it flows past the lens and through the pupil into the anterior chamber of that cavity (2). The fluid then drains into the iridocorneal angle and is removed at the scleral venous sinus (3), which is continuous with veins in the sclera.

The rods and cones, named for the shape of their outer segments, are polarized neurons. At one pole is a single photosensitive dendrite and at the other are synapses with cells of the bipolar layer. The rod and cone cells can be divided into outer and inner segments, a nuclear region, and a synaptic region (Figures 23–13 and 23–15). The **outer segments** are modified primary cilia and contain stacks of membranous saccules shaped as flattened disks. The photosensitive pigments of the retina are located in the membranes of these saccules. Both rod cells and cone cells pass through a thin layer, the **outer limiting layer**, which consists of a series of junctional complexes between the photoreceptors and the organizing glial cells of the retina called **Müller cells** (see below).

ROD CELLS

The human retina has approximately 120 million **rod cells**. They are extremely sensitive to light, responding to a single photon, and allow some vision even with light low levels, such as at dusk or nighttime. Rod cells are thin, elongated cells (50 μm × 3 μm), composed of two distinct segments (Figure 23–16). The outer segment is photosensitive; the inner segment contains the metabolic machinery for the cell's biosynthetic and energy-producing processes. The outer rod-shaped segment consists mainly of 600–1000 flattened **membranous discs** stacked like coins and surrounded by the plasma membrane. Between this outer segment and the cell's inner segment is a constriction, the **connecting stalk**, which is a modified cilium arising from a basal body (Figure 23–17). The inner segment is rich in glycogen and mitochondria near the base of this cilium (Figures 23–16 and 23–17). Abundant polyribosomes located inside the mitochondrial region produce proteins that are transported to the outer segment, where they are incorporated into the membranous discs. These proteins include the visual pigment **rhodopsin** (**visual purple**) which is bleached by light and initiates the visual stimulus.

The membranous discs form by repetitive in-folding of the plasma membrane near the connecting stalk and inserting into the lipid bilayers proteins transported there from the inner segment. The newly assembled discs in rod cells detach from the plasma membrane and are displaced distally as new discs form. Eventually discs arrive at the end of the rod, where they are shed, phagocytosed, and digested by the cells of the pigmented epithelium (Figure 23–13). Each day approximately 90 membranous discs are produced and 90 are lost from each rod, with the process of assembly, distal movement, and apical shedding taking about 10 days.

CONE CELLS

The human retina has 6 or 7 million cone cells, which are less sensitive to low light than rod cells and are specialized for color vision in bright light. Three functional types of cone cells, not distinguishable morphologically, contain variations of the visual pigment **iodopsin** with maximal sensitivities in the red, blue, or green regions of the visible spectrum, which enables these cells to detect those colors in reflected light.

Cone cells (Figure 23–16) are also elongated, with outer and inner segments, a modified cilium connecting stalk, and an accumulation of mitochondria and polyribosomes. The outer segments of cones differ from those of rods in their shorter, more conical form and in the structure of their stacked membranous disks, which in cones remain as continuous invaginations of the plasma membrane along one side (Figure 23–16). Also, newly synthesized membrane proteins are distributed uniformly throughout the outer segment of cones and although iodopsin turns over, the discs are shed much less frequently than in rods.

PHOTOTRANSDUCTION

The stacked membranous discs of the rod and cone cell outer segments are arranged in parallel with the retinal surface, which maximizes their exposure to light. The membranes in the discs of rods and cones are very densely packed with rhodopsin or one of the iodopsin proteins respectively, with each rod containing about a billion rhodopsin molecules. Each of these visual pigments contains a transmembrane protein, the **opsin**, with a bound molecule of retinal, the light-sensitive **chromophore**. Rhodopsin and each of the three iodopsins absorb light most effectively at different specific wavelengths in the visible spectrum. **Phototransduction** involves a cascade of changes in the cells triggered when light hits and activates retinal and is basically similar in both rod and cone cells.

As diagrammed for a rod in Figure 23–18, in darkness rhodopsin is not active and cation channels in the cell membrane are open. The cell is depolarized and continuously releases neurotransmitter at the synapse with the bipolar neurons. When photons of light are absorbed by the retinal of rhodopsin, the molecule changes its conformation from 11-*cis*-retinal to all-*trans*-retinal. This activates the opsin which in turn activates the adjacent molecules of **transducin**, a G protein to which opsin is coupled. The activity of transducin then indirectly stimulates many of the sodium channels to close, producing hyperpolarization which reduces the synaptic release of neurotransmitter. This change in turn depolarizes sets of bipolar neurons, which send action potentials to the various ganglion cells of the optic nerve.

The conformation change induced by light in retinal which initiates the cascade of events producing neural activity also causes the chromophore to dissociate from the opsin. This is called **bleaching** of the protein (Figure 23–18). The free all-*trans*-retinal diffuses into the pigmented epithelium where it is converted back to 11-*cis*-retinal, then transported back into a rod or cone to combine again with opsin. This cycle of retinal regeneration and rhodopsin recovery from bleaching may take a minute or more and is part of the slow adaptation of the eyes that occurs when moving from bright to dim light.

OTHER NEURONS AND GLIA

The inner nuclear layer of bipolar cells consists mainly of various **bipolar neurons** which have processes extending into the inner and outer plexiform layers and forming synaptic connections with neurons in all layers of the retina. Also having their nuclei in the inner nuclear layer are **horizontal cells** and

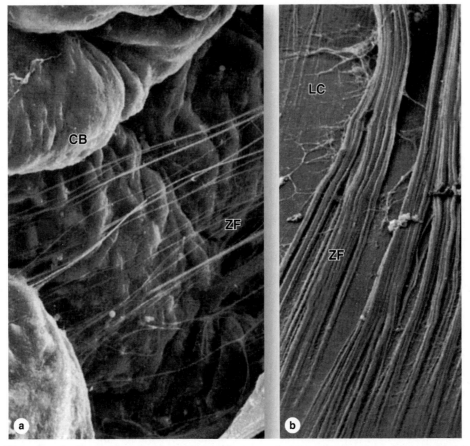

Figure 23–10. Ciliary zonule fibers. The structure of the ciliary zonule is best studied by scanning EM. **(a):** This micrograph shows the surface of the ciliary body (CB) and its projecting ciliary processes, between which emerge delicate zonular fibers (ZF). A large array of these fibers constitutes the zonule which anchors the lens in the center of the ciliary body. X400. **(b):** This micrograph shows the surface of the lens with the firm, tangential attachment of the zonular fibers to the fibrous ECM of the lens capsule (LC). X500.

amacrine cells, both with processes that mainly spread horizontally in the plexiform layers and which integrate signals from photoreceptors over a wide area of the retina.

The major supportive neuroglial cells of the retina are large, ramified **Müller cells** which provide scaffolding for the neurons of the entire retina. The nuclei of these cells are located in the inner nuclear layer and their processes extend from the inner to the outer limiting layers. The latter layer is a thin zone of adherent and occluding junctions between the photoreceptors and Müller cells. Microglial cells are also scattered throughout the retina.

The **ganglion cells** in the innermost ganglionic layer are typical nerve cells, containing large euchromatic nuclei and basophilic Nissl bodies. These cells synapse with both bipolar and amacrine cells and project their axons to the nerve fiber layer, where they come together to form the **optic nerve** (Figures 23–1 and 23–14). A subset of ganglion cells are themselves photoreceptors,

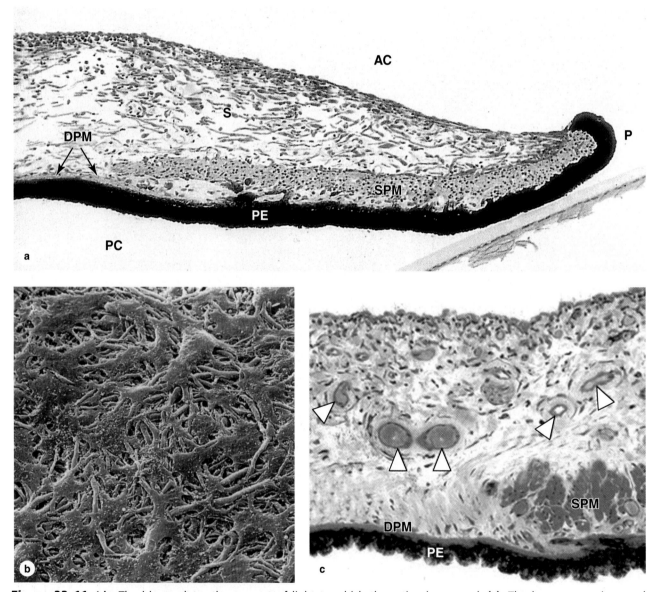

Figure 23–11. **Iris.** The iris regulates the amount of light to which the retina is exposed. **(a):** The low-power micrograph shows a section of the central iris, near the pupil (P). The anterior surface, exposed to aqueous humor in the anterior chamber (AC), has no epithelium and consists only of a matted layer of interdigitating fibroblasts and melanocytes. X140. H&E. The underlying stroma (S) contains many melanocytes with varying amounts of melanin. **(b):** The SEM shows the non-epithelial anterior surface of the iris. X900. **(c):** This micrograph shows that the deep stroma also is richly vascularized (arrowheads). The epithelium on the posterior side of the iris, adjoining the posterior chamber (PC), consists of two layers of cuboidal cells. Cells of the external pigmented epithelium (PE) are very rich in melanin granules to protect the eye's interior from an excess of light. Cells of the other layer are myoepithelial, less heavily pigmented, and comprise the dilator pupillae muscle (DPM) which extends along most of the iris. Near the pupil, fascicles of smooth muscle make up the sphincter pupillae muscle (SPM). Together the two muscles control the diameter of the pupil. X100. PT.

containing 11-*cis*-retinal bound to the protein melanopsin, involved not with vision but in detecting the changes in light quantity and quality during each 24 hr dawn/dusk cycle. Signals from these cells pass via axons of the retinohypothalamic tract to the suprachiasmatic nuclei of the hypothalamus and are important in establishing the body's physiological circadian rhythms (see Chapter 20).

SPECIALIZED AREAS OF THE RETINA

The posterior area of the retina where the optic nerve leaves the retina is devoid of photoreceptors and is known as the blind spot of the retina, or the **optic disc** (Figure 23–14).

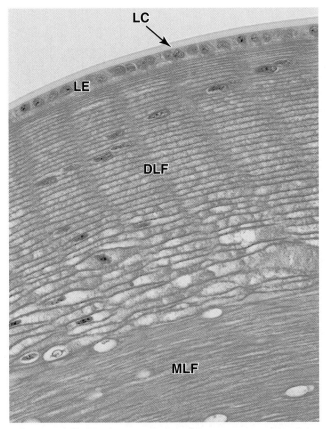

Figure 23–12. Lens. The lens is a transparent, elastic tissue that focuses light on the retina. Surrounding the entire lens, the lens capsule (LC) is a thick, homogenous external lamina formed by the epithelial cells and fibers. The anterior surface of the lens, beneath the capsule, is covered by a simple columnar lens epithelium (LE). Because of its origin as an embryonic vesicle pinching off of surface ectoderm, the basal ends of the lens epithelial cells rest on the capsule and the apical regions are directed into the lens interior. At the equator of the lens, near the ciliary zonule, the epithelial cells proliferate and give rise to cells that align parallel to the epithelium and become the lens fibers. Differentiating lens fibers (DLF) still have their nuclei, but are greatly elongating and filling their cytoplasm with proteins called crystallins. The mature lens fibers (MLF) have lost their nuclei and become densely packed to produce a unique transparent structure. The lens is difficult to process histologically and sections usually have cracks or blebs among the lens fibers. X200. H&E.

On the temporal side of the optic disc, at the posterior pole of the optical axis, lies a specialized area of the retina called the **fovea centralis** (Figure 23–14). The fovea (L. *fovea*, a small pit) is a shallow depression having only cone cells at its center, with the bipolar and ganglion cells located only at the periphery. Cone cells in the fovea are long and narrow, an adaptation that permits closer packing of the cones and thereby increases visual acuity. Blood vessels do not cross over this area and light falls directly on the cones in the central part of the fovea, which helps account for the extremely precise visual acuity of this region. Surrounding the fovea centralis is the **macula lutea** (L. *macula*, spot; *lutea*, yellow), or macula, 5.5 mm in diameter (Figure 23–14). Here all layers of the retina are present and the two plexiform layers are rich in various carotenoids, which give this area its yellowish color. The carotenoids have antioxidant properties and filter potentially damaging short-wavelength light, thus helping to protect the cone cells of the fovea.

MEDICAL APPLICATION

*A leading cause of blindness in elderly individuals of developed countries is **age-related macular degeneration**, which causes blindness in the center of the visual field. Degenerative changes in the retina around the macula include depigmentation of the posterior epithelium, focal thickening of the adjacent Bruch's membrane, major changes and blood loss in the capillaries in the choroid and retina, and eventual loss of the photoreceptor cells producing blind spots. There appears to be a genetic predisposition to the disorder, along with environmental triggers such as excessive exposure to ultraviolet radiation. Progression of the disease can be slowed by laser surgery to destroy the abnormal and excessive retinal capillaries.*

Accessory Structures of the Eye

CONJUNCTIVA

The **conjunctiva** is a thin, transparent mucosa that covers the exposed, anterior portion of the sclera and continues as the lining on the internal surface of the eyelids. It consists of a stratified columnar epithelium, with numerous small cells resembling goblet cells, supported by a thin lamina propria of loose vascular connective tissue (Figure 23–19). Mucous secretions from epithelial cells of the conjunctiva are added to the tear film coating this epithelium and the cornea.

EYELIDS

Eyelids (Figure 23–19) are pliable structures containing skin, muscle, and conjunctiva that protect the eyes. The skin is present only on the external surface. It is loose and elastic, lacking fat, and has very small hair follicles and fine hair, except at the distal edge of the eyelid, where large follicles forming eyelashes are present. Associated with the follicles of eyelashes are sebaceous glands and modified apocrine sweat glands.

Deep to the skin are fascicles of striated muscle that make up the orbicularis oculi and levator palpebrae muscles which fold the eyelids. Adjacent to the conjunctiva is a dense fibroelastic plate of connective tissue called the **tarsus** which provides support for the other tissues of the eyelids. This tissue also contains a series of 20–25 large sebaceous glands, each with many acini secreting into a long central duct that opens among the eyelashes at the eyelid's distal margin (Figure 23–19). Oils in the sebum produced by these **tarsal glands**, commonly called **Meibomian glands**, form a surface layer on the tear film, reducing its rate of evaporation, and help lubricate the ocular surface.

MEDICAL APPLICATION

Infections near an opening of the tarsal gland ducts, generally caused by Staphylococcus aureus, *are called **styes**. They are most common in infants, but can occur at any age and can be quite painful. Like certain other infections, styes can occur in periods of immunosuppression caused by poor nutrition or stress.*

LACRIMAL GLANDS

The **lacrimal glands** produce fluid continuously for the tear film that moisturizes and lubricates the cornea and conjunctiva and supplies O_2 to the corneal epithelial cells. Tear fluid also contains various metabolites, electrolytes, and proteins, including lysozyme, an enzyme that hydrolyzes the cell walls of certain

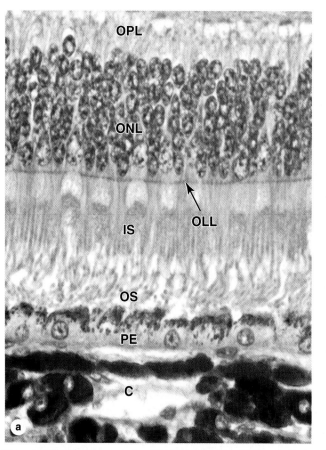
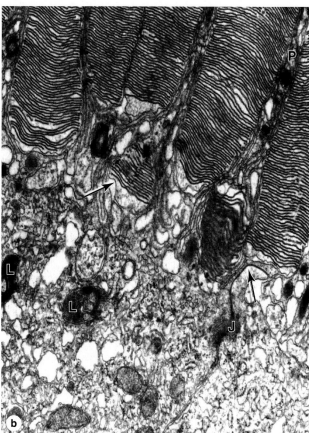

Figure 23–13. **Pigmented epithelium of retina.** The two distinct layers of the retina are the pigmented epithelium and the photosensitive layer, which are derived from the outer and inner layers of the optic cup respectively. **(a):** The light micrograph shows the interface between the two layers. The pigmented epithelium (PE) is of simple cuboidal cells resting on Bruch's membrane inside the choroid (C). Rod cells and cone cells are neurons with their nuclei collected in the outer nuclear layer (ONL) and with axons of one end forming synapses in an area called the outer plexiform layer (OPL) and modified dendrites at the other end serving as photosensitive structures. These structures have mitochondria-rich inner segments (IS) and photosensitive outer segments (OS) with stacks of folded membranes where the visual pigments are located. The inner segments of the rod and cone cells are attached to elongated glial cells called Muller cells, which are modified astrocytes of the retina. The junctional complexes of these attachments can be seen in light micrographs as the outer limiting layer (OLL). X500. H&E.

(b): The TEM shows an ultrastructural view of the interface between the pigmented epithelial cells and the outer segments of the photoreceptive cells. Junctional complexes (J) occur between lateral membranes of the epithelial cells. Above these cells are the tips of five outer segments of rod cells that interdigitate with apical processes (P) of the epithelial cells. The large vacuoles contain folded membrane stacks (arrows) that have been shed from the tips of the rods. Contents of these vacuoles are digested after fusion with secondary lysosomes (L). Also seen are mitochondria and segments of rough and smooth ER. X24,000.

species of bacteria, facilitating their destruction. The main lacrimal glands are located in the upper temporal portion of the orbit and have several lobes that drain separately through excretory ducts into the superior fornix, the conjunctiva-lined recess between the eyelids and the eye. After moving across the ocular surface, the fluid secreted by these glands collects in other parts of the bilateral **lacrimal apparatus**: flowing into two small round openings (0.5 mm in diameter) to canaliculi at the medial margins of the upper and lower eyelids and then passing into the lacrimal sac and finally draining into the nasal cavity via the nasolacrimal duct. The canaliculi are lined by stratified squamous epithelium, but the more distal sac and duct are lined by pseudostratified ciliated epithelium like that of the nasal cavity.

The lacrimal glands have tubuloalveolar acini composed of tall serous cells with basal nuclei and lightly stained secretory granules, resembling histologically the acinar cells of the parotid gland (Figure 23–20). Each acinus is surrounded by well-developed myoepithelial cells and a basal lamina and empties into a duct system leading to the excretory ducts.

EARS: THE VESTIBULOAUDITORY SYSTEM

The functions of the ear are related to both maintaining equilibrium and hearing. Ears consist of three major parts (Figure 23–21): the **external ear**, which receives sound waves; the **middle ear**, in which sound waves are transmitted from air to fluids of the internal ear via a set of small bones; and the **internal ear,** in which these fluid movements are transduced to nerve impulses that pass via the acoustic nerve to the CNS. In addition to the auditory organ, the internal ear also contains the vestibular organ which allows the body to maintain equilibrium.

External Ear

The **auricle**, or **pinna** (L. *pinna*, wing) is an irregular, funnel-shaped plate of elastic cartilage, covered by tightly adherent skin, which directs sound waves into the ear.

The waves enter the **external acoustic meatus** (L. *meatus*, passage), a canal extending from the lateral surface of the head. It is lined with stratified squamous epithelium continuous with the skin of the auricle and near its opening hair follicles, sebaceous glands, and modified apocrine sweat glands called **ceruminous glands** are found in the submucosa (Figure 23–22). **Cerumen** is the oily or waxy, yellowish material resulting from secretions of the sebaceous and ceruminous glands. It contains various proteins, saturated fatty acids, and sloughed keratinocytes and has protective, antimicrobial properties. The wall of the external auditory meatus is supported by elastic cartilage in its outer third, while the temporal bone encloses the inner part (Figure 23–21).

Across the deep end of the external acoustic meatus lies an epithelial sheet called the **tympanic membrane** or eardrum. Its external side is covered with epidermis and its inner surface is

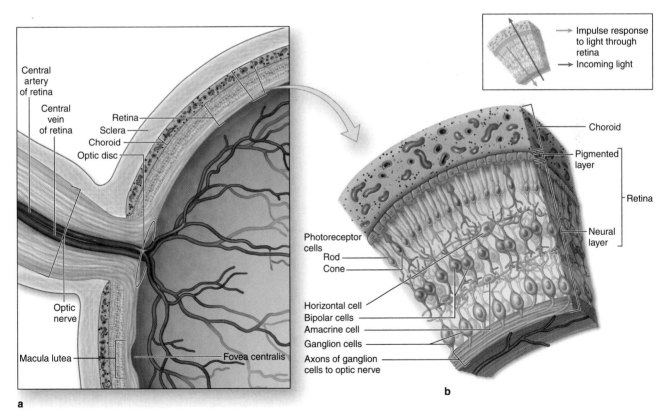

Figure 23–14. **General structure and organization of the retina.** The retina is the thick layer of the eye inside the choroid. **(a):** The diagram shows the central retinal artery and vein that pass through the optic nerve and enter the eye at the optic disc. These vessels initially lie between the vitreous body and the inner limiting layer of the retina, but their smaller lateral branches penetrate this layer and enter the retina, forming capillaries that extend as far as the inner nuclear layer. Nutrients and O₂ for the outer retinal layers diffuse from capillaries in the choroid. **(b):** This diagram illustrates the major neurons and their general organization. The long glial cells that help organize the neurons are not shown.

covered with simple cuboidal epithelium continuous with the lining of the tympanic cavity in the middle ear (see below). Between the two epithelial coverings is a thin sheet of fibrous connective tissue layer composed of collagen, elastic fibers and

fibroblasts. Vibrations of the tympanic membrane produced by sound waves transmit sound wave energy to the middle and inner ear (Figure 23–21).

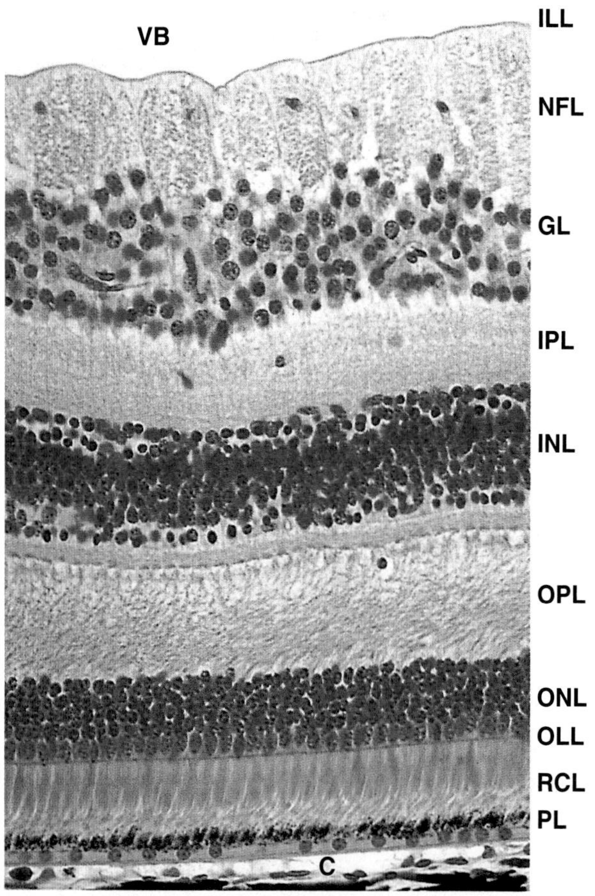

Middle Ear

The middle ear contains the air-filled **tympanic cavity**, an irregular space that lies within the temporal bone between the tympanic membrane and the bony surface of the internal ear (Figure 23–21). Anteriorly this cavity communicates with the pharynx via the **auditory tube** (**Eustachian** or **pharyngotympanic tube**) and posteriorly with the smaller, air-filled mastoid cavities of the temporal bone. The tympanic cavity is lined mainly with simple cuboidal epithelium resting on a thin lamina propria that is strongly adherent to periosteum. Near the auditory tube, this simple epithelium is gradually replaced by the ciliated pseudostratified columnar epithelium lining the tube.

Figure 23–15. Layers of the retina. Between the vitreous body (VB) and the choroid (C), the retina can usually be seen to have ten distinct layers. Following the path of the light, these are: the inner limiting layer (ILL); the nerve fiber layer (NFL), containing the ganglionic cell axons that converge at the optic disc and form the optic nerve; the ganglionic layer (GL), containing cell bodies of the ganglion cells and of somewhat variable thickness throughout the retina; the inner plexiform layer (IPL), containing fibers and synapses of the ganglion cells and the bipolar neurons of the next layer; the inner nuclear layer (INL), with the cell bodies of several types of bipolar neurons which begin to integrate signals from the rod and cone cells; the outer plexiform layer (OPL), containing fibers and synapses of bipolar neurons and rod and cone cells; the outer nuclear layer (ONL), with the cell bodies and nuclei of the photosensitive rod and cone cells; the outer limiting layer (OLL), which is a fine line formed by the junctional complexes holding the rod and cone cells to the intervening glia called Müller cells; the rod and cone cell layer (RCL), which contains the outer segments of these cells where the photoreceptors are located; and the pigmented layer (PL) which is not sensory, but has several supportive functions important for maintenance of the neural retina. X150. H&E.

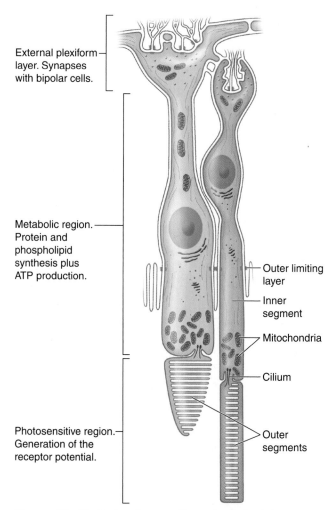

Figure 23–16. Rod and cone cells. Rod cells (right) and cone cells (left) all have the general shapes and important cytoplasmic features shown diagrammatically here. The outer limiting layer is the line of adherent junctions that attach the rod and cone inner segments to the distal ends of the Müller glial cells (not shown). The membranous discs of the cone outer segments are continuous with the cell membrane; those of the rods are not.

Although the walls of the tube are usually collapsed, it opens during the swallowing process, which serves to balance the air pressure in the middle ear with atmospheric pressure. In the medial bony wall of the middle ear are two membrane-covered regions devoid of bone: the **oval** and **round windows** (Figure 23–21).

The tympanic membrane is connected to the oval window by a series of three small bones, the **auditory ossicles**, which transmit the mechanical vibrations of the tympanic membrane to the internal ear (Figure 23–21). The ossicles are named the **malleus, incus,** and **stapes,** the Latin words respectively for "hammer," "anvil," and "stirrup," which reflect each bone's general shape. The malleus is attached to connective tissue of the tympanic membrane and the stapes to that of the membrane in the oval window. The ossicles articulate at synovial joints, which

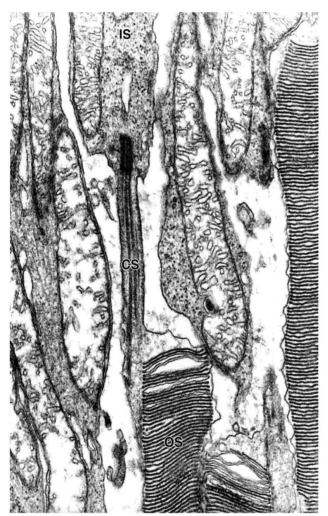

***Figure 23–17.* Connection between inner and outer segments.** A TEM of a sectioned retina shows the inner segments (IS) with mitochondria in the upper part of the figure and the outer photosensitive segment (OS) consisting of flat, parallel membranous disks. The cell in the middle of the figure shows a basal body giving rise to a cilium that forms the connecting stalk (CS) and is further modified distally as the outer segment. The stacked membranes of the discs are very distinct and electron-dense due to the high density of proteins they contain. X24,000.

along with periosteum are completely covered with simple squamous epithelium. Two small skeletal muscles insert into the malleus and stapes, restricting movement of the ossicles and helping to protect the internal ear from extremely loud noises.

Internal Ear

The internal ear is located completely within the temporal bone, where an intricate set of interconnected spaces, the **bony labyrinth,** houses a set of continuous fluid-filled, epithelium-lined tubes and chambers that make up the smaller **membranous labyrinth** (Figure 23–21). The membranous labyrinth is derived from an ectodermal vesicle, the otocyst, which invaginates into the subjacent connective tissue during the fourth week of embryonic development, loses contact with the surface ectoderm, and becomes embedded in the rudiments of the future temporal bone. During this process, the otic vesicle changes shape, giving rise to two major divisions of the membranous labyrinth:

- the **vestibular labyrinth,** which mediates the sense of equilibrium and consists of two connected sacs (the **utricle** and the **saccule**) and three **semicircular ducts** arising from the utricle, and
- the **cochlear labyrinth,** which provides for hearing and contains the **cochlear duct** connected to the saccule.

In each of these structures the epithelial lining contains large areas of columnar sensory mechanoreceptors called **hair cells** in specialized regions:

- two **maculae** of the utricle and saccule,
- three **cristae ampullaris** in the enlarged ampullary regions of each semicircular duct, and
- the long **spiral organ of Corti** in the cochlear duct.

The bony labyrinth has an irregular central cavity, the **vestibule,** where the saccule and the utricle are located. Behind this, three osseous **semicircular canals** enclose the semicircular ducts. On the other side of the vestibule, the **cochlea** (L. *cochlea,* snail, screw) contains the cochlear duct (Figure 23–23). The cochlea is about 35 mm in length and makes two-and-one-half turns around a bony core called the **modiolus.** The modiolus contains blood vessels and surrounds the cell bodies and processes of the acoustic branch of the eighth cranial nerve in the large **spiral** or **cochlear ganglion**.

All regions of the bony labyrinth are filled with **perilymph,** which is similar in ionic composition to cerebrospinal fluid and the extracellular fluid of other tissues, but contains little protein. Perilymph emerges from the microvasculature of the periosteum and is drained by a perilymphatic duct into the adjoining subarachnoid space. This fluid suspends and supports the closed membranous labyrinth, protecting it from the hard wall of the bony labyrinth. The membranous labyrinth is filled with **endolymph,** which also contains few proteins and is further characterized by a high potassium (150 mM) and low sodium (16 mM) content, similar to that of intracellular fluid. Endolymph is generated largely by capillaries in the stria vascularis in the wall of the cochlear duct and is drained from the vestibule into venous sinuses of the dura mater by the small endolymphatic duct.

SACCULE AND UTRICLE

The **saccule** and the **utricle** are composed of a very thin connective tissue sheath lined with simple squamous epithelium. The membranous labyrinth is bound to the periosteum of the bony labyrinth by strands of connective tissue that contain microvasculature supplying the tissues of the membranous labyrinth. The two maculae in the walls of the saccule and utricle are small areas of columnar neuroepithelial cells that are innervated by branches of the vestibular nerve (Figure 23–24). The macula of the saccule lies in a plane perpendicular to the macula of the utricle, but both are similar histologically. Each consists of a thickening of the wall containing several thousand mechanosensitive hair cells, along with columnar supporting cells with basal nuclei, and nerve endings.

The apical end of each hair cell has a single **kinocilium** with a basal body and a modified axoneme of microtubule doublets (Chapter 2) and a bundle of 60–100 long, rigid, unbranched **stereocilia**. The stereocilia arise from an actin-rich apical region,

the cuticular plate, which serves to return these rigid projecting structures to a normal upright position after bending. They are arranged in rows of increasing length, with the longest—about 100 μm—located adjacent to the kinocilium (Figure 23–24). The tips of the stereocilia and kinocilium are embedded in a thick, gelatinous layer of proteoglycans called the **otolithic membrane**, the outer part of which is filled with calcified structures called otoliths (or otoconia) (Figure 23–25).

At their basal ends all hair cell have synapses with afferent (to the brain) nerve endings (Figure 23–26). Some hair cells (type I) have rounded basal ends surrounded by an afferent terminal calyx (L, *calyx*, cup, chalice). The basal ends of most hair cells (type II) are cylindrical and have more typical bouton endings

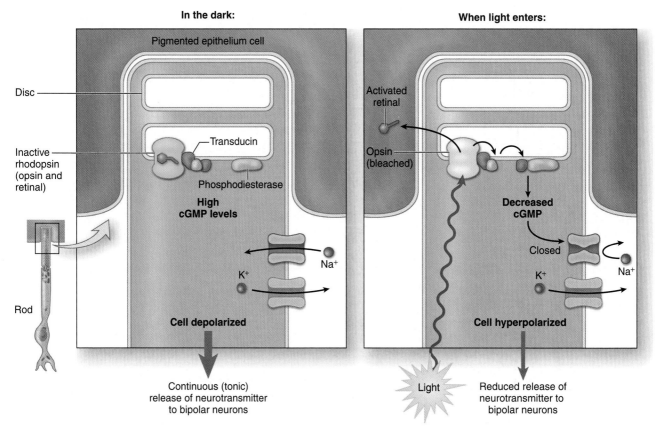

Figure 23–18. Rod cell phototransduction. Phototransduction involves a series of changes in rod and cone cells that begins when light hits the stacked membranous discs. The main parts of the process are similar in both rods and cones, but have been better studied in the more abundant rod cells, as shown here. Membranes of the discs are densely packed with proteins, although only one of each major type is shown here. In the dark rhodopsin and its 11-*cis*-retinal are inactive and the intracellular concentration of the second messenger cyclic GMP (cGMP) is high. One effect of cGMP is to keep open the abundant cation channels in the cell membrane and therefore the cell is depolarized, continuously releasing its neurotransmitter (glutamate) at the synapse with the bipolar neurons.

When photons of light are absorbed by the retinal of rhodopsin, the molecule isomerizes from 11-*cis*-retinal to all-*trans*-retinal and this change activates the opsin. This in turn activates the adjacent peripheral membrane protein transducin, a trimeric G protein, allowing it to release its α subunit, which moves laterally and stimulates another membrane protein, phosphodiesterase, to hydrolyze cGMP. With less cGMP, many of the sodium channels now close, producing hyperpolarization of the cell which decreases the release of neurotransmitter at the synapses. This change at the synapse depolarizes sets of bipolar neurons which then send action potentials to the various ganglion cells of the optic nerve that will allow the brain to produce an image. When retinal is activated by light it also dissociates from rhodopsin, leaving a more pale-colored (bleached) opsin. The free retinal moves into the surrounding pigmented epithelial cells, where the all-*trans* isomer is regenerated. It is then transported back into a rod or cone cell to again bind opsin and be used in another round of phototransduction.

from afferent nerves. Both types of hair cells, or their afferents, also have synaptic connections with efferent (from the brain) fibers that modulate the sensitivity of these mechanoreceptors (Figure 23–26). Each hair cell is also surrounded by supporting cells, which may have various functions besides providing physical support for the mechanoreceptors.

SEMICIRCULAR DUCTS

The three **semicircular ducts** are parts of the membranous labyrinth having the same general form as the semicircular canals

in the bony labyrinth. Each extends from and returns to the wall of the utricle. They lie in three different spatial planes, at approximately right angles to one another (Figure 23–23).

The enlarged **ampulla** end of each semicircular duct has an elongated ridgelike area of mechanoreceptors called the **crista ampullaris** (Figure 23–27). The ridge of each crista ampullaris is perpendicular to the long axis of the duct. Cristae are histologically similar to maculae, with hair cells, supporting cells, and nerve endings. However, the proteoglycan layer called the **cupola** attached to the sensory cells' hair bundles is thicker and does not

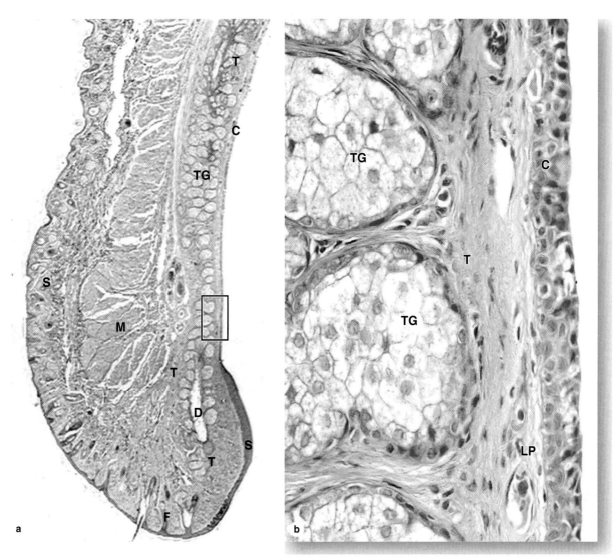

Figure 23–19. Eyelid. (a): The eyelid is a pliable tissue with skin (S) covering its external surface and smooth conjunctiva (C) lining its inner surface. At the outer rim of the eyelid are a series of large hair follicles (F) for the eyelashes. Associated with these hair follicles are small sebaceous glands and modified apocrine sweat glands. Internally eyelids contain fascicles of striated muscle (M) comprising the orbicularis oculi muscle and closer to the conjunctiva a thick plate of fibroelastic connective tissue called the tarsus (T). This tarsal plate provides structural support for the eyelid and surrounds a series of large sebaceous glands, the tarsal glands (TG) (aka Meibomian glands), with acini secreting into long central ducts (D) that empty at the free edge of the eyelids. X12.5. H&E. **(b):** At higher magnification, only the inner aspect of the eyelid is seen and it shows that the conjunctiva is a mucous membrane consisting of a stratified columnar epithelium with small cells resembling goblet cells and resting on a thin lamina propria (LP). Large cells undergoing typical holocrine secretion are shown in the tarsal gland acini (TG), and the fibrous connective tissue in the tarsus (T) surrounding the acini. Sebum from these glands is added to the tear film and helps lubricate the ocular surface. X200. H&E.

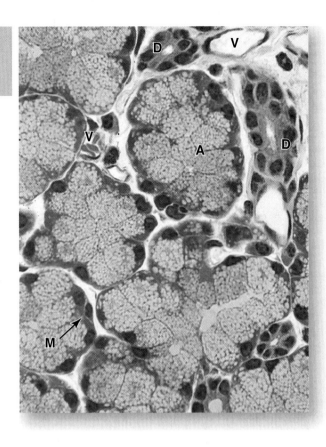

Figure 23–20. Lacrimal gland. Lacrimal glands secrete most components of the tear film that moisturizes, lubricates, and helps protect the eyes. The glands have tubuloalveolar acini (*A*) composed of secretory cells filled with small, light-staining granules and myoepithelial cells (M). Connective tissue surrounding the acini contains blood vessels (V) of the microvasculature and intra- and interlobular ducts (D) converging as excretory ducts that empty into the superior conjunctival fornix between the upper eyelid and the eye. X400. H&E.

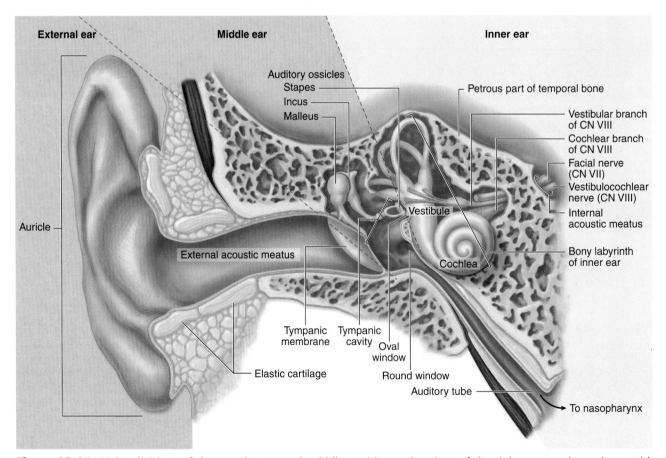

Figure 23–21. Major divisions of the ear. The external, middle, and internal regions of the right ear are shown here, with the major structures of each region.

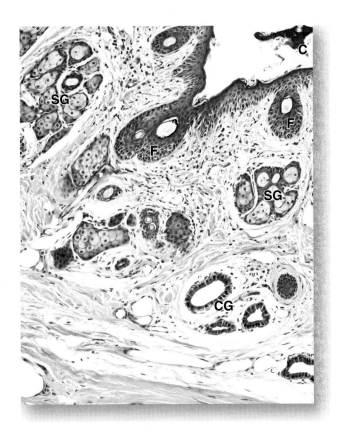

Figure 23–22. External acoustic meatus. The external acoustic meatus leads from the opening in the auricle to the tympanic membrane (eardrum). This section of the wall in the outer third of the acoustic meatus shows the lining of skin containing small hair follicles (F), sebaceous glands (SG), and modified apocrine sweat glands called ceruminous glands (CG). Secretions from these two glands form a yellowish, oily or waxy product called cerumen (C), which contains antimicrobial factors that help make the meatus uninviting for microorganisms. X50. H&E.

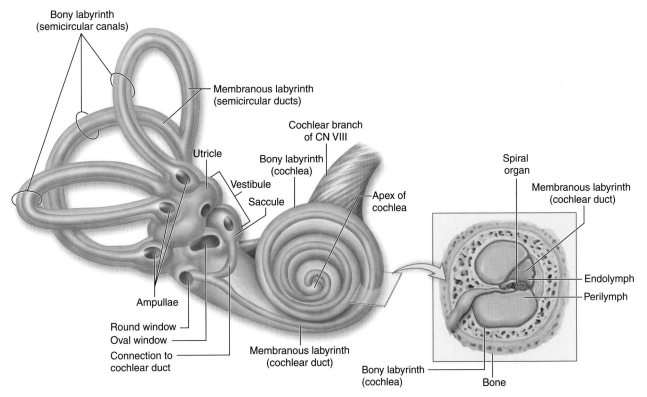

Figure 23–23. Internal ear. The internal region of the ear is composed of a cavity in the temporal bone, the bony labyrinth, which houses a fluid-filled membranous labyrinth. The membranous labyrinth includes the vestibular organs for the sense of equilibrium and balance (the saccule, utricle, and semicircular ducts) and the cochlea for the sense of hearing.

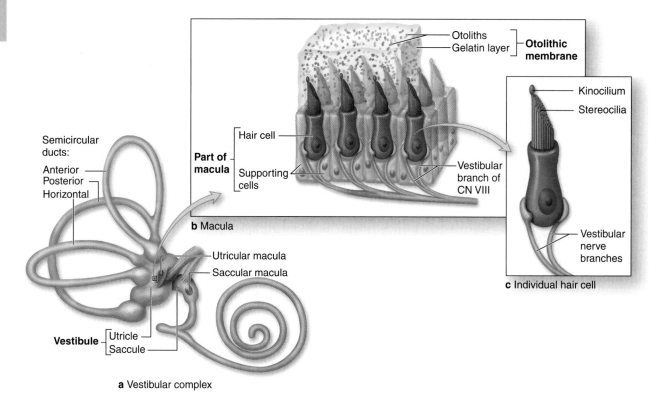

b Macula

c Individual hair cell

a Vestibular complex

Figure 23–24. **Vestibular maculae and their cells. (a):** Two sensory areas, the maculae, are located in the epithelial walls of the utricle and saccule in the vestibular complex. Both maculae are similar histologically and contain mechanoreceptor cells called hair cells which use gravity and endolymph movement to detect the orientation of the stationary head and linear acceleration of the moving head. **(b):** A detailed view of a macular wall shows that it is composed of hair cells, supporting cells, and endings of the vestibular branch of the eighth cranial nerve. The apical surface of the air cells is covered by a gelatinous otolithic layer or membrane and the basal ends of the cells have synaptic connections with the nerve fibers. **(c):** A diagram of a single generalized hair cell shows the numerous straight stereocilia, which contain bundled actin, and a longer single kinocilium, a modified cilium whose tip may be slightly enlarged.

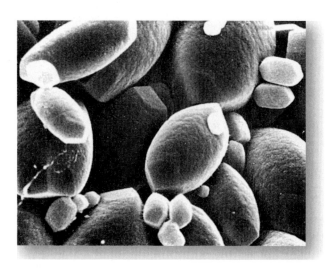

Figure 23–25. **Otoliths.** Otoliths are crystalline structures in the outer part of the otolithic membrane. Each otolith is a slightly elongated structure, up to 5 by 10 μm in size, and is composed of calcium carbonate on a matrix of proteoglycans. Their presence makes the otolithic membrane considerably heavier than endolymph alone, which facilitates bending of the kilocilia and stereocilia embedded in this membrane by gravity or movement of the head. X600. SEM. (With permission, from David J. Lim, House Ear Institute and Department of Cell & Neurobiology, University of Southern California, Los Angeles.)

contain otoliths. The cupula extends completely across the ampulla, contacting the opposite non-sensory wall (Figure 23–27).

VESTIBULAR FUNCTIONS OF THE EAR

Sensory information from the vestibular labyrinth is used, primarily in reflex mechanisms, for maintaining upright posture and balance and in allowing the eyes to stay fixed on the same point despite changes in head position. Movements of the head cause movement of the endolymph, which moves the otolith membrane of each macula and the cupula over each crista ampullaris. The sensory cell hair bundles embedded in these proteoglycan layers bend with the movement of this material, changing these cells' resting potential and their rate of neurotransmitter release to the afferent nerves. When the hair bundle is deflected *toward* the kinocilium, very small strands of protein called **tip links** that connect the stereocilia are pulled and cation channels open to allow an influx of K^+ ions (the major cation in endolymph). The resulting depolarization of the hair cell opens Ca^{2+} channels near the base of the cell and Ca^{2+} entry stimulates release of neurotransmitter (Figure 23–28). When the head stops moving, stereocilia bundles straighten to the normal position and hair cells quickly repolarize and reestablish the resting potential. Head movements that bend the stereocilia *away from* the kinocilium cause the tip links to be slack, allowing closure of the apical cation channels and hyperpolarization of the cell. This in turn closes Ca^{2+} channels at the base of the cell and reduces neurotransmitter release (Figure 23–28).

The hair cells of the cristae ampullares detect rotational or **angular movements of the head**. On each side of the head these hair cells are oriented with opposite polarity, so that turning the head causes hair cell depolarization on one side and hyperpolarization on the other. Neurons of the vestibular nuclei in the CNS receive input from the sets of semicircular ducts on each side simultaneously and interpret head rotation on the basis of the relative transmitter discharge rates of the two sides.

The hair cells in the maculae of the saccules and the utricles respond to **linear acceleration, gravity, and tilt of the head**. Because the otoliths are heavier than endolymph, the stereocilia bundles are deflected by gravity when the head is not moving, when the head is tilted with respect to gravity, and when the individual is moving in a straight line and inertia causes drag on the otolithic membrane.

Inputs from all regions of the vestibular labyrinth travel along the eighth cranial nerve to vestibular nuclei in the CNS. There

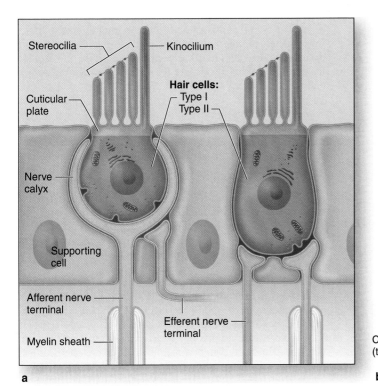

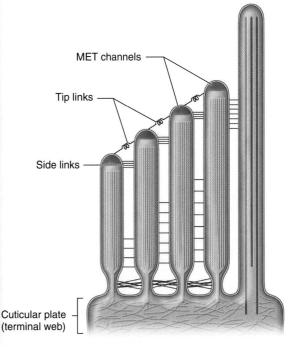

***Figure 23–26.* Hair cells and hair bundles. (a):** This diagram shows the two types of hair cells in the maculae and cristae ampullares. Basal ends of type I hair cells are rounded and enclosed within a nerve calyx on the afferent fiber. Type II hair cells are columnar and associated with typical bouton synaptic connections to their afferents. Both types are also associated with efferent fibers.

(b): A more detailed diagram of the stereocilia hair bundle of the hair cells showing that stereocilia occur in rows of increasing height, with the tallest next to the single kinocilium on one side of the cell's apical end. By TEM the end of each stereocilium shows an electron-dense region containing cation channels and proteins involved in mechanoelectric transduction (MET) that convert mechanical activity of the stereocilia to electric activity within the hair cell. Neighboring stereocilia are connected by various side links composed of proteins; the most well-understood of these are the tip links which connect the tips of stereocilia and contain very long types of cadherin proteins. Changes in the tension of the tip links caused by bending of the hair bundle open or close the adjacent cation channels and change the afferent synaptic activity of the hair cells.

they are interpreted together with inputs from mechanoreceptors of the musculoskeletal system to provide the basis for perceiving movement and orientation in space and for maintaining equilibrium or balance.

MEDICAL APPLICATION

Problems of the vestibular system can result in **vertigo**, *or dizziness, a sense of the body rotating and lack of equilibrium. This can be caused by certain infections, drugs, or tumors near the vestibular nerve. Spinning the body can produce vertigo due to overstimulation of the cristae ampullares of the semicircular ducts. Overstimulation of the maculae of the utricle caused by repetitive changes in linear acceleration and directional changes can normally lead to* **motion sickness** *(seasickness).*

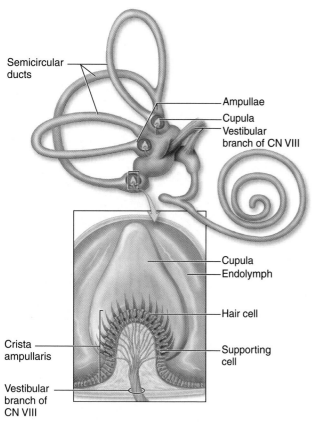

Semicircular ducts

Ampullae
Cupula
Vestibular branch of CN VIII

Cupula
Endolymph

Hair cell

Crista ampullaris

Supporting cell

Vestibular branch of CN VIII

Figure 23–27. Ampullae and cristae of the semicircular ducts. Each of the semicircular ducts has an expanded end called the ampulla. The wall of each ampulla is raised as a ridge called the crista ampullaris, a section of which is shown here diagrammatically. Hair cells of the crista epithelium resemble the two types found in the maculae, with hair bundles projecting into a dome-shape overlying layer of proteoglycan called the cupula. The cupula is attached to the wall opposite the crista and is moved by endolymph movement within the semicircular duct.

COCHLEAR DUCT AND AUDITORY FUNCTIONS

The cochlear duct, a part of the membranous labyrinth connected to the saccule, is highly specialized as a sound receptor. It is about 35 mm long, coiled two and one-half times, and is surrounded by specialized perilymphatic spaces. When observed in sections, the cochlea appears to contain three spaces: the **scala vestibuli**, the middle **cochlear duct** (or **scala media**), and the **scala tympani** (Figure 23–29). The cochlear duct contains endolymph and ends at the apex of the cochlea. The other two spaces contain perilymph and are in reality one long tube, beginning at the **oval window** and ending at the **round window** (Figure 23–23). They communicate at the apex of the cochlea via an opening known as the **helicotrema.**

Along its length, the cochlear duct is separated from the scala vestibuli by the **vestibular membrane** (Figure 23–30). This very thin structure consists of a basement membrane with simple squamous epithelium on each side: one mesothelium facing the scala vestibuli and the other part of the cochlear duct's lining. Cells of both layers have extensive tight junctions that help preserve the very high ionic gradients across this membrane between endolymph and perilymph.

In the lateral wall of the cochlear duct is the **stria vascularis** (Figure 23–30), a unique epithelium responsible for production and maintenance of the endolymph for the entire membranous labyrinth. The stria vascularis encloses a network of capillaries and consists of cells with many deep basal infoldings of their plasma membranes, where numerous mitochondria are located. Fluid and K$^+$ ions pumped from the capillaries by these epithelial cells are released in the cochlear duct as endolymph.

In the wall that separates the cochlear duct from the scala tympani is the complex structure called the **spiral organ (organ of Corti)** which contains special auditory receptors in the form of hair cells that respond to different sound frequencies. The spiral organ rests on a thick basal lamina—the basilar membrane. Two major types of hair cells are present (Figure 23–29). **Outer hair cells** (OHC) occur in three rows near the oval window, increasing to five rows near the apex of the cochlea. There is a single row of **inner hair cells** (IHC). The latter have one linear array of short stereocilia, while OHC each have a curved row of longer stereocilia (Figure 23–31). No kinocilium is present on cochlear hair cells, allowing symmetry on the cells that is important for their role in sensory transduction.

The tips of the tallest stereocilia of the OHC are embedded in the **tectorial membrane**, an acellular layer extending over the spiral organ from the modiolus (Figures 23–29 and 23–30). The tectorial membrane consists of fine bundles of collagen (types II, V, IX, and XI), associated proteoglycans and other proteins and is formed during the embryonic period from secretions of cells that come to line the adjacent region called the spiral limbus.

Both outer and inner hair cells have afferent and efferent nerve endings, with IHC much more heavily innervated. The cell bodies of the afferent bipolar neurons are located in a bony core of the modiolus and constitute the **spiral ganglion** (Figures 23–29 and 23–30).

Two major types of columnar supporting cells are associated with the hair cells of the spiral organ (Figure 23–29). **Pillar cells** are stiffened by bundles of keratin and outline a triangular, tunnel-like space between the outer and inner hair cells—another structure important in sound transduction. **Phalangeal cells** intimately surround and directly support both inner and outer hair cells, almost completely enclosing each IHC but only the basal ends of the OHC.

Stereocilia of cochlear hair cells detect movements of the spiral organ. Sound waves collected by the auricle of the external ear cause the tympanic membrane to vibrate, which causes movement of the ossicles in the middle ear (Figure 23–32). The large size of the tympanic membrane compared to the oval window and the mechanical properties of the ossicle chain connecting these two membranes allow for optimal transfer of energy between air and perilymph, from sound waves to vibrations of tissues and fluid-filled chambers.

Pressure waves within the perilymph begin at the oval window and move along the scala vestibuli. Each pressure wave causes momentary movement of the vestibular and/or basilar

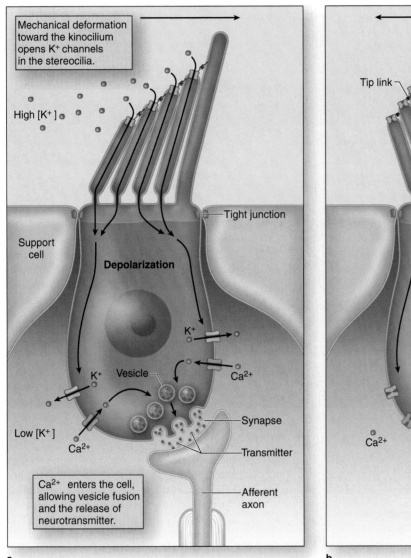
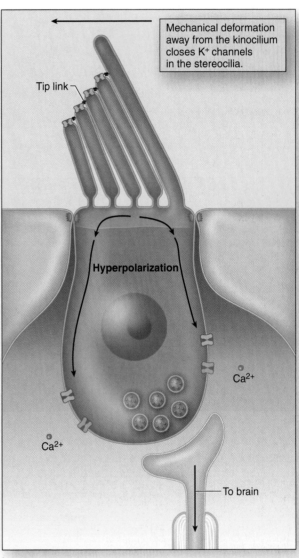

Figure 23–28. **Mechanotransduction in hair cells.** Hair cells and supporting cells are part of an epithelium with tight junctions. The apical ends of the cells are exposed to endolymph with a high concentration of K^+ and perilymph with a much lower K^+ concentration bathes their basolateral surface. At rest hair cells are polarized with a small amount of K^+ entry and a low level of neurotransmitter release to afferent nerve fibers at the basal ends of the cells. **(a):** As shown here head movements that cause the stereocilia bundle to be deflected *toward* the kinocilium produce tension in the tip links which is transduced to electrical activity by opening of adjacent cation channels. Entry of K^+ depolarizes the cell, opening Ca^{2+} at the basal end of the cell which stimulates release of neurotransmitter. When this movement stops, the cells quickly repolarize.

(b): Movements in the opposite direction, *away from* the kinocilium, produce slackness on the tip links, allowing the apical K^+ to close completely, leading to hyperpolarization and reduced transmitter release. With different numbers of afferent and efferent fibers on the hair cells and with various hair cells responding differently to endolymph movements due to their positions within the maculae and cristae ampullares, the sensory information produced collectively by these cells can be processed by the vestibular regions of the brain and used to help maintain equilibrium.

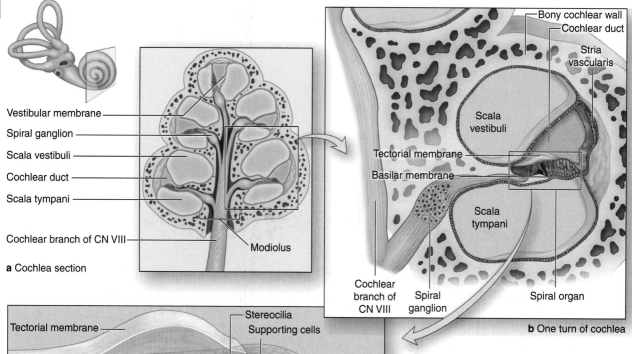

Vestibular membrane

Spiral ganglion

Scala vestibuli

Cochlear duct

Scala tympani

Cochlear branch of CN VIII

Modiolus

a Cochlea section

Bony cochlear wall

Cochlear duct

Stria vascularis

Scala vestibuli

Tectorial membrane

Basilar membrane

Scala tympani

Cochlear branch of CN VIII

Spiral ganglion

Spiral organ

b One turn of cochlea

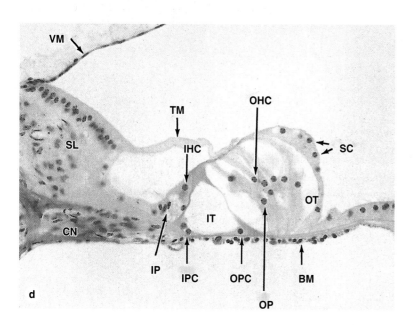

Tectorial membrane

Stereocilia

Supporting cells

Inner tunnel

Basilar membrane

Outer hair cell

Outer phalangeal cells

Inner hair cell

Pillar cells

Cochlear branch of CN VIII

Scala tympani

c Spiral organ

VM

TM

OHC

SL

IHC

SC

OT

CN

IT

IP

IPC

OPC

BM

OP

d

Figure 23–29. Cochlea and spiral organ. The auditory portion of the inner ear, the cochlea, has a snail-like spiral shape in both its bony and membranous labyrinths. **(a):** A section of the whole cochlea shows the cochlear duct cut in several places. **(b):** This diagram shows a more detailed view of one such turn of the cochlear duct and the adjacent perilymph-filled spaces, the scala vestibuli and scala tympani. Endolymph is produced in the stria vascularis, a capillary-rich area of the periosteum associated with the epithelial lining of the wall. **(c):** The lower diagram shows the spiral organ in more detail. **(d):** The micrograph shows important features, including the basilar membrane (BM) on which the spiral organ rests and the tectorial membrane (TM) which extends from cells of the spiral limbus (SL) and contacts the stereocilia of the inner (IHC) and outer hair cells (OHC). Several types of supporting cells are also present, including inner phalangeal (IP) and outer phalangeal cells (OP), which are intimately associated with the hair cells and contribute to the tight epithelium separating endolymph from perilymph in the scala tympani. Other supporting cells form various structural features of the organ important for converting vibrations into subtle stimuli to the hair cells. These include the inner (IPC) and outer pillar cells (OPC) which surround a space called the inner tunnel (IT) and other supporting cells (SC) which border the outer tunnel (OT). Afferent nerve fibers from the hair cells comprise the cochlear nerve (CN), a branch of the eighth cranial nerve. X75. H&E.

membranes and the endolymph surrounding the spiral organ. The width, rigidity, and other physical properties of the basilar membrane which supports the spiral organ vary along its length. This causes the region of maximal displacement within the vibrating spiral organ to vary with the sound waves' frequency, ie, the number of waves moving past a point per unit of time (measured in *hertz*). High-frequency sounds produce maximal movement of the spiral organ nearest the oval window. Sounds of progressively lower frequency produce pressure waves that move farther along the scala vestibuli and displace the spiral organ at a point farther from the oval window (Figure 23–32). The sounds of the lowest frequency that can be detected produce movement of the basilar membrane at the apex or helicotrema of the cochlea. After crossing the cochlear duct and spiral organ at these various points, pressure waves are transferred to the scala tympani and exit the inner ear at the round window (Figure 23–32).

The true receptors for the sense of hearing are the more heavily innervated IHC of the cochlea's spiral organ. The OHC, with the ends of their stereocilia embedded in the tectorial membrane, are depolarized when these mechanotransducers are deflected in a process similar to that of vestibular hair cells described above. Depolarization of the OHC very rapidly produces a slight shortening of these columnar cells, which is mediated by an unusual

transmembrane protein called **prestin** (It. *presto*, very fast) abundant in the lateral cell membranes. Prestin undergoes a voltage-dependent conformational change which affects the cytoskeleton, with the cells rapidly becoming shorter when the membrane is depolarized and elongating with membrane hyperpolarization. Piston-like movements of the OHC produce vibrations of the tectorial membrane against the stereocilia of the nearby IHC (Figure 23–29), amplifying the signals that these cells then send to the CNS for processing as sounds.

MEDICAL APPLICATION

*Deafness can result from many factors, which usually fall into two categories. (1) **Conductive hearing loss** involves various problems in the middle ear which can reduce conduction of vibrations by the chain of ossicles from the tympanic membrane to the oval window. A common example is otosclerosis, in which scar-like lesions develop on the bony labyrinth near the stapes which inhibit its movement of the oval window. Infection of the middle ear (otitis media) is common in young children, usually progressing from an upper respiratory infection, and can reduce sound conduction due to fluid accumulation in that cavity. (2) **Sensorineural deafness** can be congenital or acquired and due to defects in any structure or*

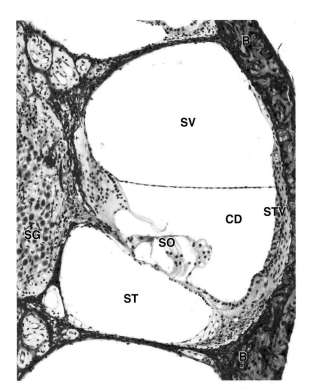

Figure 23–30. Cochlear duct and spiral ganglion. The spiral organ (SO) is located on the basal wall of the cochlear duct (CD). This duct is filled with endolymph produced in the stria vascularis (STV), an unusual association between the columnar epithelial cells which have numerous basal infoldings and the capillaries in the periosteum of the bone (B). On either side of the cochlear duct are the scala vestibuli (SV) and scala tympani (ST), which are filled with perilymph and are continuous at the apex of the cochlea. Cell bodies of bipolar neurons in the spiral ganglion (SG) send dendrites to the hair cells of the spiral organ and axons to the cochlear nuclei of the CNS. X25. H&E.

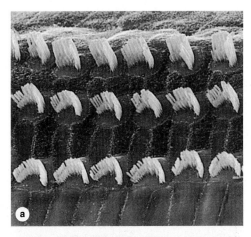

Figure 23–31. Stereocilia of cochlear hair cells. With the tectorial membrane removed, SEM shows the morphology of the three rows of outer hair cells (a), and the single row of inner hair cells (b) in the middle turn of a cochlea. X2700. (With permission, from Patricia A. Leake, Epstein Hearing Research Laboratory, University of California at San Francisco.)

cell from the cochlea to auditory centers of the brain, but commonly involves loss of hair cells or nerve degeneration.

Some hearing loss patients can be helped by **cochlear implants**. These consist of a small device worn behind the ear which contains a microphone, a sound converter, and a transmitter that sends electrical impulses to a receiver implanted under the skin of that region. The receiver is connected to a small cable with many electrodes. The cable is inserted into the internal ear and

threaded into the scala tympani along the wall containing branches of the cochlear nerve. Electrical signals produced by the transmitter in response to sounds of selected frequencies stimulate the nerve directly and are sent to the brain where they are interpreted as sounds. Cochlear implants do not restore normal hearing but can provide the deaf patient with a usable range of sounds and the potential for direct participation in speech.

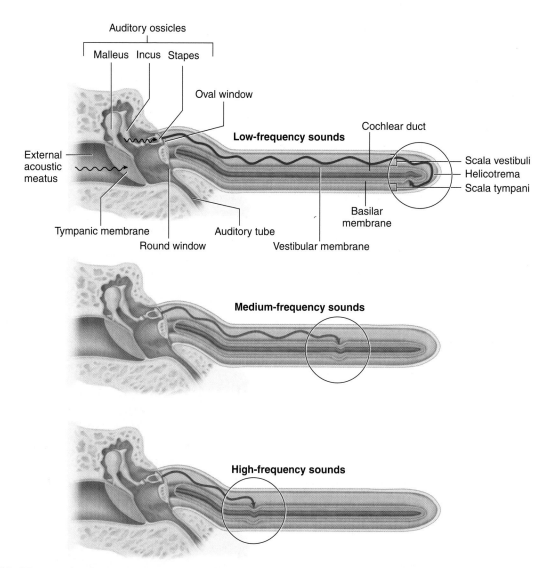

Figure 23–32. Sound waves and movements in the ear. Sound waves are funneled to the tympanic membrane by the external ear and conducted across the middle ear by movements of the three ossicles. Movements of the stapes produce pressure waves in the perilymph on the other side of the attached oval window. In these diagrams, the spiral shape of the cochlea has been straightened to better show how pressure waves affect the spiral organ. The pressure waves produce movements within the spiral organ that cause the mechanoreceptor hair cells to depolarize/hyperpolarize and release neurotransmitters to afferents of the cochlear nerve, producing signals interpreted in the CNS as sounds. Pressure waves crossing the cochlear duct are transferred to the scala tympani and dissipate at the round window. Sounds waves of different frequencies are detected by hair cells at specific sites along the spiral organ. Low frequency sounds produce pressure waves that move the spiral organ only near the end of the cochlea, near the helicotrema. High frequency sounds affect the organ close to the oval window and sounds of intermediate frequencies displace the spiral organ somewhere in between the extremes.

Appendix: Light Microscopy Stains

HEMATOXYLIN AND EOSIN (H&E)

Hematoxylin stains cellular regions rich in basophilic macromolecules (DNA or RNA) a purplish blue or blue-black color. It is the most common stain for demonstrating cell nuclei and cytoplasm rich in rough ER. Usually used as the contrasting "counterstain" with hematoxylin, eosin is an acidic stain that binds to basic macromolecules such as collagen and most cytoplasmic proteins, especially those of mitochondria. Eosin stains regions rich in such structures a pinkish red color. Tissue sections showing only structures with shades of purple and pink are stained with H&E.

PARAROSANILINE-TOLUIDINE BLUE (PT)

This dye combination stains chromatin shades of purple and cytoplasm and collagen a lighter violet. These stains penetrate plastic sections more readily than H&E and are used here primarily with acrylic resin-embedded sections to provide better detail of cell and tissue structures. Toluidine blue is also commonly used for differential staining of cellular components, particularly cytoplasmic granules.

MALLORY TRICHROME

This procedure employs a combination of stains applied in series which results in nuclei staining purple; cytoplasm, keratin, and erythrocytes staining bright red or orange; and collagen bright or light blue. Mallory trichrome is particularly useful in demonstrating cells and small blood vessels of connective tissue. Similar stains, such as Masson trichrome and Gomori trichrome, yield comparable results except that collagen stains blue-green or green.

PICRO-SIRIUS-HEMATOXYLIN (PSH)

The dye Sirius red in a solution of picric acid stains collagen red and cytoplasm a lighter violet or pink, with nuclei purple if first stained with hematoxylin. Under the polarizing microscope, collagen stained with picro-sirius red is birefringent and can be detected specifically.

PERIODIC ACID–SCHIFF REACTION (PAS)

This histochemical procedure stains complex carbohydrate-containing cell components, which become magenta (shades of purplish pink). PAS is commonly used to demonstrate cells filled with mucin granules, glycogen deposits, or the glycocalyx.

WRIGHT-GIEMSA STAIN

These are two similar combinations of stains that are widely used on fixed cells of blood or bone marrow smears to demonstrate types of blood cells. Granules in leukocytes are seen to have differential affinity for the stain components. Nuclei stain purple and erythrocytes stain uniformly pink or pinkish orange.

SILVER OR GOLD STAINS

Various procedures employing solutions of silver or gold salts have been developed to demonstrate filamentous structures in neurons and fibers of reticulin (type III collagen). By these "metal impregnation" techniques these filaments stain dark brown or black. Such stains have been largely replaced now by immunohistochemical procedures.

STAINS FOR ELASTIN

Several staining methods have been developed to distinguish elastic structures from collagen, most of which stain the elastin-rich structures brown or shades of purple. Examples of such stains are Weigert's resorcin fuchsin, aldehyde fuchsin, and orcein Van Gieson stains.

STAINS FOR LIPID

When special preparation techniques are used to retain lipids of cells, such as in frozen sections, lipophilic dyes are used to demonstrate lipid droplets and myelin. Oil red O and Sudan black stain lipid-rich structures as their names suggest. Osmium tetroxide (osmic acid), which is used as a fixative for TEM, is reduced to a black substance by unsaturated fatty acids and is also used to demonstrate lipids.

OTHER COMMON STAINS

Many basic aniline dyes, including azures, cresyl violet, brilliant cresyl blue, luxol fast blue, and light green, are used because of the permanence and brightness of the colors they impart to cellular and extracellular structures in paraffin sections. Many such stains were initially developed for use in the textile industry.

Figure Credits

Figure numbers in boldface indicate those appearing for the first time in this text; Figure numbers in lightface indicate those taken from other sources.

Berman B: *Color Atlas of Basic Histology,* 3rd ed. New York: McGraw-Hill; 2003.

Eckel C. M: *Human Anatomy Lab Manual.* New York: McGraw-Hill, Fitzpatrick, T. B. et al., *Dermatology in General Medicine.* New York: McGraw-Hill; 1971.

Lewis R, Gaffin D, Hoefnagels M, et al, *Life,* 5th ed. New York: McGraw-Hill; 2004.

McKinley M, O'Loughlin V. D: *Human Anatomy,* 2nd ed. New York: McGraw-Hill; 2008.

Widmaier E. P, Raff H, Strang K. T: *Vander's Human Physiology,* 11th ed. New York: McGraw-Hill; 2008.

Chapter 2
2-3: McKinley 2-4; **2-5:** McKinley 2-9b; **2-6:** McKinley 2-7; **2-10:** Widmaier 5-6; **2-12:** McKinley 2-12; **2-16:** McKinley 2-8; **2-20a:** McKinley 2-9a; **2-32:** McKinley 2-35.

Chapter 3
3-3: McKinley 2-17 left; **3-14:** McKinley 2-20; **3-21:** Lewis 9-10.

Chapter 4
4-4: McKinley 4-1b lower part; **4-19:** McKinley 4-4; **4-20:** McKinley 4-5; **4-21:** McKinley 4-6; **4-14d:** Berman 1-16.

Chapter 5
5-3a: Berman 2-6; **5-8a:** Berman 2-7; **5-12b:** Berman 2-24; **5-21c:** Berman 2-5.

Chapter 6
6-1a: Berman 2-18; **6-1c:** Berman 2-19; **6-1d:** Berman 2-20; **6-1e:** Berman 2-21.

Chapter 7
7-1: McKinley 6-1; **7-4:** Berman 3-3; **7-5a:** Berman 3-4.

Chapter 8
8-1: McKinley 6-8; **8-2a:** Berman 4-3; **8-9:** Berman 4-4; **8-12a-d:** McKinley 6-10; **8-13a:** Berman 5-7; **8-13b:** Berman 5-9; **8-14:** McKinley 6-11; **8-16:** McKinley 6-12a, b; **8-17a:** Berman 5-3; **8-17b:** Berman 5-4; **8-18:** McKinley 6-16; **8-19a:** McKinley 9-4.

Chapter 9
9-1: McKinley 14-1; **9-2:** McKinley 14-16; **9-3:** McKinley 14-3; **9-4:** McKinley 14-4; **9-6a:** McKinley 14-14b; **9-7:** McKinley 14-13c; **9-9b:** Eckel 4-28b; **9-10:** McKinley 14-7; **9-11a:** Berman 6-9; **9-16d:** Berman 6-8; **9-17:** Eckel 16-1c; **9-18a:** McKinley 16-2b; **9-19:** McKinley 15-4; **9-20c:** McKinley 15-7a; **9-21a:** McKinley 14-8(1); **9-21b:** McKinley 14-8(2); **9-21c:** McKinley 14-8(3); **9-21d:** McKinley 14-8(4); **9-22:** Berman 6-21; **9-23:** McKinley 14-12c; **9-25:** McKinley 14-10a; **9-26a:** McKinley 14-12a; **9-26b:** Berman 6-15; **9-27a:** McKinley 14-12b; **9-28b:** Berman 6-19; **9-28d:** Berman 6-18; **9-29a:** Berman 6-10; **9-29c:** Berman 6-12.

Chapter 10
10-01: Widmaier 9-1; **10-2:** McKinley 10-4; **10-3:** McKinley 10-1; **10-7a:** Berman 7-2; **10-7c:** Berman 7-4; **10-8:** McKinley 10-6; **10-9:** McKinley 10-5; **10-11:** McKinley 10-9; **10-12:** McKinley 10-7; **10-13a: 10-13b,c:** Widmaier 9-14; **10-14a:** Widmaier 10-4; Berman 7-6; **10-14b:** Berman 7-7; **10-15:** McKinley 10-12; **10-16:** McKinley 22-10a; **10-17a:** Berman 7-10; **10-17b:** Berman 7-11; **10-19a:** Berman 7-12; **10-21a:** McKinley 10-16.

Chapter 11
11-1: McKinley 23-3; **11-3:** McKinley 22-11, part1; **11-5:** Berman 11-7; **11-6:** Berman 11-2; **11-7:** McKinley 23-1; **11-8a:** Berman 11-11; **11-8b:** Berman 11-12; **11-13:** McKinley 23-5; **11-14a:** Berman 11-20; **11-14b:** Berman 11-22; **11-15:** Berman 11-25; **11-16:** McKinley 23-6; **11-20b:** Berman 11-21; **11-21b:** Berman 11-18; **11-21c:** Berman 11-13; **11-21d:** Berman 11-19; **11-23b:** McKinley 24-2b.

Chapter 12
12-1: McKinley 21-2; **12-3:** McKinley 21-3; **12-4a:** Widmaier 12-67; **12-4b,c:** McKinley 21-4; **12-6:** McKinley T21-3; **12-7c:** Berman 8-3; **12-9b:** Berman 8-4; **12-11b:** Berman 8-5; **12-12c:** Berman 8-6; **12-12d:** Berman 8-1; **12-13a:** Berman 8-9; **12-13c:** Berman 8-12; **12-14:** McKinley 21-10.

Chapter 13
13-1: McKinley 21-11; **13-5a:** Berman 9-6 through 9-9; **13-5b:** Berman 8-8; **13-8 top, bottom; insets:** Berman 9-2, 9-1; 9-4, 9-5; **13-11a:** Berman 9-11; **13-11b:** Berman 9-13; **13-12:** Berman 9-14.

Chapter 14
14-1: McKinley 24-1; **14-10:** Berman 10-22; **14-14:** McKinley 24-8b; **14-16a:** Berman 10-5.

Chapter 15
15-1: McKinley 26-1; **15-2:** McKinley 26-9; **15-4:** McKinley 19-6; **15-5a:** Berman 12-10; **15-5b:** Berman 12-12; **15-6:** McKinley 19-7; **15-7a:** McKinley 26-6c; **15-7b:** McKinley 26-5; **15-11a:** Berman 12-1; **15-11b:** Berman 12-4; **15-14a:** Berman 12-16; **15-15:** McKinley 26-12a; **15-16:** Berman 12-22; **15-17:** Berman 12-18; **15-25:** McKinley 26-15; **15-36:** McKinley 26-26; **15-37:** McKinley 26-17; **15-38a:** Berman 12-41; **15-38b:** Berman 12-43.

Chapter 16
16-1: McKinley 26-4a; **16-3b:** Berman 13-26; **16-6a:** Berman 13-29; **16-6b:** Berman 13-32; **16-7:** McKinley 26-20; **16-8:** Berman 13-17; **16-9a:** Berman 13-21; **16-11:** McKinley 26-19; **16-12a:** Berman 13-3; **16-12b:** Berman 13-4; **16-13a:** Berman 13-8; **16-13b:** Berman 13-7; **16-20:** McKinley 26-21; **16-21a:** Berman 13-15.

Chapter 17
17-1: McKinley 25-1; **17-3:** McKinley 19-9; **17-4:** Berman 14-1; **17-6:** McKinley 25-8; **17-7:** Berman 14-10; **17-8a:** Berman 14-11; **17-8b:** Berman 14-12; **17-9a:** Berman 14-13; **17-9c:** Berman 14-14; **17.11:** McKinley 25-9; **17-12:** Berman 14-18; **17-13:** McKinley 25-10; **17-14:** Berman 14-20; **17-18a:** McKinley 25-11.

Chapter 18
18-1: McKinley 5-1; **18-2:** McKinley 5-2; **18-3:** Berman 15-4; **18-5:** Berman 15-3; **18-6a:** Berman 15-2; **18-6b:** McKinley 5-4a; **18-7b:** Fitzpatrick 70-9; **18-8:** Fitzpatrick 7-6; **18-10:** McKinley 19-5; **18-11:** Eckel 17-2; **18-12:** McKinley 5-9; **18-13a:** Berman 15-15; **18-13b:** Berman 15-14; **18-13c:** Berman 15-13; **18-14a,b:** McKinley 5-8; **18-15a:** Berman 15-10; **18-17b:** Fitzpatrick 81-2.

Chapter 19
19-1: McKinley 27-3; **19-2:** McKinley 27-5; **19-3:** McKinley 27-4; **19-4a:** Berman 16-4; **19-5:** McKinley 27-6; **19-6b:** Berman 16-11; **19-8a:** Berman 16-8; **19-9:** McKinley 27-7; **19-13a:** Berman 16-13; **19-15:** McKinley 27-8; **19-16a,b:** McKinley 27-9b, c; **19-17a:** Berman 16-18.

Chapter 20
20-1: McKinley 20-1; **20-2:** McKinley 20-4; **20-3:** McKinley 20-15; **20-4:** Berman 17-1; **20-6:** Berman 17-3; **20-8:** McKinley 20-7; **20-9:** Berman 17-4; **20-10:** McKinley 20-10; **20-12:** McKinley 20-13a; **20-14:** McKinley 20-13c,d; **20-17c:** Berman 17-13; **20-17e:** McKinley 20-1 rt qt; **20-18a:** McKinley 20-9 up lt; **20-18b,c,d:** McKinley 20-16b; **20-19:** Berman 17-15; **20-22:** McKinley 20-11a; **20-23:** Berman 17-17.

Chapter 21
21-1: McKinley 28-11; **21-2:** McKinley 28-13; **21-3:** Berman 18-2; **21-4a:** Berman 18-5; **21-5:** McKinley 28-14; **21-6a:** Berman 18-7; **21-6b:** Berman 18-8; **21-9a:** Berman 18-10; **21-9b:** Berman 18-11; **21-10a:** Berman 18-12; **21-10c:** Berman 18-13; **21-11a:** Berman 18-14; **21-12a:** Berman 18-16; **21-13:** McKinley 28-16; **21-14a:** Berman 18-18; **21-16a:** Berman 18-20; **21-16b:** Berman 18-21; **21-17:** McKinley 28-17b; **21-18:** Berman 18-23.

Chapter 22
22-1: McKinley 28-3; **22-2:** McKinley 28-4; **22-9:** McKinley 3-7; **22-11:** Berman 19-8; **22-12:** McKinley 28-7; **22-13a:** Berman 19-16; **22-15:** McKinley 28-6; **22-17a:** Berman 19-19; **22-17b:** Berman 19-20; **22-17c:** Berman 19-21; **22-18:** McKinley 3-6; **22-21a:** Berman 19-22; **22-22a:** Berman 19-23; **22-24a:** Berman 19-24; **22-24b:** Berman 19-25; **22-24c:** Berman 19-26; **22-25a:** Berman 19-27.

Chapter 23
23-1: McKinley 19-12b; **23-2:** McKinley 19-12a; **23-3:** McKinley 19-19; **23-9:** McKinley 19-17; **23-11a:** Berman 20-4; **23-14:** McKinley 19-14a, b; **23-15:** Berman 20-9; **23-21:** McKinley 19-20; **23-23:** McKinley 19-22; **23-24:** McKinley 19-23; **23-27:** McKinley 19-25; **23-29a-c:** McKinley 19-27; **23-29d:** Berman 20-19; **23-32:** McKinley 19.29.

Index

Page numbers followed by *f* indicate figures; page numbers followed by *t* indicate tables.